Rotation, Reflection, and Frame Changes

Orthogonal tensors in computational engineering mechanics

Rotation, Reflection, and Frame Changes

Orthogonal tensors in computational engineering mechanics

R M Brannon
Department of Mechanical Engineering, University of Utah

IOP Publishing, Bristol, UK

Media content for this book is available from http://iopscience.iop.org/book/978-0-7503-1454-1.

ISBN 978-0-7503-1454-1 (ebook)
ISBN 978-0-7503-1452-7 (print)
ISBN 978-0-7503-1453-4 (mobi)

DOI 10.1016/978-0-7503-1454-1

Version: 20180401

IOP Expanding Physics
ISSN 2053-2563 (online)
ISSN 2054-7315 (print)

British Library Cataloguing-in-Publication Data: A catalogue record for this book is available from the British Library.

Published by IOP Publishing, wholly owned by The Institute of Physics, London

IOP Publishing, Temple Circus, Temple Way, Bristol, BS1 6HG, UK

US Office: IOP Publishing, Inc., 190 North Independence Mall West, Suite 601, Philadelphia, PA 19106, USA

In memory of my father, from whom I inherited a geek gene, and in grateful recognition of my mother's 'ragamuffin barefoot irreverence' for conventional wisdom. Slightly deviating from Bronowski, they cultivated in me the discipline to faithfully learn what is known before respectfully questioning it. Such is the path of artists and scholars.

Contents

Preface and acknowledgements

A review document of this size grows and improves over the years only with the assistance of a large number of technical reviewers. Rather than being the result of a single directed-research project, this book reflects numerous small lessons learned about rotation throughout many different projects, starting with Buck Schreyer at the University of New Mexico chiding me that, despite local familiarity with the name Rodriguez, the theorem about rotation was spelled (or as he once declared *spelt*) Rodrigues. My rotation education came along thereafter in graduate school at the University of Wisconsin, continuing piecemeal through my 14 years at Sandia National Laboratories, and ultimately at the University of Utah. Countless interesting conversations with my fellow grad students (especially Lee Greenler, Dallas Meyer, and Gary Gray) helped to refine my views on rotation. The subsequent support of managers during my time at Sandia National Laboratories in Albuquerque was greatly appreciated. Among those at Sandia who offered helpful suggestions and critical review, I am especially grateful to Allen Robinson, Dave Aldridge, Dan Carrol, Paul Taylor, and Erik Strack for being so thorough in their critiques of several sections. Special thanks go to Dr Ed Love, of UC Berkley when I first met him (later at Sandia), who performed the lion's share of work on the proof of the iterative fixed-point polar decomposition. Heartfelt thanks go to Grant Farnsworth for numerous interactions regarding understandability, accuracy, and efficiency of the polar-decomposition algorithms; his enthusiasm was infectious. One particularly careful and helpful reviewer, Bill Scherzinger, humorously referred to this book as 'rotation salad'. He was, of course, correct because this book aims to include as many rotation-related topics as possible, hence making it more of an archival list of facts than a coherent story leading to any particular grand conclusion. I almost changed this book's title to *Rotation Salad*, but I feared that doing so might hurt its chances of being cited in technical publications—I hope, dear reader, that you will honor Bill (and stave off our community's tendency to be ludicrously snooty) by calling this book by that name in casual conversation! In addition to catching numerous technical errors and minor grammatical typos, Bill urged me to review the flow of the material and to add more motivation at the beginning of each chapter with better transitions into subsequent chapters. The resulting book is still a rotation salad, but I hope it is now better garnished. More recently, after joining the University of Utah, I gained deep appreciation for my own graduate students and colleagues, such as K T Ramesh and Lori Brady (with whom I interacted during my 2014 sabbatical at Johns Hopkins University), Ron Smelser (who contributed indirectly to this book by giving me a chance to review his Continuum Mechanics book), and Davide Bigoni (who, in addition to giving me helpful book publishing tips, possesses an endearing wonderment of the physical world and its associated mathematics). My interactions with Dr Jim Guilkey (then with Schlumberger), John Nairn (Oregon State), Biswajit Banerjee, and Profesor Rich Regueiro (University of Colorado) improved the polar-decomposition and statistical rotation sections. A conversation with an 'academic soul mate', Peter

Mackenzie (University of Washington) was especially noteworthy: as I had been extolling my own brilliant personal insight that a Voigt compact matrix for fourth-order tensors are actually contravariant components with respect to a non-normalized basis for tensor space, Peter gently interrupted to let me know that he had published an article making the same observation. I was bummed at having been 'scooped', but he lifted my spirits by saying that my way of explaining it would be easier for more people to understand. Interactions with Professor Ashley Spear and my student Brian Leavy (who kindly contributed Python versions of the rotation tools) improved coverage of Euler angles as we worked on setting crystal orientations in mesoscale plasticity models from EBSD data. Everyone is to be thanked for driving what was added to this book based on what was *actually needed* by engineers and by talking about it in a way that is comprehensible to engineers who did not minor in mathematics. The resulting book is based on and strives to maintain a realistic perspective on the varying backgrounds and skill levels of those in professional practice. You, reader, are assumed to be familiar with multivariate calculus, linear algebra, and elementary tensor analysis. Beyond that, tutorials are provided. As I am an old gal, coding examples are in FORTRAN, and you are assumed to be able to decipher them anyway. Most of the routines are available in C++ upon request. At a more personal level, I wish to express my appreciation to Norm Johnson (from T3 division at Los Alamos National Laboratories) for introducing me to tensor analysis during the early days of operator-mounted tape drives and Tectronics displays, Sharon Vande Hei for cheering on transcriptions of my first draft to a Mac Plus (from grad school notes on engineering paper), Susan Smith for watching me tap-tap-tap my way into social oblivion and lower back pain while working on my Linux keyboards after grad school, Susan Moore for her loving—often video-conferenced—support (along with prodding to take swim breaks) as I wrapped up the production phase on my custom PC (which was built with help from my former postdoc, Seubpong Leelavanichkul, who also nagged for me to finish this cancerous book), and then Sarah 'Savory' Crawford for tolerating many travel delays while wrapping up thousands of tedious final edits (for which John Navas and Chris Benson at IOP also deserve kudos). Thank you all!

Foreword

It is a special pleasure for me to introduce the book *Rotation, Reflection and Frame Changes* by Professor Rebecca Brannon. A standout feature of this book is that it gives a clear description of rotation and how it permeates all areas of continuum and computational mechanics. From the beginning, emphasis is placed on nomenclature, notation and basic definitions such as laboratory, observer and material bases. Getting these concepts defined early and clearly allows for easier understanding of the complex topics presented in the later chapters. Who would have believed so many fundamental points can be illustrated with a simple model problem as given in her one-dimensional discussion of stress and strain?

The thorough treatment of rotations begins with the Euler–Rodrigues theorem. The author uses exceptionally clear figures to illustrate concepts of which the definition of Euler angles is a good example. The following material is necessarily intertwined with fundamental concepts of continuum and computational mechanics that is also covered in more depth than found in many standard textbooks. A small but representative set of examples of these applications include: definitions of various versions of stress and strain and their rates, polar decomposition of the deformation gradient, integration of rotation rates, rate of logarithmic strain with numerical error bounds, and remapping for computational algorithms involving large deformations. In separate chapters, Professor Brannon presents unusually comprehensive discussions of the statistics of random orientations, and of the principle of material-frame indifference with implications for valid forms of constitutive equations.

The overall depth to which Professor Brannon covers rotations is impressive and demonstrates her recognized expertise in the area. This compilation of material by a single person illustrates that the author has been involved in a wide range of research projects, and we are the beneficiaries of her experience. This book must be considered as a necessary reference, if not the textbook itself, for courses on continuum mechanics, and is a 'must have' valuable resource for those who work in the area. Professor Brannon is owed a great deal of gratitude for taking the time to compile this material and make it available to the research and engineering community associated with continuum and computational mechanics.

Howard L (Buck) Schreyer
Emeritus Professor of Mechanical Engineering
University of New Mexico
Albuquerque NM
USA

Author biography

R M Brannon

For over 25 years, first as a principal researcher (and manager) at Sandia National Laboratories and more recently as an Associate Professor of Mechanical Engineering at the University of Utah and ASME fellow, Dr Brannon has developed practical engineering constitutive models for brittle and ductile material failure at high strain rates and large strains. Her research has investigated a wide range of materials including piezoelectric ceramics, armor ceramics, geological materials, energetic materials, and metals (usually for high-rate applications). Constitutive models she has developed are used in DoD and DOE production codes such as CTH and ALEGRA. Applications have included protective structures, underground structure integrity, electroactive power supplies, artificial hip implant rapid materials ranking, shock-induced vaporization, in-vivo measurements of callus strains, and other numerous other problems in the applied sciences. Dr Brannon is particularly known for her monographs on tensor analysis, plasticity, code portability, code verification, and massive deformation kinematics in the material point method.

IOP Publishing

Rotation, Reflection, and Frame Changes
Orthogonal tensors in computational engineering mechanics
R M Brannon

Chapter 1

Introduction

Each practicing engineer must, at some point, characterize reorientations of a body or find vector and tensor components with respect to a rotated coordinate system. Additionally, rotations and reflections can contribute to overall engineering system response in combination with other predominant physical effects. Consequently, many important theorems (such as the polar decomposition and the time derivative of logarithmic strain) involve rotations as a part of a larger problem. A vast literature is available for the most common rotation-related tasks such as simple orthogonal Cartesian coordinate changes. Most reference books, however, tend to cover one or two methods for handling rotations, which makes it difficult for a researcher to translate between different methods used in different books. Resource information for less common rotation-related tasks is even scarcer. For example, mesoscale material simulations (in which crystals are explicitly modeled in a representative volume element) require methods to establish a uniform distribution of initial crystal orientations, which is a non-trivial task that is rarely explained in textbooks. Obscure rotation-related research applications can be found in disparate journal articles, but self-contained comprehensive review articles that cover *both* elementary and advanced engineering concepts *written in a manner comprehensible to engineers* are rare. Morawiec [1] offers an outstanding list of references on the topic of rotation and covers many of the topics reviewed here with greater mathematical rigor.

Available rotation review articles and textbooks seem to be aimed more at mathematicians than applications specialists. While it is true that this book will cover several mathematically advanced topics, our target audience is practicing engineers, especially those who lack a guru-level background in abstract algebra, topology, differential geometry, hyper-complex numbers, and group theory, that unfortunately seems to be a prerequisite for understanding existing resource publications. Hence, most of our discussions are limited to ordinary three-dimensional (3D) vectors of the type discussed in first calculus courses. Detailed proofs are provided only when they are non-trivial (and not readily available in the literature) or when the proof itself clarifies a concept. In this book, the assumed level of the reader's mathematical background varies considerably. Some chapters are quite elementary (first-year graduate student level), while other chapters are aimed at a target audience of mathematically adept researchers who are working on the arcane topics of contemporary interest. This book's topics and examples tend to favor materials mechanics issues.

Rather than organizing this book using increasing levels of complexity, the topics are grouped according to subject area. Each chapter (and each section within a chapter) opens with an elementary (more intuitive) approach to the subject at hand, and then it progresses into greater levels of sophistication and rigor. Consequently, if you have only limited background in mathematics, you are encouraged to drill into each chapter only to the depth appropriate to your background. Each chapter is (as much as practical) self-contained, so it is not necessary to read preceding chapters to learn about subjects in later chapters.

One aim of this book is to catalog the many different ways to define and quantify rotations and reflections. A rotation may be described mathematically by a special kind of *orthogonal* tensor. By using the 'special kind' proviso, we are implying that rotation tensors are always orthogonal, but orthogonal tensors are not necessarily rotations. A tensor $\underset{\approx}{\boldsymbol{Q}}$ is orthogonal if and only if $\underset{\approx}{\boldsymbol{Q}}^T \bullet \underset{\approx}{\boldsymbol{Q}} = \underset{\approx}{\boldsymbol{I}}$, where $\underset{\approx}{\boldsymbol{I}}$ is the identity tensor[1]. The tensor $\underset{\approx}{\boldsymbol{Q}}$ is orthogonal if and only if its component matrix $[Q]$ is orthogonal. A real matrix $[Q]$ is orthogonal if and only if

$$[Q]^T[Q] = [I], \tag{1.1}$$

where $[I]$ is the identity matrix. In indicial form,

$$Q_{ki}Q_{kj} = \delta_{ij} \tag{1.2}$$

where δ_{ij} is the Kronecker delta[2] and (using standard tensor notation conventions summarized in chapter 2) repeated indices are understood to be summed from 1 to 3. An elementary identity in tensor analysis states that a tensor $\underset{\approx}{\boldsymbol{Z}}$ is zero iff[3]

$$\underset{\sim}{\boldsymbol{u}} \bullet \underset{\approx}{\boldsymbol{Z}} \bullet \underset{\sim}{\boldsymbol{v}} = 0 \quad \forall \text{ vectors } \underset{\sim}{\boldsymbol{u}} \text{ and } \underset{\sim}{\boldsymbol{v}} \tag{1.3}$$

Accordingly, two tensors, $\underset{\approx}{\boldsymbol{A}}$ and $\underset{\approx}{\boldsymbol{B}}$, are equal iff $\underset{\sim}{\boldsymbol{u}} \bullet \underset{\approx}{\boldsymbol{A}} \bullet \underset{\sim}{\boldsymbol{v}} = \underset{\sim}{\boldsymbol{u}} \bullet \underset{\approx}{\boldsymbol{B}} \bullet \underset{\sim}{\boldsymbol{v}}$ $\forall$ vectors $\underset{\sim}{\boldsymbol{u}}$ and $\underset{\sim}{\boldsymbol{v}}$. A tensor $\underset{\approx}{\boldsymbol{Q}}$ is therefore orthogonal iff

$$\underset{\sim}{\boldsymbol{u}} \bullet \underset{\approx}{\boldsymbol{Q}}^T \bullet \underset{\approx}{\boldsymbol{Q}} \bullet \underset{\sim}{\boldsymbol{v}} = \underset{\sim}{\boldsymbol{u}} \bullet \underset{\approx}{\boldsymbol{I}} \bullet \underset{\sim}{\boldsymbol{v}} \quad \forall \text{ vectors } \underset{\sim}{\boldsymbol{u}} \text{ and } \underset{\sim}{\boldsymbol{v}} \tag{1.4}$$

---

[1] This book uses a 'self-defining' notation in which scalars, vectors and second-order tensors are written with zero, one, and two under-tildes, respectively (as in s, $\underset{\sim}{\boldsymbol{v}}$, and $\underset{\approx}{\boldsymbol{T}}$). The composition of two tensors, denoted symbolically by $\underset{\approx}{\boldsymbol{A}} \bullet \underset{\approx}{\boldsymbol{B}}$, has ij components given by $A_{ik}B_{kj}$ in which the repeated index is implicitly summed. The notation $\underset{\approx}{\boldsymbol{A}}:\underset{\approx}{\boldsymbol{B}} = A_{ij}B_{ij}$ is the tensor inner product. The notations $\underset{\approx}{\boldsymbol{A}} \bullet \underset{\sim}{\boldsymbol{u}}$ and $\underset{\sim}{\boldsymbol{v}} \bullet \underset{\approx}{\boldsymbol{B}}$ are vectors with ith components of, respectively, $A_{ik}u_k$ and v_kB_{ki}. These definitions immediately give the often-used identity $\underset{\approx}{\boldsymbol{A}} \bullet \underset{\sim}{\boldsymbol{u}} = \underset{\sim}{\boldsymbol{u}} \bullet \underset{\approx}{\boldsymbol{A}}^T$.

[2] I.e., components of the 3×3 identity matrix. Thus δ_{ij} equals 1 if $i = j$ and 0 if $i \neq j$.

[3] I.e., if and only if.

To interpret this assertion, consider the 'star' vectors in figure 1.1, which are defined to be orthogonal transformations of non-starred 'fiducial' vectors:

$$\underset{\sim}{u}^* = \underset{\approx}{Q} \bullet \underset{\sim}{u} \quad \text{and} \quad \underset{\sim}{v}^* = \underset{\approx}{Q} \bullet \underset{\sim}{v}. \tag{1.5}$$

Noting that $\underset{\sim}{u} \bullet \underset{\approx}{Q}^T = \underset{\approx}{Q} \bullet \underset{\sim}{u}$, equation (1.4) then reduces to

$$\underset{\sim}{u}^* \bullet \underset{\sim}{v}^* = \underset{\sim}{u} \bullet \underset{\sim}{v} \quad \forall \text{ vectors } \underset{\sim}{u} \text{ and } \underset{\sim}{v}. \tag{1.6}$$

As a special case, taking the two arbitrary vectors to be identical to each other proves that an orthogonal $\underset{\approx}{Q}$ must preserve lengths

$$\|\underset{\sim}{u}^*\| = \|\underset{\sim}{u}\| \quad \text{and} \quad \|\underset{\sim}{v}^*\| = \|\underset{\sim}{v}\|. \tag{1.7}$$

Denoting these magnitudes by u and v, respectively, recall from elementary vector analysis that a vector dot product satisfies $\underset{\sim}{u} \bullet \underset{\sim}{v} = uv\cos\theta$, where θ is the angle between the vectors. Consequently, equation (1.6) shows that an orthogonal transformation $\underset{\approx}{Q}$ preserves the magnitude *but not necessarily the sign* of the angle between two vectors:

$$\cos\theta^* = \cos\theta \quad \Leftrightarrow \quad \theta^* = \pm\theta \tag{1.8}$$

In two dimensions, the angle θ between $\underset{\sim}{u}$ and $\underset{\sim}{v}$ can be arbitrarily declared to be 'positive' if the vector $\underset{\sim}{u}$ must be rotated counter-clockwise by θ to become aligned with $\underset{\sim}{v}$. As indicated in figure 1.1, a 2D orthogonal transformation is called 'proper' if it preserves the angle in both magnitude and sign; it is otherwise an 'improper' transformation that will preserve the magnitude of the angle, but not the sign. An improper-orthogonal transformation is seen in the example of figure 1.1 to not only rotate the vectors but also to 'reflect' them into different relative ordering (from clockwise to counter-clockwise or vice versa). These statements may be generalized to higher dimensions based on properties of the $[Q]$ orthogonal transformation matrix itself.

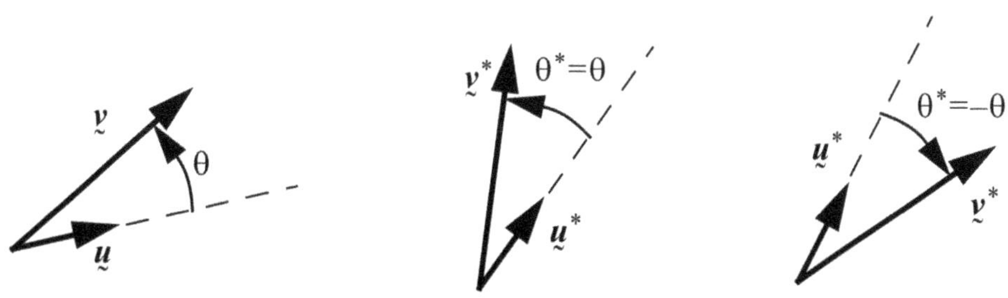

(a) fiducial vectors

(b) "star" vectors under a proper orthogonal transformation

(c) "star" vectors under an improper orthogonal transformation

Figure 1.1. Effects of an orthogonal transformation. (a) The word 'fiducial' simply means 'basis of comparison', which refers to the pre-transformation starting vectors (b) shows a change in orientation without a change in length or angle and (c) shows a change in both orientation and numerical sign of the angle.

It is fairly common for writers to refer to the set of all tensors as ‘lin’, and the subset of all orthogonal tensors is called ‘orth’ [2]. Second-order engineering tensors have nine independent components, so they are said to belong to a 9D space. Moreover, because any linear combination of tensors is itself a tensor, lin is a 9D *linear* manifold and therefore it is possible to set up a Cartesian coordinate system for all members of lin. To rigorously prove that a second-order tensor on 3D space is, in fact, also a 9D vector, one must simply confirm axioms (additive commutativity, associativity, existence of an additive inverse, etc) that must be met for something to be called a vector.

Just as the Cartesian representation for a vector $\underset{\sim}{x} = x_1\underset{\sim}{e}_1 + x_2\underset{\sim}{e}_2 + x_3\underset{\sim}{e}_3$ (in which the $\underset{\sim}{e}_i$ are orthonormal basis vectors) may be represented in array form as

$$\begin{bmatrix} x_1 \\ x_2 \\ x_3 \end{bmatrix} = x_1\begin{bmatrix} 1 \\ 0 \\ 0 \end{bmatrix} + x_2\begin{bmatrix} 0 \\ 1 \\ 0 \end{bmatrix} + x_3\begin{bmatrix} 0 \\ 0 \\ 1 \end{bmatrix}, \tag{1.9}$$

the Cartesian expansion of a tensor, $\underset{\approx}{T} = T_{11}\underset{\sim}{e}_1\underset{\sim}{e}_1 + T_{12}\underset{\sim}{e}_1\underset{\sim}{e}_2 + \ldots + T_{33}\underset{\sim}{e}_3\underset{\sim}{e}_3$ (in which the $\underset{\sim}{e}_i\underset{\sim}{e}_j$ are basis dyads), may be represented in matrix form as

$$\begin{bmatrix} T_{11} & T_{12} & T_{13} \\ T_{21'} & T_{22} & T_{23} \\ T_{31} & T_{32} & T_{33} \end{bmatrix} = T_{11}\begin{bmatrix} 1 & 0 & 0 \\ 0 & 0 & 0 \\ 0 & 0 & 0 \end{bmatrix} + T_{12}\begin{bmatrix} 0 & 1 & 0 \\ 0 & 0 & 0 \\ 0 & 0 & 0 \end{bmatrix} + \ldots + T_{33}\begin{bmatrix} 0 & 0 & 0 \\ 0 & 0 & 0 \\ 0 & 0 & 1 \end{bmatrix}. \tag{1.10}$$

This expansion has nine terms involving nine orthonormal basis tensors. The space of ordinary 3D vectors is Euclidian (and thus has a Cartesian basis), but it can contain lower-dimensional spaces, such as the surface of a sphere, that are not Euclidean—which means that a linear combination of points in that curvilinear space might not produce a point in the space. A point on the unit sphere, for example, is a vector of unit magnitude. Adding unit vectors does not generally produce a unit vector, so the set of all unit vectors is non-Euclidean. This does not mean they are not vectors. They simply belong to a 2D curvilinear space contained within the larger 3D Euclidean space of general vectors. Similarly, we will find that the set of all orthogonal tensors belongs to a 3D curvilinear space (a hyper-sphere) embedded within the larger 9D Euclidean space of general second-order tensors.

Unit vectors have only two independent components (given any two components, the third is constrained to produce a unit vector). Even though the components of a unit vector are not independent, it is nevertheless most intuitive to work with them using all three components. Similarly, even though rotations have only three independent components, it is nevertheless most intuitive to work with them in their full 3×3 form. A reduction to fewer components is useful mostly to conserve computer memory, but typically at the penalty of increased CPU cost.

The fact that orth is made of tensors having only three independent components follows because columns (or rows) of an orthogonal matrix must form an orthonormal triad. The first column has two independent components (because it is a unit vector), and the second column has only one independent component because it must be a unit vector and also have a zero inner product with the first column. Once the first and second columns of a rotation matrix are set, the third is the cross product of the first two, giving no additional independent components (other than sign if reflections are allowed). Because a linear combination of unit vectors is not generally a unit vector, it follows that a linear combination of orth tensors is not generally an orth tensor itself. Consequently, orth is a 3D *nonlinear* (curvilinear) subset of lin. As a nonlinear space, orth is not a flat higher-dimensional plane; instead, it will be seen to belong to a subset of a *hyper-sphere* in lin. Before doing any 'high-math' to prove this, you should be able to guess the radius of this sphere: just as the magnitude of a vector $\underset{\sim}{v}$ is $\sqrt{v_k v_k}$, the magnitude of a tensor $\underset{\approx}{A}$ is $\sqrt{A_{ij}A_{ij}}$. If the tensor happens to be orthogonal $\underset{\approx}{Q}$, its magnitude (and hence the radius of the hyper-sphere) is

$$\sqrt{Q_{ij}Q_{ij}} = \sqrt{Q^T_{ji}Q_{ij}} = \sqrt{\delta_{ij}} = \sqrt{3}, \tag{1.11}$$

which applies if $[Q]$ is a 3×3 orthogonal matrix. In general, an $N \times N$ orthogonal matrix has magnitude $\sqrt{N}$.

An orthogonal matrix $[Q]$ with a *positive* determinant (equal to $+1$) is called a *rotation* or *proper orthogonal* matrix [3]. The set of all proper-orthogonal tensors is often referred to as orth$^+$. An orthogonal matrix $[Q]$ with a *negative* determinant (equal to -1) is said to be an *improper-orthogonal tensor* and is often misleadingly called *reflection*—but it may actually be a reflection *in combination with* a rotation (see chapter 6). The set of improper-orthogonal tensors is often referred to as orth$^-$.

A proper-orthogonal tensor is the most useful means of characterizing a rotation even though it is inefficient to store as nine components when only three are independent. Euler angles are also popular, but of limited value. These and many other lesser-known representations will be reviewed. Orthogonal tensors emerge as the superior quantification of a rotation or reflection[4], so each definition of an alternative representation of a rotation will be accompanied by instructions for converting from that representation to an orthogonal tensor and vice versa. Algorithms and source code are provided (and may be downloaded from http://iopscience.iop.org/book/978-0-7503-1454-1).

In addition to cataloging the many ways to quantify rotation, this book covers several rotation-related applications. The topics are arranged as summarized here. For a smaller list that highlights the most significant contributions of this text, see the closing remarks in chapter 28.

[4] Having nine components, an orthogonal tensor is not superior to other methods if minimizing *storage* is the metric. Instead, it is the *structure* and ease of computation that tends to make orthogonal tensors convenient. In the same way, a unit vector $\underset{\sim}{n}$ is still most conveniently described in terms of three components (n_1, n_2, n_3) satisfying one constraint, $n_1^2 + n_2^2 + n_3^2 = 1$, even though they may be described in terms of two spherical angles as $n_1 = \cos\varphi \sin\vartheta$, $n_2 = \sin\varphi \sin\vartheta$, $n_3 = \cos\vartheta$. This representation, though efficient in storage, has an ugly imbalance as well as numerical inefficiency in rebuilding the components. Similar assertions apply to rotations. The most compelling advantage of building a full three-component unit vector is that the set of unit vectors forms a 2D curvilinear subset (surface of a sphere) contained within a larger 3D *Euclidean* space. Euclidean representations are needed to compute Jacobians (needed to quantify cluster density of unit normals on a sphere or, in a higher-dimensional analog, texture of crystallographic orientations in materials).

- Chapter 2 outlines notation and key mathematical prerequisites, as well as explaining the linear fractional transformation notation commonly used in robotics/controls.
- Chapter 3 reviews how orthogonal *direction-cosine matrices* are used to transform vector and tensor components from one orthonormal basis to another, and how these transformations are used to define engineering vectors or tensors as closure tests in the abstract mathematician's definitions that make no reference to length or direction. The distinction between a position vector's components and coordinates is made clear.
- Chapter 4 points out the distinction between *rotation* and *coordinate transformation.* Analyzing rotation demands awareness of your desired perspective. You can rotate an object, while you stay still, or you can keep the object fixed while you rotate yourself. It is important to be aware of which of these perspectives applies for your problem of interest. The distinction between these fundamentally different transformations goes beyond one being the same as the other with an opposite rotation angle.
- Chapter 5 describes nonlinear rotations (using torsion and vortex as examples), how to *convert an axis and angle* of rotation into a rotation tensor and vice versa, properties of skew tensors associated with rotation axes, and spinor vector representations of rotations.
- Chapter 6 reiterates that a rotation is characterized by a *proper*-orthogonal tensor—i.e., one having a determinant equal to +1. An orthogonal tensor with a determinant equal to −1 should be regarded (in general) as a *reflection, possibly in combination with a rotation.*
- Chapter 7 presents the representation of a rotation in terms of a unit *quaternion.* Quaternions, which (being generalizations of complex numbers) are predecessors to modern vectors, and they are enjoying a resurgent vogue in rotation-related applications. Even though a rotation tensor can be constructed uniquely from a unit quaternion (having one real component and three hyper-imaginary parts), it is explained that this does not imply that the smallest Euclidean space containing all possible rotation tensors is itself 4D.
- Chapter 8 describes how any linear vector-to-vector operator (in 3D) can be expressed as a *sum of three dyads*, and this result is specialized to rotations. This chapter provides yet another way to describe rotations that is based on a colorful *'nautical' analogy* between embedded triads in an object and embedded directions in a sailboat (mast, starboard, bow). Any orientation of these embedded directions can be specified through the boat's latitude, longitude and heading. Applications of dyadic representations of tensors include user interfaces for controlling rotation of a 3D object on a computer image, finding basis orientations that most simplify a tensor's component matrix, and constructing a rotation that will transform a specified vector into another specified vector.
- Chapter 9 shows that any general rotation may be expressed as a (non-commuting) sequence of three rotation operations. When defining a rotation in this way, it is essential to know whether subsequent rotations are defined with respect to fixed axes or follower axes. *Rotations applied sequentially*

about the fixed laboratory basis are different from rotations applied sequentially about axes of a triad that *moves with the body* as quantified *by Euler angles*, but the two descriptions are intimately related.

- Chapter 10 provides the *series expansion representation of a rotation*, which turns out to be a tensor analog to the series expansion of an exponential. This chapter also describes the Cayley transformation that constructs a rotation tensor from a skew tensor, as well as the Rayleigh transformation (which is used to apply a superimposed rotation to a tensor of any order).
- Chapter 11 derives the *eigenvalues* and *eigenvectors* of any rotation tensor.
- Chapter 12 presents the *polar-decomposition theorem*, which shows how a general deformation can be broken into two distinct (non-commuting) steps, one being a rigid rotation, and the other embodying pure material distortion and/or volume changes with no *net* material rotation.
- Chapter 13 describes how the polar decomposition can be used to strip away overall material rotation from material deformation, quantified by various *strain measures* commonly used in engineering applications. The spatial stretch which generates spatial strain measures is altered in the presence of rotation. The reference stretch which generates reference strain measures is unaffected by rotation.
- Chapter 14 discusses candidate strategies for *advecting rotations* in Eulerian physics codes or in Lagrangian codes that remap the mesh. The issues are similar to the challenge of 'mixing' or interpolating unit vectors.
- Chapter 15 shows how the *rate of a rotation tensor* is related to the conventional angular rotation vector, vorticity tensor, and vorticity vector. This chapter includes derivatives of a rotation with respect to the rotation angle and with respect to the rotation axis as well as numerical algorithms for accurately integrating a rate of rotation to update material orientation in a physics code. The famous Dienes algorithm for computing polar spin from vorticity is included, along with a debunking of the common (but erroneous) assertion that eigenvectors of the symmetric part of the velocity gradient rotate according to the vorticity tensor defined by the skew part of the velocity gradient.
- Chapter 16 shows how to compute the *time rate (or other derivative) of a principal function of a symmetric tensor*, with emphasis on finding the *time rate of logarithmic strain*. A principal function (such as the Hencky logarithmic strain used in materials modeling) requires casting the argument (a symmetric tensor) in its principal basis to apply the function to the eigenvalues, after which the result is cast back to the physical basis. Finding the rate of such functions is crucial for linearizing numerical solvers, yet is extraordinarily complicated by the fact that principal directions not only evolve but are not even unique when there are repeated eigenvalues. An algorithm for taking rates of principal functions is presented and shown to be valid even when eigenvalues are repeated. The results are furthermore used to express the rate of reference Hencky strain as a linear transform of un-rotated 'rate' of deformation (defined by a Rayleigh un-rotation of the symmetric

part of the velocity gradient and proved in this chapter to not be a path-independent rate of any measure of deformation).

- Chapter 17 shows how to generate a *uniformly random rotation* tensor, which is useful for generating grain orientations for microscale simulations. We also discuss how to find the average of a tensor over all possible (uniform) rotations of that tensor.[5] This chapter includes extensive tutorials on elementary statistical sampling methods, as well as clear explanations about how to construct a non-square Jacobian matrix for transforming random numbers to random realizations and to then use it to prove whether or not an algorithm produces a *uniform* sampling. A case study, for example, proves that uniform sampling of Euler angles will not produce a uniformly random rotation, as is needed (for example) in mesoscale polycrystalline constitutive modeling.
- Chapter 18 defines *isotropic tensors*, which are very special tensors whose components do not change upon an orthogonal change in basis. A distinction is made between *proper isotropy* and *strict isotropy*, both of which are useful concepts. The set of all isotropic tensors of a given class[6] is a linear manifold, which means that it must possess a finite number of isotropic tensors (called *primitive isotropic tensors*) that form a basis for *all* isotropic tensors of that class. In three dimensions, there is only one primitive isotropic second-order tensor—the identity tensor. There are two primitive proper-isotropic tensors in two dimensions. Fourth-order tensors in three dimensions have three primitive isotropic tensors.
- Chapter 19 provides an elementary introduction to the *principle of material-frame indifference* (PMFI), which requires that material constitutive models must be invariant under rigid rotation. An important point is that there do exist scalar invariants (which, by definition, will not change under a basis rotation) that *will* change upon a change in frame (i.e., a superimposed rotation). Likewise, a vector or tensor never changes under a basis rotation (the components change, but the sum of components times basis vectors or basis dyads remains invariant); whether or not a vector or tensor will change under a superimposed rotation depends on its physical definition. The discussion of PMFI includes a review of *objective rates* of spatial tensors, which are often employed in materials modeling (though we argue against this approach in favor of genuine rates within un-distorted or un-rotated material reference frames).
- Chapter 20 explains the difference between *tensor symmetry* and *material symmetry*, and then proceeds to prove that isotropic tensors are not always limited to mere multiples of the identity. Transverse symmetry (for which a tensor's components are unchanged upon rotation about a fixed axis) is covered in detail, including a rather brute force (hence accessible to a broader audience) explanation of why double-symmetric fourth-order transverse

[5] Spoiler alert: it is proved to be the isotropic part of that tensor.

[6] Where 'class' refers to a tensor order (also called 'rank') and the dimension of the underlying physical space.

stiffness tensors have five independent components, but only four distinct eigenvalues. The corresponding eigenprojectors are derived and a closed-form solution is derived for necessary and sufficient conditions for two such tensors to have the same eigenspaces (useful, for example, to force elastic and plastic stiffness tensors to have the same *spectral* symmetries). Application to *jointed rock* is also discussed.

- Chapter 21 defines *scalar invariants* of tensors, which are scalar-valued properties of tensors whose values do not change upon a change of basis. Not all real numbers are scalars.
- Chapter 22 returns to the topic of frame indifference as it applies to incremental constitutive models, first explaining the effect of rotation rate using an easily understood spring model, then moving towards various co-rotational rates in the literature, finishing with the obligatory derivation of oscillatory stress using certain rates (and explaining why it probably is not the rate that is the core issue) along with some guidance about how to run frame indifference testing in computational mechanics.
- Chapter 23 specializes the general continuum laws of mechanics to rigid-body mechanics. By applying the full power of tensor analysis, this presentation distinguishes itself from the derivations normally found in elementary dynamics textbooks. For example, with direct-notation tensor analysis, Euler's equations and the parallel-axis theorem can be presented in a single line.
- Chapter 24 returns to the problem of deformable media, but considers the case that bulk rigid motion is the primary motion, with relatively small deformations with respect to an appropriately defined rigidly moving observer frame (e.g., turbine blade, human bodies in a braking automobile, etc). *Phantom body forces* are derived to allow solving the problem in a non-inertial material frame.
- Chapter 25 covers some basic equations for *3D computer graphics visualization*, in particular providing equations governing mapping from a virtual 3D body and providing recommendations for 3D visualization that would *not* require a person to wear 3D goggles.
- Chapter 26 defines conventional *Voigt components* for arranging the six independent components of symmetric tensors into a 6×1 array and *demonstrates that these are in fact components of a 6D vector whose associated basis is not normalized* (which is why manipulations using Voigt components always contain unruly factors of 2 that are actually the metrics associated with the non-normalized Voigt basis). Normalization of the Voigt basis produces the Mandel basis with associated *Mandel components* that are *not* plagued by factors of 2 and which therefore obey standard linear algebra, permitting more efficient use of computer library functions such as eigenvalue solvers for fourth-order tensors. Using Mandel components, the 6×6 component matrix associated with a tensor rotation operation in physical space is derived.

- Chapter 27 completes the main text by providing a brief discussion of *rotations in higher dimensions*, which is applicable to constitutive material modeling. Ordinary physical tensor rotation operations are shown to correspond to *double-plane* rotations in stress space, where part of the tensor rotates by an amount equal to the rotation angle, another part rotates by twice this angle (as in Mohr diagrams), and a third part of the tensor does not rotate at all in tensor space.
- Appendix A provides *FORTRAN* and Python source listings that perform most of the computations presented in this book. Additional source code in these and other programming languages are available.

References

[1] Moraweic A 2004 *Orientations and Rotations: Computations in Crystallographic Textures* (Berlin: Springer)
[2] Bigoni D 2012 *Nonlinear Solid Mechanics* (New York: Cambridge University Press)
[3] Gurtin M E 1981 *An Introduction to Continuum Mechanics* (New York: Academic)

IOP Publishing

Chapter 2

Notation and tensor-analysis essentials

The reader is assumed to have a rudimentary understanding of basic vector and tensor analysis as well as multivariate calculus, along with some familiarity of numerical methods in mechanics[1]. This chapter merely outlines the notation and some essential identities.

Throughout this book, scalars are denoted in plain italics (s, r, t). Vectors are typeset with a single under-tilde ($\underset{\sim}{v}$, $\underset{\sim}{w}$, $\underset{\sim}{x}$). Second-order tensors are shown with two under-tildes ($\underset{\approx}{\sigma}$, $\underset{\approx}{S}$, $\underset{\approx}{T}$). Likewise, the order of higher-order tensors is indicated by the number of under-tildes. Throughout this book, subscripts (called indices) are permitted to vary from 1 to 3. Furthermore, any subscript that appears exactly twice in a term is implicitly understood to be summed from 1 to 3. All vectors and tensors are presumed to have real components unless otherwise specified. Accordingly, some of the identities in this section might require adjustments if they are to be generalized to complex numbers (e.g., a complex-conjugate operator would be needed on one variable for inner products, a transpose might need to be changed to a Hermitian, etc).

We will frequently refer to a triad of 'laboratory' vectors $\{\underset{\sim}{E}_1, \underset{\sim}{E}_2, \underset{\sim}{E}_3\}$, which form an orthonormal right-handed basis for three-dimensional space. A second triad $\{\underset{\sim}{e}_1, \underset{\sim}{e}_2, \underset{\sim}{e}_3\}$ will also denote an orthonormal basis that is not required to coincide with the lab basis. Any orthonormal triad of basis vectors $\{\underset{\sim}{e}_1, \underset{\sim}{e}_2, \underset{\sim}{e}_3\}$ will satisfy

$$\underset{\sim}{e}_i \bullet \underset{\sim}{e}_j = \delta_{ij}, \tag{2.1}$$

where δ_{ij} is the Kronecker delta, defined by

$$\delta_{ij} = \begin{cases} 1 & \text{if } i = j \\ 0 & \text{if } i \neq j \end{cases}. \tag{2.2}$$

[1] If not, an extensive tutorial may be found in a free downloadable resource [1]. Other good resources include [2–8] and [9, 10] for applications.

Any vector $\underset{\sim}{v}$ can always be expanded as the sum of components times basis vectors as

$$\underset{\sim}{v} = v_1\underset{\sim}{e}_1 + v_2\underset{\sim}{e}_2 + v_3\underset{\sim}{e}_3, \tag{2.3}$$

or, using the summation convention,

$$\underset{\sim}{v} = v_i\underset{\sim}{e}_i. \tag{2.4}$$

In matrix form, this expansion is equivalent to

$$\begin{bmatrix} v_1 \\ v_2 \\ v_3 \end{bmatrix} = v_1\begin{bmatrix} 1 \\ 0 \\ 0 \end{bmatrix} + v_2\begin{bmatrix} 0 \\ 1 \\ 0 \end{bmatrix} + v_3\begin{bmatrix} 0 \\ 0 \\ 1 \end{bmatrix}. \tag{2.5}$$

A single raised dot denotes the *vector inner product* defined by

$$\underset{\sim}{u} \bullet \underset{\sim}{v} = u_1v_1 + u_2v_2 + u_3v_3 = u_kv_k. \tag{2.6}$$

In publications that use matrix (rather than vector/tensor) notation, this operation might be written $\{u\}^T\{v\}$, where we are here using the notation $\{u\}$ to refer to the 3×1 matrix containing the Cartesian components of $\underset{\sim}{u}$, therefore making the transpose $\{u\}^T$ a 1×3 matrix so that $\{u\}^T\{v\}$ produces a 1×1 matrix. Of course, matrix notation is different from direct (symbolic/Gibbs) notation $\underset{\sim}{u} \bullet \underset{\sim}{v}$, which never puts transposes on vectors. Similarly, matrix notation never uses a raised dot, which is reserved for direct notation.

The magnitude of a vector is defined as

$$\|\underset{\sim}{v}\| \equiv \sqrt{\underset{\sim}{v} \bullet \underset{\sim}{v}}. \tag{2.7}$$

The *vector outer product*, also called a *dyadic product*, is herein denoted by two vectors written side-by-side without an operator between them (as in $\underset{\sim}{u}\underset{\sim}{v}$) and defined such that[2]

$$\boxed{(\underset{\sim}{u}\underset{\sim}{v})\bullet \underset{\sim}{w} = \underset{\sim}{u}(\underset{\sim}{v} \bullet \underset{\sim}{w})} \quad \forall\, \underset{\sim}{w}. \tag{2.8}$$

In matrix form, this would be written

$$[\{u\}\{v\}^T]\{w\} = \{u\}[\{v\}^T\{w\}] \quad \forall\, \{w\}. \tag{2.9}$$

In indicial form, this is written (again using implied summation for repeated indices) as

$$(u_iv_j)w_j = u_i(v_jw_j) \quad \forall\, w_j. \tag{2.10}$$

The dyad $\underset{\sim}{u}\underset{\sim}{v}$ is a rudimentary tensor having a 3×3 Cartesian component matrix whose ij component is given by u_iv_j:

[2] Boxes are used to highlight relatively important equations for quick reference.

$$[\underset{\sim}{u}\underset{\sim}{v}] = \begin{bmatrix} u_1v_1 & u_1v_2 & u_1v_3 \\ u_2v_1 & u_2v_2 & u_2v_3 \\ u_3v_1 & u_3v_2 & u_3v_3 \end{bmatrix}. \tag{2.11}$$

Unlike the matrix form $\{u\}^T\{v\}$ for the inner product, the matrix form for the outer product is $\{u\}\{v\}^T$, which has the transpose in a different location, thus producing a full 3×3 matrix. From looking at the matrix form of a dyad, it should be clear that $\underset{\sim}{u}\underset{\sim}{v} \neq \underset{\sim}{v}\underset{\sim}{u}$. Instead,

$$\underset{\sim}{u}\underset{\sim}{v} = (\underset{\sim}{v}\underset{\sim}{u})^T, \text{ which is proved indicially by noting that } u_iv_j = v_ju_i. \tag{2.12}$$

Instead of the notation $\underset{\sim}{u}\underset{\sim}{v}$, many authors typeset a dyad as $\underset{\sim}{u} \otimes \underset{\sim}{v}$. Omitting the dyadic symbol notationally establishes an associative property so that we may drop parentheses and write $\underset{\sim}{u}\underset{\sim}{v} \bullet \underset{\sim}{w}$ without ambiguity[3]. Except in the rare cases when it helps to avoid confusion, the space-hogging *circle-cross symbol* for dyadic multiplication is unnecessary and interferes with self-defining notation.

The single raised dot continues to denote the vector inner product even when acting between higher-order tensors. Therefore, instead of adopting the traditional $\boldsymbol{Ax}$ notation to denote a tensor acting on a vector, we use a *self-defining notation* for which the meaning of an expression can always be inferred by expanding each element in the expression in basis form. For example,

$$\underset{\approx}{A} \bullet \underset{\sim}{x} = (A_{ik}\underset{\sim}{e}_i\underset{\sim}{e}_k) \bullet (x_j\underset{\sim}{e}_j) = A_{ik}\underset{\sim}{e}_i(\underset{\sim}{e}_k \bullet \underset{\sim}{e}_j)x_j = A_{ij}x_j\underset{\sim}{e}_i. \tag{2.13}$$

Here, each $\underset{\sim}{e}_i\underset{\sim}{e}_k$ is a basis dyad, representing a tensor whose 3×3 Cartesian component matrix is zeros everywhere except for a 1 in the *ik* component, which is similar to the meaning of the basis vector $\underset{\sim}{e}_k$ whose 3×1 Cartesian component array is zeros everywhere except 1 in the *k*th component. Above, the last step used the orthonormal basis identity $\underset{\sim}{e}_k \bullet \underset{\sim}{e}_j = \delta_{kj}$, in which δ_{kj} is the Kronecker delta (equal to 1 if $k = j$ and 0 otherwise). The presence of $\underset{\sim}{e}_i$ in the final expression implies that $\underset{\approx}{A} \bullet \underset{\sim}{x}$ must denote a vector whose *i*th component is given by $A_{ij}x_j$. In matrix form,

$$\text{if} \quad \underset{\sim}{v} = \underset{\approx}{A} \bullet \underset{\sim}{x}, \quad \text{then} \quad \begin{bmatrix} v_1 \\ v_2 \\ v_3 \end{bmatrix} = \begin{bmatrix} A_{11} & A_{12} & A_{13} \\ A_{21} & A_{22} & A_{23} \\ A_{31} & A_{32} & A_{33} \end{bmatrix} \begin{bmatrix} x_1 \\ x_2 \\ x_3 \end{bmatrix}. \tag{2.14}$$

When a vector dots into a tensor $\underset{\approx}{B}$ from the *left*, the raised dot continues to denote contraction of adjacent indices. Specifically,

[3] Some people argue in favor of the circle-cross $\otimes$ for dyadic multiplication in order to emphasize that it is somehow 'different' from regular scalar multiplication, but logical consistency would then also require a different symbol for vector addition, as in $\underset{\sim}{u} \oplus \underset{\sim}{v}$, to distinguish it from scalar addition. As long as self-defining notation rules are adopted for this or other operations such as the dot product, our choice generalizes seamlessly to a broader range of higher-order expressions.

$$\underset{\sim}{x} \bullet \underset{\approx}{B} = x_i B_{ij} \underset{\sim}{e}_j, \tag{2.15}$$

which also implies the often-used identity,

$$\underset{\sim}{x} \bullet \underset{\approx}{B} = \underset{\approx}{B}^T \bullet \underset{\sim}{x}, \tag{2.16}$$

in which T denotes the transpose[4].

Consistent with self-defining notation, the expression $\underset{\approx}{A}\underset{\sim}{x}$, which is a tensor and vector written side-by-side without a dot between them, would have to be interpreted as a dyadic multiplication producing a third-order tensor with ijk components $A_{ij}x_j$. For clarity, writing $\underset{\approx}{A} \otimes \underset{\sim}{x}$ might be advisable if there is risk that $\underset{\approx}{A}\underset{\sim}{x}$ might be confused with what we denote by $\underset{\approx}{A} \bullet \underset{\sim}{x}$.

Any second-order tensor $\underset{\approx}{T}$ may be expanded in terms of basis dyads as

$$\underset{\approx}{T} = T_{ij} \underset{\sim}{e}_i \underset{\sim}{e}_j. \tag{2.17}$$

Applying the definition of a dyad, a *basis* dyad $\underset{\sim}{e}_i \underset{\sim}{e}_j$ represents a tensor whose component matrix has zeros everywhere except for a 1 in the ij position. Thus, in matrix form, equation (2.17) is equivalent to

$$\begin{bmatrix} T_{11} & T_{12} & T_{13} \\ T_{21} & T_{22} & T_{23} \\ T_{31} & T_{32} & T_{33} \end{bmatrix} = T_{11}\begin{bmatrix} 1 & 0 & 0 \\ 0 & 0 & 0 \\ 0 & 0 & 0 \end{bmatrix} + T_{12}\begin{bmatrix} 0 & 1 & 0 \\ 0 & 0 & 0 \\ 0 & 0 & 0 \end{bmatrix} + \ldots + T_{33}\begin{bmatrix} 0 & 0 & 0 \\ 0 & 0 & 0 \\ 0 & 0 & 1 \end{bmatrix}. \tag{2.18}$$

Comparing this with equation (2.5), it should be clear that basis dyads are the tensor generalization of basis vectors. Just as you can introduce a convenient change of basis for vectors, you can likewise change the tensor basis dyads even if you choose not to change the vector basis vectors. See, for example, the discussion of Voigt notation in chapter 26. The identity tensor $\underset{\approx}{I}$ is defined such that $\underset{\approx}{I} \bullet \underset{\sim}{x} = \underset{\sim}{x}$ for all $\underset{\sim}{x}$. Consequently,

$$\underset{\approx}{I} = \delta_{ij} \underset{\sim}{e}_i \underset{\sim}{e}_j \quad \text{with Cartesian components } [\underset{\approx}{I}] = \begin{bmatrix} 1 & 0 & 0 \\ 0 & 1 & 0 \\ 0 & 0 & 1 \end{bmatrix}. \tag{2.19}$$

Applying equation (2.18) to the identity tensor shows that

$$\underset{\approx}{I} = \underset{\sim}{e}_1 \underset{\sim}{e}_1 + \underset{\sim}{e}_2 \underset{\sim}{e}_2 + \underset{\sim}{e}_3 \underset{\sim}{e}_3 \tag{2.20}$$

or, using implied summation,

$$\boxed{\underset{\approx}{I} = \underset{\sim}{e}_k \underset{\sim}{e}_k}. \tag{2.21}$$

[4] I.e., $(\underset{\approx}{B}^T)_{ij} = B_{ji}$. Rather than the cumbersome notation $(\underset{\approx}{B}^T)_{ij}$ we will usually just write B_{ij}^T. In general, indices take the lower precedence than tensor operators. For example, B_{ij}^{-1} should be interpreted to mean $(\underset{\approx}{B}^{-1})_{ij}$, not $1/B_{ij}$. Here, $\underset{\approx}{B}^{-1}$ is the inverse of some tensor $\underset{\approx}{B}$.

The identity tensor is isotropic, which means that its basis-dyad expansion is of this form regardless of which basis is used (as long as it is orthonormal). Thus, in terms of a second orthonormal basis, denoted with captialized E, the formula is the same:

$$\boxed{\underset{\approx}{\boldsymbol{I}} = \underset{\sim}{\boldsymbol{E}}_k \underset{\sim}{\boldsymbol{E}}_k}. \tag{2.22}$$

Composition of two tensors is another example illustrating how the single raised dot continues to denote contraction of adjacent indices:

$$\underset{\approx}{\boldsymbol{A}} \bullet \underset{\approx}{\boldsymbol{B}} = A_{ik} B_{kj} \;\; \underset{\sim}{\boldsymbol{e}}_i \underset{\sim}{\boldsymbol{e}}_j. \tag{2.23}$$

This result involves two basis vectors $\underset{\sim}{\boldsymbol{e}}_i \underset{\sim}{\boldsymbol{e}}_j$ written side-by-side, indicating that $\underset{\approx}{\boldsymbol{A}} \bullet \underset{\approx}{\boldsymbol{B}}$ is a second-order tensor whose components are found via the matrix multiplication $[A][B]$. Many authors might denote this same operation by $\boldsymbol{AB}$, but our self-defining notation not only uses the tildes to indicate tensorial order, it also includes the single-dot contraction so that (2.23) could be derived directly by expanding each tensor in basis form, similar to what was done in (2.13). Consistent with the principles of self-defining notation, when we write $\underset{\approx}{\boldsymbol{A}}\underset{\approx}{\boldsymbol{B}}$ without a dot, it must represent a fourth-order tensor $A_{ij} B_{pq} \underset{\sim}{\boldsymbol{e}}_i \underset{\sim}{\boldsymbol{e}}_j \underset{\sim}{\boldsymbol{e}}_p \underset{\sim}{\boldsymbol{e}}_q$. This might also be written using an unnecessary and space-hogging (but clarity enhancing) dyadic multiplication symbol as $\underset{\approx}{\boldsymbol{A}} \otimes \underset{\approx}{\boldsymbol{B}}$.

The *isotropic* (also called 'spherical') part of a second-order tensor is denoted with a word-style operator 'iso' or perhaps a superscript 's', while the *deviatoric* part of a tensor is denoted using one of three notations: a 'prime,' a superscript d, or a word operator 'dev'. Specifically,

$$\underset{\approx}{\boldsymbol{A}} = \text{iso}\underset{\approx}{\boldsymbol{A}} + \text{dev}\underset{\approx}{\boldsymbol{A}}, \quad \text{where} \tag{2.24a}$$

$$\text{iso}\underset{\approx}{\boldsymbol{A}} = \underset{\approx}{\boldsymbol{A}}^s = \tfrac{1}{3}(\text{tr}\underset{\approx}{\boldsymbol{A}})\underset{\approx}{\boldsymbol{I}}, \tag{2.24b}$$

$$\text{dev}\underset{\approx}{\boldsymbol{A}} = \underset{\approx}{\boldsymbol{A}}' = \underset{\approx}{\boldsymbol{A}}^d = \underset{\approx}{\boldsymbol{A}} - \tfrac{1}{3}(\text{tr}\underset{\approx}{\boldsymbol{A}})\underset{\approx}{\boldsymbol{I}}. \tag{2.24c}$$

Here, $\underset{\approx}{\boldsymbol{I}}$ is the identity tensor, and 'tr' denotes the trace. Specifically,

$$\text{tr}\underset{\approx}{\boldsymbol{A}} \equiv A_{11} + A_{22} + A_{33} = A_{kk}. \tag{2.25}$$

Two other word-style operators, 'sym' and 'skw', produce the symmetric and skew-symmetric (also called anti-symmetric) parts of a tensor:

$$\underset{\approx}{\boldsymbol{A}} = \text{sym}\underset{\approx}{\boldsymbol{A}} + \text{skw}\underset{\approx}{\boldsymbol{A}}, \quad \text{where} \tag{2.26a}$$

$$\text{sym}\underset{\approx}{\boldsymbol{A}} = \tfrac{1}{2}(\underset{\approx}{\boldsymbol{A}} + \underset{\approx}{\boldsymbol{A}}^T), \tag{2.26b}$$

$$\text{skw}\underset{\approx}{\boldsymbol{A}} = \tfrac{1}{2}(\underset{\approx}{\boldsymbol{A}} - \underset{\approx}{\boldsymbol{A}}^T). \tag{2.26c}$$

A tensor $\underset{\approx}{\boldsymbol{D}}$ is *symmetric* if it equals its own symmetric part so that it equals its own transpose: $D_{ij} = D_{ji}$. A tensor $\underset{\approx}{\boldsymbol{W}}$ is *skew* (or skew-symmetric or anti-symmetric) if it equals its own skew part and hence equals the *negative* of its own transpose: $W_{ij} = -W_{ji}$.

The second-order tensor *inner product* is denoted by ':' and is defined by[5]

$$\underset{\approx}{\mathbf{A}}:\underset{\approx}{\mathbf{B}} = A_{ij}B_{ij}. \tag{2.27}$$

Similar to the vector inner product, the tensor inner product is a sum of component-by-component multiplications:

$$\begin{aligned}\underset{\approx}{\mathbf{A}}:\underset{\approx}{\mathbf{B}} = A_{11}B_{11} + A_{12}B_{12} + A_{13}B_{13}\\ + A_{21}B_{21} + A_{22}B_{22} + A_{23}B_{23}\cdot\\ + A_{31}B_{31} + A_{32}B_{32} + A_{33}B_{33}\end{aligned} \tag{2.28}$$

The double-dot operator may be written in terms of the trace by putting a transpose on either of the arguments as follows[6]:

$$\underset{\approx}{\mathbf{A}}:\underset{\approx}{\mathbf{B}} = \text{tr}(\underset{\approx}{\mathbf{A}} \bullet \underset{\approx}{\mathbf{B}}^T) = \text{tr}(\underset{\approx}{\mathbf{A}}^T \bullet \underset{\approx}{\mathbf{B}}). \tag{2.29}$$

By using the definition of the tensor inner product, it follows that

$$\underset{\approx}{\mathbf{A}}:\underset{\approx}{\mathbf{B}} = (\text{iso}\underset{\approx}{\mathbf{A}}):(\text{iso}\underset{\approx}{\mathbf{B}}) + (\text{dev}\underset{\approx}{\mathbf{A}}):(\text{dev}\underset{\approx}{\mathbf{B}}), \tag{2.30a}$$

$$\underset{\approx}{\mathbf{A}}:\underset{\approx}{\mathbf{B}} = (\text{sym}\underset{\approx}{\mathbf{A}}):(\text{sym}\underset{\approx}{\mathbf{B}}) + (\text{skw}\underset{\approx}{\mathbf{A}}):(\text{skw}\underset{\approx}{\mathbf{B}}). \tag{2.30b}$$

The double-dot is commutative (order-independent):

$$\underset{\approx}{\mathbf{A}}:\underset{\approx}{\mathbf{B}} = \underset{\approx}{\mathbf{B}}:\underset{\approx}{\mathbf{A}}. \tag{2.31}$$

Just as $\sqrt{\underset{\sim}{\mathbf{v}} \bullet \underset{\sim}{\mathbf{v}}}$ gives the magnitude of a vector, the magnitude of a second-order tensor is

$$\left\|\underset{\approx}{\mathbf{A}}\right\| \equiv \sqrt{\underset{\approx}{\mathbf{A}}:\underset{\approx}{\mathbf{A}}}. \tag{2.32}$$

Just as the single-dot vector inner product can act between tensors of any order, the double-dot tensor inner product is allowed to operate between any two tensors of *at least* second order. For example, if $\underset{\approx\approx}{\mathbf{C}}$ is a fourth-order tensor, then

$$\underset{\approx\approx}{\mathbf{C}}:\underset{\approx}{\mathbf{A}} = C_{ijkl}A_{kl}\underset{\sim}{\mathbf{e}}_i\underset{\sim}{\mathbf{e}}_j. \tag{2.33}$$

This formula can be inferred from self-defining notation by expanding both tensors in basis notation and using the fact that the double-dot acting between basis dyads is

$$\underset{\sim}{\mathbf{e}}_i\underset{\sim}{\mathbf{e}}_j:\underset{\sim}{\mathbf{e}}_p\underset{\sim}{\mathbf{e}}_q = (\underset{\sim}{\mathbf{e}}_i \bullet \underset{\sim}{\mathbf{e}}_p)(\underset{\sim}{\mathbf{e}}_j \bullet \underset{\sim}{\mathbf{e}}_q) = \delta_{ip}\delta_{jq}. \tag{2.34}$$

These are, incidentally, the *ijpq* components of the fourth-order identity tensor.

For rotation analyses, we will make extensive use of a triad of *embedded* orthonormal basis vectors $\{\underset{\sim}{\mathbf{e}}_1, \underset{\sim}{\mathbf{e}}_2, \underset{\sim}{\mathbf{e}}_3\}$ that is oriented differently than the laboratory reference basis $\{\underset{\sim}{\mathbf{E}}_1, \underset{\sim}{\mathbf{E}}_2, \underset{\sim}{\mathbf{E}}_3\}$. We will later define a 'rotation' or orientation tensor $\underset{\approx}{\mathbf{R}}$ by the sum of three dyads $\underset{\sim}{\mathbf{e}}_1\underset{\sim}{\mathbf{E}}_1 + \underset{\sim}{\mathbf{e}}_2\underset{\sim}{\mathbf{E}}_2 + \underset{\sim}{\mathbf{e}}_3\underset{\sim}{\mathbf{E}}_3$, or

---

[5] It follows that the double dot may be expressed as $\underset{\approx}{\mathbf{A}}:\underset{\approx}{\mathbf{B}} = \text{tr}(\underset{\approx}{\mathbf{A}}^T \bullet \underset{\approx}{\mathbf{B}}) = \text{tr}(\underset{\approx}{\mathbf{A}} \bullet \underset{\approx}{\mathbf{B}}^T)$. A similar-looking but fundamentally different operation $A_{ij}B_{ji} = \text{tr}(\underset{\approx}{\mathbf{A}} \bullet \underset{\approx}{\mathbf{B}})$ must *never* be referred to as an 'inner product' even though it produces a scalar result (it fails the positivity axiom for inner products).

[6] The similar-looking scalar-valued operation $\text{tr}(\underset{\approx}{\mathbf{A}} \bullet \underset{\approx}{\mathbf{B}})$ fails to satisfy the positivity axiom required for any operation to be properly referred to as an inner product. The transpose is essential!

$$\underset{\sim}{\boldsymbol{R}} = \underset{\sim}{\boldsymbol{e}}_k \underset{\sim}{\boldsymbol{E}}_k. \tag{2.35}$$

This expression involves 'mixed' dyads, not the standard 'basis dyads' associated with a single basis. If we wish to instead express the tensor $\underset{\sim}{\boldsymbol{R}}$ in the same *non-mixed* form as (2.17), we would need to find components R_{ij} such that

$$\underset{\sim}{\boldsymbol{R}} = R_{ij} \underset{\sim}{\boldsymbol{E}}_i \underset{\sim}{\boldsymbol{E}}_j. \tag{2.36}$$

Comparing the above two equations shows that

$$\underset{\sim}{\boldsymbol{e}}_j = R_{ij} \underset{\sim}{\boldsymbol{E}}_i. \tag{2.37}$$

Dotting both sides of (2.35) by $\underset{\sim}{\boldsymbol{E}}_i$ shows that

$$\underset{\sim}{\boldsymbol{e}}_i = \underset{\sim}{\boldsymbol{R}} \bullet \underset{\sim}{\boldsymbol{E}}_i. \tag{2.38}$$

Of course, this says $\underset{\sim}{\boldsymbol{R}} = \underset{\sim}{\boldsymbol{I}}$ if the $\underset{\sim}{\boldsymbol{e}}$-basis happens to coincide with the $\underset{\sim}{\boldsymbol{E}}$-basis. Equation (2.37) can be alternatively written as

$$\underset{\sim}{\boldsymbol{e}}_i = R_{ik}^T \underset{\sim}{\boldsymbol{E}}_k. \tag{2.39}$$

It might appear strange that a transpose applies here to R_{ik}^T, where there is no transpose in (2.38). These two equations are expressing the same result in two different ways. Equation (2.38) is expressed in direct notation—there is no implied summation, but there are free indices because (2.38) actually represents *three* separate equations (one for each value of the free index). Each of the three equations is of the form $\underset{\sim}{\boldsymbol{v}} = \underset{\sim}{\boldsymbol{A}} \bullet \underset{\sim}{\boldsymbol{x}}$. Just as the indicial form of $\underset{\sim}{\boldsymbol{v}} = \underset{\sim}{\boldsymbol{A}} \bullet \underset{\sim}{\boldsymbol{x}}$ is $v_k = A_{kn}x_n$, the indicial form of (2.38) *with respect to the lab basis* is $(\underset{\sim}{\boldsymbol{e}}_i)_k = R_{kn}(\underset{\sim}{\boldsymbol{E}}_i)_n$. With respect to this basis, $(\underset{\sim}{\boldsymbol{E}}_i)_n = \delta_{in}$. Therefore, $(\underset{\sim}{\boldsymbol{e}}_i)_k = R_{ki}$. In other words, the kth component of $\underset{\sim}{\boldsymbol{e}}_i$ with respect to the lab basis must equal R_{ki}, or R_{ik}^T. Equation (2.39) is expressed in basis expansion form, similar to $\underset{\sim}{\boldsymbol{v}} = v_k \underset{\sim}{\boldsymbol{E}}_k$. In this form, the kth lab component of $\underset{\sim}{\boldsymbol{v}}$ is the coefficient of $\underset{\sim}{\boldsymbol{E}}_k$. Applying this statement to (2.39) shows that the kth component of $\underset{\sim}{\boldsymbol{e}}_i$ with respect to the lab basis must equal R_{ik}^T. Hence, both (2.38) and (2.39) lead to the same result.

As a particular example, consider the special case that $i = 1$ in (2.38). Then the *laboratory component* matrix for the left-hand side of (2.38) can be written

$$\begin{bmatrix} R_{11} & R_{12} & R_{13} \\ R_{21} & R_{22} & R_{23} \\ R_{31} & R_{32} & R_{33} \end{bmatrix} \begin{bmatrix} 1 \\ 0 \\ 0 \end{bmatrix}, \tag{2.40}$$

or simply

$$\begin{bmatrix} R_{11} \\ R_{21} \\ R_{31} \end{bmatrix}. \tag{2.41}$$

This is just the first column of the $[R]$ matrix. According to (2.38) with $i = 1$, this result must be equal to the *laboratory* components of the $\underset{\sim}{\boldsymbol{e}}_1$ vector. Applying similar logic to the cases that $i = 2$ and $i = 3$, shows that …

> The ith column of the laboratory component matrix R_{ij} must contain the components of the ith embedded basis vector $\underset{\sim}{e}_i$ with respect to the lab $(\underset{\sim}{E})$ basis. Thus, $R_{ij} = \underset{\sim}{E}_i \bullet \underset{\sim}{e}_j$.

To obtain this same result using (2.39) for the case that $i = 1$, we would stay in basis notation, simply expanding out the implied sum as

$$\underset{\sim}{e}_1 = R^T_{11}\underset{\sim}{E}_1 + R^T_{12}\underset{\sim}{E}_2 + R^T_{13}\underset{\sim}{E}_3, \tag{2.42}$$

or

$$\underset{\sim}{e}_1 = R_{11}\underset{\sim}{E}_1 + R_{21}\underset{\sim}{E}_2 + R_{31}\underset{\sim}{E}_3. \tag{2.43}$$

Assembling the components of this vector into an array gives the same result as in (2.41). These and additional tensor/vector concepts will be reviewed as necessary within the main text.

2.1 Linear fractional transform

Some engineering disciplines use the notation of 4 × 4 matrix,

$$\left[\begin{array}{ccc|c} A_{11} & A_{12} & A_{13} & b_1 \\ A_{21} & A_{22} & A_{23} & b_2 \\ A_{31} & A_{32} & A_{33} & b_3 \\ \hline c_1 & c_2 & c_3 & d \end{array}\right]. \tag{2.44}$$

This notation can be regarded as a tabular presentation of four parameters parameters (a tensor $\underset{\approx}{A}$, two vectors, $\underset{\sim}{b}$ and $\underset{\sim}{c}$, as well as a scalar d) used to define a *linear fractional transform*:

$$f(\underset{\sim}{x}) = \frac{\underset{\approx}{A} \bullet \underset{\sim}{x} + \underset{\sim}{b}}{\underset{\sim}{c} \bullet \underset{\sim}{x} + d}. \tag{2.45}$$

For the special case that $\underset{\sim}{b} = \underset{\sim}{c} = \underset{\sim}{0}$ and $d = 1$, this reduces to this notation's most commonly used form

$$\left[\begin{array}{ccc|c} A_{11} & A_{12} & A_{13} & 0 \\ A_{21} & A_{22} & A_{23} & 0 \\ A_{31} & A_{32} & A_{33} & 0 \\ \hline 0 & 0 & 0 & 1 \end{array}\right], \text{ which merely represents the operation } \underset{\approx}{A} \bullet \underset{\sim}{x}. \tag{2.46}$$

2.2 Visualizing rotations

The component matrix of any tensor has the property that the kth column contains the components of the transformation of the kth-basis vector. For a rotation, the kth column contains components of $\boldsymbol{e}_k$ with respect to the laboratory (E)-basis. The rotation of any other vector $\boldsymbol{N}$ is $\boldsymbol{n} = \boldsymbol{R} \bullet \boldsymbol{N}$. Visualizations of rotations typically attempt to illustrate, directly or indirectly, the rotated positions of the six normals to the faces of a unit cube. The faces are numbered such that their outward normals are

$$\text{Face 1:} \quad \boldsymbol{E}_1 = \boldsymbol{E}_2 \times \boldsymbol{E}_3, \tag{2.47a}$$

$$\text{Face 2:} \quad \boldsymbol{E}_2 = \boldsymbol{E}_3 \times \boldsymbol{E}_1, \tag{2.47b}$$

$$\text{Face 3:} \quad \boldsymbol{E}_3 = \boldsymbol{E}_1 \times \boldsymbol{E}_2, \tag{2.47c}$$

$$\text{Face 4:} \quad -\boldsymbol{E}_1 = \boldsymbol{E}_3 \times \boldsymbol{E}_2 \ \text{(this is face 6 for the die convention)}, \tag{2.47d}$$

$$\text{Face 5:} \quad -\boldsymbol{E}_2 = \boldsymbol{E}_1 \times \boldsymbol{E}_3, \tag{2.47e}$$

$$\text{Face 6:} \quad -\boldsymbol{E}_3 = \boldsymbol{E}_2 \times \boldsymbol{E}_1 \ \text{(this is face 4 for the die convention)}. \tag{2.47f}$$

These face normals are here also expressed in terms of cross products. The vectors involved in the cross products specify local directions to be used by an observer viewing the face. For example, an observer looking at face 5 should be oriented such that $\boldsymbol{E}_1$ is pointing to the observer's right, while $\boldsymbol{E}_3$ points 'upward' relative to the observer (i.e., it points along a direction from the feet to the head). Most of the visualizations in this book show the rotated orientations of these faces using the alternative *'die' convention* for which the face numbering is changed as indicated so that opposite sides of a (fair) die always sum to 7.

Any unit vector, such as normals to the cube faces, may be regarded as the position vector extending from the origin to a point on the unit sphere. The Lambert stereographic projection maps the sphere to a disk such that the 'north pole' is in the

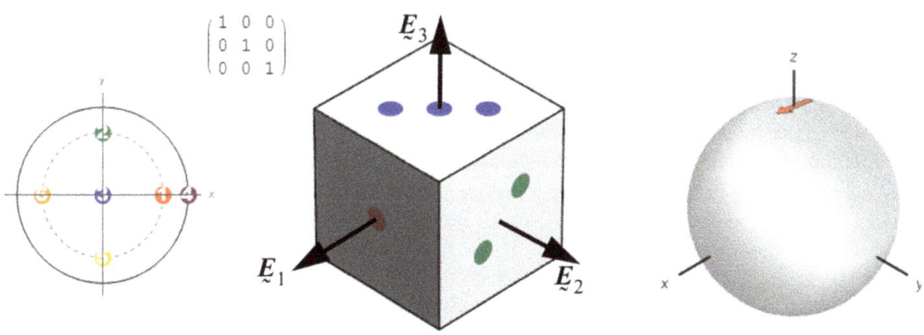

Figure 2.1. Visualization of the identity rotation (i.e., the orientation corresponding to no rotation). Note that, when the rotation matrix is the identity, the first, second, and third faces of the cube are aligned with the (x, y, z)-coordinate axes. The left-hand image is a full pole plot, which is a Lambert projection of the cube normals from the sphere to the plane (see section 17.4.4). The right-hand plot shows a 'nautical' visualization of the rotation in which the arrow represents a sailboat on the surface of a water-covered world in which the 'mast' points in the same direction as the three dots on the die and the heading (direction of the arrow) points the same as the single dot on the die.

Figure 2.2. Visualization of a rotation for which the faces remain aligned with the coordinate axes, yet the rotation is not the identity. Here, the cube face that originally pointed in the 3-direction (i.e., the face with three dots) is now pointing in the 1-direction (i.e., aligned with the x-axis). The face originally pointing in the 2-direction is now pointing in the negative 3-direction (i.e., in the negative z-direction), which we know because the face opposite face 5 must be face 2 (as the number of dots on opposite sides of a die must sum to 7). On the nautical plot, the sailboat's 'mast' is again aligned with the third face of the cube, and the arrow in the nautical plot again points in the direction of the first face of the cube (which, because opposite sides of a die always sum to 7, is opposite the side with six dots).

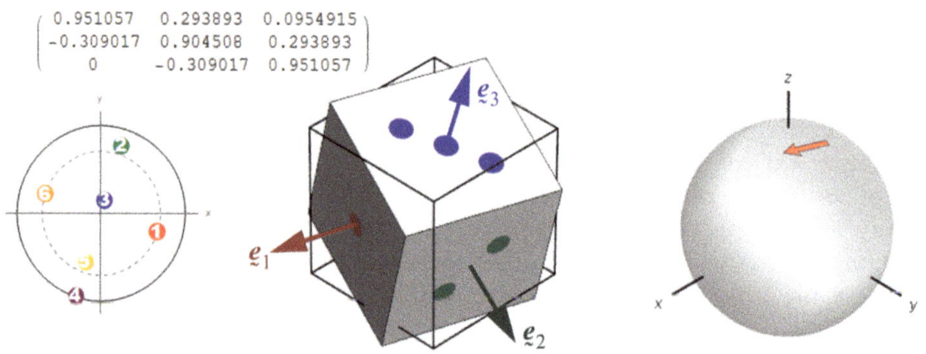

Figure 2.3. Visualization of a more general rotation. As always, the side with three dots is aligned with the 'mast' of the sailboat in the nautical plot (and the side with a single dot on the cube is again aligned with the boat's heading). The left-side plot shows that this rotation entails moving the body's 1-axis on a path around the 'equator' so that it remains perpendicular to the z-axis (as does the 6-direction because it is always opposite the 1-direction); meanwhile, the other directions on the cube are rotated about the body's 1-axis, so they show as a tilted line in the left plot. The 2-direction in that plot is outside the dashed equator line, implying that the $\underset{\sim}{e}_2$ rotated face vector points downward (towards the southern hemisphere). Similarly, the $\underset{\sim}{e}_3$ face normal is still fairly close to the north pole, so its dot in the left-side Lambert plot is near the center of the disk.

middle of the disk, the equator is the dashed line in the left side of figure 2.1, the 'southern hemisphere' is mapped to the zone outside the dashed line, and the south pole is mapped to the entire outer circle. The six unit normals to the die faces in figure 2.1 are therefore visualized as six dots in the Lambert projection. Equations defining the Lambert projections on the left sides of figures 2.1–2.3 are derived in detail in section 17.4.4, but the qualitative meaning of a Lambert projection can be discussed now. Like a stereographic projection, the Lambert projection maps points on a sphere to a disk in the plane so that all locations on the sphere, front or back, can be visualized simultaneously by 'unwrapping' or 'peeling back' the spherical surface to become 'flattened out' on the plane as if it were a deformable membrane. This plot treats the six unit normals on the cube (labeled 1 through 6) as points on a unit sphere so that their orientations may be visualized simultaneously as the labeled points in the Lambert projection. The Lambert projection is like a distorted view of the sphere looking down the z-direction. For example, face 2 points in the y-direction in figure 2.1, so the circled 2 is located where the dashed line of the Lambert projection crosses the y-axis in that plot. Points on the outer ring of the Lambert projection all correspond to negative z-direction. Thus, for example, the side opposite face 5 in figure 2.2 (which must be face 2 since opposite sides always sum to 7) is shown as a point anywhere on the south-pole outer ring in the Lambert projection.

References

[1] Brannon R M 2003 https://csmbrannon.net/2011/03/03/functional-and-structured-tensor-analysis-for-engineers/

[2] Bowen R M and Wang C C 1976 *Introduction to Vectors and Tensors* vol 1 (New York: Plenum)

[3] Elden L and Wittmeyer-Koch L 1990 *Numerical Analysis: An Introduction* (New York: Academic)

[4] Folland G B 1999 *Real Analysis: Modern Techniques and Their Applications* 2nd edn (New York: Wiley-Interscience))

[5] Friedberg S H, Insel A J and Spence L E 1996 *Linear Algebra* 3rd edn (Englewood Cliffs, NJ: Prentice-Hall)

[6] Golub G H and VanLoan C F 1996 *Matrix Computations* 3rd edn (Baltimore, MD: John Hopkins University Press)

[7] Kreyszig E 1989 *Introductory Functional Analysis with Applications* (New York: Wiley)

[8] Rudin Walter 1976 *Principles of Mathematical Analysis* (New York: McGraw-Hill)

[9] Bigoni D 2012 *Nonlinear Solid Mechanics* (New York: Cambridge University Press)

[10] Chadwick P 1999 *Continuum Mechanics: Concise Theory and Problems* (New York: Dover)

IOP Publishing

Rotation, Reflection, and Frame Changes
Orthogonal tensors in computational engineering mechanics
R M Brannon

Chapter 3

Orthogonal basis and coordinate transformations

A rigid body is an idealized collection of points (continuous or discrete) for which the distance between any two points is fixed. Consequently the angle between any two material line segments is also fixed[1]. A rigid body is capable of only uniform translation and rotation. A rotation is simply a reorientation of a body without distortion (i.e., without bending, compressing, shearing, or stretching). Most of this book focuses on fundamental theorems and concepts involving rotation without translation[2]. The body that rotates might be the physical system of interest or the 'body' of the observer. In this chapter, we focus on fundamental theorems associated with observer reorientations (i.e., changing from one orthonormal basis to another).

In this chapter, we will let $\underset{\sim}{e}_k$ denote the basis used by the observer to express components of vectors and tensors. This observer's basis may or may not coincide with the fixed laboratory basis $\underset{\sim}{E}_k$. In this chapter, we will contrast reorienting the observer (holding the material system fixed) with reorienting the material (keeping the observer stationary). Both operations can be characterized by an orthogonal matrix. An observer rotating by a certain angle will see an apparent reorientation of the body in the *opposite* direction. This duality can make the study of rotations and basis reorientations quite confusing because, as we will see, they are quantified by orthogonal matrices that are transposes of each other. Consequently, if you accidentally use the basis reorientation perspective when you should be using a body reorientation, then all of your formulas will have the orthogonal matrices off

[1] Preservation of angles follows from preservation of length only when material inversion is prohibited. When material inversion in the plane is possible, angles will remain unchanged in magnitude, but might change numerical sign (e.g., change from counter-clockwise to clockwise relative to an axis pointing 'out of the page').

[2] To include translation, the embedded material triad is merely re-interpreted as a triad that translates with a particular (convenient) point in the body, but does not change orientation. A full discussion of rigid motion (including translation) is provided in chapter 23.

doi:10.1016/978-0-7503-1454-1ch3

by a transpose. As will be explained in detail later, if you are ever unsure which perspective you need, try considering a 90° rotation as a special case. If you do not get the result you seek, simply switch perspectives by using transposes where needed.

The orientation of any rigid body can be specified by giving the orientations of three unit vectors, $\{\underset{\sim}{m}_1, \underset{\sim}{m}_2, \underset{\sim}{m}_3\}$, that characterize orthonormal material directions. We use the symbol $\boldsymbol{m}$ to emphasize that these vectors are always attached to and move with the material. For example, to characterize the orientation of a cubic crystal in a material, the three material directions could be assigned to coincide with the edges of the crystal's unit cell. For a hexagonal crystal, the unit cell is not a cube, but you can always construct three *orthonormal* material reference vectors by applying Gram–Schmidt orthonormalization on the hexagonal unit cell's edge vectors. Similar statements hold for characterizing the orientation of any body of any shape.

3.1 Superimposed rotations

Rotating a body while keeping the observer stationary (i.e., changing the $\underset{\sim}{m}_k$ vectors while holding the $\underset{\sim}{e}_k$ observer orientation equal to the lab basis $\underset{\sim}{E}_k$) is represented mathematically by finding the components of the rotated embedded $\underset{\sim}{m}_k$-basis with respect to the fixed lab $\underset{\sim}{E}_k$-basis, and assembling them into *columns* of a matrix. This matrix contains components of an orthogonal rotation tensor $\underset{\approx}{Q}$ which has the property that

$$\underset{\sim}{m}_k = \underset{\approx}{Q} \bullet \underset{\sim}{E}_k. \tag{3.1}$$

Letting Q_{ik} denote the ik laboratory components of $[Q]$, this formula may be written equivalently as[3]

$$\underset{\sim}{m}_i = Q_{ki}\underset{\sim}{E}_k = Q_{ik}^T\underset{\sim}{E}_k. \tag{3.2}$$

In terms of dyads (see equation (2.11)), the body rotation tensor can be written as[4]

$$\underset{\approx}{Q} = \underset{\sim}{m}_k\underset{\sim}{E}_k. \tag{3.3}$$

The components of any tensor $\underset{\approx}{A}$ with respect to a basis $\{\underset{\sim}{E}_1, \underset{\sim}{E}_2, \underset{\sim}{E}_3\}$ may be found by $A_{ij} = \underset{\sim}{E}_i \bullet \underset{\approx}{A} \bullet \underset{\sim}{E}_j$. Dotting (3.3) from the left by $\underset{\sim}{E}_i$ and from the right by $\underset{\sim}{E}_j$ shows that the lab components of $\underset{\approx}{Q}$ are

$$Q_{ij} = \underset{\sim}{E}_i \bullet \underset{\sim}{m}_j. \quad \text{(lab components)} \tag{3.4}$$

Similarly, the ij components of $\underset{\approx}{Q}$ with respect to the *embedded basis* may be found by

[3] Note the distinction between (3.1) and (3.2). We have *not* accidentally gotten the subscripts reversed in (3.2). See the discussion surrounding (2.39) and (2.40).

[4] Simo and Hughes [1] incorrectly refer to this expansion as a spectral expansion. The fact that this is the sum of three dyads does not make it a spectral expansion. To be a spectral decomposition, the vectors in the dyads must be *eigenvectors*. Neither $\underset{\sim}{m}_k$ nor $\underset{\sim}{E}_k$ are, in general, eigenvectors of the rotation tensor (see chapter 11).

$$\underset{\sim}{m}_i \bullet \underset{\approx}{Q} \bullet \underset{\sim}{m}_j = \underset{\sim}{m}_i \bullet (\underset{\sim}{m}_k \underset{\sim}{E}_k) \bullet \underset{\sim}{m}_j = \delta_{ik} \underset{\sim}{E}_k \bullet \underset{\sim}{m}_j = \underset{\sim}{E}_i \bullet \underset{\sim}{m}_j. \tag{3.5}$$

Comparing the right-hand sides of (3.4) and (3.5) shows that the components of $\underset{\approx}{Q}$ with respect to the embedded basis are identical to the components of $\underset{\approx}{Q}$ with respect to the laboratory basis. More broadly, the $\underset{\approx}{Q}$ tensor has proper *transverse* symmetry about its own rotation axis. Any change of basis about this axis will not affect the components of $\underset{\approx}{Q}$. In general, the components of $\underset{\approx}{Q}$ with respect to *other* bases will be different.

The fastest way to assemble the $[Q]$ matrix is to simply place the lab components of the $\underset{\sim}{m}_k$ vectors into columns of $[Q]$. We will denote this sort of column assembly by

$$[Q] = [\{\underset{\sim}{m}_1\}\{\underset{\sim}{m}_2\}\{\underset{\sim}{m}_3\}]. \tag{3.6}$$

For example, under a 30° rotation of the material about the 3-direction, we would have

$$\underset{\sim}{m}_1 = \cos 30° \underset{\sim}{E}_1 + \sin 30° \underset{\sim}{E}_2 \qquad \rightarrow \begin{Bmatrix} \cos 30° \\ \sin 30° \\ 0 \end{Bmatrix}, \tag{3.7a}$$

$$\underset{\sim}{m}_2 = -\sin 30° \underset{\sim}{E}_1 + \cos 30° \underset{\sim}{E}_2 \qquad \rightarrow \begin{Bmatrix} -\sin 30° \\ \cos 30° \\ 0 \end{Bmatrix}, \tag{3.7b}$$

$$\underset{\sim}{m}_3 = \underset{\sim}{E}_3 \qquad \rightarrow \begin{Bmatrix} 0 \\ 0 \\ 1 \end{Bmatrix}. \tag{3.7c}$$

Assembling these vectors into columns of $[Q]$ gives

$$[Q] = \begin{bmatrix} \cos 30° & -\sin 30° & 0 \\ \sin 30° & \cos 30° & 0 \\ 0 & 0 & 1 \end{bmatrix}. \tag{3.8}$$

The act of reorienting an entire system while holding the observer fixed is called a superimposed rotation. The way in which a vector transforms under a superimposed rotation depends critically on its physical definition. This transformation will look similar to an ordinary basis change except possibly when the superimposed motion varies with time. This distinction between superimposed rotation and a much simpler non-evolving basis change pertains to the related problems of non-inertial observers, associated 'phantom body forces' (needed to balance centrifugal and coriolis effects of rotating a deforming body), and material-frame indifference, all discussed in chapter 19. The situation is fundamentally different from fixed (non-evolving) basis rotations, which is the subject of this chapter. When viewed from a non-evolving but differently oriented basis (or, more precisely, when comparing what is seen by two observers that are misaligned but do not change their relative misalignment over time), no vectors (initial or embedded) will change at all, but their *components with respect to the new, rotated, basis will change*. A vector is a sum of components times basis vectors. When the basis changes, the components must change in such a manner that this sum remains unchanged.

3.2 Basis rotations

This section provides an explanation and some derivations for the basis-change formulas that are ultimately summarized in section 3.2.3.

In the remainder of this chapter, we consider the case that the material orientation vectors $\underset{\sim}{\boldsymbol{m}}_k$ (and all other system vectors or tensors) are held fixed while the *observer's* $\underset{\sim}{\boldsymbol{e}}_k$-triad is changed to a new, but still orthonormal orientation. We can again define a matrix $[Q]$ whose columns contain the components of the reoriented $\underset{\sim}{\boldsymbol{e}}_k$-basis with respect to the laboratory basis so that

$$\underset{\sim}{\boldsymbol{e}}_j = Q_{kj}\underset{\sim}{\boldsymbol{E}}_k. \tag{3.9}$$

Dotting both sides of this equation by $\underset{\sim}{\boldsymbol{E}}_i$ shows that

$$Q_{ij} = \underset{\sim}{\boldsymbol{E}}_i \bullet \underset{\sim}{\boldsymbol{e}}_j. \tag{3.10}$$

Similar to what we saw in the previous section, the columns of the $[Q]$ matrix contain the *laboratory* components of the rotated basis.

Since both the lab and reoriented bases are orthonormal, equation (3.10) shows that

$$Q_{ij} = \cos\theta_{ij}, \tag{3.11}$$

where θ_{ij} is the angle between $\underset{\sim}{\boldsymbol{E}}_i$ and $\underset{\sim}{\boldsymbol{e}}_j$. Consequently, the Q_{ij} values are often called the direction cosines.

Many authors define the direction-cosine matrix differently from our equation (3.10); you might see the direction cosines defined instead as $a_{ij} = \underset{\sim}{\boldsymbol{e}}_i \bullet \underset{\sim}{\boldsymbol{E}}_j$. Since the dot product is commutative, observe that $a_{ij} = Q_{ji}$. Consequently, both the $[a]$ and $[Q]$ matrices contain essentially the same information, but organized differently. If you prefer using $[a]$, you can replace all occurrences of $[Q]$ in the upcoming equations with $[a]^T$. We like the $[Q]$ definition because, regardless of whether you are considering basis rotation or body rotation, the fundamental orthogonal matrix is always assembled by arranging the lab components of the rotated vectors into columns of $[Q]$. If $[Q]$ appears in a body rotation equation, then $[Q]^T$ will appear in the related *basis* rotation perspective. We also like the $[Q]$ definition because, in the defining equation $Q_{ij} = \underset{\sim}{\boldsymbol{E}}_i \bullet \underset{\sim}{\boldsymbol{e}}_j$, the first vector is what you start with (the lab basis) and the second vector is what you finish with (the rotated basis).

The first index on Q_{ij} is the same as the index on the reference (lab) basis $\underset{\sim}{\boldsymbol{E}}_i$ and the second index on Q_{ij} is the same as the index on the reoriented (observer) basis $\underset{\sim}{\boldsymbol{e}}_j$. You will see that in all upcoming formulas that this relationship persists. Namely, the first index on Q_{ij} will always refer to the lab basis (or lab components) and the second index will always refer to the rotated basis or rotated components. If you

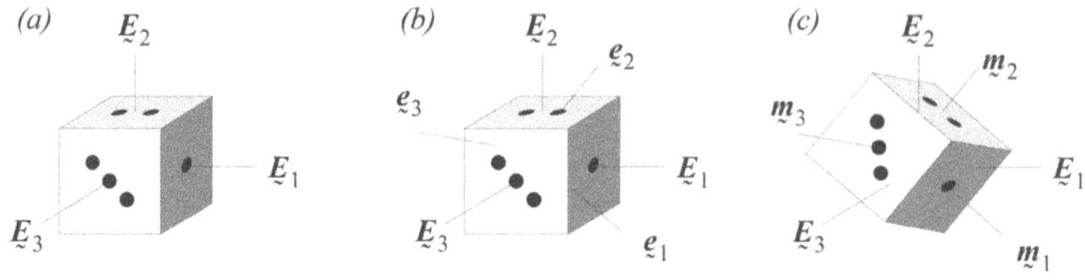

Figure 3.1. Two types of rotation transformations. (a) Initial or 'fiducial' configuration. (b) Rotation of the observer's basis keeping the body stationary. (c) Rotation of the body (moving the embedded basis along with it), keeping the observer locked to the fixed laboratory basis.

prefer to use the $[a]$ matrix instead of the $[Q]$ matrix, you will again find consistent index mapping patterns[5]. Thus, regardless of how you define your direction-cosine matrix, just remember the general structures of the equations soon to be derived. Once you have the structure written down, simply put indices onto the direction cosines in a manner consistent with how you defined it in the first place. This should become clearer as we progress.

In figure 3.1(a) we show a die that has been held stationary while the *observer* (who uses the $\{\underset{\sim}{e}_1, \underset{\sim}{e}_2, \underset{\sim}{e}_3\}$-basis) changes orientation. The coordinates of the single 'dot' (pip) on the die were originally $\{1, 0, 0\}$ with respect to the $\{\underset{\sim}{E}_1, \underset{\sim}{E}_2, \underset{\sim}{E}_3\}$ lab basis. The coordinates of *that same point* change to something unruly like $\{0.87, 0.44, -0.22\}$ with respect to the observer's new $\{\underset{\sim}{e}_1, \underset{\sim}{e}_2, \underset{\sim}{e}_3\}$-basis. In figure 3.1(b) the basis $\{\underset{\sim}{m}_1, \underset{\sim}{m}_2, \underset{\sim}{m}_3\}$ is instead *embedded*—it rotates with the body, and the observer remains fixed to the laboratory basis. In this case, the coordinates of the single 'dot' with respect to the embedded basis remain unchanged (namely, $\{1, 0, 0\}$), but the coordinates of that same point now change to new values with respect to the fixed laboratory basis. The first operation (basis rotation) changes the body's position coordinates with respect to the new $\{\underset{\sim}{e}_1, \underset{\sim}{e}_2, \underset{\sim}{e}_3\}$-basis, keeping the lab coordinates unchanged. The second operation (superimposed rotation) has the opposite effect—it changes the body's coordinates with respect to the lab basis, keeping the coordinates with respect to the $\{\underset{\sim}{m}_1, \underset{\sim}{m}_2, \underset{\sim}{m}_3\}$-triad unchanged. In figure 3.1(a), the observer changes orientation, keeping the body fixed. In figure 3.1(b), the body changes orientation by the same amount keeping the observer fixed. Intuitively, rotating a body by a certain amount should result in the same change in coordinates as rotating the *observer* about the same axis *but in the opposite direction*. Keeping track of which interpretation applies (i.e., deciding whether you are dealing with a superimposed rotation or a basis rotation) can be daunting and the mathematical difference is subtle (even when you *are* certain about the perspective, it is often not clear whether the $[Q]$ matrix or its transpose is needed because equations frequently

[5] With these operations, we comment that a nice alternative notation uses a labeling system so that, for example, q_{ij}^{AB} refers to the direction cosines between the *i*th basis vector in basis *A* and the *j*th basis vector in basis *B*. This labeling system is especially useful when dealing with more than just two bases (especially if they are not orthonormal).

pop up such as (2.39) and (2.40), which express *identical* results with one equation having a transpose and the other not having a transpose, depending on whether direct notation or basis notation is used). Consequently, we highly recommend double-checking your equations by verifying that the results turn out as expected for simpler rotations (such as 90° about a single axis) where the correct behavior is known intuitively.

As discussed in the caption of figure 3.2, no physical vector is itself changed under a basis rotation, but its *components* will generally change because the basis is changing. A vector is the sum of components times basis vectors. When the basis vectors change, the components must change in perfectly compensating manner so that the sum of components times basis vectors remains invariant. Under a material or *body* rotation, whether or not components of a vector will change depends on whether that vector is defined to move with the body. If a vector moves with the body, then (because it moved) that vector will change, but its components will not change with respect to an embedded basis that rotates with the body. If a vector (like gravity) does *not* move with the body, then the vector is unchanged while its components with respect to the rotated material basis *will* change—in fact, they will change in a manner that looks a lot like a basis rotation about the same axis but in the opposite angular direction. Under a basis transformation, all physical vectors and all physical tensors remain unchanged—forcing them to have new components. Under a superimposed body rotation, the physical definition of the vector or tensor is needed to determine if it (or its components) will change. This is the fundamental

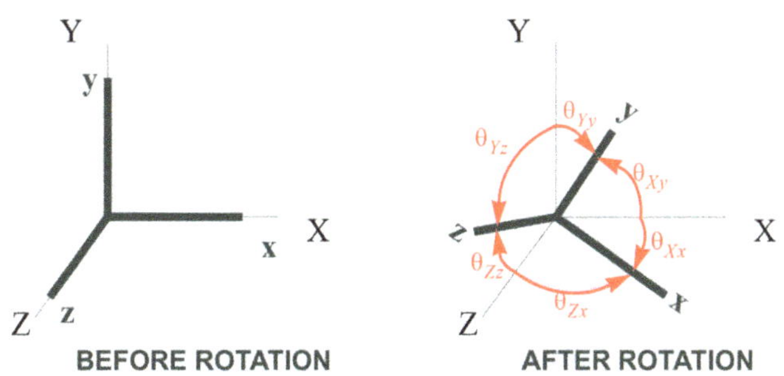

Figure 3.2. A rigid rotation. For superimposed rotation, the rotation may be described by the rotation of a triad that moves with the body, in which case (x, y, z) are body coordinates that remain unchanged if the body is rigid. Whether or not a given vector will have new components with respect to the body coordinates depends on whether that vector moves with the body. For example, the position vector from the origin to a point in the rotated body will change, but the *coordinates* of the position vector with respect to the rotated basis will not change. By contrast, if gravity points in the $-Y$-direction, then it will continue to point in that direction regardless of body rotation and therefore the gravity *vector* remains unchanged while the gravity *components* change with respect to the body basis. For basis rotation (holding the body fixed), (x, y, z) are a new set of coordinates for points in the *fixed* body. Thus, under basis rotation, the position vector stays fixed, while position *coordinates* change, and the gravity *vector* remains fixed while gravity *components* change. All transformations of vectors and tensors can be characterized by knowing the components of the rotated basis with respect to the reference (lab) basis. In either case, rotation may be fully characterized by the angles that the rotated triad axes make with the originally coincident laboratory axes. This figure shows six of these nine possible angles.

distinction between a non-evolving basis change and an evolving superimposed body rotation—evolving superimposed body rotations are the subject of the principle of frame indifference, discussed in chapter 19. Here we focus on the simpler problem of a non-evolving basis change.

Let {X, Y, Z} be a *fixed* 'laboratory' orthogonal coordinate triad, and let $\{x, y, z\}$ be a second orthonormal triad that is obtained by either rotating the observer or by rotating an 'embedded' triad with the rigid body. As indicated in figure 3.2, the orientation of the triad may be described mathematically by the angles that the re-oriented triad makes with the fixed laboratory triad. Rotation of the basis also changes the components of any physical vector or tensor in a very precise way. In fact, these precise transformation relationships are often used as a method to decide if something is a physical (engineering) vector or tensor in the first place[6].

3.2.1 Motivational discussion: principal basis

Most engineering students eventually learn that any real symmetric matrix $[A]$ can be decomposed as

$$[A] = [Q][\Lambda][Q]^T, \tag{3.12}$$

where $[\Lambda]$ is a diagonal matrix containing the eigenvalues of $[A]$, and $[Q]$ is an orthogonal matrix whose *columns* contain the lab components of the corresponding eigenvectors of $[A]$. This theorem from matrix analysis has an interpretation in tensor analysis as a change-of-basis operation. As we will soon see, the above matrix decomposition tells us the following:

$$\begin{aligned} &\text{if} \quad \underset{\sim}{\mathbf{A}} = \sum_{i=1}^{3}\sum_{j-1}^{3} A_{ij}\underset{\sim}{\mathbf{E}}_i\underset{\sim}{\mathbf{E}}_j \quad \text{and} \quad A_{ij} = A_{ji}, \\ &\text{then the spectral expansion is} \quad \underset{\sim}{\mathbf{A}} = \sum_{K=1}^{3} \lambda_K \underset{\sim}{\mathbf{p}}_K \underset{\sim}{\mathbf{p}}_K, \end{aligned} \tag{3.13}$$

where λ_K are the eigenvalues, and $\underset{\sim}{\mathbf{p}}_K$ are the orthonormal eigenvectors. Stated differently, if a tensor $\underset{\sim}{\mathbf{A}}$ has components A_{ij} with respect to the laboratory $\{\underset{\sim}{\mathbf{E}}_1, \underset{\sim}{\mathbf{E}}_2, \underset{\sim}{\mathbf{E}}_3\}$-basis, then it has components

$$[\Lambda] = \begin{bmatrix} \lambda_1 & 0 & 0 \\ 0 & \lambda_2 & 0 \\ 0 & 0 & \lambda_3 \end{bmatrix}, \tag{3.14}$$

with respect to the $\{\underset{\sim}{\mathbf{p}}_1, \underset{\sim}{\mathbf{p}}_2, \underset{\sim}{\mathbf{p}}_3\}$ principal basis. This result further motivates a general study of changes of basis. For example, if A_{ij} are the components of a tensor with

[6] Mathematicians use what might *seem* to be a very different definition of the word vector. The connection between these definitions is discussed in this chapter, just after (3.38).

respect to the laboratory $\{\underset{\sim}{E}_1, \underset{\sim}{E}_2, \underset{\sim}{E}_3\}$-basis, then what are the components $\tilde{A}_{ij}$ with respect to some *different* $\{\underset{\sim}{e}_1, \underset{\sim}{e}_2, \underset{\sim}{e}_3\}$-basis (not necessarily the principal basis)?

3.2.2 Orthonormal basis transformations

In figure 3.2, the angle that the reoriented x-axis makes with the fixed X-axis is denoted θ_{Xx}, and the angle that the reoriented x-axis makes with the fixed Z-axis is denoted θ_{Zx}, and so on. There are nine such angles, but only three are independent, as discussed below. The orientation of the rotated coordinate system can be described through the use of a *transformation matrix* $[Q]$ defined as

$$[Q] = \begin{bmatrix} \cos\theta_{Xx} & \cos\theta_{Xy} & \cos\theta_{Xz} \\ \cos\theta_{Yx} & \cos\theta_{Yy} & \cos\theta_{Yz} \\ \cos\theta_{Zx} & \cos\theta_{Zy} & \cos\theta_{Zz} \end{bmatrix}. \tag{3.15}$$

The components of $[Q]$ are called *direction cosines*. Even though we have defined this matrix in terms of cosines, it is conceptually easier to recognize that the ith *column* (not row) of $[Q]$ contains the lab components of the ith reoriented unit basis vector.

Let $\{\underset{\sim}{E}_X, \underset{\sim}{E}_Y, \underset{\sim}{E}_Z\}$ be the orthonormal set of 'laboratory' basis vectors associated with the $\{X, Y, Z\}$ system. Let $\{\underset{\sim}{e}_x, \underset{\sim}{e}_y, \underset{\sim}{e}_z\}$ be the orthonormal set of 'reoriented' basis vectors associated with the $\{x, y, z\}$ system. For any two vectors, $\underset{\sim}{u}$ and $\underset{\sim}{v}$, the vector inner product (single dot) can be defined as

$$\underset{\sim}{u} \bullet \underset{\sim}{v} = |\underset{\sim}{u}||\underset{\sim}{v}| \cos\theta_{uv}, \tag{3.16}$$

where $|\underset{\sim}{u}|$ and $|\underset{\sim}{v}|$ are the magnitudes of the vectors, $\underset{\sim}{u}$ and $\underset{\sim}{v}$, and θ_{uv} is the angle between. Therefore, recognizing that the basis vectors have unit magnitude, the direction-cosine matrix may be written as

$$[Q] = \begin{bmatrix} \underset{\sim}{E}_X \bullet \underset{\sim}{e}_x & \underset{\sim}{E}_X \bullet \underset{\sim}{e}_y & \underset{\sim}{E}_X \bullet \underset{\sim}{e}_z \\ \underset{\sim}{E}_Y \bullet \underset{\sim}{e}_x & \underset{\sim}{E}_Y \bullet \underset{\sim}{e}_y & \underset{\sim}{E}_Y \bullet \underset{\sim}{e}_z \\ \underset{\sim}{E}_Z \bullet \underset{\sim}{e}_x & \underset{\sim}{E}_Z \bullet \underset{\sim}{e}_y & \underset{\sim}{E}_Z \bullet \underset{\sim}{e}_z \end{bmatrix}. \tag{3.17}$$

Instead of using the symbols $\{X, Y, Z\}$ or $\{x, y, z\}$ as subscripts, it is common practice to identify the coordinates by *number* $\{1, 2, 3\}$, so this equation becomes

$$[Q] = \begin{bmatrix} \underset{\sim}{E}_1 \bullet \underset{\sim}{e}_1 & \underset{\sim}{E}_1 \bullet \underset{\sim}{e}_2 & \underset{\sim}{E}_1 \bullet \underset{\sim}{e}_3 \\ \underset{\sim}{E}_2 \bullet \underset{\sim}{e}_1 & \underset{\sim}{E}_2 \bullet \underset{\sim}{e}_2 & \underset{\sim}{E}_2 \bullet \underset{\sim}{e}_3 \\ \underset{\sim}{E}_3 \bullet \underset{\sim}{e}_1 & \underset{\sim}{E}_3 \bullet \underset{\sim}{e}_2 & \underset{\sim}{E}_3 \bullet \underset{\sim}{e}_3 \end{bmatrix}, \tag{3.18}$$

which can be written very compactly as

$$Q_{ij} = \underset{\sim}{E}_i \bullet \underset{\sim}{e}_j, \tag{3.19}$$

which is the same as equation (3.10).

The subscripts (called indices) vary from 1 to 3. The first subscript on Q_{ij} is the same as the subscript on the lab basis vector $\underset{\sim}{E}_i$ and the second subscript on Q_{ij} is the same as that on the reoriented basis vector $\underset{\sim}{e}_j$. Noting this, a very useful alternative notation for basis-change analysis uses a *labeling system*. Assign a label A arbitrarily to one of the basis systems and B to the other. Letting $\{\underset{\sim}{e}_1^A, \underset{\sim}{e}_2^A, \underset{\sim}{e}_3^A\}$ and $\{\underset{\sim}{e}_1^B, \underset{\sim}{e}_2^B, \underset{\sim}{e}_3^B\}$ denote the basis vectors, define two types of labeled direction cosines, q_{ij}^{AB} and q_{ji}^{BA},

$$q_{ij}^{AB} := \underset{\sim}{e}_i^A \bullet \underset{\sim}{e}_j^B = \underset{\sim}{e}_j^B \bullet \underset{\sim}{e}_i^A =: q_{ji}^{BA} \tag{3.20}$$

Thus, if we assign the labels A and B, respectively, to the $\underset{\sim}{E}$ and $\underset{\sim}{e}$ bases in this section,

$$q_{ij}^{AB} = Q_{ij} \quad \text{and} \quad q_{ij}^{BA} = Q_{ij}^T \tag{3.21}$$

or, in matrix notation,

$$[q^{AB}] = [Q] \quad \text{and} \quad [q^{BA}] = [Q]^T. \tag{3.22}$$

The remainder of this section will continue with the traditional (non-labeled) notation, but with major results stated also in the labeling system. Though seemingly cumbersome, the labeling system significantly reduces chance of error in basis-change calculations.

It will soon be shown that the $[Q]$ matrix is orthogonal[7], which means the inverse of $[Q]$ is just the transpose of $[Q]$. That is,

$$[Q][Q]^T = [Q]^T[Q] = [I]. \tag{3.23}$$

Here, $[I]$ denotes the identity matrix. The above matrix equation is written in indicial form as

$$\sum_{k=1}^{3} Q_{ik}Q_{jk} = \sum_{k=1}^{3} Q_{ki}Q_{kj} = \delta_{ij}, \tag{3.24}$$

where δ_{ij} is the Kronecker delta (equal to 1 if $i = j$ and 0 if $i \neq j$).

Because the laboratory basis $\{\underset{\sim}{E}_1, \underset{\sim}{E}_2, \underset{\sim}{E}_3\}$ is a basis, we know that there exist coefficients v_j such that any vector $\underset{\sim}{v}$ can be written as a linear combination of the lab basis vectors. Let us denote the jth component of $\underset{\sim}{v}$ with respect to the laboratory basis by v_j. This component may be computed by dotting the vector $\underset{\sim}{v}$ into the jth basis vector $\underset{\sim}{E}_j$. Stated mathematically, there exist vector components v_j such that

[7] This property holds only when both bases are orthonormal. An orthonormal basis is right-handed if crossing the first basis vector into the second basis vector gives the third basis vector. Otherwise, if the third basis vector points the opposite way, then the basis is left-handed. The determinant of $[Q]$ will be $+1$ if both bases have the same handedness, or -1 if the bases have different handedness.

$$\underset{\sim}{v} = \sum_{j=1}^{3} v_j \underset{\sim}{E}_j. \tag{3.25}$$

Importantly, note that

$$v_j = \underset{\sim}{v} \bullet \underset{\sim}{E}_j. \tag{3.26}$$

Equation (3.25) is true for *any* vector $\underset{\sim}{v}$. As a special instance, we may consider the vector $\underset{\sim}{v}$ to be one of the reoriented basis vectors $\underset{\sim}{e}_i$. Then we know that each $\underset{\sim}{e}_i$ can be expressed as a linear combination of the lab basis vectors. Let us denote the component of $\underset{\sim}{e}_i$ with respect to the jth laboratory basis vector by Q_{ji}. Then (3.25) states that each reoriented basis vector $\underset{\sim}{e}_i$ can be expressed in the form[8]

$$\boxed{\underset{\sim}{e}_i = \sum_{j=1}^{3} Q_{ji} \underset{\sim}{E}_j} \quad \text{or, with a labeling system,} \quad \underset{\sim}{e}_i^B = q_{ji}^{AB} \underset{\sim}{e}_j^A = q_{ij}^{BA} \underset{\sim}{e}_j^A. \tag{3.27}$$

This is our proof of (3.9). To remember this equation, note that the first subscript on Q_{ji} is the same as the subscript on the lab basis vector $\underset{\sim}{E}_j$ and the second subscript on Q_{ji} is the same as that on the reoriented basis vector $\underset{\sim}{e}_i$. This statement is identical to the one that occurred immediately following (3.10), so it is a powerful mnemonic —indices in transformation formulas will always be consistent with how the indices were used when you originally defined your direction cosines. Whenever an index, such as the index j in (3.27), appears exactly twice in a term (where a 'term' is defined as an expression separated by addition, subtraction, or equality symbols), then it is summed from 1 to 3.

The reoriented vectors $\{\underset{\sim}{e}_1, \underset{\sim}{e}_2, \underset{\sim}{e}_3\}$ form a basis too. Consequently, we know that there exist coefficients $\tilde{v}_j$ such that any vector $\underset{\sim}{v}$ can be written

$$\underset{\sim}{v} = \sum_{j=1}^{3} \tilde{v}_j \underset{\sim}{e}_j, \quad \text{where } \tilde{v}_j = \underset{\sim}{v} \bullet \underset{\sim}{e}_j. \tag{3.28}$$

With a labeling system and implied summation over repeated indices, this is

$$\underset{\sim}{v} = v_j^B \underset{\sim}{e}_j^B, \quad \text{where } v_j^B = \underset{\sim}{v} \bullet \underset{\sim}{e}_j^B. \tag{3.29}$$

Let us denote the component of $\underset{\sim}{E}_i$ with respect to the jth rotated basis vector by a_{ji}. In other words,

$$\underset{\sim}{E}_i = \sum_{j=1}^{3} a_{ji} \underset{\sim}{e}_j. \tag{3.30}$$

[8] This formula for the labeling system is using implied summation notation in which repeated indices are understood to be summed from 1 to 3.

Dotting both sides of this equation by $\underset{\sim}{e}_k$ and noting that $\underset{\sim}{e}_j \bullet \underset{\sim}{e}_k = \delta_{jk}$ shows that $a_{ki} = \underset{\sim}{E}_i \bullet \underset{\sim}{e}_k$. Recognizing that the right-hand side of this result is simply our Q_{ik} shows that $a_{ki} = Q_{ik}$. In other words, the $[a]$ matrix is just the transpose of the $[Q]$ matrix. We see in hindsight that it is not really necessary to introduce a new set of coefficients a_{ji}; the $[Q]$ matrix contains all of the information needed to express one basis with respect to the other. In terms of the $[Q]$ coefficients, equation (3.29) becomes

$$\boxed{\underset{\sim}{E}_i = \sum_{j=1}^{3} Q_{ij} \underset{\sim}{e}_j}, \quad \text{where } Q_{ij} = \underset{\sim}{E}_i \bullet \underset{\sim}{e}_j. \tag{3.31}$$

As before, the first subscript on Q_{ij} is the same as the subscript on the lab basis vector $\underset{\sim}{E}_i$ and the second subscript on Q_{ij} is the same as that on the reoriented basis vector $\underset{\sim}{e}_j$. Summation occurs over the repeated index j. In the future, if you ever need to recall this equation, you can start with a 'skeleton' like this

$$\underset{\sim}{E}_m = \sum_{?=1}^{3} Q_{??} \underset{\sim}{e}_n \leftarrow \text{we need to fill in the question marks.} \tag{3.32}$$

The skeleton needs to use different symbols for the subscripts on the basis vectors (here, we used m and n). You fill in the question marks by looking back at how you originally defined your direction-cosine matrix. If, like us, you originally defined it as $Q_{ij} = \underset{\sim}{E}_i \bullet \underset{\sim}{e}_j$, then (noting that this definition has the first index matching the index on the lab basis), you would fill in the first index in the above skeleton to match the index appearing on the lab basis in the skeleton, and similarly, the second index should match the rotated basis. Thus, we know that the direction cosine in the above equation would have to be Q_{mn}. In the final step, you would fill in the remaining question mark on the summation with whatever index on the right-hand side is repeated (in this case, the repeated index is n). The final result is then

$$\underset{\sim}{E}_m = \sum_{n=1}^{3} Q_{mn} \underset{\sim}{e}_n. \tag{3.33}$$

These considerations are more rapidly and reliably addressed with a labeling system, as this result would be written (again using implied summation)

$$\underset{\sim}{e}_m^A = q_{mn}^{AB} \underset{\sim}{e}_n^B. \tag{3.34}$$

The basis transformation matrix $[Q]$ is often used for relating the components of vectors with respect to different bases. Suppose $\{v_1, v_2, v_3\}$ are the components of a vector $\underset{\sim}{v}$ with respect to the fixed $\underset{\sim}{E}_i$ laboratory basis. Then $\underset{\sim}{v}$ can be written

$$\underset{\sim}{v} = \sum_{i=1}^{3} v_i \underset{\sim}{E}_i, \quad \text{where } v_i = \underset{\sim}{v} \bullet \underset{\sim}{E}_i. \tag{3.35}$$

Substituting (3.31) into (3.35) gives

$$\underset{\sim}{v} = \sum_{i=1}^{3} v_i \sum_{j=1}^{3} Q_{ij} \underset{\sim}{e}_j = \sum_{i=1}^{3}\sum_{j=1}^{3} Q_{ij} v_i \underset{\sim}{e}_j = \sum_{i=1}^{3} \tilde{v}_j \underset{\sim}{e}_j, \tag{3.36}$$

where

$$\boxed{\tilde{v}_j = \sum_{i=1}^{3} Q_{ij} v_i}; \quad \text{in matrix form,} \quad \{\tilde{v}\} = [Q]^T \{v\}. \tag{3.37}$$

With a labeling system and implied summation, this is written

$$v_j^B = q_{ij}^{AB} v_i^A = q_{ji}^{BA} v_i^A; \quad \text{in matrix form,} \quad \{v^B\} = [q^{BA}]\{v^A\}. \tag{3.38}$$

The labeling system has a clear advantage: there is no pause or chance of error in deciding whether or not a transpose is needed.

These formulas define the vector component transformation rule, which is itself frequently used in the literature as a definition of the term *vector*. Such a definition is more precise (and testable) than the vague engineering definition of a vector as 'something with length and magnitude'. This definition plays a role in the more general mathematician's definition that a vector is 'a member of a vector space'. The mathematician's definition, of course, requires a preliminary definition of a 'vector space', which is a set of objects that satisfy certain rules for multiplication and addition. Among the key, but often neglected, requirements of a vector space is the notion of *closure*. For example, the addition of two members of the set must itself be a member of the set. When an engineer uses (3.37) as the definition of a vector, that transformation property is actually serving as the discerning test for set *membership*. If, for example, two 'objects' are members of the set of engineering vectors, then they will each have three components that transform under a basis change according to (3.37). For engineering vectors, the sum of two vectors $\underset{\sim}{w} = \underset{\sim}{u} + \underset{\sim}{v}$ is defined so that $w_i = u_i + v_i$. If $\underset{\sim}{u}$ and $\underset{\sim}{v}$ transform according to (3.37), then you can prove that $\underset{\sim}{w}$ will too. Hence, this definition of addition is *closed*. Likewise, scalar multiplication $\underset{\sim}{w} = \alpha \underset{\sim}{u}$ for which $w_i = \alpha u_i$ is a closed operation under the membership test of (3.37). Hence, there is no irresolvable conflict in the difference between how engineers and mathematicians define vectors. The engineer's definition is a set membership test, and the mathematician's definition justifies the engineer's use of the word 'vector' to refer to members of the set.

By comparing (3.35) and (3.36), note that the *same* vector $\underset{\sim}{v}$ can be written as a linear combination of either the lab basis or the *reoriented* $\underset{\sim}{e}_j$-basis vectors. In other words,

$$\sum_{i=1}^{3} v_i \underset{\sim}{E}_i = \sum_{i=1}^{3} \tilde{v}_j \underset{\sim}{e}_j. \tag{3.39}$$

Whenever a basis is reoriented, then the components of a vector will change, but they will change in a perfectly compensating way so that the *sum* of the components times basis vectors remains unchanged. A vector is this sum. Without knowing the basis, you cannot know the vector. A vector (being the sum of components times basis vectors) is invariant under a basis change even though its components are not invariant. That is why we speak of a vector using the *singular* (you do not say the velocity *are* fast). A vector is a *single entity* composed of multiple parts (components and basis vectors).

We should be able to start with the expression of the vector in terms of the re-oriented basis, and recover the lab components. That is,

$$v_i = \underset{\sim}{v} \bullet \underset{\sim}{E}_i = \left(\sum_{j=1}^{3} \tilde{v}_j \underset{\sim}{e}_j \right) \bullet \underset{\sim}{E}_i. \tag{3.40}$$

Recognizing that $\underset{\sim}{e}_j \bullet \underset{\sim}{E}_i = Q_{ij}$, this may be written

$$\boxed{v_i = \sum_{j=1}^{3} Q_{ij} \tilde{v}_j}; \quad \text{in matrix form,} \quad \{v\} = [Q]\{\tilde{v}\}. \tag{3.41}$$

In the labeling system

$$v_i^A = q_{ij}^{AB} v_j^B; \quad \text{in matrix form,} \quad \{v^A\} = [q^{AB}]\{v^B\}. \tag{3.42}$$

Not only are (3.37) and (3.41) important results in their own right, they also demonstrate that the matrix $[Q]$ must be orthogonal because both matrix equations can be true only if

$$\boxed{[Q]^{-1} = [Q]^T}; \quad \text{in the labeling system, } [q^{AB}]^{-1} = [q^{BA}]. \tag{3.43}$$

All *orthogonal* vector component transformations are in the form of either (3.37) or (3.41). Component transformations involving non-orthonormal bases take a different form and fall outside the scope of this book (a discussion can be found in [2]).

Note that (3.37) and (3.41) say that the transformed components can be obtained via a matrix multiplication of either $[Q]$ or $[Q]^T$ of the starting components. It can be difficult to remember whether or not to use a transpose unless you recall the mnemonic after (3.27). For both (3.37) and (3.41), the first subscript on Q_{ij} is the same as the subscript on the lab base component v_i and the second subscript on Q_{ij} is the same as that on the reoriented component $\tilde{v}_j$. Repeated indices are summed. It is generally easier to first write the indicial expression for any transformation formula, which then helps you determine if a transpose is needed in matrix formulas. Switching to the labeling system is even better, as will be clear in upcoming matrix versions of transformation formulas.

The mnemonic rule also applies to component transformations for tensors. Let $\underset{\approx}{T}$ denote a second-order tensor. Let T_{ij} denote the components with respect to the laboratory basis so that $\underset{\approx}{T} = T_{ij} \underset{\sim}{E}_i \underset{\sim}{E}_j$. Let $\tilde{T}_{kl}$ denote the components with respect to the reoriented basis so that $\underset{\sim}{T} = \tilde{T}_{kl}\underset{\sim}{e}_k\underset{\sim}{e}_l$. There are two subscripts on the components, so the $[Q]$ matrix will appear twice. The mnemonic says that the subscripts on the laboratory components must match the *first* subscripts on the Qs, and the subscripts on the reoriented components must match the *second* subscripts on the Qs. Finally, any subscripts that appear twice in a single term must be summed. Thus, the component transformation rules for tensors are

$$\tilde{T}_{kl} = \sum_{i=1}^{3}\sum_{j=1}^{3} Q_{ik}Q_{jl}T_{ij} \quad \text{or in matrix form, } [\tilde{T}] = [Q]^T[T][Q] \tag{3.44}$$

and

$$T_{ij} = \sum_{k=1}^{3}\sum_{l=1}^{3} Q_{ik}Q_{jl}\tilde{T}_{kl}, \quad \text{or in matrix form, } [T] = [Q][\tilde{T}][Q]^T. \tag{3.45}$$

Again note that it is easier to use the mnemonic rule to *first* write the component form, and *then* construct the corresponding matrix form. Doing it that way relieves you from having to recall where the transpose goes. The labeling system is even more effective, where these equations are written

$$T_{ij}^{AA} = q_{ik}^{AB}q_{jl}^{AB}T_{kl}^{BB}, \quad \text{or in matrix form, } [T^{AA}] = [q^{AB}][T^{BB}][q^{BA}]. \tag{3.46}$$

Again, the labeling system avoids confusion about whether to use or omit a transpose on a direction-cosine matrix.

3.2.3 Summary

The key results from this section are summarized below:

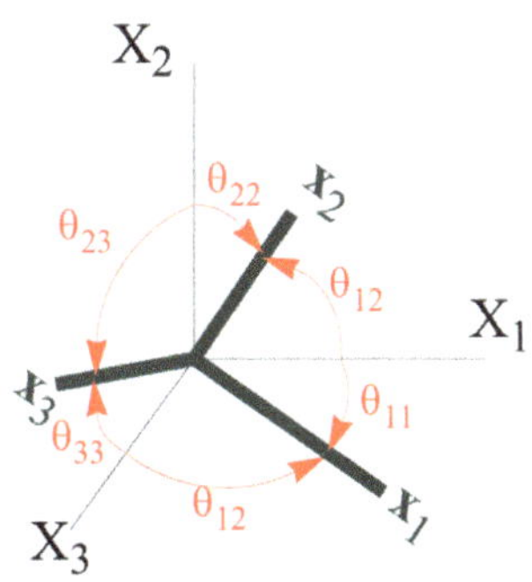

$$\text{Direction cosines: } \boxed{Q_{ij} = \underset{\sim}{E}_i \bullet \underset{\sim}{e}_j}. \tag{3.47}$$

Note that $Q_{ij} = \cos\theta_{ij}$, where θ_{ij} is the angle between $\underset{\sim}{E}_i$ and $\underset{\sim}{e}_j$. Q_{ij} is the ith component $\underset{\sim}{e}_j$ with respect to $\{\underset{\sim}{E}_1, \underset{\sim}{E}_2, \underset{\sim}{E}_3\}$-basis, and therefore the jth column of $[Q]$ contains the components of $\underset{\sim}{e}_j$ with respect to the $\{\underset{\sim}{E}_1, \underset{\sim}{E}_2, \underset{\sim}{E}_3\}$-basis. Furthermore, Q_{ij} is the jth component $\underset{\sim}{E}_i$ with respect to $\{\underset{\sim}{e}_1, \underset{\sim}{e}_2, \underset{\sim}{e}_3\}$, and therefore the ith row of

$[Q]$ contains components of $\underset{\sim}{\boldsymbol{E}}_i$ with respect to the $\{\underset{\sim}{\boldsymbol{e}}_1, \underset{\sim}{\boldsymbol{e}}_2, \underset{\sim}{\boldsymbol{e}}_3\}$-basis. Because of these relations, the direction cosines are frequently presented in tabular form as

	$\underset{\sim}{\boldsymbol{e}}_1$	$\underset{\sim}{\boldsymbol{e}}_2$	$\underset{\sim}{\boldsymbol{e}}_3$
$\underset{\sim}{\boldsymbol{E}}_1$	Q_{11}	Q_{12}	Q_{13}
$\underset{\sim}{\boldsymbol{E}}_2$	Q_{21}	Q_{22}	Q_{23}
$\underset{\sim}{\boldsymbol{E}}_3$	Q_{31}	Q_{32}	Q_{33}

For example, reading down the first column, $\underset{\sim}{\boldsymbol{e}}_1 = Q_{11}\underset{\sim}{\boldsymbol{E}}_1 + Q_{21}\underset{\sim}{\boldsymbol{E}}_2 + Q_{31}\underset{\sim}{\boldsymbol{E}}_3$. Reading across the first row, $\underset{\sim}{\boldsymbol{E}}_1 = Q_{11}\underset{\sim}{\boldsymbol{e}}_1 + Q_{12}\underset{\sim}{\boldsymbol{e}}_2 + Q_{13}\underset{\sim}{\boldsymbol{e}}_3$.

Other key results are

$$\boxed{[Q]^{-1} = [Q]^T}, \tag{3.48}$$

$$\boxed{\underset{\sim}{\boldsymbol{e}}_i = \sum_{j=1}^{3} Q_{ji}\underset{\sim}{\boldsymbol{E}}_j}, \tag{3.49}$$

$$\boxed{\underset{\sim}{\boldsymbol{E}}_i = \sum_{j=1}^{3} Q_{ij}\underset{\sim}{\boldsymbol{e}}_j}. \tag{3.50}$$

For a vector, $\underset{\sim}{\boldsymbol{v}} = v_i\underset{\sim}{\boldsymbol{E}}_i = \tilde{v}_j\underset{\sim}{\boldsymbol{e}}_j$, the components are related by

$$\boxed{\tilde{v}_j = \sum_{i=1}^{3} Q_{ij}v_i}; \quad \text{in matrix form,} \quad \{\tilde{v}\} = [Q]^T\{v\}, \tag{3.51}$$

$$\boxed{v_i = \sum_{j=1}^{3} Q_{ij}\tilde{v}_j}; \quad \text{in matrix form,} \quad \{v\} = [Q]\{\tilde{v}\}. \tag{3.52}$$

For a second-order tensor $\underset{\approx}{\boldsymbol{T}} = T_{ij}\underset{\sim}{\boldsymbol{E}}_i\underset{\sim}{\boldsymbol{E}}_j = \tilde{T}_{kl}\underset{\sim}{\boldsymbol{e}}_k\underset{\sim}{\boldsymbol{e}}_l$, the components are related by

$$\tilde{T}_{kl} = \sum_{i=1}^{3}\sum_{j=1}^{3} Q_{ik}Q_{jl}T_{ij}, \quad \text{or in matrix form, } [\tilde{T}] = [Q]^T[T][Q], \tag{3.53}$$

$$T_{ij} = \sum_{k=1}^{3}\sum_{l=1}^{3} Q_{ik}Q_{jl}\tilde{T}_{kl}, \quad \text{or in matrix form, } [T] = [Q][\tilde{T}][Q]^T. \tag{3.54}$$

These equations are restatements of (3.19), (3.43), (3.27), (3.31), (3.37), (3.41), (3.44), and (3.45), respectively. For the special case that $[T]$ is symmetric and the secondary basis is the principal basis, the second of these formulas reduces to (3.12) in which $[\tilde{T}]$ is diagonal.

Higher-order tensors transform similarly. For example, if $\underset{\approx}{\xi}$ is a third-order tensor, then

$$\tilde{\xi}_{pqr} = \sum_{i=1}^{3}\sum_{j=1}^{3}\sum_{k=1}^{3} Q_{ip}Q_{jq}Q_{kr}\xi_{ijk} \tag{3.55}$$

$$\xi_{ijk} = \sum_{p=1}^{3}\sum_{q=1}^{3}\sum_{r=1}^{3} Q_{ip}Q_{jq}Q_{kr}\tilde{\xi}_{pqr}. \tag{3.56}$$

With a labeling system, these basis-change formulas are written as follows.

$$\text{Direction cosines:}\ \boxed{q_{ij}^{AB} = \underset{\sim}{e}_i^{A} \bullet \underset{\sim}{e}_j^{B} = \underset{\sim}{e}_j^{B} \bullet \underset{\sim}{e}_i^{A} = q_{ji}^{BA}}, \tag{3.57}$$

	$\underset{\sim}{e}_1^{B}$	$\underset{\sim}{e}_2^{B}$	$\underset{\sim}{e}_3^{B}$
$\underset{\sim}{e}_1^{A}$	q_{11}^{AB}	q_{12}^{AB}	q_{13}^{AB}
$\underset{\sim}{e}_2^{A}$	q_{21}^{AB}	q_{22}^{AB}	q_{23}^{AB}
$\underset{\sim}{e}_3^{A}$	q_{31}^{AB}	q_{32}^{AB}	q_{33}^{AB}

$$\boxed{[q^{AB}]^{-1} = [q^{AB}]^{T} = [q^{BA}]}, \tag{3.58}$$

$$\boxed{\underset{\sim}{e}_i^{B} = q_{ij}^{BA}\underset{\sim}{e}_j^{A}}, \tag{3.59}$$

$$\boxed{\underset{\sim}{e}_i^{A} = q_{ij}^{AB}\underset{\sim}{e}_j^{B}}. \tag{3.60}$$

For $\underset{\sim}{v} = v_i^{A}\underset{\sim}{e}_i^{A} = v_i^{B}\underset{\sim}{e}_i^{B}$,

$$\boxed{v_i^{B} = q_{ij}^{BA}v_j^{A}};\quad \text{in matrix form,}\quad \{v^B\} = [q^{BA}]\{v^A\}, \tag{3.61}$$

$$\boxed{v_i^{A} = q_{ij}^{AB}v_j^{B}};\quad \text{in matrix form,}\quad \{v^A\} = [q^{AB}]\{v^B\}. \tag{3.62}$$

For a second-order tensor $\underset{\approx}{T} = T_{ij}^{AA}\underset{\sim}{e}_i^{A}\underset{\sim}{e}_j^{A} = T_{ij}^{BB}\underset{\sim}{e}_i^{B}\underset{\sim}{e}_j^{B}$,

$$T_{ij}^{BB} = q_{ik}^{BA}q_{jl}^{BA}T_{ij}^{AA},\quad \text{or in matrix form, } [T^{BB}] = [q^{BA}][T^{AA}][q^{AB}], \tag{3.63}$$

$$T_{ij}^{AA} = q_{ik}^{AB}q_{jl}^{AB}T_{ij}^{BB},\quad \text{or in matrix form, } [T^{AA}] = [q^{AB}][T^{BB}][q^{BA}],\ \text{etc.} \tag{3.64}$$

3.2.4 Special case: coordinate transformations

The laboratory Cartesian coordinates of a point in space are denoted $\{X, Y, Z\}$ or, using the numbered notation, $\{X_1, X_2, X_3\}$. The coordinates of that same point with respect to a reoriented orthonormal basis are denoted $\{x, y, z\}$ or $\{x_1, x_2, x_3\}$. Because both coordinate systems are Cartesian (see below), these position *coordinates* are identically equal to the position *components*. Thus, the above vector transformation relations also serve as coordinate transformation relations. Specifically, replacing the *lab* components $\{v_1, v_2, v_3\}$ by the *lab* coordinates $\{X_1, X_2, X_3\}$, and replacing the *re-oriented* components $\{\tilde{v}_1, \tilde{v}_2, \tilde{v}_3\}$ by the *reoriented* components $\{x_1, x_2, x_3\}$, equations (3.51) and (3.52) give us the coordinate transformation rules:

$$\boxed{x_j = \sum_{i=1}^{3} Q_{ij} X_i}; \quad \text{in matrix form,} \quad \{x\} = [Q]^T \{X\}, \tag{3.65}$$

$$\boxed{X_i = \sum_{j=1}^{3} Q_{ij} x_j}; \quad \text{in matrix form,} \quad \{X\} = [Q]\{x\}. \tag{3.66}$$

To understand the distinction between position components and position coordinates, consider cylindrical coordinate systems. For this system, the coordinates are (r, θ, z) whereas the position vector is given by $r\underset{\sim}{e}_r + z\underset{\sim}{e}_z$. This position vector has only two non-zero *components* with respect to the cylindrical basis. Despite having only two non-zero components, the position vector itself nevertheless still depends on all three coordinates because dependence on θ is contained implicitly in the dependence of the basis vector $\underset{\sim}{e}_r$ on θ. In this case, where the reoriented basis is not Cartesian, the *coordinate* transformation rules in (3.65) and (3.66) do *not* apply. In particular, instead of (3.66), the rule that transforms cylindrical *coordinates* to lab *coordinates* are just the familiar formulas $X = r\cos\theta$, $Y = r\sin\theta$, and $Z = z$. Keep in mind that the cylindrical system is *still* orthonormal because the three basis vectors $\{\underset{\sim}{e}_r, \underset{\sim}{e}_\theta, \underset{\sim}{e}_z\}$ form an orthonormal triad. Consequently, the vector *component* transformation rules such as (3.51) and (3.52) continue to apply. For example, we could apply these transformation rules to *prove* that the position vector must be given by $r\underset{\sim}{e}_r + z\underset{\sim}{e}_z$. To do this, we would write the position vector in two equivalent ways: $\underset{\sim}{x} = X\underset{\sim}{E}_X + Y\underset{\sim}{E}_Y + Z\underset{\sim}{E}_Z$ and $\underset{\sim}{x} = x_r\underset{\sim}{e}_r + x_\theta\underset{\sim}{e}_\theta + x_z\underset{\sim}{e}_z$. The columns of the transformation matrix contain the lab components of the cylindrical basis vectors. For example, since $\underset{\sim}{e}_r = \cos\theta\underset{\sim}{E}_X + \sin\theta\underset{\sim}{E}_Y$, the first column of the transformation matrix would contain $\cos\theta$, $\sin\theta$, and 0, respectively. Applying the usual *component* transformation rule would reveal that $x_r = r$, $x_\theta = 0$, and $x_z = z$. Note that the second *component*, x_θ, is zero even though the second *coordinate*, θ, is not generally zero. While the *position* vector always has a zero component in the $\underset{\sim}{e}_\theta$-direction, other vectors (such as gravity, velocity, force, etc) would have all three components generally non-zero.

References

[1] Simo J C and Hughes T J R 1998 *Computational Inelasticity* (Berlin: Springer)

[2] Brannon R M 2004 Curvilinear tutorial https://csmbrannon.net/2011/03/07/385/

IOP Publishing

Rotation, Reflection, and Frame Changes
Orthogonal tensors in computational engineering mechanics
R M Brannon

Chapter 4

Rotation operations

We now introduce a subtle but important distinction between basis transformations and rotation operations. The previous chapter dealt primarily with basis transformations, where the *same object* was viewed from a new perspective. In equations (3.35) and (3.36), the same vector $\underset{\sim}{v}$ was expressed from two different basis systems, and (3.37) showed how the components of $\underset{\sim}{v}$ in each system were related—that is, how one *transforms the components* of the single vector $\underset{\sim}{v}$ from one system to another. In essence, chapter 3 dealt with the kind of transformation illustrated in figure 3.1(a) where the object (vector or tensor) was held fixed while the observer changed orientation.

A similar-looking, but fundamentally different, concept is that of an *operation* that *transforms one vector to another vector*. This is the type of operation illustrated in figure 3.1(b) where the observer remains glued to the lab axes and the embedded object itself changes orientation. Now we seek to determine how the components of the rotated object change when viewed by a non-rotated observer.

The area of study called 'continuum kinematics' addresses how objects change under general deformations that include changes in size and shape. These sorts of operations on a body can be quite general. Here, however, we limit the discussion to 'right-handed linear orthogonal operations,' in which the body is permitted only to change orientation, keeping its size, shape, and handedness[1] unchanged. In particular, we will consider in this chapter the special class of homogeneous operations that are described through the use of a *rotation tensor*, $\underset{\approx}{R}$. The equation

$$\hat{\underset{\sim}{w}} = \underset{\approx}{R} \bullet \underset{\sim}{w} \tag{4.1}$$

is a symbolic way of saying that the rotation operation $\underset{\approx}{R}$ operates on a vector $\underset{\sim}{w}$, giving a new vector $\hat{\underset{\sim}{w}}$. Because $\underset{\approx}{R}$ is a rotation operation, the new vector $\hat{\underset{\sim}{w}}$ has the

[1] In other words, this chapter does not permit *reflections* where right becomes left and left becomes right. Reflections are discussed in chapter 6. This chapter also assumes that the transformation involves no rigid translation so that the point at the origin remains at the origin.

same magnitude as $\underset{\sim}{w}$, but a different orientation. The laboratory components of $\hat{\underset{\sim}{w}}$ are related to the laboratory components of $\underset{\sim}{w}$ by

$$\hat{w}_i = \sum_{j=1}^{3} R_{ij} w_j. \tag{4.2}$$

In going from the direct-notation equation (4.1) to this indicial equation (4.2), no unexpected transpose appears on the subscripts of R_{ij} as it did when we converted from (2.38) to (2.39). The distinction is that (2.38) was actually a set of *three* equations whereas (4.1) is only one equation, making (4.2) merely a direct application of (2.13).

A rotation operation must preserve length. From (4.1), note that

$$\hat{\underset{\sim}{w}} \bullet \hat{\underset{\sim}{w}} = (\underset{\approx}{R} \bullet \underset{\sim}{w}) \bullet (\underset{\approx}{R} \bullet \underset{\sim}{w}) = \underset{\sim}{w} \bullet (\underset{\approx}{R}^T \bullet \underset{\approx}{R}) \bullet \underset{\sim}{w}. \tag{4.3}$$

To preserve lengths, $\hat{\underset{\sim}{w}} \bullet \hat{\underset{\sim}{w}}$ must equal $\underset{\sim}{w} \bullet \underset{\sim}{w}$. Equivalently,

$$\hat{\underset{\sim}{w}} \bullet \hat{\underset{\sim}{w}} = \underset{\sim}{w} \bullet \underset{\approx}{I} \bullet \underset{\sim}{w}. \tag{4.4}$$

For both of these equations to hold for all $\underset{\sim}{w}$, the tensor $\underset{\approx}{R}$ must be orthogonal:

$$\underset{\approx}{R}^T \bullet \underset{\approx}{R} = \underset{\approx}{I}. \tag{4.5}$$

In the previous chapter, changing the basis meant that a vector's components would change so that the vector itself (equal to the sum of components times basis vectors) would not change. This chapter has no basis change. This chapter is different because here we actually are changing the vector into a different (rotated) vector. The components of the rotated vector will be related to those of the starting vector by formulas that look a lot like a basis-change, except that the orthogonal tensor is transposed. Recalling that the transpose of an orthogonal tensor is the same as its inverse, this observation makes sense. Rotating the observer clockwise while looking at a non-changing vector will make that vector *seem* to rotate in the opposite (inverse) direction to the eyes of the rotating observer. Conversely, leaving the observer fixed and rotating the vector will give a set of new vector components that look like a basis change in the opposite direction. This is why all engineers go bonkers trying to figure out if their orthogonal matrices need to have or omit a transpose for a given application. The fast strategy is to just guess! Then test your guess on a simple case like 90° rotation. If the answer is wrong, add a transpose and check again.

A rigid body's orientation can be fully characterized by specifying any three embedded vectors as seen by a non-rotating observer. Even for this situation, we can introduce a set of three vectors $\{\underset{\sim}{e}_1, \underset{\sim}{e}_2, \underset{\sim}{e}_3\}$ which equal the transformed locations of the laboratory basis vectors $\{\underset{\sim}{E}_1, \underset{\sim}{E}_2, \underset{\sim}{E}_3\}$, as illustrated in figure 3.1(a) and (b). In other words, *define*

$$\underset{\sim}{e}_j \equiv \underset{\approx}{R} \bullet \underset{\sim}{E}_j. \tag{4.6}$$

This triad can be regarded as a *fixed* basis that happens to coincide with the embedded basis $\{\underset{\sim}{m}_1, \underset{\sim}{m}_2, \underset{\sim}{m}_3\}$ in figure 3.1(b). The distinction between the e- and m-triads only matters in rate equations, were the material m-triad might have angular velocity whereas the instantaneously coincident e-basis is fixed (and therefore no longer aligned with the material basis an instant later).

Dotting both sides of (4.6) by $\underset{\sim}{E}_i$ shows that

$$\underset{\sim}{e}_j \bullet \underset{\sim}{E}_i = \underset{\sim}{E}_i \bullet \underset{\approx}{R} \bullet \underset{\sim}{E}_j. \tag{4.7}$$

The right-hand side of this equation represents the ij component of the rotation tensor with respect to the *laboratory* basis. The left-hand side of this equation equals the direction-cosine matrix component Q_{ij} from (3.47). Thus

$$\underset{\approx}{R} = Q_{ij}\underset{\sim}{E}_i\underset{\sim}{E}_j. \tag{4.8}$$

Since the rotation tensor is orthogonal as shown in (4.5), its inverse is simply its transpose. Hence, (4.6) can be written as

$$\underset{\sim}{E}_j = \underset{\approx}{R}^T \bullet \underset{\sim}{e}_j. \tag{4.9}$$

Dotting both sides of this equation by $\underset{\sim}{e}_i$ gives

$$\underset{\sim}{e}_i \bullet \underset{\sim}{E}_j \equiv \underset{\sim}{e}_i \bullet \underset{\approx}{R}^T \bullet \underset{\sim}{e}_j. \tag{4.10}$$

The left-hand side of this equation equals Q_{ji}, and the right-hand side is the ij component of $\underset{\approx}{R}^T$ from which it follows that

$$\underset{\approx}{R}^T = Q_{ji}\underset{\sim}{e}_i\underset{\sim}{e}_j \tag{4.11}$$

or, taking the transpose,

$$\underset{\approx}{R} = Q_{ij}\underset{\sim}{e}_i\underset{\sim}{e}_j. \tag{4.12}$$

Ordinarily, the components of a tensor change upon a change of basis. Interestingly, however, comparing (4.12) and (4.8) reveals that the components of the rotation tensor with respect to the $\{\underset{\sim}{E}_1, \underset{\sim}{E}_2, \underset{\sim}{E}_3\}$ laboratory basis are *identical* to the components with respect to the $\{\underset{\sim}{e}_1, \underset{\sim}{e}_2, \underset{\sim}{e}_3\}$ reoriented basis! This is an exceptional result that must be very carefully understood. We are *not* saying that the components of the rotation tensor are unchanged upon *any* change of basis (that would be tantamount to saying that the rotation tensor would have to be a multiple of the identity tensor, which it is not). Instead, we have merely shown that there exist two *particular* bases for which the rotation tensor components are unchanged. Henceforth, whenever we write R_{ij}, the reader may interpret these components to be components of the rotation with respect to either the lab or reoriented basis, since they are identical[2].

[2] Later, when we discuss classes of anisotropy and axes of rotation, it will become clear that a rotation tensor is always transversely isotropic about its own rotation axis, which means its components are the same in any basis that is a rotation of the laboratory basis about the same axis.

Because we have established the connection between the rotation components R_{ij} and the associated direction cosines Q_{ij}, we may write (4.2) as

$$\hat{w}_i = \sum_{j=1}^{3} Q_{ij} w_j \tag{4.13}$$

or, in matrix form

$$\{\hat{w}\} = [Q]\{w\}. \tag{4.14}$$

This expression shows how the *laboratory* components of the rotated vector $\underset{\sim}{\hat{w}}$ are related to the *laboratory* components of the un-rotated vector $\underset{\sim}{w}$. Compare this with the vector transformation rule of (3.51),

$$\{\tilde{v}\} = [Q]^T\{v\}, \tag{4.15}$$

which shows how the *reoriented* components of a *stationary* vector $\underset{\sim}{v}$ are related to the *laboratory* components of the *same* vector $\underset{\sim}{v}$. Note that (4.15) contains a transpose on the $[Q]$ matrix whereas (4.14) does not. For orthogonal matrices, the transpose is equivalent to the inverse. It will be shown later that any rotation has a unique rotation axis and rotation angle, and the inverse of a rotation is obtained by simply rotating about the same axis by the *negative* of the original rotation angle. Thus, comparing (4.15) and (4.14) verifies the intuitive understanding that rotating a vector $\underset{\sim}{w}$ by a certain angle about a given axis to obtain a new vector $\underset{\sim}{\hat{w}}$ results in laboratory components for the new vector given by a formula that is structurally the same as the transformed components resulting from viewing the original vector from a *different* basis that is rotated *in the opposite direction* about the same axis. For example, rotating a vector by 20° about the z-axis will look very similar to rotating the *observer* by −20° about the same axis. This similarity can result in considerable confusion, and it only serves to further emphasize the need for the analyst to be vigilant about which interpretation applies for the given application and which basis is being used to reference components. In one case, we have *two different vectors*, $\underset{\sim}{w}$ and $\underset{\sim}{\hat{w}}$, expressed in *one* basis. For orthogonal basis transformations, we are dealing with *one* vector expressed in *two* basis systems.

Despite the potential for confusion between body rotations and basis rotations, we will find it extremely useful to use the embedded basis associated with rigid-body rotations. In physical applications, the embedded basis is frequently referred to as the ‘material basis’ or the ‘natural basis’. In mesoscale analyses of materials, for example, the computational domain contains numerous crystals that obey a certain constitutive law that takes on a particularly simple form when expressed in terms of the natural basis for the crystal. For cubic crystals, the ‘natural’ axes are normally aligned with the crystal’s cubic lattice directions. Even for non-cubic symmetry, an orthonormal material basis can be established. These axes are not generally aligned with the laboratory axes being used by the host code in a simulation, and they certainly do not usually align with the axes for other crystals in the computation. Such computations usually make use of a rotation tensor to initialize the crystal orientation relative to the global 'lab' axes. During solution of the field equations, the velocity and stress fields are typically expressed in terms of the lab basis. However, during the evaluation of the crystal's constitutive response, these fields are typically *transformed* via a basis transformation to the local crystal basis. Thus, for such computations *both* rotations and basis transformations are used and the computer programmer must pay close attention to which operation currently applies to know whether to use the matrix $[Q]$ or its transpose $[Q]^T$. Verification testing is essential!

Now that we know that R_{ij} equals the direction-cosine matrix Q_{ij}, we can summarize the key results that apply to rotation transformations, keeping in mind that some of the proofs hark back to those provided in chapter 3:

$$\underset{\sim}{e}_i = \underset{\approx}{R} \bullet \underset{\sim}{E}_i, \tag{4.16}$$

$$R_{mn} = Q_{mn} = \underset{\sim}{E}_m \bullet \underset{\sim}{e}_n, \tag{4.17}$$

$$R_{ij} = \underset{\sim}{E}_i \bullet \underset{\approx}{R} \bullet \underset{\sim}{E}_j = Q_{ij}, \tag{4.18}$$

$$\tilde{R}_{ij} = \underset{\sim}{e}_i \bullet \underset{\approx}{R} \bullet \underset{\sim}{e}_j = Q_{ij}, \tag{4.19}$$

$$\underset{\sim}{e}_i = \sum_{j=1}^{3} R_{ji} \underset{\sim}{E}_j. \tag{4.20}$$

These results are, respectively, equivalent to (4.6), (3.47), (4.8), (4.12), and (3.49). Equation (4.17) shows that the components of a rotation tensor are the same as the direction cosines of the rotated basis vectors—the columns of $[R]$ simply contain the rotated basis components with respect to the laboratory basis. Hence, to be a rotation, the columns must form a right-handed orthonormal triad. A routine for testing whether a matrix is a rotation is in section A.1 of the appendix. In *Mathematica*, the command

```
R.Transpose[R] ==IdentityMatrix[3]
```
(4.21)

will return a value of `'True'` if the matrix is orthogonal. To further test if the matrix is a proper rotation, you would use the command `'Det[R]==1'`.

Note the distinction between (4.16) and (4.20). Readers often *wrongly* presume that the subscripts on R_{ji} are accidentally reversed in (4.16)! To obtain (4.20) from (4.16), first note that the jth component of $\underset{\sim}{e}_i$ with respect to the lab basis is obtained by dotting $\underset{\sim}{E}_j$ into $\underset{\sim}{e}_i$. Dotting both sides of (4.16) from the left by $\underset{\sim}{E}_j$ shows that the jth component of $\underset{\sim}{e}_i$ with respect to the lab basis therefore must be R_{ji}, which is exactly what appears in (4.20).

Even though R_{ji} is identically equal to Q_{ij}, we find it helpful to use R_{ji} when we know that we are working with the transformation of a vector to obtain a new vector, as in $\hat{w}_i = R_{ij} w_j$, and we use Q_{ij} when we know that we are working with transformation of the *components of a single vector* resulting from rotating the *observer* to become aligned with the rotated basis, as in $\tilde{v}_i = Q_{ji}v_j$.

Example. Rotation of an angle α about the X-axis.

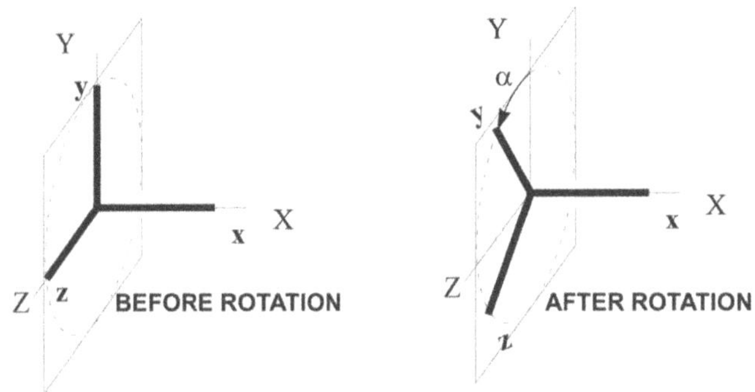

For this rotation about X, the x-axis remains aligned with the X-axis, so $\cos\theta_{xX} = 1$, $\cos\theta_{xY} = 0$, and $\cos\theta_{xZ} = 0$. The angle that the rotated y-axis makes with fixed Y-axis is θ, so $\cos\theta_{yY} = \cos\alpha$. The angle that y makes with Z is $\pi/2 - \theta$, so $\cos\theta_{yZ} = \cos(\pi/2 - \alpha) = \sin\alpha$, and so on. Thus, referring to (3.15) and (4.18), the transformation matrix and the rotation tensor are

$$[Q] = [\underset{\sim}{\boldsymbol{R}}] = \begin{bmatrix} 1 & 0 & 0 \\ 0 & \cos\alpha & -\sin\alpha \\ 0 & \sin\alpha & \cos\alpha \end{bmatrix}. \tag{4.22}$$

As always, the columns of this matrix contain the laboratory components of the rotated basis vectors. Inspecting the picture (using $\underset{\sim}{e}$ for the rotated basis and $\underset{\sim}{E}$ for the laboratory basis),

$$\underset{\sim}{e}_x = \underset{\sim}{E}_X, \tag{4.23a}$$

$$\underset{\sim}{e}_y = (\cos\alpha)\underset{\sim}{E}_Y + (\sin\alpha)\underset{\sim}{E}_Z, \tag{4.23b}$$

$$\underset{\sim}{e}_z = (-\sin\alpha)\underset{\sim}{E}_Y + (\cos\alpha)\underset{\sim}{E}_Z. \tag{4.23c}$$

The coefficients in these expansions go into *columns* of the $[Q]$ matrix. The corresponding direction-cosine table is

	$\underset{\sim}{e}_x$	$\underset{\sim}{e}_y$	$\underset{\sim}{e}_z$
$\underset{\sim}{E}_X$	1	0	0
$\underset{\sim}{E}_Y$	0	$\cos\alpha$	$-\sin\alpha$
$\underset{\sim}{E}_Z$	0	$\sin\alpha$	$\cos\alpha$

Reading off rows, this table says that the laboratory basis is expressed in terms of the rotated basis as

$$\begin{Bmatrix} \underset{\sim}{E}_X \\ \underset{\sim}{E}_Y \\ \underset{\sim}{E}_Z \end{Bmatrix} = \begin{bmatrix} 1 & 0 & 0 \\ 0 & \cos\alpha & -\sin\alpha \\ 0 & \sin\alpha & \cos\alpha \end{bmatrix} \begin{Bmatrix} \underset{\sim}{e}_x \\ \underset{\sim}{e}_y \\ \underset{\sim}{e}_z \end{Bmatrix}, \tag{4.24}$$

while the inverse relationship, giving the rotated basis in terms of the lab basis, is

$$\begin{Bmatrix} \underset{\sim}{e}_x \\ \underset{\sim}{e}_y \\ \underset{\sim}{e}_z \end{Bmatrix} = \begin{bmatrix} 1 & 0 & 0 \\ 0 & \cos\alpha & \sin\alpha \\ 0 & -\sin\alpha & \cos\alpha \end{bmatrix} \begin{Bmatrix} \underset{\sim}{E}_X \\ \underset{\sim}{E}_Y \\ \underset{\sim}{E}_Z \end{Bmatrix}. \tag{4.25}$$

Carefully note the placement of the negative sign, depending on whether the formula is expressing a rotated quantity in terms of a lab quantity or vice versa. The potential for getting this part wrong is why it is crucial to test for the right answer in special cases for which you intuitively know the answer, such as $\alpha = 90°$.

Recall that the matrix $[\underset{\approx}{R}]$ is the same as the matrix $[Q]$. Also recall that the components of position with respect to the lab system are denoted $\{X_1, X_2, X_3\}$ or $\{X, Y, Z\}$. Likewise, the components of that same location with respect to the rotated system are denoted $\{x_1, x_2, x_3\}$ or $\{x, y, z\}$. Thus, (3.65) becomes

$$\begin{Bmatrix} x \\ y \\ z \end{Bmatrix} = \begin{bmatrix} 1 & 0 & 0 \\ 0 & \cos\alpha & \sin\alpha \\ 0 & -\sin\alpha & \cos\alpha \end{bmatrix} \begin{Bmatrix} X \\ Y \\ Z \end{Bmatrix} \tag{4.26}$$

or

$$\begin{aligned} x &= X, \\ y &= (\cos\alpha)Y + (\sin\alpha)Z, \quad . \\ z &= (-\sin\alpha)Y + (\cos\alpha)Z. \end{aligned} \tag{4.27}$$

Rotations about a particular axis will always involve a sub-matrix of the form

$$\begin{bmatrix} \cos\alpha & -\sin\alpha \\ \sin\alpha & \cos\alpha \end{bmatrix}. \tag{4.28}$$

Correct placement of the 'negative sign' (tantamount to verifying that the orthogonal matrix does not need to be transposed) is checked by considering a rotation angle of exactly 90°. In this special limiting case, equation (4.27) becomes

$$\begin{aligned} x &= X, \\ y &= Z, \quad . \\ z &= -y. \end{aligned} \tag{4.29}$$

A visual inspection of the rotation sketch for a ninety degree rotation reveals that the y-axis ends up aligned with the Z-axis (so that $y = Z$) and the z-axis ends pointing opposite to the Y-axis (so that $z = -Y$). Thus, equation (4.29) gives the right result, which means the negative sign is indeed correct in (4.27). This simple '90° limiting case' test can be a lifesaver.

Example. Rotation of an angle α about the Y-axis.

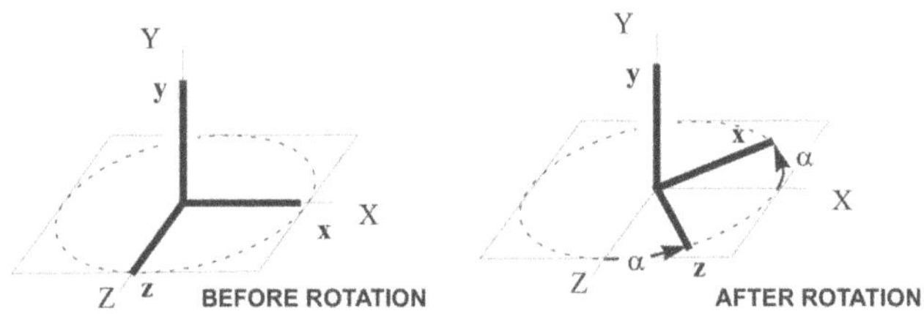

Referring to (3.15) and (4.18), the transformation matrix and the rotation matrix are

$$[Q] = [\underset{\sim}{\boldsymbol{R}}] = \begin{bmatrix} \cos\alpha & 0 & \sin\alpha \\ 0 & 1 & 0 \\ -\sin\alpha & 0 & \cos\alpha \end{bmatrix}. \tag{4.30}$$

Note the relative difference between (4.30) and (4.22). As always, the columns of this matrix contain the laboratory components of the rotated basis vectors. In the picture, it is clear that the rotated x-axis has a *negative* component, $-\sin\alpha$, in the Z-direction!

Recall that the matrix $[\underset{\sim}{\boldsymbol{R}}]$ is the same as the matrix $[Q]$. Also recall that the components of position with respect to the lab system are denoted $\{X_1, X_2, X_3\}$ or $\{X, Y, Z\}$. Likewise, the components of that same location with respect to the rotated system are denoted $\{x_1, x_2, x_3\}$ or $\{x, y, z\}$. Thus, equation (3.65) becomes

$$\begin{Bmatrix} x \\ y \\ z \end{Bmatrix} = \begin{bmatrix} \cos\alpha & 0 & -\sin\alpha \\ 0 & 1 & 0 \\ \sin\alpha & 0 & \cos\alpha \end{bmatrix} \begin{Bmatrix} X \\ Y \\ Z \end{Bmatrix} \tag{4.31}$$

or

$$\begin{aligned} x &= (\cos\alpha)X - (\sin\alpha)Z, \\ y &= Y, \\ z &= (\sin\alpha)X + (\cos\alpha)Z. \end{aligned} \tag{4.32}$$

For the special case of a 90° rotation, this becomes

$$x = -Z, \quad y = Y, \quad \text{and} \quad z = X, \tag{4.33}$$

which is consistent with visual inspection of the rotation figure. Namely, upon a ninety degree rotation, little x ends up pointing in the $-Z$-direction, little y stays equal to big Y, and little z ends up pointing in the big X-direction. Since these observations are consistent with (4.33), we can feel confident that the negative sign in (4.31) was placed in the correct position.

Example. Rotation of an angle α about the Z-axis.

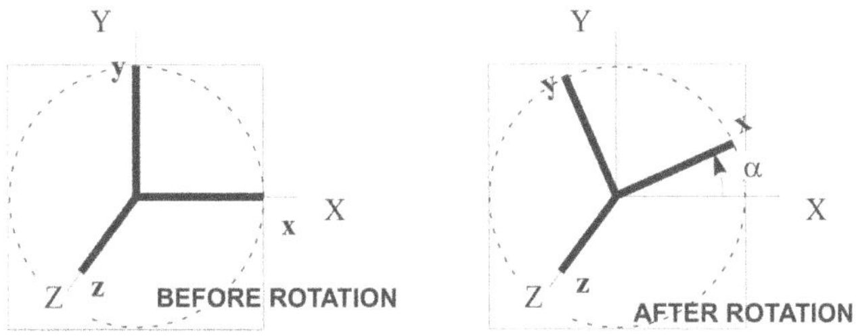

Referring to (3.15) and (4.16), the transformation matrix and the rotation matrix are

$$[Q] = [\underset{\sim}{\boldsymbol{R}}] = \begin{bmatrix} \cos\alpha & -\sin\alpha & 0 \\ \sin\alpha & \cos\alpha & 0 \\ 0 & 0 & 1 \end{bmatrix}. \tag{4.34}$$

For a 90° rotation about the Z-axis, the image specializes to

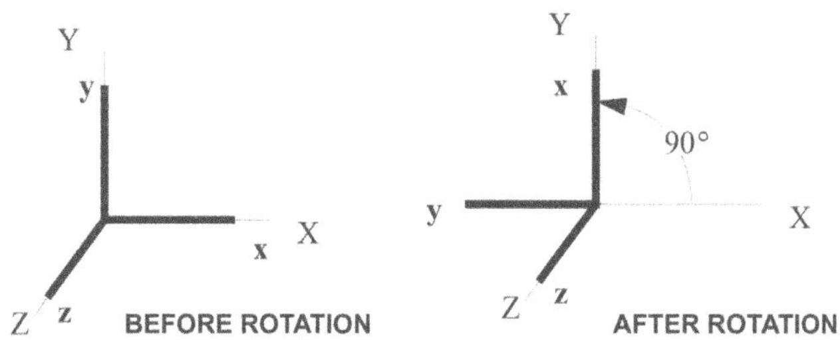

Substituting $\theta = 90°$ into (4.34) gives

$$[Q] = [\underset{\sim}{\boldsymbol{R}}] = \begin{bmatrix} 0 & -1 & 0 \\ 1 & 0 & 0 \\ 0 & 0 & 1 \end{bmatrix}. \tag{4.35}$$

The columns of this matrix contain lab components of the rotated basis, confirming correct placement of the negative!

4.1. Why apparent inconsistency in placement of the negative sign?

It might seem weird that the $-\sin\alpha$ seems to be in a 'different relative spot' in (4.22), (4.30), and (4.34). Rather than looking at the sub-matrix of sines and cosines, the negative is always placed cyclically down and left[3] from the solitary '1' on the diagonal.

[3] Moving cyclically down from the last row means looping around to the first row. Similarly, moving cyclically left from the first column means looping around to the last column.

IOP Publishing

Rotation, Reflection, and Frame Changes
Orthogonal tensors in computational engineering mechanics
R M Brannon

Chapter 5

Axis and angle of rotation

As explained above, a rotation may be described by the direction cosines of the rotation or transformation. These direction cosines are not all independent. Euler's theorem states that any rotation may be specified by an axis and angle of rotation. The axis of rotation, $\underset{\sim}{a}$, is a unit vector and the angle of rotation, α, is defined by the right-hand rule with respect to the axis $\underset{\sim}{a}$. The axis of rotation has two independent components (the third being restricted to ensure that $\|\underset{\sim}{a}\| = 1$). Hence, along with the angle of rotation, only three numbers are needed to completely specify a rotation and therefore only three of the nine direction cosines in (3.15) are independent.

'*The*' axis of rotation is not unique. A rotation angle α about a unit axis $\underset{\sim}{a}$ can be equivalently described by a rotation angle $-\alpha$ about the axis $-\underset{\sim}{a}$. Rotations depend only on the sine and cosine of the rotation angle. Consequently, the rotation angle about an axis $\underset{\sim}{a}$ can be assumed to lie in the range from 0° to 180° without loss in generality. If the rotation angle is between 180° and 360° about an axis $\underset{\sim}{a}$, then the same rotation can be described by instead using an axis $-\underset{\sim}{a}$, with the associated rotation angle then falling between 0° and 180°. Stated differently, there always exists a choice for the rotation axis that will ensure the rotation angle will be in the range from 0° to 180°. We will not necessarily invoke this reduced domain for the rotation angle in all parts of this book, and it is usually inconsequential except in certain applications (such as statistical realizations or rotations). For non-zero rotation angles, there exists a unique rotation axis $\underset{\sim}{a}$ for which $0 < \alpha \leqslant 180°$; over this domain, the sine of the angle is always non-negative and therefore it may be computed from knowledge of the cosine by $\sin\alpha = +\sqrt{1 - \cos^2\alpha}$. If the rotation angle is zero (i.e., no rotation), then the rotation axis is inconsequential and may be taken to be *any* unit vector.

Because, according to Euler's theorem, any rotation can be described by a unit vector $\underset{\sim}{a}$ and a rotation angle α, one can permissibly state that the 'pseudo'-rotation vector defined $\underset{\sim}{s} = \alpha\underset{\sim}{a}$ fully describes a rotation. The 'pseudo' modifier is *not* used to indicate that this is not a full-fledged vector—it merely serves as a reminder that rotation operations do not commute (i.e., if you apply one rotation followed by a second rotation, then the result will be different if you apply the rotations in the opposite order). Thus, even though each individual rotation can be described through individual pseudo-rotation vectors, $\underset{\sim}{s}^{(1)}$ and $\underset{\sim}{s}^{(2)}$, the pseudo-rotation vector corresponding to sequentially applied rotations will not generally equal $\underset{\sim}{s}^{(1)} + \underset{\sim}{s}^{(2)}$. This issue is discussed in much greater detail in chapter 9. For now, we will keep the rotation angle α and the unit rotation axis $\underset{\sim}{a}$ clearly separated.

5.1 Euler–Rodrigues formula

Euler's theorem states that the general displacement of a rigid body having one point fixed is a rotation about some axis that passes through that fixed point. For convenience in this section, we place our origin at the fixed point. All of the equations presented in this section are valid for any rotation angle and any rotation axis (i.e., the rotation angle does not have to be restricted to fall in the range from 0° to 180°, and the axis is not presumed to point in any particular direction).

Consider a rotation operation $\underset{\approx}{R}$ about a unit axis $\underset{\sim}{a}$. Consider an arbitrary position vector $\underset{\sim}{x}$ that points from the fixed origin to an arbitrary point on the rigid body. Letting α denote the rotation angle, the sketch in figure 5.1 shows how $\underset{\sim}{x}$ is transformed to become a new rotated vector $\underset{\approx}{R} \bullet \underset{\sim}{x}$.

The vector $\underset{\sim}{h}$ in figure 5.1 is the part of $\underset{\sim}{x}$ in the direction of $\underset{\sim}{a}$, so it must be given by

$$\underset{\sim}{h} = (\underset{\sim}{x} \bullet \underset{\sim}{a})\underset{\sim}{a}. \tag{5.1}$$

The vector $\underset{\sim}{s}$ is the part of $\underset{\sim}{x}$ that is perpendicular to $\underset{\sim}{a}$, so it must be given by

$$\begin{aligned} \underset{\sim}{s} &= \underset{\sim}{x} - \underset{\sim}{h} \\ &= \underset{\sim}{x} - (\underset{\sim}{x} \bullet \underset{\sim}{a})\underset{\sim}{a}. \end{aligned} \tag{5.2}$$

The vector $\underset{\sim}{t}$ is defined to be perpendicular to both $\underset{\sim}{a}$ and $\underset{\sim}{s}$, but with the same magnitude as $\underset{\sim}{s}$. Recalling that $\underset{\sim}{a}$ is a unit vector, we therefore conclude that

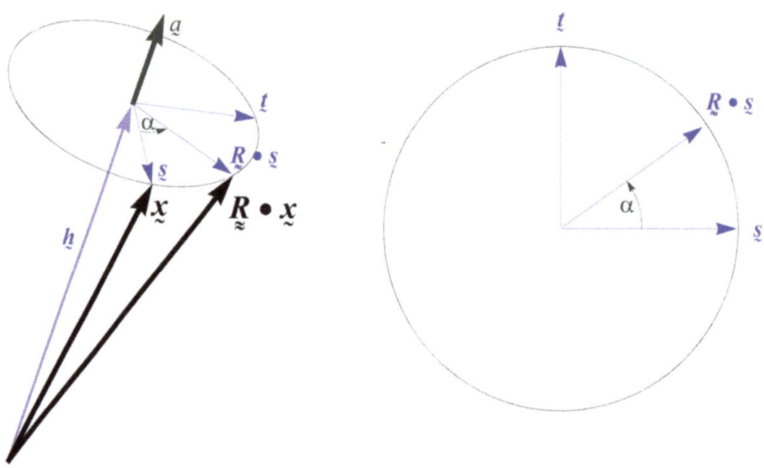

Figure 5.1. Rotation of a vector $\underset{\sim}{x}$ by a known angle α about a known axis $\underset{\sim}{a}$. In this figure, we have introduced some helper vectors: $\underset{\sim}{h}$ is the part of $\underset{\sim}{x}$ in the direction of $\underset{\sim}{a}$, while $\underset{\sim}{s}$ is the perpendicular part. The vector $\underset{\sim}{t}$ is simply defined to be perpendicular to both $\underset{\sim}{a}$ and $\underset{\sim}{x}$ as shown.

$$\begin{aligned} \underset{\sim}{t} &= \underset{\sim}{a} \times \underset{\sim}{s}, \\ &= \underset{\sim}{a} \times (\underset{\sim}{x} - \underset{\sim}{h}), \\ &= \underset{\sim}{a} \times \underset{\sim}{x}. \end{aligned} \tag{5.3}$$

In the last step, we have noted that $\underset{\sim}{a} \times \underset{\sim}{h} = \underset{\sim}{0}$ because $\underset{\sim}{h}$ is parallel to $\underset{\sim}{a}$. Looking strictly in the plane of rotation (the right side of figure 5.1), note that the transformed location of $\underset{\sim}{s}$ is given by

$$\begin{aligned} \underset{\approx}{R} \bullet \underset{\sim}{s} &= (\cos \alpha)\underset{\sim}{s} + (\sin \alpha)\underset{\sim}{t} \\ &= (\cos \alpha)[\underset{\sim}{x} - (\underset{\sim}{x} \bullet \underset{\sim}{a})\underset{\sim}{a}] + (\sin \alpha)[\underset{\sim}{a} \times \underset{\sim}{x}]. \end{aligned} \tag{5.4}$$

From the left side of figure 5.1, note that the rotated vector $\underset{\approx}{R} \bullet \underset{\sim}{x}$ can be broken into two parts as the vector sum of $\underset{\sim}{h}$ and $\underset{\approx}{R} \bullet \underset{\sim}{s}$:

$$\begin{aligned} \underset{\approx}{R} \bullet \underset{\sim}{x} &= \underset{\sim}{h} + \underset{\approx}{R} \bullet \underset{\sim}{s}, \\ &= (\underset{\sim}{x} \bullet \underset{\sim}{a})\underset{\sim}{a} + (\cos \alpha)[\underset{\sim}{x} - (\underset{\sim}{x} \bullet \underset{\sim}{a})\underset{\sim}{a}] + (\sin \alpha)[\underset{\sim}{a} \times \underset{\sim}{x}]. \end{aligned} \tag{5.5}$$

Slight rearrangement gives the so-called *Euler–Rodrigues*[1] *formula*[2]:

$$\boxed{\underset{\approx}{R} \bullet \underset{\sim}{x} = (\cos \alpha)[\underset{\sim}{x} - \underset{\sim}{a}(\underset{\sim}{a} \bullet \underset{\sim}{x})] + \underset{\sim}{a}(\underset{\sim}{a} \bullet \underset{\sim}{x}) + (\sin \alpha)[\underset{\sim}{a} \times \underset{\sim}{x}]}. \tag{5.6}$$

In the next section, we will use this formula to extract a direct-notation expression for the rotation tensor. Toward this end, recognizing that the cross product $\underset{\sim}{a} \times \underset{\sim}{x}$ is linear with respect to $\underset{\sim}{x}$, the representation theorem (from elementary tensor analysis) implies that there must exist a tensor $\underset{\approx}{A}$—called the 'axial' tensor—such that

$$\underset{\sim}{a} \times \underset{\sim}{x} = \underset{\approx}{A} \bullet \underset{\sim}{x} \;\forall\; \underset{\sim}{x}. \tag{5.7}$$

This equation may be written in (Cartesian) component form as

$$\varepsilon_{ijk} a_j x_k = A_{ik} x_k, \tag{5.8}$$

where ε_{ijk} is the permutation symbol defined by

$$\varepsilon_{ijk} = \begin{cases} +1 & \text{if} \quad ijk = 123,\ 231,\ \text{or } 312, \\ -1 & \text{if} \quad ijk = 321,\ 132,\ \text{or } 213, \\ 0 & \text{otherwise.} \end{cases} \tag{5.9}$$

Invoking the quotient rule from elementary vector analysis, the common factor of x_k in (5.8) can be eliminated to give the direct-notation and component formulas for the axial tensor, which we write below along with the indicial and matrix expressions to ensure clarity:

---

[1] In a footnote about this eponymic designation, Goldstein [1], states that Hamel (*Theoretische Mechanik*, p 103) ascribes this theorem to the French mathematician O Rodrigues (1794–1851), but Goldstein claims this might be an error, suggesting that Gibbs was the first to put it in vector form (*Vector Analysis*, p 338). The underlying formula is apparently much older still.

[2] This formula is occasionally cited unnamed and unattributed in textbooks and publications, and unfortunately, it is often cited incorrectly. For example, the version of this formula given in the first edition of [2] failed to reduce to $\underset{\approx}{R} \bullet \underset{\sim}{x} = \underset{\sim}{x}$ when the rotation angle is zero.

$$\underset{\approx}{A} \equiv -\underset{\approx}{\varepsilon} \bullet \underset{\sim}{a} = \underset{\sim}{a} \times \underset{\approx}{I} \quad \Leftrightarrow \quad A_{ij} = -\varepsilon_{ijk}a_k \quad \Leftrightarrow \quad [A] = \begin{bmatrix} 0 & -a_3 & a_2 \\ a_3 & 0 & -a_1 \\ -a_2 & a_1 & 0 \end{bmatrix}. \tag{5.10}$$

The notation $\underset{\sim}{a} \times \underset{\approx}{I}$ is self defining in the sense that

$$\underset{\sim}{a} \times \underset{\sim}{x} = \underset{\sim}{a} \times (\underset{\approx}{I} \bullet \underset{\sim}{x}) = (\underset{\sim}{a} \times \underset{\approx}{I}) \bullet \underset{\sim}{x}. \tag{5.11}$$

Noting that the identity tensor may be expanded as $\underset{\approx}{I} = \delta_{ij}\underset{\sim}{e}_i\underset{\sim}{e}_j = \underset{\sim}{e}_j\underset{\sim}{e}_j$, and also noting that the cross product of basis vectors is given by

$$\underset{\sim}{e}_k \times \underset{\sim}{e}_j = \varepsilon_{ikj}\underset{\sim}{e}_i, \tag{5.12}$$

the notation $\underset{\sim}{a} \times \underset{\approx}{I}$ is 'self-defining' when expanding each part in basis form:

$$\underset{\sim}{a} \times \underset{\approx}{I} = (a_k\underset{\sim}{e}_k) \times (\underset{\sim}{e}_j\underset{\sim}{e}_j) = a_k(\underset{\sim}{e}_k \times \underset{\sim}{e}_j)\underset{\sim}{e}_j = a_k\varepsilon_{ikj}\underset{\sim}{e}_i\underset{\sim}{e}_j = (-\varepsilon_{ijk}a_k)\underset{\sim}{e}_i\underset{\sim}{e}_j. \tag{5.13}$$

The axial tensor is skew-symmetric because the permutation symbol has the property that $\varepsilon_{ijk} = -\varepsilon_{ikj}$. Consequently, the formula $A_{ij} = \varepsilon_{ikj}a_k$ is equivalent to (5.10).

Incidentally, note the following trigonometric identity

$$\cos\alpha = 1 - 2\sin^2\left(\frac{\alpha}{2}\right). \tag{5.14}$$

Also note the vector identity

$$\underset{\sim}{a} \times (\underset{\sim}{a} \times \underset{\sim}{x}) = \underset{\sim}{a}(\underset{\sim}{a} \bullet \underset{\sim}{x}) - \underset{\sim}{x}(\underset{\sim}{a} \bullet \underset{\sim}{a}) = \underset{\sim}{a}(\underset{\sim}{a} \bullet \underset{\sim}{x})\underset{\sim}{x}. \tag{5.15}$$

In the final step, we used the fact that $\underset{\sim}{a}$ is a unit vector to write $\underset{\sim}{a} \bullet \underset{\sim}{a} = 1$. By using the above two identities, (5.6) may be written as

$$\boxed{\underset{\approx}{R} \cdot \underset{\sim}{x} = \underset{\sim}{x} + (\sin\alpha)(\underset{\sim}{a} \times \underset{\sim}{x}) + 2\sin^2\left(\frac{\alpha}{2}\right)[\underset{\sim}{a} \times (\underset{\sim}{a} \times \underset{\sim}{x})]}. \tag{5.16}$$

This is the form of the Euler–Rodrigues identity cited by Argyris [3], and it has an advantage of being written in a form that can be understood by readers who are not familiar with tensor notation.

5.2 Computing the rotation tensor given axis and angle

Given the unit axis $\underset{\sim}{a}$ and angle α, what is the rotation tensor? The answer is found by simply differentiating both sides of the *Euler–Rodrigues formula* of (5.6) by $\underset{\sim}{x}$ (or, alternatively, writing it in indicial form and using the summation rules to ensure that the index on x is the same in all terms). Performing these steps starting with equation (5.6), result is

$$\boxed{\underset{\approx}{R} = (\cos\alpha)[\underset{\approx}{I} - \underset{\sim}{a}\underset{\sim}{a}] + \underset{\sim}{a}\underset{\sim}{a} + (\sin\alpha)[\underset{\sim}{a} \times \underset{\approx}{I}]}, \quad \text{where} \quad \underset{\sim}{a} \times \underset{\approx}{I} \equiv -\underset{\approx}{\varepsilon} \bullet \underset{\sim}{a} \tag{5.17}$$

The indicial notation for this formula is

$$R_{ij} = (\cos\alpha)[\delta_{ij} - a_i a_j] + a_i a_j + (\sin\alpha)[\varepsilon_{ikj} a_k]. \tag{5.18}$$

Here, $\underset{\approx}{\boldsymbol{I}}$ is the identity tensor (represented in Cartesian systems by the 3×3 identity matrix). As explained in chapter 2, the symbol $\underset{\sim}{\boldsymbol{a}}\underset{\sim}{\boldsymbol{a}}$ denotes a vector–vector dyad (often called the *outer product*), which is simply a tensor corresponding to a 3×3 Cartesian matrix with components

$$(\underset{\sim}{\boldsymbol{a}}\underset{\sim}{\boldsymbol{a}})_{ij} = a_i a_j. \tag{5.19}$$

Written out,

$$[\underset{\sim}{\boldsymbol{a}}\underset{\sim}{\boldsymbol{a}}] = \begin{bmatrix} a_1a_1 & a_1a_2 & a_1a_3 \\ a_2a_1 & a_2a_2 & a_2a_3 \\ a_3a_1 & a_3a_2 & a_3a_3 \end{bmatrix}. \tag{5.20}$$

The second-order tensor $\underset{\approx}{\boldsymbol{A}} = \underset{\sim}{\boldsymbol{a}} \times \underset{\approx}{\boldsymbol{I}}$ is the skew-symmetric *axial tensor* associated with the axis $\underset{\sim}{\boldsymbol{a}}$, first defined in (5.10). In direct notation, the tensor $\underset{\approx}{\boldsymbol{A}}$ is defined such that $\underset{\approx}{\boldsymbol{A}} \bullet \underset{\sim}{\boldsymbol{u}} = \underset{\sim}{\boldsymbol{a}} \times \underset{\sim}{\boldsymbol{u}}$ for any vector $\underset{\sim}{\boldsymbol{u}}$. It was shown in the previous section that $\underset{\approx}{\boldsymbol{A}} = \underset{\sim}{\boldsymbol{a}} \times \underset{\approx}{\boldsymbol{I}} = -\underset{\underset{\sim}{\approx}}{\boldsymbol{\varepsilon}} \bullet \underset{\sim}{\boldsymbol{a}}$, where the symbol $\underset{\underset{\sim}{\approx}}{\boldsymbol{\varepsilon}}$ denotes the third-order alternating tensor. Expanding the operation $-\underset{\underset{\sim}{\approx}}{\boldsymbol{\varepsilon}} \bullet \underset{\sim}{\boldsymbol{a}}$ shows that the 3×3 Cartesian matrix for $\underset{\approx}{\boldsymbol{A}}$ is

$$[\underset{\approx}{\boldsymbol{A}}] = [\underset{\sim}{\boldsymbol{a}} \times \underset{\approx}{\boldsymbol{I}}] = = \begin{bmatrix} 0 & -a_3 & a_2 \\ a_3 & 0 & -a_1 \\ -a_2 & a_1 & 0 \end{bmatrix}. \tag{5.21}$$

Using the trigonometric identity $\cos\alpha = 1 - 2\sin^2(\alpha/2)$, the Euler–Rodrigues formula (equation (5.17)) can be written alternatively as

$$\underset{\approx}{\boldsymbol{R}} = \underset{\approx}{\boldsymbol{I}} + 2\sin^2\left(\frac{\alpha}{2}\right)[\underset{\sim}{\boldsymbol{a}}\underset{\sim}{\boldsymbol{a}} - \underset{\approx}{\boldsymbol{I}}] + (\sin\alpha)[\underset{\sim}{\boldsymbol{a}} - \underset{\approx}{\boldsymbol{I}}]. \tag{5.22}$$

Yet another form for the Euler–Rodrigues formula can be constructed if one introduces additional 'helper' vectors, $\underset{\sim}{\boldsymbol{b}}$ and $\underset{\sim}{\boldsymbol{c}}$, defined such $\{\underset{\sim}{\boldsymbol{a}}, \underset{\sim}{\boldsymbol{b}}, \underset{\sim}{\boldsymbol{c}}\}$ forms a right-handed orthonormal triad. In other words, $\underset{\sim}{\boldsymbol{b}}$ is any *unit* vector perpendicular to the unit axis $\underset{\sim}{\boldsymbol{a}}$, and $\underset{\sim}{\boldsymbol{c}}$ is the unit vector defined by the right-hand rule $\underset{\sim}{\boldsymbol{c}} = \underset{\sim}{\boldsymbol{a}} \times \underset{\sim}{\boldsymbol{b}}$. In this case, the identity tensor can be written as

$$\underset{\approx}{\boldsymbol{I}} = \underset{\sim}{\boldsymbol{a}}\underset{\sim}{\boldsymbol{a}} + \underset{\sim}{\boldsymbol{b}}\underset{\sim}{\boldsymbol{b}} + \underset{\sim}{\boldsymbol{c}}\underset{\sim}{\boldsymbol{c}}, \tag{5.23}$$

and the axial tensor, $\underset{\approx}{\boldsymbol{A}} = \underset{\sim}{\boldsymbol{a}} \times \underset{\approx}{\boldsymbol{I}}$, can be written

$$\underset{\approx}{\boldsymbol{A}} = \underset{\sim}{\boldsymbol{a}} \times \underset{\approx}{\boldsymbol{I}} = \underset{\sim}{\boldsymbol{a}} \times (\underset{\sim}{\boldsymbol{a}}\underset{\sim}{\boldsymbol{a}} + \underset{\sim}{\boldsymbol{b}}\underset{\sim}{\boldsymbol{b}} + \underset{\sim}{\boldsymbol{c}}\underset{\sim}{\boldsymbol{c}}) = \underset{\sim}{\boldsymbol{c}}\underset{\sim}{\boldsymbol{b}} - \underset{\sim}{\boldsymbol{b}}\underset{\sim}{\boldsymbol{c}}. \tag{5.24}$$

Thus, the Euler–Rodrigues formula given in (5.17) becomes

$$\underset{\approx}{\boldsymbol{R}} = (\cos\alpha)[\underset{\sim}{\boldsymbol{b}}\underset{\sim}{\boldsymbol{b}} + \underset{\sim}{\boldsymbol{c}}\underset{\sim}{\boldsymbol{c}}] + \underset{\sim}{\boldsymbol{a}}\underset{\sim}{\boldsymbol{a}} + (\sin\alpha)[\underset{\sim}{\boldsymbol{c}}\underset{\sim}{\boldsymbol{b}} - \underset{\sim}{\boldsymbol{b}}\underset{\sim}{\boldsymbol{c}}]. \tag{5.25}$$

Because this form of the Euler–Rodrigues formula requires construction of the two helper vectors $\underset{\sim}{\boldsymbol{b}}$ and $\underset{\sim}{\boldsymbol{c}}$, we consider it to be less efficient than (5.17) or (5.22), but it is mentioned to point out notation structures that can be converted to more elegant forms that are expressed purely in terms of the rotation axis.

5.2.1 Component form of the Euler–Rodrigues formula

Substituting (5.20) and (5.21) into (5.17) gives the component form of the Euler–Rodrigues (tensor) formula:

$$[\underset{\sim}{\boldsymbol{R}}] = \cos\alpha \begin{bmatrix} 1 & 0 & 0 \\ 0 & 1 & 0 \\ 0 & 0 & 1 \end{bmatrix} + (1 - \cos\alpha) \begin{bmatrix} a_1a_1 & a_1a_2 & a_1a_3 \\ a_2a_1 & a_2a_2 & a_2a_3 \\ a_3a_1 & a_3a_2 & a_3a_3 \end{bmatrix} + \sin\alpha \begin{bmatrix} 0 & -a_3 & a_2 \\ a_3 & 0 & -a_1 \\ -a_2 & a_1 & 0 \end{bmatrix}. \tag{5.26}$$

This equation is coded in section A.2 in the appendix. In *Mathematica*, this operation can be defined by the following function:

```
rotationMatrix[axis_, angle_] := Cos[angle] IdentityMatrix[3]
+ (1 - Cos[angle]) Outer[Times, axis, axis]
+ Sin[angle] {{0, -axis[[3]], axis[[2]]}, {axis[[3]], 0, -axis[[1]]},
{-axis[[2]], axis[[1]], 0}};
```

Example. Consider a 120° rotation about an axis that passes through the point (1,1,1). The axis of rotation is just a unit vector that points from the origin to the point (1,1,1):

$$\{\underset{\sim}{a}\} = \frac{1}{\sqrt{3}}\begin{Bmatrix} 1 \\ 1 \\ 1 \end{Bmatrix}. \tag{5.27}$$

Noting that $\sin 120° = \sqrt{3}/2$ and $\cos 120° = -1/2$, equation (5.26) gives

$$[\underset{\sim}{\boldsymbol{R}}] = \left(-\tfrac{1}{2}\right)\begin{bmatrix} 1 & 0 & 0 \\ 0 & 1 & 0 \\ 0 & 0 & 1 \end{bmatrix} + \left(\tfrac{3}{2}\right)\begin{bmatrix} 1/3 & 1/3 & 1/3 \\ 1/3 & 1/3 & 1/3 \\ 1/3 & 1/3 & 1/3 \end{bmatrix} + \frac{\sqrt{3}}{2}\begin{bmatrix} 0 & -1/\sqrt{3} & 1/\sqrt{3} \\ 1/\sqrt{3} & 0 & -1\sqrt{3} \\ -1/\sqrt{3} & 1/\sqrt{3} & 0 \end{bmatrix} \tag{5.28}$$

or

$$[\underset{\sim}{\boldsymbol{R}}] = \begin{bmatrix} 0 & 0 & 1 \\ 1 & 0 & 0 \\ 0 & 1 & 0 \end{bmatrix}. \tag{5.29}$$

This result could be obtained by using the *Mathematica* function defined above in this section. Namely, `rotationMatrix[{1, 1, 1}/Sqrt[3], 120 Degree]`. This problem is a good verification since, intuitively, this rotation should just cyclically rotate the basis vectors, making the three columns of $[\underset{\sim}{\boldsymbol{R}}]$ lab components of $\underset{\sim}{\boldsymbol{E}}_2$, $\underset{\sim}{\boldsymbol{E}}_3$, and $\underset{\sim}{\boldsymbol{E}}_1$.

Another example. Consider a 180° rotation about an axis through the point (1,1,1). Again we have the following unit vector representing the axis of rotation:

$$\{\underset{\sim}{a}\} = \frac{1}{\sqrt{3}}\begin{Bmatrix}1\\1\\1\end{Bmatrix}. \tag{5.30}$$

Noting that sin 180° = 0 and cos 180° = −1. Thus, using (5.26),

$$[\underset{\sim}{R}] = (-1)\begin{bmatrix}1&0&0\\0&1&0\\0&0&1\end{bmatrix} + 2\begin{bmatrix}1/3&1/3&1/3\\1/3&1/3&1/3\\1/3&1/3&1/3\end{bmatrix} = \begin{bmatrix}-1/3&2/3&2/3\\2/3&-1/3&2/3\\2/3&2/3&-1/3\end{bmatrix}. \tag{5.31}$$

This result could be obtained by using the *Mathematica* function defined above in this section. Namely, `rotationMatrix[{1, 1, 1}/Sqrt[3], 180 Degree]`.

This rotation tensor is distinctive because it is symmetric. It is straightforward to prove that this will happen if and only if the angle of rotation is either 0° or 180° (or an integral multiple, of course).

Another similar example. Consider 180° rotation about an axis that passes through the point (3,4,12). We have the following *unit* vector representing the axis of rotation:

$$\{\underset{\sim}{a}\} = \frac{1}{13}\begin{Bmatrix}3\\4\\12\end{Bmatrix}. \tag{5.32}$$

The sine and cosine of the angle are sin 180° = 0 and cos 180° = −1. Thus, using (5.26),

$$[\underset{\sim}{R}] = (-1)\begin{bmatrix}1&0&0\\0&1&0\\0&0&1\end{bmatrix} + \frac{2}{(13)^2}\begin{bmatrix}9&12&36\\12&16&48\\36&48&144\end{bmatrix} = \frac{1}{(13)^2}\begin{bmatrix}-151&24&72\\24&-137&96\\72&96&-119\end{bmatrix}. \tag{5.33}$$

This result could be obtained by using the Mathematica function defined above in this section by executing... `rotationMatrix[{3, 4, 12}/13, 180 Degree]`.

Note that this rotation tensor is symmetric. As previously asserted, this will happen if and only if the angle of rotation is an integral multiple of 180°.

5.2.2 Numerical example (for testing computer codes)

Consider a 76° rotation about an axis that passes through the point (1,3.2,7). The axis of rotation is a unit vector that points from the origin to the point (1,3.2,7), namely

$$a = \{0.128\,842,\ 0.412\,294,\ 0.901\,894\}. \tag{5.34}$$

The sine and cosine of the angle are

$$\begin{aligned}\sin(76°) &= 0.970\,296,\\ \cos(76°) &= 0.241\,922.\end{aligned} \tag{5.35}$$

Using (5.26),

$$\begin{aligned}[\underset{\approx}{R}] &= 0.241\,922\begin{bmatrix}1 & 0 & 0\\ 0 & 1 & 0\\ 0 & 0 & 1\end{bmatrix}\\ &+ (0.758\,078)\begin{bmatrix}0.016\,600\,3 & 0.053\,120\,8 & 0.116\,202\\ 0.053\,120\,8 & 0.169\,987 & 0.371\,846\\ 0.116\,202 & 0.371\,846 & 0.813\,413\end{bmatrix}\\ &+ 0.970\,296\begin{bmatrix}0 & -0.901\,894 & 0.412\,294\\ 0.901\,894 & 0 & -0.128\,842\\ -0.412\,294 & 0.128\,842 & 0\end{bmatrix}\end{aligned} \tag{5.36}$$

or

$$[\underset{\approx}{R}] = \begin{bmatrix}0.254\,506 & -0.834\,834 & 0.488\,138\\ 0.915\,374 & 0.370\,785 & 0.156\,873\\ -0.311\,957 & 0.406\,903 & 0.858\,552\end{bmatrix}. \tag{5.37}$$

This result is confirmed by using the Mathematica function defined in section 5.2.1:

```
rotationMatrix[{1, 3.2, 7}/Sqrt[1+3.2^2+7^2], 76 Degree].
```

5.2.3 Nonlinear rotation mappings

Let $\underset{\sim}{X}$ denote the initial location of a particle in a continuum (not necessarily rigid) body. Let the deformed location of that same particle be given by

$$\underset{\sim}{x} = \underset{\approx}{R} \bullet \underset{\sim}{X}, \tag{5.38}$$

where $\underset{\approx}{R}$ is a proper-orthogonal tensor. It might *seem* that this mapping is linear and hence corresponds to a pure rotation of a rigid body. However, we did not state that $\underset{\approx}{R}$ was independent of $\underset{\sim}{X}$, which is a necessary and sufficient additional proviso for it to be pure rotation. Otherwise, if $\underset{\approx}{R}$ is a function of $\underset{\sim}{X}$, as illustrated in figure 5.2, particles will move in circular paths, but not by the same angles, thus producing inhomogeneous deformation.

5.2.4 Lemma: a special rotation of a rotation

Consider a linear rotation operator $\underset{\approx}{R}$ that is constructed from an axis and angle using the Euler–Rodrigues formula. If we denote the Euler–Rodrigues function generically by

$$\underset{\approx}{R} = f(\underset{\sim}{a}, \alpha), \tag{5.39}$$

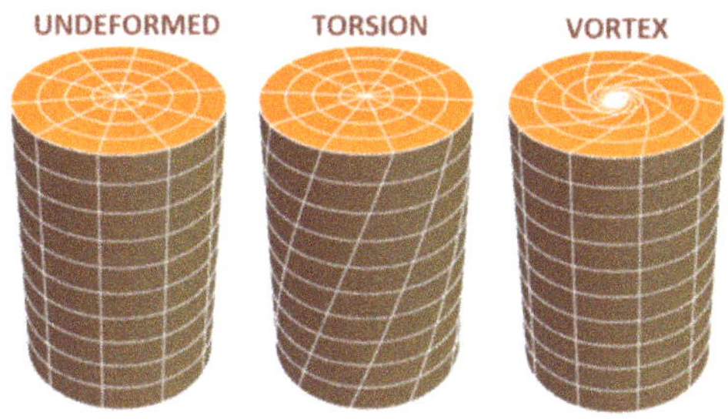

Figure 5.2. Pure circular motion of material does not necessarily imply pure rotation. For torsion, the rotation angle varies with vertical distance. For a vortex, the rotation angle varies with radius.

then, by using the indicial formula in (5.18), it follows that, for any other rotation $\underset{\approx}{Q}$, the Euler–Rodrigues formula has the following property:

$$f(\underset{\approx}{Q} \bullet \underset{\sim}{a}, \alpha) = \underset{\approx}{Q} \bullet f(\underset{\sim}{a}, \alpha) \bullet \underset{\approx}{Q}^T \quad \text{for all orthogonal } \underset{\approx}{Q}. \tag{5.40}$$

In words, changing $\underset{\sim}{a}$ to $\underset{\approx}{Q} \bullet \underset{\sim}{a}$ while keeping the rotation angle unchanged will change $\underset{\approx}{R}$ to $\underset{\approx}{Q} \bullet \underset{\approx}{R} \bullet \underset{\approx}{Q}^T$ and vice versa. This observation is used later to prove that a sequence of rotations about follower axes is identical to the same rotation angles applied to *fixed* axes in reverse order!

Additionally, considering a special case in which $\underset{\approx}{Q}$ is selected to be a rotation $\underset{\approx}{r}$ about the *same* axis $\underset{\sim}{a}$, it follows that $\underset{\approx}{r} \bullet \underset{\sim}{a} = \underset{\sim}{a}$ (i.e., rotation of the rotation axis is just the axis itself), giving the corollary

$$\underset{\approx}{R} = \underset{\approx}{r} \bullet \underset{\approx}{R} \bullet \underset{\approx}{r}^T \quad \text{for any rotations } \underset{\approx}{R} \text{ and } \underset{\approx}{r} \text{ about a shared axis.} \tag{5.41}$$

Thus, rotations about the same axis will commute with each other (i.e., be order independent) even though general rotations (about different axes) do not commute. Furthermore, a rotation is always transversely isotropic with respect to its own rotation axis. In other words, the Cartesian components of a rotation tensor will remain unchanged for any basis change that is a rotation of the laboratory basis about $\underset{\sim}{a}$. This reiterates that setting up any basis for which the 3-direction is aligned with the rotation axis will produce a component matrix of the form in (4.34), regardless of the directions chosen for the first two basis vectors (as long as the basis is right-handed and orthonormal).

5.2.5 Alternative way to construct the rotation tensor

The Euler–Rodrigues formula of (5.17) is the most elegant means of constructing the matrix of the rotation tensor, given axis and angle. However, you might encounter the following alternative algorithm (based on (5.25)), which we present here because it appears occasionally in software source code:

i. Assume the user has supplied the components $\{a_1, a_2, a_3\}$ of the unit rotation-axis vector $\underset{\sim}{a}$ with respect to the laboratory $\{\underset{\sim}{E}_1, \underset{\sim}{E}_2, \underset{\sim}{E}_3\}$-basis. First ensure that the rotation axis is a unit vector by replacing it with $\underset{\sim}{a}/|\underset{\sim}{a}|$.

ii. Determine the smallest component of $\underset{\sim}{a}$. If this component is a_K, then set $\underset{\sim}{\hat{s}} = \underset{\sim}{E}_K$ (i.e., set it equal to the Kth lab basis vector). The $\underset{\sim}{\hat{s}}$ vector will be linearly independent of $\underset{\sim}{a}$, so you can construct a unit vector $\underset{\sim}{b}$ that is perpendicular to $\underset{\sim}{a}$ by

$$\underset{\sim}{b} = \frac{\underset{\sim}{a} \times \underset{\sim}{\hat{s}}}{|\underset{\sim}{a} \times \underset{\sim}{\hat{s}}|}. \tag{5.42}$$

iii. Construct a third vector $\underset{\sim}{c}$ that is right-perpendicular to both $\underset{\sim}{a}$ and $\underset{\sim}{b}$ by

$$\underset{\sim}{c} = \underset{\sim}{a} \times \underset{\sim}{b}. \tag{5.43}$$

iv. Construct a real unitary (proper-orthogonal) matrix $[L]$ whose columns contain the vectors $\{\underset{\sim}{b}, \underset{\sim}{c}, \underset{\sim}{a}\}$:

$$[L] = \begin{bmatrix} b_1 & c_1 & a_1 \\ b_2 & c_2 & a_2 \\ b_3 & c_3 & a_3 \end{bmatrix}. \tag{5.44}$$

v. Then the matrix for the rotation tensor is

$$[\underset{\approx}{R}] = [L] \begin{bmatrix} \cos\alpha & -\sin\alpha & 0 \\ \sin\alpha & \cos\alpha & 0 \\ 0 & 0 & 1 \end{bmatrix} [L]^T. \tag{5.45}$$

In essence, this procedure:

1. constructs the transformation matrix $[L]$ that temporarily transforms to a basis for which $\underset{\sim}{a}$ is in the 3-direction, then
2. performs the rotation about this new 1-direction, and finally
3. transforms back to the laboratory configuration.

Although this approach is legitimate, direct application of the Euler–Rodrigues formula seems more elegant due to its invariant nature. Another comparative advantage of the Euler–Rodrigues formula is that one may take its rate to obtain the material time derivative $\dot{\underset{\approx}{R}}$ as a *structured (direct) notation* function of α, $\underset{\sim}{a}$, $\dot{\alpha}$, and $\dot{\underset{\sim}{a}}$ (see equation (15.16)).

5.2.6 Some properties of the axial tensor

This section (which the reader may skip without loss) summarizes several useful formulas involving the axial tensor $\underset{\approx}{A}$ that was defined in (5.10):

$$\underset{\approx}{A} \equiv -\underset{\approx}{\varepsilon} \bullet \underset{\sim}{a} = \underset{\sim}{a} \times \underset{\approx}{I} \quad \Leftrightarrow \quad A_{ij} = -\varepsilon_{ijk} a_k \quad \Leftrightarrow \quad [A] = \begin{bmatrix} 0 & -a_3 & a_2 \\ a_3 & 0 & -a_1 \\ -a_2 & a_1 & 0 \end{bmatrix}. \tag{5.46}$$

If the axial *tensor* $\underset{\approx}{\mathbf{A}}$ is known, then the axial *vector* $\underset{\sim}{\mathbf{a}}$ can be computed immediately by

$$a_1 = A_{32}, \tag{5.47a}$$
$$a_2 = A_{13}, \tag{5.47b}$$
$$a_3 = A_{21}. \tag{5.47c}$$

This operation may be written in direct notation as

$$\underset{\sim}{\mathbf{a}} = -\frac{1}{2}\underset{\approx}{\boldsymbol{\varepsilon}} : \underset{\approx}{\mathbf{A}}. \tag{5.48}$$

The indicial form is

$$a_i = -\frac{1}{2}\varepsilon_{ipq}A_{pq} = +\frac{1}{2}\varepsilon_{min}A_{mn}. \tag{5.49}$$

The axial tensor is defined to be the linear operator corresponding to the cross product. Specifically,

$$\underset{\sim}{\mathbf{a}} \times \underset{\sim}{\mathbf{x}} = \underset{\approx}{\mathbf{A}} \bullet \underset{\sim}{\mathbf{x}} \quad \text{for any vector } \underset{\sim}{\mathbf{x}}. \tag{5.50}$$

Taking $\underset{\sim}{\mathbf{x}} = \underset{\sim}{\mathbf{a}}$ as a special case shows that

$$\underset{\approx}{\mathbf{A}} \bullet \underset{\sim}{\mathbf{a}} = \underset{\sim}{0}. \tag{5.51}$$

The axial tensor is skew-symmetric. That is,

$$\underset{\approx}{\mathbf{A}}^T = -\underset{\approx}{\mathbf{A}}. \tag{5.52}$$

Axial tensors have interesting properties when they are raised to various powers. For example, even though $\underset{\approx}{\mathbf{A}}$ is skew-symmetric, its *square* $\underset{\approx}{\mathbf{A}}^2$ turns out to be symmetric[3]. Specifically:

$$\boxed{\underset{\approx}{\mathbf{A}}^2 = \underset{\sim}{\mathbf{a}}\underset{\sim}{\mathbf{a}} - \underset{\approx}{\mathbf{I}}}. \tag{5.53}$$

If applied to an arbitrary vector $\underset{\sim}{\mathbf{x}}$, this linear operator gives the *negative* projection of $\underset{\sim}{\mathbf{x}}$ onto the plane whose normal is $\underset{\sim}{\mathbf{a}}$. Replacing $\underset{\sim}{\mathbf{x}}$ by $\underset{\sim}{\mathbf{a}} \times \underset{\sim}{\mathbf{x}}$ in (5.50) shows that

$$\underset{\sim}{\mathbf{a}} \times (\underset{\sim}{\mathbf{a}} \times \underset{\sim}{\mathbf{x}}) = \underset{\approx}{\mathbf{A}}^2 \bullet \underset{\sim}{\mathbf{x}} \quad \text{for any vector } \underset{\sim}{\mathbf{x}}. \tag{5.54}$$

If we take the third power of the axial tensor, we obtain

$$\underset{\approx}{\mathbf{A}}^3 = \underset{\approx}{\mathbf{A}} \bullet \underset{\approx}{\mathbf{A}}^2 = \underset{\approx}{\mathbf{A}} \bullet (\underset{\sim}{\mathbf{a}}\underset{\sim}{\mathbf{a}} - \underset{\approx}{\mathbf{I}}) = (\underset{\approx}{\mathbf{A}} \bullet \underset{\sim}{\mathbf{a}})\underset{\sim}{\mathbf{a}} - \underset{\approx}{\mathbf{A}} \tag{5.55}$$

or, using (5.51),

$$\boxed{\underset{\approx}{\mathbf{A}}^3 = -\underset{\approx}{\mathbf{A}}}. \tag{5.56}$$

[3] This is a nice counter-example to demonstrate that the 'square root' of a symmetric tensor does not necessarily have to be symmetric!

Thus, the cube of the axial tensor is just the negative of the axial tensor itself. Higher powers are computed similarly, and they alternate between being skew-symmetric axial tensors or unit multiples of symmetric projectors.

Recall the Euler–Rodrigues expression for the rotation,

$$\underset{\approx}{\boldsymbol{R}} = \cos\alpha(\underset{\approx}{\boldsymbol{I}} - \underset{\sim}{\boldsymbol{a}}\underset{\sim}{\boldsymbol{a}}) + \underset{\sim}{\boldsymbol{a}}\underset{\sim}{\boldsymbol{a}} + \sin\alpha\underset{\approx}{\boldsymbol{A}}. \tag{5.57}$$

Applying the above identities,

$$\underset{\approx}{\boldsymbol{R}} \bullet \underset{\approx}{\boldsymbol{A}} = \underset{\approx}{\boldsymbol{A}} \bullet \underset{\approx}{\boldsymbol{R}} = \underset{\sim}{\boldsymbol{a}} \times \underset{\approx}{\boldsymbol{R}} = (\cos\alpha)\underset{\approx}{\boldsymbol{A}} + (\sin\alpha)(\underset{\sim}{\boldsymbol{a}}\underset{\sim}{\boldsymbol{a}} - \underset{\approx}{\boldsymbol{I}}), \tag{5.58}$$

$$\underset{\approx}{\boldsymbol{R}}^T \bullet \underset{\approx}{\boldsymbol{A}} = \underset{\approx}{\boldsymbol{A}} \bullet \underset{\approx}{\boldsymbol{R}}^T = \underset{\sim}{\boldsymbol{a}} \times \underset{\approx}{\boldsymbol{R}}^T = (\cos\alpha)\underset{\approx}{\boldsymbol{A}} - (\sin\alpha)(\underset{\sim}{\boldsymbol{a}}\underset{\sim}{\boldsymbol{a}} - \underset{\approx}{\boldsymbol{I}}), \tag{5.59}$$

$$\underset{\sim}{\boldsymbol{a}} \times \underset{\approx}{\boldsymbol{R}} = -(\underset{\approx}{\boldsymbol{R}}^T \times \underset{\sim}{\boldsymbol{a}})^T. \tag{5.60}$$

The axial tensor is generated from the unit axis of rotation. If we set up a basis for which the 3-direction is aligned with the axis of rotation, the component matrix for the axial tensor becomes simply

$$[\underset{\approx}{\boldsymbol{A}}] = \begin{bmatrix} 0 & -1 & 0 \\ 1 & 0 & 0 \\ 0 & 0 & 0 \end{bmatrix}. \tag{5.61}$$

In terms of this special basis, the dyad $\underset{\sim}{\boldsymbol{a}}\underset{\sim}{\boldsymbol{a}}$ becomes simply

$$[\underset{\sim}{\boldsymbol{a}}\underset{\sim}{\boldsymbol{a}}] = \begin{bmatrix} 0 & 0 & 0 \\ 0 & 0 & 0 \\ 0 & 0 & 1 \end{bmatrix}, \tag{5.62}$$

and the rotation matrix is given by

$$[\underset{\approx}{\boldsymbol{R}}] = \begin{bmatrix} c & -s & 0 \\ s & c & 0 \\ 0 & 0 & 1 \end{bmatrix}. \tag{5.63}$$

where $c = \cos\alpha$ and $s = \sin\alpha$. Many proofs are greatly simplified when the 3-direction is aligned with the rotation axis $\underset{\sim}{\boldsymbol{a}}$ Invariance properties of operators (such as trace, tensor composition, etc) then often allow asserting that the result (when cast back in direct notation) is valid for any basis.

5.2.7 Argyris's form of the Euler–Rodrigues formula

Recalling (5.50) and (5.54), a simpler (but computationally less efficient) form for the Euler–Rodrigues formula follows from (5.16),

$$\boxed{\underset{\approx}{\boldsymbol{R}} = \underset{\approx}{\boldsymbol{I}} + (\sin\alpha)\underset{\approx}{\boldsymbol{A}} + 2\sin^2\left(\frac{\alpha}{2}\right)\underset{\approx}{\boldsymbol{A}}^2}, \tag{5.64}$$

or, using the trigonometric identity $\cos\alpha = 1 - 2\sin^2(\alpha/2)$,

$$\boxed{\underset{\approx}{\boldsymbol{R}} = \underset{\approx}{\boldsymbol{I}} + (\sin\alpha)\underset{\approx}{\boldsymbol{A}} + (1 - \cos\alpha)\underset{\approx}{\boldsymbol{A}}^2}. \tag{5.65}$$

Argyris [3] introduces a *pseudo-rotation vector* defined by

$$\underset{\sim}{\boldsymbol{s}} \equiv \alpha\underset{\sim}{\boldsymbol{a}}. \tag{5.66}$$

Of course, Argyris emphasizes the caveats that two successive rotations cannot by represented by simply adding the pseudo-rotation vectors. Nonetheless, the product $\alpha\underset{\sim}{\boldsymbol{a}}$ is certainly well-defined and (in some instances) convenient. Argyris introduces a skew-symmetric tensor $\underset{\approx}{\boldsymbol{S}}$ associated with $\underset{\sim}{\boldsymbol{s}}$ defined such that

$$\underset{\sim}{\boldsymbol{s}} \times \underset{\sim}{\boldsymbol{x}} = \underset{\approx}{\boldsymbol{S}} \bullet \underset{\sim}{\boldsymbol{x}} \text{ for any vector } \underset{\sim}{\boldsymbol{x}} \ \left(\text{therefore,} \quad S_{ij} = -\varepsilon_{ijk}x_k\right). \tag{5.67}$$

Comparing this definition with (5.50), note that

$$\underset{\approx}{\boldsymbol{A}} = \frac{\underset{\approx}{\boldsymbol{S}}}{\alpha}. \tag{5.68}$$

Thus, in terms of Argyris's pseudo-rotation tensor, (5.64) can be written

$$\boxed{\underset{\approx}{\boldsymbol{R}} = \underset{\approx}{\boldsymbol{I}} + \frac{\sin\alpha}{\alpha}\underset{\approx}{\boldsymbol{S}} + \frac{1}{2}\frac{\sin^2(\alpha/2)}{(\alpha/2)^2}\underset{\approx}{\boldsymbol{S}}^2}. \tag{5.69}$$

We will later see that this expression is convenient when expanding the rotation tensor with respect to the rotation angle.

5.2.8 Spinors

In the preceding section, Argyris's form of the Euler–Rodrigues formula was based upon introduction of a new vector $\underset{\sim}{\boldsymbol{s}} = \alpha\underset{\sim}{\boldsymbol{a}}$, where (recall) $\underset{\sim}{\boldsymbol{a}}$ is the unit rotation axis and α is the rotation angle. More generally, it is not uncommon to introduce other vectors —called generalized Rodrigues pseudo-vectors (or *spinor vectors*)—of the form

$$\underset{\sim}{\boldsymbol{b}} = \beta\underset{\sim}{\boldsymbol{a}}, \tag{5.70}$$

where β is typically some function of the rotation angle α,

$$\beta = g(\alpha), \tag{5.71}$$

defined to simplify some particular application at hand. The associated axial tensor $\underset{\approx}{\boldsymbol{B}}$, called the *spinor tensor*, is defined such that

$$\underset{\sim}{\boldsymbol{b}} \times \underset{\sim}{\boldsymbol{x}} = \underset{\approx}{\boldsymbol{B}} \bullet \underset{\sim}{\boldsymbol{x}} \quad \text{for any vector } \underset{\sim}{\boldsymbol{x}}. \tag{5.72}$$

In other words, $B_{ij} = -\varepsilon_{ijk}b_k$. By using the ε - δ identity[4] in an indicial expansion, it is straightforward to show that

$$\underset{\approx}{\boldsymbol{B}}^2 = \underset{\sim}{\boldsymbol{b}}\underset{\sim}{\boldsymbol{b}} - (\underset{\sim}{\boldsymbol{b}} \bullet \underset{\sim}{\boldsymbol{b}})\underset{\approx}{\boldsymbol{I}}. \tag{5.73}$$

[4] From elementary tensor analysis, this identity is $\epsilon_{ijm}\epsilon_{pqm} = \delta_{ip}\delta_{jq} - \delta_{iq}\delta_{jp}$

Thus, any occurrence of the dyad $\underset{\sim}{b}\underset{\sim}{b}$ can be replaced by an expression involving $\underset{\approx}{B}^2$ instead.

In terms of any generalized Rodrigues pseudo-vector, the Euler–Rodrigues formula may be written

$$\underset{\approx}{R} = \underset{\approx}{I} + (\cos\alpha - 1)\left(\underset{\approx}{I} - \frac{\underset{\sim}{b}\underset{\sim}{b}}{\underset{\sim}{b} \bullet \underset{\sim}{b}}\right) + \frac{\sin\alpha}{\sqrt{\underset{\sim}{b} \bullet \underset{\sim}{b}}}\underset{\approx}{B} \tag{5.74}$$

or

$$\underset{\approx}{R} = \underset{\approx}{I} + \frac{\sin\alpha}{\beta}\underset{\approx}{B} + 2\sin^2\left(\frac{\alpha}{2}\right)\frac{\underset{\approx}{B}^2}{\beta^2}. \tag{5.75}$$

Either of these two equations may be regarded as a transformation from vectors to orthogonal tensors. A spinor fully characterizes a rotation if the rotation angle α can be determined from (5.71) by

$$\alpha = g^{-1}(\beta). \tag{5.76}$$

If so, then (5.76) can be substituted into (5.75) to give an expression for the rotation that is cast entirely in terms of the spinor vector, $\underset{\sim}{b} = \beta\underset{\sim}{a}$. Otherwise, the rotation angle must be given explicitly. Moreover, for a spinor representation to be truly useful, it must be always possible to infer the rotation axis $\underset{\sim}{a}$ from the spinor vector $\underset{\sim}{b}$. Of course, equation (5.70) implies that the rotation axis is simply a unit vector in the direction of $\underset{\sim}{b}$. However, in some upcoming spinor examples, $\beta \to \infty$ or $\beta \to 0$ at some rotation angles, which means that the representation fails at those angles.

In terms of a basis for which the 3-direction is aligned with $\underset{\sim}{a}$, the component matrices for $\underset{\approx}{B}$ and $\underset{\approx}{B}^2$ are

$$[\underset{\approx}{B}] = \beta\begin{bmatrix} 0 & -1 & 0 \\ 1 & 0 & 0 \\ 0 & 0 & 0 \end{bmatrix} \quad \text{and} \quad [\underset{\approx}{B}^2] = \beta^2\begin{bmatrix} -1 & 0 & 0 \\ 0 & -1 & 0 \\ 0 & 0 & 0 \end{bmatrix}. \tag{5.77}$$

Spinors serve as alternative representations for rotations. Because spinor tensors are derived from spinor vectors, they may be regarded as three-dimensional (3D) entities. Being skew-symmetric, spinor tensors belong to a *linear* manifold (i.e., the sum of two skew-symmetric tensors is itself a skew-symmetric tensor) whereas rotations do not form a linear manifold. It is shown later that rotations may be regarded as points on the 3D surface of a unit sphere that is embedded within higher dimensional 9D Euclidean tensor space. Spinor mappings are analogous to projections in which the 2D curved surface of an ordinary 3D sphere is mapped to a 2D flat plane. In cartography, maps of the world can be shown on 2D pieces of paper in many different ways, with the price of having to choose whether or not to preserve angles or relative areas (never both). Likewise, there are multiple ways to characterize a rotation tensor though the use of spinors. Several choices for this mapping are listed below. The entries listed below differ only by the choice for the function $g(\alpha)$ in (5.71).

If the scaling factor in (5.70) is chosen to be

$$\beta = 1, \tag{5.78}$$

then (5.75) becomes the standard Euler–Rodrigues formula

$$\underset{\approx}{\boldsymbol{R}} = \underset{\approx}{\boldsymbol{I}} + \sin\alpha\underset{\approx}{\boldsymbol{B}} + 2\sin^2\left(\frac{\alpha}{2}\right)\underset{\approx}{\boldsymbol{B}}^2. \tag{5.79}$$

Argyris's pseudo-vector is obtained by choosing

$$\beta = \alpha. \tag{5.80}$$

It was shown earlier that, with this choice, equation (5.75) becomes

$$\underset{\approx}{\boldsymbol{R}} = \underset{\approx}{\boldsymbol{I}} + \frac{\sin\alpha}{\alpha}\underset{\approx}{\boldsymbol{B}} + \frac{1}{2}\frac{\sin^2(\alpha/2)}{(\alpha/2)^2}\underset{\approx}{\boldsymbol{B}}^2. \tag{5.81}$$

It will be shown later that this choice simplifies the exponential expression for the rotation tensor as

$$\underset{\approx}{\boldsymbol{R}} = e^{\underset{\approx}{\boldsymbol{B}}}, \tag{5.82}$$

which is shorthand notation for the infinite series

$$\underset{\approx}{\boldsymbol{R}} = \underset{\approx}{\boldsymbol{I}} + \underset{\approx}{\boldsymbol{B}} + \frac{1}{2!}\underset{\approx}{\boldsymbol{B}}^2 + \frac{1}{3!}\underset{\approx}{\boldsymbol{B}}^3 + \ldots. \tag{5.83}$$

A different choice for β (used by Rodrigues to simplify the Cayley theorem) is

$$\beta = \tan\frac{\alpha}{2}, \tag{5.84}$$

in which case, (5.75) becomes

$$\underset{\approx}{\boldsymbol{R}} = \underset{\approx}{\boldsymbol{I}} + 2\cos^2\left(\frac{\alpha}{2}\right)(\underset{\approx}{\boldsymbol{B}} + \underset{\approx}{\boldsymbol{B}}^2) = (\underset{\approx}{\boldsymbol{I}} + \underset{\approx}{\boldsymbol{B}})\bullet(\underset{\approx}{\boldsymbol{I}} - \underset{\approx}{\boldsymbol{B}})^{-1}. \tag{5.85}$$

Another choice (used by DeVeubeke to eliminate singularities at $\alpha = 180°$) takes

$$\beta = \sin\frac{\alpha}{2}, \tag{5.86}$$

in which case, (5.75) becomes

$$\underset{\approx}{\boldsymbol{R}} = \underset{\approx}{\boldsymbol{I}} + 2\cos\left(\frac{\alpha}{2}\right)\underset{\approx}{\boldsymbol{B}} + 2\underset{\approx}{\boldsymbol{B}}^2. \tag{5.87}$$

The pseudo-vector associated with this spinor formulation,

$$\underset{\sim}{\boldsymbol{b}} = \sin\left(\frac{\alpha}{2}\right)\underset{\sim}{\boldsymbol{a}}, \tag{5.88}$$

is identical to the pseudo-vector $\underset{\sim}{\boldsymbol{u}}$ defined later in (7.14) in the context of *quaternions*.

A final choice (rarely explored in the engineering literature) is

$$\beta = \frac{\alpha - \sin\alpha}{\pi}, \tag{5.89}$$

where $0 < \alpha \leqslant \pi$. This choice is important because, as shown in (17.134), a uniform distribution for β leads to the proper *non-uniform* distribution of the rotation angle α that is required for a uniform distribution of random rotation tensors. A related choice [4] is

$$\beta = \left[\frac{3}{4\pi^2}(\alpha - \sin\alpha)\right]^{1/3}. \tag{5.90}$$

As will be explained later, all rotation tensors are points on the curvilinear 3D surface of a hypersphere embedded in Euclidean 9D tensor space. All invertible spinors represent projections of rotations from this abstract higher-dimensional space to ordinary vectors in ordinary Euclidean 3D space. Equation (5.90) is an 'equal-volume' mapping, analogous to the equal-area projection illustrated in figure 17.12. The associated spinor vector, $\underset{\sim}{\boldsymbol{b}} = \beta\underset{\sim}{\boldsymbol{a}}$, is an equal-volume projection of the quaternion on the 3D surface of a sphere contained within a 4D curvilinear quaternion space representation of a rotation to the 3D linear space of ordinary vectors. By this, we mean that an 'area' element on the hypersphere (which is actually a 3D curvilinear volume element) corresponds to an ordinary 3D volume element generated by this spinor. It will be explained later that a uniformly random rotation tensor corresponds to a uniform distribution on the surface of the quaternion hypersphere. Since (5.90) applies for rotation angles in the range $0 < \alpha \leqslant \pi$, it generates a set of vectors $\beta\underset{\sim}{\boldsymbol{a}}$ that fall within the interior (not just surface) of an ordinary sphere of radius $[3/(4\pi)]^{1/3}$, which is a sphere of unit volume. You can generate a uniformly random rotation tensor through the following steps: (1) first generate a uniformly random point within this ordinary unit-volume sphere in ordinary 3D space, (2) set β equal to the magnitude of the generated vector, (3) solve (5.90) numerically for α, and (4) use the Euler–Rodrigues formula in (5.26) to construct the rotation tensor. More efficient algorithms for generating uniformly random rotations will be discussed in chapter 17.

Whatever the choice for the scaling factor $\beta = g(\alpha)$ (except $\beta = 1$), the associated spinor tensor $\underset{\approx}{\boldsymbol{B}}$ is often regarded as completely characterizing the rotation. To infer both the axis and angle from a spinor tensor, $\underset{\approx}{\boldsymbol{B}}$ you would first need to compute the axial vector $\underset{\sim}{\boldsymbol{b}}$ from the formula

$$b_i = -\tfrac{1}{2}\varepsilon_{ijk}B_{jk} \quad (\text{i.e., } b_1 = B_{32},\ b_2 = B_{13},\ \text{and } b_3 = B_{21}). \tag{5.91}$$

From there, the axis and angle of rotation would be found by

$$\beta = \sqrt{\underset{\sim}{\boldsymbol{b}} \bullet \underset{\sim}{\boldsymbol{b}}}, \tag{5.92}$$

$$\alpha = g^{-1}(\beta), \tag{5.93}$$

$$\underset{\sim}{a} = \frac{\underset{\sim}{b}}{\beta}. \tag{5.94}$$

Clearly, for these operations to be valid, the scaling function $\beta = g(\alpha)$ would need to be non-zero, non-infinite, and uniquely invertible for the rotation angle $\alpha = g^{-1}(\beta)$. All but the first scaling function give a zero value for β at rotation angle of 0°, making (5.64) not applicable, but this is not a 'show-stopper' since the rotation axis is arbitrary when $\alpha = 0$. Rodrigues's function, (5.84) becomes singular at 180°, making it unacceptable for characterizing rotations.

5.3 Corollary to the Euler–Rodrigues formula: existence of a preferred basis

Recall the Euler–Rodrigues expression, equation (5.26), for expressing a rotation tensor solely in terms of the angle α and axis $\underset{\sim}{a}$ of rotation:

$$[\underset{\approx}{R}] = \cos\alpha \begin{bmatrix} 1 & 0 & 0 \\ 0 & 1 & 0 \\ 0 & 0 & 1 \end{bmatrix} + (1 - \cos\alpha) \begin{bmatrix} a_1a_1 & a_1a_2 & a_1a_3 \\ a_2a_1 & a_2a_2 & a_2a_3 \\ a_3a_1 & a_3a_2 & a_3a_3 \end{bmatrix} + \sin\alpha \begin{bmatrix} 0 & -a_3 & a_2 \\ a_3 & 0 & -a_1 \\ -a_2 & a_1 & 0 \end{bmatrix}. \tag{5.95}$$

The above expression holds no matter what basis is used. If, however, one sets up a basis such that the 3-direction is coincident with $\underset{\sim}{a}$ then

$$a_1 = a_2 = 0 \qquad \text{and} \qquad a_3 = 1. \tag{5.96}$$

In this case, the Euler–Rodrigues formula simplifies to

$$[\underset{\approx}{R}] = \begin{bmatrix} \cos\alpha & -\sin\alpha & 0 \\ \sin\alpha & \cos\alpha & 0 \\ 0 & 0 & 1 \end{bmatrix} \quad \text{when the 3-direction is aligned with } \underset{\sim}{a}. \tag{5.97}$$

In other words, it is always possible to set up a basis such that the component matrix of the rotation is given by the above expression. This can be very useful in many situations. For example, we can take the trace of (5.97) to obtain

$$\text{tr}\underset{\approx}{R} = 1 + 2\cos\alpha. \tag{5.98}$$

Even though this formula was obtained by using the expression for $\underset{\approx}{R}$ in its preferred basis, we know that it holds for *any* basis because the trace operation is an invariant. Hence, as exploited in the next section, the above formula provides a means of computing the angle of rotation when only the rotation matrix is known.

5.4 Computing axis and angle given the rotation tensor

Now consider the reverse problem. Suppose you have the rotation tensor $\underset{\approx}{R}$ and you seek the axis and angle of rotation. Actually, saying 'the axis' is misleading. More correctly, we should say 'an axis'. After all, if the rotation angle is zero, then the axis is arbitrary (there is an infinite number of solutions). If a rotation angle α is measured by right-hand-rule about a rotation direction $\underset{\sim}{a}$, then an angle of $-\alpha$ about $-\underset{\sim}{a}$ will produce the *same* rotation tensor. Therefore, the solution for the axis and angle of rotation is not unique. For non-zero rotation angles, there are always two solutions. We know that either $\underset{\sim}{a}$ or $-\underset{\sim}{a}$ serves equally well as the rotation axis. With one choice, the rotation angle ranges from 0 to π (plus or minus an integral multiple of 2π). For the other choice, the rotation angle ranges from 0 to $-\pi$ (again plus or minus an integral multiple of 2π). The *principal solution* will be the one such that

$$0 \leqslant \alpha \leqslant \pi, \quad \text{measured by right-hand-rule about } \underset{\sim}{a}. \tag{5.99}$$

Once this principal solution is found, then the family of general solutions for the axis and angle of rotation are of the form

$$\begin{aligned} &2\pi m + \alpha \quad \text{measured by right-hand-rule about } \underset{\sim}{a}, \text{ and} \\ &2\pi m - \alpha \quad \text{measured by right-hand-rule about } -\underset{\sim}{a}, \end{aligned} \tag{5.100}$$

where m is any integer.

5.4.1 Finding the principal rotation angle

Equation (5.98) gives

$$\boxed{\cos\alpha = \frac{\operatorname{tr}\underset{\approx}{R} - 1}{2} = \tfrac{1}{2}(R_{11} + R_{22} + R_{33} - 1)}. \tag{5.101}$$

The inverse cosine has *two* solutions on the interval from 0 to 2π, so simply knowing the cosine is usually not sufficient to determine an angle. However, the inverse cosine *does* uniquely determine the *principal* rotation angle of (5.99) because that angle is required to range from 0 to π. When using the principal angle of rotation, the direction of the axis ($\underset{\sim}{a}$ or $-\underset{\sim}{a}$) must consistently match with the principal rotation angle. Matching the principal rotation angle with an oppositely oriented axis will give the wrong result if you try to reconstruct the rotation tensor using the Euler–Rodrigues formula.

5.4.2 Finding the rotation axis

Recall the Euler–Rodrigues expression for the rotation

$$\underset{\approx}{R} = (\cos\alpha)\underset{\approx}{I} + (1 - \cos\alpha)\underset{\sim}{a}\underset{\sim}{a} + \sin\alpha\underset{\approx}{A}. \tag{5.102}$$

First we are going to show how to invert this formula to obtain $\underset{\sim}{a}\underset{\sim}{a}$ as a function of $\underset{\approx}{R}$. Then we are going to explain why knowing *only* $\underset{\sim}{a}\underset{\sim}{a}$ is not adequate for deducing the *principal* axis $\underset{\sim}{a}$. Then we will move on to invert the above formula for $\underset{\approx}{A}$ as a function of $\underset{\approx}{R}$, which in turn gives the correct formula for the *principal* axis.

Recalling that the axial tensor $\underset{\approx}{A}$ is skew-symmetric, note that the symmetric part of the rotation (i.e., $\operatorname{sym}\underset{\approx}{R} \equiv \frac{1}{2}(\underset{\approx}{R} + \underset{\approx}{R}^T)$) is given by

$$\operatorname{sym}\underset{\approx}{R} = (\cos\alpha)\underset{\approx}{I} + (1 - \cos\alpha)\underset{\sim}{a}\underset{\sim}{a}. \tag{5.103}$$

Solving for the dyad $\underset{\sim}{a}\underset{\sim}{a}$ gives

$$\underset{\sim}{a}\underset{\sim}{a} = \frac{\text{sym}\underset{\approx}{\boldsymbol{R}} - (\cos\alpha)\underset{\approx}{\boldsymbol{I}}}{(1 - \cos\alpha)}. \tag{5.104}$$

In the degenerate case of a zero rotation angle, we know that $\cos\alpha = 1$ and the rotation tensor is just the identity tensor. For this special case, equation (5.104) is an indeterminate 0/0 form. However, we know that the rotation axis is arbitrary whenever the rotation angle is zero. Hence, it may be set equal to $\underset{\sim}{e}_1$ without loss. The remainder of this discussion concerns the non-trivial case of non-zero rotation angles. Recall from (5.101) that

$$\cos\alpha = \frac{\text{tr}\underset{\approx}{\boldsymbol{R}} - 1}{2}. \tag{5.105}$$

Hence, for any non-zero rotation angle, we obtain an expression for the dyad $\underset{\sim}{a}\underset{\sim}{a}$:

$$\boxed{\underset{\sim}{a}\underset{\sim}{a} = \frac{\underset{\approx}{\boldsymbol{R}} + \underset{\approx}{\boldsymbol{R}}^T + (1 - \text{tr}\underset{\approx}{\boldsymbol{R}})\underset{\approx}{\boldsymbol{I}}}{3 - \text{tr}\underset{\approx}{\boldsymbol{R}}}}. \tag{5.106}$$

Recall from (5.20), that

$$[\underset{\sim}{a}\underset{\sim}{a}] = \begin{bmatrix} a_1a_1 & a_1a_2 & a_1a_3 \\ a_2a_1 & a_2a_2 & a_2a_3 \\ a_3a_1 & a_3a_2 & a_3a_3 \end{bmatrix}. \tag{5.107}$$

As seen, any non-zero column (or row) of $[\underset{\sim}{a}\underset{\sim}{a}]$ must be parallel to the rotation axis $\underset{\sim}{a}$. It is computationally most efficient to ignore the denominator in (5.106) and simply say that the rotation axis can be obtained by normalizing any non-zero column (or row) of the numerator:

$$\underset{\approx}{\boldsymbol{R}} + \underset{\approx}{\boldsymbol{R}}^T + (1 - \text{tr}\underset{\approx}{\boldsymbol{R}})\underset{\approx}{\boldsymbol{I}} \quad \leftarrow\text{see the warning below!} \tag{5.108}$$

WARNING: A disadvantage of solving the dyad $\underset{\sim}{a}\underset{\sim}{a}$ for the axis $\underset{\sim}{a}$ is that it does not give any information about the *direction* of the axis vector $\underset{\sim}{a}$. If the principal inverse cosine is to be used to find the rotation angle α from (5.105), then the result will be an angle that ranges from 0 to π. Hence, we *must* select the direction of the axial vector $\underset{\sim}{a}$ consistently. After normalizing a non-zero column of (5.108) to obtain a *trial* rotation axis $\underset{\sim}{a}$, we might be required to change the answer to $-\underset{\sim}{a}$ to be consistent with our assumption that the rotation angle lies between 0 and π. Consider, for example, a rotation obtained by a right-handed rotation of 270° about the $\underset{\sim}{e}_3$-axis:

$$[\underset{\approx}{\boldsymbol{R}}] = \begin{bmatrix} 0 & 1 & 0 \\ -1 & 0 & 0 \\ 0 & 0 & 1 \end{bmatrix}. \tag{5.109}$$

We know that 270° about the $\underset{\sim}{e}_3$-axis is *not* the principal solution for the angle and axis of rotation because 270° does not lie in the interval from 0° to 180°. However, equation (5.109) is equivalent to a 90° rotation about the $-\underset{\sim}{e}_3$-direction, and *this* is the principal solution that we seek.

First, applying (5.105) gives

$$\alpha = \mathrm{ArcCos}\left(\frac{\mathrm{tr}\underset{\approx}{\boldsymbol{R}} - 1}{2}\right) = \mathrm{ArcCos}(0) = 90°, \tag{5.110}$$

which correctly corresponds to the principal rotation angle. The expression in (5.108) becomes

$$[\underset{\approx}{\boldsymbol{R}} + \underset{\approx}{\boldsymbol{R}}^T + (1 - \mathrm{tr}\underset{\approx}{\boldsymbol{R}})\underset{\approx}{\boldsymbol{I}}] = \begin{bmatrix} 0 & 0 & 0 \\ 0 & 0 & 0 \\ 0 & 0 & 2 \end{bmatrix}. \tag{5.111}$$

Normalizing the only non-zero column gives

$$\begin{Bmatrix} 0 \\ 0 \\ 1 \end{Bmatrix}, \tag{5.112}$$

which is the wrong result. The correct answer for the *principal* rotation axis is

$$\begin{Bmatrix} 0 \\ 0 \\ -1 \end{Bmatrix}. \tag{5.113}$$

At this point, we do not have any criterion that indicates that the sign of (5.112) needs to be changed! Hence, attempting to find $\underset{\sim}{a}$ from only the dyad $\underset{\sim}{a}\underset{\sim}{a}$ is somewhat of a dead end. Of course, one could always use the Euler–Rodrigues formula to reconstruct the rotation tensor using (5.112) and then reverse the rotation axis if the result does not equal the original rotation tensor. A more elegant work-around begins by first abandoning the requirement that the rotation angle must be in the range from 0 to π. Instead, when the rotation angle is allowed to fall anywhere in the full range from 0 to 2π, then the directions, $\underset{\sim}{a}$ and $-\underset{\sim}{a}$, of the rotation axis are both acceptable if they are paired with consistent rotation angles. Multiplying the Euler–Rodrigues formula (5.17) by the axial tensor $\underset{\approx}{A}$ associated with either selected axis $\underset{\sim}{a}$ and taking the trace leads to the following formula for the *sine* of the rotation angle:

$$\sin\alpha = \tfrac{1}{2}[a_1(R_{32} - R_{23}) + a_2(R_{13} - R_{31}) + a_3(R_{21} - R_{12})]. \tag{5.114}$$

Recall that (5.105) gives

$$\cos\alpha = \tfrac{1}{2}(R_{11} + R_{22} + R_{33} - 1). \tag{5.115}$$

Thus, a two-argument arctangent operator will produce the rotation angle in the range from 0 to 2π that is consistent with the selected axis direction. This method, which permits a full 360° range for the rotation angle, is attractive for problems in which you desire rotation axes whose orientation satisfies certain properties. You might, for example, desire that the rotation axis point as closely as possible to the first octant of 3D space. In this case, if the rotation axis has a negative inner product with the vector {1,1,1} then its direction is simply reversed. For example, using this approach for the rotation matrix in (5.109) the axis direction of (5.112) has the desired positive inner product with the {1,1,1}-direction. Using this axis in (5.114) and (5.115) gives

$$\cos\alpha = 0 \quad \text{and} \quad \sin\alpha = -1. \tag{5.116}$$

A two-argument arctangent, implies a rotation angle of 270° (or, equivantly, −90°), which is correct. If the *principal* rotation angle and *principal* rotation axis are truly desired, then some adjustment might be needed. If the right-hand side of (5.114) evaluates to a positive number, then the principal solution has been found. If not, then the axis direction is merely reversed and the *principal* rotation angle can be found via an ordinary arccos operation on (5.115).

5.4.3 Finding the rotation axis: another approach

Now we will discuss an alternative method for obtaining the axis and angle of rotation that instead focuses on finding the *axial tensor* $\underset{\approx}{\boldsymbol{A}} = \underset{\sim}{\boldsymbol{a}} \times \underset{\approx}{\boldsymbol{I}}$, defined in (5.10), from which the principal rotation axis may be found by applying (5.47). Again recall the Euler–Rodrigues expression for the rotation:

$$\underset{\approx}{\boldsymbol{R}} = (\cos\alpha)\underset{\approx}{\boldsymbol{I}} + (1 - \cos\alpha)\underset{\sim}{\boldsymbol{a}}\underset{\sim}{\boldsymbol{a}} + (\sin\alpha)\underset{\approx}{\boldsymbol{A}}. \tag{5.117}$$

This time, we will take the *skew-symmetric part* to obtain

$$\text{skw}\,\underset{\approx}{\boldsymbol{R}} = \tfrac{1}{2}(\underset{\approx}{\boldsymbol{R}} - \underset{\approx}{\boldsymbol{R}}^T) = (\sin\alpha)\underset{\approx}{\boldsymbol{A}}. \tag{5.118}$$

Sometimes this result is written in terms of the 'ax', *axial vector operator*. For any tensor $\underset{\approx}{\boldsymbol{B}}$ (skew-symmetric or not), ax$\underset{\approx}{\boldsymbol{B}}$ is a vector $\underset{\sim}{\boldsymbol{b}}$ with components $b_i = -\varepsilon_{ijk}B_{jk}$. Since the permutation symbol ε_{ijk} is skew-symmetric with respect to any two of its indices, note that $\text{ax}\,\underset{\approx}{\boldsymbol{B}} = \text{ax}(\text{skw}\underset{\approx}{\boldsymbol{B}}) = \text{ax}\left(\tfrac{1}{2}[\underset{\approx}{\boldsymbol{B}} - \underset{\approx}{\boldsymbol{B}}^T]\right)$. Therefore, since the axial vector associated with $\underset{\approx}{\boldsymbol{A}}$ is the unit rotation axis $\underset{\sim}{\boldsymbol{a}}$, equation (5.118) may sometimes be written in publications as

$$\text{ax}\underset{\approx}{\boldsymbol{R}} = (\sin\alpha)\underset{\sim}{\boldsymbol{a}}. \tag{5.119}$$

Thus, as long as the rotation angle is neither 0 nor 180°,

$$\underset{\approx}{\boldsymbol{A}} = \frac{1}{2(\sin\alpha)}(\underset{\approx}{\boldsymbol{R}} - \underset{\approx}{\boldsymbol{R}}^T) \quad \left(\text{or, equivalently, } \underset{\sim}{\boldsymbol{a}} = \frac{\text{ax}\underset{\approx}{\boldsymbol{R}}}{\sin\alpha}\right). \tag{5.120}$$

Applying (5.47) gives the principal rotation axis:

$$\boxed{a_1 = \frac{1}{2(\sin\alpha)}(R_{32} - R_{23})}, \tag{5.121a}$$

$$\boxed{a_2 = \frac{1}{2(\sin\alpha)}(R_{13} - R_{31})}, \tag{5.121b}$$

$$\boxed{a_3 = \frac{1}{2(\sin\alpha)}(R_{21} - R_{12})}. \tag{5.121c}$$

The above formula gives the correct answer for the rotation axis regardless of whether the rotation angle is the *principal* angle. When we seek the *principal* axis, then we know that $\sin\alpha > 0$. Consequently, $\sin\alpha$ may be computed by the *positive* square root in the following formula

$$\sin\alpha = +\sqrt{1 - \cos^2\alpha} = +\sqrt{1 - \left(\frac{\mathrm{tr}\underset{\approx}{\mathbf{R}} - 1}{2}\right)^2}. \tag{5.122}$$

For the special case that $\sin\alpha = 0$ equation (5.121) becomes an indeterminate 0/0 form. In this case, we know that the rotation angle is either 0° (in which case the rotation axis is arbitrary) or 180°. In either case, the *pointing direction* of rotation axis is irrelevant[5]. For this special case, the method of (5.108) may be applied. The rotation tensor is symmetric whenever the rotation angle is exactly 0° or exactly 180°, so a rotation axis may be constructed in this case by normalizing any non-zero row (or column) of $\underset{\approx}{\mathbf{R}} + \underset{\approx}{\mathbf{I}}$.

Incidentally, we can always define a vector $\underset{\sim}{\boldsymbol{\gamma}}$ to be the axial vector associated with $\mathrm{skw}\underset{\approx}{\mathbf{R}}$. That is,

$$\begin{aligned} \gamma_1 &= \tfrac{1}{2}(R_{32} - R_{23}), \\ \gamma_2 &= \tfrac{1}{2}(R_{13} - R_{31}), \\ \gamma_3 &= \tfrac{1}{2}(R_{21} - R_{12}). \end{aligned} \tag{5.123}$$

Note from (5.118) that

$$\underset{\sim}{\boldsymbol{\gamma}} = (\sin\alpha)\underset{\sim}{\mathbf{a}}. \tag{5.124}$$

This vector is zero if and only if the rotation angle is 0° or 180°, which can happen if and only if the rotation tensor is symmetric.

[5] Changing the direction of an axis would normally require changing the sign of the rotation angle. However, a rotation of 180° about an axis $\underset{\sim}{\mathbf{a}}$ is the same as about $-\underset{\sim}{\mathbf{a}}$, so only the orientation, but not pointing direction, of the axis matters.

5.4.4 Method 1 algorithm for axis and angle of rotation

Given $\underset{\approx}{\mathbf{R}}$, below we provide an algorithm for finding the principal axis $\underset{\sim}{\mathbf{a}}$ and principal angle α of rotation.

STEP 1. Compute the cosine of the angle of rotation:

$$c = \tfrac{1}{2}(R_{11} + R_{22} + R_{33} - 1). \tag{5.125}$$

STEP 2. Decide if the angle is an integral multiple of 180° by checking the cosine, and follow one of these branches accordingly:

- If $c \neq \pm 1$, proceed to step 3.
- If $c = 1$ then $\underset{\approx}{\mathbf{R}} = \underset{\approx}{\mathbf{I}}$. Hence, the *angle of rotation* is 0 and the *axis of rotation* is arbitrary (so you can set it to {1,0,0} or any other convenient unit vector). Go to step 6.
- If $c = -1$, then the *angle of rotation* is 180° (=π radians). To find the *axis of rotation*, simply normalize any *non-zero* column of $\underset{\approx}{\mathbf{R}} = \underset{\approx}{\mathbf{I}}$. Go to step 6.

STEP 3. Compute the *angle of rotation*:

$$\alpha = \text{ArcCos}(c). \tag{5.126}$$

Note: in this step, we are forcing $0 \leqslant \alpha < 180°$ because that is the range of the arc-cosine function on all computing platforms. This restriction entails no loss in generality because larger or negative angles about some particular axis $\underset{\sim}{\mathbf{a}}$ can always be expressed as an angle between 0° and 180° about an *oppositely oriented* axis, $-\underset{\sim}{\mathbf{a}}$. The steps below ensure that we select the axis orientation correctly.

STEP 4. Compute the sine of the angle:

$$s = +\sqrt{1 - c^2}. \tag{5.127}$$

STEP 5. Compute the *axis of rotation*:

$$a_1 = \frac{1}{2s}(R_{32} - R_{23}) \tag{5.128a}$$

$$a_2 = \frac{1}{2s}(R_{13} - R_{31}) \tag{5.128b}$$

$$a_3 = \frac{1}{2s}(R_{21} - R_{12}). \tag{5.128c}$$

(*Note:* The check in step 2 guarantees that s will not be zero.)

STEP 6. Stop.

This algorithm is coded in section A.3 of the appendix.

Example. Let us use this algorithm to recover the axis and angle of rotation from (5.29). We are given

$$[\underset{\sim}{\boldsymbol{R}}] = \begin{bmatrix} 0 & 0 & 1 \\ 1 & 0 & 0 \\ 0 & 1 & 0 \end{bmatrix}. \tag{5.129}$$

Follow the steps of the algorithm:

STEP 1. Compute the cosine of the angle:

$$c = \tfrac{1}{2}(0 + 0 + 0 - 1) = -\tfrac{1}{2}. \tag{5.130}$$

STEP 2. Note: $c \neq \pm 1$; therefore proceed.

STEP 3. Compute the *angle of rotation*:

$$\alpha = \text{ArcCos}\left(-\tfrac{1}{2}\right) = 120°. \tag{5.131}$$

STEP 4. Compute the sine of the angle:

$$s = \sqrt{1 - \left(-\tfrac{1}{2}\right)^2} = \frac{\sqrt{3}}{2}. \tag{5.132}$$

STEP 5. Compute the *axis of rotation* $\underset{\sim}{\boldsymbol{a}}$:

$$a_1 = \frac{1}{2(\sqrt{3}/2)}(1 - 0), \tag{5.133a}$$

$$a_2 = \frac{1}{2(\sqrt{3}/2)}(1 - 0), \tag{5.133b}$$

$$a_3 = \frac{1}{2(\sqrt{3}/2)}(1 - 0), \tag{5.133c}$$

or

$$\{\underset{\sim}{\boldsymbol{a}}\} = \frac{1}{\sqrt{3}}\begin{Bmatrix} 1 \\ 1 \\ 1 \end{Bmatrix}. \tag{5.134}$$

Another example. Let us use the algorithm to recover the axis and angle of rotation from (5.31). We are given

$$[\underset{\sim}{\boldsymbol{R}}] = \begin{bmatrix} -1/3 & 2/3 & 2/3 \\ 2/3 & -1/3 & 2/3 \\ 2/3 & 2/3 & -1/3 \end{bmatrix}. \tag{5.135}$$

STEP 1. Compute the cosine of the angle:

$$c = \tfrac{1}{2}\left(-\tfrac{1}{3} - \tfrac{1}{3} - \tfrac{1}{3} - 1\right) = -1. \tag{5.136}$$

STEP 2. Note that $c = -1$, so the angle of rotation is 180°. To find the axis of rotation, we must construct the matrix $\underset{\approx}{\boldsymbol{R}} + \underset{\approx}{\boldsymbol{I}}$:

$$[\underset{\approx}{\boldsymbol{R}} + \underset{\approx}{\boldsymbol{I}}] = \begin{bmatrix} -1/3 & 2/3 & 2/3 \\ 2/3 & -1/3 & 2/3 \\ 2/3 & 2/3 & -1/3 \end{bmatrix} + \begin{bmatrix} 1 & 0 & 0 \\ 0 & 1 & 0 \\ 0 & 0 & 1 \end{bmatrix} = \begin{bmatrix} 2/3 & 2/3 & 2/3 \\ 2/3 & 2/3 & 2/3 \\ 2/3 & 2/3 & 2/3 \end{bmatrix}. \tag{5.137}$$

Thus, the axis of rotation is a unit vector in the direction of {2/3,2/3,2/3}. In other words, $\{\underset{\sim}{\boldsymbol{a}}\} = \{1, 1, 1\}/\sqrt{3}$. In this particular case *all* the columns turned out to be non-zero, but that will not be true in general, nor will they typically be identical.

5.4.5 Numerical example (for testing codes)

Now consider the numerical rotation tensor of (5.37):

$$[\underset{\approx}{\boldsymbol{R}}] = \begin{bmatrix} 0.254\,506 & -0.834\,834 & 0.488\,138 \\ 0.915\,374 & 0.370\,785 & 0.156\,873 \\ -0.311\,957 & 0.406\,903 & 0.858\,552 \end{bmatrix}. \tag{5.138}$$

We wish to recover the axis and angle of rotation.

Following the algorithm:

STEP 1. Compute the cosine of the angle:

$$c = \tfrac{1}{2}(0.254\,506 + 0.370\,785 + 0.858\,552 - 1) = 0.241\,992. \tag{5.139}$$

STEP 2. Note: $c \neq \pm 1$; therefore proceed.

STEP 3. Compute the angle:

$$\alpha = \text{acos}(0.241\,992) = 1.326\,45 \ \text{radians} = 76 \ \text{degrees}. \tag{5.140}$$

STEP 4. Compute the sine of the angle:

$$s = \sqrt{1 - (0.241\,992)^2} = 0.970\,296. \tag{5.141}$$

STEP 5. Compute the *axis of rotation*:

$$a_1 = \frac{1}{2(0.970\,296)}(0.406\,903 - 0.156\,873) = 0.128\,842, \tag{5.142a}$$

$$a_2 = \frac{1}{2(0.970\,296)}(0.488\,138 - (-0.311\,957)) = 0.412\,294, \tag{5.142b}$$

$$a_3 = \frac{1}{2(0.970\,296)}(0.915\,374 - (-0.834\,834)) = 0.901\,894. \tag{5.142c}$$

Again, we wish to emphasize that the solution for the axis is unique only within a numerical sign. Changing the sign of the axis and angle will produce the same rotation tensor. In other words, the combination $(-\alpha, -\underset{\sim}{a})$ is always a second solution. If, for example, you wish for the rotation axis to always have a positive last component (placing it in the 'northern' unit hemisphere), then a negative result for the last component is corrected by simply the reversing sign of all outputs (axis and angle).

5.4.6 Method 2 algorithm for computing axis and angle

One disadvantage of the above algorithm is that special treatment is required when the angle is 0° or 180°. One might expect some numerical round-off problems when the angle is *nearly* equal to one of these special cases. The following alternative algorithm better protects against these round-off errors and has no special cases, but it is less efficient. This algorithm computes the axis by using (5.106), and then assigns the axis orientation so that the rotation angle will range from 0° to 180°.

STEP 1. Compute the cosine of the angle of rotation:

$$c = \tfrac{1}{2}(R_{11} + R_{22} + R_{33} - 1). \tag{5.143}$$

STEP 2. Compute the *angle of rotation*:

$$\alpha = \text{ArcCos}(c). \tag{5.144}$$

STEP 3. Construct the matrix

$$\underset{\approx}{R} + \underset{\approx}{R}^T - 2c\underset{\approx}{I}. \tag{5.145}$$

STEP 4. Let $\underset{\sim}{x}$ be the column of the above matrix having the largest magnitude and set

$$x \equiv \sqrt{x_1^2 + x_2^2 + x_3^2}. \tag{5.146}$$

STEP5. Compute the *axis of rotation*:

$$a_1 = \frac{1}{x}\text{tsign}(x_1, R_{32} - R_{23}), \tag{5.147}$$

$$a_2 = \frac{1}{x}\text{tsign}(x_2, R_{13} - R_{31}), \tag{5.148}$$

$$a_3 = \frac{1}{x}\mathrm{tsign}(x_3, R_{21} - R_{12}). \tag{5.149}$$

Here, 'tsign' is the transfer of sign operation defined by

$$\mathrm{tsign}(z, w) = \begin{cases} |z| & \text{if } w \geqslant 0, \\ -|z| & \text{if } w < 0. \end{cases} \tag{5.150}$$

In FORTRAN, the 'tsign' operator is the intrinsic function '`SIGN`'.

This algorithm is coded in section A.3 of the appendix.

5.4.7 Method 3 algorithm for computing axis and angle

If you do not require the rotation angle to be the *principal* angle, then the previous algorithm can be replaced by

STEP 1. Construct the matrix of components for

$$\underset{\approx}{\mathbf{R}} + \underset{\approx}{\mathbf{R}}^T - 2c\underset{\approx}{\mathbf{I}}. \tag{5.151}$$

STEP 2. Let $\underset{\sim}{\mathbf{x}}$ be the column of the above matrix having the largest magnitude and set

$$x \equiv \sqrt{x_1^2 + x_2^2 + x_3^2}. \tag{5.152}$$

STEP 3. Compute the axis of rotation:

$$a_1 = \frac{x_1}{x}, \tag{5.153}$$

$$a_2 = \frac{x_2}{x}, \tag{5.154}$$

$$a_3 = \frac{x_3}{x}. \tag{5.155}$$

STEP 4. Compute the cosine of the angle of rotation:

$$c = \tfrac{1}{2}(R_{11} + R_{22} + R_{33} - 1). \tag{5.156}$$

STEP 5. Compute the sine of the rotation angle:

$$s = \tfrac{1}{2}[a_1(R_{32} - R_{23}) + a_2(R_{13} - R_{31}) + a_3(R_{21} - R_{12})]. \tag{5.157}$$

STEP 6. Compute the *angle of rotation* by using the *two-argument* arc tangent[6]

$$\alpha = \mathrm{ATAN2}(s, c). \tag{5.158}$$

References

[1] Goldstein H, Poole C P and Safko J L 2002 *Classical Mechanics* (Reading, MA: Addison Wesley)

[2] Lai W M, Rubin D and Krempl E 1993 *Continuum Mechanics* 3rd edn (Butterworth-Heinemann)

[3] Argyris J 1982 An excursion into large rotations *Comput. Methods Appl. Mech. Eng.* **32** 85–155

[4] Frank F C 1988 Orientation mapping *Metall. Trans.* A **19** 403–8

[6] Beware: the arguments of the double argument arctan operator are in this order for FORTRAN, but in the opposite order for many other languages such as C++ and Mathematica.

IOP Publishing

Rotation, Reflection, and Frame Changes

Orthogonal tensors in computational engineering mechanics

R M Brannon

Chapter 6

Rotations contrasted with reflections

A tensor $\mathbf{Q}$ is *orthogonal* (or 'idempotent') if and only if $\mathbf{Q}^T \bullet \mathbf{Q} = \mathbf{I}$. In other words, $\mathbf{Q}^{-1} = \mathbf{Q}^T$. These tensor operations have analogous matrix forms (with the single dot becoming matrix multiplication) when the components of these tensors are written in terms of an orthonormal basis. The determinant of a matrix product equals the products of the determinants, and the determinant of a transpose of a tensor equals the determinant of the tensor. Therefore, $\det \mathbf{Q} = \pm 1$. A *rotation* tensor $\mathbf{R}$ (also known as a *proper*-orthogonal tensor) is a special kind of orthogonal tensor for which the determinant is positive. Thus, $\mathbf{R}^T \bullet \mathbf{R} = \mathbf{I}$ and $\det \mathbf{R} = +1$. It is often wrongly claimed that if $\mathbf{Q}$ is orthogonal with a negative determinant, then it must be a reflection. Before we can rebut this claim, we need to define what *reflection* means. Thinking of mirrors, if $\mathbf{y}$ is a reflection of $\mathbf{x}$, then $\mathbf{x}$ should be a reflection of $\mathbf{y}$. Mathematically, if $\mathbf{y} = \boldsymbol{\Gamma} \bullet \mathbf{x}$, then the operator $\boldsymbol{\Gamma}$ is a reflection if and only if $\boldsymbol{\Gamma} \bullet \mathbf{y} = \mathbf{x}$; i.e., a reflection is an operation $\boldsymbol{\Gamma}$ whose inverse $\boldsymbol{\Gamma}^{-1}$ is the operation $\boldsymbol{\Gamma}$ itself:

$$\boldsymbol{\Gamma} \bullet \boldsymbol{\Gamma} = \mathbf{I}, \quad \text{or more compactly,} \quad \boldsymbol{\Gamma}^2 = \mathbf{I}. \tag{6.1}$$

Thus, a reflection is a 'square root' of the identity tensor. There are an infinite number of such tensors. Even though the identity is symmetric, its square root does not have to be symmetric. For example,

$$[\boldsymbol{\Gamma}] = \begin{bmatrix} 2 & -3 & 0 \\ 1 & -2 & 0 \\ 0 & 0 & 1 \end{bmatrix} \tag{6.2}$$

is a non-symmetric reflection tensor that satisfies (6.1). Reflection tensors are fundamentally different from orthogonal tensors. A tensor $\mathbf{Y}$ is a *proper* reflection (or an *improper*-orthogonal tensor) if and only if it is an orthogonal reflection with a negative determinant:

$$\underset{\approx}{Y} \bullet \underset{\approx}{Y} = \underset{\approx}{I} \quad \text{and} \quad \underset{\approx}{Y}^T \bullet \underset{\approx}{Y} = \underset{\approx}{I} \quad \det \underset{\approx}{Y} = -1. \tag{6.3}$$

To satisfy the first two constraints, a proper reflection $\underset{\approx}{Y}$ must be symmetric. Even if we seek only symmetric square roots of the identity tensor, there are still an infinite number of answers. For tensors in three-dimensional (3D) space, however, they are all expressible in two possible forms:

(1) a retro-reflection, or pure inversion, $\underset{\approx}{\Gamma} = -\underset{\approx}{I}$, which is a simple reversal across the origin, and

(2) a 'ricochet' reflection across a plane with unit normal $\underset{\sim}{n}$ for which $\underset{\approx}{\Gamma} = \underset{\approx}{I} - 2\underset{\sim}{n}\underset{\sim}{n}$.

For a retro-reflection, all three eigenvalues of $\underset{\approx}{\Gamma}$ equal -1, and the operation $\underset{\approx}{\Gamma} \bullet \underset{\sim}{x}$ results in simply an inversion, $-\underset{\sim}{x}$. This operation is simply a 'reversal' of space (east becomes west, north becomes south, and up becomes down)[1]. For a mirror or 'ricochet' reflection across a plane, exactly one eigenvalue of $\underset{\approx}{\Gamma}$ equals -1 while the other two equal $+1$, and $\underset{\approx}{\Gamma} \bullet \underset{\sim}{x}$ results in $\underset{\sim}{x} - 2(\underset{\sim}{x} \bullet \underset{\sim}{n})\underset{\sim}{n}$, which is the mirror image of $\underset{\sim}{x}$ across the plane perpendicular to $\underset{\sim}{n}$. Note that $\underset{\approx}{I} - 2\underset{\sim}{n}\underset{\sim}{n}$ can be written as $-\underset{\approx}{R}^{(180n)}$, where $\underset{\approx}{R}^{(180n)}$ denotes a proper rotation of exactly 180° about $\underset{\sim}{n}$.

Now that we have defined what is meant by a reflection, we can state that, for tensors in 3D space, any improper-orthogonal tensor $\underset{\approx}{Q}$ can be written in the form $-\underset{\approx}{R}$, where $\underset{\approx}{R}$ is a proper rotation. Thus any improper-orthogonal tensor (i.e., one for which $\det\underset{\approx}{Q} = -1$) can be written as a reflection across the origin *in combination with a proper rotation.*

[1] Retro-reflection is actually the mechanism by which safety 'reflectors' work (by combining three or more ricochet reflectors, they reverse an automobile's headlight illumination back to the eyes of driver). Retro-reflection is why a ball thrown into the corner of a room will always return the ball directly back regardless of the thrower's position in the room.

IOP Publishing

Rotation, Reflection, and Frame Changes

Orthogonal tensors in computational engineering mechanics

R M Brannon

Chapter 7

Quaternion representation of a rotation

Rotation matrices have three independent components. This can be readily verified by recalling that the matrix is wholly determined by a unit axis of rotation $\underset{\sim}{a}$ (two independent variables) and an angle of rotation α (the third independent variable). It can also be verified by noting that a rotation matrix has orthonormal column arrays: the first column uses two independent components (with the third constrained to produce a unit vector), the second column uses one additional independent component (with the second and third constrained to produce a unit vector that is also perpendicular to the first), and the last column has zero independent components because it must be given by a right-handed cross product of the first two columns to produce a proper-orthogonal matrix. The set of all possible rotation tensors therefore forms a 3D space, but this space is actually curvilinear. Specifically, it is a 3D hypersphere embedded in 9D tensor space. There is no lower-dimensional Euclidean (or affine) space that encloses this sphere, as can be confirmed by noting that the following set of nine orthogonal tensors is linearly independent (and remains so even when subtracting the simplest orthogonal matrix, the identity):

$\begin{pmatrix} -1 & 0 & 0 \\ 0 & 1 & 0 \\ 0 & 0 & -1 \end{pmatrix}$	$\begin{pmatrix} -1 & 0 & 0 \\ 0 & -1 & 0 \\ 0 & 0 & 1 \end{pmatrix}$	$\begin{pmatrix} 1 & 0 & 0 \\ 0 & -1 & 0 \\ 0 & 0 & -1 \end{pmatrix}$
$\begin{pmatrix} 0 & -1 & 0 \\ 1 & 0 & 0 \\ 0 & 0 & 1 \end{pmatrix}$	$\begin{pmatrix} 1 & 0 & 0 \\ 0 & 0 & -1 \\ 0 & 1 & 0 \end{pmatrix}$	$\begin{pmatrix} 0 & 0 & -1 \\ 0 & 1 & 0 \\ 1 & 0 & 0 \end{pmatrix}$
$\begin{pmatrix} 0 & -\frac{1}{\sqrt{2}} & \frac{1}{\sqrt{2}} \\ \frac{1}{\sqrt{2}} & \frac{1}{2} & \frac{1}{2} \\ -\frac{1}{\sqrt{2}} & \frac{1}{2} & \frac{1}{2} \end{pmatrix}$	$\begin{pmatrix} \frac{1}{2} & -\frac{1}{\sqrt{2}} & \frac{1}{2} \\ \frac{1}{\sqrt{2}} & 0 & -\frac{1}{\sqrt{2}} \\ \frac{1}{2} & \frac{1}{\sqrt{2}} & \frac{1}{2} \end{pmatrix}$	$\begin{pmatrix} \frac{1}{2} & \frac{1}{2} & \frac{1}{\sqrt{2}} \\ \frac{1}{2} & \frac{1}{2} & -\frac{1}{\sqrt{2}} \\ -\frac{1}{\sqrt{2}} & \frac{1}{\sqrt{2}} & 0 \end{pmatrix}$

In this chapter we present the representation of a rotation in terms of a quaternion, which treats a rotation as a point on the three-dimensional (3D) surface of a unit sphere embedded in a 4D space. Construction of the rotation tensor should be regarded as a mapping between two different spaces, somewhat analogous to drawing a map of the surface of the Earth on a flat sheet of paper (which can preserve relative areas or distances, but never both). Even though a globe can be projected to a 2D piece of paper (the map), this does not, of course, imply that the globe itself is a 'Flat Earth.' Similarly, even though a rotation is here described in terms of a 4-component object (called a quaternion), this does not imply that the set of all rotation tensors somehow belongs to a 4D subspace. The smallest enclosing Euclidean space for rotation tensors is full 9D tensor space, and this fact becomes highly pertinent later when assessing densities of rotation distributions (analogous to relative areas of countries on a globe, where an equal-area map is preferred) or when interpolating between rotations (analogous to finding shortest paths on the globe, where a conformal map is preferred).

As this book is targeted primarily at engineers, our introduction to quaternions is lightweight by design, with citations to good starting points for pursing the topic with greater rigor. A quaternion (also known as a 'hyper-complex number') is the division algebra over the real numbers generated by the elements i,j,k subject to the relations $i^2 = j^2 = k^2 = -1$, $ij = (-ji) = k$, $jk = (-kj) = i$, and $ki = (-ik) = j$. Any textbook on abstract algebra should provide a rigorous definition of 'division algebra' as well as an explanation that the elements i,j,k are essentially generalizations of the imaginary number $i = \sqrt{-1}$, and there are some alternative definitions (cf. [1]) that might be better suited to engineering applications in fluid and solid mechanics. Quaternions are predecessors to modern engineering vectors and tensors that (for many good reasons) seem to be regaining favor, especially in the computer gaming/visualization community (see [2] for a lucid overview of quaternions, including their history of `competition' with the Gibbsian tensor and vector notation that is used here in this book). For our purposes, a quaternion, $a + bi + cj + dk$, can be regarded as a 4D real vector $\{a,b,c,d\}$ in the same way that an ordinary complex number $a + bi$ may be regarded as a 2D vector $\{a, b\}$.

Like spinors, the quaternion representation of a rotation requires exactly three numbers, which is advantageous (for example) when memory or storage is limited (but the price might be higher CPU costs using rotations in calculations). In computational mechanics, speed performance is far more important than memory, motivating use of the full 9-component rotation, usually embedded in deformation tensors as only a partial contribution to bulk material movement.

7.1 Shoemake's form

Any rotation operation may be represented by a quaternion $u_0 + \boldsymbol{i}u_1 + \boldsymbol{j}u_2 + \boldsymbol{k}u_3$,[1] for which the corresponding rotation matrix may be written in the form

[1] Keep in mind: $\{\boldsymbol{i}, \boldsymbol{j}, \boldsymbol{k}\}$ are *not* unit basis vectors for a rectangular coordinate frame. They are hyper-imaginary numbers.

$$[\underset{\approx}{R}] = \begin{bmatrix} 1 - 2\left(u_2^2 + u_3^2\right) & 2u_1u_2 - 2u_0u_3 & 2u_1u_3 + 2u_0u_2 \\ 2u_2u_1 + 2u_0u_3 & 1 - 2\left(u_3^2 + u_1^2\right) & 2u_2u_3 - 2u_0u_1 \\ 2u_3u_1 - 2u_0u_2 & 2u_3u_2 + 2u_0u_1 & 1 - 2\left(u_1^2 + u_2^2\right) \end{bmatrix}. \tag{7.1}$$

The four numbers $\{u_0, u_1, u_2, u_3\}$, which are quaternion parts, must satisfy

$$u_0^2 + u_1^2 + u_2^2 + u_3^2 = 1. \tag{7.2}$$

In other words, $\{u_0, u_1, u_2, u_3\}$ is a point on the 3D surface of a 4D unit quaternion hypersphere. See [3]. Though the analytical geometry (curvilinear mathematics) of a proof is beyond scope of this book, it can be shown that the Riemann curvature tensor of the set of tensors attainable by variation of the u-components (without normalization) is not zero. Accordingly, the 3D surface of the sphere is not contained within a 4D linear manifold of 9D tensor space (yikes! what a mouthful). We know that rotation tensors belong to 9D tensor space, which is Euclidean, and they do not belong to any smaller Euclidean subspace (linear manifold) that is embedded in 9D tensor space.

A more structural direct form.

The matrix in (7.1) may be alternatively written as

$$[\underset{\approx}{R}] = \left(2u_0^2 - 1\right)\begin{bmatrix} 1 & 0 & 0 \\ 0 & 1 & 0 \\ 0 & 0 & 1 \end{bmatrix} + 2\begin{bmatrix} u_1^2 & u_1u_2 & u_1u_3 \\ u_2u_1 & u_2^2 & u_2u_3 \\ u_3u_1 & u_3u_2 & u_3^2 \end{bmatrix} + 2u_0\begin{bmatrix} 0 & -u_3 & u_2 \\ u_3 & 0 & -u_1 \\ -u_2 & u_1 & 0 \end{bmatrix}. \tag{7.3}$$

The corresponding structured (direct) notation is

$$\begin{aligned} &\underset{\approx}{R} = \left(2u_0^2 - 1\right)\underset{\approx}{I} + 2\underset{\sim}{u}\underset{\sim}{u} + 2u_0\underset{\sim}{u} \times \underset{\approx}{I} \\ &\text{i.e., } R_{ij} = \left(2u_0^2 - 1\right)\delta_{ij} + 2u_iu_j - 2u_0\varepsilon_{ijk}u_k, \\ &\text{where } \underset{\sim}{u} \equiv u_1\underset{\sim}{E}_1 + u_2\underset{\sim}{E}_2 + u_3\underset{\sim}{E}_3. \end{aligned} \tag{7.4}$$

For later use, note that

$$\underset{\sim}{u} \bullet \underset{\sim}{u} = 1 - u_0^2. \tag{7.5}$$

7.2 Relationship between quaternion and axis/angle forms

Note from (7.4) that

$$\text{tr}\underset{\approx}{R} = 4u_0^2 - 1. \tag{7.6}$$

We also know from (5.105) that

$$\text{tr}\underset{\approx}{R} = 2\cos\alpha + 1. \tag{7.7}$$

Equating these expressions gives a relationship between u_0 and α:

$$\cos\alpha = 2u_0^2 - 1 = 1 - 2\underset{\sim}{\boldsymbol{u}} \bullet \underset{\sim}{\boldsymbol{u}}. \tag{7.8}$$

We know from (5.119) that the axial vector of $\underset{\approx}{\boldsymbol{R}}$ is given by

$$ax\underset{\approx}{\boldsymbol{R}} = (\sin\alpha)\underset{\sim}{\boldsymbol{a}}. \tag{7.9}$$

Using (7.4), the axial vector associated with $\underset{\approx}{\boldsymbol{R}}$ is[2]

$$ax\underset{\approx}{\boldsymbol{R}} = 2u_0\underset{\sim}{\boldsymbol{u}}. \tag{7.10}$$

Equating the above two vectors shows that $\underset{\sim}{\boldsymbol{u}}$ is merely a multiple of the unit rotation axis $\underset{\sim}{\boldsymbol{a}}$. Equating the magnitudes of these vectors (recalling that $\underset{\sim}{\boldsymbol{a}}$ is a unit vector and using (7.5) for the magnitude of $\underset{\sim}{\boldsymbol{u}}$) shows that the rotation angle α is related to u_0 by

$$\sin\alpha = 2u_0\sqrt{\underset{\sim}{\boldsymbol{u}} \bullet \underset{\sim}{\boldsymbol{u}}} = 2u_0\sqrt{1 - u_0^2}. \tag{7.11}$$

By virtue of (7.2), we may introduce *four-dimensional spherical coordinates*[3] $\{\beta, \theta, \phi\}$ such that

$$u_0 = \cos\beta, \tag{7.12a}$$

$$u_1 = \sin\beta(\sin\theta\cos\phi), \tag{7.12b}$$

$$u_2 = \sin\beta(\sin\theta\sin\phi), \tag{7.12c}$$

$$u_3 = \sin\beta(\cos\theta). \tag{7.12d}$$

Putting (7.12*a*) into (7.8) yields $\cos\alpha = \cos 2\beta$ or, $\alpha = \pm 2\beta$. We may choose the positive solution without loss in generality (so that rotation angles in the range from 0 to π map to quaternions in the *upper* 4D hemisphere. Thus,

$$\beta = \frac{\alpha}{2}. \tag{7.13}$$

The relationship between the two vectors $\underset{\sim}{\boldsymbol{u}}$ and $\underset{\sim}{\boldsymbol{a}}$ implied by (7.9) and (7.10) can then be written

$$\boxed{\underset{\sim}{\boldsymbol{u}} = \left(\sin\frac{\alpha}{2}\right)\underset{\sim}{\boldsymbol{a}}}. \tag{7.14}$$

---

[2] Here, 'ax' is a shorthand notation for the operation that finds the axial part of a tensor. For any tensor $\underset{\approx}{\boldsymbol{B}}$ (skew-symmetric or not), ax$B$ is a vector $\underset{\sim}{\boldsymbol{b}}$ with components $b_i = -\frac{1}{2}\varepsilon_{ijk}B_{jk}$.

[3] When you have $x^2 + y^2 = R^2$, you can assert that $x = R\cos\beta$ and $y = R\sin\beta$. This basic result can be used to define higher-dimensional spherical coordinates. Equation (7.2) is of the form $x^2 + y^2 = R^2$ if we take $R = 1$, $x = u_0$, and $y^2 = u_1^2 + u_2^2 + u_3^2$. Then the equation $x = R\cos\beta$ becomes $u_0 = \cos\beta$. Noting that $R = 1$, the equation $y = R\sin\beta$, can be written $y^2 = \sin^2\beta$, or $u_1^2 + u_2^2 + u_3^2 = r^2$, where $r = \sin\beta$. Thus, for the remaining three components, we can introduce conventional 3D spherical coordinates, taking the radius r to be $\sin\beta$. The final result will be (7.12).

This vector is identical to the pseudo-vector used in the spinor formulation of (5.88). If the rotation angle α is zero, then the rotation axis $\underset{\sim}{a}$ is arbitrary, but $\underset{\sim}{u} = \underset{\sim}{0}$ and $u_0 = 1$. If the rotation angle α is 180°, then $\underset{\sim}{u} = \underset{\sim}{a}$ and $u_0 = 0$.

Equations (7.12)–(7.14) provide a means of converting from the axis/angle description to quaternion form, and vice versa. The rotation angle α may permissibly take a full range of values from $-\pi$ to $+\pi$:

$$\text{If } -\pi \leqslant \alpha \leqslant \pi, \quad \text{then } -\frac{\pi}{2} \leqslant \beta \leqslant \frac{\pi}{2}. \tag{7.15}$$

With this viewpoint, a rotation can be described as a point (unit quaternion) on a 4D hyper-sphere with coordinates (u_0, u_1, u_2, u_3).

Without loss in generality it is always possible (by changing the direction of the rotation axis) to presume that the rotation angle satisfies $\sin\alpha \geqslant 0$. Thus,

$$\text{if } 0 \leqslant \alpha \leqslant \pi, \quad \text{then } 0 \leqslant \beta \leqslant \frac{\pi}{2}. \tag{7.16}$$

In this case, a rotation is described by a unit quaternion lying on a *hemisphere* in 4D space. Both viewpoints are legitimate, and both give the same unique rotations[4]. Later, in chapter 17, it will be shown that a uniformly random distribution of rotation tensors corresponds to a uniform distribution of points on the quaternion sphere.

References

[1] Gibbon J 2007 Orthonormal quaternion frames, Lagrangian evolution equations, and the three-dimensional Euler equations *Russ. Math. Surv.* **62** 535

[2] Hanson A J 2006 Visualizing Quaternions (San Francisco, CA: Morgan-Kauffman)

[3] Shoemake K 1992 Uniform random rotations *Graphics Gems* vol 3 ed D Kirk (London: Academic) pp 124–32

[4] It might seem 'unsavory' that there is non-uniqueness in quaternion representations, but the situation should be no more disturbing than the fact that eigenvectors and associated direction cosine matrices are never unique—combining them with the diagonal eigenvalue matrix in a spectral decomposition obliterates this ambiguity to produce a unique reconstruction of the original matrix. As with spectral decompositions, ambiguity might hinder comparisons of quaternions, so the unique reconstructed rotation tensor should be used in quantitative comparisons of two different rotations.

Chapter 8

Dyadic form of an invertible linear operator

This chapter recognizes that rotations are special cases of the deformation tensors in used in Continuum Mechanics to describe both bulk material reorientation and stretching, shear, elongation, and other forms of shape change. In this more general context, at least some points in the body are presumed to have known initial and deformed locations, and therefore lines connecting points in the body can be modeled as embedded vectors, the book-keeping of which relies heavily on the notion of dyads. A dyad $\underset{\sim}{a}\underset{\sim}{b}$ is a special kind of second-order tensor with ij components given by $a_i b_j$. Any second-order tensor may be expressed as a sum of three or fewer dyads[1].

Suppose that $\{\underset{\sim}{b}_1, \underset{\sim}{b}_2, \underset{\sim}{b}_3\}$ is a set of linearly independent vectors. Suppose that some operator f is known to be linear and its action on the b vectors is known. Specifically, let

$$\underset{\sim}{c}_k = f(\underset{\sim}{b}_k). \tag{8.1}$$

Since the operation is known to be linear, we know there must exist a tensor $\underset{\approx}{F}$ such that the above expression may be written

$$\underset{\sim}{c}_k = \underset{\approx}{F} \bullet \underset{\sim}{b}_k. \tag{8.2}$$

If the six vectors $\{\underset{\sim}{b}_1, \underset{\sim}{b}_2, \underset{\sim}{b}_3\}$ and $\{\underset{\sim}{c}_1, \underset{\sim}{c}_2, \underset{\sim}{c}_3\}$ are known, then the tensor $\underset{\approx}{F}$ is expressible as the following sum of dyads:

$$\underset{\approx}{F} = \underset{\sim}{c}_k \underset{\sim}{b}^k. \tag{8.3}$$

Here, the superscripted $\{\underset{\sim}{b}^1, \underset{\sim}{b}^2, \underset{\sim}{b}^3\}$ are the so-called 'dual' vectors associated with the $\{\underset{\sim}{b}_1, \underset{\sim}{b}_2, \underset{\sim}{b}_3\}$ basis. Specifically, the dual basis is defined by

[1] The dyads in such a decomposition are not unique (different sets of dyads can sum to give the same tensor). Even though it is always possible to express a tensor as a sum of three or fewer dyads, the standard practice is to express tensors as linear combinations of the *nine* basis dyads.

doi:10.1016/978-0-7503-1454-1ch8

$$\underset{\sim}{b}^i \bullet \underset{\sim}{b}_j = \delta^i_j, \tag{8.4}$$

where δ^i_j is the Kronecker delta. It follows that

$$\underset{\sim}{b}^1 = \frac{\underset{\sim}{b}_2 \times \underset{\sim}{b}_3}{\underset{\sim}{b}_1 \bullet (\underset{\sim}{b}_2 \times \underset{\sim}{b}_3)}, \quad \underset{\sim}{b}^2 = \frac{\underset{\sim}{b}_3 \times \underset{\sim}{b}_1}{\underset{\sim}{b}_1 \bullet (\underset{\sim}{b}_2 \times \underset{\sim}{b}_3)}, \quad \text{and} \quad \underset{\sim}{b}^3 = \frac{\underset{\sim}{b}_1 \times \underset{\sim}{b}_2}{\underset{\sim}{b}_1 \bullet (\underset{\sim}{b}_2 \times \underset{\sim}{b}_3)}. \tag{8.5}$$

A faster method begins by putting the laboratory components of the b-vectors into columns of matrix. When that matrix is inverted, its ROWS contain the laboratory components of the superscripted dual vectors. If the matrix is not invertible (as will happen, for example, when the matrix is not square because of having too few or too many b-vectors, or when the b-vectors are linearly dependent), then a pseudo-inverse may be used.

8.1 Special case: lab basis

If the $\{\underset{\sim}{b}_1, \underset{\sim}{b}_2, \underset{\sim}{b}_3\}$ basis happens to be orthonormal, then the superscripted dual basis is the same as the subscripted $\{\underset{\sim}{b}_1, \underset{\sim}{b}_2, \underset{\sim}{b}_3\}$ basis. In particular, if the $\{\underset{\sim}{b}_1, \underset{\sim}{b}_2, \underset{\sim}{b}_3\}$ basis is the lab basis $\{\underset{\sim}{E}_1, \underset{\sim}{E}_2, \underset{\sim}{E}_3\}$, then

$$\underset{\approx}{F} = \underset{\sim}{c}_k \underset{\sim}{E}_k, \tag{8.6}$$

where

$$\underset{\sim}{c}_k = \underset{\approx}{F} \bullet \underset{\sim}{E}_k. \tag{8.7}$$

This means that the kth column of $[\underset{\approx}{F}]$ with respect to the lab basis is given by the lab components of $\underset{\sim}{c}_k$. Stated differently, if you know how $\underset{\approx}{F}$ transforms the three lab basis vectors, then you can immediately construct $[\underset{\approx}{F}]$ by assembling the transformations of the lab basis into the columns of $[\underset{\approx}{F}]$. Importantly, if $[F_{ij}]$ are the lab components of $\underset{\approx}{F}$, then (8.8) may be written

$$\underset{\sim}{c}_k = F_{jk} \underset{\sim}{E}_j. \tag{8.8}$$

where the index 'j' is implicitly summed. Equations (8.7) and (8.8) express the *same result* in two different forms. There is *not* a transpose error on the subscripts of F_{jk} in (8.8).

8.2 Special case: dyadic form of a rotation operation

To specialize the above results to rotation, suppose that an orthonormal triad $\{\hat{\underset{\sim}{c}}_1, \hat{\underset{\sim}{c}}_2, \hat{\underset{\sim}{c}}_3\}$ is obtained by applying a rotation $\underset{\approx}{R}$ to an orthonormal triad $\{\hat{\underset{\sim}{b}}_1, \hat{\underset{\sim}{b}}_2, \hat{\underset{\sim}{b}}_3\}$. Applying (8.3), noting that $\hat{\underset{\sim}{b}}^k = \hat{\underset{\sim}{b}}_k$ for orthonormal triads, the dyad representation of the rotation is[2]

[2] Simo and Hughes [1] *incorrectly* refer to this as a spectral expansion. However, a spectral expansion must involve eigenvectors. Neither $\hat{\underset{\sim}{c}}_k$ nor $\hat{\underset{\sim}{b}}_k$ are, in general, eigenvectors of the rotation.

$$\underset{\approx}{\boldsymbol{R}} = \hat{\underset{\sim}{\boldsymbol{c}}}_k \hat{\underset{\sim}{\boldsymbol{b}}}_k, \tag{8.9}$$

where

$$\hat{\underset{\sim}{\boldsymbol{c}}}_k = \underset{\approx}{\boldsymbol{R}} \bullet \hat{\underset{\sim}{\boldsymbol{b}}}_k \quad \Leftrightarrow \quad \hat{\underset{\sim}{\boldsymbol{c}}}_k = R_{jk} \hat{\underset{\sim}{\boldsymbol{b}}}_j. \tag{8.10}$$

This result can be used to construct the rotation relating two configurations. If, for example, the edges of a cube are defined by three unit vectors $\{\hat{\underset{\sim}{\boldsymbol{b}}}_1, \hat{\underset{\sim}{\boldsymbol{b}}}_2, \hat{\underset{\sim}{\boldsymbol{b}}}_3\}$ and that cube is rotated to a new orientation so that the original three vectors become $\{\hat{\underset{\sim}{\boldsymbol{c}}}_1, \hat{\underset{\sim}{\boldsymbol{c}}}_2, \hat{\underset{\sim}{\boldsymbol{c}}}_3\}$, then the rotation tensor can be constructed using (8.9). This construction is extremely convenient when specifying the orientation of a crystal—all that is needed is the specification of the three crystal axes in terms of the lab basis, which are then just assembled into columns to form the rotation tensor. Once the rotation tensor is known, any other orientation specification such as the axis/angle or Euler angles can be readily computed by using one of the routines in the appendix.

8.2.1 Example 1: Specifying a rotation via two key directions

Suppose you are writing a three-dimensional (3D) visualization code that will permit users to request a particular orientation of a body. We will assume that the code and the users have a common understanding of a particular embedded set of axes that can be used to define local points on the body. We will presume that the 'laboratory' axes, point respectively to the right side of the computer screen, the top of the screen, and out of the screen.

More often than not, a user will seek a particular oriented view of the body. What is an easy way to give them what they seek? One way is to permit users to specify a vector from the body's embedded origin to a particular point in the body. A re-orientation can be constructed that will put the user-specified vector 'out of the screen'. If they specify the embedded coordinates of this point as well as the coordinates of a second point that they wish to end up oriented towards the top of the screen, then that is enough information to generate a unique rotation (needed for the visualization software to provide the requested rotated view). For example, the user might wish to reorient the image of a die to get the {1,1,1} corner to point straight 'out of the page' at the observer (in Mathematica, this would be accomplished via its 'ViewVector' command); as there are many rotated views satisfying this requirement, the user would need to specify a second vector embedded in the body—perhaps pointing to another corner or face—that should be rendered to point generally upward in the rotated screen view (in Mathematica, this direction is controlled with the 'ViewVertical' command).

Let the first point (the one that will ultimately point out of the screen) be denoted by embedded vector components (u_1, u_2, u_3). Let the second point (to point toward the top of the screen) be specified by embedded components (v_1, v_2, v_3). If the user provides these coordinates, then the code can apply Gram–Schmidt orthogonalization [2] to generate an orthonormal rotated basis triad via the following sequence of array calculations:

$$\underset{\sim}{u} = \{u_1, u_2, u_3\}, \tag{8.11}$$

$$\underset{\sim}{v} = \{v_1, v_2, v_3\}, \tag{8.12}$$

$$\hat{\underset{\sim}{a}} = \frac{\{u_1, u_2, u_3\}}{\sqrt{u_1^2 + u_2^2 + u_3^2}}, \tag{8.13}$$

$$\underset{\sim}{b} = \underset{\sim}{v} - (\underset{\sim}{v} \bullet \hat{\underset{\sim}{a}})\hat{\underset{\sim}{a}}, \tag{8.14}$$

$$\hat{\underset{\sim}{b}} = \frac{\{b_1, b_2, b_3\}}{\sqrt{b_1^2 + b_2^2 + b_3^2}}, \tag{8.15}$$

$$\hat{\underset{\sim}{c}} = \hat{\underset{\sim}{b}} \times \hat{\underset{\sim}{a}}. \tag{8.16}$$

The goal now is to define a rotation that would transform these three 'hat' vectors to the laboratory basis so that, then $\hat{\underset{\sim}{a}}$ would transform to become $\underset{\sim}{E}_3$, $\hat{\underset{\sim}{b}}$ would transform to become $\underset{\sim}{E}_2$, and $\hat{\underset{\sim}{c}}$ would transform to become $\underset{\sim}{E}_1$. By the dyad representation, such a rotation would be constructed by the sum of dyads

$$\underset{\approx}{Q} = \underset{\sim}{E}_3\hat{\underset{\sim}{a}} + \underset{\sim}{E}_2\hat{\underset{\sim}{b}} + \underset{\sim}{E}_1\hat{\underset{\sim}{c}}. \tag{8.17}$$

Since the lab basis is appearing on the *left side* of each dyad, we see that associated matrix is then constructed by assembling the local vectors into *rows* of the matrix to obtain

$$[Q] = \begin{bmatrix} \hat{c}_1 & \hat{c}_2 & \hat{c}_3 \\ \hat{b}_1 & \hat{b}_2 & \hat{b}_3 \\ \hat{a}_1 & \hat{a}_2 & \hat{a}_3 \end{bmatrix}. \tag{8.18}$$

If the body is already being displayed on the screen with a known rotation matrix $[R^{\text{old}}]$, this rotation would have to be first removed before applying the above new rotation. Thus, the new orientation of the body would then be given by

$$[R^{\text{new}}] = [Q][R^{\text{old}}]^T. \tag{8.19}$$

Suppose, for example, that the user is looking at a 'die' such as the one illustrated in figure 3.1. Further suppose that they wish to specify a rotation that will cause the (1,1,1) corner to point out of the screen and the (0,0,1) face to end up oriented toward the top of the screen, then the sequence of calculations would be

$$u = \{1, 1, 1\}, \tag{8.20}$$

$$v = \{0, 0, 1\}, \tag{8.21}$$

$$\hat{a} = \left\{ \frac{1}{\sqrt{3}}, \frac{1}{\sqrt{3}}, \frac{1}{\sqrt{3}} \right\}, \tag{8.22}$$

$$b = \left\{ -\frac{1}{3}, \frac{1}{3}, \frac{2}{3} \right\}, \tag{8.23}$$

$$\hat{b} = \left\{ \frac{-1}{\sqrt{6}}, \frac{-1}{\sqrt{6}}, \frac{2}{\sqrt{6}} \right\}, \tag{8.24}$$

$$\hat{c} = \left\{ \frac{1}{\sqrt{2}}, \frac{-1}{\sqrt{2}}, 0 \right\}, \tag{8.25}$$

$$[Q] = \begin{bmatrix} \frac{1}{\sqrt{2}} & \frac{-1}{\sqrt{2}} & 0 \\ \frac{-1}{\sqrt{6}} & \frac{-1}{\sqrt{6}} & \frac{2}{\sqrt{6}} \\ \frac{1}{\sqrt{3}} & \frac{1}{\sqrt{3}} & \frac{1}{\sqrt{3}} \end{bmatrix}. \tag{8.26}$$

8.2.2 Example 2: planes of maximum normal stress

Suppose that $\underset{\approx}{\sigma}$ denotes a stress tensor. Let $\underset{\sim}{n}$ denote the normal to a plane. Then the normal and shear stresses on that plane are given by

$$p \equiv \underset{\sim}{n} \bullet \underset{\approx}{\sigma} \bullet \underset{\sim}{n} \tag{8.27}$$

and

$$\tau = \left\| \underset{\approx}{\sigma} \bullet \underset{\sim}{n} - (\underset{\sim}{n} \bullet \underset{\approx}{\sigma} \bullet \underset{\sim}{n})\underset{\sim}{n} \right\|, \tag{8.28}$$

respectively. Considering this when $\underset{\sim}{n}$ happens to coincide with a principal direction of the stress tensor, note that

$$\tau = 0 \text{ on planes whose normals are principal directions of stress.} \tag{8.29}$$

The converse is true as well: the normal of any zero-shear plane must be a principal direction of the stress tensor. For maximum generality, we will allow for the possibility that the stress tensor is not symmetric. For non-symmetric stresses, the principal stress directions are not generally orthogonal. We will demonstrate that the principal directions of a non-symmetric stress do *not* correspond to the planes of maximum or minimum normal stress. Instead, we will show that the planes of extremal normal stress correspond to principal directions of the *symmetric part* of the stress. On these planes of extremal normal stress, the shear stress is *not* generally zero.

In indicial form, the equation for the normal stress can be written

$$p \equiv n_i \sigma_{ij} n_j. \tag{8.30}$$

For a given stress tensor, the normal stress p depends on the orientation of the plane. In other words, we may regard p as a function of the plane's orientation: $p = p(\underset{\sim}{n})$. Finding the planes with the maximum and minimum normal stress is merely a matter maximizing (or minimizing) $p(\underset{\sim}{n})$ over all possible $\underset{\sim}{n}$ subject to the constraint that $\underset{\sim}{n}$ must be a unit vector. We can solve this extremum problem under the constraint that $\underset{\sim}{n} \bullet \underset{\sim}{n} = 1$ by using a Lagrange multiplier λ. Specifically, we introduce an augmented objective function,

$$f(\underset{\sim}{n},\lambda) \equiv \underset{\sim}{n} \bullet \underset{\approx}{\sigma} \bullet \underset{\sim}{n} + \lambda(1 - \underset{\sim}{n} \bullet \underset{\sim}{n}) \tag{8.31}$$

or, in indicial form,

$$f(n_1, n_2, n_3, \lambda) = n_1 \sigma_{ij} n_j + \lambda(1 - n_i n_i). \tag{8.32}$$

To minimize this function, we must set

$$\frac{\partial f}{\partial n_k} = 0 \tag{8.33}$$

or

$$n_i \sigma_{ij} \delta_{jk} + \delta_{jk} \sigma_{ij} n_j - \lambda(n_i \delta_{jk} + \delta_{jk} n_i) = 0. \tag{8.34}$$

Simplifying gives

$$\left(\frac{\sigma_{kj} + \sigma_{kj}}{2}\right) n_j = \lambda n_k. \tag{8.35}$$

In direct notation,

$$\underset{\approx}{\sigma}^{\text{sym}} \bullet \underset{\sim}{n} = \lambda \underset{\sim}{n}, \quad \text{where} \quad \underset{\approx}{\sigma}^{\text{sym}} \equiv \tfrac{1}{2}(\underset{\approx}{\sigma} + \underset{\approx}{\sigma}^T). \tag{8.36}$$

Here, $\underset{\approx}{\sigma}^{\text{sym}}$ denotes the symmetric part of the stress tensor. The above equation is merely a symmetric eigenvalue problem. Consequently, the planes corresponding to the maximum and minimum normal stresses must be principal directions of the *symmetric part* of the stress tensor. For non-symmetric stresses, the principal directions of the stress tensor do *not* correspond to planes of maximum or minimum (extremal) normal stress.

On the extremal normal stress planes, the shear stress is given by (8.28), which we may write by decomposing the stress into its symmetric and deviatoric parts

$$\tau = \left\| (\underset{\approx}{\sigma}^{\text{sym}} + \underset{\approx}{\sigma}^{\text{skew}}) \bullet \underset{\sim}{n} - \underset{\sim}{n} \bullet (\underset{\approx}{\sigma}^{\text{sym}} + \underset{\approx}{\sigma}^{\text{skew}}) \bullet \underset{\sim}{n}\underset{\sim}{n} \right\|. \tag{8.37}$$

Applying (8.36) and noting that the normal component of *any* skew-symmetric tensor is always zero this simplifies to

$$\tau = \left\| \underset{\approx}{\sigma}^{\text{skew}} \bullet \underset{\sim}{n} \right\| \text{ on planes of max/min normal stress.} \tag{8.38}$$

This result should be contrasted with (8.29).

Recall that the normals to the planes of max/min normal stress are given by principal directions of the *symmetric part* of the stress. Consequently, there always exist three orthonormal principal direction vectors $[\underset{\sim}{n}_1, \underset{\sim}{n}_2, \underset{\sim}{n}_3]$. If the corresponding eigenvalues are denoted σ_1, σ_2, and σ_3, then the symmetric part of the stress tensor may be written

$$\underset{\approx}{\sigma}^{\text{sym}} = \sigma_1 \underset{\sim}{n}_1 \underset{\sim}{n}_1 + \sigma_2 \underset{\sim}{n}_2 \underset{\sim}{n}_2 + \sigma_3 \underset{\sim}{n}_3 \underset{\sim}{n}_3, \tag{8.39}$$

which shows that the symmetric stress tensor is diagonal in its principal basis. The skew-symmetric part of the stress tensor may be fully characterized by a vector $\underset{\sim}{s}$ defined such that

$$\underset{\approx}{\sigma}^{\text{skew}} \bullet \underset{\sim}{u} = \underset{\sim}{s} \times \underset{\sim}{u} \qquad \text{for any vector } \underset{\sim}{u}. \tag{8.40}$$

In direct notation, the formula for the skew-vector is

$$\underset{\sim}{s} = -\frac{1}{2} \underset{\approx}{\varepsilon} : \underset{\approx}{\sigma}^{\text{skew}} \quad \text{or, in indicial notation} \quad s_i = \frac{1}{2} \varepsilon_{ijk} \sigma_{jk}^{\text{skew}}. \tag{8.41}$$

If the skew vector is known, then the skew-tensor may be reconstructed by the formula

$$\underset{\approx}{\sigma}^{\text{skew}} = -\underset{\approx}{\varepsilon} \bullet \underset{\sim}{s} \quad \text{or, in indicial form,} \quad \underset{\approx}{\sigma}_{ij}^{\text{skew}} = -\varepsilon_{ijk} s_k. \tag{8.42}$$

For non-symmetric stress states, this result permits the traction vector $\underset{\sim}{t}$ to be written

$$\underset{\sim}{t} = \underset{\approx}{\sigma}^{\text{sym}} \bullet \underset{\sim}{n} - (\underset{\approx}{\varepsilon} \bullet \underset{\sim}{s}) \bullet \underset{\sim}{n} \tag{8.43}$$

or

$$\underset{\sim}{t} = \underset{\approx}{\sigma}^{\text{sym}} \bullet \underset{\sim}{n} + \underset{\sim}{s} \times \underset{\sim}{n}. \tag{8.44}$$

Quite frequently, it is useful to introduce a supplemental helper basis that is aligned with the principal directions of the symmetric part of the stress. The rotation tensor that will transform the lab basis into the principal basis can be constructed by direct application of (8.9). Namely,

$$\underset{\approx}{R} = \underset{\sim}{n}_1 \underset{\sim}{E}_1 + \underset{\sim}{n}_2 \underset{\sim}{E}_2 + \underset{\sim}{n}_3 \underset{\sim}{E}_3. \tag{8.45}$$

In other words, the columns of the $[\underset{\approx}{R}]$ matrix contain the laboratory components of the principal vectors. This basis rotation tensor has the property that

$$\underset{\sim}{n}_i = \underset{\approx}{R} \bullet \underset{\sim}{E}_i. \tag{8.46}$$

Substituting this result into (8.39) gives

$$\underset{\approx}{\sigma}^{\text{sym}} = \underset{\approx}{\boldsymbol{R}} \bullet \underset{\approx}{\Sigma}^{\text{sym}} \bullet \underset{\approx}{\boldsymbol{R}}^{T}, \tag{8.47}$$

where

$$\underset{\approx}{\Sigma}^{\text{sym}} \equiv \sigma_1 \underset{\sim}{\boldsymbol{E}}_1 \underset{\sim}{\boldsymbol{E}}_1 + \sigma_2 \underset{\sim}{\boldsymbol{E}}_2 \underset{\sim}{\boldsymbol{E}}_2 + \sigma_3 \underset{\sim}{\boldsymbol{E}}_3 \underset{\sim}{\boldsymbol{E}}_3. \tag{8.48}$$

Comparing (8.39) with (8.48) shows that $\underset{\approx}{\Sigma}^{\text{sym}}$ and $\underset{\approx}{\sigma}^{\text{sym}}$ have identical components, but the components are referenced to *two different bases*. Consequently, the two tensors are not equal. One is obtained by superimposing a rotation on the other. The creation of a new tensor by a rotation of an initial tensor is intimately related to—but not equivalent to—coordinate transformation. Under a coordinate transformation, we look at the *same* tensor from two *different* orientations. Under superimposed rotation, we look at two *different* tensors holding our own orientation fixed. Using the *laboratory* basis to construct the component matrices, equation (8.47) can be written in matrix form as

$$\begin{bmatrix} \sigma_{11} & \sigma_{12} & \sigma_{13} \\ \sigma_{21} & \sigma_{22} & \sigma_{23} \\ \sigma_{31} & \sigma_{32} & \sigma_{32} \end{bmatrix}^{\text{sym}} = [R] \begin{bmatrix} \lambda_1 & 0 & 0 \\ 0 & \lambda_2 & 0 \\ 0 & 0 & \lambda_2 \end{bmatrix} [R]^T. \tag{8.49}$$

As constructed, the matrix on the left contains the *lab* components of the symmetric stress tensor, while the middle matrix on the right-hand side contains the lab components of the *different* reference symmetric stress $\underset{\approx}{\Sigma}^{\text{sym}}$. We may alternatively interpret this equation in a purely matrix context showing how the lab components of the real symmetric stress tensor can be expressed as an orthogonal coordinate transformation tensor acting on a diagonal principal value matrix. This viewpoint allows us to assert that the components of the symmetric stress *with respect to a coordinate system aligned with the principal directions* will be diagonal.

To reiterate, there are two equally valid viewpoints, and each can have its advantages. The first viewpoint tells us that we can always construct a *reference* stress tensor whose lab components are diagonal. The second viewpoint states that we can always construct a new *basis* such that the components are diagonal. Most people find the second viewpoint easiest whenever dealing with a symmetric tensor. That is because the basis transformation rule involves orthonormal bases in both cases. The merits of the *reference* tensor concept become more clear when discussing *non-symmetric* tensors. A fundamental result from matrix analysis states that, for any fully populated (permissibly non-symmetric) matrix $[A]$, there exists an invertible matrix $[L]$ and a 'canonical' matrix $[\Lambda]$ such that

$$[A] = [L][\Lambda][L]^{-1}. \tag{8.50}$$

The canonical matrix is in *Jordan form*, meaning it has eigenvalues on its diagonal, and either ones or zeros along the first upper off-diagonal. Specifically,

$$[\Lambda] = \begin{bmatrix} \lambda_1 & b & 0 \\ 0 & \lambda_2 & B \\ 0 & 0 & \lambda_3 \end{bmatrix}. \tag{8.51}$$

Here, b will equal zero if $\lambda_1 \neq \lambda_2$, and it will also equal zero if two linearly independent eigenvalues can be found when $\lambda_1 = \lambda_2$; otherwise, the geometric multiplicity of the eigenvalue is smaller than the algebraic multiplicity, and then $b = 1$. The B component is defined similarly based on algebraic versus geometric multiplicity of the eigenvalues λ_2 and λ_3. Equation (8.50) is a pure matrix analysis formula, but it does have an associated tensor-analysis interpretation. If we construct special basis-dependent tensors by

$$\underset{\approx}{\boldsymbol{L}} = L_{ij}\underset{\sim}{\boldsymbol{E}}_i\underset{\sim}{\boldsymbol{E}}_j, \tag{8.52}$$

$$\underset{\approx}{\Lambda} = \Lambda_{ij}\underset{\sim}{\boldsymbol{E}}_i\underset{\sim}{\boldsymbol{E}}_j, \tag{8.53}$$

and

$$\underset{\approx}{\boldsymbol{A}} = A_{ij}\underset{\sim}{\boldsymbol{E}}_i\underset{\sim}{\boldsymbol{E}}_j, \tag{8.54}$$

then (8.50) tells us that we can write

$$\underset{\approx}{\boldsymbol{A}} = \underset{\approx}{\boldsymbol{L}} \bullet \underset{\approx}{\Lambda} \bullet \underset{\approx}{\boldsymbol{L}}^{-1}. \tag{8.55}$$

Expanding out the canonical reference tensor in dyadic form (i.e., writing it in the form of (8.53)), the above equation may be written

$$\underset{\approx}{\boldsymbol{A}} = \underset{\approx}{\boldsymbol{L}} \bullet (\Lambda_{ij}\underset{\sim}{\boldsymbol{E}}_i\underset{\sim}{\boldsymbol{E}}_j) \bullet \underset{\approx}{\boldsymbol{L}}^{-1} \tag{8.56}$$

or

$$\underset{\approx}{\boldsymbol{A}} = \Lambda^i_{\ j}\underset{\sim}{\boldsymbol{g}}_i\underset{\sim}{\boldsymbol{g}}^j, \tag{8.57}$$

where we are using subscripted (covariant) and superscripted (contravariant) notation from analytical geometry, which defines:

$$\underset{\sim}{\boldsymbol{g}}_i \equiv \underset{\approx}{\boldsymbol{L}} \bullet \underset{\sim}{\boldsymbol{E}}_i, \quad \underset{\sim}{\boldsymbol{g}}^j = \underset{\approx}{\boldsymbol{L}}^{-T} \bullet \underset{\sim}{\boldsymbol{E}}_j, \quad \text{and} \quad \Lambda^i_{\ j} = \Lambda_{ij}. \tag{8.58}$$

By design,

$$\underset{\sim}{\boldsymbol{g}}_i \bullet \underset{\sim}{\boldsymbol{g}}^j = \delta_i^j, \tag{8.59}$$

where δ_i^j is an alternative symbol for the Kronecker delta. In general, $\underset{\sim}{\boldsymbol{g}}_i \neq \underset{\sim}{\boldsymbol{g}}^j$, so the above equation does *not* say that the 'g' vectors form an orthonormal triad. Instead, the above property is a statement of *duality* between the three 'covariant' generally non-orthonormal basis vectors $\{\underset{\sim}{\boldsymbol{g}}_1, \underset{\sim}{\boldsymbol{g}}_2, \underset{\sim}{\boldsymbol{g}}_3\}$ and their unique 'contravariant' counterparts $\{\underset{\sim}{\boldsymbol{g}}^1, \underset{\sim}{\boldsymbol{g}}^2, \underset{\sim}{\boldsymbol{g}}^3\}$. Here we have employed standard notational conventions for dealing with non-orthonormal basis in which indices can appear as both subscripts and superscripts to denote different (but closely related) vector triads.

Equation (8.57) tells us that there exists a generally non-orthonormal covariant basis (and an associated dual contravariant basis) such that the mixed component matrix for the tensor $\underset{\approx}{\boldsymbol{A}}$ will be in canonical form. Thus, just as we did for the symmetric tensor case, we have introduced two viewpoints. The first viewpoint, embodied in (8.55), tells us that we can always construct a *reference* tensor such that its *lab* components are in canonical form. The second viewpoint, shown in (8.57), tells us that we can always construct a *basis* such that the mixed components of the original tensor will be in canonical form. The first viewpoint works with two tensors using a single (lab) basis. The second viewpoint uses two bases (lab and 'g') to interpret a single tensor. Either viewpoint is legitimate. However, the first viewpoint is generally more easily understood by non-tensor experts. Furthermore, the first viewpoint permits all explicit tensor manipulations to always be performed in the nice familiar orthonormal lab basis, which lends itself better to taking rates for incremental numerical modeling (as well as using existing linear algebra numerical packages). The definitions of tensor operations for general non-orthonormal bases are considerably more complicated. That is why the first viewpoint has strong appeal to lay audiences who might never have been introduced to generalized tensor analysis.

The key point of this discussion was to illustrate yet again that you must be careful to recognize which viewpoint you are adopting. Are you using a single basis to look at two *different* objects that are related by a tensor similarity transformation or are you looking at a *single* object from two different bases that are related by a matrix similarity transformation? Either viewpoint is fine, just know which one you are adopting and be able at all times to convert to the other viewpoint so that you can communicate effectively with people who prefer different approach.

8.3 Constructing a rotation that will transform one specified vector to another specified vector

Suppose you have two unit vectors $\underset{\sim}{m}$ and $\underset{\sim}{n}$, and you wish to find a rotation tensor $\underset{\approx}{R}$ that will rotate $\underset{\sim}{m}$ into $\underset{\sim}{n}$. That is, you seek a proper-orthogonal tensor $\underset{\approx}{R}$ such that

$$\underset{\sim}{n} = \underset{\approx}{R} \bullet \underset{\sim}{m}. \tag{8.60}$$

Being unit vectors, $\underset{\sim}{m}$ and $\underset{\sim}{n}$ may be regarded as points on a unit sphere. The solution for $\underset{\approx}{R}$ is not unique. An obvious answer would be a rotation that moves from $\underset{\sim}{m}$ to $\underset{\sim}{n}$ along a major diameter of the sphere. Another answer, however, could move along a smaller circular path on the sphere. We can define two 'helper' unit vectors

$$\underset{\sim}{M} = \frac{\underset{\sim}{n} + \underset{\sim}{m}}{\|\underset{\sim}{n} + \underset{\sim}{m}\|}, \tag{8.61}$$

$$\underset{\sim}{N} = \frac{\underset{\sim}{n} - \underset{\sim}{m}}{\|\underset{\sim}{n} - \underset{\sim}{m}\|}. \tag{8.62}$$

Unlike $\underset{\sim}{m}$ and $\underset{\sim}{n}$, the vectors $\underset{\sim}{M}$ and $\underset{\sim}{N}$ are orthonormal. The angle θ separating $\underset{\sim}{m}$ and $\underset{\sim}{n}$ is given by

$$\cos\theta = \underset{\sim}{n} \bullet \underset{\sim}{m}. \tag{8.63}$$

Therefore $\underset{\sim}{m}$ and $\underset{\sim}{n}$ may be expressed in terms of $\underset{\sim}{M}$ and $\underset{\sim}{N}$ by

$$\underset{\sim}{n} = \cos\frac{\theta}{2}\underset{\sim}{M} + \sin\frac{\theta}{2}\underset{\sim}{N}, \tag{8.64}$$

$$\underset{\sim}{m} = \cos\frac{\theta}{2}\underset{\sim}{M} + \sin\frac{\theta}{2}\underset{\sim}{N}. \tag{8.65}$$

To rotate $\underset{\sim}{m}$ into $\underset{\sim}{n}$, the unit rotation axis $\underset{\sim}{a}$ must be perpendicular to $\underset{\sim}{N}$, but is otherwise unrestricted. Stated differently using the unit sphere geometry in which $\underset{\sim}{a}$ points to the 'north pole', $\underset{\sim}{N}$ would point to a location on the equator and the rotation operation would sweep $\underset{\sim}{m}$ into $\underset{\sim}{n}$ along a latitude line.

For the *shortest (major diameter) path*—i.e., a *direct* rotation of $\underset{\sim}{m}$ into $\underset{\sim}{n}$—you would define the rotation axis to be perpendicular to the plane formed by $\underset{\sim}{m}$ and $\underset{\sim}{n}$. In this case $\underset{\sim}{m}$ and $\underset{\sim}{n}$ would themselves lie on the equator, and the rotation axis also would be parallel to $\underset{\sim}{M} \times \underset{\sim}{N}$. The rotation angle would be given by $\cos\alpha = \underset{\sim}{n} \bullet \underset{\sim}{m}$. For this direct shortest-distance rotation, we can define an orthonormal basis

$$\underset{\sim}{e}_1 = \underset{\sim}{M}, \tag{8.66}$$

$$\underset{\sim}{e}_2 = \underset{\sim}{N}, \tag{8.67}$$

$$\underset{\sim}{e}_3 = \underset{\sim}{M} \times \underset{\sim}{N}. \tag{8.68}$$

Then, with respect to this basis,

$$[\underset{\approx}{R}] = \begin{bmatrix} \cos\alpha & -\sin\alpha & 0 \\ \sin\alpha & \cos\alpha & 0 \\ 0 & 0 & 1 \end{bmatrix} = \begin{bmatrix} \cos\alpha - 1 & -\sin\alpha & 0 \\ \sin\alpha & \cos\alpha - 1 & 0 \\ 0 & 0 & 0 \end{bmatrix} + \begin{bmatrix} 1 & 0 & 0 \\ 0 & 1 & 0 \\ 0 & 0 & 1 \end{bmatrix}. \tag{8.69}$$

Written in basis form,

$$\underset{\approx}{R} = (\cos\alpha - 1)(\underset{\sim}{e}_1\underset{\sim}{e}_1 + \underset{\sim}{e}_2\underset{\sim}{e}_2) + \sin\alpha(\underset{\sim}{e}_2\underset{\sim}{e}_1 - \underset{\sim}{e}_1\underset{\sim}{e}_2) + \underset{\approx}{I} \tag{8.70}$$

or

$$\underset{\approx}{R} = (\cos\alpha - 1)(\underset{\sim}{M}\underset{\sim}{M} + \underset{\sim}{N}\underset{\sim}{N}) + \sin\alpha(\underset{\sim}{N}\underset{\sim}{M} - \underset{\sim}{M}\underset{\sim}{N}) + \underset{\approx}{I}. \tag{8.71}$$

Back substituting through the definitions, it can be shown that

$$(\cos\alpha - 1)(\underset{\sim}{M}\underset{\sim}{M} + \underset{\sim}{N}\underset{\sim}{N}) = \frac{\cos\alpha(\underset{\sim}{n}\underset{\sim}{m} + \underset{\sim}{m}\underset{\sim}{n}) - (\underset{\sim}{m}\underset{\sim}{m} + \underset{\sim}{n}\underset{\sim}{n})}{1 + \cos\alpha} \tag{8.72}$$

and

$$(\sin\alpha)(\underset{\sim}{N}\underset{\sim}{M} - \underset{\sim}{M}\underset{\sim}{N}) = \underset{\sim}{n}\underset{\sim}{m} - \underset{\sim}{m}\underset{\sim}{n}. \tag{8.73}$$

8.4 Constructing a rotation from knowledge of initial and final 'marker' locations in a body

Suppose that you have computed tomography (or similar) data so that you know the initial locations, $(\underset{\sim}{x}_1, \underset{\sim}{x}_2, \ldots, \underset{\sim}{x}_N)$ of a total of N points in a body along with their final deformed locations $(\underset{\sim}{y}_1, \underset{\sim}{y}_2, \ldots, \underset{\sim}{y}_N)$. Elementary digital image correlation (DIC) seeks a tensor $\underset{\approx}{M}$ and vector $\underset{\sim}{b}$ such that.

$$\underset{\sim}{y} = \underset{\approx}{M} \bullet \underset{\sim}{x} + \underset{\sim}{b} \tag{8.74}$$

is the best affine fit to the data. We will provide (without proof) formulas for $\underset{\approx}{M}$ and $\underset{\sim}{b}$ corresponding to least-squares multi-linear regression. In general, the tensor $\underset{\approx}{M}$ will not be a pure rotation. If your goal is to find a best-fit rigid motion, then you should use the rotation from polar decomposition of $\underset{\approx}{M}$ (see chapter 12).

To find the best affine fit to the marker data, the above affine mapping is first rewritten in the alternative 'point-slope' form,

$$\underset{\sim}{y} - \underset{\sim}{y}^* = \underset{\approx}{M} \bullet (\underset{\sim}{x} - \underset{\sim}{x}^*). \tag{8.75}$$

Here, $\underset{\sim}{x}^*$ denotes any reference point in the undeformed body, while $\underset{\sim}{y}^*$ is the deformed location of that same material point. These starred reference points need not be selected from the list of marker points. In fact, for equally weighted multi-linear regression, these are simple averages:

$$\underset{\sim}{x}^* = \frac{1}{N}\sum_{\alpha=1}^{N} \underset{\sim}{x}_\alpha \quad \text{and} \quad \underset{\sim}{y}^* = \frac{1}{N}\sum_{\alpha=1}^{N} \underset{\sim}{y}_\alpha . \tag{8.76}$$

The next step in multi-linear regression forms two matrices $[X]$ and $[Y]$, each of dimension $3 \times N$ whose columns contain *relative* initial and deformed marker vectors:

$$X_{i\alpha} = \text{the } i\text{th component of } \underset{\sim}{x}_\alpha - \underset{\sim}{x}^* \tag{8.77a}$$

$$Y_{i\alpha} = \text{the } i\text{th component of } \underset{\sim}{y}_\alpha - \underset{\sim}{y}^* . \tag{8.77b}$$

The Cartesian component matrix for the best-fit $\underset{\approx}{M}$ transformation tensor is then

$$[M] = [Y][X]^{(-1)}, \tag{8.78}$$

where $[X]^{(-1)}$ is the pseudo-inverse of the $3 \times N$ non-square matrix $[X]$. The set of N marker points is sufficient to solve the problem if and only if the square 3×3 matrix $[X]\,[X]^T$ is invertible, in which case,

$$[X]^{(-1)} = [X]^T([X][X]^T)^{-1}. \tag{8.79}$$

Once $\underset{\approx}{M}$ has been found, the 'intercept' vector (if desired) is $\underset{\sim}{b} = \underset{\sim}{y}^* - \underset{\approx}{M} \bullet \underset{\sim}{x}^*$. The overall material rotation (found from polar decomposition of $\underset{\approx}{M}$) would then remove experimental error in the marker locations in an unbiased manner, where each marker is treated equally in this algorithm.

Special case: exactly four marker points. The above least-squares regression formula finds a best fit in the case of an arbitrary number of tracked points in the body, making it an overdetermined problem. The calculations simplify considerably if there are exactly four tracked points. In that case, the 'starred' reference vectors should be instead taken equal the fourth markers[3]:

$$\underset{\sim}{x}^* = \underset{\sim}{x}_4 \qquad \text{and} \qquad \underset{\sim}{y}^* = \underset{\sim}{y}_4. \tag{8.80}$$

When this choice is made, the last column of each $[X]$ and $[Y]$ is zero. Dropping that last column converts these to simple 3×3 matrices. As long as the initial four marker points are not coplanar, the $[X]$ matrix will be invertible so that an ordinary inverse (not a pseudo-inverse) may be used to evaluate $[M]$. Again, a polar decomposition of $\underset{\sim}{M}$ would provide overall material rotation.

Special case: exactly three marker points. If one is interested exclusively in finding a rotation, then only three non-colinear marker points are needed. In this case, the above method may be used by creating a fourth marker point by

$$\underset{\sim}{x}_4 = (\underset{\sim}{x}_2 - \underset{\sim}{x}_1) \times (\underset{\sim}{x}_3 - \underset{\sim}{x}_1) \tag{8.81a}$$

$$\underset{\sim}{y}_4 = (\underset{\sim}{y}_2 - \underset{\sim}{y}_1) \times (\underset{\sim}{y}_3 - \underset{\sim}{y}_1). \tag{8.81b}$$

Though not obvious, this method is roughly equivalent to a rather brute-force approach that performs Graham-Schmidt orthogonalization on $\underset{\sim}{x}_2 - \underset{\sim}{x}_1$ and $\underset{\sim}{x}_3 - \underset{\sim}{x}_1$ to obtain two unit vectors, creating a third unit vector from the cross product of these first two, and putting the now orthonormal triad into columns of a matrix $[Q_1]$. Doing a similar operation with the y-vectors to create a $[Q_2]$ matrix then allows the rotation from initial to final states to be given by $[Q_2][Q_1]^T$.

References

[1] Simo J C and Hughes T J R 1998 *Computational Inelasticity* (Berlin: Springer)

[2] Wiley C R and Barrett L C 1982 *Advanced Engineering Mathematics* (New York: McGraw-Hill)

[3] Of course, it should be clear that any marker in the set may be selected as the reference point by simply moving it to the end of the list.

Chapter 9

Sequential rotations

9.1 The distinction between fixed and follower axes

Let $\underset{\approx}{\boldsymbol{R}} = f(\underset{\sim}{\boldsymbol{a}}, \alpha)$ denote the Euler–Rodrigues formula for converting a unit axis $\underset{\sim}{\boldsymbol{a}}$ and angle α to a rotation $\underset{\approx}{\boldsymbol{R}}$. Let us consider two such rotations,

$$\underset{\approx}{\boldsymbol{R}}_1 = f(\underset{\sim}{\boldsymbol{a}}_1, \alpha_1) \qquad \text{and} \qquad \underset{\approx}{\boldsymbol{R}}_2 = f(\underset{\sim}{\boldsymbol{a}}_2, \alpha_2). \tag{9.1}$$

A *fixed-axis sequence of rotations* would involve the two axes being, well, *fixed*. Thus, if one first rotates about axis 1, and then rotates about axis 2, the resulting rotation would be

$$\underset{\approx}{\boldsymbol{R}}^{\text{fixed}} = \underset{\approx}{\boldsymbol{R}}_2^{\text{fixed}} \bullet \underset{\approx}{\boldsymbol{R}}_1^{\text{fixed}}. \tag{9.2}$$

Contrast this with a *follower-axis* sequence of rotations in which the second axis is no longer $\underset{\sim}{\boldsymbol{a}}_2$ but is instead the rotation of $\underset{\sim}{\boldsymbol{a}}_2$ under the action of the first rotation. Namely,

$$\underset{\approx}{\boldsymbol{R}}^{\text{follower}} = \underset{\approx}{\boldsymbol{R}}_2^{\text{follower}} \bullet \underset{\approx}{\boldsymbol{R}}_1^{\text{follower}}, \tag{9.3}$$

where

$$\underset{\approx}{\boldsymbol{R}}_1^{\text{follower}} = f(\underset{\sim}{\boldsymbol{a}}_1, \alpha_1) = \underset{\approx}{\boldsymbol{R}}_1^{\text{fixed}}, \tag{9.4a}$$

$$\underset{\approx}{\boldsymbol{R}}_2^{\text{follower}} = f\left(\underset{\approx}{\boldsymbol{R}}_1^{\text{follower}} \bullet \underset{\sim}{\boldsymbol{a}}_2, \alpha_2\right) = f\left(\underset{\approx}{\boldsymbol{R}}_1^{\text{fixed}} \bullet \underset{\sim}{\boldsymbol{a}}_2, \alpha_2\right). \tag{9.4b}$$

Using the lemma in (5.40),

$$\underset{\approx}{\boldsymbol{R}}_2^{\text{follower}} = \underset{\approx}{\boldsymbol{R}}_1^{\text{fixed}} \bullet f(\underset{\sim}{\boldsymbol{a}}_2, \alpha_2) \bullet \left(\underset{\approx}{\boldsymbol{R}}_1^{\text{fixed}}\right)^T = \underset{\approx}{\boldsymbol{R}}_1^{\text{fixed}} \bullet \underset{\approx}{\boldsymbol{R}}_2^{\text{fixed}} \bullet \left(\underset{\approx}{\boldsymbol{R}}_1^{\text{fixed}}\right)^T. \tag{9.5}$$

doi:10.1016/978-0-7503-1454-1ch9

Thus, there is a specific relationship between rotations using follower axes and the same rotations using fixed axes. In particular, the net rotation is

$$\underset{\approx}{\mathbf{R}}^{\text{follower}} = \underset{\approx}{\mathbf{R}}_2^{\text{follower}} \bullet \underset{\approx}{\mathbf{R}}_1^{\text{follower}} = \underset{\approx}{\mathbf{R}}_1^{\text{fixed}} \bullet \underset{\approx}{\mathbf{R}}_2^{\text{fixed}} \bullet \left(\underset{\approx}{\mathbf{R}}_1^{\text{fixed}}\right)^T \bullet \underset{\approx}{\mathbf{R}}_1^{\text{fixed}}. \tag{9.6}$$

Simplifying, using orthogonality, give the important result

$$\boxed{\underset{\approx}{\mathbf{R}}_2^{\text{follower}} \bullet \underset{\approx}{\mathbf{R}}_1^{\text{follower}} = \underset{\approx}{\mathbf{R}}_1^{\text{fixed}} \bullet \underset{\approx}{\mathbf{R}}_2^{\text{fixed}}}. \tag{9.7}$$

In words, *follower rotations are equivalent to fixed rotations in reverse order.* In this section, for example, we will define Euler angles to be a set of rotations about follower axes, and we will find that the resulting rotation matrix is the same as using fixed axes in reverse order.

9.2 Roll, pitch, yaw: sequential rotations about fixed (laboratory) axes

If a body is rigidly rotated by a rotation tensor $\underset{\approx}{\mathbf{R}}_1$ and then subsequently rotated by another rotation operation $\underset{\approx}{\mathbf{R}}_2$, then the total overall rotation tensor is given by

$$\underset{\approx}{\mathbf{R}} = \underset{\approx}{\mathbf{R}}_2 \bullet \underset{\approx}{\mathbf{R}}_1. \tag{9.8}$$

Any general rotation $\underset{\approx}{\mathbf{R}}$ can always be described in terms of three *sequential* rotations about the fixed laboratory axes $\{X, Y, Z\}$ through angles $\{\Theta_X, \Theta_Y, \Theta_Z\}$. These angles are not equivalent to Euler angles (which are discussed in the next section).

Figure 9.1 shows a simple example of sequential laboratory-referenced rotations in which a block is rotated 90° about the fixed laboratory Z-axis and then 90° about the fixed X-axis. Using (4.34) for the first rotation about Z and (4.22) for the second rotation about X, and then combining them in (9.8), the total rotation tensor is

$$[R_{ij}] = \begin{bmatrix} 1 & 0 & 0 \\ 0 & 0 & -1 \\ 0 & 1 & 0 \end{bmatrix} \begin{bmatrix} 0 & -1 & 0 \\ 1 & 0 & 0 \\ 0 & 0 & 1 \end{bmatrix} = \begin{bmatrix} 0 & -1 & 0 \\ 0 & 0 & -1 \\ 1 & 0 & 0 \end{bmatrix}. \tag{9.9}$$

The overall axis and angle of rotation may be determined by using the algorithm in section 5.4.

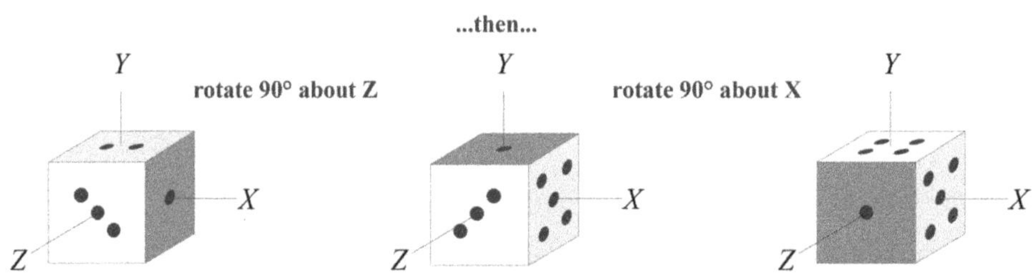

Figure 9.1. Sequential rotation about fixed axes. Note how the block rotates relative to the fixed laboratory triad. The number of dots on opposite sides of honest dice always sum to seven (hence, the side with four dots is opposite the side with three dots).

Sequential rotation operations do not commute. That is, as shown in figure 9.2, when the rotation operations are applied in a different order, the final orientation of the block will generally be different. Namely, the rotation tensor corresponding to figure 9.2 is

$$[R_{ij}] = \begin{bmatrix} 0 & -1 & 0 \\ 1 & 0 & 0 \\ 0 & 0 & 1 \end{bmatrix} \begin{bmatrix} 1 & 0 & 0 \\ 0 & 0 & -1 \\ 0 & 1 & 0 \end{bmatrix} = \begin{bmatrix} 0 & 0 & 1 \\ 1 & 0 & 0 \\ 0 & 1 & 0 \end{bmatrix}, \tag{9.10}$$

which is not equal to the rotation matrix given in (9.9). Many writers use this example as a counter-example 'proof' that rotation is not a vector. The proper conclusion is simply that sequentially applied rotation operations do not commute. We have already shown that, through the Euler–Rodrigues formula, that rotation *can* be represented by a vector. The lack of commutativity of rotations simply proves that the rotation vector of a *sequential* rotation is not in general equal to the sum of the individual rotation vectors. On the other hand, as discussed later, *simultaneously* applied rotations *do* commute.

9.3 Euler angles: sequential rotations about 'follower' axes

Euler angles are a common way to describe any general rotation. They are rather awkward for theoretical analysis or computer simulations, and their direct application in robotics made early robots move in a comically awkward 'jerky' fashion. Rather than defining the rotation relative to the fixed laboratory axes, Euler angles use the embedded triad $\{x, y, z\}$ of axes that *follow* the material as it rotates. The 'zxz' Euler angles $\{\varphi, \vartheta, \psi\}$ are defined [1] as follows. First rotate the embedded triad an angle φ about the z-axis (this causes x and y to move to new orientations while z remains unchanged). Then rotate the triad an angle ϑ about its new x-axis (this causes y and z to move to new orientations while x remains unchanged). Finally, rotate the triad an angle ψ about its new z-axis. This 'zxz' definition of Euler angles differs from a similar 'zyz' definition of Euler angles (illustrated in figure 9.3) based on which follower axis (x or y) is used in the middle step. The zxz definition, which is compared with the zyz convention in figure 9.4, is favored in materials science (to, for example, establish the orientation of a crystal).

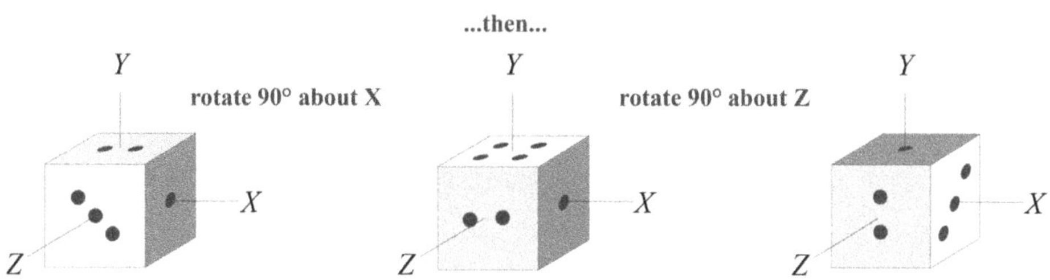

Figure 9.2. A different sequence for the rotations. Note that the final orientation of the block does not match that of figure 9.1, which proves that rotation operations do not commute. It does not prove that they are somehow not tensors, nor does it prove that rotations cannot be mapped to and from vectors (which they can via spinor vectors)—it simply means that a spinor vector associated with a sequentially applied set of rotations will not be the vector sum of the individual spinor vectors of each step. That does not make spinor vectors somehow not vectors!.

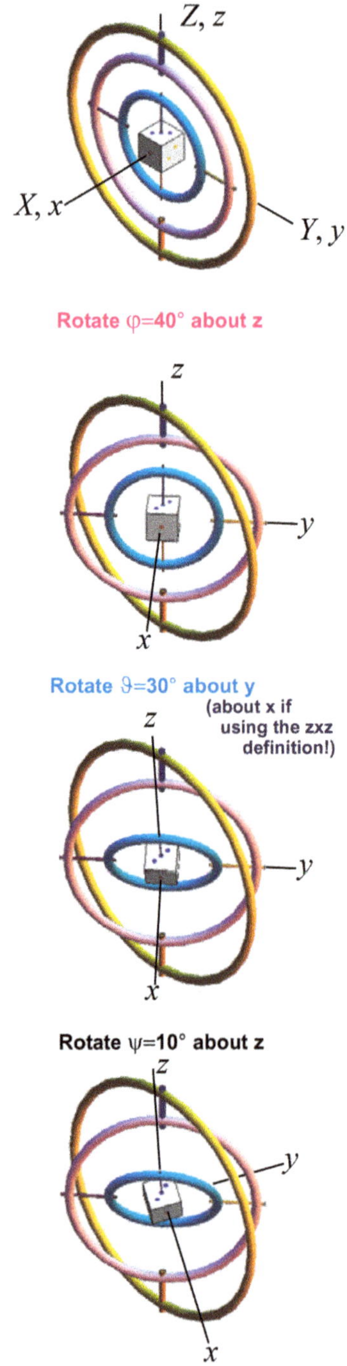

Figure 9.3. The '*zyz*' definition of Euler angles.

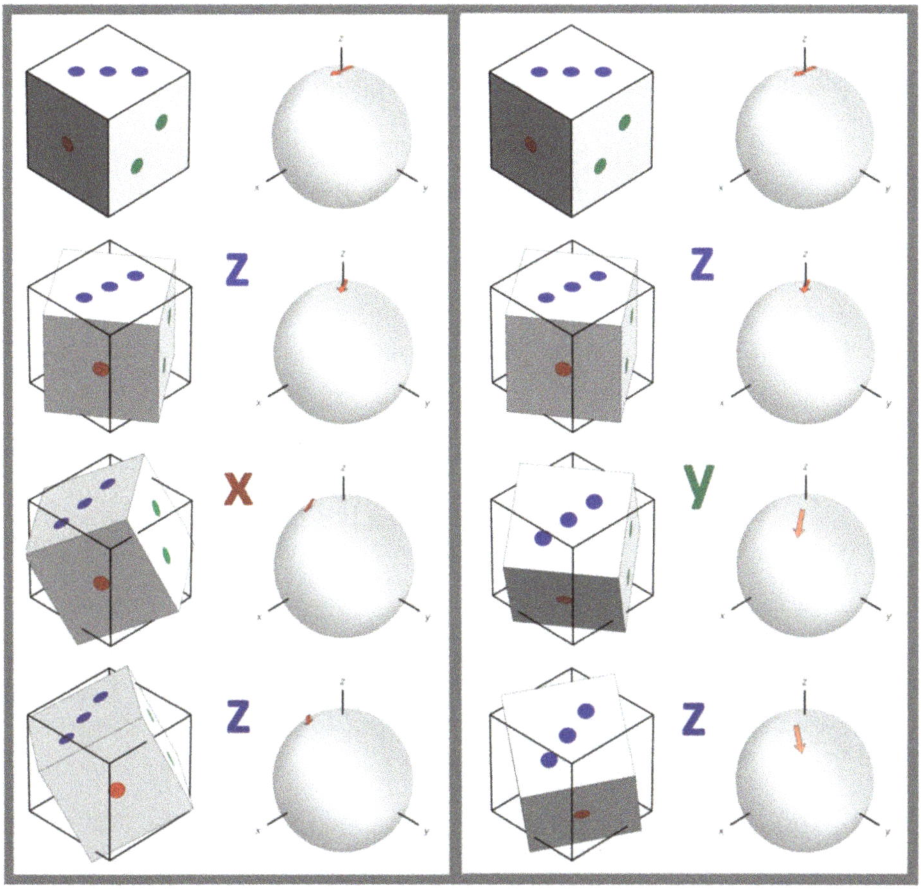

Figure 9.4. Difference between the *zxz* and *zyz* convention for Euler angles. The 'nautical' depiction of these operations is revealing. The first angle about the *z*-axis sets an initial heading for the boat (arrow) keeping it at the north pole. For the *zxz* convention, the next angle moves the boat to a different latitude by sliding it in the starboard direction down a fixed longitude line. The *zyz* convention does the same thing except the sliding direction is towards the 'bow' of the ship. The final rotation sets the heading of the boat by rotating it without a change in latitude or longitude. With these interpretations, it is easy to see why the first and last Euler angles range from 0 to 2π, whereas the middle Euler angle ranges only from 0 to π.

The second (*zyz*) definition, which appears to be favored in aerospace applications, is depicted in figure 9.3, in which the outer ring is fixed, the middle first ring rotates with angle φ (called the *precession (or yaw) angle*), carrying everything inside the ring with it. The inner ring then rotates with angle ϑ (called the *nutation (pitch) angle*) about the new (follower) *y*-axis, again carrying everything inside that ring with it. Then the object itself rotates with angle ψ (called the *spin (roll) angle*) about the new (follower) *z*-axis.

When the first (*zxz*) definition of Euler angles is used, the image looks the same except that the initially concentric rings are in the plane containing the *x*- and *z*-axes.

Unlike the previous section, which applied sequential rotations about fixed axes, Euler angles define rotations about follower axes. Recalling (9.7), this causes the sequentially multiplied rotation matrices to be *opposite of the ordering for fixed axes.*

9.3.1 Converting Euler angles to direction cosines

The matrix of components of the rotation tensor $\underset{\sim}{R}$ *with respect to the laboratory triad* is the product of the three *relative* rotation matrices as follows:

The first (zxz) Euler angle definition has opposite ordering of fixed axes:

$$[R] = \begin{bmatrix} \cos\varphi & -\sin\varphi & 0 \\ \sin\varphi & \cos\varphi & 0 \\ 0 & 0 & 1 \end{bmatrix} \begin{bmatrix} 1 & 0 & 0 \\ 0 & \cos\vartheta & -\sin\vartheta \\ 0 & \sin\vartheta & \cos\vartheta \end{bmatrix} \begin{bmatrix} \cos\psi & -\sin\psi & 0 \\ \sin\psi & \cos\psi & 0 \\ 0 & 0 & 1 \end{bmatrix}. \tag{9.11}$$

The second (zyz) Euler angle definition likewise has opposite ordering of fixed axes

$$[R] = \begin{bmatrix} \cos\varphi & -\sin\varphi & 0 \\ \sin\varphi & \cos\varphi & 0 \\ 0 & 0 & 1 \end{bmatrix} \begin{bmatrix} \cos\vartheta & 0 & \sin\vartheta \\ 0 & 1 & 0 \\ -\sin\vartheta & 0 & \cos\vartheta \end{bmatrix} \begin{bmatrix} \cos\psi & -\sin\psi & 0 \\ \sin\psi & \cos\psi & 0 \\ 0 & 0 & 1 \end{bmatrix}. \tag{9.12}$$

See section A.4 in the appendix. Unlike (9.8), the above three matrices are multiplied *in the order of operation.* This distinction arises because Euler angle matrices are defined with respect to the *embedded* follower axes whereas (9.8) is defined with respect to the fixed *laboratory* axes.

Example. Consider the particular set of zxz Euler angles $\{\varphi, \vartheta, \psi\} = \{90°, 90°, 0\}$, illustrated in figure 9.5. Euler angles are based on rotations about the *embedded* follower triad (which is different from figure 9.1 which uses the same set of rotation angles, but referenced to a *fixed* triad). Using (9.11), the rotation tensor corresponding to figure 9.5 is

$$[R_{ij}] = \begin{bmatrix} 0 & -1 & 0 \\ 1 & 0 & 0 \\ 0 & 0 & 1 \end{bmatrix} \begin{bmatrix} 1 & 0 & 0 \\ 0 & 0 & -1 \\ 0 & 1 & 0 \end{bmatrix} \begin{bmatrix} 1 & 0 & 0 \\ 0 & 1 & 0 \\ 0 & 0 & 1 \end{bmatrix} = \begin{bmatrix} 0 & 0 & 1 \\ 1 & 0 & 0 \\ 0 & 1 & 0 \end{bmatrix}. \tag{9.13}$$

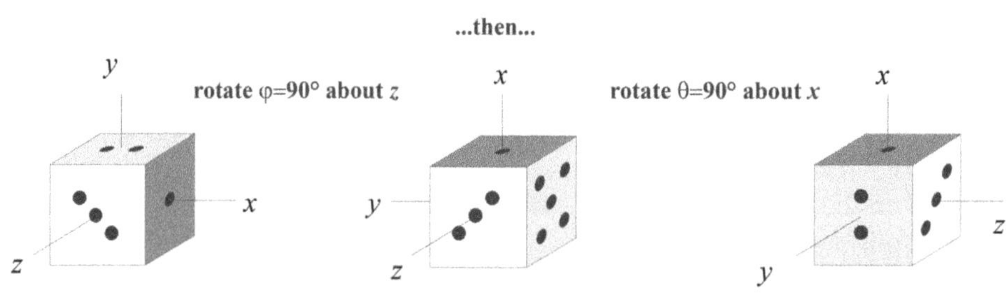

Figure 9.5. Example of the zxz definition of Euler angles, $\{\varphi, \vartheta, \psi\} = \{90°, 90°, 0\}$. The final rotation is quite different from the one defined with respect to the fixed laboratory triad.

9.3.2 Converting a rotation matrix to Euler angles

Converting an orthogonal rotation matrix back to Euler angles is relatively difficult, and the solution is not necessarily unique unless you recognize that the middle Euler angle ϑ should be limited to a smaller range than the others. Specifically, as revealed in the caption of figure 9.4,

$$0 \leqslant \vartheta < \pi \tag{9.14a}$$

whereas

$$0 \leqslant \varphi < 2\pi \qquad \text{and} \qquad 0 \leqslant \psi < 2\pi. \tag{9.14b}$$

Looking at the 'nautical' depictions in figure 9.4, it should be clear that the first two Euler angles are determined entirely from the placement of the 'mast' of the ship, which is the last column of the rotation matrix. It follows that the first two Euler angles for the *zyz* convention identically equal the spherical coordinate angles associated with the last column of the rotation matrix, which is a good argument favoring *zyz* over *zxz*! As is always the case when finding spherical coordinates from a point on the sphere (which, in this case, is the third column of the rotation matrix), a *two-argument* arctangent must be used obtain the correct value of φ over its full range (otherwise, the 'mast' of the ship would not be able to reach all octants of the sphere). An arccos may be used to find ϑ in its smaller allowable range. Once ϑ and φ have been assigned their values using only the last column of the rotation matrix, the remaining angle ψ can be determined by multiplying out the definition of the rotation (*zxz* or *zyz*) and matching components in the first two columns appropriately. A solution for the *zxz* definition is provided in section A.5 of the appendix.

9.3.3 Converting Euler angles to axis and angle

To convert from Euler angles to a single axis and angle of rotation, simply compute the rotation tensor using (9.11), and then use the algorithm in section 5.4. For example, the rotation described in figure 9.5 gave the rotation tensor of (9.13), which is identical to (5.129) for which the angle and axis were found to be 120° and $\{1, 1, 1\}/\sqrt{3}$.

Reference

[1] Desloge E A 1982 *Classical Mechanics* vol 1 and 2 (New York: Wiley)

Chapter 10

Series expression for a rotation

This chapter shows that a rotation of angle α about a unit axis $\underset{\sim}{\boldsymbol{a}}$ may be expressed as an exponential of the pseudo-rotation spinor $\underset{\approx}{\boldsymbol{S}}$, first introduced in equation (5.67). Exponentials of tensors are defined by a series expansion of the same form as the series expansion $e^x = 1 + x + \frac{1}{2!}x^2 + \frac{1}{3!}x^3 + \ldots$. Unlike exponentials of scalars, however, tensor exponentials do not (to our knowledge) constitute useful simplifications because they do not satisfy the same properties as scalar exponentials. We are presenting the series expression for a rotation here only for completeness.

Recall (5.69), which is Argyris's version of the Euler–Rodigues formula:

$$\boxed{\underset{\approx}{\boldsymbol{R}} = \underset{\approx}{\boldsymbol{I}} + \frac{\sin\alpha}{\alpha}\underset{\approx}{\boldsymbol{S}} + \frac{1}{2}\frac{\sin^2(\alpha/2)}{(\alpha/2)^2}\underset{\approx}{\boldsymbol{S}}^2}, \tag{10.1}$$

where α is the rotation angle, and Argyris's pseudo-tensor $\underset{\approx}{\boldsymbol{S}}$ is defined as

$$\underset{\approx}{\boldsymbol{S}} = \alpha\underset{\approx}{\boldsymbol{A}}. \tag{10.2}$$

Here, $A_{ij} = -\varepsilon_{ijk}a_k$, where ε_{ijk} is the permutation symbol and a_k are the Cartesian components of the unit rotation axis. Because $\underset{\approx}{\boldsymbol{S}}$ is proportional to α, note that (10.1) actually involves only sines of angles, which are bounded. Therefore, equation (10.1) is valid for all rotation angles, but the domain can be restricted to angles from 0 to 2π because of the cyclic period of the coefficients.. This chapter is devoted to expanding the above expression with respect to the rotation angle.

When forming expansions, we exploit the fact that $\underset{\approx}{\boldsymbol{S}}$ itself is proportional to (and therefore first-order with respect to) the rotation angle α. Using the indicial expression for $\underset{\approx}{\boldsymbol{A}}$, note that

doi:10.1016/978-0-7503-1454-1ch10

$$
\begin{aligned}
(\underset{\approx}{A}^2)_{ij} &= A_{ip}A_{pj} = (-\varepsilon_{ipk}a_k)(-\varepsilon_{pjs}a_s) = (\varepsilon_{ipk}\varepsilon_{pjs})a_k a_s \\
&= (\delta_{kj}\delta_{is} - \delta_{ks}\delta_{ij})a_k a_s = a_j a_j - a_s a_s \delta_{ij} \\
&= a_j a_j - \delta_{ij}.
\end{aligned} \tag{10.3}
$$

In direct notation,

$$
\boxed{\underset{\approx}{A}^2 = \underset{\sim}{a}\underset{\sim}{a} - \underset{\approx}{I}}. \tag{10.4}
$$

Dotting both sides by $\underset{\approx}{A}$ gives

$$
\underset{\approx}{A}^3 = (\underset{\approx}{A} \bullet \underset{\sim}{a})\underset{\sim}{a} - \underset{\approx}{A}. \tag{10.5}
$$

Now recall that, for *any* vector $\underset{\sim}{x}$,

$$
\underset{\approx}{A} \bullet \underset{\sim}{x} = \underset{\sim}{a} \times \underset{\sim}{x}. \tag{10.6}
$$

This holds for *any* vector $\underset{\sim}{x}$, so we are allowed to consider the case in which $\underset{\sim}{x}$ is the rotation axis $\underset{\sim}{a}$ itself. Thus

$$
\underset{\approx}{A} \bullet \underset{\sim}{a} = \underset{\sim}{0}. \tag{10.7}
$$

Consequently, equation (10.5) becomes

$$
\underset{\approx}{A}^3 = -\underset{\approx}{A}. \tag{10.8}
$$

Continuing in this manner, we find

$$
\begin{aligned}
&\underset{\approx}{A}^4 = -\underset{\approx}{A}^2, \quad \underset{\approx}{A}^6 = \underset{\approx}{A}^2, \quad \underset{\approx}{A}^8 = -\underset{\approx}{A}^2, \quad \underset{\approx}{A}^{10} = \underset{\approx}{A}^2, \\
&\underset{\approx}{A}^3 = -\underset{\approx}{A}, \quad \underset{\approx}{A}^5 = \underset{\approx}{A}, \quad \underset{\approx}{A}^7 = -\underset{\approx}{A}, \quad \underset{\approx}{A}^9 = \underset{\approx}{A}, \text{ etc.}
\end{aligned} \tag{10.9}
$$

In other words, all odd powers of $\underset{\approx}{A}$ are equal to $\pm\underset{\approx}{A}$ and all even powers of $\underset{\approx}{A}$ are equal to $\pm\underset{\approx}{A}^2$. Recalling that $\underset{\approx}{S} = \alpha\underset{\approx}{A}$, this means that

$$
\begin{aligned}
&\underset{\approx}{S}^4 = -\alpha^2\underset{\approx}{S}^2, \quad \underset{\approx}{S}^6 = \alpha^4\underset{\approx}{S}^2, \quad \underset{\approx}{S}^8 = -\alpha^6\underset{\approx}{S}^2, \quad \underset{\approx}{S}^{10} = \alpha^8\underset{\approx}{S}^2, \\
&\underset{\approx}{S}^3 = -\alpha^2\underset{\approx}{S}, \quad \underset{\approx}{S}^5 = \alpha^4\underset{\approx}{S}, \quad \underset{\approx}{S}^7 = -\alpha^6\underset{\approx}{S}, \quad \underset{\approx}{S}^9 = \alpha^8\underset{\approx}{S}, \text{ etc.}
\end{aligned} \tag{10.10}
$$

These lemmas help when expanding the rotation tensor with respect to the rotation angle. The first-order expansion of (10.1) *with respect to the rotation angle* (i.e., keeping in mind that $\underset{\approx}{S}$ itself depends on the rotation angle) is

$$
\underset{\approx}{R} = \underset{\approx}{I} + \underset{\approx}{S} + O(\alpha^2). \tag{10.11}
$$

The second-order expansion is

$$
\underset{\approx}{R} = \underset{\approx}{I} + \underset{\approx}{S} + \tfrac{1}{2}\underset{\approx}{S}^2 + O(\alpha^3). \tag{10.12}
$$

In fact, Argyris shows that the infinite series expansion is

$$
\boxed{\underset{\approx}{R} = \underset{\approx}{I} + \underset{\approx}{S} + \frac{1}{2!}\underset{\approx}{S}^2 + \frac{1}{3!}\underset{\approx}{S}^3 + \ldots + \frac{1}{n!}\underset{\approx}{S}^n + \ldots}. \tag{10.13}
$$

More compactly,

$$\boxed{\underset{\approx}{\boldsymbol{R}} = e^{\underset{\approx}{\boldsymbol{S}}} = \exp \underset{\approx}{\boldsymbol{S}}}. \tag{10.14}$$

This result may be deduced by arguments based on Lie's group theory as is done in quantum mechanics, but Argyris rightly asks 'why use a steam hammer to crack a nut?' Though intriguing and intoxicatingly compact, this result is fraught with hidden pitfalls that limit its usefulness. Identities that apply for exponentials of scalars do not generalize for exponentials of tensors. For example,

$$e^{\underset{\approx}{\boldsymbol{S}}_2} \bullet e^{\underset{\approx}{\boldsymbol{S}}1} \neq e^{\underset{\approx}{\boldsymbol{S}}_2 + \underset{\approx}{\boldsymbol{S}}_1} \neq e^{\underset{\approx}{\boldsymbol{S}}_1} \bullet e^{\underset{\approx}{\boldsymbol{S}}_2} \text{ unless } \underset{\approx}{\boldsymbol{S}}_1 \text{ and } \underset{\approx}{\boldsymbol{S}}_2 \text{ commute!} \tag{10.15}$$

This inequality is expressing the known fact that, given two rotations, $\underset{\approx}{\boldsymbol{R}}_1 = e^{\underset{\approx}{\boldsymbol{S}}_1}$ and $\underset{\approx}{\boldsymbol{R}}_2 = e^{\underset{\approx}{\boldsymbol{S}}_2}$, the *sequential* rotation $\underset{\approx}{\boldsymbol{R}}_2 \bullet \underset{\approx}{\boldsymbol{R}}_1$ is not generally equal to $\underset{\approx}{\boldsymbol{R}}_1 \bullet \underset{\approx}{\boldsymbol{R}}_2$ unless both rotations share the same axis of rotation.

10.1 Cayley transformations

Some numerical algorithms that track rotations do so by *instead* tracking a skew-symmetric tensor and then constructing the rotation when needed. The appeal of such an approach relates to round-off or solution errors, caused (for example) by material advection. Loosely speaking, advection is a *mixing* or *interpolation* operation in which a field variable is known at discrete points in space, and a value for that variable is desired at some intermediate location. For example, if a scalar field variable s is known to have values s_1 and s_2 at known locations $\underset{\sim}{\boldsymbol{x}}_1$ and $\underset{\sim}{\boldsymbol{x}}_2$, then value of s at a point in space exactly between $\underset{\sim}{\boldsymbol{x}}_1$ and $\underset{\sim}{\boldsymbol{x}}_2$ is often assigned simply as

$$s = \frac{1}{2}(s_1 + s_2). \tag{10.16}$$

The interpolation rules might be different at points that are not precisely between the known locations, but the basic idea is the same.

A mixing rule such as (10.16) can be problematic when interpolating *vectors*. Expected properties of the vector can be destroyed by the interpolation. If, for example, equation (10.16) is used to 'mix' two unit vectors $\underset{\sim}{\boldsymbol{n}}_1$ and $\underset{\sim}{\boldsymbol{n}}_2$, the final result would not be a unit vector because the average of two unit vectors is not itself a unit vector. Unit vectors do not belong to a linear manifold—they are points on a *curvilinear* space, the surface of the unit sphere. Noting that unit vectors are points on a unit sphere, one way (among many) to 'mix' unit vectors might be to employ a *stereographic projection* (see figure 17.10) that transforms points from the two-dimensional (2D) curvilinear space (sphere) to points on a 2D linear manifold (plane). Points on the plane could be 'mixed' as in (10.16), and then the inverse projection could be applied to transform the result back to a unit vector, which would naturally be sensitive to the projection.

A mixing rule such as (10.16) can also corrupt desired properties of *tensor* fields. If, for example, equation (10.16) were used to 'mix' two rotation tensors $\underset{\approx}{\boldsymbol{R}}_1$ and $\underset{\approx}{\boldsymbol{R}}_2$, the final result would not be a rotation tensor because the average of two orthogonal tensors is not itself orthogonal. Attempting to interpolate rotations leads to the same sorts of issues one encounters when mixing unit vectors—after all, every column and every row of a rotation matrix is a unit vector. Just as stereographic projections offered one possible way to mix unit vectors, an analogous tensor transformation has been used (by some researchers) to mix orthogonal tensors. You can, for example, arithmetically average rotation tensors and then set the 'mixed' rotation by extracting the polar rotation from the average.

Cayley transformations are rules that map uniquely between orthogonal tensors and skew tensors. Given an orthogonal tensor, you can compute a corresponding skew-symmetric tensor, and vice versa. Skew-symmetric tensors form a linear manifold, which means that any linear combination of skew tensors will be itself skew. Thus, a mixing operation such as (10.16) does not corrupt the property of being skew-symmetric. Consequently, a strategy for advecting or interpolating rotations is to use a Cayley transformation to interpolate the skew tensors associated with each rotation. Once done, the inverse transformation may be used to reconstruct a properly orthogonal estimate for the rotation at the interpolation location. As explained below, however, this approach does not work well for rotations that have a rotation angle of nearly 180°.

10.1.1 Constructing a rotation tensor from a skew-symmetric tensor

Given any skew-symmetric real tensor $\underset{\approx}{\mathbf{B}} = -\underset{\approx}{\mathbf{B}}^T$, the transformation

$$\underset{\approx}{\mathbf{R}} = (\underset{\approx}{\mathbf{I}} + \underset{\approx}{\mathbf{B}}) \bullet (\underset{\approx}{\mathbf{I}} - \underset{\approx}{\mathbf{B}})^{-1} \tag{10.17}$$

will produce a rotation tensor. It is straightforward to confirm that the tensors $\underset{\approx}{\mathbf{I}} + \underset{\approx}{\mathbf{B}}$ and $\underset{\approx}{\mathbf{I}} - \underset{\approx}{\mathbf{B}}$ along with the inverse $(\underset{\approx}{\mathbf{I}} - \underset{\approx}{\mathbf{B}})^{-1}$ all commute with each other. Therefore, an alternative and equivalent form of the Cayley transformation is

$$\underset{\approx}{\mathbf{R}} = (\underset{\approx}{\mathbf{I}} - \underset{\approx}{\mathbf{B}})^{-1} \bullet (\underset{\approx}{\mathbf{I}} + \underset{\approx}{\mathbf{B}}). \tag{10.18}$$

If you allow $\underset{\approx}{\mathbf{B}}$ to range over all possible skew-symmetric tensors (including those with infinite magnitude), then every possible rotation tensor can be generated. Thus, the Cayley transformation may be viewed as a projection of the 3D *linear* space of skew-symmetric tensors onto the 3D curvilinear space of rotation tensors. The set of all skew tensors is a 3D linear subspace contained within full 9D full tensor space. The set of skew tensors is 3D because only three numbers are needed to construct any skew tensor's component matrix. The set of skew tensors is a linear manifold because any linear combination of skew tensors is itself a skew tensor. Nevertheless, the Cayley transformation does not imply that the set of rotations is a linear subspace. We already know that rotations are on the 3D surface of a hypersphere. The Cayley transformation can be viewed as a mapping from a hyperplane to a curvilinear space (hypersphere). In this sense, it can be interpreted as an inverse stereographic projection.

Rather than directly using (10.17), the rotation $\underset{\approx}{\mathbf{R}}$ associated with the skew tensor $\underset{\approx}{\mathbf{B}}$ may be determined efficiently without ever performing an inverse operation by

$$\boxed{\underset{\approx}{\boldsymbol{R}} = \frac{2(\underset{\approx}{\boldsymbol{B}}^2 + \underset{\approx}{\boldsymbol{B}})}{b^2 + 1} + \underset{\approx}{\boldsymbol{I}}}, \tag{10.19}$$

which can be written in easily computed matrix form as

$$[\underset{\approx}{\boldsymbol{R}}] = \frac{2}{1 + b^2}\left\{\begin{bmatrix} b_1b_1 & b_1b_2 & b_1b_3 \\ b_2b_1 & b_2b_2 & b_2b_3 \\ b_3b_1 & b_3b_2 & b_3b_3 \end{bmatrix} + \begin{bmatrix} 0 & -b_3 & b_2 \\ b_3 & 0 & -b_1 \\ -b_2 & b_1 & 0 \end{bmatrix} + \frac{(1 - b^2)}{2}\begin{bmatrix} 1 & 0 & 0 \\ 0 & 1 & 0 \\ 0 & 0 & 1 \end{bmatrix}\right\}, \tag{10.20}$$

where $\underset{\sim}{\boldsymbol{b}}$ is the axial vector (with magnitude b) associated with $\underset{\approx}{\boldsymbol{B}}$, so it has components

$$b_1 = B_{32}, \quad b_2 = B_{13}, \quad \text{and } b_3 = B_{21}. \tag{10.21}$$

In terms of a basis having the 3-direction aligned with $\underset{\sim}{\boldsymbol{b}}$, equation (10.20) becomes

$$[R] = \begin{bmatrix} \dfrac{1 - b^2}{1 + b^2} & \dfrac{-2b}{1 + b^2} & 0 \\ \dfrac{2b}{1 + b^2} & \dfrac{1 - b^2}{1 + b^2} & 0 \\ 0 & 0 & 1 \end{bmatrix}. \tag{10.22}$$

To prove (10.17) it is helpful to first note that, because $\underset{\approx}{\boldsymbol{B}}$ is skew-symmetric,

$$(\underset{\approx}{\boldsymbol{I}} - \underset{\approx}{\boldsymbol{B}}) \bullet (\underset{\approx}{\boldsymbol{I}} + \underset{\approx}{\boldsymbol{B}}) = (\underset{\approx}{\boldsymbol{I}} + \underset{\approx}{\boldsymbol{B}}) \bullet (\underset{\approx}{\boldsymbol{I}} - \underset{\approx}{\boldsymbol{B}}). \tag{10.23}$$

In other words, $\underset{\approx}{\boldsymbol{I}} - \underset{\approx}{\boldsymbol{B}}$ and $\underset{\approx}{\boldsymbol{I}} + \underset{\approx}{\boldsymbol{B}}$ commute. With this realization, proving (10.17) becomes quite straightforward:

$$\begin{aligned} \underset{\approx}{\boldsymbol{R}}^T \bullet \underset{\approx}{\boldsymbol{R}} &= (\underset{\approx}{\boldsymbol{I}} - \underset{\approx}{\boldsymbol{B}})^{-T} \bullet (\underset{\approx}{\boldsymbol{I}} + \underset{\approx}{\boldsymbol{B}})^T \bullet (\underset{\approx}{\boldsymbol{I}} + \underset{\approx}{\boldsymbol{B}}) \bullet (\underset{\approx}{\boldsymbol{I}} - \underset{\approx}{\boldsymbol{B}})^{-1} \\ &= (\underset{\approx}{\boldsymbol{I}} + \underset{\approx}{\boldsymbol{B}})^{-1} \bullet (\underset{\approx}{\boldsymbol{I}} - \underset{\approx}{\boldsymbol{B}}) \bullet (\underset{\approx}{\boldsymbol{I}} + \underset{\approx}{\boldsymbol{B}}) \bullet (\underset{\approx}{\boldsymbol{I}} - \underset{\approx}{\boldsymbol{B}})^{-1} \end{aligned} \tag{10.24}$$

$$\begin{aligned} &= (\underset{\approx}{\boldsymbol{I}} + \underset{\approx}{\boldsymbol{B}})^{-1} \bullet (\underset{\approx}{\boldsymbol{I}} + \underset{\approx}{\boldsymbol{B}}) \bullet (\underset{\approx}{\boldsymbol{I}} - \underset{\approx}{\boldsymbol{B}}) \bullet (\underset{\approx}{\boldsymbol{I}} - \underset{\approx}{\boldsymbol{B}})^{-1} \\ &= \underset{\approx}{\boldsymbol{I}} \bullet \underset{\approx}{\boldsymbol{I}} = \underset{\approx}{\boldsymbol{I}}. \end{aligned} \tag{10.25}$$

This shows that $\underset{\approx}{\boldsymbol{R}}$ is orthogonal. To prove that it is a rotation, we need to show that $\det\underset{\approx}{\boldsymbol{R}} = 1$. To work out the determinant the following identity (which holds for any two tensors $\underset{\approx}{\boldsymbol{X}}$ and $\underset{\approx}{\boldsymbol{Y}}$ referenced to 3D space) is helpful:

$$\det(\underset{\approx}{\boldsymbol{X}} \pm \underset{\approx}{\boldsymbol{Y}}) = \det\underset{\approx}{\boldsymbol{X}} \pm \underset{\approx}{\boldsymbol{X}}^C : \underset{\approx}{\boldsymbol{Y}} + \underset{\approx}{\boldsymbol{X}} : \underset{\approx}{\boldsymbol{Y}}^C \pm \det\underset{\approx}{\boldsymbol{Y}}, \tag{10.26}$$

where the superscript C denotes the cofactor operation. The cofactor is the matrix of signed minors (see, for example, [1] or any elementary matrix analysis textbook).

For any skew tensor $\underset{\approx}{\boldsymbol{B}}$,

$$\underset{\approx}{\boldsymbol{B}}^c = \underset{\sim}{\boldsymbol{b}}\underset{\sim}{\boldsymbol{b}} \qquad \text{and} \qquad \det\underset{\approx}{\boldsymbol{B}} = 0, \tag{10.27}$$

where $\underset{\sim}{b}$ is the axial vector associated with $\underset{\approx}{B}$ and $\underset{\sim}{b}\underset{\sim}{b}$ is a dyad. For the identity tensor,

$$\underset{\approx}{I}^c = \underset{\approx}{I} \quad \text{and} \quad \det\underset{\approx}{I} = 1. \tag{10.28}$$

Thus

$$\det(\underset{\approx}{I} + \underset{\approx}{B}) = 1 + \underset{\approx}{I}:\underset{\approx}{B} + \underset{\approx}{I}:\underset{\sim}{b}\underset{\sim}{b} + 0 = 1 + \underset{\sim}{b} \bullet \underset{\sim}{b}. \tag{10.29}$$

Similarly,

$$\det(\underset{\approx}{I} - \underset{\approx}{B}) = 1 + \underset{\sim}{b} \bullet \underset{\sim}{b}. \tag{10.30}$$

The above two results therefore imply that

$$\det\left[(\underset{\approx}{I} + \underset{\approx}{B}) \bullet (\underset{\approx}{I} - \underset{\approx}{B})^{-1}\right] = \frac{1 + \underset{\sim}{b} \bullet \underset{\sim}{b}}{1 + \underset{\sim}{b} \bullet \underset{\sim}{b}} = 1. \tag{10.31}$$

This completes the proof that $\underset{\approx}{R}$ really is a proper rotation. Of course, a more straightforward way to prove that the determinant equals 1 is to recognize that the determinant is an invariant, and its value must therefore be the same for all bases. Thus, using the special aligned basis that gave rise to (10.22), we can take the determinant of that matrix to independently confirm that $\det\underset{\approx}{R} = 1$. The rotation axis associated with $\underset{\approx}{R}$ is a unit vector in the direction of $\underset{\sim}{b}$,

$$\underset{\sim}{a} = \frac{\underset{\sim}{b}}{b}, \tag{10.32}$$

and the sine and cosine of the angle are determined through use of (5.114) and (5.115) to give

$$\cos\alpha = \frac{1 - b^2}{1 + b^2} \quad \text{and} \quad \sin\alpha = \frac{2b}{1 + b^2}. \tag{10.33}$$

Again, we could have simply compared (10.22) with (5.97) to reach the same conclusion. It is impossible for $\cos\alpha$ to ever have the value -1 unless b is allowed to be infinite, which could be problematic in numerical applications. When b is infinite, the rotation axis in (10.33) becomes undefined. Thus, the Cayley transformation is incapable of producing proper-orthogonal tensors for which the rotation angle is exactly 180°. Recalling that the Cayley transformation is an inverse hyper-stereographic projection (transforming a hyper-plane of skew tensors to a hyper-sphere of rotation tensors), this mapping of infinite magnitude is analogous to the inverse mapping in an ordinary stereographic projection from infinity to one of the poles on the sphere. This singularity at infinity proves that the Cayley transformation does *not* have a one-to-one correspondence with the space of all rotations (all 180° rotations map as $b \to \infty$ with axis information lost); this point is clarified in the next section. For all other rotation angles, $b > 0$ by construction. Therefore, the sine of the rotation angle is never negative, and an ordinary arctan operation that produces angles in the range $[0, \pi)$ may be used to find the rotation angle from

$$\boxed{\tan\alpha = \frac{2b}{1-b^2}}. \tag{10.34}$$

10.1.2 Constructing a skew-symmetric tensor from a rotation tensor

The transformation in (10.17) can be easily inverted to give the forward hyper-stereographic projection from the sphere to the hyperplane of skew tensors. Specifically,

$$\underset{\approx}{\mathbf{B}} = (\underset{\approx}{\mathbf{R}} - \underset{\approx}{\mathbf{I}}) \bullet (\underset{\approx}{\mathbf{R}} + \underset{\approx}{\mathbf{I}})^{-1} \tag{10.35}$$

produces skew-symmetric matrices from any source rotation (proper-orthogonal) matrix $\underset{\approx}{\mathbf{R}}$ (proper or improper) for which the inverse is well defined. Under what conditions will the inverse not exist? By using (10.26), the determinant of $\underset{\approx}{\mathbf{R}} + \underset{\approx}{\mathbf{I}}$ is given by

$$\det(\underset{\approx}{\mathbf{R}} + \underset{\approx}{\mathbf{I}}) = 2(1 + \mathrm{tr}\underset{\approx}{\mathbf{R}}). \tag{10.36}$$

Thus, the tensor $\underset{\approx}{\mathbf{R}} + \underset{\approx}{\mathbf{I}}$ is singular if $\mathrm{tr}\underset{\approx}{\mathbf{R}} = -1$, which implies (from equation (5.101)) that the rotation angle would have to be 180° to make (10.35) ill-defined. This is not a very surprising result since we showed earlier that the Cayley transformation is incapable of producing rotations for which the rotation angle is 180°. Substituting the Euler–Rodrigues formula (5.17) for $\underset{\approx}{\mathbf{R}}$ into (10.35) produces the skew tensor

$$\underset{\approx}{\mathbf{B}} = \tan\left(\frac{\alpha}{2}\right)\underset{\approx}{\mathbf{A}}, \tag{10.37}$$

which is the spinor tensor corresponding to (5.87). The axial vector $\underset{\sim}{\mathbf{b}}$ has magnitude

$$\boxed{b = \tan\frac{\alpha}{2}}. \tag{10.38}$$

10.1.3 Similar transformations for improper-orthogonal tensors

Recall from chapter 6 that an *improper* 3D orthogonal tensor can always be written as the negative of a proper-orthogonal tensor (rotation). Thus, an *improper-*orthogonal tensor $\underset{\approx}{\mathbf{Q}}$ can be generated by changing the sign of the Cayley transformation. Given any real skew tensor $\underset{\approx}{\mathbf{B}}$, $\underset{\approx}{\mathbf{Q}} = (\underset{\approx}{\mathbf{I}} + \underset{\approx}{\mathbf{B}}) \bullet (\underset{\approx}{\mathbf{B}} - \underset{\approx}{\mathbf{I}})^{-1} = (\underset{\approx}{\mathbf{B}} - \underset{\approx}{\mathbf{I}})^{-1} \bullet (\underset{\approx}{\mathbf{I}} + \underset{\approx}{\mathbf{B}})$ gives an improper-orthogonal tensor. The inverse transformation is $\underset{\approx}{\mathbf{B}} = (\underset{\approx}{\mathbf{Q}} - \underset{\approx}{\mathbf{I}})^{-1} \bullet (\underset{\approx}{\mathbf{Q}} + \underset{\approx}{\mathbf{I}}) = (\underset{\approx}{\mathbf{Q}} + \underset{\approx}{\mathbf{I}}) \bullet (\underset{\approx}{\mathbf{Q}} - \underset{\approx}{\mathbf{I}})^{-1}$. The skew tensor generated by this formula is identical to the one produced in (10.35) if the rotation (i.e., *proper* rotation) used there is given by $\underset{\approx}{\mathbf{R}} = -\underset{\approx}{\mathbf{Q}}$.

10.1.4 Rayleigh transformations

Because of the similar sounding name, we mention here that a Rayleigh transformation is equivalent to changing each $\underset{\sim}{e}_k$-basis vector of a tensor to $\underset{\approx}{\mathbf{G}} \bullet e_k$ where

$\underset{\approx}{G}$ is a given tensor. The component formulas for Rayleigh transforms of various tensor orders should clarify what it means:

vector v_p before transform	becomes after transform: $G_{ip}v_p$,
tensor T_{pq} before transform	becomes after transform: $G_{ip}G_{jq}T_{pq}$,
tensor Y_{pqr} before transform	becomes after transform: $G_{ip}G_{jq}G_{kr}Y_{pqr}$,
tensor C_{pqrs} before transform	becomes after transform: $G_{ip}G_{jq}G_{kr}G_{ls}C_{pqrs}$.

Reference

[1] Brannon R M 2003 https://csmbrannon.net/2011/03/03/functional-and-structured-tensor-analysis-for-engineers/

IOP Publishing

Rotation, Reflection, and Frame Changes
Orthogonal tensors in computational engineering mechanics
R M Brannon

Chapter 11

Spectrum of a rotation

The spectrum (i.e., the solution to the eigenproblem) for a rotation plays a role in the exponential series expansion of a rotation (chapter 10), but it is otherwise included here only for completeness. The eigenproblem for a rotation tensor requires determination of all eigenvectors $\underset{\sim}{p}$ and eigenvalues λ for which

$$\underset{\approx}{R} \bullet \underset{\sim}{p} = \lambda \underset{\sim}{p}. \tag{11.1}$$

Suppose we set up a preferred basis $\{\boldsymbol{b}_1, \boldsymbol{b}_2, \boldsymbol{b}_3\}$ for which the 3-direction is aligned with the axis of rotation $\underset{\sim}{a}$. Then the corresponding components of $\underset{\approx}{R}$ are

$$[\underset{\approx}{R}] = \begin{bmatrix} \cos\alpha & -\sin\alpha & 0 \\ \sin\alpha & \cos\alpha & 0 \\ 0 & 0 & 1 \end{bmatrix} \text{with respect to basis } \{\underset{\sim}{b}_1, \underset{\sim}{b}_2, \text{ and } \underset{\sim}{b}_3 = \underset{\sim}{a}\}. \tag{11.2}$$

From this, we immediately see that $\underset{\sim}{p} = \underset{\sim}{a}$ is an eigenvector with associated eigenvalue $\lambda = 1$. This result reflects the intuitively clear fact that material fibers aligned with a rotation axis will experience neither rotation nor elongation. For the remaining eigen-pairs, we focus on the upper 2×2 sub-matrix. The characteristic equation for this sub-matrix is

$$\lambda^2 - 2(\cos\alpha)\lambda + 1 = 0. \tag{11.3}$$

Applying the quadratic formula gives a complex-conjugate pair of solutions:

$$\lambda = \cos\alpha \pm i \sin\alpha = e^{\pm i\alpha}, \tag{11.4}$$

where $i \equiv \sqrt{-1}$. The corresponding eigenvectors are $\{\pm i, 1\}$. Thus, the spectrum of the rotation tensor is as shown in table 11.1.

Table 11.1. Spectrum of a rotation tensor.

Eigenvalue	Eigenvector
$\lambda_1 = e^{i\alpha}$	$\underset{\sim}{p}_1 = \frac{1}{\sqrt{2}}(i\underset{\sim}{b}_1 + \underset{\sim}{b}_2)$
$\lambda_2 = e^{-i\alpha}$	$\underset{\sim}{p}_2 = \frac{1}{\sqrt{2}}(-i\underset{\sim}{b}_1 + \underset{\sim}{b}_2)$
$\lambda_3 = 1$	$\underset{\sim}{p}_3 = \underset{\sim}{b}_3(=\underset{\sim}{a})$

It can be readily verified that the rotation tensor can be reconstructed by

$$\underset{\approx}{R} = \lambda_1(\underset{\sim}{p}_1 \otimes \bar{\underset{\sim}{p}}_1) + \lambda_2(\underset{\sim}{p}_2 \otimes \bar{\underset{\sim}{p}}_2) + \lambda_3(\underset{\sim}{p}_3 \otimes \bar{\underset{\sim}{p}}_3), \tag{11.5}$$

where the superposed 'bar' denotes the complex conjugate and $\otimes$ is dyadic multiplication. Furthermore, noting that $\lambda_2 = \bar{\lambda}_1$ and $\underset{\sim}{p}_2 = \bar{\underset{\sim}{p}}_1$, the above equation may be written

$$\underset{\approx}{R} = 2\text{Re}[\lambda_1(\underset{\sim}{p}_1 \otimes \bar{\underset{\sim}{p}}_1)] + \lambda_3(\underset{\sim}{p}_3 \otimes \bar{\underset{\sim}{p}}_3). \tag{11.6}$$

Incidentally, for any non-zero rotation angle, the eigenvalues are distinct and the eigenvectors $\underset{\sim}{p}_1$ and $\underset{\sim}{p}_2$ are therefore unique to within a complex multiple even though it might appear otherwise in light of the following observation. Let $\underset{\approx}{Q}$ be an arbitrary real rotation about $\underset{\sim}{a}$. Then, because $\underset{\approx}{R}$ and $\underset{\approx}{Q}$ share the same rotation axis, we know that they commute:

$$\underset{\approx}{R} \bullet \underset{\approx}{Q} = \underset{\approx}{Q} \bullet \underset{\approx}{R}. \tag{11.7}$$

Let $\underset{\sim}{p}$ be an eigenvector of $\underset{\approx}{R}$ with eigenvalue λ. Dotting both sides of (11.7) from the right by $\underset{\sim}{p}$ gives

$$(\underset{\approx}{R} \bullet \underset{\approx}{Q}) \bullet \underset{\sim}{p} = (\underset{\approx}{Q} \bullet \underset{\approx}{R}) \bullet \underset{\sim}{p} \tag{11.8}$$

or

$$\underset{\approx}{R} \bullet (\underset{\approx}{Q} \bullet \underset{\sim}{p}) = \underset{\approx}{Q} \bullet (\underset{\approx}{R} \bullet \underset{\sim}{p}) = \underset{\approx}{Q} \bullet (\lambda \underset{\sim}{p}) = \lambda(\underset{\approx}{Q} \bullet \underset{\sim}{p}). \tag{11.9}$$

This proves that $\underset{\approx}{Q} \bullet \underset{\sim}{p}$ is an eigenvector of $\underset{\approx}{R}$ corresponding to the eigenvalue λ. Stated differently, the choice of basis vectors $\underset{\sim}{b}_1$ and $\underset{\sim}{b}_2$ is arbitrary so long as they are orthonormal and orthogonal to the rotation axis $\underset{\sim}{a}$.

Sidebar: linear independence of eigenvectors

We know from elementary matrix analysis (see [1]) that the number of linearly independent eigenvectors must be no greater than the algebraic multiplicity of the eigenvalue. The eigenvalues are all distinct as long as $\sin \alpha \neq 0$. In this case, there can be only *one single* linearly independent eigenvector associated with each eigenvalue. If $\underset{\sim}{p}$ is an eigenvector, then (11.9) states that $\underset{\approx}{Q} \bullet \underset{\sim}{p}$ is also an eigenvector associated with the same eigenvalue. Thus, equation (11.9) makes it *appear* that there are an infinite number of linearly independent eigenvectors associated with the complex eigenvalues. When a vector $\underset{\sim}{\boldsymbol{v}}$ is real and not aligned with the rotation axis, then $\underset{\approx}{\boldsymbol{Q}} \bullet \underset{\sim}{\boldsymbol{v}}$ is generally linearly independent of $\underset{\sim}{\boldsymbol{v}}$. However, when eigenvector $\underset{\sim}{\boldsymbol{p}}$ is complex, this property does not hold. We need to demonstrate that $\underset{\approx}{\boldsymbol{Q}} \bullet \underset{\sim}{\boldsymbol{p}}$ and $\underset{\sim}{\boldsymbol{p}}$ are linearly dependent for all rotations $\underset{\approx}{\boldsymbol{Q}}$ about the rotation axis $\underset{\sim}{\boldsymbol{a}}$. Denoting the rotation angle for $\underset{\approx}{\boldsymbol{Q}}$ by q, it is straightforward to demonstrate that $\underset{\approx}{\boldsymbol{Q}} \bullet \underset{\sim}{\boldsymbol{p}} = \Lambda \underset{\sim}{\boldsymbol{p}}$, where $\Lambda = \{e^{iq}, e^{-iq}, \text{ or } 1\}$. In other words, $\underset{\approx}{\boldsymbol{Q}} \bullet \underset{\sim}{\boldsymbol{p}}$ and $\underset{\sim}{\boldsymbol{p}}$ differ only by a (possibly complex) scalar multiple, which proves that they are indeed linearly dependent.

Reference

[1] Wiley C R and Barrett L C 1982 *Advanced Engineering Mathematics* (New York: McGraw-Hill)

IOP Publishing

Rotation, Reflection, and Frame Changes

Orthogonal tensors in computational engineering mechanics

R M Brannon

Chapter 12

Polar decomposition

A special tensor called the 'deformation gradient $\underset{\approx}{F}$' plays a pivotal role in the field of continuum mechanics [1, 2]. This tensor carries all information about the deformation of a material element from its known reference state to its current 'spatial' state. The polar-decomposition theorem uniquely quantifies how any general deformation can be viewed as a combination of material reorientation (i.e., rotation), material distortion (i.e., a change in shape), and material dilation (i.e., a change in size).

12.1. Difficult definition of the deformation gradient

In this section, we provide the mathematically rigorous definition of the term 'deformation gradient' normally found in textbooks. Later, we provide a more intuitive (less rigorous) definition.

Suppose that the location $\underset{\sim}{X}$ of each material particle in a body is known at some reference state (usually time zero). This reference position is regarded as the 'name' of the particle. For example, when we say 'point $\underset{\sim}{X}$', we mean 'the particle whose reference position is $\underset{\sim}{X}$'.

The complete deformation of the body may be described by a mapping function χ such that the deformed location $\underset{\sim}{x}$ of the point $\underset{\sim}{X}$ is given by

$$\underset{\sim}{x} = \chi(\underset{\sim}{X}). \tag{12.1}$$

The 'motion' of a body is a sequence of deformed states over time. Hence, the motion of a body is described by allowing the above mapping function to additionally vary with time:

$$\underset{\sim}{x} = \chi(\underset{\sim}{X}, t). \tag{12.2}$$

doi:>10.1016/978-0-7503-1454-1ch12

The deformation gradient tensor $\underset{\approx}{F}$ is mathematically defined to be the gradient of the mapping function:

$$\underset{\approx}{F} = \frac{\partial \underset{\sim}{x}}{\partial \underset{\sim}{X}} \quad \Rightarrow \quad F_{ij} = \frac{\partial x_i}{\partial X_j} \quad \Rightarrow \quad [F] = \begin{bmatrix} \frac{\partial x_1}{\partial X_1} & \frac{\partial x_1}{\partial X_2} & \frac{\partial x_1}{\partial X_3} \\ \frac{\partial x_2}{\partial X_1} & \frac{\partial x_2}{\partial X_2} & \frac{\partial x_2}{\partial X_3} \\ \frac{\partial x_3}{\partial X_1} & \frac{\partial x_3}{\partial X_2} & \frac{\partial x_3}{\partial X_3} \end{bmatrix}. \tag{12.3}$$

While this definition is rigorously precise, it offers virtually no intuitive insight into the real meaning of information contained in a matrix of F_{ij} components. If someone showed you a deformation gradient matrix, would you be able to interpret it physically? Probably not if you only understand it by the above definition.

One way to gain insight into this definition is to consider some simple specialized mapping functions. Suppose that $\underset{\sim}{X}^{(p)}$ denotes the initial location of a particle of interest. We want to gain better understanding of where that particle translates and how material in the vicinity of that particle deforms. Then it makes sense to consider a Taylor series expansion of the mapping function, centered about the particle $\underset{\sim}{X}^{(p)}$. Namely, to second-order accuracy, the mapping function near the particle can be approximated by

$$\begin{aligned} \underset{\sim}{x} = \chi(\underset{\sim}{X}, t) \approx\ & \chi(\underset{\sim}{X}^{(p)}, t) \\ & + \left[\frac{\partial \chi(\underset{\sim}{X}, t)}{\partial \underset{\sim}{X}}\right]_{\substack{\text{evaluated at} \\ \underset{\sim}{X}=\underset{\sim}{X}^{(p)}}} \bullet (\underset{\sim}{X} - \underset{\sim}{X}^{(p)}) \\ & + \frac{1}{2}\left[\frac{\partial^2 \chi(\underset{\sim}{X}, t)}{\partial \underset{\sim}{X} \partial \underset{\sim}{X}}\right]_{\substack{\text{evaluated at} \\ \underset{\sim}{X}=\underset{\sim}{X}^{(p)}}} : (\underset{\sim}{X} - \underset{\sim}{X}^{(p)})(\underset{\sim}{X} - \underset{\sim}{X}^{(p)}) \\ & + \ldots \end{aligned} \tag{12.4}$$

The first term in this expansion, $\chi(\underset{\sim}{X}^{(p)}, t)$, is the deformed location of the particle $\underset{\sim}{X}^{(p)}$, so it makes sense to denote this quantity by simply $\underset{\sim}{x}^{(p)}$:

$$\underset{\sim}{x}^{(p)} \equiv \chi(\underset{\sim}{X}^{(p)}, t). \tag{12.5}$$

Recalling (12.3), the derivative in the second term of the above expansion is the value of the deformation gradient tensor evaluated at the particle of interest. Hence, it makes sense for us to denote this derivative by a more compact symbol,

$$\underset{\approx}{F}^{(p)} \equiv \left[\frac{\partial \chi(\underset{\sim}{X}, t)}{\partial \underset{\sim}{X}}\right] \text{evaluated at } \underset{\sim}{X} = \underset{\sim}{X}^{(p)}. \tag{12.6}$$

We included the third term in the expansion of (12.4) merely to emphasize that we could continue the expansion to higher-order accuracy as long as the mapping function is adequately smooth. However, in the analysis of general nonlinear mappings, you are unlikely to ever encounter expansions that go beyond first-order[1]. Using our more compact notation defined in (12.5) and (12.6), the first-order mapping expansion in the neighborhood of a particle $\underset{\sim}{X}^{(p)}$ can be written

$$\underset{\sim}{x} \approx \underset{\sim}{x}^{(p)} + \underset{\approx}{F}^{(p)} \bullet (\underset{\sim}{X} - \underset{\sim}{X}^{(p)}) + \cdots \tag{12.7}$$

Expressions like this are similar to writing a first-order expansion of a highly nonlinear function centered about a particular point of interest. First-order expansions of *scalar* functions result in an equation of the straight line that passes through the function at the point of interest and is tangent to the curve at that point. Any good mathematician will tell you that the key to solving or interpreting a nonlinear equation is to first understand *linear* functions—that is because all but the most perverse functions can be regarded as an ensemble of linear line segments (the more you use, the better the approximation, and in the limit of an infinite number of line segments, the technique is known by the name *calculus*!).

Since understanding linear functions (straight lines) is an essential first step to understanding nonlinear functions, one might suspect that a good place to begin understanding general nonlinear mappings is to first study linear mappings of the form

$$\underset{\sim}{x} \approx \underset{\sim}{x}^{(p)} + \underset{\approx}{F}^{(p)} \bullet (\underset{\sim}{X} - \underset{\sim}{X}^{(p)}). \tag{12.8}$$

If the tensor $\underset{\approx}{F}^{(p)}$ equals the identity tensor, then this mapping describes a pure rigid translation. If the tensor $\underset{\approx}{F}^{(p)}$ is a rotation tensor, then this mapping describes translation in combination with rotation. If the tensor $\underset{\approx}{F}^{(p)}$ is not orthogonal, then this mapping is capable of describing simultaneous translation, rotation, and shape change.

Above, we said that a straight line in the plane can be described through a *linear* function. Most people would say that a linear function is expressible in the form $y = mx + b$, but the correct term is 'affine'. A linear function $y(x)$ must satisfy the property that $y(\alpha_1 x_1 + \alpha_2 x_2) = \alpha_1 y(x_1) + \alpha_2 y(x_2)$ for all scalars $\{\alpha_1, x_1, \alpha_1, x_2\}$. If this condition must hold for all choices of the scalars α_1 and α_2, then it must hold for the particular choice $\alpha_1 = \alpha_2 = 0$. Thus, if a function is linear then it must have the property that $y(0) = 0$. The function $y = mx + b$ does not satisfy this property if b is non-zero, so this function must not be linear! A truly linear function must be expressible as $y = mx$ for some scalar x. A linear function always passes through the origin. An affine function is simply a linear function plus some constant.

The function in (12.8) is not *linear*—it is *affine*. However, we can always rewrite this equation in the form $\underset{\sim}{x} - \underset{\sim}{x}^{(p)} = \underset{\approx}{F}^{(p)} \bullet (\underset{\sim}{X} - \underset{\sim}{X}^{(p)})$. Thus, the *relative* deformed position vector $\underset{\sim}{x} - \underset{\sim}{x}^{(p)}$ can always be expressed as a linear function of the *relative* initial position vector $\underset{\sim}{X} - \underset{\sim}{X}^{(p)}$. Consequently, it is sufficient for us to study the very specialized case of *linear* mapping functions:

$$\underset{\sim}{x} = \underset{\approx}{F} \bullet \underset{\sim}{X}. \tag{12.9}$$

[1] An exception is isoparametric element mappings used in finite-element analysis. With first-order mappings, squares deform to parallelograms and cubes deform to parallelepipeds. With typical isoparametric maps, squares deform to quadrilaterals, and cubes deform to quadrahedrals.

We have already explained why a general mapping can be locally approximated by a linear mapping so long as the position vectors are defined *relative* to the *local* point of interest. The hallmark of a non-affine mapping is that the local deformation gradient tensor $\underset{\approx}{F}$ varies with position. This is similar to the fact that the local slope of a tangent line varies with position along a nonlinear curve. If, as a very specialized case, the deformation gradient tensor $\underset{\approx}{F}$ does *not* vary with position, then the mapping is said to be affine, or *homogeneous*. As will be explained below, a homogeneous mapping has the property that initially straight lines deform to straight lines, flat planes deform to flat planes, squares deform to parallelograms, cubes deform to parallelepipeds, and spheres deform to ellipsoids. Since *inhomogeneous* deformations can be regarded as *locally* linear (which is like approximating a curve at a point by the local tangent at that point), we conclude that *infinitesimal* cubes in general mappings will deform to *infinitesimal* parallelepipeds, but the size, orientation, and distortion level of these deformed parallelepipeds will vary in space. These statements are illustrated in figure 12.1.

12.2. Intuitive definition of the deformation gradient

To better understand the deformation gradient, imagine a small cube-shaped element of material whose sides are aligned with the reference basis $\{\underset{\sim}{E}_1, \underset{\sim}{E}_2, \underset{\sim}{E}_3\}$. Upon deformation, the cube deforms to a parallelepiped. If the deformation is two-dimensional (2D), then a square deforms to a parallelogram, as sketched in figure 12.2. The vectors on the sides of the element deform to new vectors $\{\underset{\sim}{g}_1, \underset{\sim}{g}_2, \underset{\sim}{g}_3\}$. These so-called *convected basis vectors* are not generally orthogonal nor of unit length.

An intuitive (non-rigorous) definition of the deformation gradient tensor $\underset{\approx}{F}$ states that columns of its laboratory component matrix $[F]$ simply contains the lab components of the convected $\{\underset{\sim}{g}_1, \underset{\sim}{g}_2, \underset{\sim}{g}_3\}$ vectors. We write this statement symbolically as

$$[F] = [\{\underset{\sim}{g}_1\}\{\underset{\sim}{g}_2\}\{\underset{\sim}{g}_3\}]. \tag{12.10}$$

In direct notation, this may be expressed by a sum of dyads[2]:

$$\underset{\approx}{F} = \underset{\sim}{g}_1\underset{\sim}{E}_1 + \underset{\sim}{g}_2\underset{\sim}{E}_2 + \underset{\sim}{g}_3\underset{\sim}{E}_3. \tag{12.11}$$

Consider, for example, the deformation shown in figure 12.2, which occurs in the plane so that $\underset{\sim}{E}_3$ points out of the page. Knowing that $\underset{\sim}{E}_1$ and $\underset{\sim}{E}_2$ are of unit length, we can use a ruler to measure the vectors in figure 12.2 to obtain

[2] Here, two vectors written side-by-side $\underset{\sim}{a}\underset{\sim}{b}$ are multiplied dyadically so that $[\underset{\sim}{a}\underset{\sim}{b}]_{ij} = a_i b_j$. Therefore, the dyad $\underset{\sim}{g}_1\underset{\sim}{E}_1$ is represented by a 3×3 matrix whose components are all zero except that the first column has the same components as the lab components of the $\underset{\sim}{g}_1$ vector.

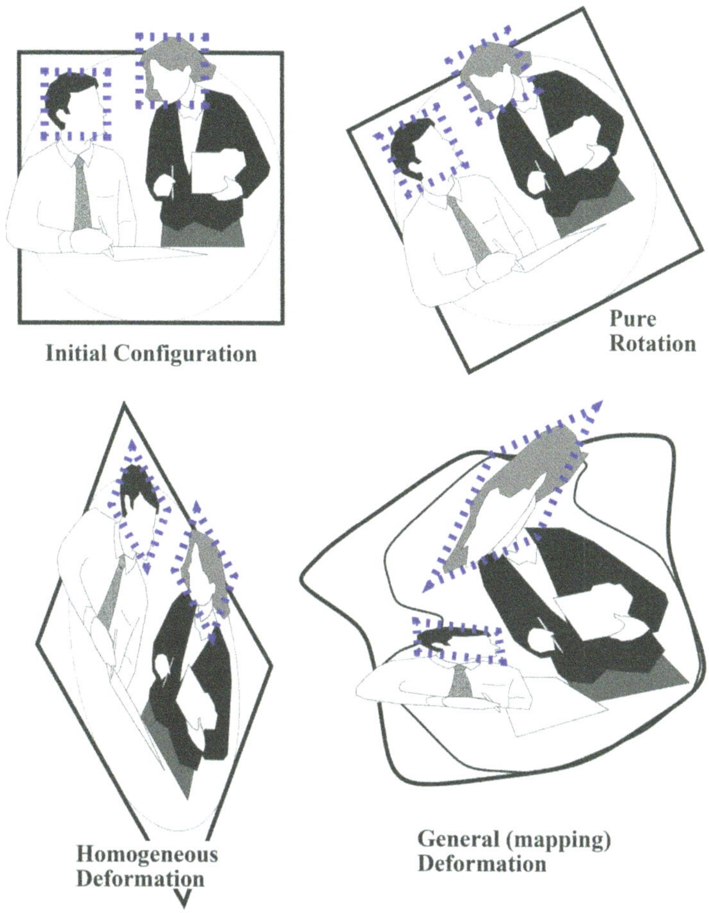

Figure 12.1. Increasingly complex deformations. The simplest deformation, rotation, does not involve any length changes—cubes remain cubes of exactly the same size but of different orientation. The next easiest deformation is 'homogeneous' deformation where the changes in shape and orientation are the same everywhere throughout the body—all reference cubes deform to parallelepipeds of the same shape and orientation. For the most general (inhomogeneous) deformation, infinitesimal cubes still deform to parallelepipeds, but the orientation and distortion of these parallelepipeds varies in space.

$$\begin{aligned} \underset{\sim}{g}_1 &\approx (0.62)\underset{\sim}{E}_1 + (0.59)\underset{\sim}{E}_2 \;\rightarrow\; \{\underset{\sim}{g}_1\} = \begin{Bmatrix} 0.62 \\ 0.59 \\ 0 \end{Bmatrix}, \\ \underset{\sim}{g}_2 &\approx (-0.71)\underset{\sim}{E}_1 + (0.13)\underset{\sim}{E}_2 \rightarrow \{\underset{\sim}{g}_2\} = \begin{Bmatrix} -0.71 \\ 0.13 \\ 0 \end{Bmatrix}. \end{aligned} \tag{12.12}$$

In addition to the deformation shown in figure 12.2, let us further suppose that the out-of-plane thickness of the material increases by a factor of 3 so that

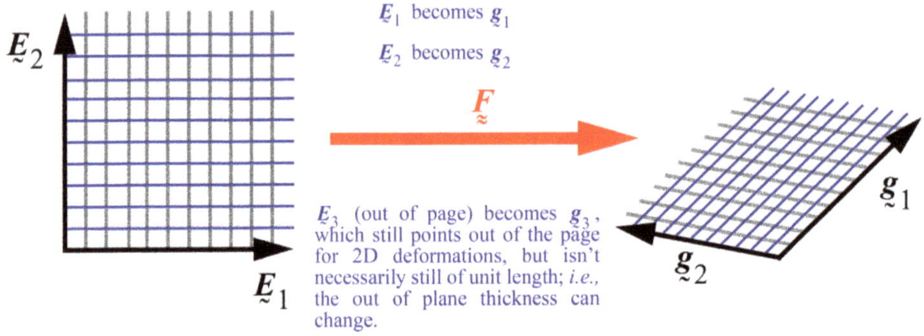

Figure 12.2. The deformation gradient tensor. A cube deforms to a parallelepiped. This deformation is shown in the plane for simplicity. For planar deformations, squares deform to parallelograms; of course, the out-of-plane thickness is allowed to change as well. Note that this deformation involves a significant amount of counter-clockwise rotation.

$$\underset{\sim}{g}_3 = (3.0)\underset{\sim}{E}_3 \rightarrow \{\underset{\sim}{g}_3\} = \begin{Bmatrix} 0 \\ 0 \\ 3.0 \end{Bmatrix}. \tag{12.13}$$

The deformation gradient component matrix[3] corresponding to figure 12.2 is constructed by assembling these three vectors into the columns of $[F]$:

$$[F] = \begin{bmatrix} 0.62 & -0.71 & 0 \\ 0.59 & 0.13 & 0 \\ 0 & 0 & 3.0 \end{bmatrix}. \tag{12.14}$$

12.2.1. Inverse problem: interpreting a deformation gradient matrix

The intuitive definition of the deformation gradient may be applied in reverse. Suppose, for example, that the deformation gradient matrix is known to be

$$[F] = \begin{bmatrix} 1.0 & 1.0 & 0 \\ 0 & 2.0 & 0 \\ 0 & 0 & 0.8 \end{bmatrix}. \tag{12.15}$$

From this, we would identify the three deformed basis vectors by the columns of this matrix. Namely,

$$\{\underset{\sim}{g}_1\} = \begin{Bmatrix} 1.0 \\ 0 \\ 0 \end{Bmatrix}, \quad \{\underset{\sim}{g}_2\} = \begin{Bmatrix} 1.0 \\ 2.0 \\ 0 \end{Bmatrix}, \quad \{\underset{\sim}{g}_3\} = \begin{Bmatrix} 0 \\ 0 \\ 0.8 \end{Bmatrix}. \tag{12.16}$$

In basis notation, these would be written

[3] With respect to the lab basis $\{\underset{\sim}{E}_1, \underset{\sim}{E}_2, \underset{\sim}{E}_3\}$.

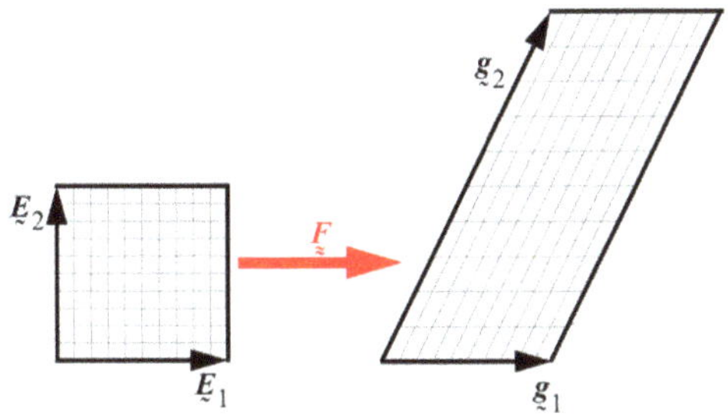

Figure 12.3. Physically interpreting a deformation gradient matrix.

$$\underset{\sim}{g}_1 = \underset{\sim}{E}_1, \qquad \underset{\sim}{g}_2 = \underset{\sim}{E}_1 + 2\underset{\sim}{E}_2, \qquad \underset{\sim}{g}_3 = (0.8)\underset{\sim}{E}_3. \tag{12.17}$$

These vectors define sides of a deformed parallelepiped as drawn in figure 12.3. Not shown in that figure is the fact that $\underset{\sim}{g}_3 = (0.8)\underset{\sim}{E}_3$, meaning that this deformation has reduced the out-of-plane thickness by 20%.

This intuitive definition of the deformation gradient has been here explained for *homogeneous* deformations in which all material elements within the body deform in exactly the same way. This means that the value of $\underset{\approx}{F}$ is the same for every point in the body. For general inhomogeneous deformations, like the last image in figure 12.1, each material point can have its own different deformation gradient tensor $\underset{\approx}{F}$. In this case, the intuitive definition of the deformation gradient must be applied to each *infinitesimal* material element individually.

12.3. The Jacobian of the deformation

A material element that deforms according to a deformation gradient $\underset{\approx}{F}$ generally changes from a reference initial volume $\mathrm{d}V_o$ to a current volume $\mathrm{d}V$. The Jacobian of the deformation is denoted J and is defined to equal the ratio

$$J = \frac{(\mathrm{d}V)}{(\mathrm{d}V_o)}. \tag{12.18}$$

Recognizing that the numerator in this ratio is the volume of formed by the three convected g-vectors, while the denominator is the volume of their initial counterparts, it follows that

$$J = \det\underset{\approx}{F}. \tag{12.19}$$

For example, for the deformation shown in figure 12.2, the Jacobian is obtained by taking the determinant of the matrix in (12.14),

$$J = \det\begin{bmatrix} 0.62 & -0.71 & 0 \\ 0.59 & 0.13 & 0 \\ 0 & 0 & 3.0 \end{bmatrix} = 1.43, \tag{12.20}$$

which means that the material element has increased in volume by 43%. Although the element appears to be smaller in size in figure 12.2, the overall volume has increased because of the large out-of-the-page expansion.

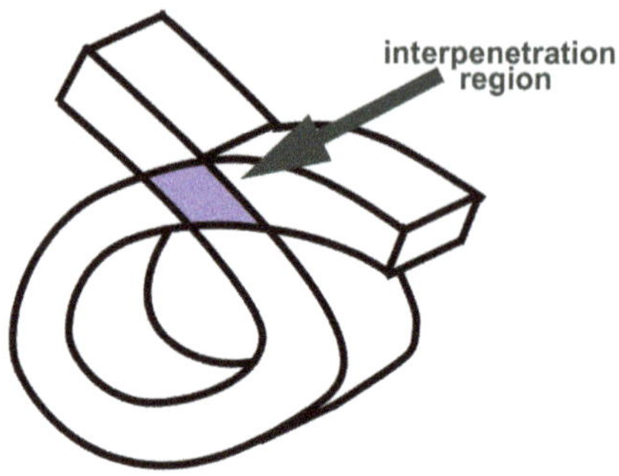

Figure 12.4. Example of a deformation that is locally invertible, but not globally invertible.

12.4. Invertibility of a deformation

For the deformation to make physical sense, the mapping function $\chi(\underset{\sim}{X}, t)$ must be 'one-to-one', meaning that any two distinct points must deform to two distinct points—they must not map to the same location (otherwise, the material will have interpenetrated itself as in figure 12.4, which is physically undesirable). Likewise, a single material point must not map to two different points—such behavior would imply the generation of new free surfaces[4]. A deformation is called 'globally invertible' if its mapping function is one-to-one.

A deformation is *locally invertible* if and only if the deformation gradient $\underset{\approx}{F}$ is invertible. In other words, the Jacobian must be non-zero: $J \neq 0$. For physically realistic deformations, both the initial and the deformed volumes in (12.18) must be positive. Hence, the condition of local invertibility may be replaced by the more restrictive condition[5],

$$J > 0. \tag{12.21}$$

This condition of local invertibility, however, is not sufficient to ensure *global invertibility* of the mapping (which requires each point $\underset{\sim}{x}$ in the deformed configuration to come from exactly one unique $\underset{\sim}{X}$ in the undeformed configuration). Global invertibility implies local invertibility, but not vice versa (see figure 12.4). This is why many numerical methods require special treatment (contact algorithms) to prevent material interpenetration.

12.5. Sequential deformations

If a material element is subjected to a deformation gradient, $\underset{\approx}{F}_1$, followed by a second deformation gradient $\underset{\approx}{F}_2$, then the total deformation gradient is given by

[4] Which would be desirable if one wishes to model, say, void nucleation or the formation of cracks. In that case, the one-to-one condition would be relaxed.

[5] We are disallowing very unusual situations, such as finding stresses in an inverted half-shell (like flipping half of a rubber racquet-ball 'inside out'), for which a negative Jacobian can be convenient to define the mapping by allowing the material to conceptually pass through itself to get to the deformed state.

$$\underset{\approx}{F} = \underset{\approx}{F}_2 \bullet \underset{\approx}{F}_1. \tag{12.22}$$

Any differential material vector $\mathrm{d}\underset{\sim}{X}$ in the reference configuration deforms to a new vector $\mathrm{d}\underset{\sim}{x}$ in the spatial configuration by the operation

$$\mathrm{d}\underset{\sim}{x} = \underset{\approx}{F} \bullet \mathrm{d}\underset{\sim}{X}. \tag{12.23}$$

If $\underset{\approx}{F}$ is given by (12.22), then $\underset{\approx}{F}_1$ first acts on $\mathrm{d}\underset{\sim}{X}$ and *then* $\underset{\approx}{F}_2$ acts on $\underset{\approx}{F}_1 \bullet \mathrm{d}\underset{\sim}{X}$. This is why the sequential deformation gradients appear in reverse order of application.

12.6. Matrix analysis version of the polar-decomposition theorem

In matrix analysis, the *singular value decomposition* (SVD) theorem states that any $M \times N$ matrix $[F]$ can be decomposed into the form

$$[F] = [q][d][r]^T, \tag{12.24}$$

for which

(i) $[q]$ and $[r]$ are square and orthogonal[6] (inverse is transpose),
(ii) $[d]$ is 'pseudo-diagonal' (i.e., $d_{ij} = 0$ whenever $i \neq j$), and
(iii) $[d]$ is positive semi-definite in the sense that its diagonal ($i = j$) components, d_i, are non-negative.

As will be seen, $[q]$ and $[r]$ are direction cosine matrices, and therefore not unique in this decomposition. Nevertheless, the product of their determinants, $(\det[q])(\det[r])$, is unique. The original matrix $[F]$ is said to be an inversion if and only if this product is negative.

The 'pseudo-diagonal' matrix $[d]$ will have the same number of rows and columns as $[F]$. If, for example, $M = 3$ and $N = 6$, then $[d]$ will be of the form

$$[d] = \begin{bmatrix} d_1 & 0 & 0 & 0 & 0 & 0 \\ 0 & d_2 & 0 & 0 & 0 & 0 \\ 0 & 0 & d_3 & 0 & 0 & 0 \end{bmatrix}, \tag{12.25}$$

where the d_i (which are called 'singular values') are all non-negative (they are strictly positive if $[F]$ is square and invertible). The generalized determinant of a non-square $[F]$ is the product of the singular values. The so-called pseudo-inverse of $[F]$ is $[F]^{(-1)} = [q][d]^{(-1)}[r]^T$, in which $[d]^{(-1)}$ is of the same form as $[d]$ with each non-zero singular value replaced by its reciprocal. A non-square $[F]$ matrix is pertinent to any transformation having a different number of outputs than inputs. The formation of a 9D rotation tensor from three inputs (Euler angles, etc) is an example that will be used later in the context of random rotations.

---

[6] This theorem can be generalized to include the possibility that $[F]$ is complex, in which case $[q]$ and $[r]$ would be *unitary*. For real matrices, unitary means the same thing as orthogonal.

The SVD implies a corollary, called polar decomposition:

$$[F] = [R][U] = [V][R]$$

where

$$\begin{aligned}[R] &= [q][i][r]^T \\ [U] &= [r][i]^T[d][r]^T . \\ [V] &= [q][d][i]^T[q]^T\end{aligned}$$

Here, $[i]$ is the $M \times N$ pseudo-diagonal identity matrix (i.e., its diagonal components are all equal to 1 with all other components zero). Unlike the SVD, the polar decomposition is unique. The next section examines the SVD and polar decomposition for the relatively intuitive case of square matrices (familiar to engineers who have experience in the field of continuum mechanics), but it should be clear that similar procedures can be used to find an SVD for non-square matrices. As done later in this book, the non-square polar decomposition is interesting when applied to a 9×3 Jacobian matrix corresponding to the derivatives of the 9 components of the rotation matrix with respect to whatever choice is made for the three coordinates (such as Euler angles). In that case, the $[r]$ direction cosine matrix might suggest an optimal change of coordinates (e.g., from Euler angles to something more computationally efficient) that would make $[U]$ diagonal. To our knowledge, such an analysis has not been performed in the literature to determine if a corresponding holonomic mapping exists (even in principle).

12.6.1. Special case: SVD for square matrices

Suppose that $M = N$ so that all matrices have the same dimensions. Since $[q]$ is orthogonal, we know that

$$[q]^T[q] = [I]. \tag{12.26}$$

Therefore, equation (12.24), once specialized to square matrices, may be written without loss in the form

$$[F] = [q][d][q]^T[q][r]^T, \tag{12.27}$$

or

$$[F] = [V][R], \quad \text{where } [V] \equiv [q][d][q]^T \text{ and } [R] \equiv [q][r]^T. \tag{12.28}$$

Similarly, since $[r]$ is orthogonal, we know that

$$[r]^T[r] = [I]. \tag{12.29}$$

Therefore, equation (12.24) may be written without loss in the form

$$[F] = [q][r]^T[r][d][r]^T, \tag{12.30}$$

or

$$[F] = [R][U], \quad \text{where } [U] \equiv [r][d][r]^T \text{ and } [R] \equiv [q][r]^T. \tag{12.31}$$

The matrix $[R]$ is itself orthogonal, while $[U]$ and $[V]$ are symmetric and positive semi-definite. Thus, the SVD guarantees existence of symmetric positive semi-definite matrices, $[U]$ and $[V]$, as well as an orthogonal matrix $[R]$ such that

$$[F] = [R][U] = [V][R]. \tag{12.32}$$

If $[F]$ is invertible, then $[U]$ and $[V]$ will be positive-*definite*. Even though the matrices $[q]$ and $[r]$ are not unique (they are direction-cosine matrices whose columns contain the eigenvectors for $[V]$ and $[U]$, respectively), their *product* $[R] \equiv [q][r]^T$ is unique. Likewise, $[U]$ and $[V]$ are unique. This specialization of the SVD theorem for square matrices will be now presented directly from the tensor-analysis perspective.

12.7. Polar decomposition—a hindsight intuitive interpretation

The essential idea behind the polar-decomposition theorem is illustrated in figure 12.5 where we have shown that the deformation $\underset{\approx}{\mathbf{F}}$ from figure 12.2 can be decomposed into a sequence of two separate steps:

- A stretch $\underset{\approx}{\mathbf{U}}$ followed by a rotation $\underset{\approx}{\mathbf{R}}$, so that $\underset{\approx}{\mathbf{F}} = \underset{\approx}{\mathbf{R}} \bullet \underset{\approx}{\mathbf{U}}$. This is the top path in figure 12.5. The tensor $\underset{\approx}{\mathbf{U}}$ is called the right stretch because it appears on the right in the expression $\underset{\approx}{\mathbf{R}} \bullet \underset{\approx}{\mathbf{U}}$.

Alternatively, the deformation may be decomposed in similar steps applied in the opposite order:

- A rotation $\underset{\approx}{\mathbf{R}}$ followed by a stretch $\underset{\approx}{\mathbf{V}}$, so that $\underset{\approx}{\mathbf{F}} = \underset{\approx}{\mathbf{V}} \bullet \underset{\approx}{\mathbf{R}}$. This is the bottom path in figure 12.5. The tensor $\underset{\approx}{\mathbf{V}}$ is called the left stretch.

In figure 12.5, a circumscribed square and an inscribed shaded circle[7] have been 'painted' onto the deforming material to help illustrate the meaning of the term *stretch*. A stretch is a special type of deformation for which the deformation gradient is both symmetric *and positive-definite*. So a stretch is any deformation for which $\underset{\approx}{\mathbf{R}} = \underset{\approx}{\mathbf{I}}$. For any pure stretch, there always exist three directions (the eigenvectors of the stretch tensor) along which the material is stretched or compressed, but not rotated. Material fibers that are originally aligned with the principal stretch directions may change length, but they do not change orientation under a pure stretch. The polar-decomposition theorem says that any general deformation can be decomposed into a pure stretch in combination with a rigid rotation.

The directions marked $\underset{\sim}{\boldsymbol{\delta}}{}_k^U$ are the principal stretch directions for the right stretch $\underset{\approx}{\mathbf{U}}$. They change length but do not change orientation during the application of $\underset{\approx}{\mathbf{U}}$. The directions marked $\underset{\sim}{\boldsymbol{\delta}}{}_k^V$ are the principal stretch directions for the left stretch $\underset{\approx}{\mathbf{V}}$. They change length but do not change orientation during the application of $\underset{\approx}{\mathbf{V}}$.

Incidentally, note for future reference that

$$\underset{\sim}{\boldsymbol{\delta}}{}_k^V = \underset{\approx}{\mathbf{R}} \bullet \underset{\sim}{\boldsymbol{\delta}}{}_k^U. \tag{12.33}$$

---

[7] Which would be a sphere for 3D deformations.

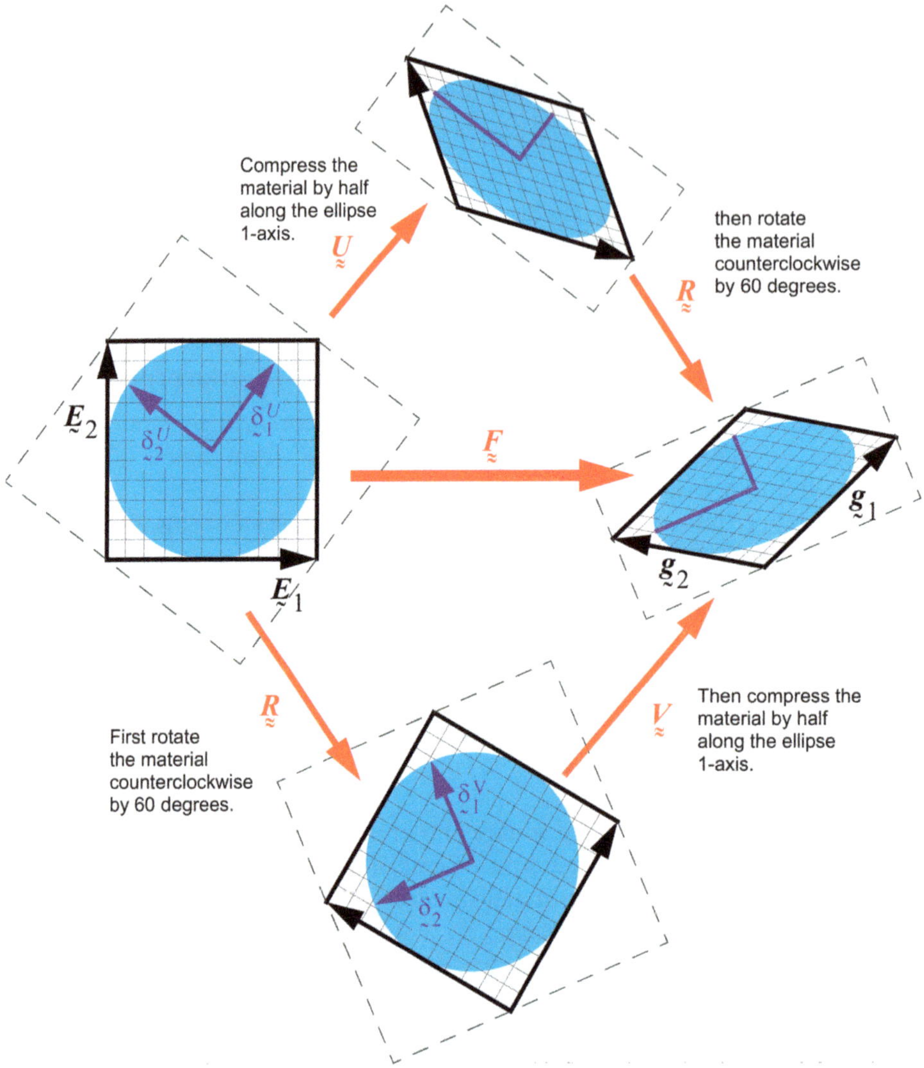

Figure 12.5. Visualization of the polar decomposition. This figure shows that the *same* deformation $\underset{\sim}{F}$ from figure 12.2 can be visualized as two different paths, each of which involves an intermediate configuration. The upper $\underset{\approx}{R} \bullet \underset{\approx}{U}$ path first compresses the material by a factor of 1/2 in the direction of the vector labeled $\underset{\sim}{\delta}_1^U$ and then rotates counter-clockwise by 60°. The same deformation is achieved on the bottom path by rotating by 60° and *then* compressing by a factor of 1/2 in the direction of the vector labeled $\underset{\sim}{\delta}_1^V$. Note that $\underset{\sim}{\delta}_1^V$ is obtained by rotating $\underset{\sim}{\delta}_1^U$ by 60°. In these figures, we have 'painted' a circle (or sphere in 3D) on the reference cube to show how it deforms into an ellipse (or ellipsoid in 3D). The vectors $\underset{\sim}{\delta}_k^U$ and $\underset{\sim}{\delta}_k^V$ lie on the major axes of the ellipse.

This section has been called a 'hindsight' interpretation because the principal directions of the stretch are not normally known *a priori.* Finding these directions is a key task for the polar-decomposition theorem.

In figure 12.5, the orthonormal triad of vectors $\underset{\sim}{\delta}_1^U$, $\underset{\sim}{\delta}_2^U$, and $\underset{\sim}{\delta}_3^U$ (which points out of the page for planar deformations) are the principal directions of the right stretch $\underset{\approx}{U}$.

These principal stretch vectors do not change orientation during the application of $\underset{\approx}{U}$. They change direction only upon application of the subsequent rotation. Material fibers that are *not* aligned with principal directions of $\underset{\approx}{U}$ generally do change orientation even before the rotation is applied. Thus, the rotation tensor describes *overall* material reorientation, but not the reorientation of individual material fibers. This observation is important in the application of ferroelectric solids, for which the transformation of embedded dipole moment vectors under a large deformation must be obtained using the deformation gradient tensor, not the polar rotation tensor [3].

For the particular example shown in figure 12.5, material fibers originally in the direction of $\underset{\sim}{\boldsymbol{\delta}}_1^U$ change length by a factor of about 1/2 (based on an estimate from looking at the picture). The fibers in the direction of $\underset{\sim}{\boldsymbol{\delta}}_2^U$ do not appear to change length. Stated differently, the ratio of their deformed length to their undeformed length is 1. If we suppose that the out-of-plane thickness increases by a factor of 3 with no out-of-plane shear, then fibers in the direction of $\underset{\sim}{\boldsymbol{\delta}}_3^U$ change length by a factor of 3 without changing orientation. The matrix for $\underset{\approx}{U}$ is diagonal in the principal stretch basis. The components are the ratios of deformed length to undeformed length. Thus, for the particular example shown in figure 12.5,

$$[\underset{\approx}{U}] = \begin{bmatrix} \frac{1}{2} & 0 & 0 \\ 0 & 1 & 0 \\ 0 & 0 & 3 \end{bmatrix} \quad \text{with respect to the } \delta_k^U \text{ triad.} \tag{12.34}$$

To express this stretch tensor in terms of the *reference* basis we must perform a coordinate transformation. To do this, first we could use a ruler to measure the components of each of the $\underset{\sim}{\boldsymbol{\delta}}_k^U$ with respect to the $\underset{\sim}{\boldsymbol{E}}_k$ reference basis (making sure to compensate for estimation errors by making sure these are orthogonal) to obtain

$$\{\underset{\sim}{\boldsymbol{\delta}}_1^U\} = \begin{Bmatrix} 0.6 \\ 0.8 \\ 0 \end{Bmatrix}, \quad \{\underset{\sim}{\boldsymbol{\delta}}_2^U\} = \begin{Bmatrix} -0.8 \\ 0.6 \\ 0 \end{Bmatrix} \quad \{\underset{\sim}{\boldsymbol{\delta}}_3^U\} = \begin{Bmatrix} 0 \\ 0 \\ 1 \end{Bmatrix}. \tag{12.35}$$

These arrays are assembled into columns to form the direction-cosine matrix $[Q]$ from (3.49):

$$[Q_{ij}] = \left[\underset{\sim}{\boldsymbol{E}}_i \bullet \underset{\sim}{\boldsymbol{\delta}}_j^U\right] = \left[\{\underset{\sim}{\boldsymbol{\delta}}_1^U\}\{\underset{\sim}{\boldsymbol{\delta}}_2^U\}\{\underset{\sim}{\boldsymbol{\delta}}_3^U\}\right] = \begin{bmatrix} 0.6 & -0.8 & 0 \\ 0.8 & 0.6 & 0 \\ 0 & 0 & 1 \end{bmatrix}. \tag{12.36}$$

Applying (3.56) gives us the components of $\underset{\approx}{U}$ with respect to the *reference* basis:

$$\begin{aligned} [\underset{\approx}{U}] &= \begin{bmatrix} 0.6 & -0.8 & 0 \\ 0.8 & 0.6 & 0 \\ 0 & 0 & 1 \end{bmatrix} \begin{bmatrix} \frac{1}{2} & 0 & 0 \\ 0 & 1 & 0 \\ 0 & 0 & 3 \end{bmatrix} \begin{bmatrix} 0.6 & -0.8 & 0 \\ 0.8 & 0.6 & 0 \\ 0 & 0 & 1 \end{bmatrix}^T \\ &= \begin{bmatrix} 0.82 & -0.24 & 0 \\ -0.24 & 0.68 & 0 \\ 0 & 0 & 3.0 \end{bmatrix} \quad \text{with respect to the } \underset{\sim}{\boldsymbol{E}}_k \text{ reference triad.} \end{aligned} \tag{12.37}$$

This tensor is symmetric. It is also positive-definite. It is therefore a stretch tensor. An alternative method to compensate for the estimation errors in identifying principal directions for $[U]$ would be to keep those errors when forming $[Q]$ (which would not generally be orthogonal when there are estimation errors) and then to replace the transpose of $[Q]$ with the inverse of $[Q]$. Either way, this process will provide a good estimate for stretch based on just looking at the picture, and this is important to serve as a back-of-the-envelope check for precise calculations.

To construct the rotation tensor, we would use a protractor to measure the angle through which the ellipse in figure (12.5) has rotated. The result appears to be about 60° and, because the deformation is planar, the axis of rotation is in the $\underset{\sim}{E}_3$-direction. Hence, the rotation tensor $\underset{\approx}{R}$ must be

$$[\underset{\approx}{R}] = \begin{bmatrix} \cos 60^\circ & -\sin 60^\circ & 0 \\ \sin 60^\circ & \cos 60^\circ & 0 \\ 0 & 0 & 1 \end{bmatrix} \text{ with respect to the } \underset{\sim}{E}_k \text{ reference triad.} \tag{12.38}$$

Multiplying the above two tensors gives

$$\begin{aligned} [\underset{\approx}{F}] = [\underset{\approx}{R} \bullet \underset{\approx}{U}] &= \begin{bmatrix} \cos 60^\circ & -\sin 60^\circ & 0 \\ \sin 60^\circ & \cos 60^\circ & 0 \\ 0 & 0 & 1 \end{bmatrix} \begin{bmatrix} 0.82 & -0.24 & 0 \\ -0.24 & 0.68 & 0 \\ 0 & 0 & 3.0 \end{bmatrix} \\ &= \begin{bmatrix} 0.618 & -0.709 & 0. \\ 0.590 & 0.132 & 0. \\ 0. & 0. & 3. \end{bmatrix} \\ &\quad \text{with respect to the } \underset{\sim}{E}_k \text{ reference triad,} \end{aligned} \tag{12.39}$$

which is the same as the matrix derived directly in (12.14). Thus, we have demonstrated very informally that the decomposition into a stretch followed by a rotation does indeed exist for the sample deformation in figure 12.5.

This section was labeled 'hindsight' because we produced the principal stretch directions by serendipity (i.e., 'out of thin air'), and then we demonstrated that they did indeed permit the decomposition into stretch followed by rotation. Section 12.9 presents the polar decomposition in a more classical manner.

12.8. Variational interpretation of the polar decomposition

Let $\underset{\approx}{F}$ be an invertible tensor. We can ask: what is the orthogonal tensor $\underset{\approx}{R}$ that is 'closest' to $\underset{\approx}{F}$ in the sense that $\|\underset{\approx}{F} - \underset{\approx}{R}\| \leqslant \|\underset{\approx}{F} - \underset{\approx}{Q}\|$ for all orthogonal $\underset{\approx}{Q}$? The polar decomposition is, in fact, the solution to this minimization problem [4]. To understand why, we will here prove that $\underset{\approx}{F} - \underset{\approx}{R}$ is orthogonal to $\dot{\underset{\approx}{R}}$, where $\dot{\underset{\approx}{R}}$ is an arbitrary variation from $\underset{\approx}{R}$. Consider the inner product of $\underset{\approx}{F} - \underset{\approx}{R}$ with an arbitrary admissible variation $\dot{\underset{\approx}{R}}$, holding the stretch constant. To be admissible, the variation must be expressible in the form $\dot{\tilde{R}} = \tilde{\Omega} \bullet \tilde{R}$ in which $\tilde{\Omega}$ is skew-symmetric. Then

$$\begin{aligned}(\tilde{F} - \tilde{R}) : \dot{\tilde{R}} &= (\tilde{F} - \tilde{R}) : (\tilde{\Omega} \bullet \tilde{R}) = \text{tr}[(\tilde{F} - \tilde{R}) \bullet (\tilde{\Omega} \bullet \tilde{R})^T] \\ &= tr[(\tilde{F} - \tilde{R}) \bullet (\tilde{R}^T \bullet \tilde{\Omega}^T)] = [(\tilde{F} - \tilde{R}) \bullet \tilde{R}^T] : \tilde{\Omega} \\ &= [\tilde{F} \bullet \tilde{R}^T - \tilde{I}] : \tilde{\Omega} = [\tilde{V} - \tilde{I}] : \tilde{\Omega} = 0.\end{aligned} \tag{12.40}$$

In the last step, we noted that $\tilde{V} - \tilde{I}$ is symmetric while $\tilde{\Omega}$ is skew-symmetric, making the inner product zero. The above derivation shows that the magnitude of $\tilde{F} - \tilde{R}$ will always be constant whenever $\tilde{R}$ is varied, holding $\tilde{F}$ (and therefore stretch) constant. Stated geometrically, the hyper-sphere of radius $\|\tilde{F} - \tilde{R}\|$ centered about $\tilde{R}$ is tangent to the convex set of all orthogonal tensors, which means that $\tilde{R}$ is closer to $\tilde{F}$ than any other orthogonal tensor. Incidentally, the set of all proper-orthogonal tensors is a subset of all tensors whose magnitude equals $\sqrt{3}$. This is true because

$$\|\tilde{R}\|^2 = \tilde{R} : \tilde{R} = \text{tr}(\tilde{R} \bullet \tilde{R}^T) = \text{tr}\tilde{I} = 3. \tag{12.41}$$

Because all rotations (i.e., proper-orthogonal tensors) have the same magnitude, they are points on a hyper-sphere in tensor space. Consequently, all rotation tensors belong to a larger convex set (the hyper-sphere) in tensor space. Even though $\tilde{F}$ is closer to $\tilde{R}$ than any to other orthogonal tensor, this should not be interpreted to mean that $\tilde{R}$ is the point on this hyper-sphere closest to $\tilde{F}$. That would be true only if the hyper-sphere of radius $\sqrt{3}$ consisted of *only* orthogonal tensors (which it does not because there exist many tensors of magnitude $\sqrt{3}$ that are *not* rotations).

12.9. A more rigorous (classical) presentation of the polar-decomposition theorem

Theorem. For each invertible tensor $\tilde{F}$ there exists a unique orthogonal tensor $\tilde{R}$, a unique symmetric positive-definite 'right stretch' $\tilde{U}$, and a unique symmetric positive-definite 'left stretch' $\tilde{V}$ such that

$$\tilde{F} = \tilde{R} \bullet \tilde{U} = \tilde{V} \bullet \tilde{R}. \tag{12.42}$$

WARNING: the descriptors 'symmetric' and 'positive-definite' are *requirements*, not consequences. Keep in mind that a stretch must be both symmetric and positive-definite—a tensor can be symmetric without being positive-definite. If only symmetry were tested, the rotation matrix of (5.31) would be wrongly identified as a stretch. Checking symmetry is not enough to declare a symmetric matrix to be a stretch—one must also prove positive-definiteness! It can be shown that the deformation gradient $\tilde{F}$ is symmetric if and only if the rotation angle is an integer multiple of 180°.

Comment: The determinant of $\underset{\approx}{\boldsymbol{R}}$ will have the same sign as the determinant of $\underset{\approx}{\boldsymbol{F}}$. Hence, the orthogonal tensor $\underset{\approx}{\boldsymbol{R}}$ will be a proper rotation if and only if det $\underset{\approx}{\boldsymbol{F}} > 0$. We are interested in second-order tensors in 3D space so that $[\underset{\approx}{\boldsymbol{F}}]$ is a 3×3 matrix. If the determinant of $\underset{\approx}{\boldsymbol{F}}$ is negative, then we can always perform the polar decomposition of $-\underset{\approx}{\boldsymbol{F}}$, which has a positive determinant in 3D space. Thus, without loss in generality, we may assume that the orthogonal tensor $\underset{\approx}{\boldsymbol{R}}$ in (12.42) is a *proper* rotation.

Theorem. A rigorous statement of polar decomposition is as follows:
GIVEN: $\underset{\approx}{\boldsymbol{F}}$ is real and invertible,
THEN: there exist real tensors $\underset{\approx}{\boldsymbol{U}}$, $\underset{\approx}{\boldsymbol{V}}$, and $\underset{\approx}{\boldsymbol{R}}$ such that

i. $\underset{\approx}{\boldsymbol{F}} = \underset{\approx}{\boldsymbol{R}} \bullet \underset{\approx}{\boldsymbol{U}} = \underset{\approx}{\boldsymbol{V}} \bullet \underset{\approx}{\boldsymbol{R}}$,
ii. $\underset{\approx}{\boldsymbol{U}}$ and $\underset{\approx}{\boldsymbol{V}}$ are symmetric,
iii. $\underset{\approx}{\boldsymbol{U}}$ and $\underset{\approx}{\boldsymbol{V}}$ are positive-definite, and
iv. $\underset{\approx}{\boldsymbol{R}}$ is orthogonal (i.e., $\underset{\approx}{\boldsymbol{R}}^{-1} = \underset{\approx}{\boldsymbol{R}}^{T}$).

Corollary. $\underset{\approx}{\boldsymbol{U}}$, $\underset{\approx}{\boldsymbol{V}}$, and $\underset{\approx}{\boldsymbol{R}}$ are unique. They are the only tensors that satisfy all of the conditions (i) through (iv), which is interesting since neither $[q]$ nor $[r]$ in the SVD is unique (they are, in fact direction-cosine matrices defining reference and spatial stretch directions, and, because eigenvector ordering and directions are never unique, direction cosine matrices are never unique.).

To prove the theorem, we first assume that $\underset{\approx}{\boldsymbol{U}}$, $\underset{\approx}{\boldsymbol{V}}$, and $\underset{\approx}{\boldsymbol{R}}$ do indeed exist. We will then derive explicit unique formulas for them as functions of $\underset{\approx}{\boldsymbol{F}}$. Since such formulas will have been proved unique by construction on the *premise* that $\underset{\approx}{\boldsymbol{U}}$, $\underset{\approx}{\boldsymbol{V}}$, and $\underset{\approx}{\boldsymbol{R}}$ exist, we will need to then verify satisfaction of all of the conditions (i)–(iv) listed above in order to confirm the premise. Assume for now that the decomposition (i) $\underset{\approx}{\boldsymbol{F}} = \underset{\approx}{\boldsymbol{R}} \bullet \underset{\approx}{\boldsymbol{U}}$ does indeed exist. Then $\underset{\approx}{\boldsymbol{F}}^{T} = \underset{\approx}{\boldsymbol{U}}^{T} \bullet \underset{\approx}{\boldsymbol{R}}^{T} = \underset{\approx}{\boldsymbol{U}} \bullet \underset{\approx}{\boldsymbol{R}}^{-1}$, where we have used the properties that (iii) $\underset{\approx}{\boldsymbol{U}}$ is symmetric and (iv) $\underset{\approx}{\boldsymbol{R}}$ is orthogonal. Consequently the product $\underset{\approx}{\boldsymbol{F}}^{T} \bullet \underset{\approx}{\boldsymbol{F}}$ must equal $\underset{\approx}{\boldsymbol{U}}^{2}$, and the stretch $\underset{\approx}{\boldsymbol{U}}$ must be given by

$$\underset{\approx}{\boldsymbol{U}} = \underset{\approx}{\boldsymbol{C}}^{1/2}, \qquad \text{where } \underset{\approx}{\boldsymbol{C}} \equiv \underset{\approx}{\boldsymbol{F}}^{T} \bullet \underset{\approx}{\boldsymbol{F}}. \tag{12.43}$$

The helper tensor $\underset{\approx}{\boldsymbol{C}}$ is positive-definite because

$$\begin{aligned} &\underset{\sim}{\boldsymbol{u}} \bullet \underset{\approx}{\boldsymbol{C}} \bullet \underset{\sim}{\boldsymbol{u}} = \underset{\sim}{\boldsymbol{u}} \bullet \underset{\approx}{\boldsymbol{F}}^{T} \bullet \underset{\approx}{\boldsymbol{F}} \bullet \underset{\sim}{\boldsymbol{u}} = (\underset{\approx}{\boldsymbol{F}} \bullet \underset{\sim}{\boldsymbol{u}}) \bullet (\underset{\approx}{\boldsymbol{F}} \bullet \underset{\sim}{\boldsymbol{u}}) \\ &= \|(\underset{\approx}{\boldsymbol{F}} \bullet \underset{\sim}{\boldsymbol{u}})\|^{2} \\ &> 0 \quad \text{for all non-zero vectors } \underset{\sim}{\boldsymbol{u}}. \end{aligned} \tag{12.44}$$

Of course the magnitude of *any* real vector is non-negative, and therefore $\|(\underset{\approx}{\boldsymbol{F}} \bullet \underset{\sim}{\boldsymbol{u}})\|^{2} \geqslant 0$. Our assertion that $\|(\underset{\approx}{\boldsymbol{F}} \bullet \underset{\sim}{\boldsymbol{u}})\|^{2} > 0$ is stronger condition. To rule out the possibility that $\|(\underset{\approx}{\boldsymbol{F}} \bullet \underset{\sim}{\boldsymbol{u}})\| = 0$, we must recall that (by premise) $\underset{\approx}{\boldsymbol{F}}$ is invertible and therefore $\underset{\approx}{\boldsymbol{F}} \bullet \underset{\sim}{\boldsymbol{u}}$ can never be zero for any non-zero $\underset{\sim}{\boldsymbol{u}}$.

Note that the helper tensor $\underset{\approx}{C}$, called the right Cauchy–Green tensor in continuum mechanics, is symmetric. It is therefore diagonal in its principal basis. By finding the eigenvalues α_i and eigenvectors $\underset{\sim}{\delta}_i^C$ of $\underset{\approx}{C}$, we may write

$$[\underset{\approx}{C}] = \begin{bmatrix} \alpha_1 & 0 & 0 \\ 0 & \alpha_2 & 0 \\ 0 & 0 & \alpha_3 \end{bmatrix} \text{ with respect to the principal } \underset{\sim}{\delta}_i^C \text{ triad} \tag{12.45}$$

where each eigenvalue α_i is strictly positive.

In basis notation,

$$\underset{\approx}{C} = \alpha_1 \underset{\sim}{\delta}_1^C \underset{\sim}{\delta}_1^C + \alpha_2 \underset{\sim}{\delta}_2^C \underset{\sim}{\delta}_2^C + \alpha_3 \underset{\sim}{\delta}_3^C \underset{\sim}{\delta}_3^C, \quad \text{where each } \alpha_i > 0. \tag{12.46}$$

We must be very careful when defining the meaning of the 'square root' $\underset{\approx}{C}^{1/2}$ in (12.43). A positive-definite and symmetric 3×3 matrix generally has an *infinite* number of square roots, of which at least eight are symmetric, and only one is both symmetric and positive-definite[8]. To ensure that our equation (12.43) defines a *unique* stretch $\underset{\approx}{U}$, we demand that the 'square root' $\underset{\approx}{C}^{1/2}$ must be interpreted as the unique positive-definite square root. Namely, the right stretch tensor $\underset{\approx}{U}$ is given by

$$\underset{\approx}{U} = \lambda_1 \underset{\sim}{\delta}_1^U \underset{\sim}{\delta}_1^U + \lambda_2 \underset{\sim}{\delta}_2^U \underset{\sim}{\delta}_2^U + \lambda_3 \underset{\sim}{\delta}_3^U \underset{\sim}{\delta}_3^U, \quad \text{where } \lambda_i \equiv +\sqrt{\alpha_i} \text{ and } \underset{\sim}{\delta}_i^U = \underset{\sim}{\delta}_1^C. \tag{12.47}$$

In matrix notation,

$$[\underset{\approx}{U}] = +\left[\underset{\approx}{C}^{1/2}\right] = \begin{bmatrix} \lambda_1 & 0 & 0 \\ 0 & \lambda_2 & 0 \\ 0 & 0 & \lambda_3 \end{bmatrix} \text{ with respect to theprincipal } \underset{\sim}{\delta}_i^C \text{ triad} \tag{12.48}$$

where $\lambda_i \equiv +\sqrt{a_i}$.

In practice, this result is phrased in terms of the laboratory basis as

$$[\underset{\approx}{U}] = [Q] \begin{bmatrix} \lambda_1 & 0 & 0 \\ 0 & \lambda_2 & 0 \\ 0 & 0 & \lambda_3 \end{bmatrix} [Q]^T \text{ with respect to the lab } \underset{\sim}{E}_i \text{ triad,} \tag{12.49}$$

where $\lambda_i \equiv +\sqrt{\alpha_i}$ and $Q_{ij} = \underset{\sim}{\delta}_i^C \bullet \underset{\sim}{E}_j$.

The ith column of the transformation matrix $[Q]$ contains the lab components of the $\underset{\sim}{\delta}_i^C$ eigenvector. There is only one square root of $\underset{\approx}{C}$ that is both symmetric and positive-definite. Hence, the stretch defined in (12.48) is the *only* stretch that satisfies the requirements of the theorem, so we have proved that if $\underset{\approx}{U}$ exists, then it is unique. Once $\underset{\approx}{U}$ is known, the rotation tensor can be found by

[8] To understand the reason for these statements, consider the following symmetric and non-symmetric square roots of the 2×2 identity matrix: $\begin{bmatrix} 1 & 0 \\ 0 & 1 \end{bmatrix}, \begin{bmatrix} -1 & 0 \\ 0 & 1 \end{bmatrix}, \begin{bmatrix} +1 & 0 \\ 0 & -1 \end{bmatrix}, \begin{bmatrix} 0 & 1 \\ 1 & 0 \end{bmatrix}, \begin{bmatrix} 2 & 3 \\ -1 & -2 \end{bmatrix}$.

$$\underset{\approx}{R} = \underset{\approx}{F} \bullet \underset{\approx}{U}^{-1}. \tag{12.50}$$

This is a single-valued operation. Hence, uniqueness of $\underset{\approx}{U}$ guarantees uniqueness of $\underset{\approx}{R}$. The tensor $\underset{\approx}{R}$ is indeed orthogonal because

$$\underset{\approx}{R}^T \bullet \underset{\approx}{R} = \underset{\approx}{U}^{-1} \bullet \underset{\approx}{F}^T \bullet \underset{\approx}{F} \bullet \underset{\approx}{U}^{-1} = \underset{\approx}{U}^{-1} \bullet \underset{\approx}{U}^2 \bullet \underset{\approx}{U}^{-1} = \underset{\approx}{I}. \tag{12.51}$$

By similar arguments, the left stretch $\underset{\approx}{V}$ must be given by

$$\underset{\approx}{V} = +\underset{\approx}{B}^{1/2}, \quad \text{where} \quad \underset{\approx}{B} \equiv \underset{\approx}{F} \bullet \underset{\approx}{F}^T. \tag{12.52}$$

As before, to ensure that $\underset{\approx}{V}$ will be a proper stretch, $\underset{\approx}{B}^{1/2}$ must be taken to be the unique symmetric and positive-definite square root. Once $\underset{\approx}{V}$ is known, an orthogonal tensor $\underset{\approx}{Q}$ can be found by

$$\hat{\underset{\approx}{R}} = \underset{\approx}{V}^{-1} \bullet \underset{\approx}{F}. \tag{12.53}$$

So far, we have proved existence of stretches, $\underset{\approx}{U}$ and $\underset{\approx}{V}$, and rotations, $\underset{\approx}{R}$ and $\hat{\underset{\approx}{R}}$, such that

$$\underset{\approx}{F} = \underset{\approx}{R} \bullet \underset{\approx}{U} = \underset{\approx}{V} \bullet \hat{\underset{\approx}{R}}. \tag{12.54}$$

All that remains is to prove that the orthogonal tensor from the left stretch is the same as the orthogonal tensor from the right stretch. In other words, we must prove that $\hat{\underset{\approx}{R}} = \underset{\approx}{R}$. Applying the definition of the left stretch, we know that $\underset{\approx}{U}^2 = \underset{\approx}{F}^T \bullet \underset{\approx}{F} = (\underset{\approx}{V} \bullet \hat{\underset{\approx}{R}})^T \bullet \underset{\approx}{V} \bullet \hat{\underset{\approx}{R}} = \hat{\underset{\approx}{R}}^T \bullet \underset{\approx}{V}^2 \bullet \hat{\underset{\approx}{R}}$ from which it follows that $\underset{\approx}{U} = \hat{\underset{\approx}{R}}^T \bullet \underset{\approx}{V} \bullet \hat{\underset{\approx}{R}}$. Substituting this result back into (12.54), shows that $\underset{\approx}{R} \bullet \hat{\underset{\approx}{R}}^T = \underset{\approx}{I}$ and therefore

$$\hat{\underset{\approx}{R}} = \underset{\approx}{R}. \tag{12.55}$$

12.9.1. Working with the inverse gradient

For some applications (such as simple shear), it is easier to work with the *inverse* of the deformation gradient. In such cases, the stretches may be computed by

$$\underset{\approx}{U} = (\underset{\approx}{F}^{-1} \bullet \underset{\approx}{F}^{-T})^{-1/2} \quad \text{and} \quad \underset{\approx}{V} = (\underset{\approx}{F}^{-T} \bullet \underset{\approx}{F}^{-1})^{-1/2}. \tag{12.56}$$

Noting that orthogonality implies $\underset{\approx}{R} = \underset{\approx}{R}^{-T}$, the rotation can be computed by

$$\underset{\approx}{R} = \underset{\approx}{F}^{-T} \bullet \underset{\approx}{U} \quad \text{and} \quad \underset{\approx}{R} = \underset{\approx}{V} \bullet \underset{\approx}{F}^{-T}. \tag{12.57}$$

As a good example of a situation where you would work with the inverse gradient, consider a Lagrangian finite-element calculation. Because the nodes move with the material, the reference position $\underset{\sim}{X}$ of each node is constant throughout time. A node's current position $\underset{\sim}{x}$ varies in time. From a computational standpoint, it is often easier to use the element shape functions to compute the gradient of $\underset{\sim}{X}$ with respect to $\underset{\sim}{x}$. This gradient is the inverse deformation gradient:

$$\underset{\approx}{F}^{-1} = \frac{\partial \underset{\sim}{X}}{\partial \underset{\sim}{x}}. \tag{12.58}$$

Equations (12.56) and (12.57) show that it is not necessary to invert (12.58) to perform a polar decomposition.

12.10. The 'fast' way to do a polar decomposition in two dimensions

For planar deformations, the deformation gradient tensor has a component matrix of the form

$$[F] = \begin{bmatrix} F_{11} & F_{12} & 0 \\ F_{21} & F_{22} & 0 \\ 0 & 0 & F_{33} \end{bmatrix}, \quad \text{where } F_{33} > 0. \tag{12.59}$$

We claim by serendipity[9] that the polar decomposition can be performed rapidly by the following formula:

$$[R] = \frac{1}{f}\begin{bmatrix} F_{11} + F_{22} & F_{12} - F_{21} & 0 \\ F_{21} - F_{12} & F_{11} + F_{22} & 0 \\ 0 & 0 & f \end{bmatrix}, \tag{12.60a}$$

$$[U] = \frac{1}{f}\begin{bmatrix} \varphi + F_{11}^2 + F_{21}^2 & F_{11}F_{12} + F_{22}F_{21} & 0 \\ F_{11}F_{12} + F_{22}F_{21} & \varphi + F_{22}^2 + F_{12}^2 & 0 \\ 0 & 0 & fF_{33} \end{bmatrix}, \tag{12.60b}$$

where

$$\begin{aligned} f &\equiv \sqrt{(F_{11} + F_{22})^2 + (F_{21} - F_{12})^2} \\ \text{and } \varphi &= \frac{\det[F]}{F_{33}} = F_{11}F_{22} - F_{12}F_{21}. \end{aligned} \tag{12.61}$$

Note that φ is simply the determinant of the upper (in-plane) 2×2 sub-matrix of $[F]$. An easier way to compute $[U]$ is to first obtain $[R]$ and then simply note that $[U] = [R]^T[F]$. Comparing (12.60*a*) with (4.34) shows that the rotation angle can be found by

$$\cos\theta = (F_{11} + F_{22})/f, \tag{12.62a}$$

$$\sin\theta = (F_{21} - F_{12})/f. \tag{12.62b}$$

[9] It is easy to verify that our formulas yield an orthogonal $[R]$ and a symmetric $[U]$. It is straightforward, but tedious, to also prove that our formula gives a *positive-definite* $[U]$ matrix. This property is essential to confirm the validity of our serendipitous formulas.

Beware! You must define $\cos\theta$ and $\sin\theta$ *separately* to uniquely determine the rotation angle. It is certainly true that

$$\tan\theta = (F_{21} - F_{12})/(F_{11} + F_{22}), \qquad \leftarrow\text{not enough!} \tag{12.63}$$

This relation does not *uniquely* define the rotation angle because *two* angles θ in the range from 0 to 2π satisfy the above equation. By contrast only *one* angle in the range from 0 to 2π satisfies (12.62).

For computer applications, the following coding will very rapidly perform a planar (2D) polar decomposition:

```
c = f(1, 1)+f(2, 2)
s = f(2, 1)-f(1, 2)
d = sqrt(c*c + s*s)
c = c/d
s = s/d

R(1, 1)=c
R(2, 2)=c
R(1, 2)=-s
R(2, 1)=s

U(1, 1)=c*F(1, 1)+s*F(2, 1)
U(1, 2)=c*F(1, 2)+s*F(2, 2)
U(2, 1)=c*F(2, 1)-s*F(1, 1)
U(2, 2)=c*F(2, 2)-s*F(1, 2)

R(3, 3)=1.
R(1, 3)=0.
R(2, 3)=0.
R(3, 1)=0.
R(3, 2)=0.

U(3, 3)=F(3, 3)
U(1, 3)=0.
U(2, 3)=0.
U(3, 1)=0.
U(3, 2)=0.
```

12.11. Scaling properties of a polar decomposition

Suppose you have the polar decomposition,

$$\underset{\approx}{\bar{F}} = \underset{\approx}{\bar{R}} \bullet \underset{\approx}{\bar{U}} = \underset{\approx}{\bar{V}} \bullet \underset{\approx}{\bar{R}}, \tag{12.64}$$

for some tensor $\underset{\approx}{\bar{F}}$. Now suppose you wish to find the polar decomposition,

$$\underset{\approx}{F} = \underset{\approx}{R} \bullet \underset{\approx}{U} = \underset{\approx}{V} \bullet \underset{\approx}{R}, \text{ for which } \underset{\approx}{F} = \alpha\underset{\approx}{\bar{F}}. \tag{12.65}$$

Here, α is a non-zero scalar multiple. The result is

$$\underset{\approx}{U} = |\alpha|\underset{\approx}{\overline{U}}, \qquad \underset{\approx}{V} = |\alpha|\underset{\approx}{\overline{V}}, \qquad \underset{\approx}{R} = \frac{|\alpha|}{\alpha}\underset{\approx}{\overline{R}}. \tag{12.66}$$

Importantly, if $\alpha > 0$, the rotation of the scaled tensor is identical to that for the original (barred) tensor. These formulas may be used to extend the fast polar decomposition algorithm to apply to the case the F(3,3) is negative. In that case, the multiplier would be -1, and therefore, neither stretch would change, while the algorithm's output for rotation would simply need to be multiplied by -1.

12.12. Classic method for obtaining a polar decomposition in 3D

The classical exact algorithm for obtaining a polar decomposition is

STEP 1. Compute $\underset{\approx}{C} \equiv \underset{\approx}{F}^T \bullet \underset{\approx}{F}$. This tensor will be both symmetric and positive-definite as long as $\underset{\approx}{F}$ is invertible (i.e., if $\det\underset{\approx}{F} \neq 0$).

STEP 2. Compute the reference (right) stretch, $\underset{\approx}{U} = \underset{\approx}{C}^{1/2}$, being sure to take the *positive-definite and symmetric* square root of $\underset{\approx}{C}$. Some methods for finding the square root of a tensor are described below.

STEP 3. Compute the polar rotation tensor by $\underset{\approx}{R} = \underset{\approx}{F} \bullet \underset{\approx}{U}^{-1}$ or by $\underset{\approx}{R} = \underset{\approx}{F}^{-T} \bullet \underset{\approx}{U}$ (the last expression follows by noting that $\underset{\approx}{R}^{-T} = \underset{\approx}{R}$).

STEP 4. If desired, the spatial (left) stretch is given by $\underset{\approx}{V} = \underset{\approx}{R} \bullet \underset{\approx}{U} \bullet \underset{\approx}{R}^T$.

If the right stretch was not desired, the left stretch can be alternatively found by $\underset{\approx}{V} = \underset{\approx}{B}^{1/2}$, where $\underset{\approx}{B} = \underset{\approx}{F} \bullet \underset{\approx}{F}^T$. Then the rotation is given by $\underset{\approx}{R} = \underset{\approx}{V}^{-1} \bullet \underset{\approx}{F}$

This classical polar-decomposition algorithm requires finding the square root $\underset{\approx}{U} = \underset{\approx}{C}^{1/2}$. The typical way to take the square root is to find the eigenvalues and eigenvectors of the component matrix $[\underset{\approx}{C}]$. If $[Q]$ is the direction-cosine matrix whose columns contain the eigenvectors of $[\underset{\approx}{C}]$ and if $[\Lambda]$ is a diagonal matrix containing the corresponding eigenvalues, then the lab components of $\underset{\approx}{U}$ may be computed by

$$[U] = [Q][\Lambda]^{1/2}[Q]^T. \tag{12.67}$$

Alternatively, if $\underset{\approx}{C} = \Lambda_1\underset{\sim}{p}_1\underset{\sim}{p}_1 + \Lambda_2\underset{\sim}{p}_2\underset{\sim}{p}_2 + \Lambda_3\underset{\sim}{p}_3\underset{\sim}{p}_3$ is the spectral expansion for $\underset{\approx}{C}$, then $\underset{\approx}{U} = \lambda_1\underset{\sim}{p}_1\underset{\sim}{p}_1 + \lambda_2\underset{\sim}{p}_2\underset{\sim}{p}_2 + \lambda_3\underset{\sim}{p}_3\underset{\sim}{p}_3$, where $\lambda_k = +\sqrt{\Lambda_k}$ (the positive square root must be taken).

While this classical method for evaluating polar decomposition is typically used in analytical work, we now review several numerical algorithms for the decomposition, which, unlike this classical method, typically first find the polar rotation (see [5]).

12.12.1. Hoger–Carlson–Sawyer (HCS) closed-form solution

The classical approach outlined above requires determination of both eigenvalues and eigenvectors. Hoger and Carlson [6] pointed out that, if the invariants of $\underset{\approx}{U}$ can be found, then $\underset{\approx}{U}$ itself may be computed directly from $\underset{\approx}{C}$ through use of the Cayley–Hamilton formula. Hoger and Carlson offered an *incorrect* method for finding the

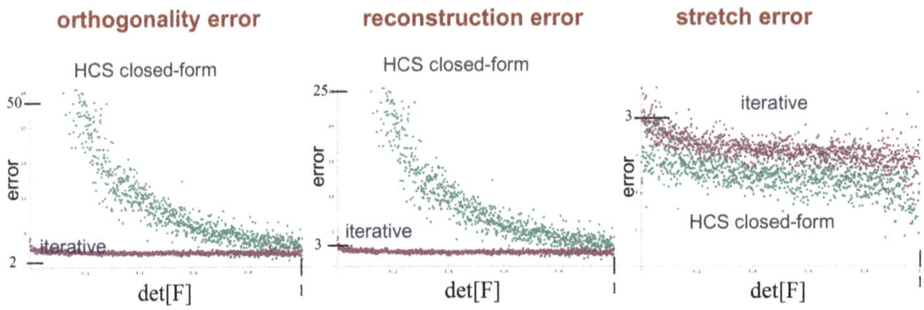

Figure 12.6. Comparison of HCS closed-form and iterative polar-decomposition algorithms. Over a million $[F]$ matrices were generated by setting the columns of $[F]$ to unit vectors of various orientations. When the unit vectors are orthonormal, then $[F] = [I]$, and both algorithms perform well. When they are highly skewed (making the determinant of $[F]$ small), the closed-form algorithm increasingly suffers from round-off error. The ordinates in these plots are the L_2 norms of $\underset{\approx}{r}^T \bullet \underset{\approx}{r} - \underset{\approx}{I}$, $\underset{\approx}{F} - \underset{\approx}{r} \bullet \underset{\approx}{u}$, and $\underset{\approx}{u} \bullet \underset{\approx}{u} - \underset{\approx}{F}^T \bullet \underset{\approx}{F}$, respectively, in units of 10^{-16}. Here $\underset{\approx}{r}$ and $\underset{\approx}{u}$are the computed solutions for $\underset{\approx}{R}$ and $\underset{\approx}{U}$, respectively.

invariants of $\underset{\approx}{U}$, which was subsequently corrected by Sawyer [7]. An algorithm for the HCS closed-form solution is may be found in [8].

It might seem that a closed-form solution like the HCS method is ideal, offering optimal accuracy. However, it is both less accurate and less efficient than iterative methods. Figure 12.6 shows the HCS closed-form solution compared with an iterative algorithm (presented soon). Both methods have comparable accuracy in computing the stretch. However, the iterative method is far more reliable than the HCS method for determining rotation for problems involving massive material distortion (e.g., taffy pulling). Moreover, the iterative algorithm is just as fast as (often faster than) the analytical method, with both run to machine precision.

12.12.2. Higham iterator for polar decomposition

A compellingly simple fixed-point algorithm exists for numerically finding a polar decomposition [9]. Defining a function that simply averages a tensor with its own inverse-transpose, $f(\underset{\approx}{A}) = \frac{1}{2}(\underset{\approx}{A} + \underset{\approx}{A}^{-T})$, the polar rotation is the limit

$$ffffffffffff\ldots fff(\underset{\approx}{F}) \rightarrow \underset{\approx}{R}. \tag{12.68}$$

The algorithm is to apply $f(\underset{\approx}{F})$ and then apply f to the result and keep on doing so until the answer stops changing, typically in only a few iterations to reach machine precision. This method is comparable in speed and accuracy to the more complicated one next. A proper (frame and basis insensitive) error measure is the Frobenius norm of the Green-Lagrange strain. The algorithm has converged to machine precision when 1.0 plus this norm evaluates to 1.0.

12.13. Another iterative polar decomposition in 3D

For any non-singular tensor $\underset{\approx}{F}$, the following recursive iteration will converge to the polar rotation tensor. This algorithm is coded in section A.9 of the appendix.

Phase 1: Compute the polar rotation tensor.

STEP 1. Compute $\mathbf{E} \equiv \mathbf{F}^T \bullet \mathbf{F}$.
STEP 2. Compute $s = 3/\mathrm{tr}(\mathbf{E})$, and then replace $\mathbf{E}$ with $\frac{1}{2}(s\mathbf{C} - \mathbf{I})$.
STEP 3. Replace s with $\sqrt{s}$ and then set $\mathbf{A} = s\mathbf{F}$.
STEP 4. Compute the initial scalar error measure $e_0 = \mathbf{E} : \mathbf{E}$.
STEP 5. If e_0 is zero to machine precision[10], go to step 10.
Otherwise continue.
STEP 6. Replace $\mathbf{A}$ with $\mathbf{A} - \mathbf{A} \bullet \mathbf{E}$.
STEP 7. Replace $\mathbf{E}$ with $\frac{1}{2}(\mathbf{A}^T \bullet \mathbf{A} - \mathbf{I})$.
STEP 8. Compute an updated scalar error measure $e = \mathbf{E} : \mathbf{E}$.
STEP 9. If $e < e_0$, replace e_0 with e and go to step 6. Otherwise, go to step 10.
STEP 10. $\mathbf{R} = \mathbf{A}$. Stop.

Phase 2: Compute the stretch.

If you seek the right stretch, it is simply $\mathbf{U} = \mathbf{R}^T \bullet \mathbf{F}$.
If you seek the left stretch, it is simply $\mathbf{V} = \mathbf{F} \bullet \mathbf{R}^T$.

The 'phase 1' algorithm converges to the polar rotation tensor. That is,

$$\mathbf{A} \rightarrow \mathbf{R}. \tag{12.69}$$

Without the scaling in step 2, convergence is ensured only if the principal stretches fall between 0 and $\sqrt{3}$ (or, equivalently, if the principal values of $\mathbf{C} = \mathbf{F}^T \bullet \mathbf{F}$ fall between 0 and 3). Scaling ensures that the algorithm will converge for *any* invertible deformation gradient.

In addition to the analytical and iterative methods we have presented here, numerous researchers have educed various other methods for obtaining a polar decomposition [9, 6, 10, 11]. However, this one does appear to be fastest and most accurate for 3×3 matrices.

12.13.1. Why does this iterative algorithm work? (A qualitative discussion)[11]

For the iterative algorithm, the fundamental fixed-point iteration expression is

$$\mathbf{A}_{n+1} = \frac{1}{2}\mathbf{A}_n \bullet \left[3\mathbf{I} - \mathbf{A}_n^T \bullet \mathbf{A}_n\right], \tag{12.70}$$

[10] A number X is zero to machine precision if (X + 1.0) − 1.0 evaluates to zero. On typical compilers, machine precision is in the neighborhood of 10^{-20}. Using the tensor L_2 norm (rather than some other matrix norm) is important to ensure the same result for all coordinate systems.

[11] This proof follows the basic ideas communicated in an email to me from Glynne Casteel, though he used the SVD from matrix analysis. Our presentation recognizes that such a decomposition is equivalent to the polar decomposition.

or

$$\underset{\approx}{\boldsymbol{A}}_{n+1} = \underset{\approx}{\boldsymbol{A}}_n \bullet [\underset{\approx}{\boldsymbol{I}} - \underset{\approx}{\boldsymbol{E}}_n], \quad \text{where} \quad \underset{\approx}{\boldsymbol{E}}_n = \tfrac{1}{2}\left[\underset{\approx}{\boldsymbol{A}}_n^T \bullet \underset{\approx}{\boldsymbol{A}}_n - \underset{\approx}{\boldsymbol{I}}\right]. \tag{12.71}$$

Physically speaking, $\underset{\approx}{\boldsymbol{E}}_n$ is the Green–Lagrange strain tensor, and the algorithm converges when this strain goes to zero.

Our goal in this section is to provide a qualitative understanding why an iterator like this should converge to the polar rotation. For understanding the *motivation* of this expression, we will first consider the case that the stretches are close to 1, in which case, we say that $\underset{\approx}{\boldsymbol{F}}$ is 'nearly orthogonal.' The concept of being nearly orthogonal is useful in computational-mechanics codes where the rotation tensor is integrated through time to obtain an updated rotation. Numerical integration naturally entails some approximation error, so the numerically integrated rotation will not generally be *exactly* orthogonal. The integration error can be removed in a basis-invariant manner by treating the nearly orthogonal updated rotation as a nearly orthogonal deformation gradient tensor, and then replacing the nearly orthogonal rotation with the precisely orthogonal rotation from a polar decomposition that removes the spurious (numerically introduced) stretch.

Our goal is to find the orthogonal tensor $\underset{\approx}{\boldsymbol{R}}$ that is closest to a given (nearly orthogonal) tensor $\underset{\approx}{\boldsymbol{F}}$. The strategy begins by considering a Taylor expansion:

$$(1 - x)^a = 1 - ax + O(x^2). \tag{12.72}$$

Substitution $x = 1 - y$ yields

$$y^a = 1 - a(1 - y) + O(1 - y)^2. \tag{12.73}$$

The above expansion was for scalar arguments, but it also extends to any symmetric tensor $\underset{\approx}{\boldsymbol{Y}}$. In particular, if we pick the exponent $a = -1/2$, we obtain

$$\underset{\approx}{\boldsymbol{Y}}^{-1/2} = \underset{\approx}{\boldsymbol{I}} + \tfrac{1}{2}(\underset{\approx}{\boldsymbol{I}} - \underset{\approx}{\boldsymbol{Y}}) + O(\underset{\approx}{\boldsymbol{I}} + \underset{\approx}{\boldsymbol{Y}})^2. \tag{12.74}$$

Let us turn our attention to the polar decomposition of $\underset{\approx}{\boldsymbol{F}}$:

$$\underset{\approx}{\boldsymbol{F}} = \underset{\approx}{\boldsymbol{R}} \bullet \underset{\approx}{\boldsymbol{U}}. \tag{12.75}$$

Define

$$\underset{\approx}{\boldsymbol{C}} = \underset{\approx}{\boldsymbol{F}}^T \bullet \underset{\approx}{\boldsymbol{F}} = \underset{\approx}{\boldsymbol{U}}^2. \tag{12.76}$$

Therefore

$$\underset{\approx}{\boldsymbol{U}}^{-1} = (\underset{\approx}{\boldsymbol{F}}^T \bullet \underset{\approx}{\boldsymbol{F}})^{-1/2}. \tag{12.77}$$

We know that

$$\underset{\approx}{\boldsymbol{R}} = \underset{\approx}{\boldsymbol{F}} \bullet \underset{\approx}{\boldsymbol{U}}^{-1} = \underset{\approx}{\boldsymbol{F}} \bullet \underset{\approx}{\boldsymbol{C}}^{-1/2}. \tag{12.78}$$

If the eigenvalues of $\underset{\approx}{U}$ are close to unity (i.e., if $\underset{\approx}{F}$ is very nearly orthogonal), then the eigenvalues of $\underset{\approx}{C}$ are also close to unity. By the Taylor expansion in (12.74),

$$\underset{\approx}{C}^{-1/2} \approx \underset{\approx}{I} + \tfrac{1}{2}(\underset{\approx}{I} - \underset{\approx}{C}) = \underset{\approx}{I} + \tfrac{1}{2}(\underset{\approx}{I} - \underset{\approx}{F}^T\underset{\approx}{F}) = \tfrac{3}{2}\underset{\approx}{I} - \tfrac{1}{2}\underset{\approx}{F}^T\underset{\approx}{F}. \tag{12.79}$$

Putting this into (12.78) shows that, to second order accuracy,

$$\underset{\approx}{R} \approx \tfrac{1}{2}\underset{\approx}{F} \bullet [3\underset{\approx}{I} - \underset{\approx}{F}^T\underset{\approx}{F}]. \tag{12.80}$$

This equation is the basis for the fixed-point iteration expression cited in (12.70). This equation represents the single-iteration estimate for $\underset{\approx}{R}$. The result after a single iteration will not generally be orthogonal, but recursive application will converge to the rotation tensor. As discussed soon, scaling is required to ensure convergence if stretches are large.

12.13.2. Proof of the algorithm for general deformation gradients[12]

Our qualitative discussion of the iterative algorithm assumed small strains (i.e., stretches nearly equal to 1). The following proof of the algorithm makes no such assumption. In fact, convergence criteria will come out as a part of the proof.

For use in the proof, define a helper function,

$$\varphi(\underset{\approx}{X}) \equiv \tfrac{1}{2}[3\underset{\approx}{X} - \underset{\approx}{X}^3], \tag{12.81}$$

where $\underset{\approx}{X}$ is any arbitrary real second-order tensor in 3D. Similarly, for scalar arguments, define $\varphi(x) = \frac{1}{2}[3x - x^3]$. Consider the fixed-point iteration

$$\underset{\approx}{A}_{n+1} = \tfrac{1}{2}\underset{\approx}{A}_n \bullet \left[3\underset{\approx}{I} - \underset{\approx}{A}_n^T \bullet \underset{\approx}{A}_n\right]. \tag{12.82}$$

Taking $\underset{\approx}{A}_0 = \underset{\approx}{F}$, consider the first iteration:

$$\begin{aligned}
\underset{\approx}{A}_1 &= \tfrac{1}{2}\underset{\approx}{A}_0 \bullet \left[3\underset{\approx}{I} - \underset{\approx}{A}_0^T \bullet \underset{\approx}{A}_0\right] \\
&= \tfrac{1}{2}\underset{\approx}{F} \bullet [3\underset{\approx}{I} - \underset{\approx}{F}^T \bullet \underset{\approx}{F}] \\
&= \tfrac{1}{2}\underset{\approx}{R} \bullet \underset{\approx}{U} \bullet [3\underset{\approx}{I} - \underset{\approx}{U}^2] \\
&= \underset{\approx}{R} \bullet \varphi(\underset{\approx}{U}) \\
&= \underset{\approx}{R} \bullet \underset{\approx}{U}_1, \text{ where } \underset{\approx}{U}_1 \equiv \varphi(\underset{\approx}{U}).
\end{aligned} \tag{12.83}$$

A standard induction argument for subsequent iterations shows that, in general,

$$\underset{\approx}{A}_k = \underset{\approx}{R} \bullet \underset{\approx}{U}_k \text{ where } \underset{\approx}{U}_k \equiv \varphi(\underset{\approx}{U}_{k-1}), \text{ with in itial condition } \underset{\approx}{U}_0 = \underset{\approx}{U}. \tag{12.84}$$

[12] This elegant and mathematically rigorous proof was provided via email from Dr Ed Love of UC Berkeley, who had seen an early draft of this book on the web. All credit should be ascribed to him.

To prove that $\underset{\approx}{A}_k \to \underset{\approx}{R}$, it suffices to prove that $\underset{\approx}{U}_k \to \underset{\approx}{I}$. Since $\underset{\approx}{U}$ is symmetric, this proof can be performed in the principal basis. In other words, it is sufficient to focus attention on a single eigenvalue (principal stretch) λ of $\underset{\approx}{U}$ and consider the fixed-point iteration $\lambda_{n+1} := \varphi(\lambda_n)$ with the initial condition $\lambda_0 = \lambda$. The convergence properties of this iteration are fully determined by the properties of the function φ:

$$\varphi(\lambda) = \tfrac{1}{2}[3\lambda - \lambda^3] \qquad \Rightarrow \varphi(1) = 1, \tag{12.85}$$

$$\varphi'(\lambda) = \tfrac{1}{2}[3 - 3\lambda^2] \qquad \Rightarrow \varphi'(1) = 0, \tag{12.86}$$

$$\varphi''(\lambda) = -3\lambda \qquad \Rightarrow \varphi''(1) = -3. \tag{12.87}$$

The iteration function $\varphi(\lambda)$ has a fixed-point at $\lambda^* := 1$. Additionally, $\varphi'(\lambda)$ is continuous and $\varphi'(\lambda^*) = 0$. By the mean-value theorem, a sufficient condition for φ to be a contraction is $|\varphi| \leqslant M < 1$. Thus, φ is a contraction in some (maximal) neighborhood of λ^*. By the contraction mapping theorem[13] in this neighborhood, the fixed-point λ^* is unique and

$$\lim_{n\to\infty} \lambda_m = \lambda^* = 1, \tag{12.88}$$

which is the desired result.

In general, the convergence of fixed-point iterations is expected to be only linear. However, since $\varphi'(\lambda^*) = \varphi'(1) = 0$, a Taylor series expansion yields

$$\begin{aligned} |\lambda_{n+1} - \lambda^*| &= |\varphi(\lambda_n) - \lambda^*| = |\varphi(\lambda_n) - \varphi(\lambda^*)| \\ &= |\varphi(\lambda^*) + \varphi'(\lambda^*)[\lambda_n - \lambda^*] + \tfrac{1}{2}\varphi''(\xi)[\lambda_n - \lambda^*]^2 - \varphi(\lambda^*)| \\ &= \tfrac{1}{2}|\varphi''(\xi)| \cdot [\lambda_n - \lambda^*]^2 \approx \tfrac{1}{2}|\varphi''(\lambda^*)| \cdot [\lambda_n - \lambda^*]^2 = \tfrac{3}{2}|\lambda_n - \lambda^*|^2. \end{aligned} \tag{12.89}$$

This shows that the asymptotic convergence rate in this case is quadratic with a limiting error constant of 3/2.

The radius of convergence of the method can be determined by examining when the iteration function φ ceases to be a contraction. As stated previously, a sufficient condition for φ to be a contraction is $|\varphi'(\lambda)| < 1$, which leads to the condition

$$\tfrac{1}{3} < \lambda^2 < \tfrac{5}{3}. \tag{12.90}$$

This shows that the iterative algorithm will converge if all eigenvalues of $\underset{\approx}{C} := \underset{\approx}{F}^T \bullet \underset{\approx}{F}$ fall in the open interval (1/3, 5/3). This is a conservative bound, and it can be shown [13] that the absolute convergence criterion is

$$0 < \lambda^2 < 3. \tag{12.91}$$

[13] See, for example, [12] or any functional analysis textbook.

Stated more physically, material fibers can stretch by as much as 73% (which corresponds to a logarithmic strain of 54%). Below, a scaling is derived that makes the algorithm converge for *any* stretch (which has been confirmed for stretches in the *thousands*).

12.13.3. Extending the convergence to arbitrary stretches via scaling

Consider a scaled tensor $\underset{\approx}{F}^*$ defined to equal the deformation gradient $\underset{\approx}{F}$ divided by a positive factor α:

$$\underset{\approx}{F}^* = \frac{1}{\alpha}\underset{\approx}{F}. \tag{12.92}$$

Just as we may write the polar decomposition

$$\underset{\approx}{F} = \underset{\approx}{R} \bullet \underset{\approx}{U}, \tag{12.93}$$

we may write one for the scaled deformation gradient

$$\underset{\approx}{F}^* = \underset{\approx}{R}^* \bullet \underset{\approx}{U}^*. \tag{12.94}$$

For any $\alpha > 0$, the polar rotation tensor for $\underset{\approx}{F}^*$ will be the same as that of $\underset{\approx}{F}$ and the stretch for $\underset{\approx}{F}^*$ will be the same as that for $\underset{\approx}{F}$ except divided by α. That is,

$$\underset{\approx}{R}^* = \underset{\approx}{R}, \tag{12.95}$$

$$\underset{\approx}{U}^* = \frac{1}{\alpha}\underset{\approx}{U}. \tag{12.96}$$

Consequently, by appropriately scaling the deformation gradient, the previously described iterative algorithm could be applied to $\underset{\approx}{F}^*$ instead of $\underset{\approx}{F}$. Convergence can be guaranteed by selecting α to make the largest eigenvalue of $\underset{\approx}{C}^* \equiv \underset{\approx}{F}^{*T} \bullet \underset{\approx}{F}^* = \underset{\approx}{U}^{*2}$ smaller than 3. An upper bound on the eigenvalues could be computed by using Gershgorin's theorem [14], but we prefer using a bound that can be computed using tensor invariant operations so that the iterative algorithm will continue to be a tensor-based operation instead of simply a matrix operation.

Let $c_{\max}$ stand for the largest eigenvalue of $\underset{\approx}{C} = \underset{\approx}{F}^T \bullet \underset{\approx}{F}$. The largest eigenvalue for $\underset{\approx}{C}^*$ is then $c^*_{\max} = c_{\max}/\alpha^2$. Convergence is guaranteed if $c^*_{\max} < 3$. Therefore, we must select the scaling factor α such that

$$\alpha^2 > c_{\max}/3. \tag{12.97}$$

The scaling condition has a geometrical interpretation in principal-value space. The scaled tensor $\underset{\approx}{C}^*$ has the same principal directions as $\underset{\approx}{C}$. In principal space, the tensor $\underset{\approx}{C}$ may be regarded as a 3D vector $\{c_1, c_2, c_3\}$. Scaling reduces the magnitude of this vector *without changing its orientation.* To guarantee convergence, the scaled vector, $\{c_1^*, c_2^*, c_3^*\}$ must lie within a box in the first octant having sides of length 3 (i.e., the box in figure 12.7 whose major diagonal extends from (0, 0, 0) to (3, 3, 3). Recall that

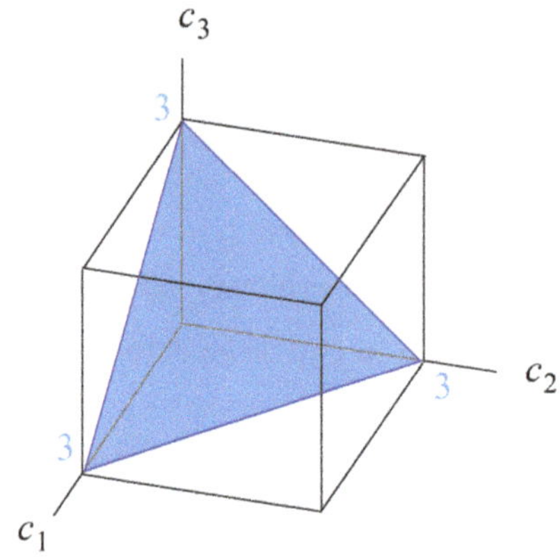

Figure 12.7. A plane that intersects each axis at a value of 3 is the target for scaling a deformation tensor to ensure and accelerate convergence of its iterative polar decomposition.

the iterative algorithm was derived from a Taylor series expansion centered at stretch equal to the identity. Therefore, to optimize convergence, the scaling strategy should be designed so that $\underset{\approx}{C}^* = \underset{\approx}{I}$ whenever $\underset{\approx}{C}$ is an isotropic tensor. A straightforward way to do this is to scale down the $\{c\}$ vector magnitude until it lies on the plane indicated in the above sketch. That plane has a normal pointing in the 111-direction and it passes through the point (1, 1, 1) which corresponds to the identity tensor. Specifically, the scaling can be defined so that $c_i^* = c_i/\alpha^2$ and $c_1^* + c_2^* + c_3^* = 3$. This strategy corresponds to the choice

$$\alpha^2 = \frac{\text{tr}\underset{\approx}{C}}{3} = \frac{\|\underset{\approx}{F}\|^2}{3}. \tag{12.98}$$

This scale factor is computationally inexpensive and has the advantage of being a tensor invariant, which means that the algorithm remains basis-independent. We have experimented with alternative scaling strategies such as scaling down the $\{c\}$ vector to non-planar surfaces within the required $3 \times 3 \times 3$ bounding box (e.g., projecting to a sphere or projecting to a $1 \times 1 \times 1$ box). While these nonlinear scaling strategies can result in fewer iterations for problems involving extremely massive distortions[14] no efficiency advantage exists for typical engineering problems.

Incidentally, based on our timing tests in which the algorithm was applied to 126 million test $\underset{\approx}{F}$ tensors spanning the range of possible principal stretch ratios, we have found that each iteration takes approximately 100 ns of computing time and, as a rule of thumb,

[14] The 'distortion' of a deformation can be here regarded as the ratio of the largest to smallest eigenvalues of the stretch. Alternatively, it can be defined as the magnitude of the stretch deviator. Distortion quantifies changes in *shape*. In typical fluid-flow problems, distortions can reach into the thousands. In solid mechanics, distortions rarely exceed 10 for even the most pliable of materials. In linear elasticity, distortions are typically very close to unity. When we speak of 'massive' distortions, we are speaking of distortions in excess of 10^6, which falls well outside values encountered in normal engineering problems. Geometrically, the closer the $\{c\}$ vector is to one of the axes in principal space (figure 12.7), the higher the material distortion.

$$\text{required number of iterations} = 3 + 5\ \ln x, \tag{12.99}$$

where x is the ratio of the largest to smallest principal stretch. For solid mechanics applications, x is very close to unity and fewer than four iterations is typical. The above rule of thumb is conservative in the neighborhood of $x = 1$. If, for example, the deformation is purely a rotation and a *size* change (i.e., no shape change and therefore all stretch eigenvalues are equal making x identically equal to 1), then the algorithm will converge in 0 iterations—that is, only the scaling operation is needed to immediately determine the rotation tensor when $x \equiv 1$.

References

[1] Chadwick P 1999 *Continuum Mechanics: Concise Theory and Problems* (New York: Dover)
[2] Malvern L E 1969 *Introduction to the Mechanics of a Continuous Medium* (Englewood Cliffs, NJ: Prentice-Hall)
[3] Brannon R M, Aidun J B, Robinson A C and Montgomery S T 2002 Macro- and mesoscale modeling of PZT ferroelectric ceramics *Shock Compression of Condensed Matter - 2001* ed M D Furnish, N N Thadhani and Y Horie (American Physical Society) pp 201–204
[4] Gurtin M E 1981 *An Introduction to Continuum Mechanics* (New York: Academic)
[5] Lu J and Papadopoulos P 1997 On the direct determination of the rotation tensor from the deformation gradient *Math. Mech. Sol.* **2** 17–26
[6] Hoger A and Carlson D E 1984 Determination of the stretch and rotation in the polar decomposition of the deformation gradient *Q. Appl. Math.* **42** 113–7
[7] Sawyer K 1986 Comment on the paper 'Determination of the stretch and rotation in the polar decomposition of the deformation gradient' by A Hoger and D E Carlson *Quart. Appl. Math.* **4** 309–11
[8] Simo J C and Hughes T J R 1998 *Computational Inelasticity* (Springer)
[9] Higham N J 1986 Computing the polar decomposition—with applications *SIAM J. Sci. Stat. Comput.* **7** 1160–74
[10] Jog C S 2002 On the explicit determination of the polar decomposition in *n*-dimensional vector spaces *J. Elasticity* **66** 159–69
[11] Dubrulle A A 1999 An optimum iteration for the matrix polar decomposition *Electron. Trans. Numer. Anal.* **8** 21–25
[12] Ramkrishna D and Amundson N R 1985 *Linear Operator Methods in Chemical Engineering with applications to Transport and Chemical Reaction Systems* (Englewood Cliffs, NJ: Prentice-Hall)
[13] Love E 2002 Personal email communication
[14] Gerald C F and Wheatley P O 1984 *Applied Numerical Analysis* (Reading, MA: Addison-Wesley)

IOP Publishing

Rotation, Reflection, and Frame Changes

Orthogonal tensors in computational engineering mechanics

R M Brannon

Chapter 13

Strain measures

In materials modeling, researchers often speaking of using the 'correct' strain measure. What do they mean by this? We will discuss this question first in the context of one-dimensional (1D) deformations, and then generalize to 3D.

13.1 One-dimensional strain measures

Consider a 1D bar of initial length L_o and current (deformed) length L. The most commonly used strain measures are members of the Seth–Hill family of 1D strain measures:

$$\varepsilon = \frac{1}{k}[\lambda^k - 1], \quad \text{where} \quad \lambda \equiv L/L_o. \tag{13.1}$$

Note that the ratio λ is the uniaxial 'stretch', and is the 1D analog of the stretch tensors $\underset{\approx}{U}$ and $\underset{\approx}{V}$ from the previous chapter. The strain definition is dictated by the user's choice for the parameter k. For example

$$\text{engineering strain:} \quad \varepsilon|_{k=1} = \frac{L - L_o}{L_0}, \tag{13.2}$$

$$\text{'true' strain:} \quad \varepsilon|_{k=-1} = \frac{L - L_o}{L}, \tag{13.3}$$

$$\text{logarithmic (Henky) strain:} \quad \varepsilon|_{k\to 0} \equiv \left(\frac{L}{L_0}\right). \tag{13.4}$$

Some people use the term 'true strain' to mean what we have here called logarithmic (Hencky) strain. A plot of the Seth–Hill strains, for various choices of the parameter k is shown in figure 13.1. Note that all of the strain measures are equivalent for small

doi:10.1016/978-0-7503-1454-1ch13

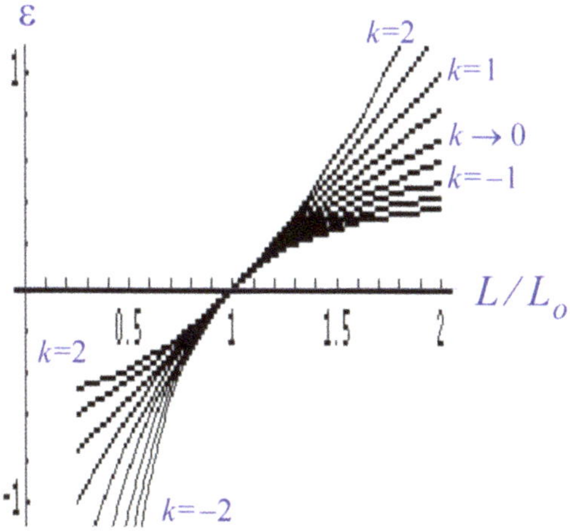

Figure 13.1. Variation of Seth–Hill strain measures with stretch.

strains (i.e., when $\lambda \approx 1$). Also note that the stretch λ must lie between 0 and ∞. Consequently, the allowable range of the Seth–Hill strain is

$$\varepsilon_{\min} = \begin{cases} -\dfrac{1}{k} & \text{if } k > 0, \\ -\infty & \text{if } k \leqslant 0, \end{cases} \tag{13.5}$$

and

$$\varepsilon_{\max} = \begin{cases} +\infty & \text{if } k \geqslant 0, \\ \dfrac{1}{|k|} & \text{if } k < 0. \end{cases} \tag{13.6}$$

Note that the only strain measure that permits a full range of strain variations is the logarithmic strain ($k = 0$).

No one strain measure is 'better' than another in the respect of being able to characterize the deformation. No matter what choice is made for k, it is always possible to solve uniquely for the ratio L/L_o. Specifically, solving (13.1) for the stretch gives

$$\lambda = (k\varepsilon + 1)^{1/k}. \tag{13.7}$$

Once this stretch is known, a different strain measure can be used without any loss of information. Consequently, any choice for the parameter k is merely a matter of convenience—no choice is better than any other as far as characterizing the deformation goes.

Suppose the bar is homogeneously deformed and subjected to a current loading force F. Because the uniaxial bar has an initial length L_o, the mass of the bar is given by $\rho_0 L_0 A_0$. Consequently, the rate of work per unit mass is

$$\dot{W} = \frac{F\dot{L}}{\rho_o L_o A_o} = \frac{F}{\rho_o A_o}\dot{\lambda} = \frac{F}{\rho_o A_o}(k\varepsilon + 1)^{\frac{1}{k}-1}\dot{\varepsilon} = \frac{F}{\rho_o A_o}\lambda^{1-k}\dot{\varepsilon}, \tag{13.8}$$

Let ρ and A denote the *current* mass density and current cross-sectional area. The product ρA is related to the initial value of this product, $\rho_o A_o$, by equating the initial and current material mass, $\rho_o L_o A_o = \rho L A$, from which it follows that

$$\rho A = \lambda(\rho_o A_o), \tag{13.9}$$

and the formula for the rate of work per unit mass can be written in one of two ways:

$$\dot{W} = \frac{F}{\rho_o A_o}\lambda^{1-k}\dot{\varepsilon} = \frac{F}{\rho A}\lambda^{-k}\dot{\varepsilon}. \tag{13.10}$$

The so-called 'conjugate' reference stress σ^r and the conjugate *spatial* stress σ^s associated with a particular choice for the strain measure are defined such that

$$\dot{W} = \frac{1}{\rho_o}\sigma^r\dot{\varepsilon} \qquad \text{and} \qquad \dot{W} = \frac{1}{\rho}\sigma^s\dot{\varepsilon}, \tag{13.11}$$

where

$$\sigma^r = \frac{F}{A_o}\lambda^{1-k} \text{ and } \sigma^s = \frac{F}{A}\lambda^{-k}. \tag{13.12}$$

Note that

$$\frac{\sigma^r}{\rho_o} = \frac{\sigma^s}{\rho}. \tag{13.13}$$

If it is desired to plot a stress–strain diagram such that the area under the curve equals the work of deformation *per unit mass*, it is essential to use one of these stress measures (divided by the appropriate mass density) on the vertical plot axis. To plot a stress–strain diagram such that the area under the curve equals the work per unit *current volume*, then σ^s should be used on the vertical axis. To make the area under the curve equal the work per unit *initial volume*, then σ^r should be used.

If the logarithmic strain ($k = 0$) is used, then the spatial stress is given by the conventional definition of stress (force per current area):

$$\sigma^s = \frac{F}{A}. \tag{13.14}$$

Suppose that the conventional stress (F/A) is known as a function of the logarithmic strain:

$$\frac{F}{A} = f(\ln(\lambda)). \tag{13.15}$$

Then the stress–strain function corresponding to a different ($k \neq 0$) strain measure would be given by

$$\sigma^s = g(\varepsilon), \quad \text{where } g(\varepsilon) = \frac{1}{k\varepsilon + 1} f\left(\frac{\ln(k\varepsilon + 1)}{k}\right). \tag{13.16}$$

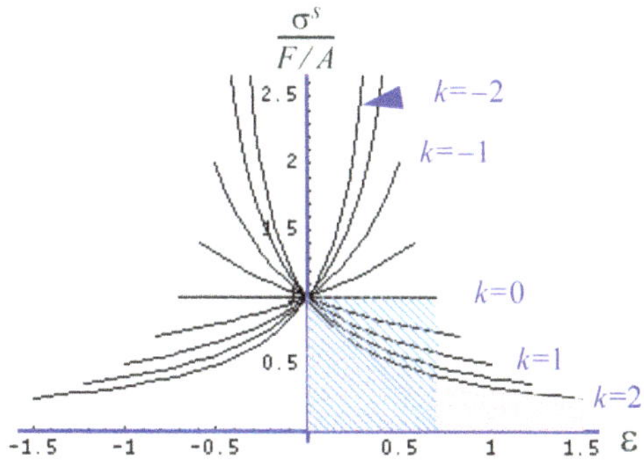

Figure 13.2. Ratio of generalized stress to 'true' stress at various strain values.

This shows that switching to a different strain measure will result in significant distortion of the stress–strain diagram. Figure 13.2 illustrates this effect by plotting the ratio of σ^s to 'true stress' as a function of the conjugate strain. This plot shows that, for $k > 0$, the generalized stress value at a given strain will become increasing smaller than the true stress (F/A) as the strain is increased in tension. This plot shows strains corresponding to a the *stretch* ranging from 0.5 to 2.0. Even though the stress value is smaller in tension when $k > 0$, note that the strain range is larger. Therefore, the shaded areas in figure 13.2 are both equal.

The logarithmic strain measure is certainly appealing in the sense that its conjugate stress happens to be spatial and identical to the familiar definition in which stress is simply force divided by area for 1D problems. However, thermodynamic formulations of constitutive laws [1] predict that it is the *reference* stress measure that is related to a derivative of an energy function. Specifically, if $u(\varepsilon, s)$ denotes the energy per unit mass, taken to depend on the logarithmic strain ε and entropy s, then a classical thermoelastic definition for the stress is

$$\sigma^s = \rho\frac{\partial u}{\partial \varepsilon}. \tag{13.17}$$

The presence of the density implies that it is the specific stress σ^s/ρ (not stress itself) that is given by the derivative of an energy potential. Stress has units of energy per volume whereas u is energy per mass. One could introduce an energy per *current* volume $\tilde{u} \equiv \rho u$, for which (13.17) would become $\sigma^s = \frac{\partial \tilde{u}}{\partial \varepsilon} - \tilde{u}$, which again shows that σ^s is not the derivative of a potential. One could introduce an alternative energy per *reference* volume $\hat{u} \equiv \rho_o u$ for which (13.17) would become

$$J\sigma^s = \frac{\partial \hat{u}}{\partial \varepsilon}, \quad \text{where } J = \frac{\rho_o}{\rho}. \tag{13.18}$$

Again, σ^s is not equal to the derivative of a potential. Touching back on the distinction between spatial and reference stress measures defined in (13.13), note that

$$\sigma^r = J\sigma^s. \tag{13.19}$$

Thus, it is the *reference* stress σ^r, not the spatial stress σ^s, that is derivable from a potential. These results arise because thermodynamics applies to closed systems where mass is constant (or, if using an Eulerian control volume, terms involving mass flux of thermodynamic state variables must be included that make the formulation functionally equivalent to a closed system). For large deformations, the density cannot be presumed constant. Therefore, the stress that appears in thermodynamic potentials is *specific* stress (equal to stress divided by density).

What do researchers mean when they speak of 'the correct' strain measure? The use of the words 'the' and 'correct' is unfortunate. Actually, a strain measure is typically considered superior if using it makes the constitutive model as simple in structure as possible. Suppose, for example, that you have some nonlinear elasticity data giving stress σ as a function of the stretch $\lambda = L/L_o$ so that $\sigma = \bar{\sigma}(\lambda)$. Naturally, for any elastic model (nonlinear or not), the stress is zero when $L = L_o$, or when $\lambda = 1$. Suppose that the stress is alternatively written in terms of one of the Seth–Hill strain measures so that $\sigma = \sigma(\varepsilon)$. Then we know that the two functions $\bar{\sigma}$ and $\hat{\sigma}$ are related by $\bar{\sigma}(\lambda) = \hat{\sigma}(\varepsilon(\lambda))$. By the chain rule,

$$\frac{d\bar{\sigma}}{d\lambda} = \frac{d\hat{\sigma}d\varepsilon}{d\varepsilon d\lambda} = \lambda^{k-1}\frac{d\hat{\sigma}}{d\varepsilon} \tag{13.20}$$

and

$$\frac{d^2\bar{\sigma}}{d\lambda^2} = k\lambda^{k-1}\frac{d^2\hat{\sigma}}{d\varepsilon^2} + (k-1)\lambda^{k-2}\frac{d\hat{\sigma}}{d\varepsilon}. \tag{13.21}$$

When evaluated at $\lambda = 1$, these give

$$\bar{\sigma}(1) = \hat{\sigma}(0), \tag{13.22}$$

$$\bar{\sigma}'(1) = \hat{\sigma}'(0), \tag{13.23}$$

$$\bar{\sigma}''(1) = k\hat{\sigma}''(0) + (k-1)\hat{\sigma}'(0). \tag{13.24}$$

Consider a Taylor series expansion for the stress as a function of the strain, centered about $\varepsilon = 0$:

$$\begin{aligned}\sigma &= \sigma(0) + \hat{\sigma}'(0)\varepsilon + \tfrac{1}{2}\hat{\sigma}''(0)\varepsilon^2 + \ldots \\ &= \bar{\sigma}(1) + \bar{\sigma}'(1)\varepsilon + \frac{1}{2}\left[\frac{\bar{\sigma}''(1) - (k-1)\bar{\sigma}'(1)}{k}\right]\varepsilon^2 + \ldots\end{aligned} \tag{13.25}$$

The coefficient $\bar{\sigma}'(1)$ is the low-strain elastic modulus. Although rarely stated explicitly in the literature, 'the correct' strain measure is implicitly regarded as the one for which the strain expansion has the highest accuracy in low-order approximations. The above expansion in strain will agree to second order with the stress–strain data only if the coefficient of ε^2 in (13.25) is zero. Setting this coefficient to zero and solving for k gives

$$k = \frac{\bar{\sigma}''(1)}{\bar{\sigma}'(1)} + 1. \tag{13.26}$$

Applying this formula to actual stress–strain data will tell you 'the best' choice for the Seth–Hill parameter k. Other choices for k remain legitimate, but they will require nonlinear elasticity to match data to the same order of accuracy as a model that uses the 'preferred' value of k given in (13.26).

13.2 Three-dimensional strain definitions

The generalization of the Seth–Hill strain measure to 3D deformation states is

$$\bar{\underset{\approx}{\varepsilon}} = \frac{1}{k}(\underset{\approx}{U}^k - \underset{\approx}{I}). \tag{13.27}$$

This is a reference strain. Of course, a *spatial* strain can also be defined,

$$\underset{\approx}{\varepsilon} = \frac{1}{k}(\underset{\approx}{V}^k - \underset{\approx}{I}). \tag{13.28}$$

Regardless of whether you use a reference or spatial strain measure, a polar rotation tensor is necessary to evaluate a constitutive law whenever the material is anisotropic. Clearly, the polar rotation would be needed when using a reference constitutive model because the output of such a model is a reference stress, and rotation is needed to translate it back to a spatial result. The polar rotation would be needed in a spatial material model to rotate the initial anisotropy directions into the spatial configuration. The rotation tensor might not appear *explicitly* if transformations of material directions are obtained by using the deformation gradient tensor $\underset{\approx}{F}$, but $\underset{\approx}{F}$ *implicitly* contains the material rotation. Our point is that knowledge of $\underset{\approx}{V}$ alone is not sufficient to evaluate an anisotropic material model. Some other tensor, which characterizes rotation, must be involved as well.

Reference

[1] McLellan A G 1980 *The Classical Thermodynamics of Deformable Materials* (London: Cambridge University Press)

IOP Publishing

Rotation, Reflection, and Frame Changes
Orthogonal tensors in computational engineering mechanics
R M Brannon

Chapter 14

Remapping, advecting, or interpolating rotations

Eulerian physics codes must routinely handle the task of 'mixing' two or more materials (or the same material in different states) in a computational cell and assigning reasonable homogenized values for the field variables of the mixture. In the absence of specific arguments to the contrary, Eulerian codes typically assign field variable values by simple mass or volume averaging. This approach, however, can be unsatisfactory when mixing rotation tensors because the sum of two orthogonal tensors is not itself generally orthogonal.

A similar problem occurs for Lagrangian codes after remapping of the mesh. Even before remeshing, physically admissible methods are needed for continuous interpolations between field values at discrete locations (nodes or element centers) to any other point in the computational domain. While most field variables are adequately approximated using standard finite-element shape functions (e.g., linear interpolation), special methods are needed to interpolate *rotations* if it is desired that the continuous approximation be a proper rotation everywhere.

In materials modeling, true 'mixing' of rotations (i.e., averages that do *not* preserve orthogonality) might be preferred. For example, single-crystal metals are anisotropic, but a large sample of polycrystalline metal is often isotropic because it contains numerous crystals at uniformly random orientations. Rolled metal has anisotropy resulting from non-uniformly distributed crystal orientations. The overall texture of a polycrystalline material can be characterized by the 'mixed' value—i.e., the arithmetic average—of the crystal orientation tensors. As a different example, the polarization of ferroelectric materials is determined by the net dipole moment per unit volume. A lab-scale sample of ferroelectric material will contain both grains of polarized material, each with the microscale dipole moment expressible in terms of a rotation or orientation tensor. In this case, the macroscale polarization can be determined by using the arithmetic mean of the orientation tensors, and this must evaluate to the zero tensor (which is definitely not itself

doi:10.1016/978-0-7503-1454-1ch14

orthogonal) to obtain the proper result of no net polarization whenever subcontinuum dipole vectors are uniformly distributed. In these and many other materials modeling applications, it is *not* desirable for a mixture of rotations to be itself a rotation. The point here is that methods for handling mixtures of rotations must be selected appropriate to the physical application.

The remainder of this chapter focuses on the advection/remapping problem in which an interpolated value of a rotation is desired to be itself a rotation. In finite-element codes, advection of a field variable ξ (scalar, vector, or tensor) is typically governed by a partial differential equation (PDE)

$$\dot{\xi} = 0, \quad \text{or} \quad \left(\frac{\partial \xi}{\partial t}\right)_{\underset{\sim}{x}^0} = 0. \tag{14.1}$$

Here, the superposed dot denotes the material 'time' rate where time t actually parameterizes the virtual movement of the grid such that a point $\underset{\sim}{x}^0$ on the pre-mapped grid moves to location $\underset{\sim}{x}(\underset{\sim}{x}^0, t)$ on the new mesh. Taking the independent variables to instead be $\underset{\sim}{x}$ and t, equation (14.1) expands via the chain rule to the following PDE:

$$\frac{\partial \xi(\underset{\sim}{x}, t)}{\partial t} + \frac{\partial \xi(\underset{\sim}{x}, t)}{\partial \underset{\sim}{x}} \bullet \frac{\partial \underset{\sim}{x}(\underset{\sim}{x}^0, t)}{\partial t} = 0, \tag{14.2}$$

which is solved numerically to remap ξ to the new grid. Equation (14.1) makes it clear that (14.2) is merely ensuring that no state variables should change at any material particle, which would be an unacceptable consequence of simply moving a mesh. Unfortunately, because this PDE is solved numerically, this approach can lead to non-orthogonal rotations and non-positive-definite stretches upon remapping. Adjustments are needed to serve the needs of material or kinematics subroutines that require a proper-orthogonal tensor as input. Below we list possible solutions to this advection problem. The list begins with the method that we believe will give the best results, and the list then works down to less attractive (but potentially still acceptable) solutions. These proposed methods have not yet been studied in the literature with sufficient depth to elucidate their advantages and disadvantages for both constrained and unconstrained mixing (where 'unconstrained' refers to an average of rotations being allowed to not be a rotation itself, as is appropriate when computing effective properties of a composite), but it seems nevertheless worthwhile avoid duplicated efforts by cataloging the many suggestions and issues that have been discussed over years of industry and research practice in computational mechanics.

14.1 Proposal 1: Map and re-compute the polar decomposition

The *initial* position vector $\underset{\sim}{X}$ may be regarded as a field in its own right. In terms of the mapping function of (12.1), it is given by

$$\underset{\sim}{X} = \chi^{-1}(\underset{\sim}{x}). \tag{14.3}$$

The vector $\underset{\sim}{x}$ may be advected just like any other field. This is particularly easy to do since the *Lagrangian* time rate of $\underset{\sim}{X}$ is always zero. Thus, the value of $\underset{\sim}{X}$ at a Lagrangian tracer particle or at a Lagrangian mesh point remains fixed over time.

For Lagrangian finite-element calculations, the value of $\underset{\sim}{X}$ at any point p may always be determined by using the shape functions. When it is determined that the mesh should be moved in a *non*-Lagrangian manner (as for remeshing or for arbitrary Lagrangian–Eulerian calculations), the value of $\underset{\sim}{X}$ can be mapped to the new locations of the nodes just as any other variable would be mapped to the nodes.

In short, the field for $\underset{\sim}{X}$ is very easy to monitor for all time and should be straightforward to remap using standard methods. With a remapped $\underset{\sim}{X}$ field available, the polar rotation and stretch can be re-computing at any location by performing a polar decomposition on the *inverse* deformation gradient tensor. Namely,

$$\underset{\approx}{F}^{-1} = \frac{\partial \underset{\sim}{X}}{\partial \underset{\sim}{x}} \Rightarrow F_{ij}^{-1} = \frac{\partial X_i}{\partial xj} \Rightarrow [F]^{-1} = \begin{bmatrix} \frac{\partial X_1}{\partial x_1} & \frac{\partial X_1}{\partial x_2} & \frac{\partial X_1}{\partial x_3} \\ \frac{\partial X_2}{\partial x_1} & \frac{\partial X_2}{\partial x_2} & \frac{\partial X_2}{\partial x_3} \\ \frac{\partial X_3}{\partial x_1} & \frac{\partial X_3}{\partial x_2} & \frac{\partial X_3}{\partial x_3} \end{bmatrix}. \tag{14.4}$$

One could always invert this matrix to obtain the new deformation gradient, but that would entail unnecessary computational cost. It is more efficient to work directly with $\underset{\approx}{F}^{-1}$. Recall that

$$\underset{\approx}{F} = \underset{\approx}{R} \bullet \underset{\approx}{U} = \underset{\approx}{V} \bullet \underset{\approx}{R}. \tag{14.5}$$

Noting that $\underset{\approx}{R}^{-1} = \underset{\approx}{R}^T$, taking the inverse-transpose gives

$$\underset{\approx}{F}^{-T} = \underset{\approx}{R} \bullet \underset{\approx}{U}^{-1} = \underset{\approx}{V}^{-1} \bullet \underset{\approx}{R}. \tag{14.6}$$

Thus, by sending $\underset{\approx}{F}^{-T}$ to a polar-decomposition routine, the rotation returned for $\underset{\approx}{F}^{-T}$ will be *the same* as the rotation for $\underset{\approx}{F}$. The stretch returned for $\underset{\approx}{F}^{-T}$ will be the inverse of the stretch for $\underset{\approx}{F}$.

This proposal is only a conjecture. It still needs to be tested. What is really needed here are 'constrained transport' algorithms for the inverse deformation gradient. This option essentially advects the material coordinates, but it is the deformation gradient that is of paramount interest. This problem is similar to the issue of advection of divergence-free magnetic flux except that here one wants to advect the curl-free inverse deformation gradient. Standard algorithms for advection of $\underset{\sim}{X}$ might work. However, algorithms that use the gradient need to be relatively insensitive to non-monotonic features in the gradients. This method might not be a good choice, for example, in a turbulence code.

14.2 Proposal 2: Discard the 'stretch' part of a mixed rotation

A second technique for directly advecting or interpolating rotations simply applies existing advection methods to advect or interpolate individual rotation tensor components as if they were scalars, thereby producing a 'not-quite-orthogonal' estimate $\underset{\approx}{A}$ for the advected rotation tensor. You could then project the non-orthogonal tensor $\underset{\approx}{A}$ to the 'nearest' orthogonal tensor. This would involve a minimization calculation. Namely, given a tensor $\underset{\approx}{A}$ (the nonorthogonal mixed rotation), find the tensor $\underset{\approx}{R}$ such that $\underset{\approx}{R}^T \bullet \underset{\approx}{R} = \underset{\approx}{I}$ and $\|\underset{\approx}{R} - \underset{\approx}{A}\|$ is minimized. As discussed in section 12.8, the polar rotation tensor $\underset{\approx}{R}$ corresponding to the tensor $\underset{\approx}{A}$ is the solution to this minimization problem [1]. Consequently, this proposal entails a conventional advection/remapping of orthogonal tensors followed by a correction step in which the polar decomposition is used to discard the spurious (numerical) stretches.

With this proposal, the tensor $\underset{\approx}{A}$ is almost (but not quite) orthogonal. Specifically, the largest eigenvalue of $\underset{\approx}{I} - \underset{\approx}{A}^T \bullet \underset{\approx}{A}$ is likely to be significantly smaller than unity. Therefore, the iterative Higham polar-decomposition algorithm section 12.12.2 is a good choice. Typically, one or two iterations will suffice.

14.3 Proposal 3: Advect the pseudo-rotation vectors

Another strategy for mixing rotation tensors is to use a conventional mass or volume average of a rotation *spinor vector* and to then convert the average rotation vector back into a rotation tensor. As discussed in section 5.2.8, there are myriad ways to define a spinor vector to characterize a rotation. The Cayley transformation described in section 15.9 is simply one instance of the 'spinor mixing' approach. The algorithm is as follows:

STEP 1. For each rotation tensor $\underset{\approx}{R}^{(k)}$ in the mixture, use any of the methods in section 5.8.2 to compute your favorite spinor vector $\underset{\sim}{b}^{(k)}$.

STEP 2. Use ordinary volume or mass averaging to compute the average of the spinors.

STEP 3. Using the averaged spinor, reconstruct the associated mixture rotation tensor $\underset{\approx}{R}$ by applying (5.74).

The above scheme for averaging rotations basically assumes that all the individual rotations are applied simultaneously, as discussed in section 15.8. The fact that angle and axis are not unique is likely to cause problems if this method is applied to a large-deformation problem.

14.4 Proposal 4: Mix the quaternions

Some researchers advocate computing the quaternions for each rotation, which may be interpreted as points on a four-dimensional unit sphere. The normalized vector sum of the rotation quaternions would probably do a decent job at assigning a mixed rotation. This method is closely related to proposal 3, and subject to similar pitfalls. Recalling that a rotation is a point on a 3D surface of a hypersphere embedded in a 4D hypercomplex space, mixing of quaternions is typically regarded as finding a point on a great hypercircle connecting the two endpoint rotations. The operation can be performed without quaternions using the method of rotating tensors in their own spanning space (see section 27.5 with the two unit tensors replaced by the two normalized rotations to be interpolated). Since a linear combination of two rotations is not generally itself a rotation, this method might require a polar decomposition after the initial mixing operation.

14.5 Advection enhancement strategy 1: Solve the compatibility equations

Regardless of which of the above proposals is adopted, further improvement on the solution is possible through various advection enhancement strategies. Instead of using (14.2), an alternative or perhaps supplemental approach might be to solve the PDEs of compatibility, which are merely conditions that must be satisfied by a strain field to ensure that it arises from a single-valued displacement field.

When phrased in terms of strain, the compatibility equations are nightmarishly complicated for large deformation problems. However, when cast in terms of the *deformation gradient*, they are intoxicatingly simple. Specifically, letting $\underset{\approx}{\boldsymbol{F}}$ denote $\frac{\partial \underset{\sim}{\boldsymbol{x}}}{\partial \underset{\sim}{\boldsymbol{X}}}$, a single-valued mapping field is ensured to exist if and only if

$$\frac{\partial F_{ij}}{\partial X_k} = \frac{\partial F_{ik}}{\partial X_j} \tag{14.7}$$

(of course, to be an invertible mapping, non-zero det $\underset{\approx}{\boldsymbol{F}}$ must still be enforced as well). By solving (14.7) numerically, a compatible mapping function can be obtained for the remeshed state. From there, the deformation gradient can be re-computed, and re-mapped rotation tensors can be obtained from a polar decomposition. When performed on an incremental level, $\underset{\sim}{\boldsymbol{X}}$ would be replaced by the pre-mapped position vector, $\underset{\sim}{\boldsymbol{x}}^0$. If it is desired to perform spatial gradients instead of reference gradients, the above equation can be replaced by

$$\frac{\partial F_{ij}}{\partial x_m} F_{mk} = \frac{\partial F_{ik}}{\partial x_n} F_{nj}. \tag{14.8}$$

Deformation gradient regularization that specifically aims to enforce compatibility (when it has been lost because of numerical truncation error) effectively advects errors in the compatibility equation out through the boundaries of the physical domain [2]!

14.6 Mixing enhancement strategy 2: Lagrangian tracers

Even if one or more of the above suggestions for mixing rotations proves to advance the state-of-the-art, none of these suggestions will continue to work well indefinitely. Remeshing (and Eulerian strategies) are used in the first place because material deformation is so severe that it produces tangled meshes (concave elements or even negative element Jacobians associate with element inversions). If the deformation continues to be severely distortional (as in fluid flow problems), then the *exact* solution for the mapping function will itself become increasingly complicated. The *exact* solution might, for example, require distortion of an initially cubic infinitesimal material element into filament-like 'tendrils' that intertwine with material from neighboring infinitesimal element as when mixing two paints of different colors. Non-infinitesimal (i.e., finite) elements cannot capture these effects. Sub-grid scale phenomena *cannot be modeled explicitly with a grid that is larger in size than the characteristic size of the material motion.* Consequently, massive material deformation would need to be accommodated through, say, turbulence models. From a practical perspective, these models are treated by the host code as if they were constitutive models.

Even if you seek only the 'bulk' effects and are willing to 'smear out' sub-grid scale deformations, there is no guarantee that proposal 1 (mapping the initial position vector $\underset{\sim}{X}$ to a new grid) or any of the other proposals will produce a deformation gradient field with a positive Jacobian everywhere. Moreover, even if a robust and acceptably accurate scheme is devised to compute mixed deformation gradients and rotations, there remains the larger and related problem of mixing internal state variables for constitutive models. Just as the rotation tensor is required to be orthogonal, constitutive internal state variables routinely have physical restrictions—especially for history dependence and anisotropy. As a simple example, an anisotropic plasticity model might employ an orthogonal tensor internal state variable to represent the polar rotation of the elastic lattice (which is different from the bulk rotation tensor) — how would the host code handle that? It is unrealistic to expect host code architects to devise, implement, and support mixing rules for each and every material constitutive variable.

Assuming you would be content with adequate 'bulk' descriptions of deformation and material constitutive response, then an attractive solution is to embed within the conventional finite-element grid a set of Lagrangian tracer particles at which all field variables (including the deformation gradient) are stored—and at which constitutive models are evaluated. The finite-element grid would be used only in an incremental sense to solve field equations. The fields used during these solution phases would be computed by mapping the Lagrangian particle information to the grid at each time step. Numerous computationally inexpensive strategies such as moving least squares [3] and the material point method [4, 5], as well as more expensive but highly accurate interpolant methods [6], already exist for such mappings and have been used in broad physics applications (mechanics, magnetohydrodynamics, etc) that would have been impossible using only traditional finite-element and finite difference methods. With a tracer-particle approach, there is no need to mix *any* quantity —not even rotations. Particle methods are a natural choice for history-dependent fluids simulations in which traditional Eulerian methods are problematic in assigning memory kernel values to mixtures. Solid mechanics problems involving severe material deformations are essentially history-dependent fluids problems from the mapping perspective and should be treated as such. Further, solid mechanics problems that are severe enough to produce fracturing and localization in unpredetermined locations are prime candidates for solution using particle methods.

References

[1] Gurtin M E 1981 *An Introduction to Continuum Mechanics* (New York: Academic)

[2] Miller G and Colella P 2001 A high-order Eulerian Godunov method for elastic-plastic flow in solids *J. Comput. Phys.* **167** 131–76

[3] Lancaster P and Salkauskas K 1981 Surfaces generated by moving least squares methods *Math. Comput.* **37** 141–58

[4] Brackbill J U 1988 A low dissipation, particle-in-cell method for fluid flow *Comput. Phys. Commun.* **48** 25–38

[5] Brackbill J U 1991 FLIP MHD: A particle-in-cell method for magnetohydrodynamics *J. Comput. Phys.* **96** 163–92

[6] Belikov V V, Ivanov V D, Kontorovich V K, Korytnik S A and Semenov A Y 1997 The non-Sibsonian interpolation: a new method of interpolation of the values of a function on an arbitrary set of points *Comput. Math. Math. Phys.* **37** 9–15

Chapter 15

Rates and other derivatives of rotation

15.1 The 'spin' tensor

Consider a vector $\underset{\sim}{b}$ that is embedded in a rigidly rotating body. Let $\bar{\underset{\sim}{b}}$ denote its initial value. Then

$$\underset{\sim}{b} = \underset{\approx}{R} \bullet \bar{\underset{\sim}{b}}. \tag{15.1}$$

Suppose that the rotation of the body is constantly changing. Then $\underset{\approx}{R} = \underset{\approx}{R}(t)$. Note that $\bar{\underset{\sim}{b}}$ is not a function of time (it is simply the *initial* value, which is forever constant). Thus, the time rate of the embedded vector $\underset{\sim}{b}$ is

$$\dot{\underset{\sim}{b}} = \dot{\underset{\approx}{R}} \bullet \bar{\underset{\sim}{b}}. \tag{15.2}$$

From (15.1), we know that $\bar{\underset{\sim}{b}} = \underset{\approx}{R}^{-1} \bullet \underset{\sim}{b} = \underset{\approx}{R}^T \bullet \underset{\sim}{b}$. Substituting this result in (15.2) gives

$$\dot{\underset{\sim}{b}} = \underset{\approx}{\Omega} \bullet \underset{\sim}{b}, \tag{15.3}$$

where the second-order tensor $\underset{\approx}{\Omega}$ is called the *spin tensor*[1] and is defined

$$\underset{\approx}{\Omega} \equiv \dot{\underset{\approx}{R}} \bullet \underset{\approx}{R}^T. \tag{15.4}$$

For later use, this implies that $\dot{\underset{\approx}{R}}$ may be expressed in terms of $\underset{\approx}{\Omega}$ by

$$\dot{\underset{\approx}{R}} = \underset{\approx}{\Omega} \bullet \underset{\approx}{R}. \tag{15.5}$$

Consider the orthogonality condition

[1] If the rotation tensor $\underset{\approx}{R}$ happens to be the *polar* rotation, the angular velocity $\underset{\approx}{\Omega}$ is sometimes called the *polar spin*. For problems involving material distortion, the polar spin tensor $\underset{\approx}{\Omega}$ must not be confused with the vorticity tensor $\underset{\approx}{W}$ (see section 15.5), which some writers call spin too.

doi:10.1016/978-0-7503-1454-1ch15

$$\underset{\approx}{R} \bullet \underset{\approx}{R}^T = \underset{\approx}{I}. \tag{15.6}$$

Taking the time rate gives:

$$\dot{\underset{\approx}{R}} \bullet \underset{\approx}{R}^T + \underset{\approx}{R} \bullet \dot{\underset{\approx}{R}}^T = \underset{\approx}{0}. \tag{15.7}$$

Substituting (15.5) into (15.7) and imposing (15.6) shows that

$$\underset{\approx}{\Omega} + \underset{\approx}{\Omega}^T = \underset{\approx}{0}. \tag{15.8}$$

In other words, the spin tensor is skew-symmetric. Consequently, it has an associated axial vector $\underset{\sim}{\omega}$, which is defined such that (15.3) becomes

$$\dot{\underset{\sim}{b}} = \underset{\sim}{\omega} \times \underset{\sim}{b}. \tag{15.9}$$

15.2 The angular velocity vector

Just as rotations are associated with an axis and angle of rotation, a spin tensor $\underset{\approx}{\Omega}$ also has an associated angular velocity vector, which is given by the so-called 'axial' or 'dual' vector:

$$\underset{\sim}{\omega} = -\tfrac{1}{2}\underset{\approx}{\varepsilon}{:}\underset{\approx}{\Omega} \qquad \text{or} \qquad \omega_i = -\tfrac{1}{2}\varepsilon_{ijk}\Omega_{jk} = \tfrac{1}{2}\varepsilon_{jik}\Omega_{jk}. \tag{15.10}$$

The operation is reversible. Given the angular velocity vector, the associated angular velocity tensor is computed by

$$\underset{\approx}{\Omega} = -\underset{\approx}{\varepsilon} \bullet \underset{\sim}{\omega} \qquad \text{or} \qquad \Omega_{ij} = -\varepsilon_{ijk}\omega_k = \varepsilon_{ikj}\omega_k. \tag{15.11}$$

The explicit relationships are

$$\omega_1 = -\Omega_{23} = \Omega_{32}, \ \omega_2 = -\Omega_{31} = \Omega_{13}, \ \omega_3 = -\Omega_{12} = \Omega_{21}. \tag{15.12}$$

The matrix relationship is

$$[\underset{\approx}{\Omega}] = \begin{bmatrix} 0 & -\omega_3 & \omega_2 \\ \omega_3 & 0 & -\omega_1 \\ -\omega_2 & \omega_1 & 0 \end{bmatrix}. \tag{15.13}$$

Dotting the spin tensor $\underset{\approx}{\Omega}$ onto an arbitrary vector $\underset{\sim}{u}$ is equivalent to crossing the angular velocity vector $\underset{\sim}{\omega}$ into $\underset{\sim}{u}$. That is,

$$\underset{\approx}{\Omega} \bullet \underset{\sim}{u} \equiv \underset{\sim}{\omega} \times \underset{\sim}{u}. \tag{15.14}$$

As a matter of fact, the angular velocity vector was constructed specifically so that this relationship would hold. By writing $\underset{\sim}{u} = \underset{\approx}{I} \bullet \underset{\sim}{u}$, the right side of this formula is $\underset{\sim}{\omega} \times \underset{\sim}{u} = \underset{\sim}{\omega} \times \underset{\approx}{I} \bullet \underset{\sim}{u}$, from which it follows that

$$\boxed{\underset{\approx}{\Omega} = \underset{\sim}{\omega} \times \underset{\approx}{I}}. \tag{15.15}$$

Final question: what relationship, if any, does the angular velocity vector have with the angular rotation vector? If the rotation axis $\underset{\sim}{a}$ is held fixed, then $\underset{\sim}{\omega} = \alpha\underset{\sim}{a}$. However, if the axis of rotation is a function of time, the relationship between the rotation vector and the angular velocity vector is highly nonlinear due to the presence of Coriolis-type terms, as discussed in the next section.

15.3 Angular velocity in terms of axis and angle of rotation

Taking the time rate of (5.17) gives $\dot{\underset{\approx}{R}}$ as

$$\dot{\underset{\approx}{R}} = -s\dot{\alpha}(\underset{\approx}{I} - \underset{\sim}{a}\underset{\sim}{a}) - c\dot{\alpha}\underset{\approx}{\varepsilon} \bullet \underset{\sim}{a} + (1 - c)[\dot{\underset{\sim}{a}}\underset{\sim}{a} + \underset{\sim}{a}\dot{\underset{\sim}{a}}] - s\underset{\approx}{\varepsilon} \bullet \dot{\underset{\sim}{a}}, \tag{15.16}$$

where $c \equiv \cos\alpha$ and $s \equiv \sin\alpha$. The transpose of (5.17) is

$$\underset{\approx}{R}^T = c(\underset{\approx}{I} - \underset{\sim}{a}\underset{\sim}{a}) + \underset{\sim}{a}\underset{\sim}{a} + s\underset{\approx}{\varepsilon} \bullet \underset{\sim}{a}. \tag{15.17}$$

After much simplification[2], the spin ($\underset{\approx}{\Omega} \equiv \dot{\underset{\approx}{R}} \bullet \underset{\approx}{R}^T$) is given by

$$\underset{\approx}{\Omega} = \dot{\alpha}\underset{\approx}{A} + c(1 - c)[\underset{\sim}{a}\dot{\underset{\sim}{a}} - \dot{\underset{\sim}{a}}\underset{\sim}{a}] + s\dot{\underset{\approx}{A}}, \tag{15.18}$$

where

$$\underset{\approx}{A} = -\underset{\approx}{\varepsilon} \bullet \underset{\sim}{a}, \qquad \text{and} \qquad \dot{\underset{\approx}{A}} = -\underset{\approx}{\varepsilon} \bullet \dot{\underset{\sim}{a}}. \tag{15.19}$$

Taking the axial vector of (15.18) gives the angular velocity vector as

$$\boxed{\underset{\sim}{\omega} = \dot{\alpha}\underset{\sim}{a} + c(1 - c)(\dot{\underset{\sim}{a}} \times \underset{\sim}{a}) + s\dot{\underset{\sim}{a}}}, \tag{15.20}$$

where (recall) $c = \cos\alpha$ and $s = \sin\alpha$.

15.4 Derivatives of rotation with respect to angle and axis

By the chain rule,

$$\dot{\underset{\approx}{R}} = \left(\frac{\partial \underset{\approx}{R}}{\partial \alpha}\right)_{\underset{\approx}{A}} \dot{\alpha} + \left(\frac{\partial \underset{\approx}{R}}{\partial \underset{\approx}{A}}\right)_{\alpha} : \dot{\underset{\approx}{A}}. \tag{15.21}$$

Recall the version of the Euler–Rodrigues formula given in (5.65):

$$\underset{\approx}{R} = \underset{\approx}{I} + s\underset{\approx}{A} + (1 - c)\underset{\approx}{A}^2, \tag{15.22}$$

where s and c are the sine and cosine, respectively. The rate of this equation is

$$\dot{\underset{\approx}{R}} = \dot{\alpha}(c\underset{\approx}{A} + s\underset{\approx}{A}^2) + s\dot{\underset{\approx}{A}} + (1 - c)(\underset{\approx}{A} \bullet \dot{\underset{\approx}{A}} + \dot{\underset{\approx}{A}} \bullet \underset{\approx}{A}). \tag{15.23}$$

From which it follows that

---

[2] It is critical to note that $\underset{\sim}{a} \bullet \underset{\sim}{a} = 1$ and therefore $\dot{\underset{\sim}{a}} \bullet \underset{\sim}{a} = 0$.

$$\left(\frac{\partial \underset{\approx}{\mathbf{R}}}{\partial \alpha}\right)_{\underset{\approx}{\mathbf{A}}} = (\cos\alpha)\underset{\approx}{\mathbf{A}} + (\sin\alpha)\underset{\approx}{\mathbf{A}}^2 \tag{15.24}$$

and

$$\begin{aligned}\left(\frac{\partial R_{ij}}{\partial A_{mn}}\right)_{\alpha} &= \frac{s}{2}(\delta_{im}\delta_{jn} - \delta_{in}\alpha_{jm}) \\ &+ \frac{(1-c)}{2}[\delta_{im}A_{nj} + A_{im}\delta_{jn} - \delta_{in}A_{mj} - A_{in}\delta_{jm}].\end{aligned} \tag{15.25}$$

Recalling that $\underset{\approx}{\mathbf{A}}$ is skew-symmetric and $\underset{\approx}{\mathbf{A}}^2$ is symmetric, note that

$$\underset{\approx}{\mathbf{R}}^T = \underset{\approx}{\mathbf{I}} - s\underset{\approx}{\mathbf{A}} + (1-s)\underset{\approx}{\mathbf{A}}^2. \tag{15.26}$$

Recall that $\underset{\approx}{\mathbf{A}}^3 = -\underset{\approx}{\mathbf{A}}$ and $\underset{\approx}{\mathbf{A}}^4 = -\underset{\approx}{\mathbf{A}}^2$. Furthermore, it is straightforward to show that

$$\underset{\approx}{\mathbf{R}} \bullet \underset{\approx}{\mathbf{A}} = \underset{\approx}{\mathbf{A}} \bullet \underset{\approx}{\mathbf{R}}, \tag{15.27}$$

and therefore

$$\left(\frac{\partial \underset{\approx}{\mathbf{R}}}{\partial \alpha}\right)_{\underset{\approx}{\mathbf{A}}} = \underset{\approx}{\mathbf{A}} \bullet \underset{\approx}{\mathbf{R}} = \underset{\approx}{\mathbf{R}} \bullet \underset{\approx}{\mathbf{A}}. \tag{15.28}$$

The derivative of $\underset{\approx}{\mathbf{R}}$ with respect to $\underset{\approx}{\mathbf{A}}$ does not appear to lend itself to such a compact final result. However, if a_k denotes the kth component of the unit axis vector, it can be shown that

$$\left(\frac{\partial \underset{\approx}{\mathbf{R}}}{\partial a_k}\right)_{\alpha} = \underset{\approx}{\mathbf{Z}}_k \bullet \underset{\approx}{\mathbf{R}}, \tag{15.29}$$

where, with respect to a basis having the 3-direction aligned with $\underset{\sim}{\mathbf{a}}$,

$$[\underset{\approx}{\mathbf{Z}}_1] = \begin{bmatrix} 0 & 0 & 1-\cos\alpha \\ 0 & 0 & -\sin\alpha \\ -(1-\cos\alpha) & \sin\alpha & 0 \end{bmatrix}, \tag{15.30a}$$

$$[\underset{\approx}{\mathbf{Z}}_2] = \begin{bmatrix} 0 & 0 & \sin\alpha \\ 0 & 0 & 1-\cos\alpha \\ -\sin\alpha & -(1-\cos\alpha) & 0 \end{bmatrix}, \text{ and} \tag{15.30b}$$

$$\underset{\approx}{\mathbf{Z}}_3 = \underset{\approx}{\mathbf{0}}. \tag{15.30c}$$

15.5 Difference between vorticity and polar spin

In continuum mechanics, the velocity gradient tensor $\underset{\approx}{\mathbf{L}}$ is the spatial derivative of the velocity field. Thus, if $\underset{\sim}{\mathbf{v}}$ is velocity and $\underset{\sim}{\mathbf{x}}$ is the spatial position vector, then

$$L_{ij} = \frac{\partial v_i}{\partial x_j}. \tag{15.31}$$

The so-called 'rate' of deformation tensor $\underset{\approx}{D}$ is defined to equal the symmetric part of the velocity gradient:

$$D_{ij} = \frac{1}{2}(L_{ij} + L_{ji}) = \frac{1}{2}\left(\frac{\partial v_i}{\partial x_j} + \frac{\partial v_j}{\partial x_i}\right). \tag{15.32}$$

The word 'spin' is used inconsistently in the literature. We use the phrase *polar spin* to refer to what many authors also call the *angular velocity tensor* $\underset{\approx}{\Omega} = \dot{\underset{\approx}{R}} \bullet \underset{\approx}{R}^T$ in which $\underset{\approx}{R}$ is the polar rotation tensor. Some writers use the word 'spin' to refer to what we call the *vorticity tensor* $\underset{\approx}{W}$, defined to be the skew-symmetric part of the velocity gradient tensor:

$$W_{ij} = \frac{1}{2}(L_{ij} - L_{ji}) = \frac{1}{2}\left(\frac{\partial v_i}{\partial x_j} - \frac{\partial v_j}{\partial x_i}\right). \tag{15.33}$$

The vorticity vector $\underset{\sim}{w}$ is the axial vector associated with $\underset{\approx}{W}$:

$$\underset{\sim}{w} = -\frac{1}{2}\underset{\approx}{\varepsilon} : \underset{\approx}{W}. \tag{15.34}$$

Written out in Cartesian components,

$$\begin{aligned} w_1 &= \frac{1}{2}\left(\frac{\partial v_3}{\partial x_2} - \frac{\partial v_2}{\partial x_3}\right), \\ w_2 &= \frac{1}{2}\left(\frac{\partial v_1}{\partial x_3} - \frac{\partial v_3}{\partial x_1}\right), \\ w_3 &= \frac{1}{2}\left(\frac{\partial v_2}{\partial x_1} - \frac{\partial v_1}{\partial x_2}\right). \end{aligned} \tag{15.35}$$

Therefore,

$$\boxed{\underset{\sim}{w} = \frac{1}{2}\nabla \times \underset{\sim}{v}}. \tag{15.36}$$

Strangely, the name 'vorticity' is often defined by this formula without the factor of 1/2, but that factor is essential to ensure that the vorticity vector becomes identical to the angular velocity vector in the case of rigid rotation.

A fundamental theorem from continuum mechanics (proved via the chain rule) states that the velocity gradient is related to the deformation gradient tensor $\underset{\approx}{F}$ by

$$\underset{\approx}{L} = \dot{\underset{\approx}{F}} \bullet \underset{\approx}{F}^{-1}. \tag{15.37}$$

Recall the polar decomposition

$$\underset{\approx}{F} = \underset{\approx}{V} \bullet \underset{\approx}{R}, \tag{15.38}$$

where $\underset{\approx}{R}$ is the orthogonal rotation tensor and $\underset{\approx}{V}$ is the symmetric positive-definite left stretch. Substituting (15.38) into (15.37) and noting that $\underset{\approx}{L} = \underset{\approx}{D} + \underset{\approx}{W}$ gives

$$\underset{\approx}{D} + \underset{\approx}{W} = \dot{\underset{\approx}{V}} \bullet \underset{\approx}{V}^{-1} + \underset{\approx}{V} \bullet \underset{\approx}{\Omega} \bullet \underset{\approx}{V}^{-1}, \tag{15.39}$$

where the polar spin $\underset{\approx}{\Omega}$ is defined as

$$\underset{\approx}{\Omega} \equiv \dot{\underset{\approx}{R}} \bullet \underset{\approx}{R}^T. \tag{15.40}$$

The polar angular velocity vector $\underset{\sim}{\omega}$ is defined as

$$\underset{\sim}{\omega} = -\frac{1}{2}\underset{\approx}{\varepsilon}:\underset{\approx}{\Omega}. \tag{15.41}$$

The purpose of the remainder of this section is to summarize Dienes's derivation [1] of the difference between $\underset{\sim}{w}$ and $\underset{\sim}{\omega}$. Physically, the negative of this difference is the material vorticity seen by a polar observer (whose orientation is set by the polar rotation).

The transpose of (15.39) is

$$\underset{\approx}{D} - \underset{\approx}{W} = \underset{\approx}{V}^{-1} \bullet \dot{\underset{\approx}{V}} - \underset{\approx}{V}^{-1} \bullet \underset{\approx}{\Omega} \bullet \underset{\approx}{V}. \tag{15.42}$$

Post multiplying (15.39) by $\underset{\approx}{V}$ gives

$$\underset{\approx}{D} \bullet \underset{\approx}{V} + \underset{\approx}{W} \bullet \underset{\approx}{V} = \dot{\underset{\approx}{V}} + \underset{\approx}{V} \bullet \underset{\approx}{\Omega}. \tag{15.43}$$

Taking the skew-symmetric part of both sides gives

$$\boxed{\underset{\approx}{Z} = \underset{\approx}{V} \bullet \underset{\approx}{H} + \underset{\approx}{H} \bullet \underset{\approx}{V}}, \tag{15.44}$$

where

$$\boxed{\underset{\approx}{Z} \equiv \underset{\approx}{D} \bullet \underset{\approx}{V} - \underset{\approx}{V} \bullet \underset{\approx}{D}} \tag{15.45}$$

and

$$\boxed{\underset{\approx}{H} \equiv \underset{\approx}{\Omega} - \underset{\approx}{W}}. \tag{15.46}$$

We are interested in the difference between the vorticity and the polar spin. In other words, we seek to solve (15.44) for the tensor $\underset{\approx}{H}$, which is the negative of material vorticity (i.e., vorticity seen by an observer rotating with the polar rotation so that the deformation (from such an observer's perspective) always appears to be a pure stretch, thus producing vorticity strictly from rotation of the principal directions of reference stretch). Below, we first explain why a solution exists, and then we proceed to illustrate Dienes' approach to efficiently finding the solution.

Incidentally, $\underset{\approx}{H}$ (the negative of which is called 'material vorticity' seen by a polar observer) is zero if and only if the reference stretch $\underset{\approx}{U}$ commutes with its own rate $\dot{\underset{\approx}{U}}$, which means these two tensors share a common set of eigenvectors; that is actually a very common situation in engineering and otherwise a good approximation for many applications[3]. Material vorticity is zero if and only if the reference stretch tensor's eigenvectors are stationary; non-zero material vorticity is essential to substantiate any

[3] An exception would be the type of loading that forms, say, a long cylinder by rolling it between your hands. That problem is solved exactly for a simple von Mises model in [2].

claim of predictive superiority of large-deformation kinematics formulations in comparison to generally cheaper models that tacitly neglect material vorticity.

15.5.1 Existence of a solution

Equations (15.44) and (15.45) may be written in standard linear form as

$$\underset{\sim}{\boldsymbol{Z}} = \underset{\sim}{\boldsymbol{V}} \bullet \underset{\sim}{\boldsymbol{H}} \bullet \underset{\sim}{\boldsymbol{I}} + \underset{\sim}{\boldsymbol{I}} \bullet \underset{\sim}{\boldsymbol{H}} \bullet \underset{\sim}{\boldsymbol{V}} = \underset{\approx}{\boldsymbol{Y}} \colon \underset{\sim}{\boldsymbol{H}}, \tag{15.47a}$$

$$\underset{\sim}{\boldsymbol{Z}} = \underset{\sim}{\boldsymbol{I}} \bullet \underset{\sim}{\boldsymbol{D}} \bullet \underset{\sim}{\boldsymbol{V}} - \underset{\sim}{\boldsymbol{V}} \bullet \underset{\sim}{\boldsymbol{D}} \bullet \underset{\sim}{\boldsymbol{I}} = \underset{\approx}{\boldsymbol{\psi}} \colon \underset{\sim}{\boldsymbol{D}} \tag{15.47b}$$

where we have introduced two new fourth-order tensors with components given by

$$Y_{ijkl} = V_{ik}\delta_{jl} - V_{il}\delta_{jk}, \tag{15.48a}$$

$$\psi_{ijkl} = \delta_{ik}V_{jl} + \delta_{il}V_{jk}. \tag{15.48b}$$

Noting that $\underset{\sim}{\boldsymbol{D}}$ is symmetric, while $\underset{\sim}{\boldsymbol{Z}}$ and $\underset{\sim}{\boldsymbol{H}}$ are skew, appropriate symmetries could be imposed on these formulas, but it turns out that they are easier to work with as stated. With respect to the principal basis for $\underset{\sim}{\boldsymbol{V}}$, the standard[4] component matrices for $[Y]$ and $[\psi]$ are, respectively,

$$\begin{bmatrix} \alpha_{11} & 0 & 0 & 0 & 0 & 0 & 0 & 0 & 0 \\ 0 & \alpha_{22} & 0 & 0 & 0 & 0 & 0 & 0 & 0 \\ 0 & 0 & \alpha_{33} & 0 & 0 & 0 & 0 & 0 & 0 \\ 0 & 0 & 0 & \alpha_{23} & 0 & 0 & 0 & 0 & 0 \\ 0 & 0 & 0 & 0 & \alpha_{31} & 0 & 0 & 0 & 0 \\ 0 & 0 & 0 & 0 & 0 & \alpha_{12} & 0 & 0 & 0 \\ 0 & 0 & 0 & 0 & 0 & 0 & \alpha_{32} & 0 & 0 \\ 0 & 0 & 0 & 0 & 0 & 0 & 0 & \alpha_{13} & 0 \\ 0 & 0 & 0 & 0 & 0 & 0 & 0 & 0 & \alpha_{21} \end{bmatrix} \text{ and } \begin{bmatrix} \lambda_{11} & 0 & 0 & 0 & 0 & 0 & 0 & 0 & 0 \\ 0 & \lambda_{22} & 0 & 0 & 0 & 0 & 0 & 0 & 0 \\ 0 & 0 & \lambda_{33} & 0 & 0 & 0 & 0 & 0 & 0 \\ 0 & 0 & 0 & \lambda_{23} & 0 & 0 & 0 & 0 & 0 \\ 0 & 0 & 0 & 0 & \lambda_{31} & 0 & 0 & 0 & 0 \\ 0 & 0 & 0 & 0 & 0 & \lambda_{12} & 0 & 0 & 0 \\ 0 & 0 & 0 & 0 & 0 & 0 & \lambda_{32} & 0 & 0 \\ 0 & 0 & 0 & 0 & 0 & 0 & 0 & \lambda_{13} & 0 \\ 0 & 0 & 0 & 0 & 0 & 0 & 0 & 0 & \lambda_{21} \end{bmatrix}, \tag{15.49}$$

where, in terms of the three eigenvalues $\{\lambda_1, \lambda_2, \lambda_3\}$ of $\underset{\sim}{\boldsymbol{V}}$,

$$\alpha_{IJ} = \lambda_J + \lambda_I \quad \text{(suspend indicial notation convention)}, \tag{15.50a}$$

$$\lambda_{IJ} = \lambda_J - \lambda_I \quad \text{(suspend indicial notation convention)}. \tag{15.50b}$$

These formulas define 3 × 3 components of matrices, not tensors. Given that $\underset{\sim}{\boldsymbol{V}}$ is positive-definite, it follows that each α_{IJ} is positive, and the matrix $[Y]$ is therefore invertible. Accordingly, setting (15.47*a*) and (15.47*b*) equal to each other implies that

$$\underset{\sim}{\boldsymbol{H}} = \underset{\approx}{\boldsymbol{T}} \colon \underset{\sim}{\boldsymbol{D}}, \quad \text{where} \quad \underset{\approx}{\boldsymbol{T}} = \underset{\approx}{\boldsymbol{Y}}^{-1} \colon \underset{\approx}{\boldsymbol{\psi}} \quad (\text{and thus } [T] = [Y]^{-1}[\psi]). \tag{15.51}$$

---

[4] See chapter 26.

Letting $\underset{\approx}{\boldsymbol{P}}_I$ denote the eigenprojector associated with eigenvalue λ_I, we note that

$$\underset{\approx\approx}{\boldsymbol{T}} = \sum_I^3 \sum_J^3 \beta_{IJ} \left(\underset{\approx}{\boldsymbol{P}}_I \underset{\approx}{\boldsymbol{P}}_J \right)^L, \text{ where } \boxed{\beta_{IJ} = \frac{\lambda_J - \lambda_I}{\lambda_J + \lambda_I}} \text{ (no index rules)}, \tag{15.52}$$

where $\underset{\approx}{\boldsymbol{P}}_I \underset{\approx}{\boldsymbol{P}}_J$ are multiplied dyadically and the superscript L denotes a 'leafing' operator such that, for any fourth-order tensor with components A_{ijkl}, the components of its leaf are $A_{ijkl}^L = A_{ikjl}$. Since each eigenprojector is formed as the dyadic product of two eigenvectors,

$$\underset{\approx}{\boldsymbol{P}} = \underset{\sim}{\boldsymbol{p}}_I \underset{\sim}{\boldsymbol{p}}_I \quad \text{(no implied sum)}. \tag{15.53}$$

It follows that

$$\underset{\approx\approx}{\boldsymbol{T}} = \sum_I^3 \sum_J^3 \beta_{IJ} (\underset{\sim}{\boldsymbol{p}}_I \underset{\sim}{\boldsymbol{p}}_J \underset{\sim}{\boldsymbol{p}}_I \underset{\sim}{\boldsymbol{p}}_J) \text{ and } \boxed{\underset{\approx}{\boldsymbol{H}} = \sum_1^3 \beta_{IJ} D_{IJ}^{<P>} \underset{\sim}{\boldsymbol{p}}_I \underset{\sim}{\boldsymbol{p}}_J}, \tag{15.54}$$

where $D_{IJ}^{<P>} = \underset{\sim}{\boldsymbol{p}}_I \bullet \underset{\approx}{\boldsymbol{D}} \bullet \underset{\sim}{\boldsymbol{p}}_J$ are the components of $\underset{\approx}{\boldsymbol{D}}$ with respect to the orthonormalized principal basis of $\underset{\approx}{\boldsymbol{V}}$. In words, this result for $\underset{\approx}{\boldsymbol{H}}$ would be implemented via the following algorithm.

STEP 1. Obtain eigenvalues $\{\lambda_1, \lambda_2, \lambda_3\}$ and corresponding orthonormalized eigenvectors $\{\underset{\sim}{\boldsymbol{p}}_1, \underset{\sim}{\boldsymbol{p}}_2, \underset{\sim}{\boldsymbol{p}}_3\}$ of $\underset{\approx}{\boldsymbol{V}}$. Put the lab components of the eigenvectors into the columns of an orthogonal matrix $[Q]$.
STEP 2. Obtain the components $D_{IJ}^{<P>} = \underset{\sim}{\boldsymbol{p}}_I \bullet \underset{\approx}{\boldsymbol{D}} \bullet \underset{\sim}{\boldsymbol{p}}_J$ of $\underset{\approx}{\boldsymbol{D}}$ with respect to the principal basis of $\underset{\approx}{\boldsymbol{V}}$. If $[D^{<lab>}]$ contains the lab components of $\underset{\approx}{\boldsymbol{D}}$, then $[D^{<P>}] = [Q]^T [D^{<lab>}][Q]$.
STEP 3. Compute $H_{IJ}^{<P>} = \beta_{IJ} D_{IJ}^{<P>}$, where $\beta_{IJ} = \dfrac{\lambda_J - \lambda_I}{\lambda_J + \lambda_I}$ (and tensor indicial rules are suspended).
STEP 4. Compute the lab components of $\underset{\approx}{\boldsymbol{H}}$ by $[H^{<lab>}] = [Q][H^{<P>}][Q]^T$.

This algorithm is fairly involved. As outlined below, Dienes found an alternative approach that does not require an eigen-analysis of the stretch.

15.5.2 Dienes' solution

The tensor $\underset{\approx}{\boldsymbol{H}}$ is skew-symmetric, so there exists an axial vector $\underset{\sim}{\boldsymbol{h}}$ such that

$$H_{ij} = \varepsilon_{ikj} h_k. \tag{15.55}$$

Similarly, there exists an axial vector $\underset{\sim}{\boldsymbol{z}}$ such that

$$Z_{ij} = \varepsilon_{ikj} z_k. \tag{15.56}$$

Thus, the component form of (15.44) can be written

$$\varepsilon_{ikj} z_k = V_{im} \varepsilon_{mpj} h_p + \varepsilon_{iqn} h_q V_{nj}. \tag{15.57}$$

Multiplying both sides by ε_{irj} and using the identity $\varepsilon_{irj} \varepsilon_{ikj} = 2\delta_{rk}$ gives

$$2\delta_{rk}z_k = \varepsilon_{irj}V_{im}\varepsilon_{mpj}h_p + \varepsilon_{irj}\varepsilon_{iqn}h_q V_{nj}. \tag{15.58}$$

Noting that $\varepsilon_{irj}\varepsilon_{mpj} = \delta_{im}\delta_{rp} + \delta_{ip}\delta_{rm}$ and $\varepsilon_{irj}\varepsilon_{iqn} = \delta_{rq}\delta_{jn} + \delta_{rn}\delta_{jq}$, this becomes

$$2\delta_{rk}z_k = (\delta_{im}\delta_{rp} + \delta_{ip}\delta_{rm})V_{im}h_p + (\delta_{rq}\delta_{jn} + \delta_{rn}\delta_{jq})h_q V_{nj} \tag{15.59}$$

or, eliminating all Kronecker deltas,

$$2z_r = (V_{mn}h_r + V_{pr}h_p) + (h_r V_{jj} + h_j V_{rj}). \tag{15.60}$$

This may be written in direct notation as

$$\underset{\sim}{z} = [(\mathrm{tr}\,\underset{\approx}{V})\underset{\approx}{I} + \underset{\approx}{V}] \bullet \underset{\sim}{h}. \tag{15.61}$$

Recalling (15.46), we note that $\underset{\sim}{h} = \underset{\sim}{\omega} - \underset{\sim}{w}$ and therefore the above equation may be inverted to give

$$\boxed{\underset{\sim}{\omega} = \underset{\sim}{w} + [(\mathrm{tr}\,\underset{\approx}{V})\underset{\approx}{I} + \underset{\approx}{V}]^{-1} \bullet \underset{\sim}{z}}, \tag{15.62}$$

where $\underset{\sim}{z}$ is the axial vector associated with $\underset{\approx}{Z} \equiv \underset{\approx}{D} \bullet \underset{\approx}{V} - \underset{\approx}{V} \bullet \underset{\approx}{D}$.

From this we see that the polar spin equals the vorticity if and only if $\underset{\sim}{z} = \underset{\sim}{0}$. Recalling (15.45), this will occur if and only if $\underset{\approx}{D}$ and $\underset{\approx}{V}$ commute[5], which implies that there must exist a triad of orthonormal vectors that are principal directions of *both* $\underset{\approx}{D}$ and $\underset{\approx}{V}$. This is naturally the case for rigid motion where $\underset{\approx}{D} = \underset{\approx}{0}$ and $\underset{\approx}{V} = \underset{\approx}{I}$. Another (less trivial) way for $\underset{\approx}{D}$ and $\underset{\approx}{V}$ to commute occurs when the principal directions of $\underset{\approx}{V}$ remain constant throughout all time. In this case, both the vorticity and the spin are zero. More generally, $\underset{\approx}{D}$ and $\underset{\approx}{V}$ to commute if and only if the principal directions of the *reference* stretch $\underset{\approx}{U}$ are stationary. Finally, note that $\underset{\approx}{D} \bullet \underset{\approx}{V} - \underset{\approx}{V} \bullet \underset{\approx}{D} = \underset{\approx}{D}' \bullet \underset{\approx}{V}' - \underset{\approx}{V}' \bullet \underset{\approx}{D}'$ where a prime denotes the deviatoric part. This identity implies that there is a difference between the vorticity and spin only if the deformation involves significant distortion—isotropic expansion or contraction has no influence on the distinction between vorticity and spin. For this reason, Dienes [1] considered simple shear as a counter-example to demonstrate that the Jaumann stress rate (based on vorticity) predicts anomalous oscillatory shear stresses when used with a constant modulus hypoelastic constitutive law. The polar rate predicts an intuitive monotonically increasing shear stress. These differences in material response require significantly large deformations, well beyond the elastic range for typical materials, and therefore these conclusions are more theoretical than of practical engineering importance, except perhaps for cyclic loading.

15.6 The (commonly mis-stated) Gosiewski's theorem

In this section we prove the following theorem:

If $\underset{\sim}{d}$ *is a unit* material *fiber that is instantaneously coincident with a principal direction of the 'rate' of deformation* $\underset{\approx}{D}$*, then its instantaneous rate of rotation is given by the vorticity; i.e., at this instant,* $\dot{\underset{\sim}{d}} = \underset{\approx}{W} \bullet \underset{\sim}{d}$.

[5] I.e., $\underset{\sim}{z} = \underset{\sim}{0}$ if and only if $\underset{\approx}{D} \bullet \underset{\approx}{V} = \underset{\approx}{V} \bullet \underset{\approx}{D}$.

This theorem is often mis-interpreted to mean that the principal directions of $\underset{\approx}{\boldsymbol{D}}$ rotate according to the vorticity $\underset{\approx}{\boldsymbol{W}}$. To the contrary, it is only the *material fibers* parallel to the principal directions of $\underset{\approx}{\boldsymbol{D}}$ that rotate according to the vorticity. These material fibers do not generally remain aligned with the principal directions of $\underset{\approx}{\boldsymbol{D}}$. Hence, the above theorem (i) applies only to material fibers and (ii) only at the instant when they pass over principal directions of $\underset{\approx}{\boldsymbol{D}}$—any time thereafter, they rotate at a speed that is *not* related to the vorticity.

Before proceeding with the proof of the correct theorem, let us first provide a counter-example that discredits the common false interpretation. We wish to prove that the principal directions of $\underset{\approx}{\boldsymbol{D}}$ do *not* rotate according to the vorticity. To do this, we use simple shear as a counter-example. For this canonical motion, the vorticity is a non-zero constant tensor, but the $\underset{\approx}{\boldsymbol{D}}$ tensor is also constant, so its principal directions cannot possibly rotate. Under simple shear, the deformation gradient is of the form

$$[\underset{\approx}{\boldsymbol{F}}] = \begin{bmatrix} 1 & 2\varepsilon & 0 \\ 0 & 1 & 0 \\ 0 & 0 & 1 \end{bmatrix}, \tag{15.63}$$

where ε is a time-varying measure of the amount of shear. The corresponding rate of deformation and vorticity tensors are

$$[\underset{\approx}{\boldsymbol{D}}] = \begin{bmatrix} 0 & \dot{\varepsilon} & 0 \\ \dot{\varepsilon} & 0 & 0 \\ 0 & 0 & 0 \end{bmatrix} \quad \text{and} \quad [\underset{\approx}{\boldsymbol{W}}] = \begin{bmatrix} 0 & \dot{\varepsilon} & 0 \\ -\dot{\varepsilon} & 0 & 0 \\ 0 & 0 & 0 \end{bmatrix}. \tag{15.64}$$

Note that the principal directions of the rate of deformation $\underset{\approx}{\boldsymbol{D}}$ are $\{1, 1, 0\}$, $\{1, -1, 0\}$, and $\{0, 0, 1\}$ which *never change with time*. According to the mis-interpretation of Gosiewski's theorem, this would be possible only if the vorticity were zero, which is false. This completes the proof that $\underset{\approx}{\boldsymbol{W}}$ does *not* govern the rotation rate of the principal directions of $\underset{\approx}{\boldsymbol{D}}$.

15.6.1 Proof of the correct theorem

For clarity, we consider a homogeneous deformation field. Let $\underset{\sim}{\boldsymbol{f}}$ denote a material fiber. By this, we mean that there exists a vector $\underset{\sim}{\boldsymbol{f}}_o$ that is 'painted' on the material in the reference configuration such that it deforms to

$$\underset{\sim}{\boldsymbol{f}} = \underset{\approx}{\boldsymbol{F}} \bullet \underset{\sim}{\boldsymbol{f}}_o. \tag{15.65}$$

The reference (initial) material fiber is fixed in time, so taking rates gives

$$\dot{\underset{\sim}{\boldsymbol{f}}} = \dot{\underset{\approx}{\boldsymbol{F}}} \bullet \underset{\sim}{\boldsymbol{f}}_o. \tag{15.66}$$

From continuum mechanics, we know that $\dot{\mathbf{F}} = \mathbf{L} \bullet \mathbf{F}$, where $\mathbf{L}$ is the spatial velocity gradient. Therefore, the above equation becomes

$$\dot{\mathbf{f}} = \mathbf{L} \bullet \mathbf{f}. \tag{15.67}$$

Decomposing $\mathbf{L}$ into its symmetric part $\mathbf{D}$ plus its skew-symmetric part $\mathbf{W}$ gives

$$\dot{\mathbf{f}} = \mathbf{D} \bullet \mathbf{f} + \mathbf{W} \bullet \mathbf{f}. \tag{15.68}$$

This relationship holds for *any* material fiber $\mathbf{f}$. We wish to determine what part of $\dot{\mathbf{f}}$ is caused by change in length and what part is caused by change in orientation. Towards this end, we write

$$\mathbf{f} = f\mathbf{n}, \tag{15.69}$$

where f is the magnitude of $\mathbf{f}$ defined by

$$f \equiv \sqrt{\mathbf{f} \bullet \mathbf{f}} \tag{15.70}$$

and $\mathbf{n}$ is a unit vector in the direction of $\mathbf{f}$. In other words, $\mathbf{n} = \mathbf{f}/f$ so that

$$\mathbf{n} \bullet \mathbf{n} = 1. \tag{15.71}$$

Taking rates of the above three equations, we note that

$$\dot{\mathbf{f}} = \dot{f}\mathbf{n} + f\dot{\mathbf{n}}, \tag{15.72}$$

$$\dot{f} = \frac{\mathbf{f} \bullet \dot{\mathbf{f}}}{f}, \tag{15.73}$$

$$\mathbf{n} \bullet \dot{\mathbf{n}} = 0. \tag{15.74}$$

Hence, equation (15.68) may be written

$$\dot{f}\mathbf{n} + \dot{f}\mathbf{n} = f\mathbf{D} \bullet \mathbf{n} + f\mathbf{W} \bullet \mathbf{n}. \tag{15.75}$$

Dotting both sides by $\mathbf{n}$ gives

$$\boxed{\dot{f} = fD_{nn}}, \quad \text{where } D_{nn} \equiv \mathbf{n} \bullet \mathbf{D} \bullet \mathbf{n}. \tag{15.76}$$

Here, we have noted that $\mathbf{n} \bullet \mathbf{W} \bullet \mathbf{n} = 0$ because $\mathbf{W}$ is skew-symmetric. Substituting (15.76) back into (15.75) and dividing by f gives

$$\boxed{\dot{\mathbf{n}} = \mathbf{D} \bullet \mathbf{n} - D_{mn}\mathbf{n} + \mathbf{W} \bullet \mathbf{n}}. \tag{15.77}$$

The above boxed equations hold for *any* material fiber $\mathbf{f}$ regardless of its instantaneous orientation.

If the material fiber *happens* to instantaneously coincide with a principal direction of $\underset{\approx}{D}$ then *at this instant*, $\underset{\approx}{D} \bullet \underset{\sim}{n} = \lambda \underset{\sim}{n}$, and therefore $D_{nn} = \underset{\sim}{n} \bullet \underset{\approx}{D} \bullet \underset{\sim}{n} = \lambda$. Consequently, *at this instant in time,* the first two terms in (15.77) cancel leaving only $\dot{\underset{\sim}{n}} = \underset{\approx}{W} \bullet \underset{\sim}{n}$, which proves the theorem that the vorticity governs the angular velocity of material fibers at the moment when they are coincident with a principal direction of $\underset{\approx}{D}$. This special theorem seems of little practical use since the same material fiber will *not* be rotating according to the vorticity an instant later. We have included the theorem here only for the purpose of disproving its common misinterpretation.

15.7 Rates of sequential rotations

Consider three rotations, $\underset{\approx}{R}_1$ about the X-axis, $\underset{\approx}{R}_2$ about the Y-axis, and $\underset{\approx}{R}_3$ about the Z-axis. If these rotations are applied sequentially (i.e., the first followed by the second and then the third), we know that the total rotation is given by

$$\underset{\approx}{R} = \underset{\approx}{R}_3 \bullet \underset{\approx}{R}_2 \bullet \underset{\approx}{R}_1. \tag{15.78}$$

Less obvious is the fact that, for any rotation $\underset{\approx}{R}$, there exist three rotations $\{\underset{\approx}{R}_1, \underset{\approx}{R}_2,$ and $\underset{\approx}{R}_3\}$ about the laboratory axes such that (15.78) holds.

If the three axis rotations are known for all time, how is the spin tensor $(\underset{\approx}{\Omega} \equiv \dot{\underset{\approx}{R}} \bullet \underset{\approx}{R}^T)$ related to the individual spin tensors for each of the axis rotations?

The time rate of (15.78) gives

$$\dot{\underset{\approx}{R}} = \dot{\underset{\approx}{R}}_3 \bullet \underset{\approx}{R}_2 \bullet \underset{\approx}{R}_1 + \underset{\approx}{R}_3 \bullet \dot{\underset{\approx}{R}}_2 \bullet \underset{\approx}{R}_1 + \underset{\approx}{R}_3 \bullet \underset{\approx}{R}_2 \bullet \dot{\underset{\approx}{R}}_1. \tag{15.79}$$

Substituting (15.5) for $\dot{\underset{\approx}{R}}$ and making analogous substitutions for $\dot{\underset{\approx}{R}}_k$ shows that

$$\underset{\approx}{\Omega} = \underset{\approx}{\Omega}_3 + \underset{\approx}{R}_3 \bullet \underset{\approx}{\Omega}_2 \bullet \underset{\approx}{R}_3^T + \underset{\approx}{R}_3 \bullet \underset{\approx}{R}_2 \bullet \underset{\approx}{\Omega}_1 \bullet \underset{\approx}{R}_2^T \bullet \underset{\approx}{R}_3^T. \tag{15.80}$$

This formulation shows that the spin tensor also depends on the order of application of the sequentially applied rotation tensors. In the limit of *infinitesimally small* rotations (but not necessarily small rotation rates), each $\underset{\approx}{R}_k$ is approximately equal to the identity tensor and we obtain the result that

$$\underset{\approx}{\Omega} = \underset{\approx}{\Omega}_3 + \underset{\approx}{\Omega}_2 + \underset{\approx}{\Omega}_1. \tag{15.81}$$

Thus, for infinitesimal sequential rotations,

$$\underset{\sim}{\omega} = \underset{\sim}{\omega}_3 + \underset{\sim}{\omega}_2 + \underset{\sim}{\omega}_1. \tag{15.82}$$

For the *special case* that the rotation axis of each $\underset{\approx}{R}_k$ is constant, we have

$$\underset{\sim}{\omega} = \dot{\alpha}_1 \underset{\sim}{a}_1 + \dot{\alpha}_2 \underset{\sim}{a}_2 + \dot{\alpha}_3 \underset{\sim}{a}_3. \tag{15.83}$$

Even for this special case where each rotation axis is constant, this shows that the rotation axis of the *total* rotation is not necessarily constant—it can be varied by changing the individual rotation rates $\{\dot{\alpha}_1, \dot{\alpha}_2,$ and $\dot{\alpha}_3\}$. This result motivates the study of *simultaneously* applied rotations.

15.8 Rates of simultaneous rotations

Consider the rotation illustrated in figure 9.5. By visual inspection, the rotation axis passes through the point (1,1,1), and the rotation angle is 120°. One could imagine directly rotating the body about the axis so that rotations would be simultaneously occurring about all axes. For a rotation that is applied directly from zero angle to final angle without changing the axis of rotation, the angular velocity vector is

$$\underset{\sim}{\omega} = \dot{\alpha}\underset{\sim}{a}. \tag{15.84}$$

Writing the axis as $\underset{\sim}{a} = a_1\underset{\sim}{e}_1 + a_2\underset{\sim}{e}_2 + a_3\underset{\sim}{e}_3$, this equation becomes

$$\underset{\sim}{\omega} = \dot{\alpha}_1\underset{\sim}{e}_1 + \dot{\alpha}_2\underset{\sim}{e}_2 + \dot{\alpha}_3\underset{\sim}{e}_3, \tag{15.85}$$

where

$$\dot{\alpha}_k \equiv \dot{\alpha}a_k. \tag{15.86}$$

Once the rotation is complete, the final rotation tensor is given by the Euler–Rodrigues formula, (5.17). In view of (15.85), this final rotated configuration can be visualized as the simultaneous application of three rotations about the three coordinate axes with the respective rotation angles given by

$$\alpha_k \equiv \alpha a_k. \tag{15.87}$$

Thus, for the rotation in figure 9.5, the simultaneously applied angles of rotation about each coordinate axis are $120°/\sqrt{3} = 69.28°$.

Given the rotation angles $\{\alpha_1, \alpha_2, a_3\}$, the total rotation vector is given by

$$\alpha\underset{\sim}{a} = \alpha_1\underset{\sim}{e}_1 + \alpha_2\underset{\sim}{e}_2 + \alpha_3\underset{\sim}{e}_3. \tag{15.88}$$

Note, however, that α_k is not equivalent to $\alpha_k + 2\pi$. In other words, it is not admissible to add an integral multiple of 2π to a coordinate rotation angle. Referring to (15.86), the individual coordinate angles must always be less than or equal to the total rotation angle. Furthermore,

$$\dot{\alpha}_k \neq \omega_k. \tag{15.89}$$

Equality holds only if the total rotation axis never changes with time.

In practice, angular velocity generally changes in both magnitude and orientation. Hence rotations are not generally achieved by rotations about a constant axis. Nevertheless, given any rotation tensor $\underset{\approx}{R}$, there always exist a rotation axis and angle regardless of whether the rotation was actually achieved directly through the axis and angle of rotation. There are corresponding coordinate rotation angles $\{\alpha_1, \alpha_2, \alpha_3\}$.

For simultaneous rotations about the coordinate axes, the relationship

$$\underset{\sim}{\omega} = \omega_1\underset{\sim}{e}_1 + \omega_2\underset{\sim}{e}_2 + \omega_3\underset{\sim}{e}_3 \tag{15.90}$$

always holds, but because the magnitudes of the angular velocities may vary, the total axis of rotation generally varies with time and is not coincident with the direction of the angular velocity vector.

To find the total rotation axis, the total rotation tensor is first found by solving (15.4) as a differential equation. Once the total rotation is found, the angle and axis of rotation may be determined via (5.131) and (5.128). Hence, the axis and angle of rotation are always well defined, and they may be decomposed into coordinate rotations, but they are related to the angular velocity in a highly nonlinear way.

A principal point of this section is to show that it *is* possible to define rotation vectors, but they are of limited value because their time rates are not related to the angular velocity in a straightforward way. This concept of simultaneous rotations is perhaps most useful in robotics to determine how to simultaneously actuate gimbal motors to obtain a smooth and natural-looking rotation about a fixed axis (rather than the jerky 'robot dance' motion that results from applying each Euler angle sequentially). The mathematical problem, addressed above, determines the non-linear path through 3D Euler-angle space required to move along an efficient 'shortest path' on the hypersphere of rotations in embedded in 9D tensor space: first find the axis and angle of the total rotation, and then use (15.87) to find the individual simultaneously-applied rotation angles. We now discuss this problem more abstractly to apply to any choice of the three parameters selected for controlling rotation.

15.8.1. Avoiding the robot dance: wobble-free rotation for abstract control systems!

Early robots had an unnatural looking motion (resulting in thousands of entertaining YouTube 'robot dance' videos). Early control algorithms clearly moved the three rotation parameters (usually Euler angles) in a straight line through parameter space, producing a 'wobbly' sequence of rotations. Why? Because an efficient path through coordinate space is not generally an efficient path through tensor space[6]. A natural motion, on the other hand, needs to minimize the kinetic energy required to rotate a body from one orientation $\underset{\approx}{\boldsymbol{Q}}_1$ to another $\underset{\approx}{\boldsymbol{Q}}_2$. Defining the orientation change by $\underset{\approx}{\boldsymbol{Q}} = \underset{\approx}{\boldsymbol{Q}}_2 \bullet \underset{\approx}{\boldsymbol{Q}}_1^T$, the inverse Euler–Rodrigues formula (see section 5.4) determines the corresponding axis $\underset{\sim}{\boldsymbol{a}}$ and total angle α. The transition from one orientation to another must progress through a sequence of admissible rotations so that the time-varying rotation is

$$\underset{\approx}{\boldsymbol{R}}(t) \bullet \underset{\approx}{\boldsymbol{Q}}_1. \tag{15.91}$$

A natural-looking (energy-efficient) rotation keeps the rotation axis fixed so that only the angle changes through time. Then, equation (5.17) gives

$$\underset{\approx}{\boldsymbol{R}}(t) = \cos(\alpha[t])\ (\underset{\approx}{\boldsymbol{I}} - \underset{\sim}{\boldsymbol{a}}\underset{\sim}{\boldsymbol{a}}) + \underset{\sim}{\boldsymbol{a}}\underset{\sim}{\boldsymbol{a}} + \sin(\alpha[t])\ (\underset{\sim}{\boldsymbol{a}} \times \underset{\approx}{\boldsymbol{I}}). \tag{15.92}$$

Because the axis is not rotating, it follows that

$$\dot{\underset{\approx}{\boldsymbol{R}}}(t) = \dot{\alpha}[-\sin(\alpha[t])\ (\underset{\approx}{\boldsymbol{I}} - \underset{\sim}{\boldsymbol{a}}\underset{\sim}{\boldsymbol{a}}) + \cos(\alpha[t])\ (\underset{\sim}{\boldsymbol{a}} \times \underset{\approx}{\boldsymbol{I}})]\alpha. \tag{15.93}$$

[6] If you are not sure what this means, think about an analogous problem in which the (x, y) location of a point in space is given in polar coordinates as $(r\cos\theta, r\sin\theta)$. An efficient straight line in (r, θ) space will produce an inefficient curved path in physical space. Conversely, a 'natural' efficient straight path in physical (x, y) space generally requires an 'inefficient' curvy path in (r, θ) space.

In terms of the natural (fixed) basis having the 3-direction aligned with the axis, the component matrix for the rate of rotation is

$$[\dot{\underset{\approx}{R}}] = \dot{\alpha}\begin{bmatrix} -\sin(\alpha) & -\cos(\alpha) & 0 \\ \cos(\alpha) & -\sin(\alpha) & 0 \\ 0 & 0 & 0 \end{bmatrix}. \tag{15.94}$$

The function $\underset{\approx}{R}(t)$ may be interpreted as a parametric description of a path through nine-dimensional (9D) tensor space. In robotics, a specified rotation might be achieved by actuators. Since three numbers are required to specify a rotation, it follows that there must be three actuators. Let the nine components of the rotation tensor be denoted R_I, where the capitalized index I varies from 1 to 9. We must consider *all nine* components in order to consider $\underset{\approx}{R}$ as being a vector in 9D Euclidean space[7]. Because rotations actually belong to a 3D curvilinear space embedded within the larger 9D space, we know that a rotation can be constructed from three numbers, $\{\eta^1, \eta^1, \eta^3\}$, which we will refer to as 'coordinates' and denote with superscripts to be consistent with notation used in computational geometry. The coordinates might be Euler angles, displacements of linear actuators, Arvo algorithm inputs, etc. For whatever coordinate choice is made, you can construct a 9×3 Jacobian matrix:

$$F_{Ik} = \frac{\partial R_I}{\partial \eta^k}. \tag{15.95}$$

The kth column of this matrix is a 9×1 array containing the nine components of the kth so-called covariant basis tensor,

$$\underset{\approx}{G}_k = \frac{\partial \underset{\approx}{R}}{\partial \eta^k}. \tag{15.96}$$

These covariant basis tensors may be regarded as known because (by premise) you already have formulas that give the rotation tensor as a function of your chosen coordinates (such as actuator displacements in robotics). Being second-order tensors, these are 9D objects, but the total number of basis tensors is three because only three numbers are required to specify a rotation. Associated with this set of three basis tensors is a corresponding 3×3 set of 'covariant metrics' defined as

$$G_{ij} = \underset{\approx}{G}_i : \underset{\approx}{G}_j. \tag{15.97}$$

Recalling that R_I are the nine Cartesian components of the rotation in Euclidean 9D tensor space, it follows that a double-dot operation in conventional tensor notation becomes a single-dot operator in 9D vector notation. Accordingly, the 3×3 covariant metric matrix $[G_{ij}]$ may be evaluated from the operation $[F]^T[F]$. The

[7] This is similar to the needing to consider all three components of a unit vector in order to regard it as a member of a Euclidean space even though it also is constrained to be always on the unit sphere, which is a curvilinear subset within the larger Euclidean space.

inverse of the covariant metric matrix has components denoted G^{ij} and referred to as contravariant metrics. Once the contravariant metrics have been found, you can compute the corresponding contravariant basis tensors by

$$\underset{\approx}{\boldsymbol{G}}^i = G^{ik}\underset{\approx}{\boldsymbol{G}}_k, \quad \text{where the index } k \text{ is summed implicitly from 1 to 3.} \tag{15.98}$$

Using computational geometry, these contravariant basis vectors may be interpreted as the gradient of the coordinate constrained to rotation space:

$$\underset{\approx}{\boldsymbol{G}}^i = \frac{\partial \eta^i}{\partial \underset{\approx}{\boldsymbol{R}}}. \tag{15.99}$$

The most important property of the contravariant basis tensor is

$$\underset{\approx}{\boldsymbol{G}}^i : \underset{\approx}{\boldsymbol{G}}_j = \delta^i_j, \tag{15.100}$$

where δ^i_j is the Kronecker delta.

What does all this junk have to do with achieving a natural-looking rotation? Recall that we have decided that a natural rotation should have a fixed rotation axis so that only the rotation angle changes with time as the body is reoriented. This assertion (or any other opinion about what makes a rotation look natural) gives a specific path to follow through tensor space, defined parametrically by $\underset{\approx}{\boldsymbol{R}}(t)$. The task at hand is therefore to determine the path through 3D coordinate space (actuator control space in robotics) that will lead to the desired path through 9D tensor space. If, for example, the coordinates are Euler angles, moving them linearly through time will *not* move the rotation as desired—it will give the robot dance in which the robot moves through a bizzare sequence of inefficient rotations before finishing at the correct one. To find the correct path through parameter (actuator) space, note that

$$\dot{\underset{\approx}{\boldsymbol{R}}} = \sum_{i=1}^{3} \frac{\partial \underset{\approx}{\boldsymbol{R}}}{\partial \eta^i} \dot{\eta}^i = \sum_{i=1}^{3} \underset{\approx}{\boldsymbol{G}}_i \dot{\eta}^i. \tag{15.101}$$

Double-dotting both sides of this expression by the contravariant basis tensor $\underset{\approx}{\boldsymbol{G}}^k$ and using (23.112) gives a formula for the rate of the actuator coordinates from the desired rate of rotation:

$$\boxed{\dot{\eta}^k = \dot{\underset{\approx}{\boldsymbol{R}}} : \underset{\approx}{\boldsymbol{G}}^k}. \tag{15.102}$$

For any specified path of the rotation tensor in 9D Euclidean tensor space, this formula can be integrated through time to infer the 'pre-image path' (i.e., a parametric description of motion through actuator parameter space). Broadly speaking, the method outlined here applies to any output vector or tensor that is described using actuator control coordinates.

Equation (15.102) provides a way to numerically determine the value of the coordinates (such as actuator stimuli) when an analytical inverse mapping from a rotation to the coordinates is not available. Such high-powered mathematics is not necessary if you have an analytical inverse mapping that can convert the nine components of a rotation to its associated three control coordinates. If you have $\underset{\approx}{\boldsymbol{R}}(t)$, you can evaluate it at any instant in time to get the rotation tensor at that moment. With an analytical inverse mapping, you would then have the coordinates at that moment. You might, for example, find the rotation tensor's axis $\underset{\sim}{\boldsymbol{a}}$ and angle α, which is then sufficient to determine its quaternion (u_1, u_2, u_3) coordinates via (7.14). For a natural (non-wobbly) motion of a steady rotation rate about a fixed rotation axis, this means that the quaternion spinor vector would be

$$\underset{\sim}{\boldsymbol{u}} = \left[\sin\left(\frac{\alpha}{2}\frac{t - t_1}{t_2 - t_1}\right)\right]\underset{\sim}{\boldsymbol{a}}. \tag{15.103}$$

15.9 Integration of rotation rates

Let $\underset{\approx}{\boldsymbol{R}}$ denote a time-varying rotation tensor. Its associated skew-symmetric spin (angular velocity) tensor is

$$\underset{\approx}{\boldsymbol{\Omega}} = \dot{\underset{\approx}{\boldsymbol{R}}} \bullet \underset{\approx}{\boldsymbol{R}}^T. \tag{15.104}$$

Conversely, if spin is known, the time rate of the rotation tensor is given by

$$\dot{\underset{\approx}{\boldsymbol{R}}} = \underset{\approx}{\boldsymbol{\Omega}} \bullet \underset{\approx}{\boldsymbol{R}}. \tag{15.105}$$

Suppose that the angular velocity tensor $\underset{\approx}{\boldsymbol{\Omega}}$ is known[8] as a function of time t throughout some interval from an 'old' initial time $(t = t_O)$ to a 'new' final time $(t = t_N)$. Given the 'old' value of the rotation tensor $\underset{\approx}{\boldsymbol{R}}_O$ at the beginning of the time interval $(t = t_O)$, we wish to integrate (15.105) over the time interval $\Delta t = t_N - t_O$ to obtain the 'new' rotation tensor $\underset{\approx}{\boldsymbol{R}}_N$ at the 'new' time $(t = t_N)$. Using conventional finite differencing is unwise because it leads to loss of orthogonality of the predicted rotation tensor. Specifically, standard first-order accurate finite differencing begins by multiplying the above equation by a small time increment Δt to obtain $\underset{\approx}{\boldsymbol{\Omega}}\Delta t = (\Delta\underset{\approx}{\boldsymbol{R}}) \bullet \underset{\approx}{\boldsymbol{R}}^T$. Then the updated rotation is found to first-order accuracy by $\underset{\approx}{\boldsymbol{R}}_N = \underset{\approx}{\boldsymbol{R}}_O + \Delta\underset{\approx}{\boldsymbol{R}} = \underset{\approx}{\boldsymbol{R}}_O + \underset{\approx}{\boldsymbol{\Omega}} \bullet \underset{\approx}{\boldsymbol{R}}_O\Delta t$. Here, the subscript O stands for the 'old' value at the beginning of the step and N stands for the desired 'new' value at the end of the step. This formula is indeed first-order accurate, but it has the extremely unsavory property of predicting updated rotations that are nearly, but *not quite* orthogonal. Repeated application of a standard finite difference scheme can therefore result in accumulated errors, making the spurious 'stretch' part of the rotation increase over time. The root of the problem is that *the sum of two orthogonal tensors is not itself generally orthogonal.*

[8] If, instead, the angular velocity *vector* $\underset{\sim}{\boldsymbol{\omega}}$ is known, then the angular velocity *tensor* (i.e., the spin) $\underset{\approx}{\boldsymbol{\Omega}}$ can be regarded as known because it can be readily constructed through use of (15.13).

If (15.105) were only a simple *scalar* equation, $\dot{r} = wr$ with w being a known function of time, then its exact solution would be $r_N = qr_O$, where $q \equiv \exp\left(\int_{t_O}^{t_N} w(t)\mathrm{d}t\right)$; furthermore, if w happened to be *constant* over the time interval, then q would be simply $e^{w\Delta t}$. The exact solution to the *tensor* equation, (15.105), is more complicated, but it does share some similar features. Specifically, since we know that $\underset{\approx}{\boldsymbol{R}}_O$ is an orthogonal tensor and, of course, the exact solution for $\underset{\approx}{\boldsymbol{R}}$ at any later time must also be orthogonal, then we know there must exist an orthogonal tensor $\underset{\approx}{\boldsymbol{Q}}(t)$ that represents the *change* in rotation that occurs between $t = t_O$ and some later time t:

$$\underset{\approx}{\boldsymbol{R}}(t) = \underset{\approx}{\boldsymbol{Q}}(t) \bullet \underset{\approx}{\boldsymbol{R}}_O. \qquad (15.106)$$

Thus, solution of (15.105) can be accomplished by solving the following initial value problem that governs the multiplicative change of rotation tensor:

$$\dot{\underset{\approx}{\boldsymbol{Q}}} = \underset{\approx}{\Omega} \bullet \underset{\approx}{\boldsymbol{Q}} \text{ subject to } \underset{\approx}{\boldsymbol{Q}} = \underset{\approx}{\boldsymbol{I}} \text{ when } t = t_O. \qquad (15.107)$$

Recall that (15.20) expresses the angular velocity vector in terms of the axis and angle of rotation

$$\underset{\sim}{\omega} = \dot{\alpha}\underset{\sim}{a} + c(1 - c)(\dot{\underset{\sim}{a}} \times \underset{\sim}{a}) + s\dot{\underset{\sim}{a}}, \quad \text{where } c = \cos\alpha \text{ and } s = \sin\alpha. \qquad (15.108)$$

If we could extract from this formula individual equations governing the rotation angle α and the rotation axis $\underset{\sim}{a}$ separately, then these equations could be integrated to obtain time-varying values for α and $\underset{\sim}{a}$, from which the tensor $\underset{\approx}{\boldsymbol{Q}}$ could be reconstructed using any of the many methods discussed in chapter 5. Note that

$$\underset{\sim}{\omega} \bullet \underset{\sim}{a} = \dot{\alpha}. \qquad (15.109)$$

This equation provides a means of finding the rate of the rotation angle as a function of time *if the total rotation axis* $\underset{\sim}{a}$ *is known as a function of time*. The instantaneous *total rotation* axis $\underset{\sim}{a}$ is not the same as the instantaneous axis of angular velocity, which is simply a unit vector in the direction of $\underset{\sim}{\omega}$. In light of (15.108), the angular velocity vector is *not* parallel to $\underset{\sim}{a}$, $\dot{\underset{\sim}{a}}$, or $\dot{\underset{\sim}{a}} \times \underset{\sim}{a}$; it is a linear combination of these vectors in general. Furthermore α is *not* the magnitude of the angular velocity vector. Instead,

$$\underset{\sim}{\omega} \bullet \underset{\sim}{\omega} = (\dot{\alpha})^2 + [1-2c^3 + c^4](\dot{\underset{\sim}{a}} \bullet \dot{\underset{\sim}{a}}). \qquad (15.110)$$

Note that $\dot{\alpha}$ equals the magnitude of the angular velocity vector only at the beginning of the rotation interval (when $c = 1$). To our knowledge, no progress has been made toward optimizing integration algorithms when the direction of the angular velocity vector varies through the time interval. We will now develop an *exact* algorithm for the special case that the spin tensor (or, equivalently, the angular velocity vector $\underset{\sim}{\omega}$) is constant throughout the entire interval. In this case,

$$\underset{\sim}{\omega} = \dot{\alpha}\underset{\sim}{a}. \qquad (15.111)$$

Since the axis is constant, this may be integrated over the time step to obtain the rotation angle associated with $\underset{\approx}{\boldsymbol{Q}}$ at any time $t \geqslant t_O$:

$$\alpha \equiv \int_{t_O}^{t} \omega \, \mathrm{d}t = \dot{\alpha}\Delta t = \omega \Delta t \text{ for a constant spin tensor, where } \omega \equiv \sqrt{\underset{\sim}{\omega} \bullet \underset{\sim}{\omega}}. \quad (15.112)$$

The rotation axis $\underset{\sim}{\boldsymbol{a}}$ associated with $\underset{\approx}{\boldsymbol{Q}}$ is simply the unit vector in the direction of $\underset{\sim}{\omega}$. Thus, with both the rotation angle and axis known, the incremental rotation tensor $\underset{\approx}{\boldsymbol{Q}}$ can be constructed by applying the Euler–Rodrigues formula of (5.26). Specifically,

Given a rotation $\underset{\approx}{\boldsymbol{R}}_O$ that is known at some time $t = t_O$, and also given an angular velocity (spin) tensor $\underset{\approx}{\Omega}$ that is *constant* over an interval $t_O \leqslant t \leqslant t_N$, the exact expression for the rotation $\underset{\approx}{\boldsymbol{R}}_N$ at $t = t_N$ is found by the following sequence of calculations:

$$\Delta t = t_N - t_O,$$

$$\omega_1 = \Omega_{32},\ \omega_2 = \Omega_{13},\ \omega_3 = \Omega_{21},$$

$$\omega = \sqrt{\omega_1^2 + \omega_2^2 + \omega_3^2},$$

$$a_k = \omega_k/\omega,$$

$$c = \cos(\omega\Delta t),$$

$$s = \sin(\omega\Delta t),$$

$$[Q] = c\begin{bmatrix} 1 & 0 & 0 \\ 0 & 1 & 0 \\ 0 & 0 & 1 \end{bmatrix} + (1-c)\begin{bmatrix} a_1a_1 & a_1a_2 & a_1a_3 \\ a_2a_1 & a_2a_2 & a_2a_3 \\ a_3a_1 & a_3a_2 & a_3a_3 \end{bmatrix} + s\begin{bmatrix} 0 & -a_3 & a_2 \\ a_3 & 0 & -a_1 \\ -a_2 & a_1 & 0 \end{bmatrix},$$

$$[R_N] = [Q][R_O].$$

This algorithm is implemented in section A.6 of the appendix. It is *exact* (infinite-order accuracy) if the spin tensor is constant over the step. It does not matter how far t_N is ahead in time from t_O. Nor does it matter how large the angular velocity is. Thus, this algorithm is especially well suited to problems that involve extremely fast rotation rates (e.g., turbine blades) or extremely large time steps.

This use of the Euler–Rodrigues formula is computationally efficient and, we believe, the superior method for reconstructing the incremental rotation tensor $\underset{\approx}{\boldsymbol{Q}}$ from its angle α and axis $\underset{\sim}{\boldsymbol{a}}$. However, you might encounter in the literature *spinor*-based reconstruction methods (discussed in section 5.2) used to obtain the same result. For example, the incremental rotation alternatively can be found by applying 5.85 as the *Cayley transformation*,

$$\underset{\approx}{\mathbf{Q}} = (\underset{\approx}{\mathbf{I}} + \underset{\approx}{\mathbf{B}}) \bullet (\underset{\approx}{\mathbf{I}} - \underset{\approx}{\mathbf{B}})^{-1}, \tag{15.113}$$

where, recalling (10.38) and (10.38), $\underset{\approx}{\mathbf{B}}$ is the skew-symmetric tensor associated with an axial spinor vector $\underset{\sim}{\mathbf{b}}$ that is related to the axis and angle of rotation by

$$\underset{\sim}{\mathbf{b}} = \left(\tan \frac{\alpha}{2}\right)\underset{\sim}{\mathbf{a}}. \tag{15.114}$$

Recalling 10.19, note that (15.113) can be computed without performing an inverse operation by

$$\underset{\approx}{\mathbf{Q}} = \frac{2(\underset{\approx}{\mathbf{B}}^2 + \underset{\approx}{\mathbf{B}})}{b^2 + 1} + \underset{\approx}{\mathbf{I}}, \quad \text{where } b^2 \equiv \underset{\sim}{\mathbf{b}} \bullet \underset{\sim}{\mathbf{b}}. \tag{15.115}$$

For constant spins, the rotation axis is parallel to the angular velocity vector and the rotation angle is given by $\alpha = \omega\Delta t$. Thus,

$$\underset{\sim}{\mathbf{b}} = \zeta\underset{\sim}{\boldsymbol{\omega}}, \quad b = \zeta\underset{\sim}{\omega} \quad \text{and} \quad \underset{\approx}{\mathbf{B}} = \zeta\underset{\approx}{\boldsymbol{\Omega}}, \quad \text{where } \zeta = \frac{1}{\omega}\tan\left(\frac{\omega\Delta t}{2}\right). \tag{15.116}$$

This formula for $\underset{\sim}{\mathbf{b}}$ is *exact* (infinite-order accuracy) for arbitrarily large time increments, provided that the spin tensor is constant. For small time steps, note that

$$\zeta = \frac{\Delta t}{2} + \frac{\omega^2}{3}\left(\frac{\Delta t}{2}\right)^3 + O\left[\left(\frac{\Delta t}{2}\right)^5\right]. \tag{15.117}$$

Thus, ζ can be approximated to second-order accuracy for small time steps by

$$\zeta \approx \frac{\Delta t}{2}, \tag{15.118}$$

which is used by Flannagan and Taylor [3] in their numerical integration scheme. The half step, $\Delta t/2$ appears in these formulas only as a by-product of the Taylor series expansion—they do not imply a prediction for the rotation at the half step. These factors will lead to a prediction of the rotation change at the end of the *entire* Δt step. Incidentally, Trangenstein [4] uses the spinor reconstruction method described here, but with a different (and, it would seem, excessively complicated) approximating formula for the ζ factor. It is not clear why any *approximate* formula for the ζ factor would be attractive in comparison to the *exact* formula of (15.116), which is valid for arbitrarily large time steps as long as the spin is constant. Appealing to computational efficiency is not a valid justification for using the approximate formulas for ζ. If you seek efficiency *and* accuracy, then the very idea of even using the Cayley transformation in (15.113) or (15.115) should be abandoned. The Euler–Rodrigues approach is far more efficient. Moreover the Cayley transformation method is highly susceptible to round-off errors for time steps or angular velocities that are large enough to permit $\omega\Delta t$ approach π, making ζ, $\underset{\sim}{\mathbf{b}}$, and $\underset{\approx}{\mathbf{B}}$ all become infinite. This is a realistic concern for, say, turbine blade analyses where large material rotations are superimposed on small structural deformations.

We have already intimated that, at the very least, a rotation integration algorithm should be designed so that it is exact if the spin (angular velocity) is constant over a time step. Now suppose that the angular velocity vector has a constant *direction*, but varying *magnitude*. In this case, the incremental rotation can be computed *exactly* if it is possible to evaluate the integral in equation (15.112) exactly. If this is not possible, then any algorithm that optimizes accuracy of that integral is superior. For second-order accuracy, the integral can be approximated by taking $\alpha \approx \omega_H \Delta t$, where the subscript H denotes evaluation at the half step. When the angular velocity vector can change in both magnitude *and direction* over a time interval, then optimal integration strategies remain the subject of debate and, to our knowledge, there exists no exact analytical solution. Using the spin at the half step appears to be a reasonable approach lacking other ideas that might be optimized for a given application. However, even this is subject to potential pitfalls. Suppose, for example, that the actual angular velocity is oriented about the 1-direction at the beginning of the step and transitions linearly to an angular velocity of the same magnitude but oriented about the 2-direction at the end of the step. Recalling the contrast between figures 9.1 and 9.2, this motion will produce a rotation increment that is quite different from the motion from rotating initially about the 2-direction and then about the 1-direction. An integration scheme that uses the spin at the half step, however, would not be able to distinguish between these two scenarios. Thus, it appears that obtaining second-order accuracy for arbitrary angular velocity histories requires the rotation integration scheme to make *direct* use of the angular *acceleration* as well as the angular velocity—it is unlikely that an 'effective' angular velocity vector can be defined that will lead to second-order accuracy when sent as input to an integration routine that treats angular velocity to be constant. In mechanics calculations, the angular acceleration could be obtained through appropriate continuum mechanics formulas that involve the spatial gradient of the material *acceleration* field. The acceleration field is available after solution of the linear momentum equation, so why not use it?

References

[1] Dienes J K 1979 On the analysis of rotation and stress rate in deforming bodies *Acta Mech.* **32** 217–32

[2] Brannon R M and Leelavanichkul S 2010 A multi-stage return algorithm for solving the classical damage component of constitutive models for rocks, ceramics, and other rock-like media *Int. J. Fracture* **163** 133–49

[3] Flanagan D P and Taylor L M 1987 An accurate numerical algorithm for stress integration with finite rotations *Comput. Methods Appl. Mech. Eng.* **62** 305–20

[4] Trangenstein J A and Colella P 1991 A higher-order Godunov method for modeling finite deformation in elastic-plastic solids *Commun. Pure Appl. Math.* **44** 41–100

IOP Publishing

Rotation, Reflection, and Frame Changes
Orthogonal tensors in computational engineering mechanics
R M Brannon

Chapter 16

Variations of tensor-valued functions of scalars and vectors

In mechanics, many material models are first formulated in terms of material-dependent functions and fundamental physical laws that are typically nonlinear. Numerical methods must often be used to solve these equations. Typically, the values of all material and physical variables are known at some given point in time, and it is desired to determine the state at some later time. Any differentiable function, no matter how nonlinear, will always become linear with respect to rates (or increments). For example, a nonlinear function $y = f(x)$ becomes $\dot{y} = s\dot{x}$ in rate form, where $s \equiv f'(x)$. Since s depends only on the state variable x, but not on its rate, the equation $\dot{y} = s\dot{x}$ is linear with respect to rates. Any *system* of nonlinear equations can similarly be transformed into a linear system in rate form. The updated state at the end of a time interval is found by integrating these rates through time. Thus, an ability to determine variations (slopes) of nonlinear functions with respect to variations in the independent variables is of fundamental importance in mechanics. The mathematics required to determine variations of tensor-to-tensor nonlinear transformations is the subject of this chapter. This topic is related to the subject of rotations and orthogonal matrices because tensor-to-tensor functions are often defined in terms of principal space. Variations of such functions are extraordinarily complicated because not only do principal values change, but so do principal *directions*. A change in orientation of a principal triad is accomplished through rotation operations.

In this chapter, we consider a function whose independent variables are (i) a set of scalars and (ii) a set of vectors arranged in such a manner that the output of the function is a tensor. The fact that any tensor $\underset{\approx}{\boldsymbol{T}}$ can be expressed as the sum of lab components (T_{ij} for i and j taking values from 1 to 3) times *dyads* of basis vectors ($\underset{\sim}{\boldsymbol{E}}_i\underset{\sim}{\boldsymbol{E}}_j$ for i and j taking values from 1 to 3) tells us that a tensor itself may be regarded as a function of single numbers (the components) and vectors (the basis vectors). Thus, in a purely functional sense, we may write

$$\underset{\approx}{T} = f(T_{11}, T_{12}, \ldots, T_{33}, \underset{\sim}{E}_1, \underset{\sim}{E}_2, \underset{\sim}{E}_3). \tag{16.1}$$

By this, we mean that having the basis vectors and the components is sufficient to reconstruct the tensor $\underset{\approx}{T}$. Specifically,

$$f(T_{11}, T_{12}, \ldots, T_{33}, \underset{\sim}{E}_1, \underset{\sim}{E}_2, \underset{\sim}{E}_3) = T_{ij}\underset{\sim}{E}_i\underset{\sim}{E}_j. \tag{16.2}$$

When this function is applied using a different set of arguments it gives

$$f\left(T_{11}^*, T_{12}^*, \ldots, T_{33}^*, \underset{\sim}{E}_1^*, \underset{\sim}{E}_2^*, \underset{\sim}{E}_3^*\right) = T_{ij}^*\underset{\sim}{E}_i^*\underset{\sim}{E}_j^*. \tag{16.3}$$

For the function f given in (16.1) to correspond to a valid *tensor construction* rule, it must satisfy the important constraint that

$$T_{ij}\underset{\sim}{E}_i\underset{\sim}{E}_j = T_{ij}^*\underset{\sim}{E}_i^*\underset{\sim}{E}_j^*. \tag{16.4}$$

In other words, whenever the basis vectors are changed, the components must change in a perfectly compensating manner so that the sum of components times basis dyads remains unchanged, which is the basis for the component transformation rules discussed in chapter 3. If this constraint is to hold for all time, then it must also hold in incremental or rate form even when the basis varies in time. Taking rates of this constraint (noting that the lab basis is constant) gives

$$\dot{T}_{ij}\underset{\sim}{E}_i\underset{\sim}{E}_j = \dot{T}_{ij}^*\underset{\sim}{E}_i^*\underset{\sim}{E}_j^* + T_{ij}^*\dot{\underset{\sim}{E}}_i^*\underset{\sim}{E}_j^* + T_{ij}^*\underset{\sim}{E}_i^*\dot{\underset{\sim}{E}}_j^*. \tag{16.5}$$

If the two basis vectors are related by[1]

$$\underset{\sim}{E}_i^* = Q_{ik}^{-1}\underset{\sim}{E}_k, \tag{16.6}$$

then (16.4) implies the component transformation rule

$$T_{mn}^* = Q_{im}Q_{jn}T_{ij}. \tag{16.7}$$

Taking rates of this equation gives

$$\dot{T}_{ij}^* = \dot{Q}_{im}Q_{jn}T_{ij} + Q_{im}\dot{Q}_{jn}T_{ij} + Q_{im}Q_{jn}\dot{T}_{ij}. \tag{16.8}$$

To take rates of (16.6), we need to keep in mind that the rate of the inverse of a matrix is *not* equal to the inverse of the rate of that matrix. Instead, $\frac{d}{dt}(Q_{ik}^{-1}) = -Q_{im}^{-1}\dot{Q}_{mn}Q_{nk}^{-1}$ (which is the matrix analog to the scalar identity that the rate of $1/x$ is given by $-\dot{x}/x^2$). With this in mind, we see that

[1] If both bases are orthonormal, then the $[Q]$ matrix will be an orthogonal direction-cosine matrix and therefore $Q_{ik}^{-1} = Q_{ik}^T = Q_{ki}$. Although not a major emphasis of this section, the statements made in this section also hold for non-orthonormal basis changes, in which case the $[Q]$ matrix would not be orthonormal, which is why we have kept the inverse typeset as Q_{ik}^{-1} instead of Q_{ki}.

$$\dot{\underset{\sim}{E}}_i^* = Q_{ik}^{-1}\dot{\underset{\sim}{E}}_k - Q_{im}^{-1}\dot{Q}_{mn}Q_{nk}^{-1}\dot{\underset{\sim}{E}}_k. \tag{16.9}$$

The relationships derived in this section are rather abstract, but they are the key to making the connection between the rate of a tensor computed using one basis to the rate of that same tensor computed using a different basis. In most engineering applications, at least one basis is held stationary, but that is only a convenience, not a necessity.

An important application of these concepts will be explored in section 16.3, where we explain how to compute the time rate of a function that is defined in terms of the principal basis of another tensor. In that case, the $\underset{\sim}{E}$-basis will be the fixed (laboratory) basis while the $\underset{\sim}{E}^*$-basis will be the principal basis, which might vary in time. Some subtleties will occur in that section associated with the fact that eigenvectors are not unique whenever eigenvalues are repeated.

16.1 A motivational example

The reader may wonder why rates of tensors should receive extensive attention in a book that is focusing on *rotations*. Sometimes, the rate of a *tensor* shares strong qualitative properties with the counterpart formulas for scalar-valued rates. For example,

$$\frac{\mathrm{d}}{\mathrm{d}t}(\underset{\approx}{A}^{-1}) = -\underset{\approx}{A}^{-1} \bullet \dot{\underset{\approx}{A}} \bullet \underset{\approx}{A}^{-1} \text{ is similar to } \frac{\mathrm{d}(1/s)}{\mathrm{d}t} = \frac{-\dot{s}}{s^2}. \tag{16.10}$$

Other times, however, the properties of a tensor rate can differ significantly from analogous rates of scalars. The distinction is directly related to the fact that the underlying principal directions associated with tensors are themselves capable of rotating. For symmetric tensors, this rotation is quantified by the orthogonal matrix whose columns contain the lab components of the orthonormal eigentriad, and the rate of rotation is quantified by an associated skew-symmetric spin tensor. Below we discuss a topic that is intimately connected to the consequences of rotating principal directions.

16.1.1 Does the sym part of the velocity gradient equal the rate of any tensor?

Given a spatially varying velocity field, the symmetric part of the velocity gradient (typically denoted $\underset{\approx}{D}$ and erroneously referred to as the 'rate' of deformation) is related to the deformation gradient $\underset{\approx}{F}$, and therefore to the polar rotation and stretch tensors according to

$$\underset{\approx}{D} = \tfrac{1}{2}[\dot{\underset{\approx}{F}} \bullet \underset{\approx}{F}^{-1} + \underset{\approx}{F}^{-T} \bullet \dot{\underset{\approx}{F}}^T], \tag{16.11}$$

$$\underset{\approx}{D} = \tfrac{1}{2}[\underset{\approx}{R} \bullet (\dot{\underset{\approx}{U}} \bullet \underset{\approx}{U}^{-1} + \underset{\approx}{U}^{-1} \bullet \dot{\underset{\approx}{U}}) \bullet \underset{\approx}{R}^T]. \tag{16.12}$$

The so-called 'un-rotated' symmetric part of the velocity gradient is defined by

$$\overline{\underset{\approx}{D}} = \underset{\approx}{R}^T \bullet \underset{\approx}{D} \bullet \underset{\approx}{R} = \tfrac{1}{2}(\dot{\underset{\approx}{U}} \bullet \underset{\approx}{U}^{-1} + \underset{\approx}{U}^{-1} \bullet \dot{\underset{\approx}{U}}). \tag{16.13}$$

If this were a scalar equation, it would read simply $\overline{D} = \dot{U}/U$, which is just the time derivative of the logarithmic strain ln(U). Many researchers treat $\overline{\underset{\approx}{D}}$ as if it *were* a logarithmic strain rate even though it is not. As will be discussed later, this is only an approximation, albeit a very good one for many engineering applications.

A perennial question in continuum mechanics addresses whether or not the full tensor $\underset{\approx}{D}$ is a 'true rate'. By this, people are asking whether or not there exists a tensor $\underset{\approx}{\varsigma}$ that is uniquely determined from $\underset{\approx}{F}$ such that $\underset{\approx}{D} = \dot{\underset{\approx}{\varsigma}}$. Most researchers are aware the answer is 'no', but they are not always able to provide a *proof*. Whenever you wish to disprove a conjecture[2], all you have to do is come up with a counter-example for which the conjecture is false. This is done by assuming the conjecture is true, and then demonstrating that your counter-example results in a contradiction. If $\underset{\approx}{\varsigma}$ is solely a function of $\underset{\approx}{F}$, then any path that begins and ends with the same $\underset{\approx}{F}$ should also begin and end with the same $\underset{\approx}{\varsigma}$. Stated more mathematically,

$$\text{if } \underset{\approx}{\varsigma} = \underset{\approx}{\varsigma}(\underset{\approx}{F}), \text{ then every path satisfying} \\ \int_{t_1}^{t_2} \dot{\underset{\approx}{F}}\, dt = 0 \text{ will also satisfy} \int_{t_1}^{t_2} \dot{\underset{\approx}{\varsigma}}\, dt = 0. \tag{16.14}$$

Stated in the logically equivalent contrapositive,

$$\text{if there exists } \textit{any} \text{ path such that} \int_{t_1}^{t_2} \dot{\underset{\approx}{F}}\, dt = 0 \text{ but} \int_{t_1}^{t_2} \underset{\approx}{\varsigma}\, dt \neq 0, \\ \text{then } \underset{\approx}{\varsigma} \text{ is } \textit{not} \text{ expressible as a function of } \underset{\approx}{F} \\ \text{and therefore } \dot{\underset{\approx}{\varsigma}} \text{ is not really a 'true rate'}. \tag{16.15}$$

Below, we are going to prove that neither $\underset{\approx}{D}$ nor $\overline{\underset{\approx}{D}}$ is a true rate, which is why careful writers do not typeset them as rates.

To demonstrate that $\underset{\approx}{D}$ is *not* the time rate of any proper function of $\underset{\approx}{F}$, all we need to do is devise a path that is closed in $\underset{\approx}{F}$ for which $\int \underset{\approx}{D}\, dt \neq 0$. For this purpose, consider the following four-stage deformation (illustrated in figure 16.1(a)):

$$[F] = \begin{bmatrix} 1 & 0 \\ 0 & 1 \end{bmatrix}(1-t) + \begin{bmatrix} 1 & 0 \\ 0 & 2 \end{bmatrix}t \qquad \text{for } 0 < t < 1, \tag{16.16}$$

$$[F] = \begin{bmatrix} 1 & 0 \\ 0 & 2 \end{bmatrix}(2-t) + \begin{bmatrix} 1 & 1 \\ 0 & 2 \end{bmatrix}(t-1) \text{ for } 1 < t < 2, \tag{16.17}$$

[2] In this case, the conjecture is that $\underset{\approx}{\varsigma}$ exists such that it is purely a function of $\underset{\approx}{F}$ and $\underset{\approx}{D} = \dot{\underset{\approx}{\varsigma}}$. Incidentally, note that alternatively supposing that $\underset{\approx}{\varsigma}$ is a function of $\underset{\approx}{V}$ is really a special case of our more general conjecture. After all, if $\underset{\approx}{\varsigma}$ depends solely on $\underset{\approx}{V}$, then we know that $\underset{\approx}{\varsigma}$ also depends solely on $\underset{\approx}{F}$ because, from the polar-decomposition theorem, we know that $\underset{\approx}{V}$ itself depends solely on $\underset{\approx}{F}$.

$$[F] = \begin{bmatrix} 1 & 1 \\ 0 & 2 \end{bmatrix}(3 - t) + \begin{bmatrix} 1 & 1 \\ 0 & 1 \end{bmatrix}(t - 2) \text{ for } 2 < t < 3, \tag{16.18}$$

$$[F] = \begin{bmatrix} 1 & 1 \\ 0 & 1 \end{bmatrix}(4 - t) + \begin{bmatrix} 1 & 0 \\ 0 & 1 \end{bmatrix}(t - 3) \text{ for } 3 < t < 4. \tag{16.19}$$

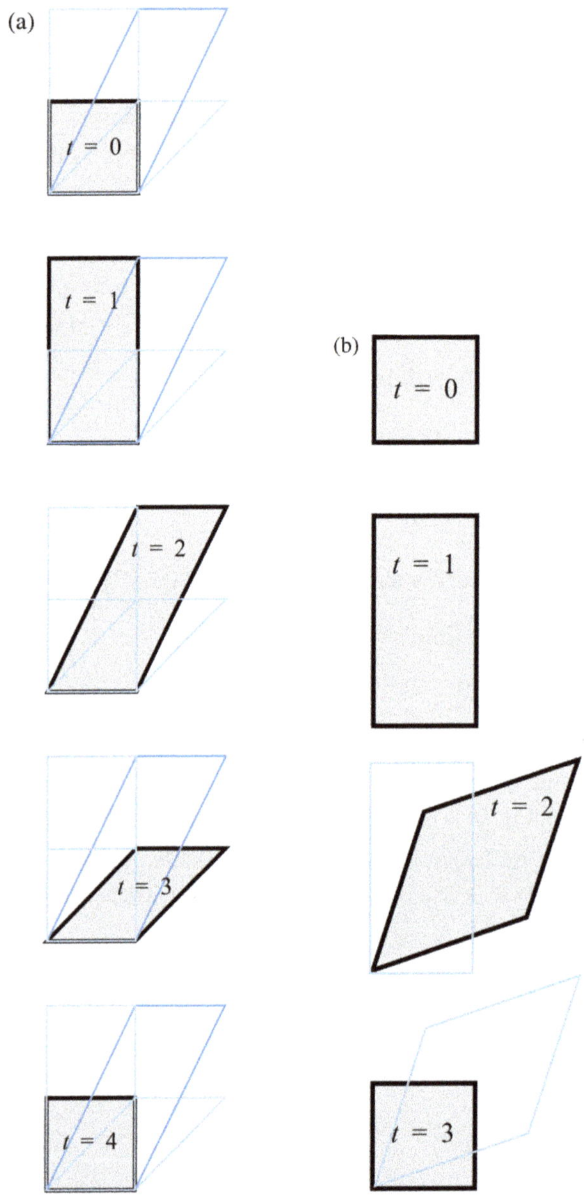

Figure 16.1. Cyclic deformations start and end in the same state (a) has four simple stages but the shearing parts involve both stretch and rotation (b) has three.

Note that $[F]$ at $t = 0$ equals $[F]$ at $t = 4$. Therefore, the path is closed in $\underset{\approx}{\mathbf{F}}$. That is, $\int_0^4 \dot{\underset{\approx}{\mathbf{F}}}\, \mathrm{d}t = 0$. The corresponding symmetric part of the velocity gradient can be computed using (16.11) to give

$$[D] = \begin{bmatrix} 0 & 0 \\ 0 & \dfrac{1}{1+t} \end{bmatrix} \text{ for } 0 < t < 1, \tag{16.20}$$

$$[D] = \begin{bmatrix} 0 & 1/4 \\ 1/4 & 0 \end{bmatrix} \text{ for } 1 < t < 2, \tag{16.21}$$

$$[D] = \begin{bmatrix} 0 & 0 \\ 0 & \dfrac{1}{t-4} \end{bmatrix} \text{ for } 2 < t < 3, \tag{16.22}$$

$$[D] = \begin{bmatrix} 0 & 1/2 \\ 1/2 & 0 \end{bmatrix} \text{ for } 3 < t < 4. \tag{16.23}$$

The time integral of $\underset{\approx}{\mathbf{D}}$ is

$$\begin{aligned} \int_0^4 \underset{\approx}{\mathbf{D}}\, \mathrm{d}t &= \int_0^1 \underset{\approx}{\mathbf{D}}\, \mathrm{d}t + \int_1^2 \underset{\approx}{\mathbf{D}}\, \mathrm{d}t + \int_2^3 \underset{\approx}{\mathbf{D}}\, \mathrm{d}t + \int_3^4 \underset{\approx}{\mathbf{D}}\, \mathrm{d}t \\ &= \begin{bmatrix} 0 & 0 \\ 0 & \ln 2 \end{bmatrix} + \begin{bmatrix} 0 & 1/4 \\ 1/4 & 0 \end{bmatrix} + \begin{bmatrix} 0 & 0 \\ 0 & -\ln 2 \end{bmatrix} + \begin{bmatrix} 0 & -1/2 \\ -1/2 & 0 \end{bmatrix} \\ &= \begin{bmatrix} 0 & -1/4 \\ -1/4 & 0 \end{bmatrix}. \end{aligned} \tag{16.24}$$

The integral is non-zero, which, applying equation (16.15), completes the proof that $\underset{\approx}{\mathbf{D}}$ is not a true rate of any quantity that can be expressed as a function of the deformation gradient.

The preceding derivation used a four-stage deformation sequence: (1) elongate vertically, (2) shear to the right, (3) compress vertically, and (4) shear to the left. This path was selected to simplify the equations as much as possible for the counter-example. Now that we know that $\underset{\approx}{\mathbf{D}}$ is not a true rate, a natural follow-up activity is to determine what *subclass* of deformations, if any, will permit $\underset{\approx}{\mathbf{D}}$ to be a true rate. In other words, we are now exploring the time integral of $\underset{\approx}{\mathbf{D}}$ associated with closed paths in $\underset{\approx}{\mathbf{F}}$ for which all values of $\underset{\approx}{\mathbf{F}}$ are restricted to belong to some particular subclass.

In the preceding counter-example, the non-zero contributions to the time integral of $\underset{\approx}{\mathbf{D}}$ might have appeared (at first glance) to arise solely from the simple shearing stages, so it is natural to question whether the time integral of $\underset{\approx}{\mathbf{D}}$ might come out to be zero whenever we limit ourselves to deformations that are always pure stretches (i.e., deformations where the polar rotation tensor $\underset{\sim}{\boldsymbol{R}}$ always equals the identity $\underset{\sim}{\boldsymbol{I}}$). For this class of deformations, there is no distinction between $\underset{\sim}{\boldsymbol{D}}$ and $\underset{\sim}{\overline{\boldsymbol{D}}}$. Hence, this question is the same as asking whether or not $\underset{\sim}{\overline{\boldsymbol{D}}}$, defined in (16.13) is a true rate. The answer again turns out to be 'no' and the following three-stage loading, illustrated in Figure 16.1(b), is a counter-example:

$$[F] = \begin{bmatrix} 1 & 0 \\ 0 & 1 \end{bmatrix}(1 - t) + \begin{bmatrix} 1 & 0 \\ 0 & 2 \end{bmatrix}t \qquad \text{for } 0 < t < 1, \tag{16.25}$$

$$[F] = \begin{bmatrix} 1 & 0 \\ 0 & 2 \end{bmatrix}(2 - t) + \begin{bmatrix} 3/2 & 1/2 \\ 1/2 & 3/2 \end{bmatrix}(t - 1) \text{ for } 1 < t < 2, \tag{16.26}$$

$$[F] = \begin{bmatrix} 3/2 & 1/2 \\ 1/2 & 3/2 \end{bmatrix}(3 - t) + \begin{bmatrix} 1 & 0 \\ 0 & 1 \end{bmatrix}(t - 2) \text{ for } 2 < t < 3. \tag{16.27}$$

This problem can be analyzed in a manner similar to what was done in the four-stage loading counter-example, but the math is messy (so we recommend using a symbolic math program such as Mathematica). The time integral of $\underset{\sim}{\overline{\boldsymbol{D}}}$ again turns out to be non-zero even though the deformation consists only of pure stretches.

As will be clarified later, $\underset{\sim}{\overline{\boldsymbol{D}}}$ is a true rate if and only if principal directions of the reference (right) stretch do not rotate. The above example illustrates that the principal directions of stretch can change *without* having any net material rotation (i.e., the polar rotation can remain $\underset{\sim}{\boldsymbol{R}} = \underset{\sim}{\boldsymbol{I}}$). Also, the condition does not exclude rotation of spatial (left) stretch directions.

16.2 A comment about rates of proper functions

Whenever a tensor $\underset{\sim}{\boldsymbol{Y}}$ depends solely on a second tensor $\underset{\sim}{\boldsymbol{A}}$, then the chain rule tells us that its rate can be expressed as

$$\dot{\underset{\sim}{\boldsymbol{Y}}} = \underset{\approx}{\boldsymbol{L}}:\dot{\underset{\sim}{\boldsymbol{A}}}, \text{ where } \underset{\approx}{\boldsymbol{L}} = \frac{\mathrm{d}\underset{\sim}{\boldsymbol{Y}}}{\mathrm{d}\underset{\sim}{\boldsymbol{A}}}. \tag{16.28}$$

The fourth-order tangent tensor $\underset{\approx}{\boldsymbol{L}}$ depends on $\underset{\sim}{\boldsymbol{A}}$, but not on $\dot{\underset{\sim}{\boldsymbol{A}}}$. If the laboratory components of $\underset{\sim}{\boldsymbol{Y}}$ can be expressed in terms of the laboratory components of $\underset{\sim}{\boldsymbol{A}}$, then the indicial form of the first expression in (16.28) is $\dot{Y}_{ij} = L_{ijkl}\dot{A}_{kl}$ with the lab components of the fourth-order tangent $\underset{\approx}{\boldsymbol{L}}$ tensor given by

$$L_{ijkl} = \frac{\partial Y_{ij}}{\partial A_{kl}} \quad \text{possibly with adjustments, as explained below.} \tag{16.29}$$

Suppose, as an extremely trivial example, that

$$\underset{\sim}{\boldsymbol{Y}} = \underset{\sim}{\boldsymbol{A}}. \tag{16.30}$$

Then

$$Y_{ij} = A_{ij}, \tag{16.31}$$

and therefore

$$L_{ijkl} = \delta_{ik}\delta_{jl}. \tag{16.32}$$

If the function is to be applied in situations where $\underset{\approx}{Y}$ will always be symmetric, then the first two indices of L_{ijkl} will *by necessity* be symmetric (i.e., $L_{ijkl} = L_{jikl}$). If the function arises in a physical model for which the independent variable $\underset{\approx}{A}$ will also be always symmetric, then it is customary to assume 'without loss in generality' that L_{ijkl} has the property that $L_{ijkl} = L_{ijlk}$. *This property will not be satisfied automatically—the analyst must force it upon the* $\underset{\approx\approx}{L}$ *tensor*. That is, once a nominal expression for L_{ijkl} has been obtained, then

$$\text{replace} \quad L_{ijkl} \text{ by } \tfrac{1}{2}(L_{ijkl} + L_{ijlk}). \tag{16.33}$$

Thus, for our example that $\underset{\approx}{Y} = \underset{\approx}{A}$, we would assert that

$$L_{ijkl} = \tfrac{1}{2}(\delta_{ik}\delta_{jl} + \delta_{il}\delta_{jk}). \tag{16.34}$$

This result is *not* identical to (16.32). This adjustment for symmetric tensors must be done manually after a nominal result is obtained. When the tensor $\underset{\approx\approx}{L}$ in (16.32) operates on an arbitrary—permissibly non-symmetric—tensor $\underset{\approx}{B}$, then the result is $\underset{\approx\approx}{L} : \underset{\approx}{B} = \underset{\approx}{B}$, whereas when the tensor in (16.33) operates on $\underset{\approx}{B}$ the result is $\underset{\approx\approx}{L} : \underset{\approx}{B} = \frac{1}{2}(\underset{\approx}{B} + \underset{\approx}{B}^T)$. If the tensor $\underset{\approx}{B}$ happens to be symmetric, then both results are identical, which is why researchers assert that they may impose minor symmetry in the last two indices without loss in generality. The tensor in (16.34) is the identity for the subspace of symmetric tensors.

Equation (16.29) was derived on the premise that it was possible to express the lab components of $\underset{\approx}{Y}$ in terms of the lab components of $\underset{\approx}{A}$. There are some functions, however, for which this is possible in principal, but not especially clean in practice. We discuss one important class of such functions now.

16.3 The time rate of a principal function of a symmetric tensor[3]

*Note:*Throughout this section, we will suspend the summation conventions so that *there are no implied summations*. All equations that require a summation will show the summation symbol explicitly. Any non-summed indices that appear in an equation should be regarded as free indices. We will resume using the summation conventions after this section is completed.

[3] Several results and techniques described in this section were contributed by Bill Scherzinger via personal discussions. Most of the information in this section has appeared periodically in *specialized* contexts in the literature [1], so the pedigree for credit is unclear. Our own new contributions include: (1) addressing issues associated with multiple eigenvalue roots, (2) illustrating the mathematics via an example of computing the rate of the square of a tensor, (3) introducing the function in (16.50), (4) the compact formula given in section 16.3, (5) specific application to compute the logarithmic strain rate via a principal component adjustment of the 'rate' of deformation tensor, and (6) a practical run-time estimate of the error associated with approximating the log strain rate by the symmetric part of the velocity gradient (section 16.4).

Suppose that a tensor $\underset{\approx}{Y}$ depends solely on a second-order symmetric tensor $\underset{\approx}{A}$. In this section, we will consider one particular class of this type of function in which it is not always possible to express the lab components Y_{ij} as an *algebraic* function of the lab components A_{ij}. *Principal functions* are defined with respect to the eigenvalues $\{\lambda_1, \lambda_2, \lambda_3\}$ and *orthonormal* eigenvectors $\{\underset{\sim}{p}_1, \underset{\sim}{p}_2, \underset{\sim}{p}_3\}$ of $\underset{\approx}{A}$ in the form

$\underset{\approx}{Y} = \sum_{i=1} f(\lambda_i)\underset{\sim}{p}_i\underset{\sim}{p}_i$ which implies the matrix formula

$$[Y] = \begin{bmatrix} f(\lambda_1) & 0 & 0 \\ 0 & f(\lambda_2) & 0 \\ 0 & 0 & f(\lambda_3) \end{bmatrix}_{<A>} \quad . \tag{16.35}$$

The subscript <A> on the matrix indicates that the components are with respect to the principal basis $\{\underset{\sim}{p}_1, \underset{\sim}{p}_2, \underset{\sim}{p}_3\}$ for $\underset{\approx}{A}$. An equation $\underset{\approx}{Y} = \underset{\approx}{A}$ is a 'trivial' example of a principal function corresponding to $f(\lambda_i) = \lambda_i$. Similarly, if $\underset{\approx}{Y} = \underset{\approx}{A} \bullet \underset{\approx}{A}$, then $f(\lambda_i) = \lambda_i^2$. More generally, if $\underset{\approx}{Y} = \underset{\approx}{A}^k$ for some integer k, then $f(\lambda_i) = \lambda_i^k$. For all of these examples, it is possible to find the rate of $\underset{\approx}{Y}$ through classical means. For example, if $\underset{\approx}{Y} = \underset{\approx}{A} \bullet \underset{\approx}{A}$, then $\dot{\underset{\approx}{Y}} = \dot{\underset{\approx}{A}} \bullet \underset{\approx}{A} + \underset{\approx}{A} \bullet \dot{\underset{\approx}{A}}$. The classical method for finding rates of $\underset{\approx}{A}^k$ becomes impractical when the exponent is large. For example, finding the rate of $\underset{\approx}{Y} = \underset{\approx}{A}^{100}$ would require one hundred terms, with each term involving 100 matrix multiplications. When the function f is not a simple polynomial (e.g., if $f(\lambda_i) = \ln \lambda_i$), then the classical methods of taking rates do not apply, and alternative methods are needed.

If $\underset{\approx}{Y}$ is defined according to (16.35) then its rate must be

$$\dot{\underset{\approx}{Y}} = \sum_{i=1}^{3}\left[f'(\lambda_i)\dot{\lambda}_i\underset{\sim}{p}_i\underset{\sim}{p}_i + f(\lambda_i)\dot{\underset{\sim}{p}}_i\underset{\sim}{p}_i + f(\lambda_i)\underset{\sim}{p}_i\dot{\underset{\sim}{p}}_i\right], \tag{16.36}$$

where

$$f'(\lambda_i) \equiv \frac{df}{d\lambda_i}. \tag{16.37}$$

The remainder of this section will focus on proving (and then illustrating via example) the following theorems:

If $\underset{\approx}{Y} = \sum_{i=1}^{3} f(\lambda_i)\underset{\sim}{p}_i\underset{\sim}{p}_i$, where $\{\lambda_1, \lambda_2, \lambda_3\}$ are the eigenvalues of some symmetric tensor $\underset{\approx}{A}$ and $\{\underset{\sim}{p}_1, \underset{\sim}{p}_2, \underset{\sim}{p}_3\}$ are the associated orthonormalized eigenvectors, then the *laboratory components* of $\dot{\underset{\approx}{Y}}$ can be computed from the *laboratory components* of $\dot{\underset{\approx}{A}}$ by

$$\boxed{\dot{Y}_{mn}^{<lab>} = \sum_{r=1}^{3}\sum_{s=1}^{3}\sum_{k=1}^{3}\sum_{l=1}^{3}(Q_{mk}Q_{nl}Q_{rk}Q_{sl}\gamma_{kl})\dot{A}_{pq}^{<lab>}}, \tag{16.38}$$

where, letting λ_i denote the ith eigenvalue of $\underset{\approx}{A}$

$$\text{and}\quad \boxed{\gamma_{ij} = \begin{cases} \dfrac{f(\lambda_i) - f(\lambda_j)}{\lambda_i - \lambda_j} & \text{if } \lambda_i \neq \lambda_j \\ f'(\lambda_i) & \text{if } \lambda_i = \lambda_j \end{cases}}. \tag{16.39}$$

Here, Q_{ij} denotes the ith lab component of the jth *orthonormalized* eigenvector of $\underset{\approx}{A}$ (i.e., $Q_{ij} \equiv \underset{\sim}{e}_i \bullet \underset{\sim}{p}_j$ so that the columns of $[Q]$ contain the lab components of the eigenvectors). *This result holds even if the tensor* $\underset{\approx}{A}$ *has repeated eigenvalues*; that is, this formula still gives the correct answer no matter what choice is made for the orthonormalized eigenvectors despite the fact they are not unique when there are repeated eigenvalues!

Commands for implementing this function in Mathematica are

```
prinFn[A_, f_] := Module[{eval, evecs, Q, Y},
  {eval, evecs} = Eigensystem[A]; Q = Transpose[Orthogonalize[evecs]];
  Y = Q.DiagonalMatrix[f[eval]].Q^T;
  {Y, {eval, Q}}
 ]
```

```
rateOfprinFn[A_, Adot_, f_] := Module[{eval, evecs, Q, G, γ, Y, Ydot, m, n, p, q, k, l},
  {Y, {eval, Q}} = prinFn[A, f];
  G[x_, y_] := If[x == y,  f'[x],  (f[x] - f[y])/(x - y)];
  γ = Table[G[eval[[i]], eval[[j]]], {i, 3}, {j, 3}];
  Ydot = Table[
    Sum_{p=1}^{3} Sum_{q=1}^{3} Sum_{k=1}^{3} Sum_{l=1}^{3} Q[[m, k]] Q[[n, l]] Q[[p, k]] Q[[q, l]] Adot[[p, q]] γ[[k, l]]
    , {m, 3}, {n, 3}];
  {Y, Ydot, {eval, Q}}]
```

The formula for the rate may be written entirely in direct notation as follows:

If $\underset{\approx}{Y} = \sum_{i=1}^{3} f(\lambda_i)\underset{\approx}{P}_i$, where $\{\lambda_1, \lambda_2, \lambda_3\}$ are the eigenvalues of some symmetric tensor $\underset{\approx}{A}$ and $\{\underset{\approx}{P}_1, \underset{\approx}{P}_2, \underset{\approx}{P}_3\}$ are the associated eigen*projectors*, then

$$\boxed{\dot{\underset{\approx}{Y}} = \sum_{i=}^{3}\sum_{j=1}^{3}\gamma_{ij}\underset{\approx}{P}_i \bullet \underset{\approx}{A} \bullet \underset{\approx}{P}_j, \quad \text{where} \quad \gamma_{ij} = \begin{cases} \dfrac{f(\lambda_i) - f(\lambda_j)}{\lambda_i - \lambda_j} & \text{if } \lambda_i \neq \lambda_j \\ f'(\lambda_i) & \text{if } \lambda_i = \lambda_j \end{cases}}.$$

After outlining the proof, an example will be provided section 16.3.2 for a verification problem in which a more classical analytical solution is available. Then, in section 16.4, this general result will be applied to the physical problem of evaluating the material time rate, the objective time rate, and the reference time rate of the logarithmic strain, where we will provide an even simpler formula showing that the reference logarithmic strain rate can be obtained directly from a simple modification of the *off-diagonal* components of the symmetric part of the velocity gradient expressed in the principal stretch basis.

Proof. The following proof of the boxed statement (equation (16.38)) employs more than one basis at a time, so we use labels to distinguish which basis is associated with components. In this chapter, a subscript or superscript <lab> will be used to indicate that components are referenced to the laboratory basis $\{\mathbf{E}_1, \mathbf{E}_2, \mathbf{E}_3\}$. Thus, for any tensor $\mathbf{Z}$,

$$Z_{ij}^{\langle\text{lab}\rangle} \equiv \mathbf{E}_i \bullet \mathbf{Z} \bullet \mathbf{E}_j \text{ and therefore } \mathbf{Z} = Z_{ij}^{\langle\text{lab}\rangle}\mathbf{E}_i\mathbf{E}_j. \tag{16.40}$$

For our proof of equation (16.38), we are interested in finding the lab components of $\dot{\mathbf{Y}}$ when $\mathbf{Y}$ is given by a principal function of some symmetric tensor $\mathbf{A}$. We will need to discuss the steps of the proof using components with respect to the principal basis for $\mathbf{A}$. Such components will be called the 'A-space' components. A superscript <A> will be used to indicate that components of a tensor are referenced to this principal basis for $\mathbf{A}$ (on matrices, a *subscript* <A> will be used to avoid confusion with matrix operators and exponents). Of course, with respect to its own principal basis, the A-space component matrix for $\mathbf{A}$ will be diagonal:

$$[\mathbf{A}] = \begin{bmatrix} \lambda_1 & 0 & 0 \\ 0 & \lambda_2 & 0 \\ 0 & 0 & \lambda_3 \end{bmatrix}_{\langle A\rangle}, \tag{16.41}$$

where $\{\lambda_1, \lambda_2, \lambda_3\}$ are the eigenvalues of $\mathbf{A}$. Even though the A-space component matrix for $\mathbf{A}$ is diagonal, the A-space component matrices for *other* tensors (including the rate $\dot{\mathbf{A}}$) can, in general, be fully populated with up to nine non-zero components. For any tensor $\mathbf{Z}$, the 'A-space components' are defined

$$Z_{ij}^{\langle A\rangle} \equiv \mathbf{p}_i \bullet \mathbf{Z} \bullet \mathbf{p}_j \quad \text{and therefore} \quad \mathbf{Z} = Z_{ij}^{\langle A\rangle}\mathbf{p}_i\mathbf{p}_j, \tag{16.42}$$

where $\{\mathbf{p}_1, \mathbf{p}_2, \mathbf{p}_3\}$ are the orthonormalized eigenvectors of $\mathbf{A}$, ordered consistently with the eigenvalues $\{\lambda_1, \lambda_2, \lambda_3\}$. If the A-space components of any tensor $\mathbf{Z}$ are known, then the lab components of $\mathbf{Z}$ are obtained by the usual transformation rule:

$$Z_{ij}^{\langle\text{lab}\rangle} = \sum_m^3 \sum_n^3 Q_{im}Q_{jn}Z_{mn}^{\langle A\rangle}, \quad \text{where} \quad Q_{rs} = \mathbf{E}_r \bullet \mathbf{p}_s. \tag{16.43}$$

Here, $[Q]$ is a matrix whose columns contain the lab components of the eigenvectors.

With this A-space notation now defined, we are ready to proceed with the proof of the boxed statement in section 16.3.

If the eigenvalues of $\underset{\approx}{A}$ are all distinct, then the eigenvectors themselves are unique to within a scalar multiple. If, on the other hand, two eigenvalues are equal, then the associated two eigenvectors are non-unique in magnitude *and direction*—they are merely constrained to lie within a given plane. If three eigenvalues are equal (i.e., if $\underset{\approx}{A}$ is proportional to the second-order identity tensor), then *any* vector is an eigenvector of $\underset{\approx}{A}$. Whenever eigenvalues are repeated, then one must look at the *rate* of the tensor to determine if the eigenvectors will continue to be arbitrary an instant later.

To first order, the tensor $\underset{\approx}{A}$ will vary such that

$$\underset{\approx}{A}(t) = \underset{\approx}{A}(t_o) + \dot{\underset{\approx}{A}}(t_o)(t - t_o). \tag{16.44}$$

It might happen that the tensor $\underset{\approx}{A}$ has eigenvalues with multiplicity two at $t = t_o$, in which case the associated eigenvectors are arbitrary, so long as they lie in a given plane. However, the *rates* of the eigenvalues and their associated eigenvectors might *not* be arbitrary—it all depends on the nature of $\dot{\underset{\approx}{A}}$. If two eigenvalues are repeated at an instant in time, then they will remain equal to first order in time only if their rates are also equal. Otherwise, if the rates of two instantaneously equal eigenvalues are different, then the eigenvectors can potentially jump discontinuously from their arbitrary orientations into new *non-arbitrary* directions to align themselves with the new principal directions required at $t = t_o^{(+)}$. The only way to prevent a discontinuous jump in the principal directions is to *choose* them from the outset to be consistent with the new directions implied $\dot{\underset{\approx}{A}}$. Equation (16.91) will later provide an example of how to compute these continuously varying eigenvectors. Fortunately, we will find that it is *not* necessary to actually compute them—we only need to assert that *there exists* at least one choice for the orthonormal triad that will be continuous in time from $t = t_o$ to $t = t_o^{(+)}$. We will now proceed with the derivation on the assumption that the principal directions are so-aligned. Then, at the end of the proof, we will explain why the principal directions can be replaced with their original arbitrary in-plane directions at $t = t_o^{(-)}$ without affecting the final result.

Given that the eigenvectors are orthonormal, and given that we have presumed that they occupy orientations that make them continuously differentiable through time, we know that their time rate must be expressible in the form of a skew-symmetric tensor $\underset{\approx}{W}$ operating on the eigenvector itself:

$$\dot{\underset{\sim}{p}}_i = \underset{\approx}{W} \bullet \underset{\sim}{p}_i, \quad \text{where} \quad \underset{\approx}{W}^T = -\underset{\approx}{W}. \tag{16.45}$$

Substituting this expression into (16.36) gives

$$\dot{\underset{\approx}{Y}} = \sum_{i=1}^{3} [f'(\lambda_i)\dot{\lambda}_i \underset{\sim}{p}_i \underset{\sim}{p}_i + f(\lambda_i)(\underset{\approx}{W} \bullet \underset{\sim}{p}_i)\underset{\sim}{p}_i + f(\lambda_j)\underset{\sim}{p}_i(\underset{\approx}{W} \bullet \underset{\sim}{p}_i)]. \tag{16.46}$$

The A-space *rs* component of $\dot{\underset{\approx}{Y}}$, which is the component with respect to the principal basis of $\underset{\approx}{A}$, will be denoted by $(\dot{\underset{\approx}{Y}})_{rs}^{<A>}$. Rates have higher precedence than other operations. Specifically, $(\dot{\underset{\approx}{Y}})^{<A>}_{rs}$ is *not* the time derivative of the A-space *rs* component of $\underset{\approx}{Y}$,

$$(\dot{\underset{\approx}{Y}})^{<A>}_{rs} \neq \frac{\mathrm{d}}{\mathrm{d}t}(Y^{<A>}_{rs}). \tag{16.47}$$

Recalling that the superscript <A> indicates that these components are referenced to the principal basis for $\underset{\approx}{A}$, the above equation may be written equivalently as

$$\underset{\sim}{p}_r \bullet \frac{\mathrm{d}\underset{\approx}{Y}}{\mathrm{d}t} \bullet \underset{\sim}{p}_s \neq \frac{\mathrm{d}}{\mathrm{d}t}(\underset{\sim}{p}_r \bullet \underset{\approx}{Y} \bullet \underset{\sim}{p}_s). \tag{16.48}$$

Eigenvectors of $\underset{\approx}{A}$ might themselves vary with time, so the time derivative on the left side of (16.48) cannot be 'pulled to the outside'. For this same reason, the A-space components of the time rate $\dot{\underset{\approx}{A}}$ will not generally be diagonal even though the A-space components of $\underset{\approx}{A}$ *are* diagonal. In fact, the A-space components of $\dot{\underset{\approx}{A}}$ will be diagonal only if the principal directions of $\underset{\approx}{A}$ do not vary in time. We are interested in the case where these principal directions *do* vary in time.

According to (16.48), the *rs* A-space component of $\dot{\underset{\approx}{Y}}$ can be found by dotting (16.46) from the left by $\underset{\sim}{p}_r$ and from the right by $\underset{\sim}{p}_S$, giving

$$\left(\dot{\underset{\approx}{Y}}\right)^{<A>}_{rs} = \sum_{i=1}^{3}[f'(\lambda_i)\dot{\lambda}_i\delta_{ir}\delta_{is} + f(\lambda_i)(\underset{\approx}{W})^{<A>}_{ri}\delta_{is} + f(\lambda_i)\delta_{ir}(\underset{\approx}{W})^{<A>}_{si}], \tag{16.49}$$

which (because of the Kronecker deltas in the summations) simplifies to

$$(\dot{\underset{\approx}{Y}})^{<A>}_{rs} = f'(\lambda_r)\dot{\lambda}_r\delta_{rs} + f(\lambda_s)(\underset{\approx}{W})^{<A>}_{rs} + f(\lambda_r)(\underset{\approx}{W})^{<A>}_{sr}, \tag{16.50}$$

or, noting that $W_{rs} = -W_{sr}$,

$$(\dot{\underset{\approx}{Y}})^{<A>}_{rs} = f'(\lambda_r)\dot{\lambda}_i\delta_{rs} + (\underset{\approx}{W})^{<A>}_{sr}[f(\lambda_r) - f(\lambda_S)]. \tag{16.51}$$

At this point, we would like to introduce a 'helper' function $G(x,\ \mathrm{y})$ defined as follows:

$$G(x, y) = \begin{cases} \dfrac{f(x) - f(y)}{x - y} & \text{if } x \neq y, \\ f'(x) & \text{if } x = y. \end{cases} \tag{16.52}$$

Even though we have provided a special definition of this function for the special case when $x = y$, note that

$$\lim_{x \to y} \frac{f(x) - f(y)}{x - y} = f'(x). \tag{16.53}$$

Therefore $G(x, y)$ is a continuous function. Referring to (16.53), and keeping in mind that there are no implied sums in this chapter, note that

$$f'(\lambda_r)\delta_{rs} = \gamma_{rs}\delta_{rs}, \quad \text{where} \quad \gamma_{rs} \equiv G(\lambda_r, \lambda_s). \tag{16.54}$$

Also

$$f(\lambda_r) - f(\lambda_s) = \gamma_{rs}[\lambda_r - \lambda_S]. \tag{16.55}$$

Therefore, equation (16.53) may be written

$$(\dot{\underset{\approx}{Y}})_{rs}^{<A>} = \gamma_{rs}[\delta_{rs}\dot{\lambda}_r + (\lambda_r - \lambda_s)(\underset{\sim}{W})_{sr}^{<A>}]. \tag{16.56}$$

This result must hold for any differentiable function f. Therefore it must hold for the special case that $f(x) = x$. This particular choice for the function f corresponds to $\underset{\approx}{Y} = \underset{\approx}{A}$. Furthermore, when $f(x) = x$, equation (16.52) shows that $G(x, y) = 1$ and therefore all nine of the γ_{rs} components happen to equal 1 when $f(x) = x$. Thus, when applied to the special case $\underset{\approx}{Y} = \underset{\approx}{A}$, equation (16.56) gives

$$(\dot{\underset{\approx}{A}})_{rs}^{<A>} = [\delta_{rs}\dot{\lambda}_r + (\lambda_r - \lambda_s)(\underset{\sim}{W})_{sr}^{<A>}]. \tag{16.57}$$

Comparing this result with the general formula in (16.56) shows that, for the general case of a principal function, $\underset{\approx}{Y} = f(\underset{\approx}{A})$, each A-space component of $\dot{\underset{\approx}{Y}}$ is simply a multiple of the corresponding A-space components of $\dot{\underset{\approx}{A}}$, with this proportionality factor being $G(\lambda_r, \lambda_s)$: $\boxed{(\dot{\underset{\approx}{Y}})_{rs}^{<A>} = \gamma_{rs}(\dot{\underset{\approx}{A}})_{rs}^{<A>}}$, no sum on r or s,

$$\text{where} \quad \boxed{\gamma_{ij} = \begin{cases} \dfrac{f(\lambda_i) - f(\lambda_j)}{\lambda_i - \lambda_j} & \text{if } \lambda_i \neq \lambda_j \\ f'(\lambda_i) & \text{if } \lambda_i = \lambda_j \end{cases} = G(\lambda_i, \lambda_j)}. \tag{16.58}$$

Of course, with the A-space components known, the lab components can, at any time, be obtained through the simple component transformation in (16.42). Staying in A-space for now, the matrix form of the above result can be written

$$\begin{bmatrix} (\dot{\underset{\approx}{Y}})_{11} & (\dot{\underset{\approx}{Y}})_{12} & (\dot{\underset{\approx}{Y}})_{13} \\ & (\dot{\underset{\approx}{Y}})_{22} & (\dot{\underset{\approx}{Y}})_{23} \\ \text{sym} & & (\dot{\underset{\approx}{Y}})_{33} \end{bmatrix}_{<A>} = \begin{bmatrix} \gamma_{11}(\dot{\underset{\approx}{A}})_{11} & \gamma_{12}(\dot{\underset{\approx}{A}})_{12} & \gamma_{13}(\dot{\underset{\approx}{A}})_{13} \\ & \gamma_{22}(\dot{\underset{\approx}{A}})_{22} & \gamma_{23}(\dot{\underset{\approx}{A}})_{23} \\ \text{sym} & & \gamma_{33}(\dot{\underset{\approx}{A}})_{33} \end{bmatrix}_{<A>}. \tag{16.59}$$

The subscript <A> emphasizes that these are A-space components. Even though *principal* functions of $\underset{\approx}{A}$ are a special category of functions, they are nevertheless still valid *and differentiable* functions. Therefore, we may assert there exists a fourth-order tensor $\underset{\approx\approx}{L}$ such that

$$\dot{\underset{\approx}{Y}} = \underset{\approx\approx}{L} : \dot{\underset{\approx}{A}}. \tag{16.60}$$

When expressed in the A-space basis, equation (16.58) shows that

$$
\begin{aligned}
(\dot{\underset{\approx}{\mathbf{Y}}})_{ij}^{<A>} &= \sum_{k=1}^{3}\sum_{l=1}^{3}(\underset{\approx\approx}{\mathbf{L}})_{ijkl}^{<A>} : (\dot{\underset{\approx}{\mathbf{A}}})_{kl}^{<A>}, \\
\text{where } (\underset{\approx\approx}{\mathbf{L}})_{ijkl}^{<A>} &= \gamma_{ij}\left(\frac{\delta_{ik}\delta_{jl} + \delta_{il}\delta_{jk}}{2}\right) \text{ (no sums implied).}
\end{aligned}
\tag{16.61}
$$

These are the A-space components of $\underset{\approx\approx}{\mathbf{L}}$. If the A-space components for $\dot{\underset{\approx}{\mathbf{A}}}$ and $\dot{\underset{\approx}{\mathbf{Y}}}$ are stored as 6×1 arrays, then the above result may be written alternatively with respect to the A-space as[4]

$$
\begin{bmatrix}
(\dot{\underset{\approx}{\mathbf{Y}}})_{11}^{<A>} \\
(\dot{\underset{\approx}{\mathbf{Y}}})_{22}^{<A>} \\
(\dot{\underset{\approx}{\mathbf{Y}}})_{33}^{<A>} \\
(\dot{\underset{\approx}{\mathbf{Y}}})_{23}^{<A>} \\
(\dot{\underset{\approx}{\mathbf{Y}}})_{31}^{<A>} \\
(\dot{\underset{\approx}{\mathbf{Y}}})_{12}^{<A>}
\end{bmatrix}
=
\begin{bmatrix}
\gamma_{11} & 0 & 0 & 0 & 0 & 0 \\
0 & \gamma_{22} & 0 & 0 & 0 & 0 \\
0 & 0 & \gamma_{33} & 0 & 0 & 0 \\
0 & 0 & 0 & \gamma_{23} & 0 & 0 \\
0 & 0 & 0 & 0 & \gamma_{31} & 0 \\
0 & 0 & 0 & 0 & 0 & \gamma_{12}
\end{bmatrix}
\begin{bmatrix}
(\dot{\underset{\approx}{\mathbf{A}}})_{11}^{<A>} \\
(\dot{\underset{\approx}{\mathbf{A}}})_{22}^{<A>} \\
(\dot{\underset{\approx}{\mathbf{A}}})_{33}^{<A>} \\
(\dot{\underset{\approx}{\mathbf{A}}})_{23}^{<A>} \\
(\dot{\underset{\approx}{\mathbf{A}}})_{31}^{<A>} \\
(\dot{\underset{\approx}{\mathbf{A}}})_{12}^{<A>}
\end{bmatrix}.
\tag{16.62}
$$

This result can be transformed back to give the components of $\underset{\approx\approx}{\mathbf{L}}$ with respect to the *laboratory* basis

$$
L_{mnpq}^{<lab>} = \sum_{i=1}^{3}\sum_{j=1}^{3}\sum_{k=1}^{3}\sum_{l=1}^{3} Q_{mi}Q_{nj}Q_{pk}Q_{ql}\gamma_{ij}\left(\frac{\delta_{ik}\delta_{jl} + \delta_{il}\delta_{jk}}{2}\right),
\tag{16.63}
$$

which, noting that $\gamma_{ij} = \gamma_{ji}$, simplifies to

$$
L_{mnpq}^{<lab>} = \frac{1}{2}\sum_{k=1}^{3}\sum_{l=1}^{3}(Q_{mk}Q_{nl} + Q_{ml}Q_{nk})(Q_{pk}Q_{ql})\gamma_{kl}.
\tag{16.64}
$$

In basis notation,

$$
\underset{\approx\approx}{\mathbf{L}} = \sum_{i=1}^{3}\sum_{j=1}^{3}\gamma_{ij}\left(\frac{\underset{\sim}{\mathbf{p}}_i\underset{\sim}{\mathbf{p}}_j + \underset{\sim}{\mathbf{p}}_j\underset{\sim}{\mathbf{p}}_i}{2}\right)\left(\frac{\underset{\sim}{\mathbf{p}}_i\underset{\sim}{\mathbf{p}}_j + \underset{\sim}{\mathbf{p}}_j\underset{\sim}{\mathbf{p}}_i}{2}\right),
\tag{16.65}
$$

This tensor represents the derivative of $\underset{\approx}{\mathbf{Y}}$ with respect to $\underset{\approx}{\mathbf{A}}$. Thus, the calculation of $\dot{\underset{\approx}{\mathbf{Y}}}$ can be performed directly in the lab basis by applying the dyadic form of the expression

[4] The 6×6 matrix in this equation is the *Mandel* matrix (see chapter 26), which means that the L_{11}, L_{22}, and L_{33} diagonal components equal the L_{1111}, L_{2222}, and L_{3333} components of the fourth-order tensor, respectively, whereas the L_{44}, L_{55}, and L_{66} diagonal components equal $2L_{2323}$, $2L_{3131}$, and $2L_{1212}$, respectively.

$$\underset{\approx}{\mathbf{Y}} = \underset{\approx\approx}{\mathbf{L}} : \underset{\approx}{\mathbf{A}} \tag{16.66}$$

to give

$$\dot{\underset{\approx}{\mathbf{Y}}} = \frac{1}{2}\sum_{i=1}^{3}\sum_{j=1}^{3}\gamma_{ij}(\underset{\sim}{\mathbf{p}}_i\underset{\sim}{\mathbf{p}}_j + \underset{\sim}{\mathbf{p}}_j\underset{\sim}{\mathbf{p}}_i)(\underset{\sim}{\mathbf{p}}_i \bullet \dot{\underset{\approx}{\mathbf{A}}} \bullet \underset{\sim}{\mathbf{p}}_j), \tag{16.67}$$

where we have noted that, because the operand $\dot{\underset{\approx}{\mathbf{A}}}$ is symmetric,

$$\left(\frac{\underset{\sim}{\mathbf{p}}_i\underset{\sim}{\mathbf{p}}_j + \underset{\sim}{\mathbf{p}}_j\underset{\sim}{\mathbf{p}}_i}{2}\right) : \dot{\underset{\approx}{\mathbf{A}}} = \underset{\sim}{\mathbf{p}}_i \bullet \dot{\underset{\approx}{\mathbf{A}}} \bullet \underset{\sim}{\mathbf{p}}_j. \tag{16.68}$$

In pure component form, one way to compute the *lab* components of $\dot{\underset{\approx}{\mathbf{Y}}}$ in terms of the lab components of $\dot{\underset{\approx}{\mathbf{A}}}$ is

$$\boxed{\dot{Y}_{mn}^{\text{<lab>}} = \sum_{r=1}^{3}\sum_{s=1}^{3}\sum_{k=1}^{3}\sum_{l=1}^{3}(Q_{mk}Q_{nl}Q_{rk}Q_{sl}\gamma_{kl})\dot{A}_{pq}^{\text{<lab>}}}. \tag{16.69}$$

In this equation, which is the key result quoted in equation (16.38), the components of all tensors are the *laboratory* components. Neither $[Q]$ nor $[\gamma]$ are tensors—they are coefficients that must be determined by spectral analysis of $\underset{\approx}{\mathbf{A}}$. The precise values for Q_{ij} and γ_{ij} permute depending on the ordering of the eigenvalues, but this permutation has no impact on the final result—the ordering of eigenvalues and eigenvectors is inconsequential. Moreover, as explained below, the final result is also unaffected by arbitrariness of eigenvectors even when there are double or triple eigenvalue roots. All that matters is that $[Q]$ and $[\gamma]$ be constructed consistently.

A possibly more efficient way to compute the time rate of a principal function is to use eigenprojectors. If $\underset{\approx}{\mathbf{P}}_k$ are the eigenprojectors of $\underset{\approx}{\mathbf{A}}$,[5] then (16.65) may be written

$$\underset{\approx\approx}{\mathbf{L}} = \sum_{i=1}^{3}\sum_{j=1}^{3}\gamma_{ij}(\underset{\approx}{\mathbf{P}}_i \otimes \underset{\approx}{\mathbf{P}}_j)^{\lambda}, \tag{16.70}$$

where the superscript λ denotes the symmetric 'leaf' operation[6]. In light of this result, equation (16.67) may be written

$$\dot{\underset{\approx}{\mathbf{Y}}} = \underset{\approx\approx}{\mathbf{L}} : \dot{\underset{\approx}{\mathbf{A}}} = \sum_{i=1}^{3}\sum_{j=1}^{3}\gamma_{ij}\underset{\approx}{\mathbf{P}}_i \bullet \dot{\underset{\approx}{\mathbf{A}}} \bullet \underset{\approx}{\mathbf{P}}_j. \tag{16.71}$$

---

[5] That is, $\underset{\approx}{\mathbf{P}}_k = \underset{\sim}{\mathbf{p}}_k\underset{\sim}{\mathbf{p}}_k$ (no sum).

[6] For a fourth-order tensor with components T_{ijkl}, the 'leaf' of the tensor (which is a generalized transpose) has components $T^{L}_{ijkl} = T_{ikjl}$. The symmetric leaf is the same except with symmetry imposed on the first and last index pairs: $T^{\lambda}_{ijkl} = \frac{1}{4}(T_{ikjl} + T_{jkil} + T_{iljk} + T_{jlik})$. As a special case, $(\underset{\approx}{\mathbf{A}} \otimes \underset{\approx}{\mathbf{A}})L_{ijkl} = A_{ik}A_{jl}$, and $(\underset{\approx}{\mathbf{A}} \otimes \underset{\approx}{\mathbf{A}})^{\lambda}_{ijkl} = \frac{1}{2}(A_{ik}A_{jl} + A_{il}A_{jk})$.

If the projectors used in this formulation are the unique eigen*space* projectors[7] for the tensor $\underset{\approx}{A}$, then the number of terms can be reduced. Specifically:

$$\begin{aligned}&\text{If } \underset{\approx}{A} \text{ has exactly one distinct root } \lambda, \text{ then } \dot{\underset{\approx}{Y}} = f'(\lambda)\dot{\underset{\approx}{A}}.\\ &\text{If } \underset{\approx}{A} \text{ has exactly two distinct roots } \lambda_1 \text{ and } \lambda_2, \text{ then}\end{aligned} \tag{16.72}$$

$$\begin{aligned}\dot{\underset{\approx}{Y}} = {} & f'(\lambda_1)\underset{\approx}{P}_1 \bullet \dot{\underset{\approx}{A}} \bullet \underset{\approx}{P}_1\\ & + f'(\lambda_2)\underset{\approx}{P}_2 \bullet \dot{\underset{\approx}{A}} \bullet \underset{\approx}{P}_2 + 2\left\{\frac{f(\lambda_2) - f(\lambda_1)}{\lambda_2 - \lambda_1}\right\}\operatorname{sym}\,(\underset{\approx}{P}_1 \bullet \dot{\underset{\approx}{A}} \bullet \underset{\approx}{P}_2).\end{aligned} \tag{16.73}$$

If $\underset{\approx}{A}$ has exactly three distinct roots, then

$$\begin{aligned}\dot{\underset{\approx}{Y}} = {} & f(\lambda_1)\underset{\approx}{P}_1 \bullet \dot{\underset{\approx}{A}} \bullet \underset{\approx}{P}_1 + f'(\lambda_2)\underset{\approx}{P}_2 \bullet \dot{\underset{\approx}{A}} \bullet \underset{\approx}{P}_2 + f'(\lambda_3)\underset{\approx}{P}_3 \bullet \dot{\underset{\approx}{A}} \bullet \underset{\approx}{P}_3\\ & + \left\{\frac{f(\lambda_1) - f(\lambda_2)}{\lambda_1 - \lambda_2}\right\}\underset{\approx}{P}_1 \bullet \dot{\underset{\approx}{A}} \bullet \underset{\approx}{P}_2 + \left\{\frac{f(\lambda_2) - f(\lambda_3)}{\lambda_2 - \lambda_3}\right\}\underset{\approx}{P}_2 \bullet \dot{\underset{\approx}{A}} \bullet \underset{\approx}{P}_3\\ & + \left\{\frac{f(\lambda_3) - f(\lambda_1)}{\lambda_3 - \lambda_1}\right\}\underset{\approx}{P}_1 \bullet \dot{\underset{\approx}{A}} \bullet \underset{\approx}{P}_2.\end{aligned} \tag{16.74}$$

16.3.1 Effect of double or triple multiplicity eigenvalues

Our derivation of (16.59) was built around the assertion that a triad of principal directions always exists such that these principal directions vary continuously through time. This assertion meant that, in principle, one would need to compute the eigenvectors of

$$\underset{\approx}{A}(t) = \underset{\approx}{A}(t_o) + \dot{\underset{\approx}{A}}(t_o)(t - t_o) \tag{16.75}$$

and then use the eigenvectors corresponding to the limit as $t \to t_o^+$. In this limit, the eigenvectors of $\underset{\approx}{A}(t)$ will always coincide with those of $\underset{\approx}{A}(t_o)$ if $\underset{\approx}{A}(t_o)$ has distinct eigenvalues. However, if $\underset{\approx}{A}(t_o)$ has a double root, then there is a degree of arbitrariness in the two associated eigenvectors—they are restricted to lie within a given plane, but their directions are otherwise arbitrary. Of this family of in-plane orthonormal eigenvectors, only one choice will generally *also* be eigenvectors of $\underset{\approx}{A}(t)$

[7] When an eigenvalue is a double root, its eigenvectors are arbitrary in direction—they are only constrained to lie within a given plane. The plane is unique, not the vectors themselves. However, the *sum* of the two non-unique eigenprojectors results in a *unique* projector, the eigenspace projector. When an eigenvalue is a triple root, the sum of its three eigenprojectors equals the identity tensor, which is the unique eigenspace projector in this case.

in the limit as $t \to t_o^+$. The situation is even clearer when $\underset{\approx}{A}(t_o)$ has a triple root. In this case, $\underset{\approx}{A}(t_o)$ is isotropic so that *any* direction is an eigenvector of $\underset{\approx}{A}(t_o)$. To determine the eigenvectors of $\underset{\approx}{A}$ at an instant later—that is, the eigenvectors of $\underset{\approx}{A}(t_o^+)$—one would need to look at the rate $\dot{\underset{\approx}{A}}(t_o)$. Thankfully, this specially aligned triad that corresponds to eigenvectors of $\underset{\approx}{A}(t_o^+)$ does *not* have to be computed in practice. We wish to demonstrate that boxed statement in section 16.3 holds true even if *any* eigenvector triad of $\underset{\approx}{A}(t_o)$ is used. When double or triple roots implies arbitrary eigenvector directions, you can use *any* choice for the eigenvectors, not necessarily the special triad that is also a principal triad for $\underset{\approx}{A}(t_o^+)$.

Consider first the case of triple multiplicity: $\lambda_1 = \lambda_2 = \lambda_3 = \Lambda$. In this case, all nine values of γ_{ij} simply equal $f'(\Lambda)$, and therefore (16.59) becomes simply

$$\begin{bmatrix} (\dot{\underset{\approx}{Y}})_{11} & (\dot{\underset{\approx}{Y}})_{12} & (\dot{\underset{\approx}{Y}})_{13} \\ & (\dot{\underset{\approx}{Y}})_{22} & (\dot{\underset{\approx}{Y}})_{23} \\ \text{sym} & & (\dot{\underset{\approx}{Y}})_{33} \end{bmatrix} = f'(\Lambda) \begin{bmatrix} (\dot{\underset{\approx}{A}})_{11} & (\dot{\underset{\approx}{A}})_{12} & (\dot{\underset{\approx}{A}})_{13} \\ & (\dot{\underset{\approx}{A}})_{22} & (\dot{\underset{\approx}{A}})_{23} \\ \text{sym} & & (\dot{\underset{\approx}{A}})_{33} \end{bmatrix}. \tag{16.76}$$

For a triple eigenvalue, the orientation of the principal triad is completely arbitrary. Recall that (16.59) was derived using the eigenvector triad that would *not* be forced to jump discontinuously through time to align itself with the principal directions at t_o^+. If we had used a different triad, then the orthogonal $[Q]$ matrix would naturally change, but so would the principal components $(\dot{\underset{\approx}{A}})_{pq}$. These two changes would be self-compensating so that the *final* laboratory components predicted using (16.69) would come out to be identical. That is, no matter what eigentriad is used, the final result would be

$$\dot{\underset{\approx}{Y}} = f'(\Lambda)\dot{\underset{\approx}{A}}, \quad \text{whenever } \lambda_1 = \lambda_2 = \lambda_3 = \Lambda. \tag{16.77}$$

Similar conclusions hold for double multiplicity eigenvalues. If, for example, $\lambda_2 = \lambda_3$, then (16.58) would give

$$\gamma_{22} = \gamma_{33} = \gamma_{23} = \gamma_{32} \text{ and } \gamma_{12} = \gamma_{13}. \tag{16.78}$$

Consequently, the result of (16.69) is unchanged even when expressed in terms of the *any* (even non-ideally aligned) eigenvectors. To further illustrate this point, we will now proceed to an example that has double roots.

16.3.2 Example: rate of the square of a tensor

To serve as a verification example for which there exists a straightforward classical solution method, we will illustrate equation (16.38) by taking the time rate of the square of a tensor. Results were checked using a FORTRAN test program cited in section A.10 of the appendix, and double-checked using the Mathematica commands in section A.11. Of course, different codes might order the eigenvalues differently, so intermediate 'helper' quantities such as the Q_{ij} and γ_{ij} coefficients might be different in two codes, but they should all predict the same final result!

Suppose that the time-varying *lab* components of $\underset{\approx}{A}$ are given by

$$[A]_{\langle\text{lab}\rangle} = \begin{bmatrix} 41e^{t-1} & -12t & 0 \\ -12t & 34t^3 & \ln t \\ 0 & \ln t & 50t^2 \end{bmatrix}_{\langle\text{lab}\rangle} . \tag{16.79}$$

Differentiating this matrix gives the *lab* components of $\dot{\underset{\approx}{A}}$:

$$[\dot{A}]_{\langle\text{lab}\rangle} = \begin{bmatrix} 41e^{t-1} & -12 & 0 \\ -12 & 102t^2 & \frac{1}{t} \\ 0 & \frac{1}{t} & 100t \end{bmatrix}_{\langle\text{lab}\rangle} . \tag{16.80}$$

For illustration, let us suppose that we wish to compute the rate of $\underset{\approx}{Y} = \underset{\approx}{A}^2$, evaluated at $t = 1$. In this particular case, we know that the answer can be computed without using the specialized formulas derived in the previous section. We know the answer is simply

$$\dot{\underset{\approx}{Y}} = \dot{\underset{\approx}{A}} \bullet \underset{\approx}{A} + \underset{\approx}{A} \bullet \dot{\underset{\approx}{A}}. \tag{16.81}$$

At $t = 1$, equations (16.79) and (16.80) evaluate to

$$[A]_{\langle\text{lab}\rangle} = \begin{bmatrix} 41 & -12 & 0 \\ -12 & 34 & 0 \\ 0 & 0 & 50 \end{bmatrix}_{\langle\text{lab}\rangle} \quad \text{and} \quad [\dot{A}] = \begin{bmatrix} 41 & -12 & 0 \\ -12 & 102 & 1 \\ 0 & 1 & 100 \end{bmatrix}_{\langle\text{lab}\rangle} \tag{16.82}$$

from which the traditional method of (16.82) gives

$$[\dot{Y}] = \begin{bmatrix} 41 & -12 & 0 \\ -12 & 34 & 0 \\ 0 & 0 & 50 \end{bmatrix} \begin{bmatrix} 41 & -12 & 0 \\ -12 & 102 & 1 \\ 0 & 1 & 100 \end{bmatrix} + \begin{bmatrix} 41 & -12 & 0 \\ -12 & 102 & 1 \\ 0 & 1 & 100 \end{bmatrix} \begin{bmatrix} 41 & -12 & 0 \\ -12 & 34 & 0 \\ 0 & 0 & 50 \end{bmatrix}$$
$$= \begin{bmatrix} 3650 & -2616 & -12 \\ -2616 & 7224 & 84 \\ -12 & 84 & 10000 \end{bmatrix}. \tag{16.83}$$

Now we wish to demonstrate that the methods of the previous section will give *this same result*. At the instant of interest, the eigenvalues of the laboratory matrix $[A]_{\langle\text{lab}\rangle}$ are {25, 50, 50} and the corresponding orthonormal eigenvectors are assembled into columns to create the Q matrix:

$$[Q] = \begin{bmatrix} 3/5 & 0 & -4/5 \\ 4/5 & 0 & 3/5 \\ 0 & 1 & 0 \end{bmatrix}. \tag{16.84}$$

This is a direction-cosine matrix, so it does not represent components of a tensor in lab-space, A-space, or any other tensor space. For this example, $f(x) = x^2$. Therefore the function in (16.52) is given by

$$G(x, y) = \begin{cases} \dfrac{x^2 - y^2}{x - y} & \text{if } x \neq y \\ 2x & \text{if } x = y \end{cases}, \quad \text{or simply} \quad G(x, y) = x + y. \tag{16.85}$$

For this example, the function $G(x, y)$ turned out to be reducible to a form that no longer required a special-case definition when $x = y$. This happened because we chose a polynomial function $f(\lambda)$ for our example, corresponding to a problem having a straightforward classical solution.

The matrix of γ_{ij} coefficients is computed by

$$[\gamma_{ij}] = [G(\lambda_i, \lambda_j)] = \begin{bmatrix} 50 & 75 & 75 \\ 75 & 100 & 100 \\ 75 & 100 & 100 \end{bmatrix}. \tag{16.86}$$

The components of $\dot{\underset{\approx}{A}}$ with respect to the principal basis are

$$[\dot{A}]_{<A>} = [Q]^T[\dot{A}]_{<lab>}[Q] = \begin{bmatrix} 3/5 & 4/5 & 0 \\ 0 & 0 & 1 \\ -4/5 & 3/5 & 0 \end{bmatrix} \begin{bmatrix} 41 & -12 & 0 \\ -12 & 34 & 0 \\ 0 & 0 & 50 \end{bmatrix} \begin{bmatrix} 3/5 & 0 & -4/5 \\ 4/5 & 0 & 3/5 \\ 0 & 1 & 0 \end{bmatrix}$$
$$= \frac{1}{25}\begin{bmatrix} 1713 & 20 & 816 \\ 20 & 2500 & 15 \\ 816 & 15 & 1862 \end{bmatrix}_{<A>}. \tag{16.87}$$

Multiplying each component of $[\dot{A}]_{<A>}$ by the corresponding component of $[\gamma_{ij}]$ gives the components of $\dot{\underset{\approx}{Y}}$ with respect to the principal basis:

$$[\dot{Y}]_{<A>} = \begin{bmatrix} 3426 & 60 & 2448 \\ 60 & 10000 & 60 \\ 2448 & 60 & 7448 \end{bmatrix}. \tag{16.88}$$

Transforming this result back to the lab basis gives the lab components of $\dot{\underset{\approx}{Y}}$:

$$[\dot{Y}]_{<lab>} = [Q][\dot{Y}]_{<A>}[Q]^T$$
$$= \begin{bmatrix} 3/5 & 0 & -4/5 \\ 4/5 & 0 & 3/5 \\ 0 & 1 & 0 \end{bmatrix} \begin{bmatrix} 3426 & 60 & 2448 \\ 60 & 10000 & 60 \\ 2448 & 60 & 7448 \end{bmatrix} \begin{bmatrix} 3/5 & 4/5 & 0 \\ 0 & 0 & 1 \\ -4/5 & 3/5 & 0 \end{bmatrix} \tag{16.89}$$

or

$$[\dot{Y}]_{<lab>} = \begin{bmatrix} 3650 & -2616 & -12 \\ -2616 & 7224 & 84 \\ -12 & 84 & 10000 \end{bmatrix}_{<lab>}, \tag{16.90}$$

which agrees with the result derived in the traditional manner, (16.83). The calculations needed to obtain this result using Mathematica [2] are given in section A.11. For this example, the second and third eigenvalue both equal 50. We used Mathematica to verify that the *final* result is unaffected if the second and third eigenvectors used in the analysis are replaced by any in-plane rotation of the original second and third eigenvectors. Of course, intermediate principal component matrices vary, but the final *laboratory* result is unchanged.

Even though we have shown that the continuous eigenvectors are *not* needed to compute principal function rates, we would like to record for completeness what the continuous eigenvectors were for this example. At $t_o = 1$, equation (16.75) becomes

$$\underset{\approx}{A}(t) = \begin{bmatrix} 41 & -12 & 0 \\ -12 & 34 & 0 \\ 0 & 0 & 50 \end{bmatrix}_{\text{<lab>}} + \begin{bmatrix} 41 & -12 & 0 \\ -12 & 102 & 1 \\ 0 & 1 & 100 \end{bmatrix}_{\text{<lab>}} (t-1). \tag{16.91}$$

At $t = 1$, the eigenvalues are $\{25, 50, 50\}$. The double root makes the last two eigenvectors are arbitrary. In other words, the last two columns of the $[Q]$ matrix are arbitrary as long as they are perpendicular to each other and to the first column. Even though there are double roots at $t = 1$, all three roots are distinct an instant later, which means there are suddenly three unique eigenvector orientations an instant after $t = 1$. In the limit as t approaches 1 from the right, these three distinct eigenvectors of $\underset{\approx}{A}(t)$ are the columns of

$$[Q] = \begin{bmatrix} 0.6 & -0.018\,7933 & 0.799\,779 \\ 0.8 & 0.014\,095 & -0.599\,834 \\ 0 & 0.999\,724 & 0.023\,4916 \end{bmatrix}. \tag{16.92}$$

If we had used this continuous direction-cosine matrix instead of the simpler one in (16.84), we *still* would have obtained the same final result in (16.90).

16.4 Time rate of the logarithmic strain

The reference 'logarithmic strain' $\underset{\approx}{\bar{\varepsilon}}$ is a special kind of principal function defined by natural log of the *reference* stretch from the polar decomposition of the deformation gradient, $\underset{\approx}{F} = \underset{\approx}{R} \bullet \underset{\approx}{U}$:

$$\underset{\approx}{\bar{\varepsilon}} = \ln \underset{\approx}{U} = \sum_{i=1}^{3} \ln(\lambda_i) \underset{\sim}{\bar{p}}_i \underset{\sim}{\bar{p}}_j , \tag{16.93}$$

where $\{\lambda_1, \lambda_2, \lambda_3\}$ are the eigenvalues of $\underset{\approx}{U}$, and $\{\underset{\sim}{\bar{p}}_1, \underset{\sim}{\bar{p}}_2, \underset{\sim}{\bar{p}}_3\}$ are the corresponding eigenvectors. The *spatial* logarithmic strain $\underset{\approx}{\varepsilon}$ (which is *not* recommended for use in constitutive models) is defined similarly, but using the *spatial* stretch from the polar decomposition $\underset{\approx}{F} = \underset{\approx}{V} \bullet \underset{\approx}{R}$:

$$\underset{\approx}{\varepsilon} = \ln \underset{\approx}{V} = \sum_{i=1}^{3} \ln(\lambda_i) \underset{\sim}{p}_i \underset{\sim}{p}_j . \tag{16.94}$$

The eigenvalues $\{\lambda_1, \lambda_2, \lambda_3\}$ for $\underset{\approx}{\boldsymbol{V}}$ are identical to the eigenvalues for $\underset{\approx}{\boldsymbol{U}}$, and the eigenvectors $\{p_1, p_2, p_3\}$ for $\underset{\approx}{\boldsymbol{V}}$ are related to the eigenvectors for $\underset{\approx}{\boldsymbol{U}}$ by

$$\underset{\sim}{\boldsymbol{p}}_i = \underset{\approx}{\boldsymbol{R}} \bullet \underset{\sim}{\bar{\boldsymbol{p}}}_i \,. \tag{16.95}$$

Consequently, the two logarithmic strains are related to each other by

$$\underset{\approx}{\bar{\varepsilon}} = \underset{\approx}{\boldsymbol{R}}^T \bullet \underset{\approx}{\varepsilon} \bullet \underset{\approx}{\boldsymbol{R}}. \tag{16.96}$$

In other words, the reference log strain is the *un-rotated* version of the spatial log strain. In general, for any spatial tensor $\underset{\approx}{\boldsymbol{A}}$ the 'un-rotation' operation will be denoted by an 'overbar' operation:

$$\underset{\approx}{\overline{\boldsymbol{A}}} = \underset{\approx}{\boldsymbol{R}}^T \bullet \underset{\approx}{\boldsymbol{A}} \bullet \underset{\approx}{\boldsymbol{R}}. \tag{16.97}$$

Incidentally, note that the reference stretch $\underset{\approx}{\boldsymbol{U}}$ is just the un-rotation of the spatial stretch. That is,

$$\underset{\approx}{\boldsymbol{U}} \equiv \underset{\approx}{\overline{\boldsymbol{V}}}. \tag{16.98}$$

Recall that the rate of the rotation tensor is given by $\underset{\approx}{\dot{\boldsymbol{R}}} = \underset{\approx}{\Omega} \bullet \underset{\approx}{\boldsymbol{R}}$, where $\underset{\approx}{\Omega}$ is the skew-symmetric polar spin. Therefore, taking rates of (16.96), and making use of our 'overbar' definition reveals that

$$\underset{\approx}{\dot{\bar{\varepsilon}}} = \underset{\approx}{\overline{\mathring{\varepsilon}}}, \quad \text{where} \quad \underset{\approx}{\mathring{\varepsilon}} \equiv \underset{\approx}{\dot{\varepsilon}} - \underset{\approx}{\Omega} \bullet \underset{\approx}{\varepsilon} + \underset{\approx}{\varepsilon} \bullet \underset{\approx}{\Omega}. \tag{16.99}$$

Even though $\underset{\approx}{\bar{\varepsilon}}$ is the un-rotation of $\underset{\approx}{\varepsilon}$, note that $\underset{\approx}{\dot{\bar{\varepsilon}}}$ is *not* the un-rotation of $\underset{\approx}{\dot{\varepsilon}}$. Instead, it is the un-rotation of the co-rotational objective polar rate $\underset{\approx}{\mathring{\varepsilon}}$. Fortunately, material constitutive models that are cast in spatial rate form typically require a co-rotational rate, and the above equation shows that the co-rotational polar rate is easily obtained by simply rotating the ordinary reference rate into the spatial configuration. For this reason, we will proceed with the ensuing analysis using the reference stretch and the reference log strain. In the upcoming equations, an annotation <U> or the phrase 'U-space' will be used to indicate that components are referenced to the principal basis for $\underset{\approx}{\boldsymbol{U}}$ and <lab> will be used to show that components are laboratory components.

This strain definition in (16.93) is a special case of (16.35) in which $\underset{\approx}{\boldsymbol{A}}$ is now the stretch $\underset{\approx}{\boldsymbol{U}}$ and $f(\lambda_i) = \ln \lambda_i$, where $\{\lambda_1, \lambda_2, \lambda_3\}$ are the principal stretches. Therefore, the γ_{ij} matrix in (16.58) is given by

$$\gamma_{ij} = \begin{cases} \dfrac{\ln(\lambda_i) - \ln(\lambda_j)}{\lambda_i - \lambda_j} & \text{if } \lambda_i \neq \lambda_j \\ \dfrac{1}{\lambda_i} & \text{if } \lambda_i = \lambda_j \end{cases}, \tag{16.100}$$

and the U-space components of the rate of log strain are therefore given by

$$(\dot{\underset{\approx}{\varepsilon}})_{pq}^{<U>} = \begin{cases} \left(\dfrac{\ln(\lambda_p) - \ln(\lambda_q)}{\lambda_p - \lambda_q}\right)(\dot{\underset{\approx}{U}})_{pq}^{<U>} & \textit{if } \lambda_p \neq \lambda_q \\ \dfrac{(\dot{\underset{\approx}{U}})_{pq}^{<U>}}{\lambda_p} & \textit{if } \lambda_p = \lambda_q \end{cases}. \tag{16.101}$$

Frequently, the rate of the un-rotated (reference) logarithmic strain is approximated by the un-rotated symmetric part of the velocity gradient, defined by

$$\underset{\approx}{\overline{D}} \equiv \tfrac{1}{2}(\dot{\underset{\approx}{U}} \bullet \underset{\approx}{U}^{-1} + \underset{\approx}{U}^{-1} \bullet \dot{\underset{\approx}{U}}). \tag{16.102}$$

Writing this equation in terms of the principal basis for $\underset{\approx}{U}$ reveals that the components of $\underset{\approx}{\overline{D}}$ are given by

$$(\underset{\approx}{\overline{D}})_{pq}^{<U>} = \tfrac{1}{2}\left(\frac{1}{\lambda_p} + \frac{1}{\lambda_q}\right)(\dot{\underset{\approx}{U}})_{pq}^{<U>}, \quad \text{no sum on } p \text{ or } q. \tag{16.103}$$

Note that

$$(\underset{\approx}{\overline{D}})_{pq}^{<U>} = \begin{cases} \tfrac{1}{2}\left(\dfrac{1}{\lambda_p} + \dfrac{1}{\lambda_1}\right)(\dot{\underset{\approx}{U}})_{pq}^{<U>} & \text{if } \lambda_p \neq \lambda_q \\ \dfrac{(\dot{\underset{\approx}{U}})_{pq}^{<U>}}{\lambda_p} & \text{if } \lambda_p = \lambda_q \end{cases}. \tag{16.104}$$

Therefore, the time rate of log strain can be computed *directly* from the U-space components of the un-rotated symmetric part of the velocity gradient by

$$\boxed{(\dot{\underset{\approx}{\varepsilon}})_{pq}^{<U>} = \begin{cases} \left(\dfrac{r_{pq}}{\sinh r_{pq}}\right)(\underset{\approx}{\overline{D}})_{pq}^{<U>} & \text{if } r_{pq} \neq 0 \\ (\underset{\approx}{\overline{D}})_{pq}^{<U>} & \text{if } \lambda_{pq} = 0 \end{cases}}, \quad \text{where } r_{pq} \equiv \ln \frac{\lambda_p}{\lambda_q}. \tag{16.105}$$

An example verifying this formula is given near the end of this section. Note that

$$(\dot{\underset{\approx}{\varepsilon}})_{pq}^{<U>} = (\underset{\approx}{\overline{D}})_{pq}^{<U>} \quad \text{when} \quad \lambda_p = \lambda_q. \tag{16.106}$$

In words, the diagonal U-space components of $\underset{\approx}{\overline{D}}$ identically equal the counterpart components of the log strain. Stated in direct notation,

$$\underset{\sim}{\delta}_k^U \bullet \dot{\underset{\approx}{\varepsilon}} \bullet \underset{\sim}{\delta}_k^U = \underset{\sim}{\delta}_k^U \bullet \underset{\approx}{\overline{D}} \bullet \underset{\sim}{\delta}_k^U \tag{16.107}$$

(keep in mind: no implied sums in this chapter)

where $\{\underset{\sim}{\delta}_1^U, \underset{\sim}{\delta}_2^U, \underset{\sim}{\delta}_3^U\}$ are the principal directions of $\underset{\approx}{U}$. Equation (16.105) also shows that

$$(\dot{\underset{\approx}{\varepsilon}})_{pq}^{<U>} = (\underset{\approx}{\bar{\mathbf{D}}})_{pq}^{<U>}, \quad \text{when} \quad (\underset{\approx}{\bar{\mathbf{D}}})_{pq}^{<U>} = 0. \tag{16.108}$$

In other words, if an off-diagonal U-space component of $\underset{\approx}{\bar{\mathbf{D}}}$ is zero, then the corresponding U-space component of the log strain will also be zero. Otherwise, if a U-space component of $\underset{\approx}{\bar{\mathbf{D}}}$ is non-zero and if that component is not on the U-space diagonal, then the corresponding component of the log strain will be different by an multiple given by[8]

$$\boxed{\frac{(\underset{\approx}{\bar{\mathbf{D}}})_{pq}^{<U>}}{(\dot{\underset{\approx}{\varepsilon}})_{pq}^{<U>}} = \frac{\sinh r_{pq}}{r_{pq}}}, \quad \text{where} \quad r_{pq} \equiv \ln\left(\frac{\lambda_p}{\lambda_q}\right). \tag{16.109}$$

This result quantifies the amount by which $\underset{\approx}{\bar{\mathbf{D}}}$ differs from the logarithmic strain rate. A plot of this function is shown in figure 16.1.

A Taylor series expansion of the function in (16.109) shows that

$$\boxed{\frac{(\underset{\approx}{\bar{\mathbf{D}}})_{pq}}{(\dot{\underset{\approx}{\varepsilon}})_{pq}} = 1 + \frac{1}{6}r_{pq}^2 + O(r_{pq}^4)}, \quad \text{where} \quad r_{pq} \equiv \ln\left(\frac{\lambda_p}{\lambda_q}\right). \tag{16.110}$$

Thus, an excellent approximation to the 'percent error' associated with approximating the log strain rate by the symmetric part of the deformation gradient is

$$\%\ \text{error} \approx 50 r_{pq}^2/3. \tag{16.111}$$

If it is desired that the error should never exceed 1%, then the largest eigenvalue ratio must never exceed 1.3, which corresponds to an engineering strain of 30% (well above the failure strain for a large number of engineering materials)!.

Of course, this error corresponds to the error in the off-diagonal components associated with non-equal eigenvalues (the other components have *zero* error). Equation (16.105) tells us that the difference between $\underset{\approx}{\bar{\mathbf{D}}}$ and $\dot{\underset{\approx}{\varepsilon}}$ is given by

$$(\underset{\approx}{\bar{\mathbf{D}}})_{pq}^{<U>} - (\dot{\underset{\approx}{\varepsilon}})_{pq}^{<U>} = \begin{cases} \left(1 - \dfrac{r_{pq}}{\sinh r_{pq}}\right)(\underset{\approx}{\bar{\mathbf{D}}})_{pq}^{<U>} & \text{if } r_{pq} \neq 0 \\ 0 & \text{if } r_{pq} = 0 \end{cases}, \quad \text{where} \quad r_{pq} \equiv \ln \frac{\lambda_p}{\lambda_p} \tag{16.112}$$

Note that $r_{pq} = 0$ whenever $p = q$. Consequently, there is *zero* error in the diagonal components, when those components are expressed in the principal basis of the reference stretch. It follows that *there is zero error in computing the volumetric part of the logarithmic strain rate by the trace of* $\underset{\approx}{\bar{\mathbf{D}}}$. That is,

$$\text{tr}\underset{\approx}{\bar{\mathbf{D}}} = \text{tr}\dot{\underset{\approx}{\varepsilon}}. \tag{16.113}$$

[8] This is, to our knowledge, a new result. We would appreciate being informed if this is not the case.

All error associated with approximating $\dot{\underset{\approx}{\varepsilon}}$ by $\underset{\approx}{\bar{D}}$ comes exclusively from contributions of the *deviators*, $\dot{\underset{\approx}{\varepsilon}}^d$ and $\underset{\approx}{\bar{D}}^d$, of these tensors. In particular, equation (16.105) implies that the difference between these deviators is given by

$$\|\underset{\approx}{\bar{D}}^d - \dot{\underset{\approx}{\varepsilon}}^d\| = \left\{2\left(h_{12}^2\left[(\bar{D})_{12}^d\right]^2 + h_{23}^2\left[(\bar{D})_{23}^d\right]^2 + h_{31}^2\left[(\bar{D})_{31}^d\right]^2\right)\right\}^{1/2},$$
$$\text{where } h_{pq} = \begin{cases} \left(1 - \dfrac{r_{pq}}{\sinh r_{pq}}\right) & \text{if } r_{pq} \neq 0 \\ 0 & \text{if } r_{pq} = 0 \end{cases}. \tag{16.114}$$

Here, the factor of 2 appears inside the square root because the off-diagonal components contribute *twice* to the inner product. Note that

$$h_{pq} \leqslant 1 - \frac{R}{\sinh R} \leqslant \frac{R^2}{6}, \quad \text{where} \quad R = \ln\left(\frac{\lambda^{max}}{\lambda^{min}}\right). \tag{16.115}$$

Here, $\lambda^{max}/\lambda^{min}$ is the ratio of the largest to the smallest eigenvalue of $\underset{\approx}{U}$. The expression $R^2/6$ was introduced because it is computationally simpler to compute and (for eigenvalue ratios as high as 2.8) it is an extremely tight bound (within 2%) on the expression '$1 - R/(\sinh R)$'.

In practice, it is generally easier to compute the *square* stretch, $\underset{\approx}{U}^2 = \underset{\approx}{F}^T \bullet \underset{\approx}{F}$. If we denote the eigenvalues of $\underset{\approx}{F}^T \bullet \underset{\approx}{F}$ by $\{\Lambda^{min}, \Lambda^{mid}, \Lambda^{max}\}$, then we can define

$$\rho = \ln\left(\frac{\Lambda^{max}}{\Lambda^{min}}\right) = 2R \quad \text{and therefore} \quad h_{pq} \leqslant \frac{\rho^2}{24}. \tag{16.116}$$

Consequently, equation (16.114) tells us that

$$\|\underset{\approx}{\bar{D}}^d - \dot{\underset{\approx}{\varepsilon}}^d\| \leqslant \frac{\sqrt{2}}{24}\rho^2\left\{\left[(\bar{D})_{12}^d\right]^2 + \left[(\bar{D})_{23}^d\right]^2 + \left[(\bar{D})_{31}^d\right]^2\right\}^{1/2}. \tag{16.117}$$

Equivalently,

$$\|\underset{\approx}{\bar{D}}^d - \dot{\underset{\approx}{\varepsilon}}^d\| \leqslant \frac{\rho^2}{24}\left\{\|\underset{\approx}{\bar{D}}^d\|^2 - \left(\left[(\bar{D})_{11}^d\right]^2 + \left[(\bar{D})_{22}^d\right]^2 + \left[(\bar{D})_{32}^d\right]^2\right)\right\}^{1/2}. \tag{16.118}$$

This equation should be a reasonably tight bound, but it has the disadvantage that it still requires determination of the principal stretch directions to evaluate the diagonal components. A far looser—but probably more useful—error bound results from observing that

$$\|\underset{\approx}{\bar{D}}^d - \dot{\underset{\approx}{\varepsilon}}^d\| \leqslant \frac{\rho^2}{24}\|\underset{\approx}{\bar{D}}^d\|. \tag{16.119}$$

Recalling that there is *zero* error in the isotropic part, we obtain our final result:

Error bound for approximating log strain rate by the symmetric part of the velocity gradient:

$$\boxed{\|\underset{\sim}{\bar{\mathbf{D}}} - \underset{\approx}{\dot{\bar{\varepsilon}}}\| \leqslant \frac{\rho^2}{24}\|\underset{\sim}{\bar{\mathbf{D}}}^d\|},$$

where ρ is the natural logarithm of the ratio of the largest to smallest eigenvalue of $\underset{\sim}{\mathbf{F}}^T \bullet \underset{\sim}{\mathbf{F}}$, and $\underset{\sim}{\bar{\mathbf{D}}}^d$ is the deviatoric part of $\underset{\sim}{\bar{\mathbf{D}}}$.

This is, to our knowledge a new result that will permit straightforward and inexpensive runtime monitoring of the error because any of several theorems (such as the Gershgorin bound [3]) that provide an upper limit on the eigenvalue ratio can be used, thereby permitting an error bound *without* suffering the expense of computing the stretch or its eigensystem. Though this is a loose bound on $\|\underset{\sim}{\bar{\mathbf{D}}} - \underset{\approx}{\dot{\bar{\varepsilon}}}\|$, it should be an extremely tight bound on at least one off-diagonal principal component of $\underset{\sim}{\bar{\mathbf{D}}} - \underset{\approx}{\dot{\bar{\varepsilon}}}$. The *overall* bound quoted above is much looser because we know that the diagonal principal components will have zero error. We were forced to disregard this information to obtain an expression for the error bound that could be computed without having to determine the principal stretch directions themselves.

As shown below, a similar bound exists for the difference between the spatial co-rotational polar log strain rate $\underset{\approx}{\mathring{\varepsilon}}$ and the spatial 'rate' of deformation $\underset{\sim}{\mathbf{D}}$.

16.4.1 Example: simple shear

In simple shear, the deformation gradient is of the form

$$\left[F_{ij}^{\text{<lab>}}\right] = \begin{bmatrix} 1 & 2\sinh\varepsilon \\ 0 & 1 \end{bmatrix}, \tag{16.120}$$

where the superscripted label <lab> indicates lab components. As the parameter ε varies in time, the material is increasingly sheared. The shear component is written in this odd form because ε turns out to equal the largest principal Hencky strain. Using the fast two-dimensional (2D) formulas in (12.60), the polar rotation and reference stretch corresponding to (16.120) are

$$\left[R_{ij}^{\text{<lab>}}\right] = \begin{bmatrix} \operatorname{sech}\varepsilon & \tanh\varepsilon \\ -\tanh\varepsilon & \operatorname{sech}\varepsilon \end{bmatrix}, \tag{16.121}$$

$$\left[U_{ij}^{\text{<lab>}}\right] = \begin{bmatrix} \operatorname{sech}\varepsilon & \tanh\varepsilon \\ \tanh\varepsilon & (\cosh 2\varepsilon)(\operatorname{sech}\varepsilon) \end{bmatrix}. \tag{16.122}$$

The rate of stretch therefore has laboratory components given by

$$\left[\dot{U}_{ij}^{\text{<lab>}}\right] = \begin{bmatrix} -(\operatorname{sech}\varepsilon)(\tanh\varepsilon) & \operatorname{sech}^2\varepsilon \\ \operatorname{sech}^2\varepsilon & (2 + \cosh 2\varepsilon)(\operatorname{sech}\varepsilon)(\tanh\varepsilon) \end{bmatrix}. \tag{16.123}$$

The eigensystem of $[U_{ij}^{<\text{lab}>}]$ is

$$\textit{eigenvalues } e^{\pm\varepsilon} \text{ with } \textit{eigenvectors} \begin{bmatrix} \sqrt{\frac{1}{2}(1 \mp \tanh\varepsilon)} \\ \pm\sqrt{\frac{1}{2}(1 \pm \tanh\varepsilon)} \end{bmatrix}, \tag{16.124a}$$

$$\text{with corresponding } \textit{eigenprojectors} \begin{bmatrix} \frac{1}{2}(1 \mp \tanh\varepsilon) & \pm\operatorname{sech}\varepsilon \\ \pm\operatorname{sech}\varepsilon & \frac{1}{2}(1 \pm \tanh\varepsilon) \end{bmatrix}. \tag{16.124b}$$

Defining $c = (1 + e^{-2\varepsilon})^{-1/2}$ and $s = (1 + e^{2\varepsilon})^{-1/2}$,

$$\text{the stretch } \textit{direction cosine matrix} \text{ is} \begin{bmatrix} c & s \\ -s & c \end{bmatrix}. \tag{16.12c}$$

The *reference Hencky logarithm strain in simple shear* is therefore

$$\boxed{\left[\varepsilon_{ij}^{<\text{lab}>}\right] = \begin{bmatrix} -a & b \\ b & a \end{bmatrix} \text{ where } \begin{matrix} a = \varepsilon(\tanh\varepsilon) \\ b = \varepsilon(\operatorname{sech}\varepsilon) \end{matrix}}. \tag{16.125}$$

Because this is a 2D problem, we have found an analytical solution for the lab components of the Hencky strain, which allows direct evaluation of its rate, giving the *rate of the reference Hencky strain in simple shear*:

$$\boxed{\left[\dot{\varepsilon}_{ij}^{<\text{lab}>}\right] = \begin{bmatrix} -\dot{a} & \dot{b} \\ \dot{b} & \dot{a} \end{bmatrix} \text{ where } \begin{matrix} \dot{a} = \dot{\varepsilon}[\varepsilon(\operatorname{sech}^2\varepsilon) + \tanh\varepsilon) \\ \dot{b} = \dot{\varepsilon}[(\operatorname{sech}\varepsilon)(1 - \varepsilon\tanh\varepsilon)] \end{matrix}}. \tag{16.126}$$

We now wish to use these directly derived results to confirm the general formulas derived in this chapter. The process begins by first finding the lab components of the rate of $\underset{\approx}{\boldsymbol{F}}$

$$\left[\dot{F}_{ij}^{LL}\right] = \dot{\varepsilon}\begin{bmatrix} 0 & 2\cosh\varepsilon \\ 0 & 0 \end{bmatrix}. \tag{16.127}$$

The velocity gradient, $\underset{\approx}{\boldsymbol{L}} = \dot{\underset{\approx}{\boldsymbol{F}}} \bullet \underset{\approx}{\boldsymbol{F}}^{-1}$, happens to be given by the same formula:

$$\left[L_{ij}^{LL}\right] = \dot{\varepsilon}(\cosh\varepsilon)\begin{bmatrix} 0 & 2 \\ 0 & 0 \end{bmatrix}. \tag{16.128}$$

The rate-of-deformation and vorticity tensors, respectively defined to be the symmetric and skew parts of the velocity gradient, are then

$$\left[D_{ij}^{LL}\right] = \dot{\varepsilon}(\cosh\varepsilon)\begin{bmatrix} 0 & 1 \\ 1 & 0 \end{bmatrix} \text{ and } \left[W_{ij}^{LL}\right] = \dot{\varepsilon}(\cosh\varepsilon)\begin{bmatrix} 0 & 1 \\ -1 & 0 \end{bmatrix}. \tag{16.129}$$

The un-rotated rate of deformation, defined $\underset{\approx}{\overline{\boldsymbol{D}}} = \underset{\approx}{\boldsymbol{R}}^T \bullet \underset{\approx}{\boldsymbol{D}} \bullet \underset{\approx}{\boldsymbol{R}}$, has laboratory components

$$\left[\overline{D}_{ij}^{LL}\right] = \dot{\varepsilon}\begin{bmatrix} -2\tanh\varepsilon & \operatorname{sech}\varepsilon - (\sinh\varepsilon)(\tanh\varepsilon) \\ \operatorname{sech}\varepsilon - (\sinh\varepsilon)(\tanh\varepsilon) & 2\tanh\varepsilon \end{bmatrix}. \tag{16.130}$$

This same result can optionally be found by using $\underset{\approx}{\overline{\boldsymbol{D}}} = \text{sym}(\underset{\approx}{\dot{\boldsymbol{U}}} \bullet \underset{\approx}{\boldsymbol{U}}^{-1})$. Using the stretch direction cosines, the component matrices for $\underset{\approx}{\boldsymbol{U}}$, its rate $\underset{\approx}{\dot{\boldsymbol{U}}}$, and the un-rotated rate of deformation $\underset{\approx}{\overline{\boldsymbol{D}}}$ are written in the principal basis of stretch as

$$\left[U_{ij}^{<U>}\right] = \begin{bmatrix} e^{-\varepsilon} & 0 \\ 0 & e^{\varepsilon} \end{bmatrix} \quad \text{and} \quad \left[\dot{U}_{ij}^{<U>}\right] = \dot{\varepsilon}\begin{bmatrix} -e^{-\varepsilon} & -\tanh\varepsilon \\ -\tanh\varepsilon & e^{\varepsilon} \end{bmatrix}, \tag{16.131a}$$

$$\left[\overline{D}_{ij}^{<U>}\right] = \dot{\varepsilon}\begin{bmatrix} -1 & -\sinh\varepsilon \\ -\sinh\varepsilon & 1 \end{bmatrix}. \tag{16.131b}$$

Using (16.100) with $\lambda_1 = e^{-\varepsilon}$ and $\lambda_2 = e^{\varepsilon}$,

$$[\gamma_{ij}] = \begin{bmatrix} e^{\varepsilon} & \varepsilon(\operatorname{csch}\varepsilon) \\ \varepsilon(\operatorname{csch}\varepsilon) & e^{-\varepsilon} \end{bmatrix}. \tag{16.132}$$

Using (16.101), the U-space components of the rate of Hencky strain are therefore

$$\left[\dot{\varepsilon}_{ij}^{<U>}\right] = \dot{\varepsilon}\begin{bmatrix} -1 & -\varepsilon(\operatorname{sech}\varepsilon) \\ -\varepsilon(\operatorname{sech}\varepsilon) & 1 \end{bmatrix}. \tag{16.133}$$

When this is transformed back to the laboratory basis, the result agrees with (16.126), which was found by direct rates of the analytical solution for Hencky strain, thus verifying the formula in (16.101) in this case of simple shear.

A major disadvantage of (16.101) is that it requires the rate of reference stretch, which is difficult to find in numerical simulations. We therefore wish to verify the alternative formula for rate of Hencky strain given in (16.105), for which the eigenvalues $\lambda_1 = e^{-\varepsilon}$ and $\lambda_2 = e^{\varepsilon}$ give the following r_{pq} values:

$$[r_{pq}] = \begin{bmatrix} 0 & -2\varepsilon \\ 2\varepsilon & 0 \end{bmatrix}. \tag{16.134}$$

The formula in (16.105) is verified because it does indeed confirm (16.133).

16.4.2 Rate of reference stretch and material frame in general

Using the polar decomposition, $\underset{\approx}{\boldsymbol{F}} = \underset{\approx}{\boldsymbol{R}} \bullet \underset{\approx}{\boldsymbol{U}}$, it follows that the velocity gradient is

$$\underset{\approx}{\boldsymbol{L}} = \underset{\approx}{\dot{\boldsymbol{F}}} \bullet \underset{\approx}{\boldsymbol{F}}^{-1} = \underset{\approx}{\Omega} + \underset{\approx}{\boldsymbol{R}} \bullet \underset{\approx}{\dot{\boldsymbol{U}}} \bullet \underset{\approx}{\boldsymbol{U}}^{-1} \bullet \underset{\approx}{\boldsymbol{R}}^T, \tag{16.135}$$

where $\underset{\approx}{\Omega} = \dot{\underset{\approx}{R}} \bullet \underset{\approx}{R}^T$ is the polar spin. The un-rotated velocity gradient is

$$\overline{\underset{\approx}{L}} = \underset{\approx}{R}^T \bullet \underset{\approx}{L} \bullet \underset{\approx}{R} = \overline{\underset{\approx}{\Omega}} + \dot{\underset{\approx}{U}} \bullet \underset{\approx}{U}^{-1}, \qquad (16.136)$$

where $\overline{\underset{\approx}{\Omega}} = \underset{\approx}{R}^T \bullet \underset{\approx}{\Omega} \bullet \underset{\approx}{R} = \underset{\approx}{R}^T \bullet \dot{\underset{\approx}{R}}$ is the un-rotated polar spin. Formulas for fields as they are perceived from the perspective of a transiently rotating observer are later derived in section 19.3.1, where the tensor $\underset{\approx}{q}$ is the *transpose* of the observer's rotation. Therefore, taking $\underset{\approx}{q} = \underset{\approx}{R}^T$, the velocity gradient in the material frame (i.e., one tracking with the polar rotation) is, according to (19.53*b*)

$$\underset{\approx}{L}^* = \underset{\approx}{q} \bullet \underset{\approx}{L} \bullet \underset{\approx}{q}^T + \dot{\underset{\approx}{q}} \bullet \underset{\approx}{q}^T = \overline{\underset{\approx}{L}} - \overline{\underset{\approx}{\Omega}} = \dot{\underset{\approx}{U}} \bullet \underset{\approx}{U}^{-1}. \qquad (16.137)$$

In other words, the velocity gradient in the material frame is given by a formula that is the same as it would be if the deformation gradient were simply $\underset{\approx}{F}^* = \underset{\approx}{U}$. This also follows from (19.48) when we recognize that $\underset{\approx}{R} = \underset{\approx}{I}$ in the initial state, and therefore $\underset{\approx}{q}_0^T = \underset{\approx}{I}$. This observation implies that constitutive models may be designed without loss in generality by assuming that the deformation is always a pure stretch, and then large-deformation considerations arise exclusively from large stretches and/or rotation of stretch eigenvectors in the absence of bulk material rotation.

The un-rotated rate of deformation and un-rotated vorticity tensors are

$$\overline{\underset{\approx}{D}} = \text{sym}\overline{\underset{\approx}{L}} = \tfrac{1}{2}(\dot{\underset{\approx}{U}} \bullet \underset{\approx}{U}^{-1} + \underset{\approx}{U}^{-1} \bullet \dot{\underset{\approx}{U}}), \qquad (16.138a)$$

$$\overline{\underset{\approx}{W}} = \text{skw}\overline{\underset{\approx}{L}} = \overline{\underset{\approx}{\Omega}} + \tfrac{1}{2}(\dot{\underset{\approx}{U}} \bullet \underset{\approx}{U}^{-1} - \underset{\approx}{U}^{-1} \bullet \dot{\underset{\approx}{U}}). \qquad (16.138b)$$

The rate of deformation and vorticity as seen by a material-frame observer are

$$\underset{\approx}{W}^* = \overline{\underset{\approx}{W}} - \overline{\underset{\approx}{\Omega}} = \tfrac{1}{2}(\dot{\underset{\approx}{U}} \bullet \underset{\approx}{U}^{-1} - \underset{\approx}{U}^{-1} \bullet \dot{\underset{\approx}{U}}). \qquad (16.139)$$

This shows, incidentally, that a material-frame observer perceives no vorticity if the principal directions of reference stretch are stationary[9]. At least for initially isotropic media, such conditions can be used to justify using constitutive models that employ additive strain rates instead of a more complicated multiplicative decomposition of the deformation gradient. Moreover, results of this section demonstrate that this is a very good approximation over a very large strain range (beyond which many materials would have failed, hence reducing to relatively trivial response models).

[9] At least for an initially isotropic medium, plasticity theories that use a multiplicative decomposition of the deformation gradient reduce to additive decompositions of strain rate in the case of stationary directions of reference stretch. Therefore, any claim of superiority of multiplicative theories over additive ones needs to demonstrate no advantage in cases of vortex-free pure stretch deformation (i.e., when reference stretch directions are stationary) and then demonstrate compelling advantage with a vortex. In our engineering experience, there is no advantage of the multiplicative models for real materials subjected to severe loading such as explosive compaction and fragmentation. An advantage might exist, however, for cyclic loading.

Of interest at this point is the fact that $\overline{\underset{\approx}{D}}$ might be easier to compute than $\underset{\approx}{\dot{U}}$. In that case, equation (16.138*a*) could represent a formula solvable for $\underset{\approx}{\dot{U}}$. Specifically,

$$\overline{D}_{ij} = K_{ijrs}\dot{U}_{rs}, \quad \text{where} \quad K_{ijrs} = \tfrac{1}{2}(\delta_{ir}U_{js}^{-1} + U_{ir}^{-1}\delta_{js}). \tag{16.140}$$

Formally, this represents a linear system solvable for $\underset{\approx}{\dot{U}}$. To solve the problem more efficiently, however, we note that (16.103) expresses the un-rotated rate of deformation in terms of the rate of reference stretch. Thus, for cases where the un-rotated rate of deformation is actually easier to compute, equation (16.103) may be used in inverse form as:

$$(\underset{\approx}{\dot{U}})_{pq}^{\langle U \rangle} = \frac{(\overline{\underset{\approx}{D}})_{pq}^{\langle U \rangle}}{\frac{1}{2}\left(\frac{1}{\lambda_p} + \frac{1}{\lambda_q}\right)}, \quad \text{no sum on } p \text{ or } q \quad . \tag{16.141}$$

This gives the components of the *rate* of stretch in the principal basis for stretch, which may be then transformed back to the laboratory frame to obtain the laboratory components. This process effectively (and efficiently) applies the inverse of the 6 × 6 matrix for K_{ijrs} in (16.140).

For finite time stepping, we comment that perhaps a better way of computing the rate of stretch (in a manner more consistent with the numerical method) might be to compute the deformation at the beginning and end of the step, and simply use the corresponding polar decompositions at the beginning and end of the step to set the best value for the rates of rotation and stretch.

16.4.3 The stress that is conjugate to logarithmic strain

The stress power per unit mass is defined by

$$\frac{1}{\rho}\sum_{i=1}^{3}\sum_{j=1}^{3}\sigma_{ij}D_{ij}, \tag{16.142}$$

where σ_{ij} is the Cauchy stress and D_{ij} is the symmetric part of the velocity gradient. In terms of the un-rotated Cauchy stress and the un-rotated symmetric part of the velocity gradient, the stress power is

$$\frac{1}{\rho}\sum_{i=1}^{3}\sum_{j=1}^{3}\overline{\sigma}_{ij}\overline{D}_{ij}. \tag{16.143}$$

This holds for any basis, so it must also hold in the U-space basis. Thus, using (16.105), the stress power may be written

$$\frac{1}{\rho}\sum_{i=1}^{3}\sum_{j=1}^{3}\overline{\sigma}_{ij}^{\langle U \rangle}\overline{D}_{ij}^{\langle U \rangle} = \frac{1}{\rho}\sum_{i=1}^{3}\sum_{j=1}^{3}\overline{\Sigma}_{ij}^{\langle U \rangle}\dot{\varepsilon}_{ij}^{\langle U \rangle}, \tag{16.144}$$

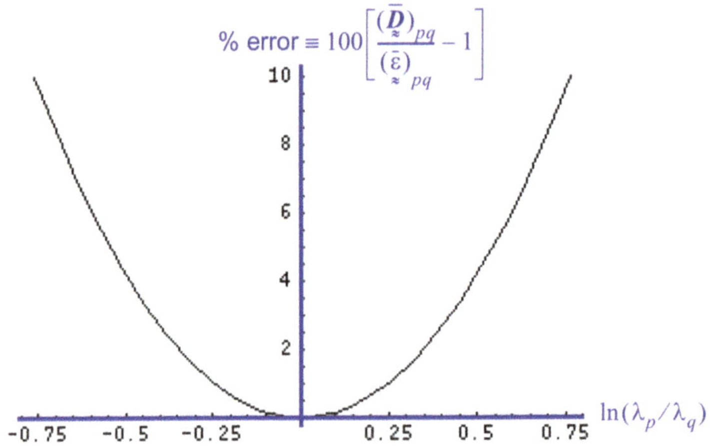

Figure 16.2. Percent error in approximating log strain rate by the symmetric part of the deformation gradient. Note that the error in any U-space diagonal component is zero. Only off-diagonals can have error.

where

$$\boxed{\bar{\Sigma}_{pq}^{\langle U\rangle} = \begin{cases} \left(\dfrac{\sinh r_{pq}}{r_{pq}}\right)\bar{\sigma}_{pq}^{\langle U\rangle} & \text{if } r_{pq} \neq 0 \\ \bar{\sigma}_{pq}^{\langle U\rangle} & \text{if } r_{pq} = 0 \end{cases}}. \tag{16.145}$$

This is the stress that is work conjugate to the logarithmic strain rate. The diagonal components of this stress (in the <U> basis) always coincide with the Cauchy stress component. Differences appear only in the off-diagonals. Consequently, laboratory testing that is limited to axial deformations and similar tests that do not involve rotation of the reference stretch directions might fail to distinguish between Cauchy stress and this truly conjugate stress. In plasticity theory, this class of deformation (with rotating reference stretch directions) is required to distinguish models that use a multiplicative decomposition of the deformation gradient and those that use an additive decomposition of Hencky (logarithmic) strain. Without such validation data, any claim of superiority of multiplicative decompositions is conjecture.

16.4.4 Spatial co-rotational logarithmic strain rate

Although we do not recommend its use in material models, some researchers employ the co-rotational logarithmic strain rate, so we will discuss it here for completeness. Equation (16.105) implies there must exist a fourth-order tensor $\underset{\approx\approx}{\bar{\mathbf{Z}}}$ such that

$$\overset{\circ}{\underset{\approx}{\varepsilon}} = \underset{\approx\approx}{\bar{\mathbf{Z}}} : \underset{\approx}{\bar{\mathbf{D}}}. \tag{16.146}$$

As indicated in the previous section, the fourth-order tensor is never actually computed because it is easier to compute the strain rate directly. Recalling (16.99), the above equation may be written as

$$\overline{\overset{\circ}{\underset{\approx}{\varepsilon}}} = \overline{\underset{\approx}{Z}} : \overline{\underset{\approx}{D}}. \tag{16.147}$$

Here, the 'overbar' operation is generalized to fourth-order tensors as

$$\overline{Z}_{ijkl} = R^T_{ip} R^T_{jq} R^T_{kr} R^T_{ls} Z_{pqrs}, \text{ with implied summations.} \tag{16.148}$$

Generalizations of the overbar operation to tensors of any order is similar. So defined, the overbar operation can be readily shown to be distributive over multiplication. Therefore, equation (16.147) can be immediately converted into a spatial result for the co-rotational log strain rate as

$$\overset{\circ}{\underset{\approx}{\varepsilon}} = \underset{\approx}{Z} : \underset{\approx}{D}, \tag{16.149}$$

from which it follows that the components of the co-rotational log strain rate with respect to the *spatial* principal directions of the *spatial stretch* (i.e., V-space) are given by an equation very similar to (16.105). Namely

$$(\overset{\circ}{\underset{\approx}{\varepsilon}})^{<V>}_{pq} = \begin{cases} \left(\dfrac{r_{pq}}{\sinh r_{pq}}\right)(\underset{\approx}{D})^{<V>}_{pq} & \text{if } r_{pq} \neq 0 \\ (\underset{\approx}{D})^{<V>}_{pq} & \text{if } r_{pq} = 0 \end{cases}, \quad \text{where } r_{pq} \equiv \ln\left(\frac{\lambda_p}{\lambda_q}\right). \tag{16.150}$$

The *material time rate* of the spatial log strain (i.e., the rate that also includes additional contributions to the angular velocity of principal stretch directions resulting from material rotation) is given by

$$(\dot{\underset{\approx}{\varepsilon}})^{<V>}_{pq} = \begin{cases} \left(\dfrac{\ln(\lambda_p) - \ln(\lambda_q)}{\lambda_p - \lambda_p}\right)(\dot{\underset{\approx}{V}})^{<V>}_{pq} & \text{if } \lambda_p \neq \lambda_q \\ \dfrac{(\dot{\underset{\approx}{V}})^{<V>}_{pq}}{\lambda_p} & \text{if } \lambda_p = \lambda_q \end{cases}. \tag{16.151}$$

This expression does not have the same appealing connection to the symmetric part of the deformation gradient that is seen in (16.150), and it is probably of very limited use anyway because the material constitutive laws using it would probably violate the principle of material-frame indifference (see chapter 19). If a material model is built from series expansions of a governing scalar-valued energy function, then the model will be frame-indifferent, but the *qualitative* behavior of the model is still affected by the choice of objective strain rate measure or, equivalently, by the choice of the physical reference configuration. This fact was demonstrated by Dienes in the case of a two-term series expansion of the hyperelastic energy function. The first term in this expansion equals the stress that is conjugate to the chosen strain measure. The next term equals the elastic stiffness times the strain increment. The applicability range of a two-term series expansion depends critically on the choice of strain measure. For example, a fourth-order expansion of the function

$f(x) = 3\sin x$ is $f(x) \sim 3x - \frac{x^3}{2} + O(x^5)$. If x is retained as the independent variable, then a one term expansion of $f(x)$ will be second-order accurate. However, if we switch to a different independent variable, $y \equiv 3x - x^3/2$, and rewrite $f(x)$ in terms of y, then a one-term expansion of $f(y)$ will become *fourth-order* accurate. This is why strain measures make a difference—it is not that any one strain measure characterizes the deformation any better than another strain measure. Instead optimal choices for the strain measure provide higher accuracy to series expansions of the energy function when evaluated at larger strains. Even when an optimal strain measure has been identified, it is *still* important to carefully select the best co-rotational strain rate to use in frame-indifferent rate-forms of spatial material law formulations. As will be discussed later (see section 22.4), three popular choices for the co-rotational rates are

$$\text{polar rate:} \quad \overset{\circ\text{P}}{\underset{\approx}{\varepsilon}} \equiv \dot{\underset{\approx}{\varepsilon}} - \underset{\sim}{\Omega} \bullet \underset{\approx}{\varepsilon} + \underset{\approx}{\varepsilon} \bullet \underset{\sim}{\Omega}, \tag{16.152}$$

$$\text{Jaumann rate:} \quad \overset{\circ\text{J}}{\underset{\approx}{\varepsilon}} \equiv \dot{\underset{\approx}{\varepsilon}} - \underset{\sim}{\boldsymbol{W}} \bullet \underset{\approx}{\varepsilon} + \underset{\approx}{\varepsilon} \bullet \underset{\sim}{\boldsymbol{W}}, \tag{16.153}$$

$$\text{convected rate:} \quad \overset{\circ\text{C}}{\underset{\approx}{\varepsilon}} \equiv \dot{\underset{\approx}{\varepsilon}} - \underset{\approx}{\boldsymbol{L}} \bullet \underset{\approx}{\varepsilon} - \underset{\approx}{\varepsilon} \bullet \underset{\approx}{\boldsymbol{L}}^T. \tag{16.154}$$

Recall that $\overline{\overset{\circ\text{P}}{\underset{\approx}{\varepsilon}}} = \dot{\overline{\underset{\approx}{\varepsilon}}}$, where, for any tensor $\underset{\approx}{\boldsymbol{A}}$, the overbar is defined $\overline{\underset{\approx}{\boldsymbol{A}}} \equiv \underset{\approx}{\boldsymbol{R}}^T \bullet \underset{\approx}{\boldsymbol{A}} \bullet \underset{\approx}{\boldsymbol{R}}$. Thus, formulating the rate form of the constitutive equation in terms of the *unrotated* reference configuration is equivalent to solving the spatial equation using the polar objective rate. Similarly, it can be shown that $\overline{\overset{\circ\text{C}}{\underset{\approx}{\varepsilon}}} = \dot{\overline{\underset{\approx}{\varepsilon}}}$ if the overbar operation is redefined to mean $\overline{\underset{\approx}{\boldsymbol{A}}} \equiv \underset{\approx}{\boldsymbol{F}}^{-1} \bullet \underset{\approx}{\boldsymbol{A}} \bullet \underset{\approx}{\boldsymbol{F}}^{-T}$. Thus, formulating the constitutive law in terms of the *unconvected* reference configuration is equivalent to using the convected rate in the spatial configuration. Either of these choices are special cases of what are called Lie derivatives, in which an objective rate expressible as the 'push forward' of the ordinary rate of a 'pulled back' tensor. The Jaumann rate is not expressible as a Lie derivative, and hence lacks the numerical appeal of performing all constitutive calculations in a pulled back reference configuration.

References

[1] Chadwick P and Ogden R W 1971 A theorem of tensor calculus and its applications to isotropic elasticity *Arch. Ration. Mech. Anal.* **44** 54–68

[2] Wolfram Steven 1999 *The Mathematica Book* 4th edn (Cambridge: Wolfram Media/ Cambridge University Press)

[3] Gerald C F and Wheatley P O 1984 *Applied Numerical Analysis* 4th edn (Reading, MA: Addison-Wesley)

Chapter 17

Statistics of random orientation

For problems at the mesoscopic scale where, for example, crystal anisotropy is modeled, simulations often require a means of assigning a random rotation of the crystal from its unit cell reference orientation to its actual orientation [1]. For a macroscopically isotropic material of statistically uniform grain size, a uniformly random distribution of grain orientations is desired. This chapter discusses how to achieve realizations of orientation and texture distributions. We have already mentioned one algorithm for generating a uniformly random rotation (in the text following (5.90)), but that algorithm requires a numerical solver to obtain the rotation angle. In this chapter, we will develop the proof of why that algorithm works, and we will present alternative algorithms that can be applied in closed form (i.e., not requiring a solver). This chapter begins with an overview of how to generate non-uniformly distributed scalars from a uniformly random number. This capability is then applied to generate uniformly random unit vectors. This prerequisite work leads up to the discussion of random rotations starting in section 17.6.

17.1 Elementary probability and statistics refresher

Let X denote a realization of a continuously random variable, and let x denote possible values for this realization. The probability distribution function (PDF) is denoted $p(x)$ and defined such that the probability that X falls between a and b is

$$\int_a^b p(x')\mathrm{d}x' = g(b) - g(a). \tag{17.1}$$

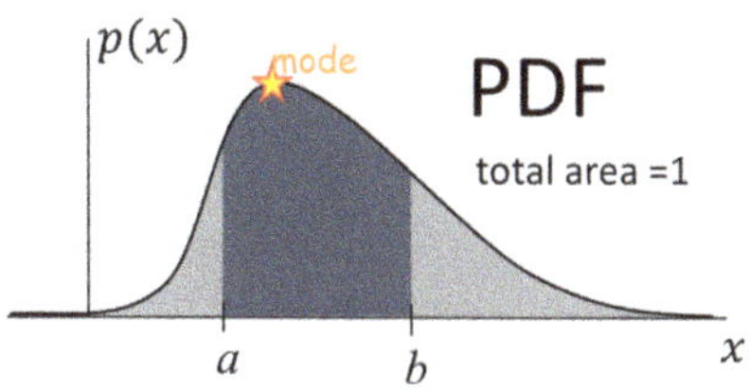

doi:10.1016/978-0-7503-1454-1ch17

The corresponding cumulative distribution function (CDF) is denoted $g(x)$ and defined as

$$g(x) = \int_{-\infty}^{x} p(x')\mathrm{d}x'. \tag{17.2}$$

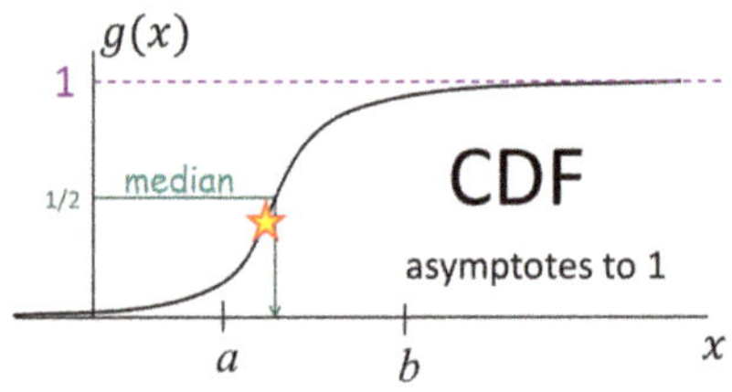

Thus, $g(x)$ is the probability that $X > x$ (i.e., it is the likelihood that the actual realization of X is larger than x). Key attributes of these functions are:

- The PDF always starts and ends at zero. The area under the PDF is unity. The physical units of $p(x)$ are therefore the reciprocal of the units for x.
- The CDF always starts at 0 and grows monotonically to 1 (so it *never* has a negative slope). The value of the CDF at x is just the area under the PDF to the left of x.
- It helps to think of the CDF in terms of probability that some 'significant event' will have occurred at or below the value of x. Suppose, for example, that x is the stress in a material at the point that it fails. After testing hundreds of specimens, it is observed that stress at failure is statistically distributed. Now suppose that you are given a fresh specimen. There is zero chance of failure at zero stress, and the failure probability continuously increases toward 100% as the stress is increased—that is the CDF! The CDF for failure stress shows how increasing the applied stress increases likelihood of failure.
- The *mode* (marked with a star in the sketches) is the value of x at the peak of the PDF, so it is therefore the point on the CDF with the steepest slope. If you had to bet that the outcome would fall within a small neighborhood of a selected value, then put your money on the mode. Multiple peaks in a PDF, if present, represent the most likely outcome in that neighborhood.
- The *mean* (also called the *expected value*) is the density-weighted average,

$$\bar{x} = \int_{-\infty}^{+\infty} x' p(x')\mathrm{d}x'. \tag{17.3}$$

  The notation $E(X)$ is also used for the mean

- The *median* (also called the *50% quantile*, marked by the endpoint of the arrow originating at 1/2 on the ordinate of the CDF) is the value of x for which X turns out to be smaller than x half of the time. While the mean might be a more mathematically useful concept, accounting for human nature (namely a tendency to remove outliers even when they are actually legitimate) makes the median a far more reliable measure to report in experimental data.

Removal of a strong outlier can affect the mean significantly while having relatively little influence on the median.

- The mode, mean, and median are not generally equal to each other. When they are different, it is a sign that there is 'skewness' (non-symmetry) in the distribution.
- Mean values receive the glory in probability theory, as they are most useful in decision theory. However, the mean often gives distorted impressions of what is 'typical'. Suppose, for example, that professors in a university are 'scored' for their citation rate, normalized in comparison to the mean citation rate. If there is one sassy professor getting thousands of citations from a fluke popular article, everyone else's scores will plummet. The median, on the other hand, is insensitive to such outliers. As mentioned, experimentalists tend to remove what they judge to be outliers in their data, spuriously causing the mean to be closer to the median.
- When dealing with more than one random variable at a time, a subscript label might be added to the p and g functions to identify their associated random variable. For example, $p_X()$ is the PDF for X, while $p_Y()$ is for some other variable Y.
- If $Y = f(X)$ is a random variable computed from X, then the expected value of Y is

$$E(Y) = \int_{-\infty}^{+\infty} f(x')p(x')\mathrm{d}x'. \tag{17.4}$$

- If $Y = f(X)$ is a random variable computed from X, and if the function $f(X)$ is invertible (i.e., not changing in slope), then the CDF for Y is $g_Y(y) = g_X(f^{-1}(y))$, which (by the chain rule) implies that the PDF for Y is $p_Y(y) = \frac{p_X(f^{-1}(y))}{f'(f^{-1}(y))}$. In this case of an invertible relationship between Y and X, the median for Y is $f(\hat{X})$, where $\hat{X}$ is the median for X.

To illustrate these statements, consider an example from damage mechanics, where the compliance of a cracked medium depends on the expected value of $y = a^3$, where a is a randomly distributed crack size. Suppose that the crack size is exponentially distributed so that its PDF is $p_a(a) = \frac{e^{-a/\bar{a}}}{\bar{a}}$, and therefore its CDF is $g_a(a) = 1 - e^{-a/\bar{a}}$, where $\bar{a}$ is the mean crack size. Then (using $f(a) = a^3$ and $f^{-1}(y) = y^{1/3}$), the CDF for Y is $g_Y(y) = 1 - e^{-y^{1/3}/\bar{a}}$ and the PDF is $p_Y(y) = \frac{e^{-y^{1/3}/\bar{a}}}{3\bar{a}y^{2/3}}$. In this example, the median crack size is $\hat{a} = \bar{a}\ln 2$. The median of crack size cubed is simply the cube of the median crack size: $\hat{y} = \hat{a}^3$. The mean of the cube, on the other hand, is $\bar{y} = 6\bar{a}^3$, which is larger than the cube of the mean by a factor of 6. These results imply that the median and mean of the cube are related by $\hat{y} = \frac{1}{6}(\ln 2)^3\,\bar{y}$. Thus, while the median $\hat{a}$ is about 40% larger than the mean $\bar{a}$, a much larger discrepancy exists for $y = a^3$, where the median $\hat{y}$ is about 18 times larger than the mean $\bar{y}$. This skewness of the distribution makes it difficult to model a cracked medium using an effective single crack.

- From here forward, no further effort is expended to always distinguish the actual realization values X from possible values x unless the distinction is essential. Otherwise, intended meaning can be inferred from context.

17.1.1 How to generate a random realization

Consider any non-uniform distribution $p(x)$ on the interval $[a, b)$[1]. We seek a mapping function $x = f(r)$ such that x will have the desired $p(x)$ distribution whenever r has a uniform distribution on $[0, 1)$. We will further require the function f to have a correspondence such that $f(0) = a$ and $f(1) = b$. Let us first define the *cumulative distribution*:

$$\boxed{g_x(x) \equiv \int_{-\infty}^{x} p_x(x')\mathrm{d}x'}. \tag{17.5}$$

The density function $p(x)$ is non-negative, so the CDF $g(x)$ must be a non-decreasing function, and therefore it has an inverse except possibly at discrete points. In other words, g^{-1} may permissibly contain discrete bounded jumps.

The distribution function p is defined such that the probability that x will lie between x_1 and x_2 is

$$\int_{x_1}^{x_2} p(x')\mathrm{d}x' = \int_{f(r_1)}^{f(r_2)} p(x')\mathrm{d}x' = g(f(r_2)) - g(f(r_1)). \tag{17.6}$$

Since r is uniformly distributed on the interval $[0, 1)$, the probability that $r_1 < r < r_2$ is

$$\int_{r_1}^{r_2} \mathrm{d}r = r_2 - r_1. \tag{17.7}$$

Because we are seeking an invertible mapping function, these last two probabilities must be equal:

$$g_x(f(r_2)) - g_x(f(r_1)) = r_2 - r_1. \tag{17.8}$$

This must hold for any values of r_1 and r_2. Let us take $r_1 = 0$ and $r_2 = r$. Recall that we have required the point $r = 0$ to map to the point $x = a$. Therefore $f(0) = a$, and (17.8) implies

$$g(f(r)) - g(a) = r. \tag{17.9}$$

Recalling that $p(x) = 0$ for all $x < a$, we note that $g_x(a) = 0$, and the above equation becomes

$$g(f(r)) = r, \tag{17.10}$$

[1] The mismatched [...) delimiters here are not typos. The notation $x \in [a, b)$ means $a \leqslant x < b$.

or

$$f(r) = g^{-1}(r). \tag{17.11}$$

In summary, if r has a uniform distribution on [0,1), then $f(r)$ will have the distribution $p_x(x)$ on the interval $[a, b)$. This was a lot of work leading to the following simple result: to get a random realization distributed according to the PDF $p(x)$, first get the CDF $g(x)$. Then set the CDF equal to r and solve for x. That is it!

$$\boxed{\begin{array}{l}\text{random realization}\\ X = g^{-1}(r)\end{array}} \tag{17.12}$$

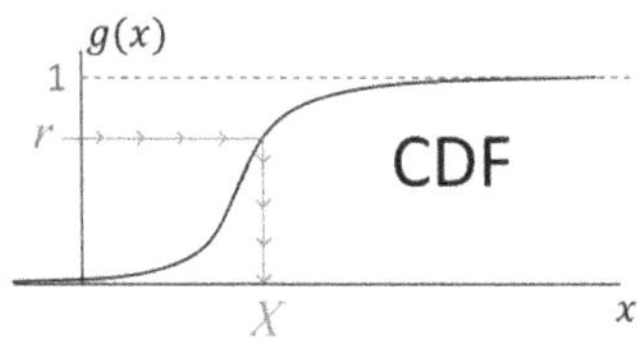

As an example, suppose that the PDF is

$$p(x) = \tfrac{1}{2}\sin x \text{ for } 0 < x < \pi \text{ (and zero outside this interval).} \tag{17.13}$$

For x in this interval, the CDF is

$$g(x) = \int_0^x \tfrac{1}{2}\sin x'\ \mathrm{d}x' = \tfrac{1}{2}(1 + \cos x). \tag{17.14}$$

Replacing $g(x)$ with r and solving for x gives the formula for a random realization: $X = \cos^{-1}(2r - 1)$. Easy!

17.1.2 Finite sampling errors

Suppose that you take 100 measurements of a random variable. Is that enough to discern its distribution to a good approximation? How about 1000 measurements? How many is good enough? This question pertains to the topic of 'confidence', covered in any good book on statistics (see e.g., 'Student's t-test' on Wikipedia). Here, we simply show how to create empirical CDFs (called eCDFs) from measured data, and then we give you a sense of the frighteningly large errors seen when fitting distributions from only a finite number of realizations. As a result, even if your model is an exact description of reality, your statistical predictions can have errors exceeding 15%–30% simply because it is not practical to take enough laboratory measurements to fit your model with high confidence[2].

[2] Incidentally, this means that you should look askance at any researcher's claims of high accuracy with a statistical prediction; by its nature, high confidence (low error) is impractical within ordinary engineering budgets. Challenge these model-tuning braggarts to show you the 1000+ data points!

For our case study, consider a positive variable F that is Weibull distributed. This distribution is defined by two parameters: the shape parameter m (often called the Weibull modulus) and a scale parameter α. In the classical application, the random variable F represents the force at which a chain of length L breaks. The strength of the chain is determined entirely by the strength of its weakest link. Let $g(F)$ denote the probability that a single link is safe from failure when subjected to a force F. Since each link in a serial chain is subjected to the same force, the probability that the chain is safe from failure is the single-failure probability multiplied by itself as many times as there are links. If there are n links per unit length, the total number of links in the chain is nL, so that

$$P^{\text{safe}} = [g(F)]^{nL}. \tag{17.15}$$

Unsurprisingly, this has a size effect: longer chains are less likely to be safe than short ones. That makes sense: the longer the chain, the more likely it will be that it contains a very weak link from a random selection of links uses to create the chain.

Taking the logarithm gives

$$\ln P^{\text{safe}} = (nL)\ln[g(F)]. \tag{17.16}$$

To further separate the effect of chain link strength from the intrinsic length effect, we would like to take another logarithm. However, because $0 < P^{\text{safe}} < 1$, it follows that $\ln P^{\text{safe}} < 0$. Therefore, before we can take another logarithm, we must adjust the equation by multiplying both sides by -1 to obtain

$$\ln \frac{1}{P^{\text{safe}}} = (nL)\ln \frac{1}{g(F)}. \tag{17.17}$$

Now we can take another logarithm to write

$$\ln \ln \frac{1}{P^{\text{safe}}} = \ln(nL) + \ln \ln \frac{1}{g(F)}. \tag{17.18}$$

Recognizing that P^{safe} is actually the complementary CDF, equal to $1 - P^{\text{fail}}$ where P^{fail} is the CDF, let us make $\bar{F}$ denote the *median* breaking force for a chain of length L. Then, by definition of median (which corresponds to the CDF equal to 1/2), evaluating the above equation using $P^{\text{safe}} = 1/2$ and $F = \bar{F}$ gives

$$\ln \ln 2 = \ln(nL) + \ln \ln \frac{1}{g(\bar{F})}. \tag{17.19}$$

Subtracting the above two equations eliminates the explicit presence of the chain length (hiding it instead in size dependence of median chain strength) to give

$$\ln \ln \frac{1}{P^{\text{safe}}} - \ln \ln 2 = \ln \ln \frac{g(\bar{F})}{g(F)}. \tag{17.20}$$

In classical Weibull theory, the single-chain function $g(F)$ is assumed to be of a particular (power-law) form such that

$$\ln \ln \frac{1}{P^{\text{safe}}} - \ln \ln 2 = m \ln\left(\frac{F}{\bar{F}}\right). \tag{17.21}$$

where m is a material parameter called the *Weibull modulus*. Recalling that $P^{\text{safe}} = 1 - P^{\text{fail}}$ and that P^{fail} is actually the CDF for the failure force, this equation may be solved for the CDF to give the Weibull CDF

$$\boxed{P^{\text{fail}} = 1 - e^{-(F/\beta)m}}, \qquad \text{where} \qquad \beta = \frac{\bar{F}}{(\ln 2)^{1/m}}. \tag{17.22}$$

The parameter m, is also called the *shape parameter* because it controls the amount of spread (i.e., variance in the CDF). The parameter β is called the *scale parameter* because it is directly proportional to the median strength (which itself depends on the size of the chain). Note that (17.21) can be written in the compact form

$$y = mx, \qquad \text{where} \quad y = \ln \ln \frac{1}{P^{\text{safe}}} - \ln \ln 2 \qquad \text{and} \qquad x = \ln\left(\frac{F}{\bar{F}}\right). \tag{17.23}$$

Note that y is merely a transformation of the CDF ordinate, while x is a transformation of the abscissa. Thus, a plot of y versus x should be viewed as merely a distorted CDF plot. Such a transformation can be applied to *any* CDF plot. The formula $y = mx$ simply says that the data are Weibull distributed if they map to a straight line in this transformed plot. Fitting the data to a Weibull distribution is a simple matter of linear regression in the Weibull-transformed plot. The slope of that fit is the Weibull modulus. Of course, the scale parameter is determined directly from the median, which can be evaluated from the population.

Given a total of N sample failure forces, $F_1, F_2, \ldots, F_N$, arranged from smallest to largest, the median $\bar{F}$ is the middle value. Associated with each F_k value is an eCDF value given by an estimator that (for the realistically small numbers of experiments usually available to a laboratory budget) is best taken as

$$C_k = \frac{k - \frac{1}{2}}{N}. \tag{17.24}$$

With this estimator for the eCDF ordinate values, the eCDF steps up by an amount $1/N$ at each measured F_k value. The subtraction of 1/2 puts the eCDF estimator at the middle of this step. A table of (F_k, C_k) values can then be transformed to a set of (x_k, y_k) data pairs to be depicted in the Weibull-transformed space.

Figure 17.1 shows, however, that even if data are sampled directly from a known Weibull distribution, the eCDF fits to the data will not closely match the exact Weibull source CDF unless a fairly large set of measurements is taken. Typical testing budgets are do not allow enough experiments to reach a 5%–10% confidence, so funding agencies must be taught that they should not expect high accuracy from low-budget testing[3].

[3] Even more funding would be needed to determine if fitting to a Weibull distribution is appropriate!

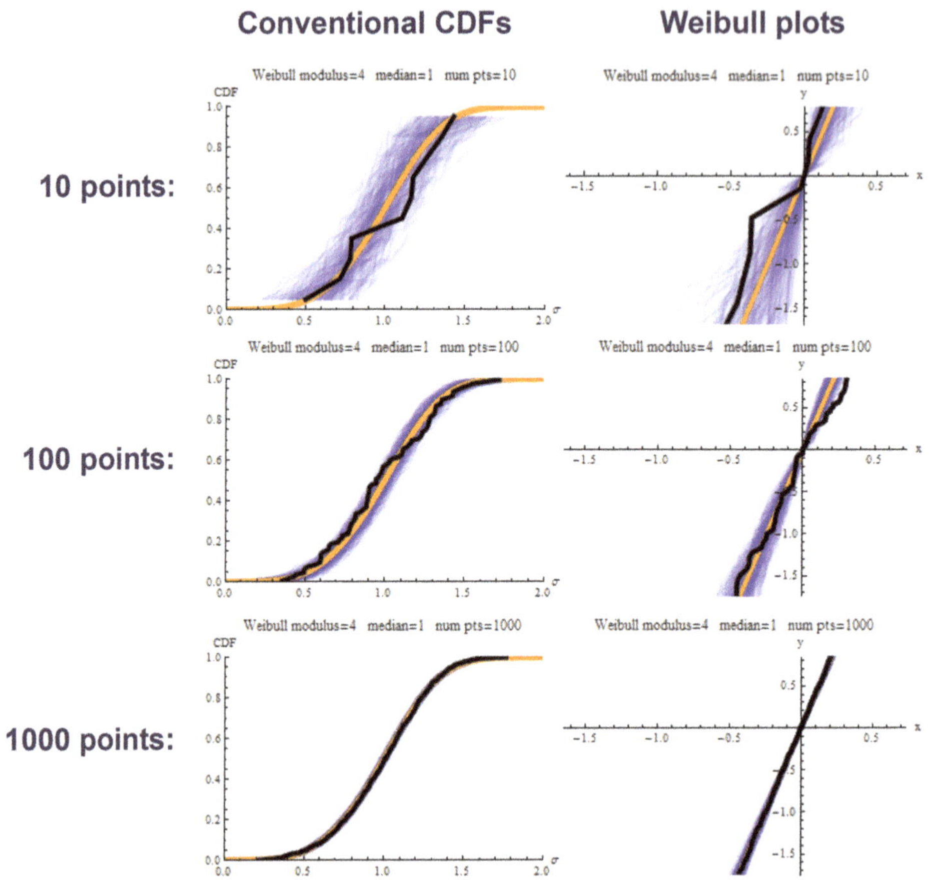

Figure 17.1. A finite number of exactly Weibull-distributed σ stress values were generated and analyzed as if they were experimental data to produce the black plot, shown here with the exact Weibull curve in orange. The process was repeated 200 times, each time producing a different version of the black plot, except shown in blue. This figure gives a general 'sense' of how much error might exist simply from finite sampling even when the data are exactly Weibull-distributed. Even a sampling of 1000 points shows visible error!

17.1.3 Two random variables

For two random variables, ν and η, let $p_{\nu,\eta}(\nu, \eta)$ denote the joint distribution defined such that the probability that $\nu_1 < \nu < \nu_1$ and $\eta_1 < \eta < \eta_1$ is

$$\int_{\nu_1}^{\nu_2} \int_{\eta_1}^{\eta_2} p_{\nu,\eta}(\nu', \eta') \mathrm{d}\nu' \mathrm{d}\eta'. \tag{17.25}$$

The two random variables are said to be *independent* if the joint PDF turns out to be a separable function: $p_{\nu,\eta}(\nu, \eta) = p_\nu(\nu) p_\eta(\eta)$. Of course, the above integral must evaluate to unity when the integration domain includes the full range of possible values of the random variables; for impossible values (e.g., when the random variables are always positive), simply treat the impossible regions to have PDF values of zero. For readers familiar with mechanics, it often helps to think of a PDF as mass density so that the above integral represents the mass in a region contained within a 'unit-mass universe', so it is the fraction of total mass in the integration domain.

When considering the entire range of possibilities for both ν and η, the PDF for ν is found by integrating away the other random variable:

$$p_\nu(\nu) = \int_{-\infty}^{+\infty} p_{\nu,\eta}(\nu, \eta)\mathrm{d}\eta, \qquad \text{and similarly} \qquad p_\eta(\eta) = \int_{-\infty}^{+\infty} p_{\nu,\eta}(\nu, \eta)\mathrm{d}\nu. \quad (17.26)$$

When a PDF is seen as analogous to density (= mass per area for two-dimensional problems), this integral creates a lineal density (= mass per length).

Now consider a different, far more narrow, range of possibilities. Specifically, suppose that the value of η is known. In that case, the 'conditional' PDF for the other random variable ν is denoted $p(\nu|\eta)$, which is read as *the probability of v given* η, and is given by

$$p(\nu|\eta) = \frac{p_{\nu,\eta}(\nu, \eta)}{p_\eta(\eta)}, \qquad \text{and similarly} \qquad p(\eta|\nu) = \frac{p_{\nu,\eta}(\nu, \eta)}{p_\nu(\nu)}. \quad (17.27)$$

Together, these imply a useful relation, called *Bayes theorem*, which allows you to transform conditional probabilities to swap which variable is known:

$$p(\nu|\eta) = \frac{p(\eta|\nu)p_\nu(\nu)}{p_\eta(\eta)}. \quad (17.28)$$

To create a random realization for the pair (ν^*, η^*), you must first generate two random numbers (r_1, r_2), each uniformly distributed on the interval from 0 to 1. Letting $g_\eta(\eta)$ denote the CDF associated with the PDF $p_\eta(\eta)$, the first step is to set $g_\eta(\eta^*) = r_1$ and solve for η^* to generate the random realization for η, as discussed previously for a single-variable distribution. Now that a realization has been found for one of the random variables, $\eta = \eta^*$, the next step is to find a realization ν^* for the other one. Letting $g_\nu(\nu|\eta^*)$ denote the CDF associated with the conditional PDF $p(\nu|\eta^*)$,

$$g_\nu(\nu|\eta^*) = \int_{-\infty}^{\nu} \frac{p_{\nu,\eta}(\nu', \eta^*)}{p_\eta(\eta^*)}\mathrm{d}\nu, \quad (17.29)$$

you similarly create a realization for ν by setting $g_\nu(\nu^*|\eta^*) = r_2$ and solving for ν^*.

17.1.4 Special case: random position vector

If $\underset{\sim}{x}$ is a random vector in 3D space, then the distribution function $p_{\underset{\sim}{x}}(\underset{\sim}{x})$ is defined such that the probability that $\underset{\sim}{x}$ lies in a region of space B is

$$\int_B p_{\underset{\sim}{x}}(\underset{\sim}{x}')\mathrm{d}V, \quad (17.30)$$

where $\mathrm{d}V$ is the volume element. To actually evaluate this integral, coordinates must be introduced. If ordinary (x, y, z) Cartesian coordinates are used, so that $\underset{\sim}{x} = \underset{\sim}{x}(x, y, z)$, then the volume element is $\mathrm{d}V = \mathrm{d}x\mathrm{d}y\mathrm{d}z$, and the above integral becomes

$$\int_B p_{\underset{\sim}{x}}(\underset{\sim}{x}')\mathrm{d}V = \iiint p_{x,y,z}(x, y, z)\mathrm{d}x\ \mathrm{d}y\ \mathrm{d}z. \tag{17.31}$$

In this case, the joint distribution for the Cartesian coordinates is intuitively given by equating the integrands:

$$p_{x,y,z}(x, y, z) = p_{\underset{\sim}{x}}(\underset{\sim}{x}(x, y, z)). \tag{17.32}$$

Thus, for example, if a point is uniformly distributed in B, then $p_{\underset{\sim}{x}}(\underset{\sim}{x})$ is a constant in B and zero outside B. To ensure that the integral of $p_{\underset{\sim}{x}}(\underset{\sim}{x})$ over all possibilities is unity, the value of the constant must be the reciprocal of the volume of B.

Now consider spherical (r, ϑ, φ) coordinates, related to Cartesian coordinates by

$$x = r \sin\vartheta \cos\varphi, \qquad y = r \sin\vartheta \sin\varphi, \qquad z = r\cos\vartheta. \tag{17.33}$$

With these relationships, the Cartesian PDF function $p_{x,y,z}(x, y, z)$ may be expressed in terms of the spherical coordinates. However,

$$p_{r,\vartheta,\varphi}(r, \vartheta, \varphi) \neq p_{\underset{\sim}{x}}(\underset{\sim}{x}(r, \vartheta, \varphi)) \ \text{ in general.} \tag{17.34}$$

That is, you cannot obtain $p_{r,\theta,\varphi}(r, \theta, \varphi)$ by simply substituting (17.33) into $p_{x,y,z}(x, y, z)$. Doing that would simply be expressing $p_{x,y,z}$ in terms of spherical coordinates, which is not the same as expressing $p_{r,\vartheta,\varphi}$ in terms of spherical coordinates.

Given the joint distribution $p_{\underset{\sim}{x}}$ or, equivalently, given $p_{x,y,z}(x, y, z)$, what is the joint distribution $p_{r,\vartheta,\varphi}(r, \vartheta, \varphi)$ for the spherical coordinates? In this case, noting that $\mathrm{d}V = \sin\vartheta\ \mathrm{d}r\ \mathrm{d}\vartheta\ \mathrm{d}\varphi$,

$$\int_B p_{\underset{\sim}{x}}(\underset{\sim}{x}')\mathrm{d}V = \iiint p_{\underset{\sim}{x}}(\underset{\sim}{x}'(r, \vartheta, \varphi)\sin\vartheta\ \mathrm{d}r\ \mathrm{d}\vartheta\ \mathrm{d}\varphi. \tag{17.35}$$

Identifying the entire integrand of the right-hand side (including the crucial $\sin\vartheta$ part) to be $p_{r,\vartheta,\varphi}(r, \vartheta, \varphi)$ and using (17.32), this implies that

$$p_{r,\vartheta,\varphi}(r, \vartheta, \varphi) = p_{x,y,z}(x, y, z)\Big|_{\substack{x\to r\sin\vartheta\cos\varphi \\ y\to r\sin\vartheta\sin\varphi \\ z\to r\cos\vartheta}} \sin\vartheta. \tag{17.36}$$

More generally, given a generic joint PDF $p_{u_1,u_2,u_3}(u_1, u_2, u_3)$, the PDF under a change of variables to (v_1, v_2, v_3) is given by

$$p_{v_1,v_2,v_3}(v_1, v_2, v_3) = p_{u_1,u_2,u_3}(u_1, u_2, u_3)\Big|_{\substack{u_1\to u_2(v_1,v_2,v_3) \\ u_2\to u_2(v_1,v_2,v_3) \\ u_3\to u_3(v_1,v_2,v_3)}} J(v_1, v_2, v_3), \tag{17.37}$$

where J is the *Jacobian determinant*, found by

$$J = \det[\mathbf{J}], \qquad \text{in which} \qquad [\mathbf{J}] = \begin{bmatrix} \frac{\partial u_1}{\partial v_1} & \frac{\partial u_1}{\partial v_2} & \frac{\partial u_1}{\partial v_3} \\ \frac{\partial u_2}{\partial v_1} & \frac{\partial u_2}{\partial v_2} & \frac{\partial u_2}{\partial v_3} \\ \frac{\partial u_3}{\partial v_1} & \frac{\partial u_3}{\partial v_2} & \frac{\partial u_3}{\partial v_3} \end{bmatrix}. \tag{17.38}$$

The number of random variables (three in the above example) is not always the same. If, for example, we had only v_1 and v_2, as would be the case when generating a random (three-component) position vector $\underset{\sim}{x}$ on the surface of a sphere by using random ϑ and φ spherical coordinates, then the above formulas should be revised as

$$p_{v_1,v_2,v_3}(v_1, v_2, v_3) = p_{u_1,u_2,u_3}(u_1, u_2, u_3)\Big|_{\substack{u_1 \to u_1(v_1,v_2) \\ u_2 \to u_2(v_1,v_2) \\ u_3 \to u_3(v_1,v_2)}} J(v_1, v_2), \tag{17.39}$$

$$J = \sqrt{\det([\mathbf{J}]^T[\mathbf{J}])}, \qquad \text{in which} \qquad [\mathbf{J}] = \begin{bmatrix} \frac{\partial u_1}{\partial v_1} & \frac{\partial u_1}{\partial v_2} \\ \frac{\partial u_2}{\partial v_1} & \frac{\partial u_2}{\partial v_2} \\ \frac{\partial u_3}{\partial v_1} & \frac{\partial u_3}{\partial v_2} \end{bmatrix}. \tag{17.40}$$

This is a simple way to find the product of the non-zero singular values of the non-square matrix. An excellent example is given in the theory of random unit vectors. This result is used in the next section for a conditional random distribution of spherical coordinates subject to the constraint that the radial coordinate is constant (i.e., we will find random points on a sphere).

The Jacobian J has a useful geometrical interpretation. Imagine laying out a rectangular grid (or any other tesselation) covering the domain of possible values for $\{v_1, v_2\}$. Thinking of this grid as being painted on a flat deformable membrane, the change of variables that gives $\{u_1, u_3, u_3\}$ in terms of $\{v_1, v_2\}$ is like picking up that membrane and deforming it into a 3D shape (like a saddle or surface of a sphere). The Jacobian is the ratio of area of the deformed grid cell (in the limit of infinitesimal cells, of course) to the initial area. Accordingly, any mapping that has a Jacobian that does not depend on position will have mapped cells that are perhaps distorted in different ways at different locations, but all of them would have the same factor defining change in area! Constant-Jacobian mappings are therefore useful to visualize distributions of random points because a clustering in one space will produce a similar clustering in the mapped space. If a Jacobian varies with position, then a clustering might simply be caused by the mapping itself (such as pinching of grid lines near the North and South Poles when mapping from spherical coordinates to the globe, making grid cells there smaller than those at the equator).

17.2 Uniformly random unit vectors—the theory

Before considering the case of a random rotation, first consider how to randomly select a unit vector such that no unit vector orientation is more likely than any other. Unit vectors may be regarded as points on the unit sphere. Therefore a unit vector $\underset{\sim}{n}$ is completely specified by the spherical coordinates ϑ and φ (not to be confused with Euler angles of the same symbols):

$$\underset{\sim}{n} = \sin\vartheta \cos\varphi \underset{\sim}{e}_1 + \sin\vartheta \cos\varphi \underset{\sim}{e}_2 + \cos\vartheta \underset{\sim}{e}_3. \tag{17.41}$$

We want any point on the unit sphere to be equally likely—we seek a *uniform* distribution. Hence, the distribution function $p_{\underset{\sim}{n}}(\underset{\sim}{n})$ must be a constant. In fact, requiring that the integral of this PDF over the entire sphere be unity, the constant must be the inverse of the *area* of the unit sphere: $p_{\underset{\sim}{n}}(\underset{\sim}{n}) = 1/(4\pi)$. Recalling (17.34), this does not imply that the joint distribution for ϑ and φ are also constant. To the contrary, using the definition of joint distributions,

$$p_{\vartheta,\varphi}(\vartheta, \varphi) = J p_{\underset{\sim}{n}}(\underset{\sim}{n}) = \frac{J}{4\pi}, \qquad \text{where} \qquad J = \frac{d(n_1, n_2, n_3)}{d(\vartheta, \varphi)}. \tag{17.42}$$

Here, J is the Jacobian of the transformation, denoted here using a notation common on multivariate calculus to refer to the following process: First use (17.41) to obtain the 3×2 Jacobian *matrix* as

$$[\mathbf{J}] = \begin{bmatrix} \frac{\partial n_1}{\partial\vartheta} & \frac{\partial n_1}{\partial\varphi} \\ \frac{\partial n_2}{\partial\vartheta} & \frac{\partial n_2}{\partial\varphi} \\ \frac{\partial n_3}{\partial\vartheta} & \frac{\partial n_3}{\partial\varphi} \end{bmatrix} = \begin{bmatrix} \cos\vartheta\cos\varphi & -\sin\vartheta\sin\varphi \\ \cos\vartheta\sin\varphi & \sin\vartheta\cos\varphi \\ -\sin\vartheta & 0 \end{bmatrix}. \tag{17.43}$$

Then, using the theory of singular values, the Jacobian of the transformation from spherical coordinates to a point on the unit sphere is

$$J = \sqrt{\det([\mathbf{J}]^T[\mathbf{J}])} = \ldots = \sin\vartheta. \tag{17.44}$$

These results may be computed in Mathematica as follows, where the variable names emphasize that the Jacobian matrix is the derivative of the `outputs` (Cartesian components of a point on the sphere) with respect to the `inputs` (spherical coordinates) defined by the mapping in (17.33):

```
input = {θ, ϕ};
output = {Cos[ϕ] Sin[θ], Sin[θ] Sin[ϕ], Cos[θ]};
Jmat = Table[D[output[[i]], input[[j]]], {i, 3}, {j, 2}];
J = Simplify[√Det[Jmatᵀ.Jmat], {0 < θ < π, 0 < ϕ < 2 π}];
Print["matrix = ", Jmat // MatrixForm];
Print["Jacobian = ", J];
```

```
         ( Cos[θ] Cos[ϕ]  -Sin[θ] Sin[ϕ] )
matrix = ( Cos[θ] Sin[ϕ]   Cos[ϕ] Sin[θ] )
         (   -Sin[θ]            0        )

Jacobian = Sin[θ]
```

The Jacobian is the ratio of an area $\mathrm{d}A$ on the sphere to a notional area element $\mathrm{d}\vartheta\mathrm{d}\varphi$ in coordinate space. These areas are infinitesimal versions of the small areas enclosed by grid lines in the upper-right image of figure 17.2. This result is the familiar one for an area element on the sphere:

$$\mathrm{d}A = \sin\vartheta\,\mathrm{d}\vartheta\,\mathrm{d}\varphi. \tag{17.45}$$

Using (17.45) in (17.42) gives the joint distribution

$$\boxed{p_{\vartheta,\varphi}(\vartheta, \varphi) = \frac{\sin\vartheta}{4\pi}}. \tag{17.46}$$

This function is separable. That is, $p_{\vartheta,\varphi}(\vartheta, \varphi) = p_\vartheta(\vartheta)p_\varphi(\varphi)$, where

$$\boxed{p_\vartheta(\vartheta) = \frac{1}{2}\sin\vartheta} \quad \text{and} \quad \boxed{p_\varphi(\varphi) = \frac{1}{2\pi}}. \tag{17.47}$$

The divisors in these formulas are determined from the requirement that the integral of each PDF must be unity. Note that the zenith angle φ is uniformly distributed on the range from 0 to 2π, but (as illustrated in figure 17.2) *the azimuthal angle ϑ must be non-uniformly distributed in order to achieve a uniform distribution on the sphere*. In that figure, it is noted that it is actually the axial z coordinate, $z = \cos\vartheta$, that is uniformly distributed. This is seen most readily by writing (17.45) as

$$\mathrm{d}A = \sin\vartheta\,\mathrm{d}\vartheta\,\mathrm{d}\varphi = -\mathrm{d}(\cos\vartheta)\mathrm{d}\varphi = -\mathrm{d}z\,\mathrm{d}\varphi. \tag{17.48}$$

The negative merely indicates that area on the sphere decreases if z increases. This result shows that a uniform distribution on the sphere requires uniform and independently distributed z and φ:

$$\boxed{p_z(z) = \frac{1}{2}} \quad \text{and} \quad \boxed{p_\varphi(\varphi) = \frac{1}{2\pi}}. \tag{17.49}$$

Again, the divisors have been set to give unity when the PDF is integrated over the range of the variable. Each PDF is implicitly understood to equal zero outside of the

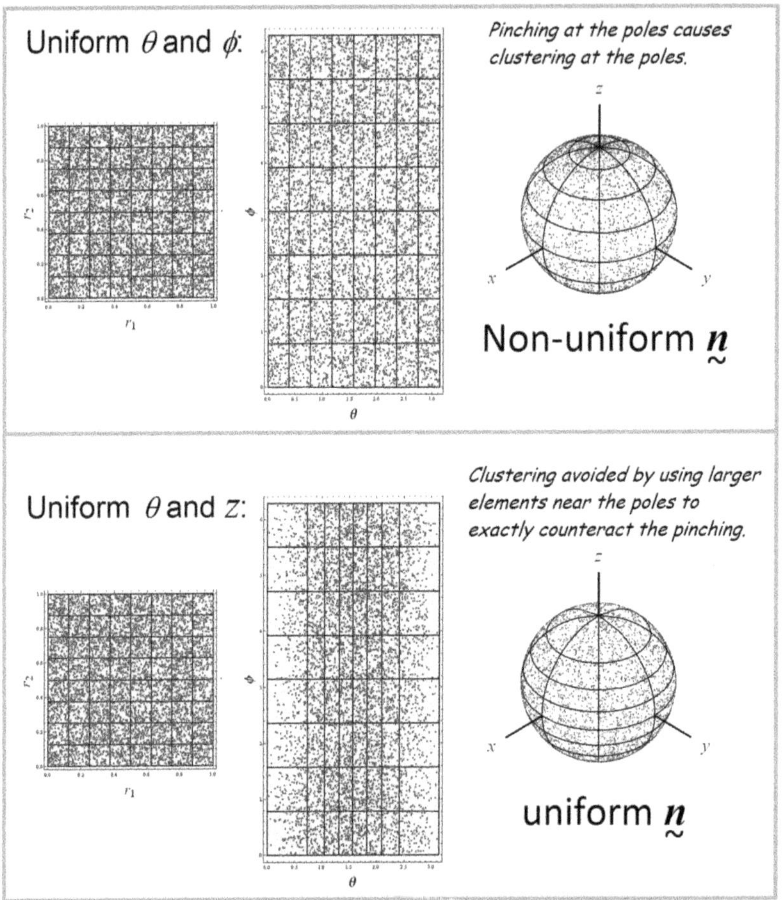

Figure 17.2. The process of generating uniform normals. Two uniformly random numbers on $[0,1)$ are generated. These are uniform in the unit square. If these are mapped to uniform spherical coordinates (middle panel of the top row), then the points will cluster at the poles because the grid 'pinches' at the poles. This pinching is counteracted in the bottom row to ensure that each element is of constant area.

admissible coordinate range. The z coordinate ranges from −1 to 1, so its total range is 2. As illustrated in the bottom half of figure 17.2, slicing the unit sphere into patches of equal area requires uniform slicing in φ and uniform slicing in thickness z, which implies *non-uniform* slicing of ϑ to prevent grid cells near the poles from being smaller than those near the equator.

Equation (17.49) gives the following algorithm to generate a uniformly random unit vector (repeated for each dot):

STEP 1. Generate two random numbers, r_1 and r_2, each in the range from 0 to 1.
STEP 2. Generate uniform realizations of z and φ, by

$$\boxed{z = 2r_1 - 1} \text{ (or, alternatively, } z = 1 - 2r_1\text{)}, \tag{17.50}$$

$$\boxed{\varphi = 2\pi r_2}. \tag{17.51}$$

STEP 3. Noting that $z = \cos \vartheta$, compute the realization for ϑ by

$$\boxed{\vartheta = \cos^{-1} z}. \tag{17.52}$$

STEP 4. Construct the random unit vector:

$$\boxed{\underset{\sim}{n} = \sin \vartheta \cos \varphi \underset{\sim}{e}_1 + \sin \vartheta \sin \varphi \underset{\sim}{e}_2 + \cos \vartheta \underset{\sim}{e}_3}. \tag{17.53}$$

The algorithm is illustrated (with proof of uniformity) below:

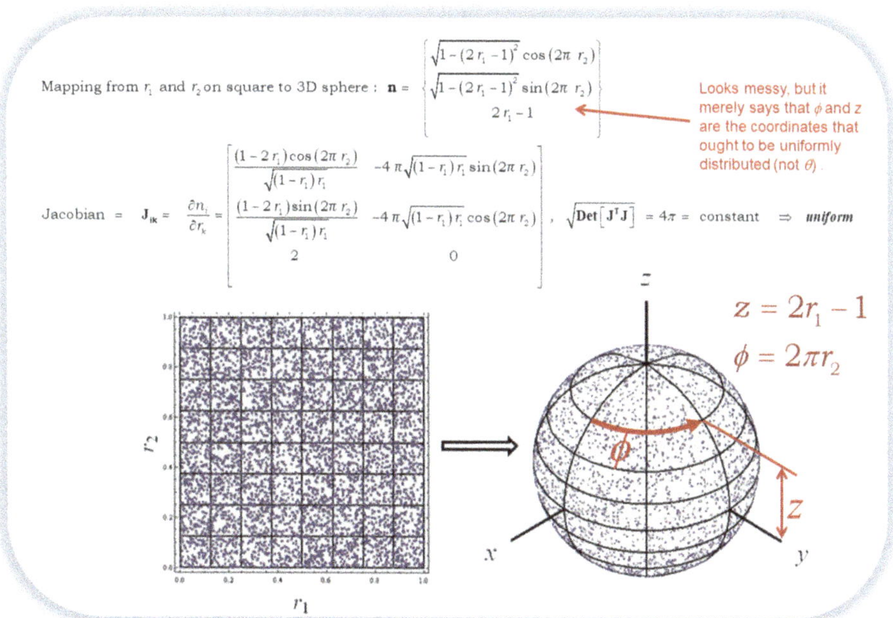

This algorithm is implemented in section A.7 of the appendix. Incidentally, the alternative formula mentioned in (17.50) allows for two options: one makes z range from +1 to −1 as r_1 varies from 0 to 1. The other option ranges in the opposite direction, but still uniformly.

17.2.1 The expected value of a uniformly random unit vector is zero

The expected value of a randomly generated unit vector $\underset{\sim}{n}$ is defined by

$$\bar{\underset{\sim}{n}} = \int_{A^{\text{sphere}}} \underset{\sim}{n}\ p(\underset{\sim}{n}) \mathrm{d}A, \tag{17.54}$$

where $p(\underset{\sim}{n})$ is the distribution function, which is $p(\underset{\sim}{n}) = \frac{1}{A^{\text{sphere}}} = \frac{1}{4\pi}$ for a uniform distribution so that

$$\bar{\underset{\sim}{n}} \equiv \frac{1}{4\pi}\int_{A^{\text{sphere}}} \underset{\sim}{n}\; \mathrm{d}A = \underset{\sim}{0} \text{ (by the divergence theorem).} \tag{17.55}$$

17.3 Uniformly random unit vectors—alternative implementation

The techniques outlined in the preceding section followed classical methods of probability theory. Specifically, a uniformly random unit vector was generated by first sampling two random numbers, r_1 and r_2, each uniformly distributed on the interval [0, 1). Then (17.50) through (17.53) were applied to construct the uniformly random unit vector. Unfortunately, numerical implementations of that procedure require calling trigonometric intrinsic functions, which can be expensive.

In this section we describe a much more straightforward way to generate a uniformly random unit vector, and this method may be computationally faster than the method of the preceding section, as long as the computer's pseudo-random number generator is well-optimized (or, if possible and appropriate, you have a lookup table of random numbers). The idea behind the alternative algorithm is to first generate a (non-unit) vector $\underset{\sim}{r}$ using a uniform distribution inside the unit sphere. Then this vector is simply normalized to a unit vector $\underset{\sim}{n}$ by applying the formula

$$\underset{\sim}{n} = \frac{\underset{\sim}{r}}{\sqrt{\underset{\sim}{r} \bullet \underset{\sim}{r}}}. \tag{17.56}$$

Generating the vector $\underset{\sim}{r}$ from a uniformly distributed sample space inside the unit sphere is algorithmically trivial. First, three pseudo-random numbers, $\{r_1, r_2, r_3\}$ on the interval [0, 1) are generated. Then, a scalar $\mu = r_1^2 + r_2^2 + r_3^2$ is computed. If $\mu > 1$, the three random numbers must be discarded and a new set of three random numbers is generated. This process continues until a set of three random numbers is generated for which $\mu \leqslant 1$, in which case, the unit vector $\underset{\sim}{n}$ is then given by (17.56). As explained later in figure 17.14, you must *not* simply normalize the vector if it turns out to fall outside the sphere. You must discard it.

17.4 'Centroidally random' unit vectors

So far, we have described only methods for producing a truly probabilistic uniform distribution of points on a sphere. For this kind of distribution, it is not at all uncommon for the distance between neighboring points to vary considerably. A different kind of uniform distribution, here called a centroidally uniform distribution, is a uniform distribution subject to the constraint that the mean distance between two points must never fall below a certain value. More precisely, the region on the sphere surrounding each point should be as close to uniform as possible.

To understand the distinction between *truly* random distributions and centroidally random distributions, consider the pattern that ten raindrops make on a small courtyard. The raindrops can fall anywhere, and it is quite natural that some raindrops might fall in very close vicinities to others. This is a classical uniformly random distribution.

Compare the raindrop situation with dropping ten (non-helium) spherical balloons onto the courtyard. In this case, the centers of two balloons can never be any closer than a balloon diameter, and the center of a balloon is never any closer to the courtyard boundary than one balloon radius. Aside from these constraints, the distribution is otherwise uniformly random.

An even more restrictive case involves distributing ten *people* on the courtyard, all of whom are antisocial introverts with less-than-stellar hygiene. In addition to being repelled by each other, further suppose that these people are equally repelled by some nasty poisonous thorn bushes at the boundary of the courtyard. Then you would expect these people to distribute themselves in a manner that gives them maximum personal 'safe' space. Unlike the balloon example, not only would the individual centroids be no closer than a 'person diameter', those centroids maximize their distance from other centroids and from the dangerous courtyard wall. In this book, we call this kind of distribution 'centroidal Voronoi' (CV). These three different kinds of distributions are illustrated in figure 17.3.

In this section, we outline how to generate a CV distribution of unit vectors. This distribution would be similar to the distribution expected for a set of electrons on a metal sphere. We use the term 'centroidal Voronoi' because, for this type of distribution, the centroids of the associated Voronoi[4] cells are coincident with the particle itself. See figure 17.4. Even though we will speak of the Voronoi regions associated with particles, the algorithm presented here is especially attractive because it does *not* require ever generating the Voronoi diagram. We merely assert that it *exists*. The CV-tesselation (CVT) techniques described in this section were pioneered by Du, Faber, and Gunzburger [2].

For the remainder of this section, we generalize the analysis to consider centroidal particle distributions in N-dimensions. The N-dimensional volume element will be denoted $\mathrm{d}V$ and the $(N-1)$-dimensional 'area' element will be denoted $\mathrm{d}A$. Of course, the N-dimensional analysis can be simplified to 2D, in which case the 2D 'volume' element would really be an area, and the 1D 'area element' would actually be a line.

From a mechanics perspective, generating a CV point distribution on a domain Ω is basically the same as trying to minimize the average rotational moment of inertia of each associated Voronoi cell. If we let ω_p denote the Voronoi cell associated with the pth particle, then

$$\Omega = \bigcup_{p=1}^{N} \omega_p, \tag{17.57}$$

[4] The Voronoi cell associated with a particle is the region of space that is closer to that particle than to any other particle.

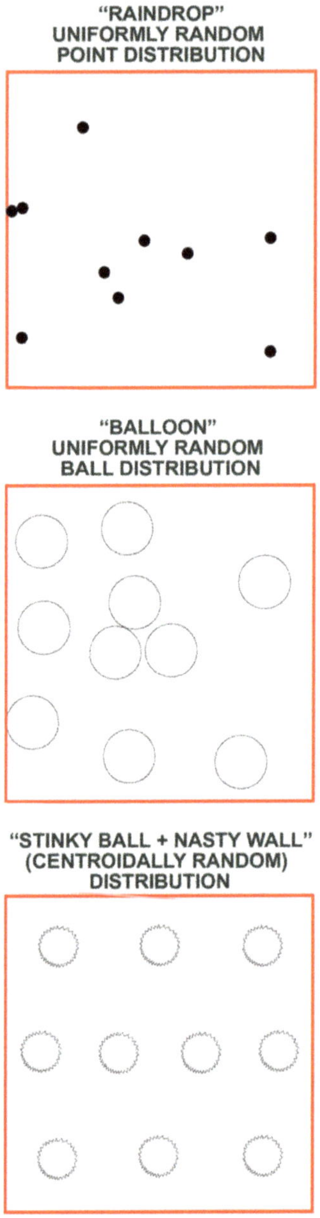

Figure 17.3. Different types of uniform distributions.

and the total domain volume is the sum of the cell volumes:

$$V = \sum_{p=1}^{N} V_p, \tag{17.58}$$

where N is the number of particles. The size and shape of a Voronoi cell can be characterized in a manner that is basis-independent [3] by using moments. The

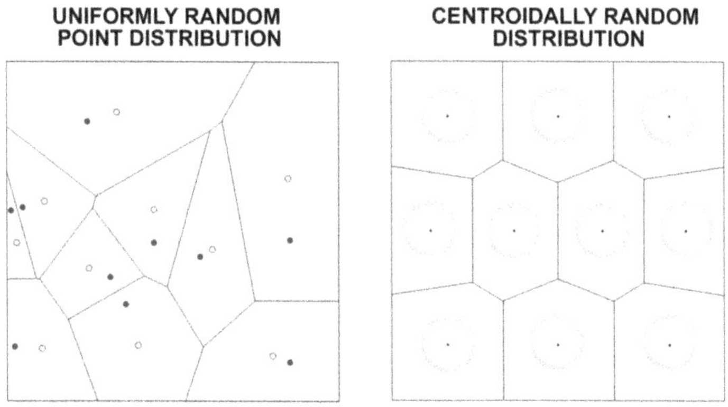

Figure 17.4. Voronoi diagrams for uniformly random and uniformly centroidal distributions. In the left figure, the particle locations are shown as solid dots and the hollow dots represent the Voronoi cell centroids.

zeroth moment of a Voronoi cell is simply its volume (or area in 2D), The first moment is its centroid. The second moment of the Voronoi cell is a second-order tensor $\underset{\approx}{y}_p$ defined by the average *dyadic* distance from the particle:

$$\underset{\approx}{y}_p \equiv \frac{1}{V_p}\int_{\omega_p}(\underset{\sim}{x} - \underset{\sim}{x}_p)(\underset{\sim}{x} - \underset{\sim}{x}_p)\mathrm{d}V. \tag{17.59}$$

The scalar moment of inertia is defined by $\psi_p \equiv \mathrm{tr}\underset{\approx}{y}_p$, or simply

$$\psi_p \equiv \frac{1}{V_p}\int_{\omega_p}(\underset{\sim}{x} - \underset{\sim}{x}_p)\bullet(\underset{\sim}{x} - \underset{\sim}{x}_p)\mathrm{d}V. \tag{17.60}$$

We can define the ensemble 'energy' (or, in the lexicon of the mathematics literature, the 'cost' or variance) of the system by

$$E \equiv \tfrac{1}{2}\sum_{p=1}^{N}\int_{\omega_p}(\underset{\sim}{x} - \underset{\sim}{x}_p)\bullet(\underset{\sim}{x} - \underset{\sim}{x}_p)\mathrm{d}V. \tag{17.61}$$

Our goal is to select particle locations that will minimize E. The analysis ultimately leads to the following iterative algorithm:

STEP 0. Insert N particles into the domain, perhaps using a random distribution, but any distribution method will suffice. This initial distribution comprises the initial set of generating points.

STEP 1. Use a disposable helper grid (as explained in section 17.4.3) to estimate the centroids of the Voronoi cells corresponding to the generating points.

STEP 2. Move the particles to these centroids to establish a new distribution of generating points for the next iteration.

STEP 3. Check for convergence (e.g., check if particle 'motion' has stopped to machine accuracy or if all new cell centroids are tolerably close to generating points). If not converged, go to STEP 1.

Figure 17.5 shows the result of applying step 2 to the particle distribution shown in figure 17.4. Clearly, the revised particle placement has 'pushed apart' the particles that were initially close together. Let us now proceed with an explanation of why having the particles located at centroids of Voronoi cells results in a stationary energy (i.e., cost or variance). Centroidal configurations might, in principle, correspond to a saddle point in energy, but a by-product of the iterative scheme with Richardson extrapolation (detailed below) is that the solutions do tend to converge to true energy *minimums*.

Let $\dot{E}$ denote the variation in E resulting from an arbitrary small variation, $\dot{\underset{\sim}{x}}_q$, of a single particle $\underset{\sim}{x}_q$, while holding the other particles are held fixed. When a particle $\underset{\sim}{x}_q$ changes location, then the Voronoi domain ω_q associated with that particle will change, *as will the domains of all of its nearest neighbors of* $\underset{\sim}{x}_q$. Therefore, applying the Leibniz formula for differentiating an integral whose domain itself is affected by variation,

$$\dot{E} = -\sum_{p=1}^{N} \int_{\omega_p} (\underset{\sim}{x} - \underset{\sim}{x}_p) \bullet \dot{\underset{\sim}{x}}_p \, \mathrm{d}V + \frac{1}{2} \sum_{p=1}^{N} \int_{\partial\omega_p} [(\underset{\sim}{x} - \underset{\sim}{x}_p) \bullet (\underset{\sim}{x} - \underset{\sim}{x}_p)](\underset{\sim}{v} \bullet \underset{\sim}{n}) \mathrm{d}A. \quad (17.62)$$

Here, $\partial\omega_p$ denotes the boundary of ω_p, $\underset{\sim}{n}$ denotes the local outward unit normal of this boundary, and $\underset{\sim}{v}$ denotes the velocity of the boundary. The integral over $\partial\omega_p$ can be broken into separate terms for each planar surface that forms the whole boundary. For each of these planar surfaces, the contribution coming from one Voronoi cell is exactly cancelled by the second contribution of equal magnitude, but opposite sign, from the adjoining cell. This occurs because all of the arguments in the second integral of (17.62) are continuous across the shared interface, *except* the

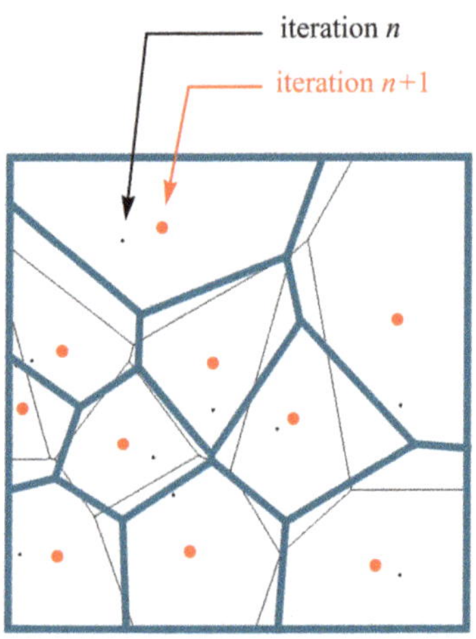

Figure 17.5. Improved Voronoi diagram resulting from moving particles to the centroids of the old Voronoi cells. The initial Voronoi diagram is shown in the left panel of figure 17.4 and as thin black lines here. The bold lines in this figure show the improved diagram after the first iteration. The final converged Voronoi diagram is shown in the right panel of figure 17.4.

outward normal $\underset{\sim}{n}$ of each Voronoi cell changes sign when crossing from one cell to another. Consequently, equation (17.62) reduces to

$$\dot{E} = -\sum_{p=1}^{N} \int_{\omega_p} (\underset{\sim}{x} - \underset{\sim}{x}_p) \bullet \dot{\underset{\sim}{x}}_p \, \mathrm{d}V = -\sum_{p=1}^{N} (\bar{\underset{\sim}{x}}_p - \underset{\sim}{x}_p) \bullet \dot{\underset{\sim}{x}}_p V_p,$$
$$\text{where } \bar{\underset{\sim}{x}}_p \equiv \frac{1}{V_q} \int_{\omega_q} \underset{\sim}{x} \, \mathrm{d}V. \tag{17.63}$$

In the last step, we have noted that neither $\underset{\sim}{x}_p$ nor $\dot{\underset{\sim}{x}}_p$ varies spatially, so they may be taken outside the integral. A *necessary* condition to minimize the energy is that the particles must be placed such that the energy is 'stationary' ($\dot{E} = 0$) with respect to infinitesimal perturbations of the particle locations. Thus, taking $\underset{\sim}{x}_q$ to be the only moving particle, and recognizing that we must require that $\dot{E} = 0$ for all possible variations, $\dot{\underset{\sim}{x}}_p$, leads to the conclusion that the particle must reside at the cell centroid:

$$\underset{\sim}{x}_q = \frac{1}{V_q} \int_{\omega_q} \underset{\sim}{x} \, \mathrm{d}V. \tag{17.64}$$

If this condition is satisfied, then we will call the distribution 'centroidal'. There exist particle distributions that are centroidal, but which are not global minimizers of energy. For example, when particles in a square are distributed in a perfectly square array, then they will be centroidal, but the energy will not be minimized (it is an unstable saddle point). When the number of particles is quite large, and when the energy is truly minimized, the Voronoi cells tend approach perfect hexagons (in 2D), with some geometrically necessary distortion near the boundary.

Allowing for a spatially varying particle density is quite straightforward—all that is needed is a particle density field $\rho(x)$, which equals the number of particles per unit volume (reciprocal of the volume per particle). The energy function is then generalized as

$$E \equiv \frac{1}{2M} \sum_{p=1}^{N} \int_{\omega p} (\underset{\sim}{x} - \underset{\sim}{x}_p) \bullet (\underset{\sim}{x} - \underset{\sim}{x}_p) \rho(\underset{\sim}{x}) \mathrm{d}V, \text{ where } M \equiv \int_{\Omega} \rho(\underset{\sim}{x}) \mathrm{d}V. \tag{17.65}$$

In this form, the energy is clearly analogous to kinetic energy ($\frac{1}{2}mv^2$) from mechanics or to linear spring energy ($\frac{1}{2}kx^2$). Strictly speaking, if $\rho(x)$ represents the desired number of particles per unit volume, then M would have to equal (to the nearest integer) the total number of particles N within the domain Ω. The energy function may be re-written in the form

$$E \equiv \frac{1}{2} \sum_{p=1}^{N} \int_{\omega_p} (\underset{\sim}{x} - \underset{\sim}{x}_p) \bullet (\underset{\sim}{x} - \underset{\sim}{x}_p) \hat{\rho}(\underset{\sim}{x}) \mathrm{d}V, \quad \text{where} \quad \hat{\rho}(\underset{\sim}{x}) = \frac{\rho(\underset{\sim}{x})}{M}. \tag{17.66}$$

In this form, $\hat{\rho}(\underset{\sim}{x})$ is a *relative* density field that may be regarded as a probability density. Because our goal is to minimize the energy E, the final particle placements will be unaffected by any constant scalar multiple of the density field.

The previous algorithm corresponded to particles being placed at the Voronoi cell centroids, while this one requires placement at the cell 'center of mass'. Consequently, particles will pack more closely in regions where the density $\rho(\underset{\sim}{x})$ is high.

The energy function can be generalized to permit particle *texture*[5] so that the Voronoi cells will generally look like *elongated* hexagons (in 2D). This is done by writing the *energy function for textured particle distributions* in the form

$$\boxed{E \equiv \frac{1}{2}\sum_{p=1}^{N}\int_{\omega_p}(\underset{\sim}{x} - \underset{\sim}{x}_p) \bullet \underset{\approx}{A}^{-2} \bullet (\underset{\sim}{x} - \underset{\sim}{x}_p)\mathrm{d}V}, \tag{17.67}$$

where $\underset{\approx}{A}$ is a user-specified, permissibly spatially varying, positive-definite second-order tensor. Note that (17.66) is a special case of (17.67) in which

$$\underset{\approx}{A}^{-2} = \hat{\rho}(x)\underset{\approx}{I}. \tag{17.68}$$

The new generalized energy function may be written

$$E = \frac{1}{2}\sum_{p=1}^{N}\int_{\omega_p}(\underset{\sim}{y} - \underset{\sim}{y}_p) \bullet (\underset{\sim}{y} - \underset{\sim}{y}_p)\mathrm{d}V, \quad \text{where } \underset{\sim}{y} = \underset{\approx}{A}^{-1} \bullet \underset{\sim}{x},\ \underset{\sim}{y}_p = \underset{\approx}{A}^{-1} \bullet \underset{\sim}{x}_p. \tag{17.69}$$

If the $\underset{\approx}{A}$ field satisfies compatibility conditions for a deformation mapping, then this integral could be transformed to a Euclidean $\underset{\sim}{y}$-space where *isotropic* CVT could be applied, with the results transformed back to $\underset{\sim}{x}$-space to obtain the final *anisotropic* CVT particle placements. However, what if the $\underset{\approx}{A}$ field is not compatible? In this case, let us go back to the conditions for minimizing energy. For minimization of the new E defined in (17.67), the generalization of (17.62) is

$$\dot{E} = -\sum_{p=1}^{N}\int_{\omega_p}(\underset{\sim}{x} - \underset{\sim}{x}_p) \bullet \underset{\approx}{A}^{-2} \bullet \dot{\underset{\sim}{x}}_p\mathrm{d}V + \frac{1}{2}\sum_{p=1}^{N}\int_{\partial\omega_p}[(\underset{\sim}{x} - \underset{\sim}{x}_p) \bullet \underset{\approx}{A}^{-2} \bullet (\underset{\sim}{x} - \underset{\sim}{x}_p)](\underset{\sim}{v} \bullet \underset{\sim}{n})\mathrm{d}A. \tag{17.70}$$

[5] An enhancement that is, to our knowledge, a new advance.

The volume integral may be written

$$-\sum_{p=1}^{N}\int_{\omega_p}(\underset{\sim}{x}-\underset{\sim}{x}_p)\bullet\underset{\approx}{A}^{-2}\bullet\dot{\underset{\sim}{x}}_p\,\mathrm{d}V=\sum_{p=1}^{N}(\underset{\sim}{x}_p\bullet\underset{\approx}{B}_p-\underset{\sim}{c}_p)\bullet\dot{\underset{\sim}{x}}_p,$$
$$\text{where } \underset{\approx}{B}_p=\int_{\omega_p}\underset{\approx}{A}^{-2}\,\mathrm{d}V \text{ and } \underset{\sim}{c}_p=\int_{\omega_p}\underset{\approx}{A}^{-2}\bullet\underset{\sim}{x}\,\mathrm{d}V. \tag{17.71}$$

Unlike the previous *isotropic* distribution case, the boundary integral does not vanish. Consider the common Voronoi boundary between two particles, $\underset{\sim}{x}_p$ and $\underset{\sim}{x}_q$:

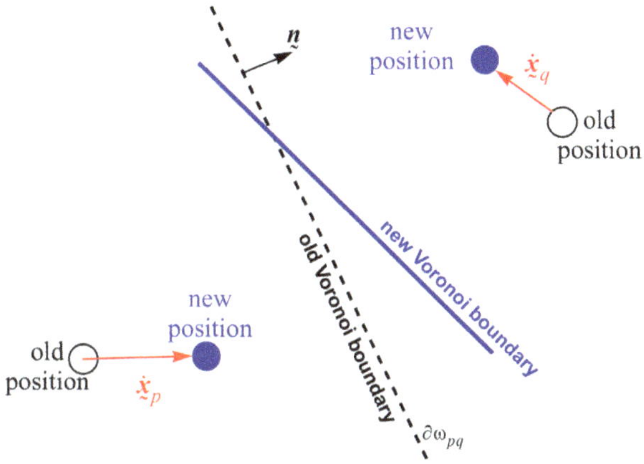

Along the common Voronoi boundary, $\partial\omega_{pq}$, the boundary integral has contributions from $\underset{\sim}{x}_p$ and $\underset{\sim}{x}_q$ as follows:

$$\tfrac{1}{2}\int_{\partial\omega_{pq}}[(\underset{\sim}{x}-\underset{\sim}{x}_p)\bullet\underset{\approx}{A}^{-2}\bullet(\underset{\sim}{x}-\underset{\sim}{x}_p)](\underset{\sim}{v}\bullet\underset{\sim}{n}_p)\mathrm{d}A$$
$$+\int_{\partial\omega_{pq}}[(\underset{\sim}{x}-\underset{\sim}{x}_q)\bullet\underset{\approx}{A}^{-2}\bullet(\underset{\sim}{x}-\underset{\sim}{x}_q)](\underset{\sim}{v}\bullet\underset{\sim}{n}_q)\mathrm{d}A \tag{17.72}$$

or, noting from the figure that $\underset{\sim}{n}_p=\underset{\sim}{n}$ and $\underset{\sim}{n}_q=\underset{\sim}{n}$,

$$\tfrac{1}{2}\int_{\partial\omega}([(\underset{\sim}{x}-\underset{\sim}{x}_p)\bullet\underset{\approx}{A}^{-2}\bullet(\underset{\sim}{x}-\underset{\sim}{x}_p)]\;-\;[(\underset{\sim}{x}-\underset{\sim}{x}_q)\bullet\underset{\approx}{A}^{-2}\bullet(\underset{\sim}{x}-\underset{\sim}{x}_q)])(\underset{\sim}{v}\bullet\underset{\sim}{n})\mathrm{d}A. \tag{17.73}$$

We can define two helper vectors:

$$\underset{\sim}{s}=\frac{\underset{\sim}{x}_q-\underset{\sim}{x}_p}{2}\quad\text{and}\quad\underset{\sim}{t}=\frac{\underset{\sim}{x}_q-\underset{\sim}{x}_p}{2}, \tag{17.74}$$

so that

$$\underset{\sim}{x}_q=\underset{\sim}{t}+\underset{\sim}{s}\quad\text{and}\quad\underset{\sim}{x}_p=\underset{\sim}{t}-\underset{\sim}{s}. \tag{17.75}$$

This change of variables allows (17.73) to simplify to

$$2\int_{\partial\omega_{pq}} [\underset{\sim}{s} \bullet \underset{\approx}{A}^{-2} \bullet (\underset{\sim}{x} - \underset{\sim}{t})](\underset{\sim}{v} \bullet \underset{\sim}{n})\mathrm{d}A. \tag{17.76}$$

The vector $\underset{\sim}{x} - \underset{\sim}{t}$ is always perpendicular to $\underset{\sim}{s}$. Therefore, if the tensor $\underset{\approx}{A}^{-2}$ is isotropic (making its off-diagonal components zero with respect to *any* basis), the integrand is zero, which is why the *isotropic* CVT formulations had no contribution from the boundary integral. In this anisotropic case, however, the boundary contribution is non-zero, so we need to work out the formula for the normal component of velocity of the moving Voronoi boundary. At all times, a point $\underset{\sim}{x}$ on the moving Voronoi boundary satisfies

$$(\underset{\sim}{x} - \underset{\sim}{t}) \bullet \underset{\sim}{s} = 0. \tag{17.77}$$

Taking rates gives

$$(\dot{\underset{\sim}{x}} - \dot{\underset{\sim}{t}}) \bullet \underset{\sim}{s} + (\underset{\sim}{x} - \underset{\sim}{t}) \bullet \dot{\underset{\sim}{s}} = 0 \tag{17.78}$$

or

$$\dot{\underset{\sim}{x}} \bullet \underset{\sim}{s} = \dot{\underset{\sim}{t}} \bullet \underset{\sim}{s} - (\underset{\sim}{x} - \underset{\sim}{t}) \bullet \dot{\underset{\sim}{s}}. \tag{17.79}$$

The normal component of the velocity of the Voronoi boundary is

$$\underset{\sim}{v} \bullet \underset{\sim}{n} = \dot{\underset{\sim}{x}} \bullet \frac{\underset{\sim}{s}}{\|\underset{\sim}{s}\|} = \frac{\dot{\underset{\sim}{t}} \bullet \underset{\sim}{s} - (\underset{\sim}{x} - \underset{\sim}{t}) \bullet \dot{\underset{\sim}{s}}}{\|\underset{\sim}{s}\|}. \tag{17.80}$$

Therefore, equation (17.76) becomes

$$2\int_{\partial\omega_{pq}} \frac{[\underset{\sim}{s} \bullet \underset{\approx}{A}^{-2} \bullet (\underset{\sim}{x} - \underset{\sim}{t})](\dot{\underset{\sim}{t}} \bullet \underset{\sim}{s} - (\underset{\sim}{x} - \underset{\sim}{t}) \bullet \dot{\underset{\sim}{s}})}{\|\underset{\sim}{s}\|}\mathrm{d}A. \tag{17.81}$$

Using (17.74), the term in this expression that is linear with respect to $\dot{\underset{\sim}{x}}_p$ is

$$\underset{\sim}{z}_{pq} \bullet \dot{\underset{\sim}{x}}_p, \tag{17.82}$$

where

$$\underset{\sim}{z}_{pq} = \int_{\partial\omega_{pq}} \frac{[\underset{\sim}{s} \bullet \underset{\approx}{A}^{-2} \bullet (\underset{\sim}{x} - \underset{\sim}{t})](\underset{\sim}{s} + (\underset{\sim}{x} - \underset{\sim}{t}))}{\|\underset{\sim}{s}\|}\mathrm{d}A. \tag{17.83}$$

To minimize the energy (i.e., to make $\dot{E} = 0$) for all variations $\dot{\underset{\sim}{x}}_p$, the coefficient of $\dot{\underset{\sim}{x}}_p$ in the expansion of (17.70) must vanish. Therefore, using (17.71) the condition for a stationary energy is

$$\underset{\approx}{B}_p \bullet \underset{\sim}{x}_p = \underset{\sim}{c}_p - \sum_q \underset{\sim}{z}_{pq}. \tag{17.84}$$

The sum over q extends over all points $\underset{\sim}{x}_q$ that generate the boundary of the Voronoi cell for $\underset{\sim}{x}_p$. For an *isotropic* particle distribution, each $\underset{\sim}{z}_{pq}$ is zero, implying a particle

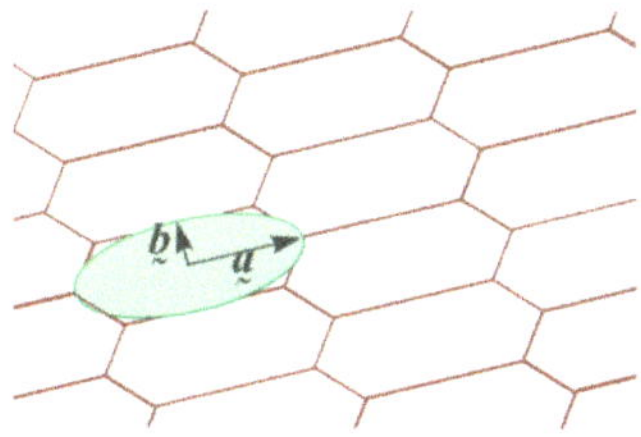

Figure 17.6. A textured CVT. Each Voronoi cell is elongated such that its second moment is the same as the second moment of an ellipse defined by two 'directed' mean free path vectors, $\underset{\sim}{a}$ and $\underset{\sim}{b}$.

placement rule $\underset{\sim}{x}_p = \underset{\approx}{B}_p^{-1} \bullet \underset{\sim}{c}_p$, which is consistent with the isotropic results derived earlier[6]. The unresolved research question here is how to evaluate the last term in (17.84) for an anisotropic $\underset{\approx}{A}$ tensor.

17.4.1 Clarification of the anisotropy tensor

Recall that the cells for a 2D isotropic particle distribution tend to take hexagonal shapes. We can always construct a circle having the same moment of inertia as the hexagon. For the 2D *anisotropic* case, the tensor $\underset{\approx}{A}$ is defined to produce Voronoi cells that look like *elongated* hexagons, and we can always construct an *ellipse* that has the same second moment tensor $\underset{\approx}{y}_p$ as the Voronoi cell. Similarly, for 3D or higher dimensions, we can construct ellipsoids that characterize the overall shape eccentricity of the Voronoi cells. This observation can guide you in determining the tensor $\underset{\approx}{A}$ that will produce a desired texture.

The tensor $\underset{\approx}{A}$ can be regarded as a generalization of a 'mean free path'. If, for example, it is desired for the cells to be elongated hexagons that are of the same general shape as an ellipse having mutually orthogonal major and minor axes given by two vectors $\underset{\sim}{a}$ and $\underset{\sim}{b}$ (see figure 17.6), then the tensor $\underset{\approx}{A}$ (in two dimensions) should be specified equal to

$$\underset{\approx}{A}^2 = \underset{\sim}{a}\underset{\sim}{a} + \underset{\sim}{b}\underset{\sim}{b}, \quad \text{or, in component form,} \quad (\underset{\approx}{A}^2)_{ij} = a_i a_j + b_i b_j. \tag{17.85}$$

Suppose, for example, that you wished to construct a particle placement for which the texture (i.e., the shape eccentricity of the Voronoi cells) was similar to the grid eccentricity for the ordinary polar coordinates in figure 17.7, which has grid cells elongated in the radial direction near the origin but in the transverse direction far from the origin.

With polar coordinates, grid cells far from the origin are longer in the hoop direction than in the radial direction. The opposite is true near the origin, where grid cells are narrow in the hoop direction and elongated in the radial direction. To imitate the spatially varying size, orientation, and eccentricity of a polar grid, the vector $\underset{\sim}{a}$ in (17.85) would be the radial increment of the position vector ($\Delta r \underset{\sim}{e}_r$), while the vector $\underset{\sim}{b}$ would be angular increment of the position vector $r\Delta\theta \underset{\sim}{e}_\theta$. Hence, with respect to the polar basis,

---

[6] Given that the sum of two positive-definite tensors is itself positive-definite, it follows that the volume integral of the positive-definite tensor $\underset{\approx}{A}^{-2}$ must also be positive-definite. Therefore, the tensor $\underset{\approx}{B}\underset{\approx}{B}$ is invertible.

$$[\underset{\sim}{A}] = \begin{bmatrix} \Delta r & 0 \\ 0 & r\Delta\theta \end{bmatrix}. \tag{17.86}$$

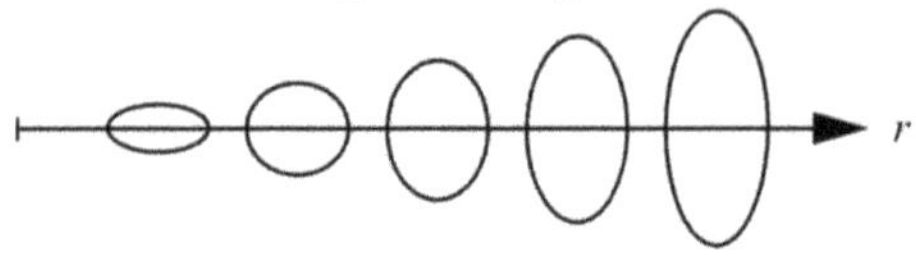

As desired, the spatially varying ellipses in the sketch have eccentricities just like the polar coordinate grid. Generalizing,

$$[\underset{\sim}{A}^{-2}] = \begin{bmatrix} \dfrac{1}{(\Delta r)^2} & 0 \\ 0 & \dfrac{1}{(r\Delta\theta)^2} \end{bmatrix} \quad \text{with respect to the polar basis.} \tag{17.87}$$

In terms of a Cartesian basis,

$$[\underset{\sim}{A}^{-2}] = \begin{bmatrix} \beta + \left(\dfrac{\alpha - \beta}{r^2}\right)x_1^2 & \left(\dfrac{\alpha - \beta}{r^2}\right)x_1x_2 \\ \left(\dfrac{\alpha - \beta}{r^2}\right)x_1x_2 & \beta + \left(\dfrac{\alpha - \beta}{r^2}\right)x_2^2 \end{bmatrix}, \tag{17.88}$$

$$\text{where} \quad \alpha = \frac{1}{(\Delta r)^2}, \quad \beta = \frac{1}{(r\Delta\theta)^2}, \quad \text{and} \quad r^2 = x_1^2 + x_2^2.$$

The physical interpretation of the $\underset{\sim}{A}$ tensor as an oriented mean free path extends naturally to higher dimensions. For example, to generate a 3D particle placement with Voronoi cells having the same moment of inertia as an ellipsoid with axes $\underset{\sim}{a}$, $\underset{\sim}{b}$, and $\underset{\sim}{c}$, then the 'mean free path' tensor would be specified as

$$\underset{\sim}{A}^2 = \underset{\sim}{a}\underset{\sim}{a} + \underset{\sim}{b}\underset{\sim}{b} + \underset{\sim}{c}\underset{\sim}{c} \text{ (the '}A\text{' tensor is a directed mean-free-path).} \tag{17.89}$$

17.4.2 Isotropic CV distributions on a curvilinear surface (sphere)

To apply the CVT method to place particles on a curvilinear surface defined by $f(\underset{\sim}{x}) = 0$, we introduce a Lagrange multiplier λ into the energy function to account for the constraint that $\underset{\sim}{x}_p$ must lie on the surface:

$$E \equiv \sum_{p=1}^{N}\left[\int_{\omega_p} [(\underset{\sim}{x} - \underset{\sim}{x}_p) \bullet (\underset{\sim}{x} - \underset{\sim}{x}_p)]\hat{\rho}(\underset{\sim}{x})\mathrm{d}V + \lambda f(\underset{\sim}{x}_p)\right]. \tag{17.90}$$

To illustrate, suppose we wish to place particles onto a *sphere*. Then $f(\underset{\sim}{x}) = \underset{\sim}{x} \bullet \underset{\sim}{x} - 1$ and the energy functional becomes

$$E \equiv \sum_{p=1}^{N}\left[\int_{\omega_p} [(\underset{\sim}{x} - \underset{\sim}{x}_p)\bullet(\underset{\sim}{x} - \underset{\sim}{x}_p)]\hat{\rho}(\underset{\sim}{x})\mathrm{d}V + \lambda(\underset{\sim}{x}_p \bullet \underset{\sim}{x}_p - 1)\right]. \tag{17.91}$$

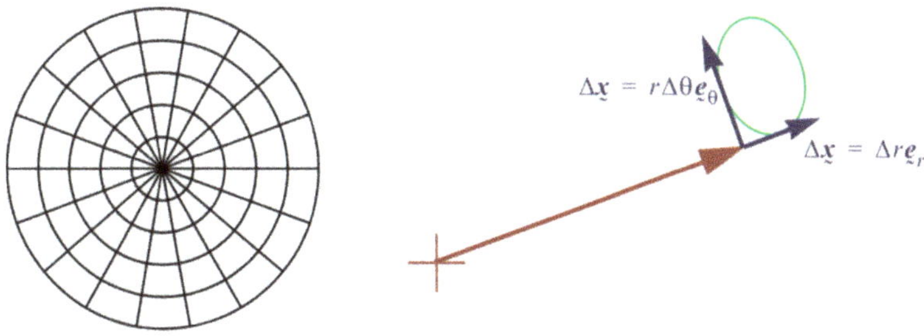

Figure 17.7. Defining ellipses that have the same eccentricity as polar grid cells.

Taking the derivative with respect to $\underset{\sim}{x}_q$ (again recognizing that Leibniz contributions from variations of $\partial\omega_p$ cancel each other out) gives

$$\frac{\partial E}{\partial \underset{\sim}{x}_q} \equiv -2\int_{\omega_p} (\underset{\sim}{x} - \underset{\sim}{x}_q)\hat{\rho}(\underset{\sim}{x})\mathrm{d}V + 2\lambda\underset{\sim}{x}_q. \tag{17.92}$$

Setting this equal to zero and solving for $\underset{\sim}{x}_q$ gives

$$\underset{\sim}{x}_q \equiv \frac{\displaystyle\int_{\omega_p} \underset{\sim}{x}\hat{\rho}(\underset{\sim}{x})\mathrm{d}V}{\lambda + \displaystyle\int_{\omega_p} \hat{\rho}(\underset{\sim}{x})\mathrm{d}V}. \tag{17.93}$$

The constraint that gives the value of λ is $\underset{\sim}{x}_q \bullet \underset{\sim}{x}_q = 1$. Therefore $\underset{\sim}{x}_q$ is simply a unit vector in the direction of

$$\int_{\omega_p} \underset{\sim}{x}\hat{\rho}(\underset{\sim}{x})\mathrm{d}V. \tag{17.94}$$

A sample 2562-point CVT tesselation of the sphere is in figure 17.8. This CVT spherical tesselation could be used to 'bin' flaw orientations for modeling micro-cracked material or for dipole orientations in a ferroelectric.

17.4.3 Statistical algorithm for finding planar CV particle distributions

All CV algorithms require an initial set of candidate starting points. These can be generated from a uniformly random distribution, or from some more 'savvy' distribution that is expected to be as close as possible to the desired result. To avoid the cost and inconvenience of having to construct Voronoi cells, a statistical method was pioneered by Du *et al* [2], and the idea of simplifying their approach by using a regular sampling grid is newly presented here. The domain is broken into an $m \times n$ regular grid, and the approximate, grid-constrained, Voronoi tesselation is generated. Specifically, if the center of grid cell K is closer to particle p than to any other particle, then that grid cell is assigned a Voronoi integer identifier p. Once an identifier has been assigned to each grid cell, the centroid of all grid cells with identifier p is computed, and particle p is moved to this location for the next iteration. This procedure is repeated until no further movement of the particles

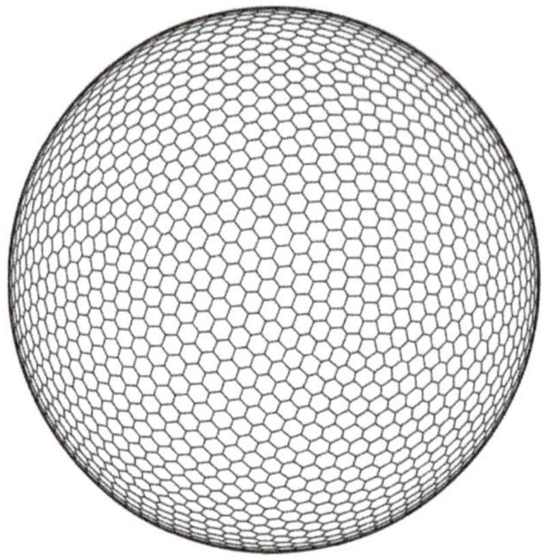

Figure 17.8. Voronoi cells associated with a centroidally uniform distribution of 2562 points on the sphere. Figure courtesy of Qiang Du, Max Gunzburger, and Lili Ju.

occurs. At that point, a new grid of dimensions $\frac{m}{2} \times \frac{n}{2}$ is generated and the procedure is repeated. In principle, if this process is continued using successively smaller grid sizes then, eventually, the particle locations will be centroidal. Of course, in practice, this is not possible so the solution is only approximate. However, the convergence of the method can be enhanced by using Richardson extrapolation. If $\underset{\sim}{x}_p^h$ is the location of the particle obtained using a grid spacing h and if $\underset{\sim}{x}_p^H$ is the location found using a spacing H, then we can approximate the exact solution by

$$\underset{\sim}{x}_p = \underset{\sim}{x}_p^h - \frac{\underset{\sim}{x}_p^H - \underset{\sim}{x}_p^h}{\frac{H}{h} - 1}. \tag{17.95}$$

If we commit to always using a factor of two in grid reduction, then the prediction of the limit as the grid is infinitely refined is

$$\underset{\sim}{x}_p = \underset{\sim}{x}_p^h + \underset{\sim}{\delta}_p, \quad \text{where } \underset{\sim}{\delta}_p = \underset{\sim}{x}_p^h + \underset{\sim}{x}_p^H. \tag{17.96}$$

The grid should continue to be refined until the magnitude of $\underset{\sim}{\delta}_p$ falls below some acceptable error tolerance, at which point this formula would be used to set the final locations.

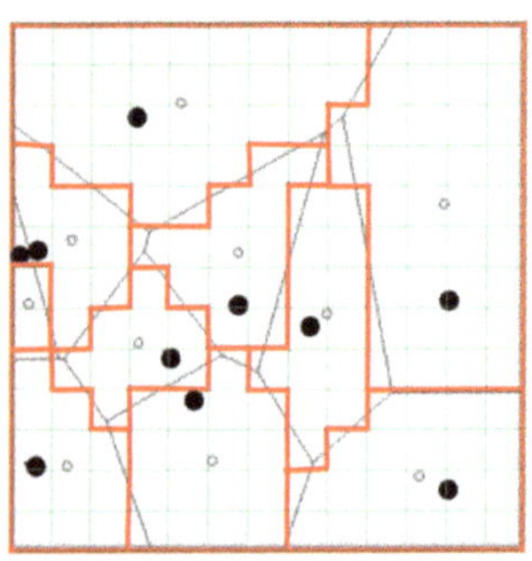

17.4.4 Visualizing unit vectors (angle or area-preserving projections)

A unit vector $\underset{\sim}{n}$ may be regarded as a point on the unit sphere. For visualization, points on the unit sphere must be mapped to the plane (this sheet of paper). Any such mapping is called a *stereographic projection* if it is conformal (i.e., preserves angles at the cost of distorting relative areas), although this phrase if often used by engineers to also refer to area-preserving projections. The angle-preserving ones tend to be favored for visualization (e.g., a circle on the sphere maps to a circle on the projection), while the equal-area projections are crucial to provide an accurate image of point density for statistical analysis. The most common projection is what you would get if you were to take a photograph of a sphere, which is neither angle nor area-preserving but instead best depicts the 3D nature of the original ball. The surface of the sphere maps to the planar surface of the photo. Multiple photos of the sphere (showing the front and the rear) would be needed for this type of projection. There are numerous alternative ways to define the projection, and the best choice depends on the application. The equations for a photograph projection are derived later in (25.22). This section is focused on explaining why an 'area-preserving' projection is the most natural choice for visualizing statistical distributions (or any other quantity defined per solid angle—i.e., 'per-area' on the sphere).

Before discussing the 'sphere–plane' problem, first consider the simpler circle–line problem of visualizing points on a simple circle defined by the equation $x^2 + y^2 = R^2$. This circle is a 1D entity existing in two dimensions. As sketched in figure 17.9, we can 'snip' the circle at the point $(0, -1)$ and then 'unwrap' or 'flatten' it into a line. This is an easy mapping to visualize because both the circle and the line are Euclidean spaces (both the circle and the line can be formed by a strip of paper).

In crystallography, a different kind of stereographic projection (shown in figure 17.10) is commonly used. For mapping a circle to a line, a disadvantage of this kind of projection is that arc length is not preserved. Thus, if a classical stereographic

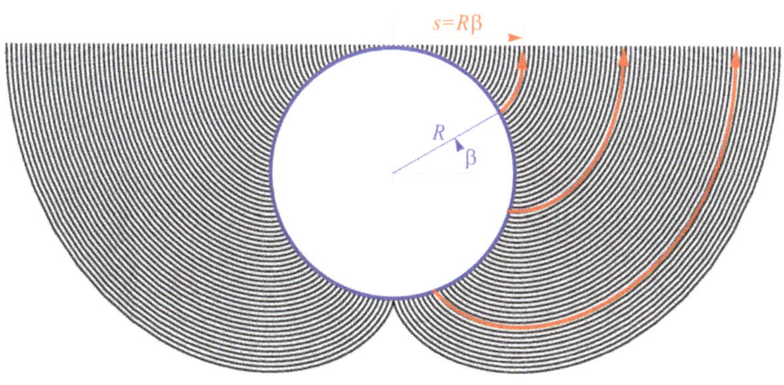

Figure 17.9. Length-preserving mapping of a circle to a 1D straight line. Each point on the circle maps to a point on a straight line. The polar coordinate θ maps to a linear coordinate s, which is just the arc length. Note that the lower hemisphere maps to a discontinuous region defined by $|s| > R\pi/2$. The mapping is one-to-one everywhere except that the point $(0, -1)$ maps to both $s = 1$ and $s = -1$.

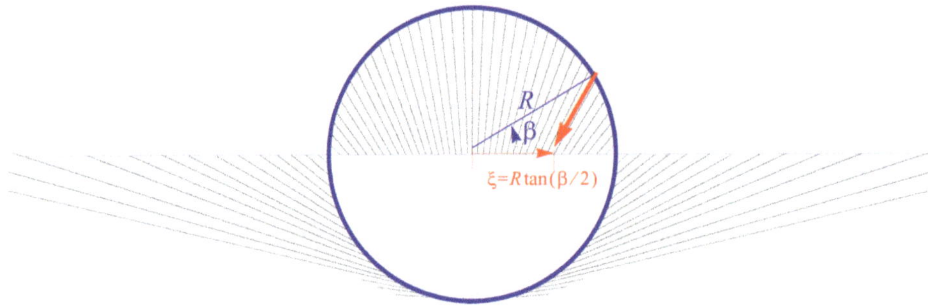

Figure 17.10. Classical stereographic mapping of a circle to a 1D straight line. The polar coordinate β maps to a linear coordinate ξ, which is defined by where a line from the south pole to the point intersects the equator plane. Note that the upper semicircle maps to points for which $\xi < R$ and points on the lower semicircle map to points for which $\xi > R$.

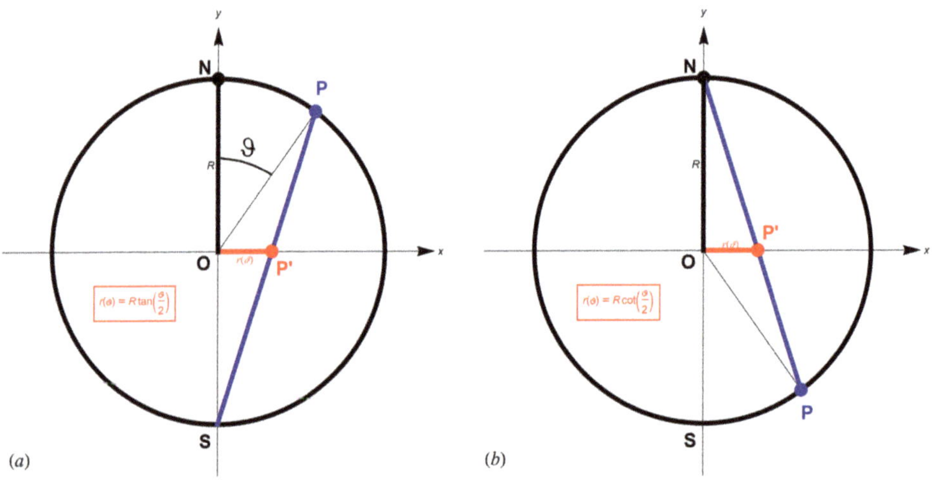

Figure 17.11. Side view of a sphere with zenith angle ∠NOP denoted by ϑ, and ranging from 0 to π. Choices for stereographic projection. (a) A south-pole projection maps the upper hemisphere to the center of the projected disk. (b) A north-pole projection maps the southern hemisphere to the center of the disk.

projection is used, uniformly distributed dots on the circle will map non-uniformly to the line.

Figure 17.11 shows how formulas for the projected radius in an azimuthal projection change slightly depending on whether or not the projection anchor is the north or south pole.

Now let's examine issues associated with the sphere-to-plane projection. Mapping the surface of a *sphere* embedded in 3D space to the 2D plane (defined by this sheet of paper) must involve area distortion since the surface of a sphere is non-Euclidean. In other words, an area element on the sphere will map to an area element on the plane, but the two elements will generally have different aspect ratios or different sizes. Imagine introducing a tiny hole at the 'south pole' of a racquetball. To bend the racquetball into the plane, the area elements that were originally near the south pole would have to be significantly stretched out. Unlike the circle–line mapping of figure 17.9 (which involved zero change in arc length), the sphere–plane mapping necessarily involves aspect ratio distortion and or relative element sizes. As we soon show, it is possible to define a projection such that the magnitude of the area is preserved despite the fact that the *shape* of the area element necessarily changes. It is also possible to define a stereographic projection that preserves shape at the cost of not preserving area. No mapping from the sphere to a plane can preserve both shape and area.

The left side of figure 17.12 shows the conventional (photographic) rendering of the sphere as seen when looking down the {1, 1, 1}-axis passing through the first octant. In such an illustration, the sphere 'looks like a sphere', which is aesthetically appealing in most situations. However, when illustrating statistical distributions on a sphere, it is more useful to employ an *area-preserving* 'Lambert' mapping of the sphere to the 2D plane. In this way, a probability function defined on the surface of the sphere will have the same value when mapped to the 2D plane. With equal-area mapping, a uniform distribution on the sphere will map to a uniform distribution on the plane—any clustering in the 2D map will correspond to real clusters on the sphere.

We are ultimately interested in mapping a *unit* sphere to the plane, but let us derive the equations for a sphere of radius R, and then later set $R = 1$. Equating the area elements shown in figure 17.12 gives

$$R^2(\sin \vartheta)\mathrm{d}\vartheta\, \mathrm{d}\varphi = r\mathrm{d}r\, \mathrm{d}\psi. \tag{17.97}$$

We *choose* to limit to attention to *azimuthal mapping*, which means $\psi = \varphi$, so that (17.97) becomes an ordinary differential equation,

$$R^2(\sin \vartheta)\mathrm{d}\vartheta = r\, \mathrm{d}r, \tag{17.98}$$

which is solved (subject to $r = 0$ when $\vartheta = 0$) to give r as a function of ϑ. Specifically,

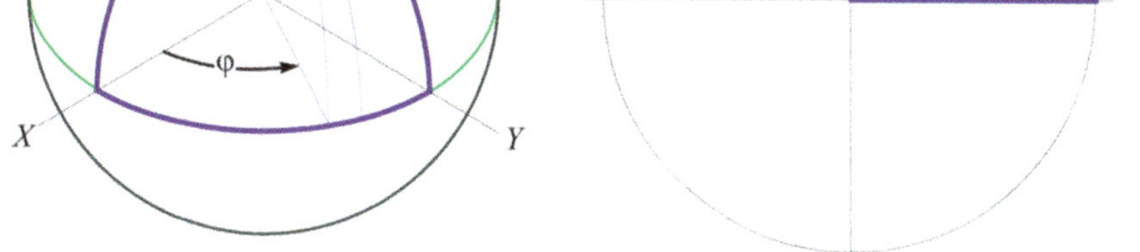

Figure 17.12. Mapping points on a sphere of radius R to a 2D disk. The 2D disk shown on the right-hand side of the above figure may be regarded as a *distorted* view of the sphere as seen looking down the Z-axis.

Lambert equal-area mapping: If a point on a sphere has spherical coordinates $\{R, \vartheta, \varphi\}$, then the (unit Jacobian) mapped point on the 2D disk has *polar* (r, ψ) coordinates

$$\boxed{\psi = \varphi} \quad \text{and} \quad \boxed{r \equiv R\sqrt{2(1 - \cos\vartheta)} = 2R\sin\left(\frac{\vartheta}{2}\right)}. \tag{17.99}$$

In some textbooks, you might see a cosine instead of a sine, which means they are using the angle measured from the south pole instead of north pole (which would map the southern hemisphere to the center of the Lambert disk).

As shown in figure 17.13, a point on the equator of the sphere (where $\theta = \pi/2$) maps to a radius $r = \sqrt{2}R$. All points in the upper hemisphere have mapped radial coordinates smaller than $\sqrt{2}R$. As seen in the graph in figure 17.13, points in the upper hemisphere have a mapped radial coordinate r that is *roughly* equal to the spherical arc length $R\theta$. The equal-area mapping is well defined for the larger values of θ in the lower hemisphere (shown in grey in figure 17.13). The south-pole ($\theta = \pi$) maps to the outer boundary of the disk where $r = 2R$. To preserve area, points mapped from the lower hemisphere require a greater amount of distortion (i.e., Δr is significantly different from $R\Delta\theta$).

17.4.5 Jacobian for azimuthal mappings

Spherical coordinates involve two angles. The *zenith angle* is the angle that the position vector forms with the z-axis, so it ranges from 0 to π. A constant zenith angle corresponds to constant latitude. The *azimuthal angle* is the other angle that ranges from 0 to 2π, so a constant azimuthal angle corresponds to constant longitude. Engineers use (r, ϑ, φ) for (radius, zenith, azimuth), while mathematicians usually reverse the letters used for the angles. We use the engineering convention, so an area element on the sphere is

$$\mathrm{d}A^{\text{sphere}} = R^2(\sin\vartheta)\mathrm{d}\vartheta\,\mathrm{d}\varphi. \tag{17.100}$$

An *azimuthal mapping* preserves the azimuthal angle φ (i.e., the one that varies from 0 to 2π) so that the radius r in the projection is a function of the zenith angle: $r = r(\vartheta)$. The area element on the azimuthal projection is

$$\mathrm{d}A^{\text{disk}} = r\mathrm{d}r\,\mathrm{d}\varphi = r(\vartheta)\frac{\mathrm{d}r(\vartheta)}{\mathrm{d}\vartheta}\mathrm{d}\vartheta\,\mathrm{d}\varphi. \tag{17.101}$$

The Jacobian of this mapping is the ratio of the mapped area to the original area:

$$J(\vartheta) = \frac{(\mathrm{d}A^{\text{disk}})}{(\mathrm{d}A^{\text{sphere}})} = \frac{r(\vartheta)}{R^2(\sin\vartheta)}\frac{\mathrm{d}r(\vartheta)}{\mathrm{d}\vartheta}. \tag{17.102}$$

By design, the Lambert mapping in (17.99) has a Jacobian $J = 1$ indicating equal areas. The classical stereographic projection in figure 17.11 preserves angles but not areas. That projection has a Jacobian that increases monotonically with ϑ.

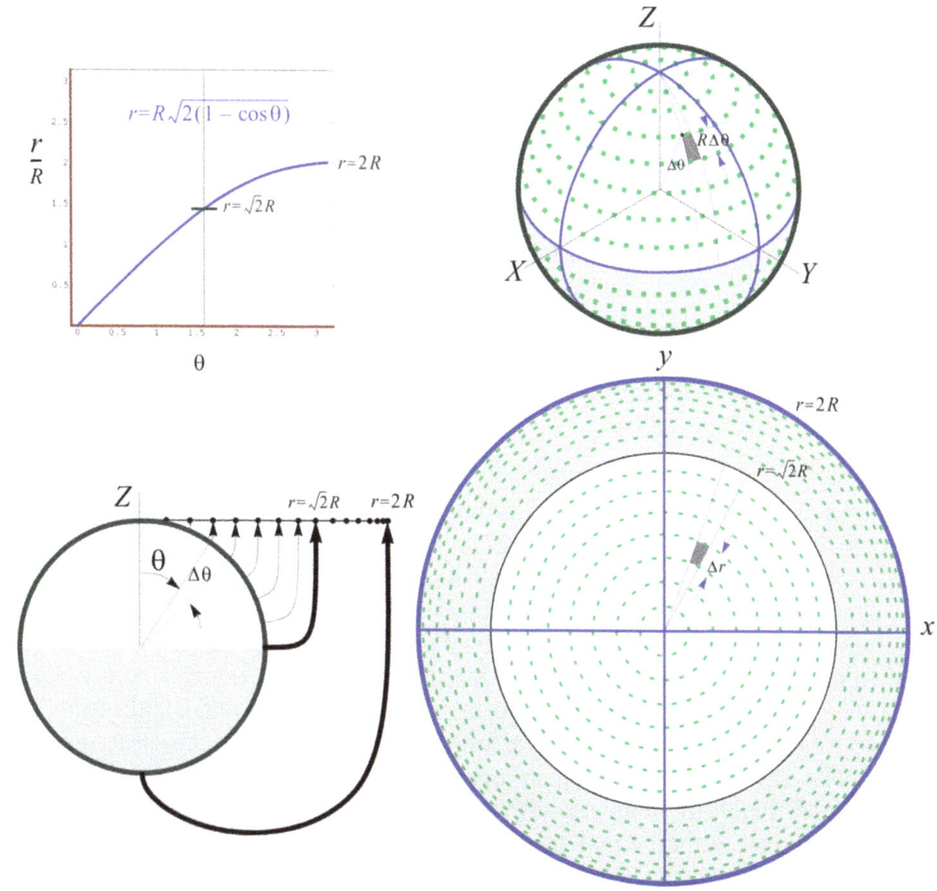

Figure 17.13. Equal-area (Lambert) mapping. The upper-left graph of r versus ϑ shows that $r \approx R\theta$ for points in the upper hemisphere. The lower-left plot shows a side view of the sphere to illustrate how the ϑ coordinate maps to the disk radial coordinate r. The upper-right sketch shows a conventional '3D' rendering of the sphere as seen down the {1,1,1}-axis. The lower-right plot is the equal-area mapping of the sphere as seen down the Z-axis. The relative sizes of the plots are shown to correct scale (i.e., the disk has twice the radius of the sphere.

Consequently, a uniform sprinkling of points on the sphere would map with the points relatively concentrated at the center of a stereographic projection, becoming increasingly spaced apart from each other moving towards the equator. The Lambert projection has no such mapping-induced clustering.

17.4.6 Why is equal-area (Lambert) mapping important?

For visually assessing distributions of dots, the point density on the projection plane needs to be the same as on the sphere so that there will be a cluster in the projection if and only if there is a cluster on the sphere. With a simple orthogonal or photographic projection, a uniform distribution on a sphere will have clustering of points near the outer boundary of the projection of the sphere onto a 2D disk (this sheet of paper). Because of its non-uniform Jacobian, the traditional stereographic projection will show relative clustering at the center of the projection that is an artifact of the projection, not indicating a clustering that is actually near the pole on the sphere. The Lambert equal-area mapping removes spurious clustering. If a distribution of points is uniform on the sphere, then it will also be uniform on the 2D equal-area mapping. Any clustering in an equal-area map is indicative of clustering on the sphere.

Figure 17.14(a) shows a plot of uniformly random unit vectors displayed using the Lambert equal-area stereographic projection. Because the mapping preserves the relative magnitudes of area elements, a uniform distribution of points on the sphere corresponds to a uniform distribution of points on the stereographic projection disk. Points on the interior of the white circle correspond to points on the upper half of the sphere. Points outside the white circle correspond to points in the lower hemisphere.

17.4.7 Other (non-equal-area) mappings to the sphere

Equal-area mapping helps to expose faulty statistical sampling methods. Suppose for example, that one attempts (incorrectly) to generate a set of uniform unit normals by simply using a uniform distribution on [0, 1) to generate three reals $\{r_1, r_2, r_3\}$, from which a unit vector is obtained by normalization. Essentially, this technique is equivalent to generating a set of uniformly distributed points in a *cube*, and then normalizing them all. The flaw in this scheme is that a larger percentage of points will be generated from points near the cube corners. Upon normalization, this produces an over-sampling of unit vectors that 'echo' their origins near the cube corners. The equal-area mapping in figure 17.14(b) clearly shows this spurious clustering and therefore demonstrates visually the inadequacy of the method.

We defined the Lambert equal-area mapping in such a way that the north pole on the sphere maps to the center of the mapped disk, and this makes the equator of the sphere map to the white circle in figure 17.14. However, any point P can be used as the pole for the projection. For example, by choosing the point P to be a location on the equator of the sphere, the equator will then map to a straight line cutting through

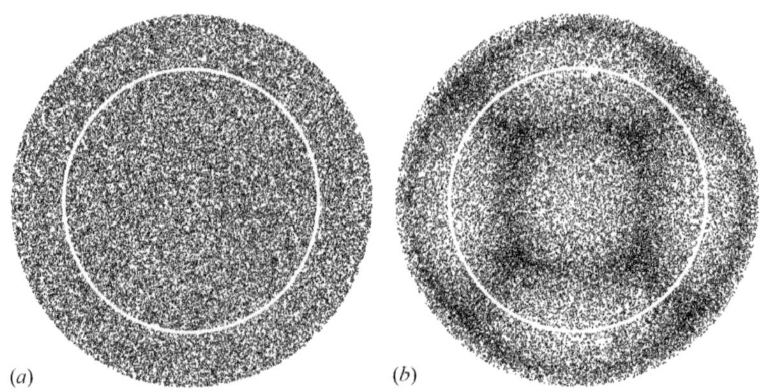

Figure 17.14. Lambert projections of 50000 random unit vectors. (a) A uniformly random unit vector distribution found by normalizing a uniform sampling of points *inside* the sphere. (b) A non-uniform set of unit vectors resulting from normalizing a uniform sampling of points inside a unit cube.

the center of the disk. In cartography, a minor variant of the Lambert mapping (called the *Hammer–Aitoff projection*) puts P on the equator and then uniformly stretches the Lambert disk to a 2:1 ellipse, which better shows the equator while preserving the equal-area property.

Figure 17.14 was created by a mapping that started with a 3D vector $\underset{\sim}{r}$,

$$\underset{\sim}{r}(r_1, r_2, r_3) = r_1\underset{\sim}{e}_1 + r_2\underset{\sim}{e}_2 + r_3\underset{\sim}{e}_3, \tag{17.103}$$

contained inside the unit cube so that each component was uniformly random over the range $-1 \leqslant r_k < 1$. Then a point $\underset{\sim}{x}$ on the unit sphere was found by normalizing $\underset{\sim}{r}$:

$$\underset{\sim}{x}(r_1, r_2, r_3) = \frac{r_1\underset{\sim}{e}_1 + r_2\underset{\sim}{e}_2 + r_3\underset{\sim}{e}_3}{\sqrt{r_1^2 + r_2^2 + r_3^2}}. \tag{17.104}$$

The point $\underset{\sim}{x}$ was mapped via Lambert projection to the disk for visualization. A related mapping, often used to study crystal orientations, starts instead with a point on the 2D surface of a unit cube:

$$\underset{\sim}{y}(u, v) = u\underset{\sim}{e}_1 + v\underset{\sim}{e}_2 + \underset{\sim}{e}_3 \quad \text{for} - 1 \leqslant u < 1 \text{ and} - 1 \leqslant v < 1. \tag{17.105}$$

This vector is then normalized to be projected on the sphere so that

$$\underset{\sim}{x}(u, v) = \frac{u\underset{\sim}{e}_1 + v\underset{\sim}{e}_2 + \underset{\sim}{e}_3}{\sqrt{u^2 + v^2 + 1}}. \tag{17.106}$$

If an evenly spaced grid is painted on the upper face of a unit cube, then its projection onto the sphere is as shown as the blue regions in figure 17.15, which shows similar projections for the other faces. The bottom (south pole) face maps to the dark outer ring in the Lambert projection.

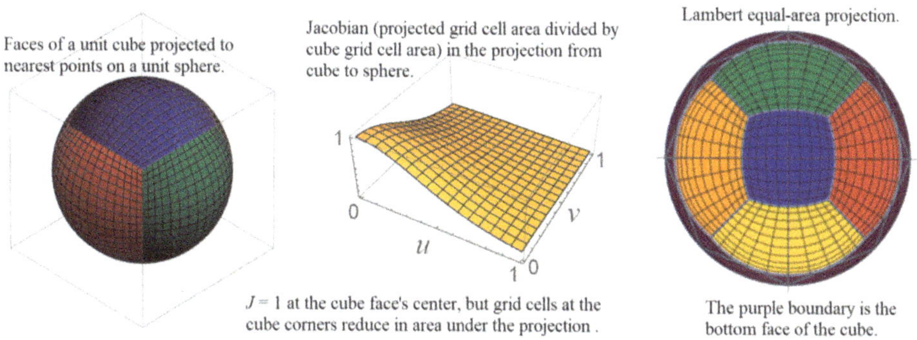

Figure 17.15. Projection of a cube to the sphere. The 3D image on the left is viewed from the {1,1,1} direction. The image on the right is the Lambert projection of the entire sphere, including the southern hemisphere (and therefore showing the bottom face as the purple ring on the boundary). The middle image shows that the Jacobian is non-uniform. In particular, an evenly spaced grid on the unit cube deforms to be preserved in size ($J = 1$) near $u = v = 0$ but is severely 'pinched down' ($J < 1$) for grid cells near cube corners.

17.5 'Nautical' visualization of a rotation

A rotation can be characterized by making an analogy with the orientation of a sailboat on the surface of the earth after some climate catastrophe makes the entire surface water. A right-handed system of embedded directions that characterize the orientation of the boat can be defined to point in the following directions:

1: heading direction (towards the bow) = first column of [*R*],

2: port direction (to the left) = second column of [*R*],

3: up the sailboat's mast = third column of [*R*].

Any possible orientation of the boat's embedded triad can be achieved by specifying the boat's latitude, longitude, and heading. To make the mast point 'downward', for example, the boat would be in the southern hemisphere. The convention for triad directions is rather arbitrary as long as it is right-handed. We might, for example, take the first and second columns of [*R*] to instead point starboard and in the heading direction, respectively.

The nautical representation of rotation is occasionally used in the materials science community, and it is useful to visually check if a random rotation is, in fact, *uniformly* random. For example, figure 17.16 (where the sailboat is depicted as an arrow) shows that you will *not* get a uniformly random rotation if you create a random rotation by from uniformly random Euler angles. Correct (non-biased) algorithms for generating uniformly random rotations are given in chapter 17, where the simplest one places the 'mast' of the sailboat at a uniformly random point on the sphere, with the sailboat's heading set to be a uniformly random rotation about the mast.

17.6 Uniformly random rotations

In this section, we discuss several methods for generating a uniformly random rotation. There are many correct (and incorrect) ways to do this, but the most elegant and efficient one is *Arvo's algorithm*, summarized in the box below

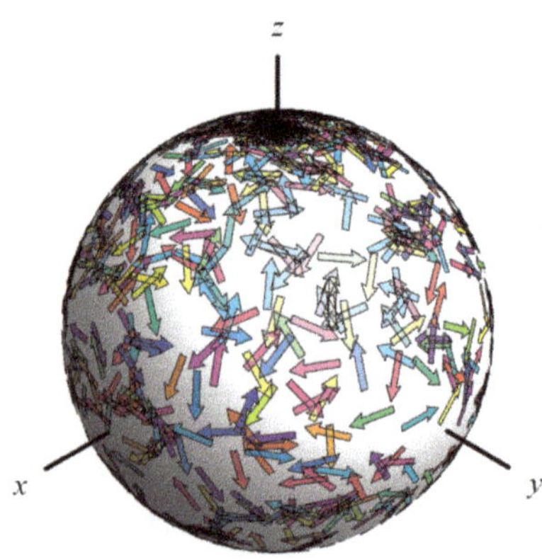

Figure 17.16. North-pole clustering. Nonuniform rotations are generated from uniform Euler angles.

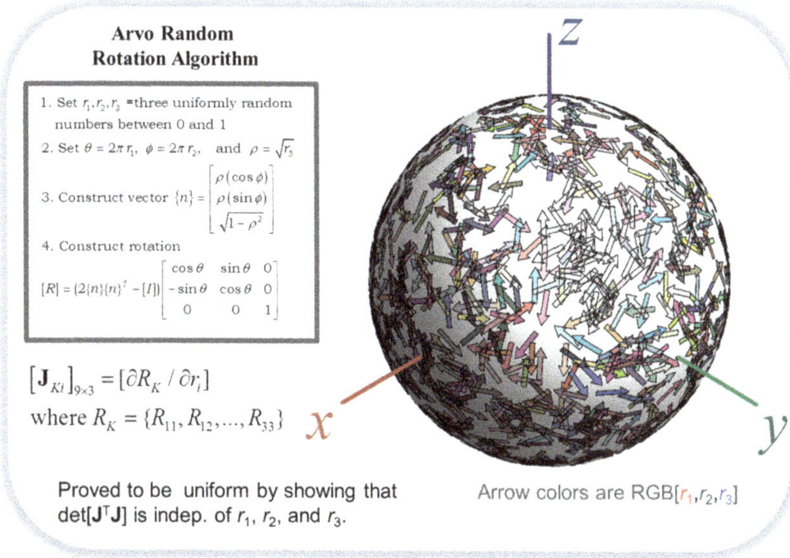

Like any engineering tensor, a rotation tensor has nine components, but only three of them are independent[7]. Thus it makes sense that the algorithm takes three inputs (the random numbers[8]) and gives nine outputs (components of the rotation tensor). The corresponding 9×3 Jacobian matrix $\mathbf{J}$ is, as indicated in the figure, defined by derivatives of the outputs with respect to the inputs. Being non-square, the Jacobian determinant of the transformation (from random numbers to rotation components) must be found by $J = \det\sqrt{J^T J}$. The algorithm is proved to be uniform by noting that the Jacobian determinant is the same at all points, meaning that it is independent of (r_1, r_2, r_3). While this is a rigorous mathematical definition of uniformity of a rotation distribution, it helps to understand the consequence of uniformity by knowing how to detect cluster bias, as we now explain.

Let $\underset{\sim}{\boldsymbol{N}}$ denote an arbitrarily selected unit *reference vector*. The *sample vector* $\underset{\sim}{\boldsymbol{n}}$ generated from a random rotation realization $\underset{\approx}{\boldsymbol{R}}$ is defined as

$$\underset{\sim}{\boldsymbol{n}} \equiv \underset{\approx}{\boldsymbol{R}} \bullet \underset{\sim}{\boldsymbol{N}}. \tag{17.107}$$

The rotation tensor $\underset{\approx}{\boldsymbol{R}}$ is uniformly distributed if and only if $\underset{\sim}{\boldsymbol{n}}$ is uniformly distributed for any reference direction $\underset{\sim}{\boldsymbol{N}}$. Equivalently (see [4]), a rotation tensor $\underset{\approx}{\boldsymbol{R}}$ is uniformly distributed if and only if $\underset{\approx}{\boldsymbol{Q}} \bullet \underset{\approx}{\boldsymbol{R}}$ and $\underset{\approx}{\boldsymbol{R}} \bullet \underset{\approx}{\boldsymbol{Q}}$ have the same distribution

[7] This number makes sense when you think about constructing a rotation matrix. Two numbers are needed to form the first column (because there is one constraint: it must be a unit vector). One number is needed for the second column (because it is subject to two constraints: it must be a unit vector and it must be perpendicular to the first column). Zero numbers are needed for the last column (as it is found by the cross product of the first two columns).

[8] Each between 0 and 1, as is standard for pseudo-random number generators in computers.

as $\underset{\approx}{\boldsymbol{R}}$ for any fixed rotation $\underset{\approx}{\boldsymbol{Q}}$. As with uniformly random unit normals, the mean of a uniformly random rotation is zero!

Cluster bias in a distribution typically reveals itself by using (17.107) to generate rotated orientations of the laboratory basis vectors. Treating these rotated vectors as points on the unit sphere, any clustering in their Lambert projections will (because the Lambert projection preserves area) indicate bias in the random rotation algorithm, as shown in figure 17.17. If no clustering is evident, the next step is to rigorously prove non-bias by showing that the Jacobian is uniform (i.e., not a function of the random numbers that are input to the rotation function).

After discussing a few more rotation algorithms (both good and bad), we finish this section by showing how to average a tensor over all possible uniformly

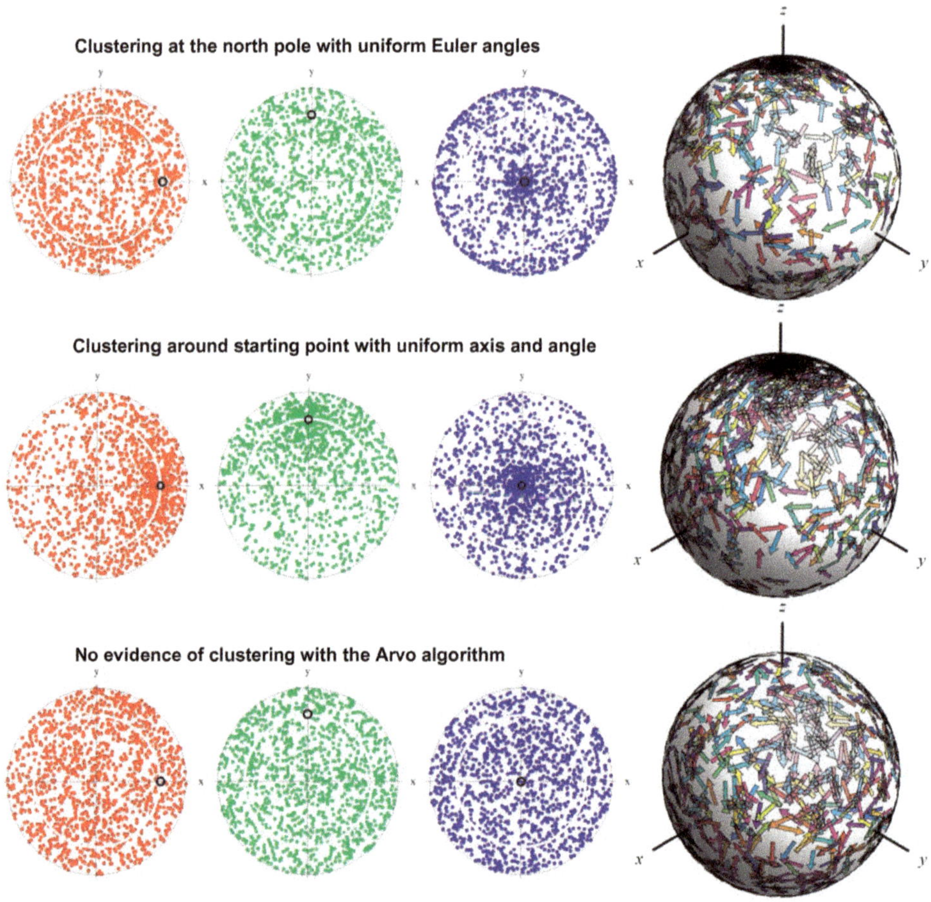

Figure 17.17. Clustering (with Euler angles or uniform axis/angle) and non-clustering (with Arvo method) of rotated basis vectors. The small black hollow circle shows the Lambert projection of the non-rotated first-, second-, and third-basis vectors, with all other dots being their rotations. With uniform Euler angles, clustering is quite clear in the Lambert projection of the third rotated basis vector. With a uniform axis/angle, rotated vectors are biased towards their initial positions, resulting in clusters around the black circles. Lack of clustering in the Arvo approach motivates *proving* uniformity by demonstrating homogeneity of the Jacobian.

distributed reorientations of that tensor. This procedure will be seen to produce the isotropic part of the tensor even in higher dimensions.

17.6.1 Inadequacy of using uniformly random angle and axis

It is clear from the clustering in the middle row of figure 17.17 that a uniform rotation *cannot* be obtained by using a uniform angle and axis of rotation. This section explains why uniform rotations require the distribution for the rotation angle to be biased toward 180°.

A uniformly random rotation tensor must have an expected value of $\underset{\approx}{\mathbf{0}}$ so that the rotated sample vector $\underset{\sim}{\boldsymbol{n}}$ is equally likely to point in any direction. As discussed by Shoemake [4], one might *guess* (incorrectly) that using a uniformly random rotation axis with a uniformly random rotation angle would provide a uniformly random rotation tensor (computed via equation (5.17)). However, as shown in the numerical experiment of figure 17.18, such a scheme makes the rotated sample vectors cluster about their original orientation, and the rotation is therefore not uniform. As a matter of fact, the expected value of the rotation turns out to be $\underset{\approx}{\boldsymbol{I}}/6\pi$, which is not zero as required for a uniform rotation.

To create figure 17.18, we generated roughly 4000 random rotations by taking the axis $\underset{\sim}{\boldsymbol{a}}$ and angle α to be uniform. The rotation tensor itself was then constructed via the Euler–Rodrigues formula of (5.17): $\underset{\approx}{\boldsymbol{R}} = \cos\alpha(\underset{\approx}{\boldsymbol{I}} - \underset{\sim}{\boldsymbol{a}}\underset{\sim}{\boldsymbol{a}}) + \underset{\sim}{\boldsymbol{a}}\underset{\sim}{\boldsymbol{a}} + \sin\alpha\underset{\sim}{\boldsymbol{a}} \times \underset{\approx}{\boldsymbol{I}}$. A version using even more sample points is shown in figure 17.18. In all cases, the rotated location of $\underset{\sim}{\boldsymbol{E}}_k$ tends to cluster about $\underset{\sim}{\boldsymbol{E}}_k$ itself!

To better understand why a uniform rotation must have a *non*-uniform rotation angle, consider any reference vector $\underset{\sim}{\boldsymbol{N}}$ to be rotated about a given rotation axis. If the rotation axis is uniformly distributed, then a significant fraction[9] of the rotation tensors will have a rotation axis that forms an angle with $\underset{\sim}{\boldsymbol{N}}$ of less than 45°. As indicated in figure 17.19, these 'polar' axes will *always* produce a sample vector $\underset{\approx}{\boldsymbol{R}} \bullet \underset{\sim}{\boldsymbol{N}}$ that lies in the upper hemisphere closest to $\underset{\sim}{\boldsymbol{N}}$. These polar axes are incapable of producing rotated sample points in the lower hemisphere.

Only 'equatorial' rotation axes that lie farther than 45° from $\underset{\sim}{\boldsymbol{N}}$ can *possibly* generate points in the lower hemisphere. Biasing the rotation angle α toward 180° allows the equatorial axes to produce a compensating number of sample points in the lower hemisphere. We later show that the rotation angle α has a probability density function given by

$$\begin{aligned} p_\alpha(\alpha) &= \frac{1}{\pi}\sin^2\frac{\alpha}{2} \quad \text{for rotation angles ranging from } -\pi \text{ to } \pi, \\ p_\alpha(\alpha) &= \frac{2}{\pi}\sin^2\frac{\alpha}{2} \quad \text{for rotation angles ranging from } 0 \text{ to } \pi, \end{aligned} \tag{17.108}$$

which is minimum at 0° and maximum at 180°.

---

[9] Namely, about 21% (=1 − $1/\sqrt{2}$).

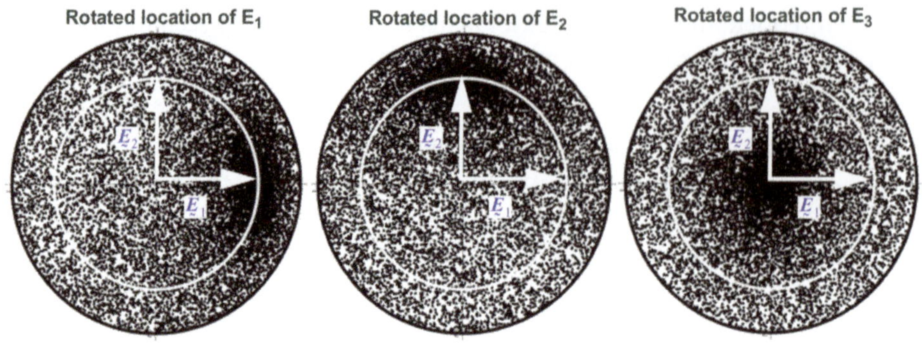

Figure 17.18. Uniform axis and angle do not lead to uniform rotation. Roughly 10^4 random rotations were generated using uniform axis and angle. For each sample rotation, the rotated basis vector, $\underset{\approx}{R} \bullet \underset{\sim}{E}_k$, was computed and (regarding it as a point on the unit sphere) mapped to the plane via Lambert equal-area mapping. The white circle demarks the boundary between the upper and lower hemispheres (see figure 17.13). The rotated positions of the three basis vectors tend to cluster about their original orientations.

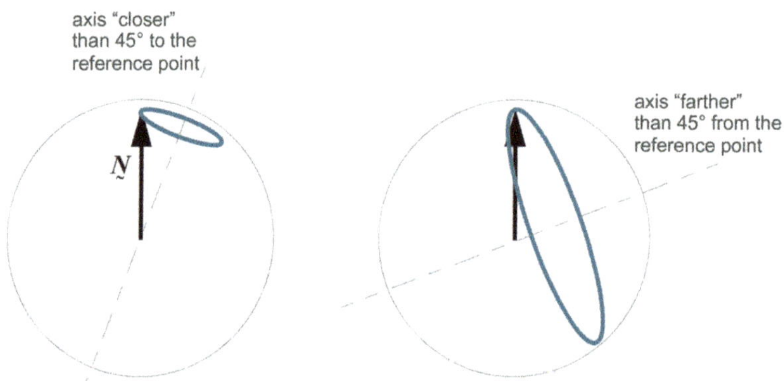

Figure 17.19. Qualitative reason why a uniform rotation must have a non-uniform rotation angle. Suppose that the rotation axis forms an angle smaller than 45° with the reference vector $\underset{\sim}{N}$. Then regardless of the distribution for the rotation angle, the rotated sample vector $\underset{\approx}{R} \bullet \underset{\sim}{N}$ will lie in the upper hemisphere closest to $\underset{\sim}{N}$. Only axes farther than 45° away from the reference vector even have the capability of producing sample points in the lower hemisphere. Consequently, the rotation angle must be biased towards 180° to allow the equatorial axes to generate enough sample points in the lower hemisphere to balance those in the upper hemisphere.

17.6.2 Inadequacy of using uniform Euler angles

Above, we showed that a uniformly random rotation must have a rotation angle that is biased towards 180°. Similarly, this section explains that a uniformly random rotation must have its second Euler angle biased toward 90°. A uniformly random rotation tensor must have an expected value of zero. This is a necessary, but not sufficient, condition for uniformity. Suppose that you generate three uniformly random Euler angles (φ, ϑ, ψ) and then construct the rotation tensor by applying (9.11),

$$[R] = \begin{bmatrix} \cos\varphi & -\sin\varphi & 0 \\ \sin\varphi & \cos\varphi & 0 \\ 0 & 0 & 1 \end{bmatrix} \begin{bmatrix} 1 & 0 & 0 \\ 0 & \cos\vartheta & -\sin\vartheta \\ 0 & \sin\vartheta & \cos\vartheta \end{bmatrix} \begin{bmatrix} \cos\psi & -\sin\psi & 0 \\ \sin\psi & \cos\psi & 0 \\ 0 & 0 & 1 \end{bmatrix}. \tag{17.109}$$

The resulting rotation tensor will have a zero expected value (as desired), but it still will not be *uniformly* random. A rotation is deemed uniformly random if the distribution for $\underset{\sim}{\mathbf{R}} \bullet \underset{\sim}{\mathbf{N}}$ is uniform for any unit vector $\underset{\sim}{\mathbf{N}}$. The sample points plotted in figure 17.17 showed north-pole clustering of the rotated locations of $\underset{\sim}{\mathbf{R}} \bullet \underset{\sim}{\mathbf{E}}_3$. That plot showed immediately apparent bias of the $\underset{\sim}{\mathbf{R}} \bullet \underset{\sim}{\mathbf{E}}_1$ and $\underset{\sim}{\mathbf{R}} \bullet \underset{\sim}{\mathbf{E}}_2$ plots. There is, in fact, also clustering in those plots near the equator. To better see this, figure 17.20 shows 10^4 sample points generated using uniform Euler angles viewed by Lambert projection down the y-axis (i.e., down the $\underset{\sim}{\mathbf{E}}_2$-axis). The equatorial clustering is then clearly visible from this different viewpoint, hence showing the importance of looking at dot plots from multiple angles.

To produce a truly uniform rotation, the Euler angle distributions must be

$$\begin{aligned} p_\varphi(\varphi) &= \frac{1}{2\pi} && \text{on } -\pi < \varphi < \pi, \\ p_\vartheta(\vartheta) &= \frac{1}{2}\sin\vartheta && \text{on } 0 < \vartheta < \pi, \\ p_\psi(\psi) &= \frac{1}{2\pi} && \text{on } -\pi < \psi < \pi. \end{aligned} \tag{17.110}$$

Note that φ and ψ are uniform, but ϑ must be biased toward values of 90°.

17.7 A basic algorithm for generating a uniformly random rotation

Any rotation tensor $\underset{\sim}{\mathbf{R}}$ can be defined by how it transforms the three laboratory basis vectors, $\{\underset{\sim}{\mathbf{E}}_1, \underset{\sim}{\mathbf{E}}_2, \underset{\sim}{\mathbf{E}}_3\}$. Upon a rigid rotation, this triad becomes $\{\underset{\sim}{\mathbf{e}}_1, \underset{\sim}{\mathbf{e}}_2, \underset{\sim}{\mathbf{e}}_3\}$. Then the

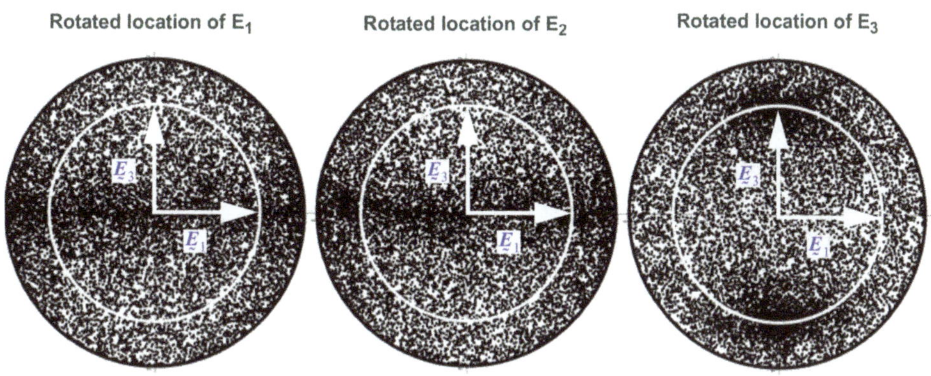

Figure 17.20. Uniform Euler angles produce non-uniform rotations. Equatorial clustering is more visible when viewing down the $\underset{\sim}{\mathbf{E}}_2$-axis.

rotation tensor is $\underset{\approx}{R} = \underset{\sim}{e}_1\underset{\sim}{E}_1 + \underset{\sim}{e}_2\underset{\sim}{E}_2, \underset{\sim}{e}_3\underset{\sim}{E}_3$, meaning that the columns of the component matrix are the rotated basis vectors:

$$[\underset{\approx}{R}] = [\{\underset{\sim}{e}_1\}\{\underset{\sim}{e}_2\}\{\underset{\sim}{e}_3\}]. \tag{17.111}$$

To generate a uniformly random rotation tensor $\underset{\approx}{R}$, we set $\underset{\sim}{e}_1$ to be a uniformly random unit vector (constructed via (17.52), (17.51), and (17.53)). We then construct a second uniformly random vector $\underset{\sim}{m}$ and subtract away its part in the direction of $\underset{\sim}{e}_1$. In other words, we compute $\underset{\sim}{e}_2$ as follows:

$$\underset{\sim}{e}_2 = \underset{\sim}{\mu}/\|\underset{\sim}{\mu}\|, \text{ where } \underset{\sim}{\mu} = \underset{\sim}{m} - (\underset{\sim}{m} \bullet \underset{\sim}{e}_1)\underset{\sim}{e}_1. \tag{17.112}$$

The third rotated unit vector is given by the right-hand cross product: $\underset{\sim}{e}_3 = \underset{\sim}{e}_1 \times \underset{\sim}{e}_2$. With the three rotated basis vectors known, the uniformly random rotation $\underset{\approx}{R}$ is given by (17.111).

The numerical implementation of this technique is given in section A.8 of the appendix. Using this algorithm to generate 10^4 uniformly random rotations, we called the routine DC2AA (section A.3) to 'experimentally' measure the distributions for axis and angle shown in figure 17.21.

Figure 17.21 illustrates that the rotation axis is distributed uniformly. The rotation angle α clearly is not uniform. As previously explained, the rotation angle must be biased toward values of 180°. We later prove that a uniform rotation must have a distribution for the rotation angle given by

$$p_\alpha(\alpha) = \frac{1}{\pi}\sin^2\frac{\alpha}{2} \quad \text{for } \alpha \in [-\pi, \pi], \tag{17.113}$$

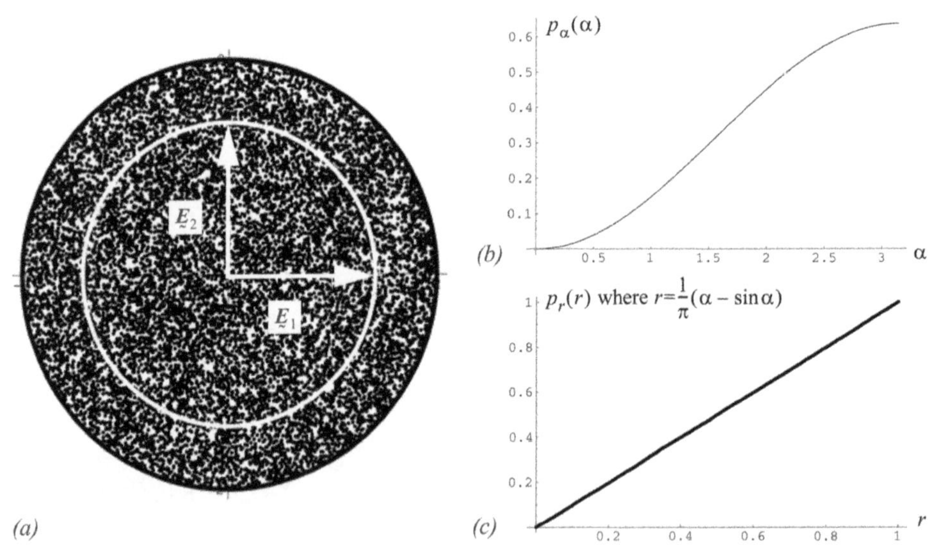

Figure 17.21. Numerical measurement of rotation axis and angle. The left side (a) shows a dot plot of the rotation axis from a truly uniform rotation algorithm (e.g., the basic one in this section or the previously provided Arvo algorithm). The graph (b) is the cumulative distribution for the rotation angle. The graph (c) is the cumulative distribution for the variable $r = (\alpha - \sin\alpha)/\pi$.

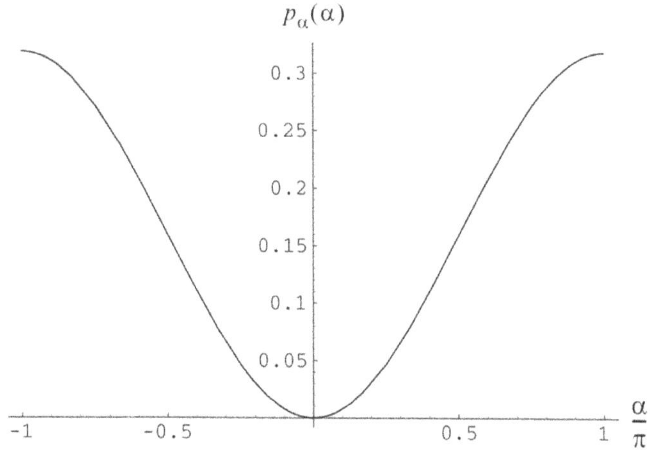

Figure 17.22. Distribution for the angle of rotation.

which is plotted in figure 17.22. Note that the rotation angle is weighted toward ±180°, as expected.

Without loss in generality, the rotation angle may be alternatively taken to be always positive (as is presumed in the routine DC2AA), in which case the distribution is

$$p_\alpha(\alpha) = \frac{2}{\pi}\sin^2\frac{\alpha}{2} \qquad \text{for } \alpha \in [0, \pi]. \tag{17.114}$$

This means that the CDF must be given by

$$g_\alpha(\alpha) = \int_0^\alpha p_\alpha(\alpha')\mathrm{d}\alpha' = \frac{1}{\pi}(\alpha - \sin\alpha). \tag{17.115}$$

Consequently, the quantity $(\alpha - \sin\alpha)/\pi$ must be uniformly distributed on the interval from 0 to 1, which is verified in the numerical experiment of figure 17.21. Since we know that

$$r = (\alpha - \sin\alpha)/\pi \tag{17.116}$$

is uniformly distributed on the interval from 0 to 1, then we know we can generate a representative sampling of the rotation angle by generating a number r that is uniformly random on the interval from 0 to 1, and then the rotation angle α would then be obtained by numerically solving (17.116) for α.

17.8 Generalization to generate transversely isotropic orientation distributions

When metals are rolled to form sheets, not only do their crystals deform from nearly regular polyhedra to 'flattened' polyhedra, their orientations also change from being uniform to having bias such that the material direction triads (which define orientation tensors) tend to have preference for one triad basis vector being biased in the direction of the sheet's normal. The other two directions are uniformly distributed about the biased normal. The biased crystal orientation distribution can be obtained using the CVT method to generate the first basis vector with a bias toward the north and south poles, and then using a uniform distribution for the other basis vector directions. From there, the biased crystal shape can be obtained by using the shape tensor from (17.89) with the directions of the $\underset{\sim}{a}$, $\underset{\sim}{b}$, and $\underset{\sim}{c}$ vectors being determined from the biased orientation triad and their magnitudes being set to reflect the desired crystal stretch magnitudes.

17.9 Alternative algorithm for uniformly random rotation

The above method for generating a uniformly random rotation simply applied Gram–Schmidt orthogonalization to two uniformly random unit vectors, with the third unit vector being given by a cross product. The resulting right-handed orthogonal set of unit vectors was then assembled to form columns of a 3 × 3 rotation matrix. Algorithmically this is a trivial procedure so long as a function is already available to generate a uniformly random unit vector. However, an unsavory numerical issue is that the second vector must be re-generated if it is parallel (or nearly parallel) to the first vector.

An alternative algorithm begins in the same manner by generating a uniformly random unit vector $\underset{\sim}{n}_1$. If the kth laboratory component of $\underset{\sim}{n}_1$ is the smallest (or tied for smallest), then we know that the lab basis vector $\underset{\sim}{E}_k$ will be linearly independent of $\underset{\sim}{n}_1$ (i.e., $\underset{\sim}{n}_1 \bullet \underset{\sim}{E}_k \neq 0$). Consequently, we can construct a vector $\underset{\sim}{m}$ that is perpendicular to $\underset{\sim}{n}_1$ by the formula

$$\underset{\sim}{m} = \frac{\underset{\sim}{E}_k - \underset{\sim}{n}_1(\underset{\sim}{E}_k \bullet \underset{\sim}{n}_1)}{\|\underset{\sim}{E}_k - \underset{\sim}{n}_1(\underset{\sim}{E}_k \bullet \underset{\sim}{n}_1)\|}. \tag{17.117}$$

A third unit vector can be created by

$$\underset{\sim}{t} = \underset{\sim}{n}_1 \times \underset{\sim}{m}. \tag{17.118}$$

The two unit vectors $\underset{\sim}{m}$ and $\underset{\sim}{t}$ form a plane perpendicular to $\underset{\sim}{n}_1$. If R is a random number on the interval [0, 1), then the vector

$$\underset{\sim}{n}_2 = \cos(2\pi R)\underset{\sim}{m} + \sin(2\pi R)\underset{\sim}{t} \tag{17.119}$$

is uniformly random in the plane perpendicular to $\underset{\sim}{n}_1$. The uniformly random rotation tensor would then be constructed by

$$[\underset{\sim}{R}] = [\{\underset{\sim}{n}_1\}\{\underset{\sim}{n}_2\}\{\underset{\sim}{n}_3\}], \quad \text{where } \underset{\sim}{n}_3 = \underset{\sim}{n}_1 \times \underset{\sim}{n}_2. \tag{17.120}$$

This algorithm is provided here only because we have seen it used in research, and we can confirm that it gives correct results.

17.10 Shoemake's algorithm for uniformly random rotations

Recall that a rotation tensor may always be expressed in terms of a point (u_o, u_1, u_2, u_3) lying on the surface of a sphere in 4D quaternion space:

$$\underset{\approx}{R} = \left(2u_o^2 - 1\right)\underset{\approx}{I} + 2\underset{\sim}{u}\underset{\sim}{u} - 2u_o\underset{\approx}{\varepsilon} \bullet \underset{\sim}{u}, \qquad \text{where } \underset{\sim}{u} \equiv u_1\underset{\sim}{E}_1 + u_2\underset{\sim}{E}_2 + u_3\underset{\sim}{E}_3. \tag{17.121}$$

The mapping from the 4-component quaternion to 9-component rotation tensor can be shown to have a uniform Jacobian (making it effectively like a higher dimensional version of a Lambert projection). Accordingly, a uniform rotation distribution corresponds to a uniform quaternion distribution and vice versa. The surface area of the unit quaternion hyper-sphere is proved below to equal $2\pi^2$. Therefore, for a uniform distribution on the 4D this hyper-sphere, the joint distribution function must be defined such that

$$p_{\vec{u}}(\vec{u}) = \frac{1}{2\pi^2}. \tag{17.122}$$

The symbol $\vec{u}$ denotes the 4D vector $\{u_o, u_1, u_2, u_3\}$. By virtue of (7.2), we may introduce 4D spherical coordinates $\{\beta, \theta, \phi\}$ such that

Coordinate choice 'a':

$$\begin{aligned} u_1 &= \sin\beta(\sin\theta\cos\phi), \\ u_2 &= \sin\beta(\sin\theta\sin\phi), \\ u_3 &= \sin\beta(\cos\theta), \\ u_o &= \cos\beta, \\ &\text{for which} \quad 0 \leqslant \beta \leqslant \pi,\ 0 \leqslant \theta \leqslant \pi,\ -\pi \leqslant \phi \leqslant \pi. \end{aligned} \tag{17.123a}$$

The area element is $dA = [\sin\theta(\sin^2\beta)]d\beta\, d\theta\, d\phi$.

This is not the only choice for 4D spherical coordinates. Another option is

Coordinate choice 'b':

$$\begin{aligned} u_1 &= \sin\psi_1\cos\psi_3, \\ u_2 &= \cos\psi_1\cos\psi_3, \\ u_3 &= \sin\psi_2\sin\psi_3, \\ u_o &= \cos\psi_2\sin\psi_3, \end{aligned} \tag{17.123b}$$

$$\text{for which,} \quad -\pi \leqslant \psi_1 \leqslant \pi,\ -\pi \leqslant \psi_2 \leqslant \pi,\quad 0 \leqslant \psi_3 \leqslant \frac{\pi}{2}.$$

The area element is $dA = \left[\frac{1}{2}\sin(2\psi_3)\right]d\psi_1 d\psi_2 d\psi_3$.

The area element on the 4D unit hyper-sphere is found using standard methods of differential geometry, evaluating the Jacobian in a manner similar to the derivation of equation (17.45). The coordinate ranges cited above cover the *entire* 4D hyper-sphere. Integrating the area element over the coordinate domains gives the area of the entire 4D hyper-sphere as $2\pi^2$. For a uniform distribution on the 4D hyper-sphere, the joint distribution function for the spherical coordinates must be defined such that

$$p_{\vec{u}}(\vec{u})\mathrm{d}A = p_{\beta,\theta,\phi}(\beta, \theta, \phi)\mathrm{d}\beta\, \mathrm{d}\theta\, \mathrm{d}\phi, \tag{17.124a}$$

$$p_{\vec{u}}(\vec{u})\mathrm{d}A = p_{\psi_1,\psi_2,\psi_3}(\psi_1, \psi_2, \psi_3)\mathrm{d}\psi_1\, \mathrm{d}\psi_2\, \mathrm{d}\psi_3. \tag{17.124b}$$

Substituting (17.122) and (17.123*a*) into (17.124*a*) gives the joint distribution for coordinate choice 'a':

$$p_{\beta,\theta,\phi}(\beta, \theta, \phi) = \frac{\sin\theta(\sin^2\beta)}{2\pi^2}. \tag{17.125a}$$

Similarly, using coordinate choice 'b',

$$p_{\psi_1,\psi_2,\psi_3}(\psi_1, \psi_2, \psi_3) = \frac{\sin(2\psi_3)}{4\pi^2}. \tag{17.125b}$$

Not surprisingly, these joint distributions are separable, allowing us to define individual distributions for each coordinate:

$$p_\beta(\beta) = \tfrac{2}{\pi}\sin^2(\beta), \quad p_{\psi_1}(\psi_1) = \tfrac{1}{2\pi}, \tag{17.126}$$

$$p_\theta(\theta) = \tfrac{1}{2}\sin\theta, \quad p_{\psi_1}(\psi_2) = \tfrac{1}{2\pi}, \tag{17.127}$$

$$p_\phi(\phi) = \tfrac{1}{2\pi}, \quad p_{\psi_3}(\psi_3) = \sin(2\psi_3). \tag{17.128}$$

The coefficients for these independent distributions ensure that the integral of each distribution function over the full domain of its argument will equal unity, as required for any distribution function.

17.10.1 Axis/angle formulation for generating a uniformly random rotation

The Euler–Rodrigues formula,

$$\underset{\approx}{R} = \cos\alpha(\underset{\approx}{I} - \underset{\sim}{a}\underset{\sim}{a}) + \underset{\sim}{a}\underset{\sim}{a} - \sin\alpha\ \underset{\approx\approx}{\varepsilon} \bullet \underset{\sim}{a}, \tag{17.129}$$

may be used to generate a uniformly random rotation if one has functions to generate properly distributed realizations of the unit rotation axis $\underset{\sim}{a}$ and rotation angle α. The rotation axis and rotation angle are related to choice 'a' for the quaternion spherical coordinates defined in (17.123*a*). Specifically, θ and ϕ are the spherical coordinates that define the rotation axis, and $\beta = \alpha/2$.

Comparing the distributions $p_\theta(\theta)$ and $p_\phi(\phi)$ in (17.127*a*) and (17.128*a*) with those in (17.47) shows that the rotation axis is uniformly distributed. Listing 7 in section A.8 could be used to generate the uniformly random rotation axis $\underset{\sim}{a}$. The rotation angle, on the other hand, is non-uniformly distributed. Because the rotation angle α equals 2β, the density function for the rotation angle α must be given by using (17.126*a*):

$$p_\alpha(\alpha) = \tfrac{1}{2} p_\beta\left(\frac{\alpha}{2}\right) = \tfrac{1}{\pi}\sin^2\left(\frac{\alpha}{2}\right). \tag{17.130}$$

The above distribution corresponds to a full domain in the rotation angle

$$-\pi \leqslant \alpha \leqslant \pi. \tag{17.131}$$

Without loss in generality it is always possible (by changing the direction of the rotation axis) to presume that the rotation angle satisfies $\sin\alpha \geqslant 0$, in which case,

$$0 \leqslant \alpha \leqslant \pi, \qquad \text{and} \qquad p_\alpha(\alpha) = \tfrac{2}{\pi}\sin^2\left(\frac{\alpha}{2}\right). \tag{17.132}$$

In this case, a rotation is described by a unit quaternion lying on a spherical *hemisphere* in 4D space (so the factor of $2\pi^2$ in (17.125) would become π^2 and all of the subsequent arguments would go through unchanged to ultimately lead to (17.132)).

The function needed to convert from a pseudo-random number on [0, 1) to a rotation angle $\alpha \in (0, \pi]$ with distribution of (17.132) is determined by applying (17.5):

$$g(\alpha) \equiv \int_0^\alpha p_\alpha(\alpha')\mathrm{d}\alpha' = \int_0^\alpha \tfrac{2}{\pi}\sin^2\left(\frac{\alpha'}{2}\right)\, \mathrm{d}\alpha' = \frac{\alpha - \sin\alpha}{\pi}. \tag{17.133}$$

Setting $g(\alpha) = r$ gives

$$\boxed{\alpha - \sin\alpha = \pi r}. \tag{17.134}$$

If r is uniformly distributed on the interval from 0 to 1, then α is distributed from 0 to π according to (17.114), and vice versa. Figure 17.21(b) verifies the converse statement. Specifically, we generated 10^4 random rotations, and for each rotation, we used the function in section A.8 to compute the rotation angle α. Figure 17.21(b) shows that the cumulative distribution for $r = (\alpha - \sin\alpha)/\pi$ is linear, indicating that r is uniform.

Unfortunately, equation (17.134) cannot be solved to give α as an analytical function of r, so a numerical root solver would be needed to obtain a realization of α. Once this is done, the Euler–Rodrigues formula from (5.17) may be used to build the uniformly random rotation realization. The numerical complexity of this approach again shows the superiority of the Arvo algorithm.

17.10.2 Shoemake's quaternion algorithm for directly computing a uniform rotation

Coordinate choice 'a' in (17.123*a*) is appealing in the sense that it is intimately related to the angle and axis representation for a rotation. However, as seen in the previous section, generating a realization for a rotation angle requires a numerical solver.

Consider now coordinate choice ‘b’ in (17.123*b*). Coordinates ψ_1 and ψ_2 are uniformly distributed on $[-\pi, \pi)$ or, without loss, on the interval $[0, 2\pi]$, so realizations are very easy to generate for them. Specifically, if r_1 and r_2 are two independently generated random numbers on [0, 1), then

$$\psi_1 = 2\pi r_1, \tag{17.135}$$

$$\psi_2 = 2\pi r_2. \tag{17.136}$$

The third coordinate ψ_3 has the distribution cited in (17.128b):

$$p_{\psi_3}(\psi_3) = \sin(2\psi_3). \tag{17.137}$$

Integrating this to obtain the CDF and setting the result equal to a random number r on [0, 1) gives

$$\sin^2 \psi_3 = r_3. \tag{17.138}$$

Recalling that ψ_3 falls in the range from 0 to $\pi/2$, it follows that

$$\sin \psi_3 = \sqrt{r_3} \qquad \text{and} \qquad \cos \psi_3 = \sqrt{1 - r_3}\,. \tag{17.139}$$

Equations (17.135), (17.136), and (17.139) may be substituted into (17.123*b*) to obtain the quaternion. Then the rotation tensor is constructed by applying (17.121). To summarize, Shoemake states that the following algorithm will result in a uniform rotation:

STEP 1. Generate three independent random numbers $\{r_1, r_2, r_3\}$ that are uniformly distributed between 0 and 1.
STEP 2. Compute two uniformly distributed angles, $\psi_1 = 2\pi r_1$ and $\psi_2 = 2\pi r_2$, and their sines and cosines, s_1, c_1, s_2, c_2.
STEP 3. Compute $s_3 = \sqrt{r_3}$ and $c_3 = \sqrt{1 - r_3}$.
STEP 4. Compute the unit quaternion with components

$$\begin{aligned} u_1 &= s_1 c_3, \\ u_2 &= c_1 c_3, \\ u_3 &= s_2 s_3, \\ u_0 &= c_2 s_3. \end{aligned} \tag{17.140}$$

STEP 5. Compute the rotation by applying (17.121):

$$\underset{\approx}{R} = \left(2u_o^2 - 1\right)\underset{\approx}{I} + 2\underset{\sim}{u}\underset{\sim}{u} - 2u_o \underset{\approx}{\varepsilon} \bullet \underset{\sim}{u}, \quad \text{where } \underset{\sim}{u} \equiv u_1\underset{\sim}{E}_1 + u_2\underset{\sim}{E}_2 + u_3\underset{\sim}{E}_3. \tag{17.141}$$

Shoemake’s algorithm has been verified by us and is given in the routine called ‘RANDC_QUATERN’ in section A.8.

17.10.3 The expected value (average) of a uniformly random rotation is zero

Expected values often fail to share properties with the thing being averaged. For example, the average of all points on a sphere is the origin, which is not on the sphere. Likewise, the average of uniform rotations is *zero*, not the identity tensor. We can write the rotation axis in terms of spherical coordinates as

$$\underset{\sim}{a} = \sin\theta(\cos\phi\,\underset{\sim}{e}_1 + \sin\phi\,\underset{\sim}{e}_2) + \cos\theta\,\underset{\sim}{e}_3. \qquad (17.142)$$

If the rotation axis is uniformly distributed, then the distribution functions for θ and ϕ are

$$p_\theta(\theta) = \tfrac{1}{2}\sin\theta \ \text{ and } \ p_\phi(\phi) = \tfrac{1}{2\pi}. \qquad (17.143)$$

Let $\underset{\sim}{N}$ be any unit vector. We wish to determine whether the distribution for $\underset{\approx}{R} \bullet \underset{\sim}{N}$ is uniform. We can always set up a basis such that $\underset{\sim}{e}_3 = \underset{\sim}{N}$. Then

$$\begin{aligned}
\underset{\approx}{R} \bullet \underset{\sim}{N} &= \cos\alpha(\underset{\sim}{e}_3 - \underset{\sim}{a}\cos\theta) + \underset{\sim}{a}\cos\theta + \sin\alpha(\sin\theta(-\cos\phi\,\underset{\sim}{e}_2 + \sin\phi\,\underset{\sim}{e}_1)) \\
&= \underset{\sim}{e}_1[(\sin\theta)(\cos\phi)(\cos\theta)(1 - \cos\alpha) + (\sin\alpha)(\sin\theta)(\sin\phi)] \\
&\quad + \underset{\sim}{e}_2[(\sin\theta)(\sin\phi)(\cos\theta)(1 - \cos\alpha) - (\sin\alpha)(\sin\theta)(\cos\phi)] \\
&\quad + \underset{\sim}{e}_3[\cos^2\theta + (\cos\alpha)\sin^2\theta].
\end{aligned}$$

Let us verify that the distributions of (17.113) and (17.143) correspond to a zero expected value of $\underset{\approx}{R} \bullet \underset{\sim}{e}_3$:

$$E[\underset{\approx}{R} \bullet \underset{\sim}{N}] = \int_{-\pi}^{\pi}\int_0^{\pi}\int_0^{(2\pi)} (\underset{\approx}{R} \bullet \underset{\sim}{e}_3)(p_\phi(\phi)(p_\theta(\theta)p_\alpha(\alpha)))\mathrm{d}\phi \ \mathrm{d}\theta \ \mathrm{d}\phi. \qquad (17.145)$$

Performing this integral does indeed show that the distribution has a zero expected value. This is a necessary (not sufficient[10]) condition for $\underset{\approx}{R} \bullet \underset{\sim}{N}$ to be uniformly distributed on the unit sphere. To prove that $\underset{\approx}{R} \bullet \underset{\sim}{N}$ is uniformly distributed, we write

$$\underset{\approx}{R} \bullet \underset{\sim}{N} = n_1\underset{\sim}{e}_1 + n_2\underset{\sim}{e}_2 + n_3\underset{\sim}{e}_3. \qquad (17.146)$$

The distributions of (17.113) and (17.143) correspond to a uniform distribution for the rotation if and only if the coordinates (n_1, n_2, n_3) are uniformly distributed on the unit sphere. Thus

$$p_{\underset{\sim}{n}}(\underset{\sim}{n}) = \tfrac{1}{4\pi}, \qquad (17.147)$$

which is indeed the distribution implied by (17.113) and (17.143).

[10] Uniform Euler angles will produce a non-uniform distribution of $\underset{\approx}{R} \bullet \underset{\sim}{N}$ that nevertheless has a zero expected value.

17.10.4 The average of a rotated tensor over a distribution of rotations

Let $\underset{\approx}{S}$ be some known tensor and let $\underset{\approx}{\bar{S}}$ denote a second tensor that is generated by superimposing a rotation on the first tensor:

$$\underset{\approx}{\bar{S}} = \underset{\approx}{R}^T \bullet \underset{\approx}{S} \bullet \underset{\approx}{R}. \tag{17.148}$$

If $\underset{\approx}{R}$ is uniformly random, then (without proof) the expected value of $\underset{\approx}{\bar{S}}$ is

$$E[\underset{\approx}{\bar{S}}] = \tfrac{1}{3}(\text{tr}\ \underset{\approx}{S})\underset{\approx}{I}. \tag{17.149}$$

In other words, the expected value of a tensor over all possible uniform rotations of that tensor is just the *isotropic part* of that tensor. This result also holds for higher-order tensors. In general, the isotropic part of a tensor is its projection to the space of isotropic tensors. For example, a second-order tensor $\underset{\approx}{A}$ is isotropic if and only if

$$\underset{\approx}{R}^T \bullet \underset{\approx}{A} \bullet \underset{\approx}{R} = \underset{\approx}{A} \qquad \text{or} \qquad R_{pi}R_{qj}A_{pq} = A_{ij} \text{ for all rotations } \underset{\approx}{R}. \tag{17.150}$$

Writing this definition in component form eventually reveals that any isotropic tensor (in 3D space) must be a multiple of the identity tensor $\underset{\approx}{I}$. Therefore, the identity tensor must be a *basis* for the space of isotropic second-order tensors referenced to 3D physical space. By 'referenced to 3D physical space' we simply mean that the tensor is expressible in terms of a 3×3 matrix. Because all isotropic 3×3 matrices are expressible as some scalar times one matrix (the identity matrix), the space of isotropic second-order tensors referenced to 3D physical space must be 1D. A 3D second-order tensor is projected to its 1D isotropic space by applying the following projection operation:

$$\text{iso}\underset{\approx}{S} = (\underset{\approx}{S} * \underset{\approx}{\hat{I}})\underset{\approx}{\hat{I}} \qquad \text{for 3D second-order tensors.} \tag{17.151}$$

Here, the * denotes the inner product, which for second-order tensors is defined $\underset{\approx}{A}*\underset{\approx}{B} = A_{ij}B_{ij}$. The 'hat' over the identity tensor denotes normalization of the identity tensor. Specifically,

$$\underset{\approx}{\hat{I}} = \frac{\underset{\approx}{I}}{\sqrt{\underset{\approx}{I}*\underset{\approx}{I}}} = \frac{\underset{\approx}{I}}{\sqrt{3}}. \tag{17.152}$$

Noting that $\underset{\approx}{S}*\underset{\approx}{I}$ is the same thing as tr $\underset{\approx}{S}$, equation (17.151) becomes the traditional expression for the isotropic part of a 3D second-order tensor:

$$\text{iso}\underset{\approx}{S} = \tfrac{1}{3}(\text{tr}\underset{\approx}{S})\underset{\approx}{I} \qquad \text{for 3D second-order tensors.} \tag{17.153}$$

We have emphasized that this result holds for second-order tensors referenced to *three-dimensional* physical space. To see why this is important, consider *two-dimensional* space. A 2D second-order tensor is expressible in terms of a 2×2 matrix and the tensor is isotropic if and only if

$$\begin{bmatrix} \cos\alpha & \sin\alpha \\ -\sin\alpha & \cos\alpha \end{bmatrix} \begin{bmatrix} A_{11} & A_{12} \\ A_{21} & A_{22} \end{bmatrix} \begin{bmatrix} \cos\alpha & -\sin\alpha \\ \sin\alpha & \cos\alpha \end{bmatrix} = \begin{bmatrix} A_{11} & A_{12} \\ A_{21} & A_{22} \end{bmatrix} \text{ for all } \alpha. \tag{17.154}$$

Analyzing this definition reveals two restrictions:

$$A_{11} = A_{22} \qquad \text{and} \qquad A_{12} = -A_{21}. \tag{17.155}$$

Therefore, any isotropic tensor in *two-dimensions* is expressible in the form

$$\begin{bmatrix} \gamma & \beta \\ -\beta & \gamma \end{bmatrix} \text{for some scalars } \gamma \text{ and } \beta. \tag{17.156}$$

In other words, any isotropic tensor in *two dimensions* can always be written as a linear combination of the 2D identity tensor $\underset{\approx}{\boldsymbol{I}}$ and the 2D alternating tensor $\underset{\approx}{\varepsilon}$ (defined such that ε_{ij} is zero if $i = j$, +1 if $ij = 12$, and −1 if $ij = 21$). Thus, the space of isotropic tensors in *two dimensions* is 2D. The matrix expressions for the two basis tensors are

$$[\underset{\approx}{\boldsymbol{I}}] = \begin{bmatrix} 1 & 0 \\ 0 & 1 \end{bmatrix} \qquad \text{and} \qquad [\underset{\approx}{\varepsilon}] = \begin{bmatrix} 0 & 1 \\ -1 & 0 \end{bmatrix}. \tag{17.157}$$

The orthonormalized basis is

$$\hat{\underset{\approx}{\boldsymbol{I}}} = \frac{\underset{\approx}{\boldsymbol{I}}}{\sqrt{\underset{\approx}{\boldsymbol{I}} * \underset{\approx}{\boldsymbol{I}}}} = \frac{\underset{\approx}{\boldsymbol{I}}}{\sqrt{2}} \qquad \text{and} \qquad \hat{\underset{\approx}{\varepsilon}} = \frac{\underset{\approx}{\varepsilon}}{\sqrt{\underset{\approx}{\varepsilon} * \underset{\approx}{\varepsilon}}} = \frac{\underset{\approx}{\varepsilon}}{\sqrt{2}}. \tag{17.158}$$

Here, the * operator is the 2D inner product, defined by $\underset{\approx}{\boldsymbol{A}} * \underset{\approx}{\boldsymbol{B}} = A_{ij}B_{ij}$ with repeated indices summed from 1 to 2.

The projection of a 2D second-order tensor $\underset{\approx}{\boldsymbol{S}}$ to its isotropic part is obtained by the projection operation

$$\text{iso}\underset{\approx}{\boldsymbol{S}} = (\underset{\approx}{\boldsymbol{S}} * \hat{\underset{\approx}{\boldsymbol{I}}})\hat{\underset{\approx}{\boldsymbol{I}}} + (\underset{\approx}{\boldsymbol{S}} * \hat{\underset{\approx}{\varepsilon}})\hat{\underset{\approx}{\varepsilon}} \text{ for 2D second-order tensors.} \tag{17.159}$$

Expanding this out in component form gives

$$[\text{iso}\underset{\approx}{\boldsymbol{S}}] = \begin{bmatrix} \frac{1}{2}(S_{11} + S_{22}) & \frac{1}{2}(S_{12} - S_{21}) \\ \frac{1}{2}(S_{21} - S_{12}) & \frac{1}{2}(S_{11} + S_{22}) \end{bmatrix} \text{for 2D second-order tensors.} \tag{17.160}$$

This formula is significantly different from the corresponding operation for 3D tensors! The polar rotation, incidentally, is found by simply normalizing the columns of this 2D isotropic part!

The techniques described so far for second-order tensors generalize similarly to higher-order tensors. For example, a 3D third-order tensor $\underset{\approx}{\boldsymbol{A}}$ is isotropic if and only if

$$R_{pi}R_{qj}R_{sk}A_{pqs} = A_{ijk} \qquad \text{for all rotations } \underset{\sim}{R}. \tag{17.161}$$

It turns out that this condition can be satisfied if and only if $\underset{\approx}{A}$ is proportional to the third-order alternating tensor. Therefore, the isotropic part of a third-order tensor is obtained by the projection operation

$$\operatorname{iso}\underset{\approx}{\zeta} = (\underset{\approx}{\zeta} * \underset{\approx}{\hat{\varepsilon}})\underset{\approx}{\hat{\varepsilon}}, \tag{17.162}$$

where * is now the third-order inner product ($\underset{\approx}{A} * \underset{\approx}{B} = A_{ijk}B_{ijk}$) and

$$\underset{\approx}{\hat{\varepsilon}} = \frac{\underset{\approx}{\varepsilon}}{\sqrt{\underset{\approx}{\varepsilon} * \underset{\approx}{\varepsilon}}} = \frac{\underset{\approx}{\varepsilon}}{\sqrt{6}}. \tag{17.163}$$

As before, the isotropic part of a third-order tensor is identical to the expected value of that tensor over the set of uniform superimposed rotations. Thus, the expected value of a 3D third-order tensor over a uniformly random rotation is given by

$$E[\underset{\approx}{\bar{\zeta}}] = \operatorname{iso}\underset{\approx}{\zeta} = \tfrac{1}{6}(\zeta_{mnp}\varepsilon_{mnp})\varepsilon_{ijk}\underset{\sim}{e}_i\underset{\sim}{e}_j\underset{\sim}{e}_k. \tag{17.164}$$

Continuing in this same vein, the expected value of a fourth-order tensor over uniform rotations is given by the isotropic part of the tensor:

$$E[\underset{\approx}{\bar{C}}] = \operatorname{iso}\underset{\approx}{\bar{C}}. \tag{17.165}$$

Again, to compute the isotropic part, we must project the tensor to the space of isotropic fourth-order tensors. It turns out that an isotropic 3D fourth-order tensor is always expressible as a linear combination of three isotropic basis tensors having *ijkl* components given by

$$\delta_{ij}\delta_{kl}, \qquad \delta_{ik}\delta_{jl}, \qquad \text{and} \qquad \delta_{il}\delta_{jk}. \tag{17.166}$$

These three tensors form a basis for isotropic 3D fourth-order tensors. However, the basis is not orthonormalized, as it must be if we wish to define the projection operation is the usual way. An alternative orthogonal set of base-tensors for the space of isotropic fourth-order tensors is

$$\begin{aligned} P^{\text{ISO}}_{ijkl} &= \tfrac{1}{3}\delta_{ij}\delta_{kl}, \\ P^{\text{SD}}_{ijkl} &= \tfrac{1}{2}(\delta_{ik}\delta_{jl} + \delta_{il}\delta_{jk}) - \tfrac{1}{3}\delta_{ij}\delta_{kl}, \\ P^{\text{SKM}}_{ijkl} &= \tfrac{1}{2}(\delta_{ik}\delta_{jl} - \delta_{il}\delta_{jk}). \end{aligned} \tag{17.167}$$

The tensors in this alternative basis are projectors themselves. When P^{ISO}_{ijkl} acts on a second-order tensor, the result is the isotropic part of that tensor. When P^{SD}_{ijkl} acts on a second-order tensor, the result is the symmetric-deviatoric part of that tensor. When P^{SKW}_{ijkl} acts on a tensor, the result is the skew-symmetric part of that tensor.

We note that

$$P^{\mathrm{ISO}}_{ijkl}P^{\mathrm{ISO}}_{ijkl} = 1, \qquad P^{\mathrm{SD}}_{ijkl}P^{\mathrm{SD}}_{ijkl} = 5, \qquad \text{and} \qquad P^{\mathrm{SKW}}_{ijkl}P^{\mathrm{SKW}}_{ijkl} = 3. \tag{17.168}$$

Therefore, the isotropic part of a fourth-order tensor C_{ijkl} is obtained by

$$(\mathrm{iso}\,\underset{\approx}{C})_{ijkl} = \left(C_{pqrs}P^{\mathrm{ISO}}_{pqrs}\right)P^{\mathrm{ISO}}_{ijkl} + \frac{\left(C_{pqrs}P^{\mathrm{SD}}_{pqrs}\right)P^{\mathrm{SD}}_{ijkl}}{5} + \frac{\left(C_{pqrs}P^{\mathrm{SKW}}_{pqrs}\right)P^{\mathrm{SKW}}_{ijkl}}{3}. \tag{17.169}$$

An algorithm for quickly evaluating this tensor and for computing a scalar measure of the degree of anisotropy is derived in section 20.5.

References

[1] Schindler W 1997 On the efficient simulation of a particular class of random rotations relevant to computer graphics *J. Comput. Appl. Math.* **81** 107–14

[2] Du Q, Faber V and Gunzburger M 1999 Centroidal Voronoi tesselations: applications and algorithms *SIAM Rev.* **41** 637–77

[3] Burkardt J, Gunzburger M, Peterson J and Brannon R M 2002 User manual and supporting information for library of codes for centroidal Voronoi point placement and associated zeroth, first, and second moment determination. *Sandia National Laboratories Technical Report* SAND2002-0099.

[4] Shoemake K 1992 Uniform random rotations *Graphics Gems* vol 3 ed D Kirk (London: Academic) 124–32 vol 3

IOP Publishing

Rotation, Reflection, and Frame Changes
Orthogonal tensors in computational engineering mechanics
R M Brannon

Chapter 18

Introduction to material and tensor symmetries

Anisotropy is the macroscopic (homogenized) manifestation of organized or patterned microscale heterogeneity, called *texture*. Examples might include a tendency for microcracks to be aligned, layering in rock deposits, etc. Texture can be modeled directly in heterogeneous mesoscale simulations (e.g., by explicitly modeling each fiber in a composite or each layer in rock), but it is accommodated through anisotropy in homogenized macroscale models, which necessitates quantifying material orientations not aligned with laboratory basis directions[1].

18.1 Anisotropy classification via group theory

This chapter introduces the concept of material symmetry by assessing changes in the constitutive response (e.g., the stress prediction) caused by applying a member of a specified 'group' of transformations to the material *prior* to applying an arbitrary *fiducial*[2] deformation[3]. The material is said to have symmetry *with respect to the given transformation group* if its constitutive response is unaffected by applying any member of the group prior to application of the fiducial deformation.

[1] Another (often unwisely neglected) effect of sub-continuum heterogeneity is *aleatory uncertainty*, which is the macroscale influence of natural imperfections, such as irregularity in the texture pattern or other microscale fluctuations that make material property measurements statistically variable from sample to sample and which therefore strongly influence bifurcations (see [1]).

[2] The word 'fiducial' (which may be a noun or adjective) means 'something compared against'. In this case, we compare the fiducial material response (e.g., shearing or pulling horizontally) to the response observed under the same fiducial deformation except preceded with the extra transformation (such as rotating before shearing or pulling).

[3] A set of transformations (such as specific types of deformations or rotations) is a *group* if for any $\underset{\approx}{A}$, $\underset{\approx}{B}$, and $\underset{\approx}{C}$ in the set the following hold: (1) closed under multiplication (i.e., $\underset{\approx}{A} \bullet \underset{\approx}{B}$ is also in the set); (2) associative $(\underset{\approx}{A} \bullet \underset{\approx}{B}) \bullet \underset{\approx}{C} = \underset{\approx}{A} \bullet (\underset{\approx}{B} \bullet \underset{\approx}{C})$; (3) a transformation $\underset{\approx}{I}$ (the identity) exists so that $\underset{\approx}{A} \bullet \underset{\approx}{I} = \underset{\approx}{I} \bullet \underset{\approx}{A} = \underset{\approx}{A}$; (4) each $\underset{\approx}{A}$ has an inverse $\underset{\approx}{A}^{-1}$such that $\underset{\approx}{A} \bullet \underset{\approx}{A}^{-1} = \underset{\approx}{A}^{-1} \bullet \underset{\approx}{A} = \underset{\approx}{I}$. The set is called *Abelian* or *commutative* if, for any members of the set, $\underset{\approx}{A} \bullet \underset{\approx}{B} = \underset{\approx}{B} \bullet \underset{\approx}{A}$ (this is not typical). The set of all rotations, for example, is a non-Abelian group. Rotations about a fixed axis form an Abelian group.

doi:10.1016/978-0-7503-1454-1ch18

18.1.1 Elementary symmetry groups

A *properly isotropic material* (also known as *hemitropic* [2]), is invariant under the group of proper orthotropic tensors, which means that the constitutive response is unaffected by arbitrarily rotating the material prior to application of an arbitrary fiducial deformation[4]. A *strictly isotropic material* extends the symmetry group to include improper-orthogonal tensors (e.g., inversions[5]).

A fiber-reinforced composite having all of the fibers laid out horizontally will require a large applied stress to deform the material horizontally (in the fiber direction). However, if the composite is first rotated by 90° to reorient the fibers vertically, then the same horizontal deformation will no longer require such a large stress, because that deformation is now loading the material in the more compliant direction (transverse to the fibers). In general, if material response to a deformation noticeably changes when the material is rotated *prior to* applying the deformation, then the material is said to be *intrinsically* or *inherently anisotropic*. Even isotropic materials become anisotropic in their loaded states, which is a condition called *deformation-induced anisotropy* caused by texturing (such as the flattening of spheres to ellipses in figure 18.1) under load; the *material* in that figure is still isotropic, because the induced anisotropy of its *stiffness* would not change if the material is first rotated before loading—equivalently, an isotropic material's response (and its induced anisotropy) is the same relative to the loading directions no matter which way those loading directions point. The key here is that the symmetry group (in this case the set of all proper-orthogonal tensors) is applied *before* loading to test if the resulting stress is unchanged. There is no implication of insensitivity of subsequent changes in stress under the same symmetry group *after* loading. In fact, for the porous isotropic material in figure 18.1, it should be clear that axial loading of an isotropic material induces a loaded state that is transversely isotropic (invariant under rotations about the loading axis). When limited data (such as axial stress–strain response) are fitted to a model with an isotropic stiffness, induced anisotropy might make it seem that a highly compressed material has a pressure-dependent shear modulus, but that is a spurious consequence of interpreting the data in the context of an inappropriate material model. The effect is more properly interpreted as an induced isotropy—to merely adjust an isotropic stiffness to have a pressure-sensitive shear modulus would violate thermodynamic admissibility [3].

A *transversely isotropic material* has a constitutive response that is unchanged if the material is first rotated about a given axis before applying a fiducial deformation; examples include a deck of cards or fiber-reinforced composite having all fibers

[4] Frame indifference constraints (described later) require the constitutive model to predict specific *(mostly intuitive)* changes in the material response (relative to a fiducial response) when an arbitrary rotation is superimposed *after* the fiducial deformation. Even isotropic material responses will change when rotation is applied *after* a fiducial deformation.

[5] An *inversion* in 3D is either the negative of the identity tensor, $-\underset{\sim}{I}$ (causing any vector $\underset{\sim}{v}$ to transform to $-\underset{\sim}{v}$), or a reflection, $\underset{\sim}{I} - 2\underset{\sim}{n}\underset{\sim}{n}$ (causing a vector to be transformed to its mirror image across a plane with unit normal $\underset{\sim}{n}$). Improper-orthogonal transformations are orthogonal tensors having a negative determinant, so they might be inversions in combination with rotations.

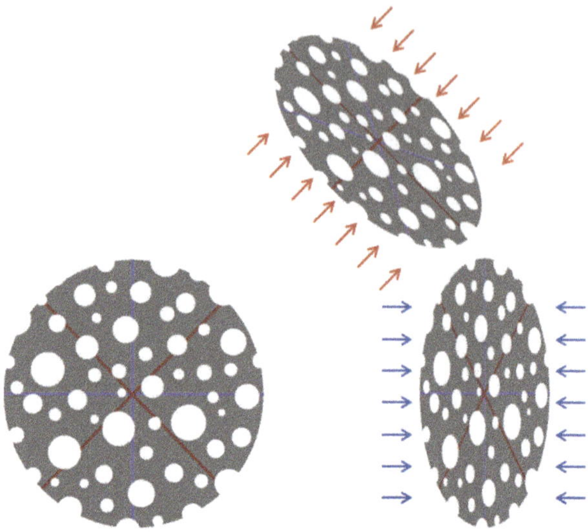

Figure 18.1. An approximately isotropic porous sponge acquiring anisotropy under load. In the loaded state, the initially spherical pores become ellipsoids, which *induces* anisotropy, but this degree of anisotropy would not be changed if the sponge had been first rotated prior to deformation. Thus, it is an isotropically induced anisotropy!

pointing in a common direction. If a material has a set of three mutually orthogonal directions (called *material* or *preferred directions*) for which the constitutive response is insensitive to 180° rotations about these directions (applied before any other deformation), then the material is said to be *orthotropic*[6]. If the constitutive response is insensitive to any 90° rotations about a mutually orthogonal set of directions, then it is said to be *cubic*. Implicit in these definitions of various types of material symmetry is an unstated assertion that the response generally *will be* affected by any pre-applied transformation not belonging to the material's symmetry group. Also, as evident in these examples, anisotropy typically is described in the context of a special frame of orientation, called the *material frame*. Even if material directions are naturally aligned with a set of skewed axes (as for hexahedral crystal morphology), you can still refer to a mutually orthonormal and right-handed material basis so that the skewed directions are defined relative to that material frame. Later in this chapter, methods of measuring and reporting material orientation distributions are discussed in the context of cubic symmetry, but it should be clear that the same ideas would apply to any form of material symmetry.

An isotropic material is analogous to a sphere, whose appearance is unaffected by any rotation. A transversely isotropic material is analogous to a cylinder, which is

[6] Some researchers allow the orthotropy group to include reflections across the material planes, which requires the constitutive response to be unchanged if a material's microstructure is replaced with an exact mirror image of the original. We choose not to include reflections; we use the word 'polar' to indicate that response is altered by reflections (and non-polar is otherwise assumed).

unchanged (in appearance) if it is rotated about its symmetry axis. An orthotropic medium is analogous to a shoebox shape, whose appearance insensitive to 180° rotations about the box axes. A cubic material is analogous to, well, a cube. Material symmetries can be attributed, in fact, to geometrical symmetries like these at the sub-continuum scale—the micromorphology of the atoms (i.e., their geometrical arrangement) and/or the distribution of crystals in a polycrystal exhibit these geometrical shape properties.

18.1.2 Other symmetries

The above examples of different types of material symmetry refer to symmetry groups that always consist of orthogonal tensors, but this is not required. For example, a *fluid* might be defined as having a constitutive response that is unaltered if the material is first subjected to any quasistatically applied *isochoric* (*unimodular*) transformation (i.e., any deformation having shape change without volume change). The degree to which a deformation is 'quasi-static' is loosely quantified by the so-called *Deborah number*, which is a ratio of the characteristic time over which a deformation is applied (hours, days, microseconds, etc) divided by the a characteristic material response time (which, for example, could be the time required for a shear stress relax to half of its initial value induced by a suddenly applied deformation that is then held fixed).

18.1.3 The distinction between material and tensor symmetry

Aside from *material* symmetries (which refer to insensitivity of a constitutive response to certain transformations applied before application of fiducial deformation), there are also *tensor* symmetries that go by the same names, but which refer to insensitivity of a tensor's components to basis transforms under the same transformation groups. For example, a *properly isotropic tensor* is one whose components do not change under any proper-orthogonal transformation (i.e., pure rotation) of the basis vectors. A *strictly isotropic tensor* additionally has components unaffected by improper-orthogonal basis changes (e.g., from a right-handed to a left-handed system). As explained in chapter 20, an isotropic tensor is *not* necessarily a simple multiple of the identity tensor; that assertion is true for second-order tensors in a three-dimensional (3D) space, but it does not generally apply to different tensorial orders or to differently dimensioned spaces. A *transversely isotropic tensor* has components that do not change under rotation about a given axis. An orthotropic tensor has components that stay the same for 180° rotations about a fixed orthonormal set of material directions. Again, an implicit part of these definitions is that the tensor components *would* generally change for transformations that are not in the symmetry group. For example, a transversely isotropic tensor's components would change for a rotation axis that is not aligned with the symmetry axis.

In its *undeformed* state, a material of a given symmetry will generally have constitutive tensors (such as elastic stiffness) of the same symmetries (e.g., a cubic material has a cubic stiffness in the undeformed state). However, most materials will *acquire* evolving deformation-induced anisotropy as a result of directionally dependent changes in the material microstructure under load, and this makes the constitutive tensors (such as the *tangent* elastic stiffness) potentially change their symmetry properties with deformation while possibly not having any effect on the material symmetry. For example, the functionally isotropic poroelastic material in figure 18.1 acquires orthotropy (aligned with the principal axes of the ellipse) under deformation. This assertion is later proved rigorously in the case of isotropic hyperelasticity for which the tangent stiffness of an isotropic tensor is not itself isotropic. Even though an isotropic material might become anisotropic under load, it should still be classified as an isotropic material because the induced anisotropy is the same if the material is first rotated arbitrarily before application of loads.

While material symmetry is analyzed by considering the effect of a superimposed rotation *before* application of loads, the topic of the next chapter, frame indifference, broadens this analysis to derive constraints that must be satisfied by any and all governing equations (including constitutive models) based on how their predictions ought to change in response to superimposed rotation that is applied during or at the end of a deformation. Simple requirements of basis-change transformations imply that governing equations must be well-formed tensor equations, but frame-change analysis takes this much further by deriving additional constraints that must be satisfied for a change in frame that might evolve through time (not fixed as it is in basis-change analysis).

18.2 Quantifying and visualizing orientations

Cubic symmetry arises from cubic geometry of a material's atomic lattice. Such symmetry is usually also preserved if the atoms are swapped under a reflection transformation. Accordingly, when a cubic crystal's orientation is characterized by a rotation tensor $\underset{\approx}{\boldsymbol{Q}}$, a different crystal will be 'equivalent' (i.e., have identical constitutive material response) if its orientation is $\underset{\approx}{\boldsymbol{R}} \bullet \underset{\approx}{\boldsymbol{Q}} \bullet \underset{\approx}{\boldsymbol{R}}^T$ in which $\underset{\approx}{\boldsymbol{R}}$ is a 90° rotation or a reflection across the one of the original crystal's axes. An axis direction for a cubic crystal is just the unit normal to one of its faces. Hence, the lab components of the crystal's six axis directions are merely the components of the three columns of its orientation matrix $[\underset{\approx}{\boldsymbol{Q}}]$. For every instance of a face with normal (n_1, n_2, n_3), there will be a face with normal $(-n_1, -n_2, -n_3)$, $(-n_1, n_2, n_3)$, etc. Hence, data are needed only for the distribution of normals with *positive* components, reducing the population down to points on the first octant of the unit sphere. For cubic symmetry, invariance under 90° rotations and reflections further implies that the components of the unit normal can be permuted. Accordingly, when examining statistical distributions of cubic face normals (as is done in *electron backscatter diffraction* (EBSD)), one only needs to track the face normals with canonically ordered components:

$$n_1 \leqslant n_2 \leqslant n_3. \tag{18.1}$$

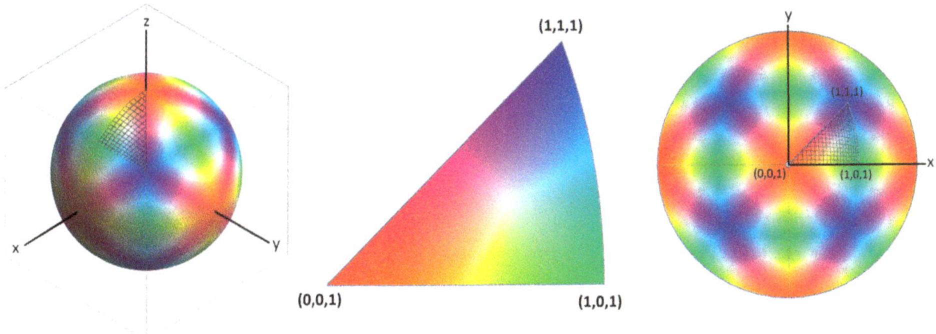

Figure 18.2. Inverse pole color legends. The zones on the sphere (left image) or its Lambert projection (right image) have coloring associated with cubic symmetry. EBSD or similar experimental procedures might measure the orientations of a population of crystals, each quantified by a rotation tensor $\underset{\approx}{Q}$. The *k*th column of each $[Q]$ matrix defines the *k*th cube face normal, which is a unit vector and can therefore be drawn as a two black dots on the sphere (on polar opposite sides). Drawing dots for each column of $[Q]$ and for each crystal orientation would generate a *pole plot*. Rather than rotate the crystal, one could alternatively imagine the rotation of the laboratory basis vectors relative to the cube, defined by the transpose (i.e., inverse) of $\underset{\approx}{Q}$. Plotting corresponding dots would generate an *inverse pole plot*. These dots would have cubic symmetry in their distributions, so the entire population of dots could be depicted visually on just the 'triangular' sector shown in the middle of this figure.

This canonically ordered zone is convenient in computer codes because any unit normal $\underset{\sim}{n}^*$ that does not satisfy this component-ordering property can be converted to an equivalent canonical normal $\underset{\sim}{n}$ by the operation[7]

$$\{n_1, n_2, n_3\} = \text{sort}\{|n_1^*|, |n_2^*|, |n_3^*|\}. \tag{18.2}$$

Once this canonical sort order is obtained, all other equivalent zones on the sphere can be reached by permutation and/or sign changes of these values. Of these, the most common choice is a swapping of the first two components to give

$$\{n_2, n_1, n_3\}. \tag{18.3}$$

Recalling figure 17.15, the surface of a sphere can be broken into zones associated with projection of a cube's faces onto the sphere. Of these zones, the one satisfying (18.1) but also with the first two components swapped (as in equation (18.3)) is shown in figure 18.2 as a 'triangular patch', called the *fundamental sector* [4].

Any given cube orientation has a set of six face normals, all of which can be drawn as dots on the sphere or on the Lambert projection of the sphere, as is done on the left sides of figures 2.1–2.3. Such figures are called a *pole plots*. Rather than depicting the rotation of a body relative to the laboratory basis, one might imagine the perspective of an observer who co-rotates with the body, and therefore sees the world rotating in the opposite (inverse) direction (i.e., from the perspective of a

[7] Here, 'sort' arranges values from smallest to largest. In Mathematica, the function can be defined `toCubicZone[n_]:=Sort[Abs@n,Less]`, where 'Less' forces Mathematica to use a proper numerical sort even if the components are exact square roots.

Figure 18.3. EBSD stitched scans of aluminum oxynitride colored by the average grain orientation from the [001] inverse pole figure. Since an un-rotated cube has all three laboratory axes piercing the middle of the cube faces (where that piercing location on the sphere is colored red), it follows that the red zones in this plot would have a crystal orientation closely aligned with the laboratory axes. Data from [5], visualized using [6].

material-frame observer, the laboratory basis vectors appear to rotate in the opposite direction). If this body-aligned material observer were to draw dots corresponding to the inversely rotated laboratory basis vectors, the plot would be similar except that it would depict the inverse (i.e., the transpose) of the original rotation. Such an image is called an *inverse pole plot*.

If a crystal grain is rotated relative to the laboratory axis according to a rotation tensor $\underset{\approx}{\boldsymbol{Q}}$, it follows that an observer in the material frame (i.e., one who is always aligned with the crystal would perceive the laboratory axes to have rotated in the opposite direction according to $\underset{\approx}{\boldsymbol{Q}}^T$). For each distinct crystal orientation, an observer in the material frame would see a distinct laboratory axes orientation. Each laboratory axis is a unit vector, whose components can be transformed to canonical ordering according to (18.2), optionally then permuted by (18.3). For each laboratory axis, this process will generate three 'dots' in the fundamental sector, which we name **x**, **y**, and **z**. If the cubic crystal is aligned with the laboratory axes, then all three laboratory axes, (1, 0, 0), (0, 1, 0), and (0, 0, 1), will map to the same canonically ordered point (0,0,1) in the inverse pole figure. If the cube is rotated counter-clockwise about the x-axis, then the laboratory axes would rotate clockwise relative to the cube. Looking at the colored sphere in figure 18.2, this motion would cause the y- and z-axes (and hence the **y**- and **z**-points in the inverse pole plot) to move away from the red zone towards the green zone while keeping the z-axis (and hence the **z**-point in the inverse pole figure) stationary. Similarly, a pure rotation about the y-axis would move the **z**- and **x**-points in the inverse pole plot from red towards green. The only way to move one of the **x**-, **y**-, or **z**-points away from the red–green bottom of the pole 'triangle' (fundamental sector) would be to use a rotation axis that is not aligned with the laboratory axes. Only out-of-plane rotation would move one of the **x**-, **y**-, or **z**-dots into or towards the blue zone of the fundamental sector. For example, by imagining how the axes on the colored sphere in figure 18.2 would move under a rotation about the axis parallel to $\{-1, 1, 0\}$, the

z-axis (and hence the **z**-point) would move from its initial location in the red zone directly towards the blue zone, while the **x**- and **y**-points would move along an arcing path through the green and then white zones.

A single-crystal orientation corresponds to exactly three (**x**, **y**, **z**) dots on the inverse pole plot, one for each coordinate axis. A population of crystal orientations then gives a population of (**x**, **y**, **z**) dots, as shown in figure 18.3 for a measured set of **x**-dots. Strong clustering of each type of dot (**x**, **y**, or **z**) would indicate that the crystals are tending to prefer a single orientation (corresponding to strong texture in the polycrystal composite). If the dots are clustered in a line, then the polycrystal would be tending towards transverse isotropy. Recalling that the inverse pole plot is a projection of the sphere, it would be crucial to use an equal-area (Lambert) mapping to avoid spurious clustering that is merely an artifact of the mapping. Unlike a stereographic projection, the Lambert mapping ensures that a uniformly random set of crystal orientations would produce a uniformly random set of dots in the inverse pole plot, thus indicating isotropy.

18.2.1 Coloring orientation plots

Color provides a legend for other plots, such as the one in figure 18.3. As long as there is a unique correspondence between color and orientation, the color scheme is a matter of aesthetic taste. This section discusses two color mapping problems:

- Mapping *two* parameters defining a unit vector $\underset{\sim}{x}$ to three RGB color weights.
- Mapping *three* parameters defining a rotation tensor $\underset{\approx}{Q}$ to three RGB color weights.

The inverse pole plot in figure 18.3, for example, adopts an aesthetically pleasing broadening of the yellow, magenta, and cyan regions in comparison to the more narrow bands of those ranges in our figure 18.2

The inverse pole plot in figure 18.2 applies a simple RGB scheme that associates the following colors with the corners of the 'triangular' fundamental sector:

$$\text{RED:} \qquad f[\{0, 0, 1\}] = \{0, 0, 1\}, \tag{18.4a}$$

$$\text{GREEN:} \qquad f[\{1, 0, 1\}] = \frac{\{0, 1, 1\}}{\sqrt{2}}, \tag{18.4b}$$

$$\text{BLUE:} \qquad f[\{1, 1, 1\}] = \frac{\{1, 1, 1\}}{\sqrt{3}}, \tag{18.4c}$$

where the function f first transforms the argument to a point on the sphere and then applies (18.2) to produce canonical sorting of the components from small to large. Let $\underset{\sim}{x}$ denote a point on the unit sphere that has also been moved to the fundamental sector via this function. If the above three vectors are denoted $\{\underset{\sim}{g}_1, \underset{\sim}{g}_2, \underset{\sim}{g}_3\}$, then we can always find coefficients[8] (X_1, X_2, X_3) such that

[8] For readers familiar with analytical geometry, these are contravariant components of the vector.

$$\underset{\sim}{x} = X_1\underset{\sim}{g}_1 + X_2\underset{\sim}{g}_2 + X_3\underset{\sim}{g}_3. \tag{18.5}$$

Expanding this in component form produces a linear system solvable for the X_k coefficients. Namely,

$$\begin{bmatrix} X_1 \\ X_2 \\ X_3 \end{bmatrix} = \begin{bmatrix} 0 & -1 & 1 \\ -\sqrt{2} & \sqrt{2} & 0 \\ \sqrt{3} & 0 & 0 \end{bmatrix} \begin{bmatrix} x_1 \\ x_2 \\ x_3 \end{bmatrix}. \tag{18.6}$$

Because $\underset{\sim}{x}$ falls in the fundamental sector, these X_k values all fall in the range from 0 to 1 and can therefore be used as RGB weights. When $\underset{\sim}{x}$ coincides with the red, green, or blue directions defined above, this mapping produces $\underset{\sim}{X}$ vectors {1, 0, 0}, {0, 1, 0}, and {0, 0, 1}, respectively, so this transformation reduces the problem of assigning color on a distorted curvilinear triangle in $\underset{\sim}{x}$-space to assigning color on a symmetric triangular domain $\underset{\sim}{X}$, as is done in figure 18.4. That figure includes an option to 'push' the center (white) point away from the RGB corners. Specifically, letting $\{p_1, p_2, p_3\}$ denote 'push' weights, the final RGB inputs are

$$\frac{\{p_1 X_1, p_2 X_2, p_3 X_3\}}{\max[p_1 X_1, p_2 X_2, p_3 X_3]}. \tag{18.7}$$

The divisor ensures full color saturation by making the largest component of the final array unity. In Mathematica, this color assignment is programmed as

```
maykColor[x_, rgbPUSH_] := RGBColor@@Normalize[rgbPUSH
{{0,-1,1}, {-Sqrt[2],Sqrt[2],0}, {Sqrt[3],0,0}}.x), Max]
```

Broadly speaking, equation (18.6) provides a good starting point to assign relative weights of the three corners of the triangle, which (if desired) can be then adjusted to 'stretch' the intermediate zones of yellow, magenta, and cyan, move the relative location of the center point using push factors, etc. In Mathematica, the color scheme for RGB weights can be applied to a sphere plot or inverse pole plot using the 'ColorFunction' command. The sphere plot, for example, is a parametric description of 3D position vectors, defined by

$$\underset{\sim}{x} = \cos\varphi \sin\vartheta \underset{\sim}{e}_1 + \sin\varphi \sin\vartheta \underset{\sim}{e}_2 + \cos\vartheta \underset{\sim}{e}_3. \tag{18.8}$$

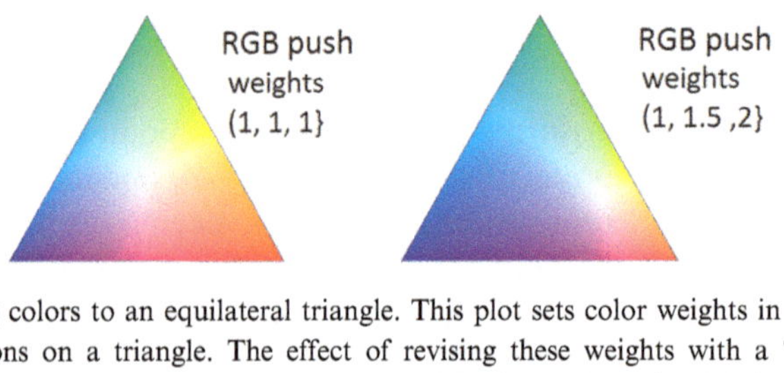

Figure 18.4. Mapping colors to an equilateral triangle. This plot sets color weights in (18.7) to equal finite-element shape functions on a triangle. The effect of revising these weights with a 'push' specification is illustrated. The largest third push value expands the zone of the third RGB color (blue), thus pushing the white point away from that corner.

Mathematica's `'ColorFunction'` command is an option in its 3D parametric plots to support coloring by the parameter values, which would be the spherical coordinate angles in this case. To set the color, $\underset{\sim}{x}(\vartheta, \varphi)$, would need to be sent as the first argument to the previously defined `maykColor` color command, which would convert it to an RGB directive.

For the inverse pole plot, the fundamental sector on the sphere can be defined parametrically by

$$\underset{\sim}{x} = \frac{u\underset{\sim}{e}_1 + v\underset{\sim}{e}_2 + \underset{\sim}{e}_3}{\sqrt{u^2 + v^2 + 1}}. \tag{18.9}$$

To ensure that this point lies in the fundamental sector where $x_1 \leqslant x_2 \leqslant x_3$, the first parameter u varies from 0 to 1, while the second parameter v varies from u to 1. A location in the inverse pole plot would be assigned by applying an equal-area Lambert projection to convert the 3D vector $\underset{\sim}{x}(u, v)$ to a 2D vector in the plane. The color for that location would be found by applying the `maykColor` command using the 3D vector $\underset{\sim}{x} = \underset{\sim}{x}(u, v)$ as the first argument.

Rotation color mapping. The color schemes discussed above have converted any choice of two parameters, such as (ϑ, φ) or (u, v), to a 3D vector $\underset{\sim}{x}$ on the unit sphere, which was then converted to RGB values. These schemes could, of course, be applied to assign a color to each column of a rotation tensor's component matrix, which is what is conventionally done by providing three separate inverse pole plots, one for each rotated location of each laboratory axis relative to the material frame. However, if an entire rotation tensor is known, a far better strategy would be to realize that rotation tensors are constructed from three input parameters in the first place, so it would make sense to simply map those three input parameters to three RGB weights. The scheme used to do this, however, must be carefully selected to avoid misleading, discontinuous, or unevenly weighted coloring. As pointed out in [4], for example, coloring by setting relative RGB weights according to Euler angles can produce confusing discontinuous jumps in color for nearly equal orientations, and coloring by Euler angles furthermore has a bias inherited from the fact that a uniform distribution of rotations does not correspond to a uniform distribution of Euler angles. This bias can be eliminated by inverting formulas for uniformly random rotations, discussed in section 17.6. Those algorithms take a set of input parameters (r_1, r_2, r_3), each ranging from 0 to 1, and map them uniformly to rotation tensors. One such method maps the first two parameters to create a uniformly distributed unit axis of rotation, according to

$$\underset{\sim}{a} = \sqrt{1 - r_1^2}\,[\cos(2\pi r_2)\underset{\sim}{e}_1 + \sin(2\pi r_2)\underset{\sim}{e}_2] + r_1\underset{\sim}{e}_3. \tag{18.10}$$

Recalling (17.134) non-uniformly distributed angle of rotation is found by solving

$$\alpha - \sin\alpha = \pi r_3.$$

Therefore, if a given rotation tensor's axis $\underset{\sim}{a}$ and angle α are first found using the methods in sections 5.4.1 and 5.4.2,[9] then corresponding (r_1, r_2, r_3) values can be set according to

$$r_1 = a_3, \tag{18.11a}$$

$$r_2 = \frac{1}{2\pi}\text{ArcTan}[a_1, a_2], \tag{18.11b}$$

$$r_3 = \frac{\alpha - \sin\alpha}{\pi}. \tag{18.11c}$$

A collection of measured orientations could then be visualized using the 'nautical' representation in figure 17.17 with each arrow assigned a unique color by using RGB$[r_1, r_2, r_3]$. If the collection of rotations is uniformly distributed, then the collection of mapped (r_1, r_2, r_3) dots will be uniformly distributed in a unit cube. This scheme does not eliminate the possibility of discontinuity in the mapping from rotation to r-values, which is resolved in [4], though it is not clear that those methods preserve uniformity.

References

[1] Strack O E, Leavy R B and Brannon R M 2015 Aleatory uncertainty and scale effects in computational damage models for failure and fragmentation *Int. J. Numer. Meth. Eng.* **102** 468–95

[2] Spencer A J M 2004 *Continuum Mechanics* (New York: Dover)

[3] Fuller T J and Brannon R M 2012 On the effects of deformation-induced anisotropy on classical elastic-plastic materials *Int. J. Numer. Anal. Methods Geomech.* **37** 1079–94

[4] Nolze Gert and Hielscher R 2016 Orientations—perfectly colored *J. Appl. Crystallogr.* **49** 1786–802

[5] Sano T 2009 AlON EBSD reconstruction, unpublished research.

[6] Wright S I and Nowell M 2009 A review of *in situ* EBSD studies *Electron Backscatter Diffraction in Materials Science* (Boston, MA: Springer)

[9] If the third axis component is negative ($a_3 < 0$), then both the axis and angle should be multiplied by -1, which will produce the same rotation. The rotation angle should be shifted appropriately to always fall in the range from 0 to 2π.

Chapter 19

Frame indifference

Material symmetries are identified by examining the effect of a transformation applied *before* application of a fiducial deformation. A 'fiducial' motion is any deformation that serves as a starting point for investigating how material response is expected to change under deformations that are different from the fiducial one in some specific way. Previous discussions of well-formed tensor equations merely asserted how material response must change if a rotation is applied before the fiducial deformation, leading only to the requirement that governing equations must be well formed tensor equations. This chapter expands focus to include transformations that are superimposed any time during or after the fiducial one (e.g., superimposing extra rotation during a shearing motion, like performing a shear experiment while on a turntable). We now discuss the broader concept of material-frame indifference, which mandates certain constraints on all governing equations (including but not limited to constitutive models) to ensure that they respond appropriately to a superimposed rotation (or translation) applied *after* (or during) the fiducial deformation[1]. If, for example, you horizontally pull a rubber band (defining a fiducial deformation) and then rotate it (defining a new deformation that is the same as the fiducial case other than this superimposed rotation), then Cauchy stress in the new problem should have the same magnitude as in the fiducial problem, but with directions that have also rotated with the material by *exactly* the same amount as the superimposed rotation. Such a response is called *co-rotational.* At first blush, it might seem that all vectors and all tensors would be co-rotational, but they are not (and that makes frame-change analysis differ from simple basis-change analysis). The manner in which something changes under superimposed rotation and/or translation depends on how

[1] As will be soon explained, superimposed rotation and/or translation is an alternative viewpoint of what must be understood to actually be a frame change, where two differently oriented and differently located observers are looking at the *same* event. If one observer is moving left relative to a stationary observer, then the left-moving observer will see the same thing as the first observer except that all material is perceived to be subjected to superimposed translation to the *right*. Similarly, an observer who rotates clockwise will see the same thing as the other observer except with superimposed rotation in the opposite (counter-clockwise) direction.

doi:10.1016/978-0-7503-1454-1ch19

it is defined. If, for example, a vector is defined to be velocity at time zero, then it would be naturally unaffected by any superimposed rotation that happens later. The velocity at the current time, on the other hand would be whatever it was in the fiducial problem plus extra velocity coming from superimposed motion. This chapter is organized to systematically step through the following topics, each producing a surprisingly tangled web of mathematics for what really is a simple principle:

- A motivational introduction shows that even intuitive expectations (such as what a rubber-band pulling model should predict in response to translating or rotating the band) can be infuriatingly difficult to achieve in a dumb computer implementation that does only what it is told.
- A frame change (i.e., translation and/or rotation of an observer relative to another 'fiducial' observer) is explained to be equivalent to superimposing a rigid motion *in the opposite direction* to whatever event is observed by the fiducial observer. This perspective is useful for running frame-change simulations.
- Kinematics of frame change is worked out in excruciating detail to prove mostly intuitive results (such as the fact that a left-moving observer will perceive the world to be moving to the right in comparison to what is seen by a non-moving observer).
- A broad statement of the principle of material-frame indifference (PMFI) is provided so that (in chapter 24) the concept can be used to determine 'phantom' body forces that are felt by non-inertial observers, such as a backward 'gravity' felt by an observer riding in a forward accelerating automobile or the Coriolis effect for rotating observers.
- Understanding that superimposed rigid motion is actually a frame change helps to address the most common confusion that students have about frame indifference: *why does a superimposed rigid rotation not induce centrifugal forces?* The answer is that it would, but such forces would elongate the material and therefore violate the assertion that the superimposed motion is rigid. Hence, implicit in an assumption of superimposed rigid motion is that a properly implemented actual simulation of such motion is required to include body forces that exactly counter-balance the phantom body forces derived for non-inertial kinematics! Such simulations are best regarded as consistency checks of the Method of Manufactured Solutions (c.f. [2]) that (in our opinion) should be part of any well-done regression testing suite for any computational mechanics code.
- Tensors commonly used in constitutive modeling are classified based on how they are affected by superimposed motion.
- Several examples of the PMFI are provided, with emphasis on the meaning of 'co-rotational rates' and how to address their computational burdens (including avoidable numerical error) by re-casting them as Lie derivatives.

Details about frame indifference are tedious, allowing one to get bogged down in the mathematics, so it is important to 'step back' and remember that we are addressing the very intuitive topic of how a given event should appear to unfold from the perspectives of two observers who have different positions, orientations, velocities, or even accelerations. It is the potential for accelerations (i.e., non-inertial frame change) that causes the greatest mathematical burden, though the results are usually intuitive. If, for example, one observer is watching a ball drop from a high-altitude balloon while the second observer is a skydiver falling alongside the ball, then the stationary ground-based observer will perceive both the ball and skydiver to be simultaneously accelerating downward towards a stationary earth. The skydiving observer, on the other hand, will perceive the ball to be stationary relative to his frame (while the earth rushes upward). The two observers perceive forces differently. For example, neglecting air drag, the falling observer feels weightless and thus perceives gravity to be zero. If drag is included, the falling observer perceives gravity to be smaller than that for a stationary observer; upon reaching terminal velocity, however, the two observers perceive gravity to be the same while velocities continue to be seen differently. Frame indifference requires the same governing equations to be used by all observers. The ground-based one will perceive normal downward-pointing gravity and thus apply Newton's second law to predict that the ball falls with the acceleration of gravity. The drag-free skydiver, perceiving no gravity, applies Newton's second law to predict zero acceleration, and *this is correct* because the ball is falling right alongside the skydiver and thus has no acceleration in that observer's frame. Thus, Newton's law does *not* require an inertial frame to predict valid results—it does, however, require all equation inputs to be whatever they seem to be to the observer (and these are not related to inputs of a different observer by mere basis-change transformations).

Frame indifference is an admissibility condition to be applied to all governing equations. It requires that all observers (whether stationary or non-inertial) should be able to use *exactly* the same governing equations even though they will provide different inputs and thus generate different outputs. Frame indifference requires that these predictions be consistent with changes intuitively expected in a frame change. In our previous example, the stationary observer should be able to use his own observation of ball motion and his own observation of the movement of the skydiver to predict what is perceived, measured, and computed by the skydiver. A change in observer frame must produce a very particular change in perception of variables. When applied to material constitutive models, this concept is called the *principle of material-frame indifference*. As long as each observer uses input data consistent with how it appears to be from his own frame, then his predictions using frame-indifferent governing equations should provide outputs that are *consistent with* similar outputs computed by another person in a different frame using the *same* governing equations. This chapter builds systematically towards a testable definition of 'consistency' under frame change, and then applies it to infer admissibility constraints on constitutive models. It will be proved, for example, that the only way that Cauchy stress can be expressed as a pure function of the deformation gradient would be if the material is isotropic—anisotropic media would require additional inputs, such as directions of material fibers.

At a fundamental level, think of a constitutive model as being a cell-phone app. PMFI requires that all observers should be able to use that app *with inputs as they seem to be from their perspective*. If inputs of two observers are related as they should be under a frame change, then the cell-phone app's predicted stress ought to be consistently different as well. Suppose, for example, that the first observer has a 'rubber-band app' that predicts a certain horizontal force is required to stretch a rubber band horizontally. Now suppose that a second observer is rotated *clockwise* relative to the first (fiducial) one. Then the second observer sees the same physical event as being the same amount of stretching in combination with what appears to be a rotation of the band in the *opposite* (counter-clockwise) direction. Accordingly, the second observer gives the app these *consistently different* inputs. The app passes PMFI testing if it predicts the required force to be the same magnitude but rotated *clockwise* in comparison to the fiducial prediction from the *same* app. In other words, both observers need to predict that the force is aligned with the rubber band even if the rubber band appears to be rotating. PMFI is just a consistency condition. If a constitutive model fails PMFI testing, then it is definitely a bad model. However, passing PMFI does not ensure that the model is good. After all, it might predict that it takes 50 kN to elongate the rubber band by one millimeter, which is a ludicrous prediction. PMFI merely requires that the other observer's result should be equally as bad.

As the mathematics of frame indifference is covered systematically, we will use examples in constitutive modeling to clarify the concepts. Frame indifference constraints might require adding strange complications to the governing equations. From the perspective of a software developer who is creating code that must be universally applicable for observers, moving or not, fixing a PMFI issue might be as simple as changing the code so that something that was previously programmed as a universal (built-in/hard-coded) constant would instead need to become a required new input argument[2]. At other times, the conditions of frame indifference can be unexpectedly gnarly, so it is crucial to learn the rigorous mathematical procedures for deriving frame indifference admissibility constraints. To gain a sense of these procedures, we will now start with a simple example.

19.1. A 3D spring—who expected it to be this hard!?

Let us suppose that you are a graduate student who has been assigned the task of writing computer code that generalizes a simple spring model, $f = k\delta$, into vector form so that the resulting model can be used to connect a collection of springs of different orientations in a three-dimensional (3D) lattice. Here in the original formula, k is a constant, while f is the force in a spring that is attached to a wall on one end and is elongated a distance δ on the other end. We will use this hypothetical modeling and programming task as a way to illustrate the mathematics of frame indifference, along with the irritation of implementing material models into computers that do exclusively what they are programmed to do without the courtesy of doing what we *obviously intended* even if we did not program them that way. We start with an obviously flawed spring model, and use rigorous frame indifference

[2] Recall, for example, our promise to prove that a function predicting Cauchy stress as a pure function of deformation gradient will not pass frame indifference unless the function (and hence the material) is isotropic. This does not mean anisotropic media cannot be modeled. It just means that the stress for an anisotropic medium must also depend on material fiber directions, and these must never be programmed internally as software parameters. Recognizing that they change in response to superimposed material rotation, they must be supplied as an input to the model.

mathematics to prove that it is flawed. We use the analysis to figure out how to fix it. The work is done in stages, by first considering rigid translation and later rotation.

19.1.1. Translational frame indifference

In this section, we present a simple and obviously flawed 1D spring model that is *not* frame indifferent. The idea here is to pretend that you have programmed your model into a dumb computer that is capable of *only* applying the model literally—the computer does not assess the validity of the model. The computer does not automatically make adjustments to the model inputs that we human beings would do without even consciously thinking about it. Any rational human being would have no problem discerning the *intended* meaning of the flawed model. The point of this section is that the *computer* (which always applies the model exactly as you program it) will get the wrong result. Frame analysis is a systematic tool for rooting out bugs in algorithms and/or their implementations. More often than not, the shortcoming is that the code fails to take in enough information so that different observers can get consistent predictions when they run the code using inputs as they perceive them to be despite differences in their observer location or orientation. Here, we consider only a translating observer, which is equivalent to superimposing motion on the spring in the opposite direction.

For a classical (non-vector) spring model, the force f in the spring equals a spring constant k times change in length δ,

$$f = k\delta. \tag{19.1}$$

We wish to generalize this model so that it applies in vector form. Our first attempt might introduce the following variables:

$$\begin{aligned} \underset{\sim}{x} &= \text{current location of the tip of the spring,} \\ \underset{\sim}{x}_o &= \text{original (unstressed) location of the tip of the spring.} \end{aligned} \tag{19.2}$$

An undergraduate physics student might be tempted to propose the following as a vector generalization of (19.1):

$$\underset{\sim}{f} = k(\underset{\sim}{x} - \underset{\sim}{x}_o) \quad \leftarrow \text{(violates PMFI)}. \tag{19.3}$$

This model violates invariance under rigid translation because *in its literal sense*, it does not account for the possibility that the *tail* of the spring might move from where

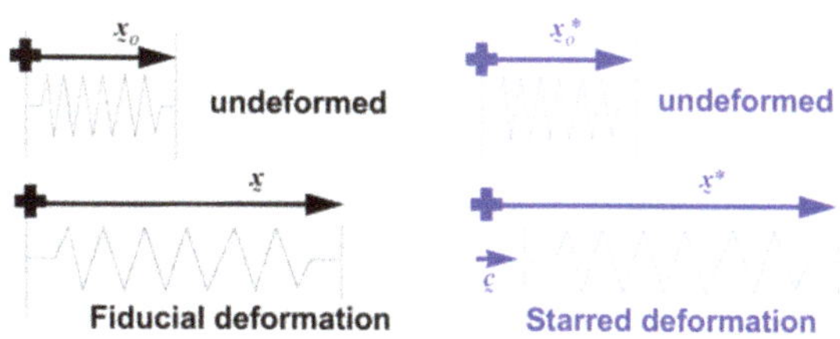

Figure 19.1. Two deformations involving the same spring stretch for which the faulty model wrongly predicts different spring forces. Each dark plus sign is the location of the origin. On the left is a deformation in which the tail of the spring does not move. On the right is a deformation that is identical to the fiducial deformation except that the spring is also rigidly translated. The model of (19.3) predicts the wrong answer because it does not include the position of the *tail*.

it was initially. If we were to implement (19.3) on a dumb computer, the resulting program would wrongly predict a *larger* spring force for the star deformation in Figure 19.1 than for the fiducial deformation, even though both deformations involve the same initial configuration and the same amount of spring stretch.

As human beings, we know *intuitively* that the deformation on the right side of Figure 19.1 should involve the same spring force since it involves the same amount of spring elongation. However, a computer program using (19.1) has no intuition—it just gives us the wrong answer (probably a trillion times per second). For Galilean frame indifference (i.e., invariance under translation), our spring model must have the property that it will predict exactly the same force whenever we rigidly translate the spring in space. This is an intuitively simple concept, but its mathematical expression is, unfortunately, less obvious. The idea here is that we compare two deformations:

- *Fiducial deformation:* stretch the spring by some known amount with the left end anchored to not move.
- *Starred deformation:* stretch the spring by the same amount and also translate it by an amount $\underset{\sim}{c}$. All variables here will be marked with an asterisk to compare them with the fiducial case.

Both scenarios involve the same starting tip location, so the initial location of the spring's right-side end $\underset{\sim}{x}_o$ is the same in both cases:

$$\underset{\sim}{x}_o^* = \underset{\sim}{x}_o. \tag{19.4}$$

Let $\underset{\sim}{x}$ denote the deformed location of the tip (right-side end) for the fiducial deformation. Let $\underset{\sim}{x}^*$ denote the deformed location of the tip for the starred deformation. Denoting the amount of translation by $\underset{\sim}{c}$, the location of the spring tip for the starred deformation is given by

$$\underset{\sim}{x}^* = \underset{\sim}{x} + \underset{\sim}{c}. \tag{19.5}$$

These two equations have different forms because the amount of superimposed translation is different in the initial and deformed configurations. This is an essential trait of frame analysis—you must consider problems for which differences between observers (or, equivalently, differences in superimposed motion) evolve to be different at different times during the problem. There was no translation vector in the initial state of equation (19.4) because there had been no superimposed translation applied at that time. The PMFI under translation says that both scenarios are *supposed to involve the same spring force.* Thus, we require that our spring model should predict $\underset{\sim}{f}$ to be the same for both the starred and fiducial deformations:

$$\underset{\sim}{f}^*_{\text{desired}} = \underset{\sim}{f}. \tag{19.6}$$

We demand that this condition should hold no matter how much translation we use. To demonstrate that (19.3) violates translational frame indifference, we simply apply it in both scenarios:

$$\text{for the fiducial deformation } \underset{\sim}{f} = k(\underset{\sim}{x} - \underset{\sim}{x}_o). \tag{19.7}$$

For starred deformation, the spring model predicts

$$\underset{\sim}{f}^{*}_{\text{predicted}} = k\left(\underset{\sim}{x}^{*} - \underset{\sim}{x}^{*}_{o}\right). \tag{19.8}$$

To check whether or not this prediction for $\underset{\sim}{f}^{*}$ agrees with the desired result given in (19.6), we substitute (19.4) and (19.5) into (19.8) to obtain

$$\underset{\sim}{f}^{*}_{\text{predicted}} = k(\underset{\sim}{x} + \underset{\sim}{c} - \underset{\sim}{x}_{o}) = \underset{\sim}{f}^{*}_{\text{desired}} + k\underset{\sim}{c}. \tag{19.9}$$

We want our spring model to give the correct answer for both the fiducial and starred deformations. However, the only way that $\underset{\sim}{f}^{*}_{\text{predicted}}$ will equal $\underset{\sim}{f}^{*}_{\text{desired}}$ is if the translation $\underset{\sim}{c}$ is zero. Thus, our undergraduate spring model of (19.3) should be applied only under the restriction of zero spring translation. For general motions, it violates the principle of translational frame indifference.

The above discussion captures the strange flavor of invariance analysis. We used a simple and obviously wrong spring equation to illustrate the precise (but stilted and often non-intuitive) *proof* that the model was wrong. Flaws in real-life models are not always so intuitively apparent, so the abstract mathematical analysis might be the only way to check a model for translational frame indifference.

To summarize, analysis of translational frame indifference goes as follows:

- Use kinematics and/or careful application of definitions to deduce how the independent variables in the translated 'starred' deformation must be related to those of the fiducial deformation. For our spring model, for example, we asserted that $\underset{\sim}{x}^{*} = \underset{\sim}{x} + \underset{\sim}{c}$. Superimposing translation affects only the deformed state—we do not impose a translation on the initial state. That is why we stated that $\underset{\sim}{x}^{*}_{o} = \underset{\sim}{x}_{o}$ to reflect the fact that both deformations started from the same initial condition. The fact that the fiducial and starred deformations have the same initial state is the key difference that distinguishes frame indifference from mere coordinate invariance[3]. If the model is written in rate form, then the kinematical condition that relates the starred velocity to the fiducial velocity would be $\underset{\sim}{v}^{*} = \underset{\sim}{v} + \dot{\underset{\sim}{c}}$.
- Apply the PMFI to decide how the dependent variables for the starred deformation should be related to those of the fiducial deformation. For our spring model, we asserted the physical condition that the spring force should be unchanged by rigid translation ($\underset{\sim}{f}^{*} = \underset{\sim}{f}$).
- Apply the material constitutive model for both the fiducial and starred deformations. The model is translationally invariant if consistency of the fiducial and starred model predictions requires no restrictions on the frame

[3] The principle of translational coordinate invariance says that we should be able to translate the origin to a new location without affecting the prediction of the model. If the origin were moved by an amount $\underset{\sim}{C}$, then, $\underset{\sim}{x}_{o}$ would become $\underset{\sim}{x}'_{o} = \underset{\sim}{x}_{o} - \underset{\sim}{C}$ and $\underset{\sim}{x}$ would become $\underset{\sim}{x}' = \underset{\sim}{x} - \underset{\sim}{C}$. Consequently, $\underset{\sim}{x}' - \underset{\sim}{x}'_{o} = \underset{\sim}{x} - \underset{\sim}{x}_{o}$, thereby demonstrating that the model would give the same answer for spring force even if the origin were to be moved. Thus, our spring model *does* satisfy translational coordinate invariance even though it violates translational frame indifference, which effectively requires invariance even if the origin moves to different places *during the application of the model*. Later, this distinction will be used to consider invariance under a stationary frame change as being necessary but not sufficient for invariance under an evolving frame change.

> translation vector $\underset{\sim}{c}$. For our undergraduate spring model, we proved that the fiducial and starred spring forces were consistent only if $\underset{\sim}{c} = \underset{\sim}{0}$, which therefore proved that the model was not translationally frame invariant. To be translationally invariant, the fiducial force needed to be equal to the starred force regardless of the value of $\underset{\sim}{c}$.

Again, keep in mind that we knew *intuitively* that something was wrong with (19.3) just by considering figure 19.1. The hard part was *proving it mathematically*! Frame indifference analysis is really about enforcing what we intuitively know should be true. Computers do exactly what we tell them to do—they lack intuition, so we must be very careful when developing our numerical models to ensure they pass frame indifference.

19.1.2. Rotational frame indifference (spring case study)

To satisfy *translation* invariance, a masters-level student might propose a new improved version of (19.3). Namely,

$$\underset{\sim}{f} = k(\underset{\sim}{L} - \underset{\sim}{L}_o) \leftarrow \text{better, but still bad.} \tag{19.10}$$

Here, $\underset{\sim}{L}$ is the vector pointing from the tail of the spring to the tip of the spring. Since this vector represents the difference between two points on the spring, it will be invariant under a superimposed translation. Thus, under translation, $\underset{\sim}{L}^* = \underset{\sim}{L}$. We again have the restriction that $\underset{\sim}{L}_o^* = \underset{\sim}{L}_o$ and again PMFI requires that $\underset{\sim}{f}^* = \underset{\sim}{f}$. If (19.10) is assumed to be true, then the equation $\underset{\sim}{f}^* = k(\underset{\sim}{L}^* - \underset{\sim}{L}_o^*)$ is also true without having to impose any assumptions about the translation vector $\underset{\sim}{c}$. Consequently, equation (19.10) is indeed *translationally* frame invariant. That is very good, but not enough—this model fails frame rotation indifference!

Unfortunately, the spring model of (19.10) still has problems because it is not *rotationally* frame invariant. To understand the issue, consider the carefully contrived counter-example in figure 19.2 where we show a problem with the same amount of spring elongation, but a different amount of spring rotation. For convenience, we

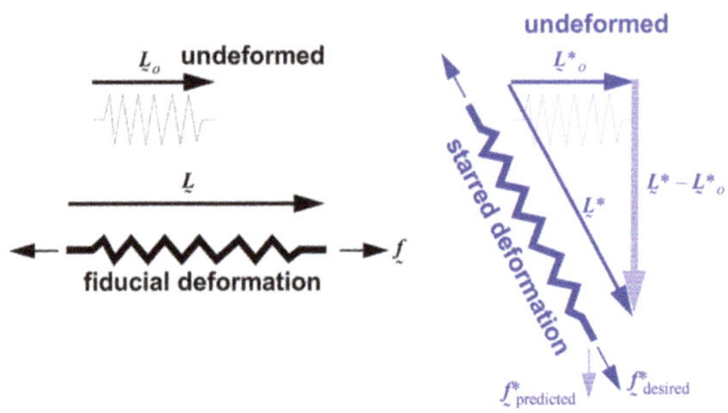

Figure 19.2. A rotationally faulty model. On the left is a deformation in which the orientation of the spring remains fixed. On the right is a deformation in which the spring experiences the *same elongation* but it is also rotated. To be rotationally consistent with the fiducial deformation, the starred force vector $\underset{\sim}{f}^*$ needs to be the same as the fiducial $\underset{\sim}{f}$ except rotated with the spring. For the picture on the right, we contrived the amount of rotation so that the vector $\underset{\sim}{L} - \underset{\sim}{L}_o$ would point straight down. That means our faulty masters-level spring equation predicts a spring force that is vertical instead of lined up with the spring the way it should be!

artificially selected a rotation that would make the vector $\underset{\sim}{L} - \underset{\sim}{L}_o$ point straight down. If we were to blindly program (19.10) into a dumb computer, then it would predict that the force in the spring is oriented vertically. As human beings, we know *intuitively* that this is an absurd result—the force in the spring should point along the axis of the spring for *both* the fiducial and starred deformations. Following the same sort of logic as we used for translational invariance, we will now prove *mathematically* that (19.10) is not rotationally invariant. The idea is to again to check whether or not the predictions of the model for the starred rotated deformation are physically consistent with the prediction in the fiducial deformation. Again consider two scenarios:

- *Fiducial deformation*: stretch the spring without rotation and without motion of the left endpoint.
- *Starred deformation*: starting with the same initial configuration as the fiducial problem (i.e., a horizontal unstretched spring), apply the same spring elongation as the fiducial deformation, but also rigidly rotate the spring while holding the left endpoint fixed.

Both scenarios involve the same starting length vector. Thus,

$$\underset{\sim}{L}_o^* = \underset{\sim}{L}_o. \tag{19.11}$$

Let $\underset{\sim}{L}$ denote the deformed length vector for the fiducial deformation. Let $\underset{\sim}{L}^*$ denote the deformed length vector for the starred deformation. Let $\underset{\approx}{Q}$ denote the rotation tensor. Then simple kinematics reveals that

$$\underset{\sim}{L}^* = \underset{\approx}{Q} \bullet \underset{\sim}{L}. \tag{19.12}$$

The principle of invariance under rotation imposes the intuitively 'obvious' requirement that the spring force predicted for the starred deformation should be the same as the fiducial force except rotated with the material. Stated differently, the force in the spring should form the same angle *relative* to the spring for both scenarios. Some authors would state this property by saying that, under rigid rotation, the force must 'co-rotate' with the spring. Mathematically, this requirement is expressed by asserting that the new 'starred' force should be related to the fiducial force by

$$\underset{\sim}{f}^*_{\text{desired}} = \underset{\approx}{Q} \bullet \underset{\sim}{f}. \tag{19.13}$$

To demonstrate that (19.10) violates rotational frame indifference, we will apply it in both scenarios and demonstrate that $\underset{\sim}{f}^*_{\text{predicted}} \neq \underset{\sim}{f}^*_{\text{desired}}$. The spring model says that $\underset{\sim}{f} = k(\underset{\sim}{L} - \underset{\sim}{L}_o)$ and therefore (19.13) becomes

$$\underset{\sim}{f}^*_{\text{desired}} = k(\underset{\approx}{Q} \bullet \underset{\sim}{L} - \underset{\approx}{Q} \bullet \underset{\sim}{L}_o). \tag{19.14}$$

For the starred (rotated) deformation, the spring model predicts $\underset{\sim}{f}^*_{\text{predicted}} = k(\underset{\sim}{L}^* - \underset{\sim}{L}_o^*)$. Hence, substituting (19.11) and (19.12) gives

$$\text{for the starred deformation } \underset{\sim}{f}^*_{\text{predicted}} = k(\underset{\approx}{Q} \bullet \underset{\sim}{L} - \underset{\sim}{L}_o). \tag{19.15}$$

To satisfy rotational frame indifference, $\underset{\sim}{f}^*_{\text{predicted}}$ must be the same as $\underset{\sim}{f}^*_{\text{desired}}$. However, subtracting (19.14) from (19.15) gives

$$\underset{\sim}{f}^*_{\text{predicted}} - \underset{\sim}{f}^*_{\text{desired}} = k(\underset{\approx}{Q} \bullet \underset{\sim}{L}_o - \underset{\sim}{L}_o). \tag{19.16}$$

The only way this result will equal zero (as required for rotational frame indifference) is if the rotation tensor equals the identity tensor. In other words, there must be no rotation. We therefore conclude that our masters-level spring model violates rotational frame indifference. We desire our spring model to give the correct answer for both the fiducial and the starred scenarios, so we now look into a PhD modification of the spring model.

19.1.3. A spring model that satisfies PMFI

To satisfy both translational and rotational frame indifference, a PhD-level student might propose another new-and-improved version of (19.10). Namely,

$$\underset{\sim}{f} = k\delta\underset{\sim}{n},$$
$$\text{where } \delta \equiv \|\underset{\sim}{L}\| - \|\underset{\sim}{L}_o\|, \quad \text{and } \underset{\sim}{n} = \frac{\underset{\sim}{L}}{\|\underset{\sim}{L}\|}. \tag{19.17}$$

To prove that this equation is indeed both translationally and rotationally invariant, we again consider how the prediction would change if we were to consider a 'starred' deformation that superimposes a rigid translation and rotation on the deforming material. Keep in mind the fiducial and starred motions have identical initial configurations with the only difference between them being superimposed rigid motion (translation and rotation). Therefore, deformed lengths are the same in both cases. Hence, $\|\underset{\sim}{L}^*\| = \|\underset{\sim}{L}\|$. Of course, as before, the initial state is the same for both cases, so $\|\underset{\sim}{L}_o^*\| = \|\underset{\sim}{L}_o\|$. Consequently,

$$\delta^* = \delta. \tag{19.18}$$

Under a superimposed rigid motion, the 'starred' length vector $\underset{\sim}{L}^*$ is given by $\underset{\approx}{Q} \bullet \underset{\sim}{L}$. Thus, the 'starred' unit vector $\underset{\sim}{n}^*$ is related to the fiducial (un-starred) unit vector $\underset{\sim}{n}$ by a simple rotation

$$\underset{\sim}{n}^* = \underset{\approx}{Q} \bullet \underset{\sim}{n}. \tag{19.19}$$

As before, we require that the force for the starred deformation should be related to the fiducial force by

$$\underset{\sim}{f}^*_{\text{desired}} = \underset{\approx}{Q} \bullet \underset{\sim}{f} = k\delta\underset{\approx}{Q} \bullet \underset{\sim}{n}. \tag{19.20}$$

If we apply the improved spring model of (19.17) to the starred deformation, we obtain

$$\underset{\sim}{f}^*_{\text{predicted}} = k\delta^*\underset{\sim}{n}^*. \tag{19.21}$$

In light of (19.18) and (19.19), we conclude that this new spring model is frame-indifferent because we finally have consistent results. Namely,

$$\underset{\sim}{f}^{*}_{\text{predicted}} = \underset{\sim}{f}^{*}_{\text{desired}}. \tag{19.22}$$

Incidentally, the spring equation in (19.17) is equivalent to an energy-based spring model defined by

$$e = \tfrac{1}{2}k\delta^2, \quad \delta \equiv \|\underset{\sim}{L}\| - \|\underset{\sim}{L}_o\|, \quad \underset{\sim}{f} = \frac{\mathrm{d}e}{\mathrm{d}\underset{\sim}{L}}. \tag{19.23}$$

Here, e is the internal energy of the spring. As a rule of thumb, constitutive models that can be cast in terms of scalar potentials like this are likely to satisfy PMFI because most physical scalars are unaffected by superimposed rigid motion. However, even if a model passes PMFI in theory, its numerical implementation should also be verified to pass PMFI in order to detect bugs.

19.1.4. Is frame indifference enough to be good for applications? No!

The PhD-level model in (19.17) satisfies both translational and rotational frame indifference requirements, which is necessary but not sufficient for it to be predictive. For example, perhaps the spring is actually nonlinear. Even if linearity is an adequate assumption, a professional research scientist would soon realize that k should not be a constant. Putting two springs in series, for example, would reduce the effective spring constant by a factor of 2. In fact, the effective spring constant should be inversely proportional to the initial spring length, so a better model is

$$\underset{\sim}{f} = K\varepsilon\underset{\sim}{n},$$
$$\text{where } \varepsilon \equiv \frac{\|\underset{\sim}{L}\| - \|\underset{\sim}{L}_o\|}{\|\underset{\sim}{L}_o\|}, \quad \underset{\sim}{n} = \frac{\underset{\sim}{L}}{\|\underset{\sim}{L}\|}, \text{ and } K \text{ is a new constant.} \tag{19.24}$$

This point generalizes to broader contexts. In the continuum mechanics literature, it is surprising how often constitutive modelers to declare their model to be valid for large deformations merely because it satisfies PMFI admissibility restrictions. To declare a model to be 'valid' for large deformations, however, it must be also agree with experimental data (or, if such data are impractical, higher-fidelity simulations[4]) (see [1, 2]).

Frame indifference analysis of material models follows similar (albeit more complicated) procedures as used in this spring example. When you want to check if a material constitutive equation satisfies frame indifference, you must first note what variables are involved in the constitutive equation. You must use physical reasoning with rigorous application of definitions to assert how those quantities are *supposed to* change upon a superimposed rigid motion. Then you must demonstrate that the

[4] In the case of this spring model, higher-fidelity *in silico* (virtual) validation might be a simulation of the actual spring geometry. Such simulations, for example, would reveal the need to have the spring constant vary inversely with spring length even if no real experiments were conducted.

constitutive model gives consistent results with and without superimposed rigid motion, both fixed and evolving. This self-consistency must be achieved without imposing *any* conditions on the superimposed rigid motion or on the *rate* of the rigid motion. We have illustrated these concepts in the simple setting of a spring model where the nature of each problem was quite obvious. Real material constitutive laws, however, have some more subtle issues that demand careful mathematics.

19.1.5. Frame indifference testing in codes

Even once a model is proved to *theoretically* satisfy frame indifference, any implementation of that model in a code must also be verified to indeed give predictions consistent with theory[5]. A single-element test is the simplest form of numerical PMFI testing in which the user runs a series of basic homogeneous fiducial tests, such as simple shear, uniaxial strain, hydrostatic compression, etc. Then those very same deformations would need to be run again with superimposed rotation, ideally a random rotation one (see chapter 17). The formulas derived in this chapter mandate how the code's output *ought* to change in comparison to the fiducial tests. The test passes if it actually does change as it should. Given how common it is for constitutive models to fail even simple pull-and-rotate PMFI testing, it is remarkable that commercial mechanics codes do not (to our knowledge) include automated random superimposed rotations in all single-element regression tests.

As will be explained here in this chapter, full frame-indifference testing would supplement single-element PMFI tests by running full-field multi-element tests, which is complicated because (as will be shown), they require the host code to add extra body forces (derived in chapter 24) that were not present in the fiducial test and which are not determined exclusively from the amount of superimposed rotation (but instead depend on the location and motion of the material as well).

The MMS[6] deformation in figure 19.3 provides a simple, yet surprisingly challenging, way to simultaneously check a code's momentum solver and PMFI in simple shear. Using body forces derived in [3], the aim is to induce the deformation field depicted in figure 19.3, which is colored by initial angular coordinate to easily see material motion. That deformation is designed to have zero displacement and zero strain at the boundaries and can therefore be easily run in codes lack support for traction boundary conditions and/or that require all domains to be rectangular (this one can be extended to simply hold stationary material) including the interior and exterior of the ring as stationary material points. Inside the ring, every material point is subjected to simple shear of varying intensities subjected to superimposed rotation,

---

[5] And, oh my heck, to check against mistakes in the theoretical proof.

[6] MMS stands for the 'method of manufactured solutions', which is a standard method of verifying momentum solvers in mechanics codes (see [2]). With this method, one works out analytically the boundary and body forces required to induce a 'manufactured' displacement field (in this case, a vortex ring). Then the code is run with those body forces. The predicted displacement can then be verified against the original manufactured displacement field to quantify simulation error. Frame indifference testing falls under the heading of *validation* since it is based on a principle that has been universally observed in laboratory testing, but its enforcement is mostly mathematical rather than physical.

Figure 19.3. MMS vortex ring. All points move in circles at different speeds designed to give zero strain (and hence zero traction) at the boundaries [3]. Here, hue is proportional to the initial polar angle.

making this problem also serve as a PMFI test. The solution quality is quantified by errors in the predicted displacement field compared to the MMS field.

19.2. Introduction to frame indifference

The effect of an observer's position and orientation is crucial in physics. In order to report what is seen, an observer must necessarily introduce an origin and reference directions (typically *to my right*, *above my head*, and *behind my back*). You as an observer might be able to see other observers using different origins and reference directions. Different observers will *perceive* different position vectors, different velocities, etc, even though they are actually looking at the same thing[7]. In order for the thing being observed to be physically meaningful irrespective of the *perceived* differences coming from the artifice of the observer's location and orientation, one observer must be able to look at the position and orientation of a second observer relative *to his own* and be able to successfully predict not only the second observer's perceptions, but also the result of the second observer's calculations using the same physical equations (like Newton's second law) as I am using.

If you are sitting in an automobile while it breaks hard, then (relative to a stationary observer standing on the sidewalk) you will perceive an 'extra' component of gravity pointing in the opposite direction of your acceleration (which points rearward relative to the stationary observer outside the car) that seems to be throwing you towards the dashboard. If you hold a glass of water in this automobile, the surface of the water will tilt from this extra gravity that is not perceived by the stationary observer who instead sees and predicts the same tilt as a consequence of inertia. If you are in a state of free-fall in a vacuum, then (relative to a stationary observer) you will feel an 'extra' component of gravity pointing upward in the opposite direction of your downward acceleration (relative to a stationary observer on the ground), and this upward component combines with regular downward gravity to give you a sensation of zero gravity. If you are on a spinning amusement park ride, you will perceive 'extra' gravity

[7] For example, you might say a ball is thrown upward, while I might say it is thrown to the right if I happen to be watching the same event while I am lying on my left side. You might say a rock is stationary, but I might say it is wobbling if I am watching it from a massage chair.

throwing you outward (in the opposite direction of your actual acceleration relative to an observer on the ground). As long as all observers use inputs to governing equations that are consistent with their own *perceptions* of those inputs, any model satisfying frame indifference will give them outputs that are also consistent with their own frame and predictably different from what might be seen in another frame. For example, you might have been told that Newton's second law, $F = ma$, should never be applied in a non-inertial (accelerating) frame, but that is wrong! Consider a free-falling ball being observed by someone on the ground, who sees the force from gravity as mg and hence predicts acceleration is $a = g$. Now consider the same free-falling ball being observed by someone who is falling along with it. Neglecting drag, the free-falling observer perceives zero gravity, and therefore (applying Newton's second law) zero force, thus predicting zero acceleration (from his perspective), which is correct! Including drag, the falling observer's acceleration changes with time, but the conclusions of consistency in predictions will not change—that's because Newton's second law is properly frame indifferent and therefore applicable in any frame, accelerating or not.

The underlying requirement of frame indifference is that the governing equations must be identical in form for all observers. All of them should be able to use exactly the same computer programs that implement the governing equations. They should enter inputs as they perceive them to be, and the resulting outputs should be consistent with (not identical to) outputs from the same code run by a different observer using inputs as he perceives them to be. In the case of the falling ball, a stationary observer would enter $F = mg$ into Newton's law to predict that $a = g$. The free-falling observer would enter a different perception of force ($F^* = 0$ from perceived weightlessness) to predict that acceleration in that frame is $a^* = 0$. These two predictions used the same governing equation, $F = ma$. They gave two different results for the acceleration in two different frames, but those results are consistent with a frame change. If the ground-based observer knows, from his own observations that the second observer is accelerating along with the ball, then he infers that the falling observer must be seeing acceleration as zero. These statements are consistent. Frame indifference never requires answers to be correct—it only requires consistency. An equation that has 10% error in one frame, for example, ought to have exactly the same error in any other frame. Now let us start on the mathematics of frame change.

19.3. Kinematics changes under superimposed rigid motion

Kinematics is the mathematics of motion. Recalling that a frame change is tantamount to superimposed rigid motion, this section is dedicated to working out how kinematical variables will change upon superimposing translation and or rotation that might even exist initially and which might furthermore evolve over time. As another motivation, suppose that you have already painstakingly analyzed a motion $\underset{\sim}{x} = \chi(\underset{\sim}{X}, t)$ to derive values for all of kinematic variables (like the deformation gradient $\underset{\approx}{F}$, velocity gradient $\underset{\approx}{L}$, particle acceleration $\underset{\sim}{a}$, displacement $\underset{\sim}{u}$, left stretch $\underset{\approx}{V}$, etc). Furthermore assume that you have worked out all of the forces required to induce this motion. Now imagine that your boss walks into your office to assign a new job: derive all of these same kinematics variables and associated forces for a different (possibly transiently non-inertial) observer (which is equivalent to superimposing rigid motion with appropriate additional body forces on the fiducial motion). Being transient means it might change as the deformation proceeds, so this task is much more complicated than a simple case of a non-evolving basis change.

This new problem with superimposed motion is *somewhat* like running a new experiment as the one you just analyzed, but now conducted in the bed of a moving truck. This is not exactly that scenario, however, if the truck is accelerating (either by giving it gas, braking, or going around a curve). Not only would the forces in this new experiment need to be the same as were used in the original (fiducial) experiment, but you would also need extra body forces (derived in chapter 24) to apply while in the moving truck to exactly offset any 'phantom' body forces it might induce through its own motion such as braking for a stop-light causing 'phantom' gravity to push things towards the front of the truck—that would need to be exactly neutralized if you want this new, non-inertial experiment to be identical to the old one but now relative to the frame of the truck. Likewise, you would need to add exactly offsetting body forces to counteract the phantom gravity coming from the truck taking a corner. What is worse, you would even need to figure out 'extra' gravity to add to your moving experiment to exactly offset Coriolis effects caused by material motion moving radially relative to the truck's curving path. What a mess! While this imagined scenario might not happen in real life, similar ones are very common in numerical analysis of deformations that occur in a moving frame. If, for example you want to analyze cyclic damage induced in a tire from rolling, a simulation that actually rotated the tire would quickly accumulate spurious integration errors from the bulk non-inertial material motion. The solution is to run the simulation from the perspective of a non-inertial observer who is rotating and translating with the moving wheel's hubcap. To do that, however, the simulation would need to include new radial body forces from frame rotation to provide the centrifugal body forces in the actual problem. The simulation would also need phantom forces from the Coriolis effect associated with compression of the tire towards the center of rotation. Similar revisions in the apparent body forces would be required to correctly analyze deformations of windmill or turbine blades, rocket sleds, or any other system experiencing large accelerations with respect to a stationary frame, but small accelerations in a well-chosen accelerating frame.

For analyzing a tire, the real system is spinning, and a non-inertial frame is needed to foster numerical accuracy. This section considers the inverse problem that would correspond to converting the tire solution found in the moving frame back to the laboratory frame. In that case, the inertial body forces would have to be removed and solution variables would need to be transformed back as well. At a fundamental level, we seek formulas for how kinematics variables in a 'fiducial' problem will change if that problem is viewed from the perspective of an observer who is translating and rotating, perhaps transiently (and thus with acceleration) over time. All that matters is *relative* motion. As mentioned, moving and rotating an observer is equivalent to keeping the observer stationary while superimposing translation and rotation in the opposite directions. Such a perspective is especially useful to set up simulations of the view as seen by a moving observer. If those simulations are defined properly, then frame indifference (which requires the same governing equations to be used in both the fiducial and new problem) demands that the predicted results in the code should transform relative to the fiducial problem in the ways derived here in this section. If they fail to transform properly, there is something wrong with the algorithm, its implementation, or the code user's understanding of how to run problems with superimposed rotation. The most common mistake of code users is a failure to recognize that the problem with rotation needs to include extra body forces that were not present in the fiducial simulation!

As long as two motions really have deformations that differ only by a rigid motion, then we can derive relationships between kinematic quantities in the two problems. Later, we invoke the PMFI to argue that the loadings required to induce these motions must likewise have specific relationships to each other. A simpler example might be that you worked out variables for pulling horizontally on a bar, but now your boss wants to know what the same deformation would look like from the perspective of an observer riding on a roller coaster. The deformation is applied on the ground as in the previously solved problem, but you now seek what that same deformation looks like from the perspective of a moving observer. If the roller coaster is accelerating forward, then you reason (correctly) that the moving observer will perceive the deformation on the ground to be exactly what you solved previously except that the material is accelerating backward. If the roller coaster is turning left, it will seem to the moving observer that the deformation is identical to the previously solved problem except also rigidly rotating to the right. In short, to figure out what is seen by the observer on the roller coaster, you can simply work out formulas for the previously solved problem subjected to a pure superimposed rigid motion. Do you have to start from scratch, or can you convert your old results to new results that include the superimposed rotation? The purpose of this section is to show how to work out those formulas, the most important of which are shown as boxed equations. While this work is tedious, it is rewarding to step through all of it before attempting to use any of it.

Let us use the word *fiducial* to refer to anything you might perceive when you watch an event from a particular (fiducial) choice of origin and orientation. Let a 'star' be used on variables as they would seem to be if you were to watch the same event from a different (potentially evolving) location and orientation. If you get to this 'star' state by rotating *yourself* clockwise, that is the same as perceiving that the *entire world* (and all events happening in it) to be rotated counter-clockwise relative to what was seen in the fiducial perspective. If you move yourself to the left, that is the same as perceiving the entire world to shift to the right. Generally, if $\underset{\sim}{x}$ denotes the position vector before you move, the perceived position vector $\underset{\sim}{x}^*$ in the new frame is

$$\boxed{\underset{\sim}{x}^* = \underset{\approx}{q} \bullet \underset{\sim}{x} + \underset{\sim}{c}}. \tag{19.25}$$

Here $\underset{\approx}{q}$ is the inverse of your change in rotation relative to the fiducial orientation (if you rotated clockwise relative to the fiducial frame, then $\underset{\approx}{q}$ is a counter-clockwise rotation). Similarly, $\underset{\sim}{c}$ is the negative of your translation relative to the fiducial frame (if you moved to the left, this vector points to the right). Neither $\underset{\approx}{q}$ nor $\underset{\sim}{c}$ varies with position, but they both might change with time, making it possible that the star observer is accelerating relative to the fiducial one. Frame indifference analysis, discussed in an upcoming section, dictates how physical *perceptions* must change so the thing being observed does not change. Specifically, frame indifference requires that everyone must use the same governing equations. They must provide inputs to these equations as they perceive them to be. The outputs will be predictions of what they perceive, and these predictions must be related in a particular way to the same predictions found by a different observer using inputs as he perceives them to be. Again, the inputs and outputs will be different but related in a particular predictable way, while the governing equations themselves must be the same for everyone! More on this topic later.

For now, we derive what must be seen and predicted by a different 'star' observer based exclusively on what we see and predict in the fiducial frame and knowledge of $\underset{\approx}{q}$ and $\underset{\sim}{c}$ as functions of time. This discussion could be generalized to also include a shift in time (often called different 'clocks', like observers on the east and west coast looking at the same comet even though they are in different time zones, or a compression in time associated with moving near the speed of light). However, for simplicity, we presume Newtonian relativity with both observers using the same clocks so that any difference in their perceptions comes exclusively from differences in their origin and/or orientation[9].

If we let $\underset{\approx}{q}_0$ and $\underset{\sim}{c}_0$ denote the values of $\underset{\approx}{q}$ and $\underset{\sim}{c}$ at time zero, then a point's *initial* position vector $\underset{\sim}{X}^*$ seen by the star observer is related to the fiducial initial position vector $\underset{\sim}{X}$ of the same point by simply evaluating (19.25) at time zero:

$$\boxed{\underset{\sim}{X}^* = \underset{\approx}{q}_0 \bullet \underset{\sim}{X} + \underset{\sim}{c}_0}. \tag{19.26}$$

This formula is the frame transformation relationship evaluated at time zero. Therefore this formula has no time dependence. Accordingly, we note for future reference that

$$\left(\frac{\partial \underset{\sim}{X}}{\partial t}\right)_{\underset{\sim}{X}^*} = \underset{\sim}{0}. \tag{19.27}$$

Displacement is $\underset{\sim}{u} = \underset{\sim}{x} - \underset{\sim}{X}$. Therefore, since all observers must use the same governing equations (including definitions of physical variables), the star observer is required to define displacement by the same formula, except with star labels (meaning that the formula must use the variable values perceived in that frame): $\underset{\sim}{u}^* = \underset{\sim}{x}^* - \underset{\sim}{X}^*$. Recalling that we have already derived formulas for how the star position vectors are related to the non-starred (fiducual) ones, we obtain

$$\underset{\sim}{u}^* = (\underset{\approx}{q} \bullet \underset{\sim}{x} + \underset{\sim}{c}) - (\underset{\approx}{q}_0 \bullet \underset{\sim}{X} + \underset{\sim}{c}_0). \tag{19.28}$$

---

[9] This section considers a rigid relationship between $\underset{\sim}{x}$, and $\underset{\sim}{x}^*$, $\underset{\sim}{x}^*(\underset{\sim}{x}, t) = \underset{\approx}{q}(t) \bullet \underset{\sim}{x} + \underset{\sim}{c}(t)$. To extend the analysis to a more general *affine* relationship for shifts and distortions of both space and time, similar methods would be applied using $\underset{\sim}{x}^*(\underset{\sim}{x}, t) = \underset{\approx}{G}(t) \bullet \underset{\sim}{x} + \underset{\sim}{k}(t)$ and $t^*(\underset{\sim}{x}, t) = \alpha(\underset{\sim}{x})t + c(\underset{\sim}{x})$. If you are up for that, then you might as well change your name to Einstein since that is what he and his wife did!

Writing $\underset{\sim}{x} = \underset{\sim}{X} + \underset{\sim}{u}$, this becomes

$$\boxed{\underset{\sim}{u}^* = \underset{\approx}{q} \bullet \underset{\sim}{u} + (\underset{\approx}{q} - \underset{\approx}{q}_0) \bullet \underset{\sim}{X} + (\underset{\sim}{c} - \underset{\sim}{c}_0)}. \tag{19.29}$$

This result is our first chance to see that an evolving frame change is different from a mere basis change. This formula doesn't look like a vector basis change unless observer displacement and orientation don't change with time (knocking out the last two terms). The displacement connects the positions of a single material point at two different moments in time. Contrast this with a material line, $\underset{\sim}{\mu}$, that connects two different points, $\underset{\sim}{x}_1$ and $\underset{\sim}{x}_2$ at any given moment in time: $\underset{\sim}{\mu} = \underset{\sim}{x}_2 - \underset{\sim}{x}_1$. Therefore

$$\underset{\sim}{\mu}^* = \underset{\sim}{x}_2^* - \underset{\sim}{x}_1^* = (\underset{\approx}{q} \bullet \underset{\sim}{x}_2 + \underset{\sim}{c}) - (\underset{\approx}{q} \bullet \underset{\sim}{x}_1 + \underset{\sim}{c}). \tag{19.30}$$

Since each of the two points are at the same moment in time, the $\underset{\sim}{c}$ vectors cancel and the $\underset{\approx}{q}$ tensor may be factored out to give

$$\boxed{\underset{\sim}{\mu}^* = \underset{\approx}{q} \bullet \underset{\sim}{\mu}}. \tag{19.31}$$

The material rate is defined in the fiducial problem as

$$\dot{\#} = \left(\frac{\partial \#}{\partial t}\right)_{\underset{\sim}{X}}. \tag{19.32}$$

Recalling that we are considering no clock shift, so that $t^* = t$, the chain rule implies that the material rate in the star problem is related to the rate in the fiducial problem by

$$\textit{Material rate:} \quad \left(\frac{\partial \#}{\partial t^*}\right)_{\underset{\sim}{X}^*} = \left(\frac{\partial \#}{\partial t}\right)_{\underset{\sim}{X}}. \tag{19.33}$$

In other words, because we have taken the clock to be the same in both cases, the material rate in the star problem equals the material rate in the fiducial problem. This result proves there is no ambiguity in using a 'dot' to denote the material time rate as $\dot{\#}^*$. In other words, you can compute the starred version of a quantity and take its rate in the starred frame, and this will be the same as computing the rate in the fiducial problem and transforming it to the starred frame. Spatial gradients, on the other hand, are affected by frame rotation. Applying the chain rule,

$$\left(\frac{\partial \#}{\partial \underset{\sim}{X}^*}\right)_{t^*} = \left(\frac{\partial \#}{\partial \underset{\sim}{X}}\right)_t \bullet \left(\frac{\partial \underset{\sim}{X}}{\partial \underset{\sim}{X}^*}\right)_t. \tag{19.34}$$

Using (19.26),

$$\textit{Reference gradient:} \quad \left(\frac{\partial \#}{\partial \underset{\sim}{X}^*}\right)_{t^*} = \left(\frac{\partial \#}{\partial \underset{\sim}{X}}\right)_t \bullet \underset{\approx}{q}_0. \tag{19.35}$$

Keep in mind why we are deriving formulas like these: it is to be able to quickly convert already computed fiducial results to the new 'star' problem without having to redo all of the calculations. This result is saying that, if the reference gradient in the fiducial problem is known, then it is immediately converted to the reference gradient in the 'star' problem by merely post multiplying by whatever observer rotation already existed at time zero. A spatial gradient is similarly found to satisfy

$$\textit{Spatial gradient:} \qquad \left(\frac{\partial \#}{\partial \underset{\sim}{x}^*}\right)_{t^*} = \left(\frac{\partial \#}{\partial \underset{\sim}{x}}\right)_t \bullet \underset{\approx}{q}. \tag{19.36}$$

The two derivatives of the fiducial deformation mapping are

$$\textit{Fiducial deformation gradient:} \qquad \underset{\approx}{F} = \left(\frac{\partial \underset{\sim}{x}}{\partial \underset{\sim}{X}}\right)_t. \tag{19.37a}$$

$$\textit{Fiducial material velocity:} \qquad \underset{\sim}{v} = \left(\frac{\partial \underset{\sim}{x}}{\partial t}\right)_{\underset{\sim}{X}}. \tag{19.37b}$$

Frame indifference requires everyone to use the same governing equations, including definitions. Therefore, the 'star' observer is required to use form-invariant definitions:

$$\textit{Star deformation gradient:} \qquad \underset{\approx}{F}^* = \left(\frac{\partial \underset{\sim}{x}^*}{\partial \underset{\sim}{X}^*}\right)_{t^*}. \tag{19.38a}$$

$$\textit{Star material velocity:} \qquad \underset{\sim}{v}^* = \left(\frac{\partial \underset{\sim}{x}^*}{\partial t^*}\right)_{\underset{\sim}{X}^*}. \tag{19.38b}$$

Equations (19.25) and (19.26) give us the information we need to express these star derivatives in terms of their fiducial counterparts. Starting with (19.25), which gives $\underset{\sim}{x}^*$ as a function of $\underset{\sim}{x}$ and t, the chain rule gives the first step in finding the star $\underset{\sim}{v}^*$:

$$\underset{\sim}{v}^* = \left(\frac{\partial \underset{\sim}{x}^*}{\partial t^*}\right)_{\underset{\sim}{X}^*} = \left(\frac{\partial \underset{\sim}{x}^*}{\partial t}\right)_{\underset{\sim}{x}}\left(\frac{\partial t}{\partial t^*}\right)_{\underset{\sim}{X}^*} + \left(\frac{\partial \underset{\sim}{x}^*}{\partial \underset{\sim}{x}}\right)_t \bullet \left(\frac{\partial \underset{\sim}{x}}{\partial t^*}\right)_{\underset{\sim}{X}^*}. \tag{19.39}$$

Assuming no time shift (so that $t^* = t$) it is trivial that

$$\left(\frac{\partial t}{\partial t^*}\right)_{\underset{\sim}{X}^*} = 1.$$

Recalling that $\underset{\approx}{q}$ and $\underset{\sim}{c}$ do not vary with position,

$$\left(\frac{\partial \underset{\sim}{x}^*}{\partial t}\right)_{\underset{\sim}{x}} = \left(\frac{\partial[\underset{\approx}{q} \bullet \underset{\sim}{x} + \underset{\sim}{c}]}{\partial t}\right)_{\underset{\sim}{x}} = \dot{\underset{\approx}{q}} \bullet \underset{\sim}{x} + \dot{\underset{\sim}{c}}. \tag{19.41}$$

where the superposed dots in $\dot{\underset{\approx}{q}}$ and $\dot{\underset{\sim}{c}}$ denote the derivative with respect to time. This is the part of the star velocity resulting from motion of the star frame relative to the fiducial frame. The remaining terms in the chain rule expansion will be now shown to come from velocity in the fiducial frame. Differentiating (19.25), keeping in mind that $\underset{\approx}{q}$ and $\underset{\sim}{c}$ depend only on time,

$$\left(\frac{\partial \utilde{x}^*}{\partial \utilde{x}}\right)_t = \left(\frac{\partial[\utilde{q} \bullet \utilde{x} + \utilde{c}]}{\partial \utilde{x}}\right)_t = \utilde{q}. \tag{19.42}$$

Again recalling that $t = t^*$, the chain rule, along with (19.27), gives

$$\left(\frac{\partial \utilde{x}}{\partial t^*}\right)_{\utilde{X}^*} = \left(\frac{\partial \utilde{x}}{\partial t}\right)_{\utilde{X}^*} = \underbrace{\left(\frac{\partial \utilde{x}}{\partial t}\right)_{\utilde{X}}}_{\utilde{v}} \underbrace{\left(\frac{\partial t}{\partial t}\right)_{\utilde{X}^*}}_{1} + \underbrace{\left(\frac{\partial \utilde{x}}{\partial \utilde{X}}\right)_t}_{\utilde{F}} \bullet \underbrace{\left(\frac{\partial \utilde{X}}{\partial t}\right)_{\utilde{X}^*}}_{\utilde{0}} = \utilde{v} \tag{19.43}$$

The above four equations then provide the final result for the prediction of perceived velocity in the star frame, based on knowledge of the event in the fiducial frame (and motion of the star observer relative to the fiducial observer):

$$\boxed{\utilde{v}^* = \utilde{q} \bullet \utilde{v} + \dot{\utilde{q}} \bullet \utilde{x} + \dot{\utilde{c}}}. \tag{19.44}$$

This result might be found by directly taking the rate of (19.25), which is what we do in upcoming equations, but the details have been shown here to rigorously justify the steps, especially to hint how to generalize this analysis to include time shifts.

The deformation gradient perceived by the star observer is found similarly. Recalling an assumption of no time shift,

$$\utilde{F}^* = \left(\frac{\partial \utilde{x}^*}{\partial \utilde{X}^*}\right)_t = \left(\frac{\partial \utilde{x}^*}{\partial \utilde{X}}\right)_t \bullet \left(\frac{\partial \utilde{X}}{\partial \utilde{X}^*}\right)_t. \tag{19.45}$$

From (19.25) and (19.26),

$$\left(\frac{\partial \utilde{x}^*}{\partial \utilde{X}}\right)_t = \left(\frac{\partial[\utilde{q} \bullet \utilde{x} + \utilde{c}]}{\partial \utilde{X}}\right)_t = \utilde{q} \bullet \utilde{F}, \tag{19.46}$$

$$\left(\frac{\partial \utilde{X}^*}{\partial \utilde{X}}\right)_t = \left(\frac{\partial[\utilde{q}_0 \bullet \utilde{X} + \utilde{c}_0]}{\partial \utilde{X}}\right)_t = \utilde{q}_0 \quad \Rightarrow \quad \left(\frac{\partial \utilde{X}}{\partial \utilde{X}^*}\right)_t = \utilde{q}_0^T. \tag{19.47}$$

Thus,

$$\boxed{\utilde{F}^* = \utilde{q} \bullet \utilde{F} \bullet \utilde{q}_0^T}. \tag{19.48}$$

From this, note that

$$(\utilde{F}^*)^{-1} = \utilde{q}_0 \bullet \utilde{F}^{-1} \bullet \utilde{q}^T, \tag{19.49a}$$

$$(\utilde{F}^*)^T = \utilde{q}_0 \bullet \utilde{F}^T \bullet \utilde{q}^T, \tag{19.49b}$$

$$(\utilde{F}^*)^{-T} = \utilde{q} \bullet \utilde{F}^T \bullet \utilde{q}_0^T. \tag{19.49c}$$

The left Cauchy–Green tensor $\underset{\approx}{\boldsymbol{B}} = \underset{\approx}{\boldsymbol{F}} \bullet \underset{\approx}{\boldsymbol{F}}^T$ and right Cauchy–Green tensor, $\underset{\approx}{\boldsymbol{C}} = \underset{\approx}{\boldsymbol{F}}^T \bullet \underset{\approx}{\boldsymbol{F}}$ (frequently appearing in continuum mechanics) therefore transform as

$$\boxed{\underset{\approx}{\boldsymbol{B}}^* = \underset{\approx}{\boldsymbol{q}} \bullet \underset{\approx}{\boldsymbol{B}} \bullet \underset{\approx}{\boldsymbol{q}}^T}, \tag{19.50a}$$

$$\boxed{\underset{\approx}{\boldsymbol{C}}^* = \underset{\approx}{\boldsymbol{q}}_0 \bullet \underset{\approx}{\boldsymbol{C}} \bullet \underset{\approx}{\boldsymbol{q}}_0^T}. \tag{19.50b}$$

We now introduce some new terms:

$$\text{Tensor } \underset{\approx}{\boldsymbol{A}} \text{ is a } \textit{two-point push tensor} \text{ iff } \underset{\approx}{\boldsymbol{A}}^* = \underset{\approx}{\boldsymbol{q}} \bullet \underset{\approx}{\boldsymbol{A}} \bullet \underset{\approx}{\boldsymbol{q}}_0^T. \tag{19.51a}$$

$$\text{Tensor } \underset{\approx}{\boldsymbol{A}} \text{ is a } \textit{two-point pull tensor} \text{ iff } \underset{\approx}{\boldsymbol{A}}^* = \underset{\approx}{\boldsymbol{q}}_0 \bullet \underset{\approx}{\boldsymbol{A}} \bullet \underset{\approx}{\boldsymbol{q}}^T. \tag{19.51b}$$

$$\text{Tensor } \underset{\approx}{\boldsymbol{A}} \text{ is a } \textit{spatial (corotational) tensor} \text{ iff } \underset{\approx}{\boldsymbol{A}}^* = \underset{\approx}{\boldsymbol{q}} \bullet \underset{\approx}{\boldsymbol{A}} \bullet \underset{\approx}{\boldsymbol{q}}^T. \tag{19.51c}$$

$$\textit{Tensor } \underset{\approx}{\boldsymbol{A}} \text{ is a } \textit{reference tensor} \text{ iff } \underset{\approx}{\boldsymbol{A}}^* = \underset{\approx}{\boldsymbol{q}}_0 \bullet \underset{\approx}{\boldsymbol{A}} \bullet \underset{\approx}{\boldsymbol{q}}_0^T. \tag{19.51d}$$

Therefore, $\underset{\approx}{\boldsymbol{F}}$ and $\underset{\approx}{\boldsymbol{F}}^{-T}$ are two-point push tensors (the first one pushes a reference material line element $\mathrm{d}\underset{\sim}{\boldsymbol{X}}$ to its 'covariantly' deformed state $\mathrm{d}\underset{\sim}{\boldsymbol{x}} = \underset{\approx}{\boldsymbol{F}} \bullet \mathrm{d}\underset{\sim}{\boldsymbol{X}}$, whereas the inverse-transpose operation corresponds to a 'contravariant' push that tracks with the normals of surface areas). Both $\underset{\approx}{\boldsymbol{F}}^{-1}$ and $\underset{\approx}{\boldsymbol{F}}^T$ are two-point pull tensors (they act on vectors in the deformed state to transform them—like running a movie backwards—to the reference state). The left Cauchy–Green tensor $\underset{\approx}{\boldsymbol{B}}$ is a spatial tensor (it characterizes the deformed size and shape of a material element after application of any rotation), and $\underset{\approx}{\boldsymbol{C}}$ is a reference tensor (it characterizes deformed size and shape of an element before application of any rotation beyond the initial rotation, and it is therefore unsurprising that it is unaffected by any evolution of superimposed rotation).

Now let us work on the following second-order derivatives:

$$\textit{Fiducial:} \quad \dot{\underset{\approx}{\boldsymbol{F}}} = \left(\frac{\partial^2 \underset{\sim}{\boldsymbol{x}}}{\partial \underset{\sim}{\boldsymbol{X}} \partial t}\right) = \left(\frac{\partial \underset{\sim}{\boldsymbol{v}}}{\partial \underset{\sim}{\boldsymbol{X}}}\right)_t \tag{19.52a}$$

$$\textit{Fiducial:} \quad \underset{\approx}{\boldsymbol{L}} = \left(\frac{\partial \underset{\sim}{\boldsymbol{v}}}{\partial \underset{\sim}{\boldsymbol{x}}}\right)_t = \dot{\underset{\approx}{\boldsymbol{F}}} \bullet \underset{\approx}{\boldsymbol{F}}^{-1} \tag{19.52b}$$

$$\textit{Fiducial:} \quad \underset{\sim}{\boldsymbol{a}} = \left(\frac{\partial^2 \underset{\sim}{\boldsymbol{x}}}{\partial t^2}\right)_{\underset{\sim}{\boldsymbol{X}}}. \tag{19.52c}$$

The star definitions are identical except with stars on all variables, as before. Performing chain rule analysis gives

$$\boxed{\dot{\underset{\approx}{F}}^* = \underset{\approx}{q} \bullet \dot{\underset{\approx}{F}} \bullet \underset{\approx}{q}_0^T + \dot{\underset{\approx}{q}} \bullet \underset{\approx}{F} \bullet \underset{\approx}{q}_0^T}, \tag{19.53a}$$

$$\boxed{\underset{\approx}{L}^* = \underset{\approx}{q} \bullet \underset{\approx}{L} \bullet \underset{\approx}{q}^T + \dot{\underset{\approx}{q}} \bullet \underset{\approx}{q}^T}, \tag{19.53b}$$

$$\boxed{\underset{\sim}{a}^* = \underset{\approx}{q} \bullet \underset{\sim}{a} + \ddot{\underset{\approx}{q}} \bullet \underset{\sim}{x} + 2\dot{\underset{\approx}{q}} \bullet \underset{\sim}{v} + \ddot{\underset{\sim}{c}}}. \tag{19.53c}$$

Recall that an accelerating frame change will cause the 'star' observer to perceive so-called phantom body forces. These (derived in detail in chapter 24) are reflected in the trailing three terms for the acceleration, showing that the acceleration in the 'star' problem is not merely a rotation of the fiducial acceleration, nor is it a simple matter of adding intuitive radial centrifugal force for a spinning drum (which is part of the $\ddot{\underset{\approx}{q}} \bullet \underset{\sim}{x}$ term) but also counter-intuitive Coriolis effects from the $2\dot{\underset{\approx}{q}} \bullet \underset{\sim}{v}$ term.

Using (19.53*b*), and also noting that the frame spin $\dot{\underset{\approx}{q}} \bullet \underset{\approx}{q}^T$ is skew-symmetric, the vorticity tensor, $\underset{\approx}{W} = \text{skw}\underset{\approx}{L}$, and the 'rate' of deformation, $\underset{\approx}{D} = \text{sym}\underset{\approx}{L}$, transform according to

$$\boxed{\underset{\approx}{W}^* = \underset{\approx}{q} \bullet \underset{\approx}{W} \bullet \underset{\approx}{q}^T + \dot{\underset{\approx}{q}} \bullet \underset{\approx}{q}^T}, \tag{19.54}$$

$$\boxed{\underset{\approx}{D}^* = \underset{\approx}{q} \bullet \underset{\approx}{D} \bullet \underset{\approx}{q}^T}. \tag{19.55}$$

At this point, it should be clear that an accelerating frame change is generally more than just a basis change. A simple basis change is fixed and therefore non-accelerating, giving non-changing rotation ($\underset{\approx}{q} = \underset{\approx}{q}_0$) and non-changing origin shift ($\underset{\sim}{c} = \underset{\sim}{c}_0$), making $\dot{\underset{\approx}{q}}$, $\ddot{\underset{\approx}{q}}$, $\dot{\underset{\sim}{c}}$, and $\ddot{\underset{\sim}{c}}$ all zero. Accordingly, a non-evolving frame requires no extra body forces to counteract inertia. The next few sections apply a decomposition to $\underset{\approx}{q}$ and $\underset{\sim}{c}$ in order to conceptually separate a non-evolving frame change from an evolving one.

19.3.1. Decomposing superimposed rigid motion

To clarify the distinct notions of a fixed and evolving frame change, let us now introduce a rotation tensor $\underset{\approx}{Q}$ that is the amount of rotation superimposed in addition to whatever frame rotation existed in the initial state. Likewise, let $\underset{\sim}{k}$ correspond to any *evolving* shift in the star frame's origin, beyond the initial shift. In other words, these two new symbols are defined such that

$$\underset{\sim}{c}(t) = \underset{\sim}{c}_0 + \underset{\sim}{k}(t) \qquad \text{and} \qquad \underset{\approx}{q}(t) = \underset{\approx}{Q}(t) \bullet \underset{\approx}{q}_0. \tag{19.56}$$

Therefore, also introducing $\underset{\approx}{\Upsilon} = \dot{\underset{\approx}{q}} \bullet \underset{\approx}{q}^T = \dot{\underset{\approx}{Q}} \bullet \underset{\approx}{Q}^T$ to denote the frame spin tensor,

$$\dot{\underset{\sim}{c}} = \dot{\underset{\sim}{k}} \quad \text{and} \quad \dot{\underset{\approx}{q}} = \dot{\underset{\approx}{Q}} \bullet \underset{\approx}{q}_0 = \dot{\underset{\approx}{Q}} \bullet \underset{\approx}{Q}^T \bullet \underset{\approx}{q} = \underset{\approx}{\Upsilon} \bullet \underset{\approx}{q}. \tag{19.57}$$

With these substitutions, the boxed equations in the preceding section are now

$$\boxed{\underset{\sim}{x}^* = \underset{\approx}{Q} \bullet (\underset{\approx}{q}_0 \bullet \underset{\sim}{x}) + \underset{\sim}{c}_0 + \underset{\sim}{k}}, \tag{19.58a}$$

$$\boxed{\underset{\sim}{X}^* = (\underset{\approx}{q}_0 \bullet \underset{\sim}{X}) + \underset{\sim}{c}_0}, \tag{19.58b}$$

$$\boxed{\underset{\sim}{u}^* = \underset{\approx}{Q} \bullet (\underset{\approx}{q}_0 \bullet \underset{\sim}{u}) + (\underset{\approx}{Q} - \underset{\approx}{I}) \bullet (\underset{\approx}{q}_0 \bullet \underset{\sim}{X}) + \underset{\sim}{k}}, \tag{19.58c}$$

$$\boxed{\underset{\sim}{\mu}^* = \underset{\approx}{Q} \bullet (\underset{\approx}{q}_0 \bullet \underset{\sim}{\mu})}, \tag{19.58d}$$

$$\boxed{\underset{\sim}{v}^* = \underset{\approx}{Q} \bullet (\underset{\approx}{q}_0 \bullet \underset{\sim}{v}) + \underset{\approx}{\Upsilon} \bullet \underset{\approx}{Q} \bullet (\underset{\approx}{q}_0 \bullet \underset{\sim}{x}) + \dot{\underset{\sim}{k}}}, \tag{19.58e}$$

$$\boxed{\underset{\approx}{F}^* = \underset{\approx}{Q} \bullet (\underset{\approx}{q}_0 \bullet \underset{\approx}{F} \bullet \underset{\approx}{q}_0^T)}, \tag{19.58f}$$

$$\boxed{\underset{\approx}{B}^* = \underset{\approx}{Q} \bullet \left(\underset{\approx}{q}_0 \bullet \underset{\approx}{B} \bullet \underset{\approx}{q}_0^T\right) \bullet \underset{\approx}{Q}^T}, \tag{19.58g}$$

$$\boxed{\underset{\approx}{C}^* = \underset{\approx}{q}_0 \bullet \underset{\approx}{C} \bullet \underset{\approx}{q}_0^T}, \tag{19.58h}$$

$$\boxed{\dot{\underset{\approx}{F}}^* = \underset{\approx}{Q} \bullet \left(\underset{\approx}{q}_0 \bullet \dot{\underset{\approx}{F}} \bullet \underset{\approx}{q}_0^T\right) + \underset{\approx}{\Upsilon} \bullet \left(\underset{\approx}{q}_0 \bullet \underset{\approx}{F} \bullet \underset{\approx}{q}_0^T\right)}, \tag{19.58i}$$

$$\boxed{\underset{\approx}{L}^* = \underset{\approx}{Q} \bullet \left(\underset{\approx}{q}_0 \bullet \underset{\approx}{L} \bullet \underset{\approx}{q}_0^T\right) \bullet \underset{\approx}{Q}^T + \underset{\approx}{\Upsilon}}, \tag{19.58j}$$

$$\boxed{\underset{\approx}{W}^* = \underset{\approx}{Q} \bullet \left(\underset{\approx}{q}_0 \bullet \underset{\approx}{W} \bullet \underset{\approx}{q}_0^T\right) \bullet \underset{\approx}{Q}^T + \underset{\approx}{\Upsilon}}, \tag{19.58k}$$

$$\boxed{\underset{\approx}{D}^* = \underset{\approx}{Q} \bullet \left(\underset{\approx}{q}_0 \bullet \underset{\approx}{D} \bullet \underset{\approx}{q}_0^T\right) \bullet \underset{\approx}{Q}^T}, \tag{19.58l}$$

$$\boxed{\underset{\sim}{a}^* = \underset{\approx}{Q} \bullet (\underset{\approx}{q}_0 \bullet \underset{\sim}{a}) + (\dot{\underset{\approx}{\Upsilon}} + \underset{\approx}{\Upsilon}^2) \bullet \underset{\approx}{Q} \bullet (\underset{\approx}{q}_0 \bullet \underset{\sim}{x}) + 2\underset{\approx}{\Upsilon} \bullet \underset{\approx}{Q} \bullet (\underset{\approx}{q}_0 \bullet \underset{\sim}{v}) + \ddot{\underset{\sim}{k}}}. \tag{19.58m}$$

19.3.2. Case 1: non-evolving frame change

The previous section summarized the 'star' formulas using decomposition of the frame change to be its initial change supplemented with additional change (if any) over time. That helps to distinguish between a general evolving frame change and a simpler *fixed* (non-changing) basis change. Let us consider a frame change that does *not* evolve in time. In that case, there is no extra frame displacement beyond the initial shift (so $\underset{\sim}{k} = \underset{\sim}{0}$), and there is no extra frame rotation beyond the initial one associated with the fixed-basis change (so $\underset{\approx}{Q} = \underset{\approx}{I}$ and $\underset{\approx}{\Upsilon} = \underset{\approx}{0}$). In this special case of a non-evolving frame change, the above equations reduce as follows:

$$\underset{\sim}{x}^* = \underset{\approx}{q}_0 \bullet \underset{\sim}{x} + \underset{\sim}{c}_0 \quad \text{origin-dependent.} \tag{19.59a}$$

$$\underset{\sim}{X}^* = \underset{\approx}{q}_0 \bullet \underset{\sim}{X} + \underset{\sim}{c}_0 \quad \text{origin-dependent.} \tag{19.59b}$$

From here forward, everything else (in the case of non-evolving frame change) transforms just like a basic basis change:

$$\underset{\sim}{u}^* = \underset{\approx}{q}_0 \bullet \underset{\sim}{u} \tag{19.59c}$$

$$\underset{\sim}{\mu}^* = \underset{\approx}{q}_0 \bullet \underset{\sim}{\mu} \tag{19.59d}$$

$$\underset{\sim}{v}^* = \underset{\approx}{q}_0 \bullet \underset{\sim}{v} \tag{19.59e}$$

$$\underset{\approx}{F}^* = \underset{\approx}{q}_0 \bullet \underset{\approx}{F} \bullet \underset{\approx}{q}_0^T \tag{19.59f}$$

$$\underset{\approx}{B}^* = \underset{\approx}{q}_0 \bullet \underset{\approx}{B} \bullet \underset{\approx}{q}_0^T \tag{19.59g}$$

$$\underset{\approx}{C}^* = \underset{\approx}{q}_0 \bullet \underset{\approx}{C} \bullet \underset{\approx}{q}_0^T \tag{19.59h}$$

$$\dot{\underset{\approx}{F}}^* = \underset{\approx}{q}_0 \bullet \dot{\underset{\approx}{F}} \bullet \underset{\approx}{q}_0^T \tag{19.59i}$$

$$\underset{\approx}{L}^* = \underset{\approx}{q}_0 \bullet \underset{\approx}{L} \bullet \underset{\approx}{q}_0^T \tag{19.59j}$$

$$\underset{\approx}{W}^* = \underset{\approx}{q}_0 \bullet \underset{\approx}{W} \bullet \underset{\approx}{q}_0^T \tag{19.59k}$$

$$\underset{\approx}{D}^* = \underset{\approx}{q}_0 \bullet \underset{\approx}{D} \bullet \underset{\approx}{q}_0^T \tag{19.59l}$$

$$\underset{\sim}{a}^* = \underset{\approx}{q}_0 \bullet \underset{\sim}{a}. \tag{19.59m}$$

Looking at the transformations under a non-evolving frame change allows variables to be identified based on their transformation formulas under a fixed (proper) frame change:

$$\textit{Free scalars:} \quad s^* = s \tag{19.60a}$$

$$\textit{Free vectors:} \qquad \underset{\sim}{v}^* = \underset{\approx}{q}_0 \bullet \underset{\sim}{v} \tag{19.60b}$$

$$\textit{Free second-order tensors:} \qquad \underset{\approx}{T}^* = \underset{\approx}{q}_0 \bullet \underset{\approx}{T} \bullet \underset{\approx}{q}_0^T. \tag{19.60c}$$

The pattern extends to higher-order tensors. In general, a *free tensor* (of any order) transforms under a non-evolving frame change in the same way components would change under a non-evolving basis change in the opposite direction of the frame rotation. In the previous list, we identify velocity and acceleration to be free vectors under fixed-frame change, while the position vector $\underset{\sim}{x}$ is an *origin-tied* vector because of its dependence on $\underset{\sim}{c}_0$. An example of an origin-tied *scalar* would be the magnitude of an origin-tied vector. More broadly, coordinates for the position vector are generally origin tied. Vectors constructed from origin-tied vectors might not be themselves origin tied. For example, the displacement $\underset{\sim}{u} = \underset{\sim}{x} - \underset{\sim}{X}$ is a free vector under a non-evolving frame change because, in that special case, $\underset{\sim}{u}^* = \underset{\sim}{x}^* - \underset{\sim}{X}^* = \underset{\approx}{q}_0 \bullet \underset{\sim}{u}$.

Keep in mind that frame-change analysis aims to ensure that governing equations are universal. The same set of governing equations (i.e., the same computer programs) should be used by each observer. Each observer will use inputs as they perceive them to be (which are different for different observers), but the equations themselves need to be the same. The governing equations are declared to be properly frame-indifferent[10] if their predictions are self-consistent with a frame change. This means that you should be able to write each governing equation with stars on each variable, and then substitute formulas for starred variables in terms of unstarred variables to obtain the exact same governing equation in the unstarred (fiducial) frame without needing any restriction on (potentially non-constant) origin translation and/or basis orientation. This section considered only a special subclass of possible frame change—those that do not change over time. The resulting transformation formulas ended up looking a lot like simple basis-change formulas. In fact, that is exactly how the equations should be interpreted if the vectors in the formulas are all polar. A necessary condition for frame indifference in this case is merely that the governing equations be well-formed tensor equations! Thus, a non-moving frame change is not very interesting. It just says that everything must transform properly under a basis change. Since this considered only non-moving frame change, this condition is merely necessary, not sufficient for frame indifference. The next section considers further restrictions that arise to satisfy frame indifference under evolving frame changes.

Frame changes can be extended to consider observer inversion, in which the superimposed tensor $\underset{\approx}{q}$ is improper orthogonal, as in the reflection of figure 19.4. A reflection implies that the 'handedness' of the two observers (fiducial and star) is reversed. If the fiducial system is right-handed, the star one is left-handed or vice versa. In this case, a vector (or, more broadly, any odd-order tensor) is said to be *polar* if, like the velocity vector in figure 19.4, the transformations in (19.60) continue

[10] The term *objective* is also used to refer to a frame-indifferent equation.

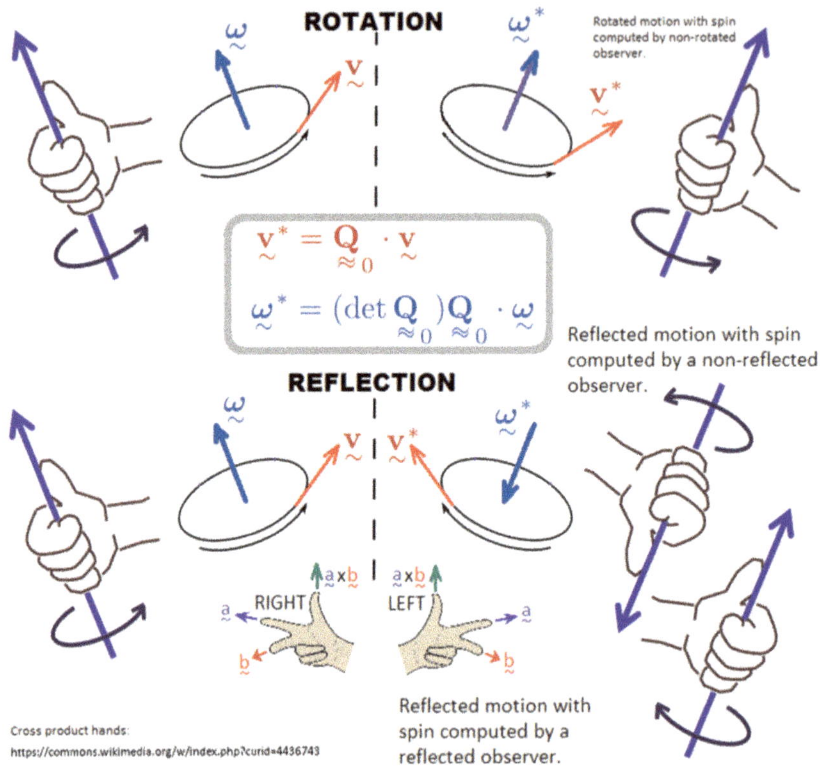

Figure 19.4. Superimposed rotation (top) compared to reflection (bottom). Axial vectors (like angular velocity) would be both reflected and inverted if a right-handed observer were to compute them directly from other data in the reflected motion.

to apply even under an improper (still fixed) frame change, but it is *axial* if, like the angular velocity vector in figure 19.4, the result also changes direction. Stated mathematically, for any non-evolving frame change (proper or not),

$$\textit{Polar vectors:} \quad \underset{\sim}{v}^* = \underset{\approx 0}{q} \bullet \underset{\sim}{v} \tag{19.61a}$$

$$\textit{Axial vectors:} \quad \underset{\sim}{w}^* = (\det \underset{\approx 0}{q})(\underset{\approx 0}{q} \bullet \underset{\sim}{w}). \tag{19.61b}$$

As an example, suppose that $\underset{\sim}{a}$ and $\underset{\sim}{b}$ are polar vectors in the plane of this page. If $\underset{\approx 0}{q}$ is a reflection in the horizontal direction, this means these vectors will transform into their mirror images:

What does this imply about cross products, such as those used in spinning motion and electromagnetism? Should an inversion change the right-handed cross product to a left-handed one? On the left (fiducial) side of this image, the cross product,

$\boldsymbol{w} = \boldsymbol{a} \times \boldsymbol{b}$, points out of the page, but the cross product $\boldsymbol{w}^* = \boldsymbol{a}^* \times \boldsymbol{b}^*$ points into the page, making it the *negative* of the mirror image and thus not a polar vector. This behavior persists for any choice of improper non-evolving frame change, so the cross product of polar vectors is an axial vector. There is some ambiguity in this assessment, however, because we are implicitly applying a right-handed definition of a cross product to a frame having a left-handed observer, who would apply the left-handed cross product and therefore obtain a reflected angular velocity vector, just like the polar vectors. This now comes down to pure choice: when the fiducial observer performs calculations to predict what is observed by the left-handed observer, should the fiducial observer change the sign convention of a cross product? A purist would say no, but it is more practical to say yes to the sign change simply to avoid having to revise an untold number of software libraries to take extra information about observer handedness into consideration before supplying output involving a cross-product operation. Since frame inversions are extremely rare, it makes more sense to just let the person calling this software make appropriate sign changes to the output so that these libraries can continue to run without revision. Let us examine a tensor formula for the cross product,

$$\boldsymbol{a} \times \boldsymbol{b} = \boldsymbol{\xi} : (\boldsymbol{a}\boldsymbol{b}), \tag{19.62}$$

where $\boldsymbol{\xi}$ is the permutation tensor whose components with respect to an orthonormal basis are $\xi_{ijk} = (\boldsymbol{e}_i \times \boldsymbol{e}_j) \bullet \boldsymbol{e}_k = \pm\varepsilon_{ijk}$. Here, ε_{ijk} denotes the permutation symbol, and the positive is used for a right-handed system and a negative is used for a left-handed system so that the volume of the cube with edges given by the basis vectors will change sign if the cube is inverted. Since frame indifference requires all observers to use the same governing equations,

$$\boldsymbol{a}^* \times \boldsymbol{b}^* = \boldsymbol{\xi}^* : (\boldsymbol{a}^*\boldsymbol{b}^*) = \boldsymbol{\xi}^* : [(\boldsymbol{Q}_0 \bullet \boldsymbol{a})(\boldsymbol{Q}_0 \bullet \boldsymbol{b})] = \boldsymbol{Q}_0 \bullet \left[\left(\boldsymbol{Q}_0^T > \boldsymbol{\xi}^*\right) : (\boldsymbol{a}\boldsymbol{b})\right]. \tag{19.63}$$

The last formula, which can be easily derived using indicial notation, uses the Raleigh product[11] for compactness. This brings us to the question: how should the permutation tensor itself change under a fixed improper frame change? We know that, because this tensor is isotropic, it will not change at all under a proper basis change. The question here is whether or not it should change sign under an improper frame change. It is proved later, in equation (20.40), that the permutation symbol's component matrix must change sign under a differently-handed basis change, but this is done to ensure that the permutation tensor itself (formed not just from components but also from the the underlying basis) does not change. Think of frame-change analysis as being constrained by an over-arching need for all observers to use exactly the same computer programs in their computations. A cross-product program that works for a right-handed system gives the third component of the cross product by

$$a_2 b_1 - a_1 b_2. \tag{19.64}$$

[11] If $\boldsymbol{T}$ is a third-order tensor and $\boldsymbol{A}$ is a second-order tensor, the *Rayleigh product* $\boldsymbol{B} = \boldsymbol{A} > \boldsymbol{T}$ is a third-order tensor defined indicially by $B_{ijk} = A_{ip}A_{jq}A_{kr}T_{pqr}$.

When this program is sent components in the star system, the starred third component will be $a_2^* b_1^* - a_2^* b_2^*$. The other components will be treated similarly. The results will imply that the cross-product operator produces results consistent with an axial vector. That is only because the computer program uses a right-handed convention for the cross product—it has no information about the handedness of the frame. At this point, you can choose to rewrite all cross-product programs to take the components of the permutation tensor as an additional input so that its output becomes a left-handed cross product when the observer is left-handed, or you can choose the more practical approach of leaving these programs unchanged so that you would apply the sign change as a wrapper to the output, making the program itself axial and the wrapper polar. Frame analysis is rooted in a desire for universality in the governing equations, which means universality of the computer programs that use those equations. To keep the cross-product operator in countless existing programs, it is simply more practical to recognize that these programs might have a sign change that must be accounted for after the fact when a result is sought in a left-handed frame. This choice is effectively defining

$$\underset{\approx}{\boldsymbol{q}}_0^T > \underset{\approx}{\boldsymbol{\xi}}^* = (\det \underset{\approx}{\boldsymbol{q}}_0)\underset{\approx}{\boldsymbol{\xi}} \tag{19.65}$$

so that, when the input vectors are polar, the output is axial:

$$\underset{\sim}{\boldsymbol{a}}^* \times \underset{\sim}{\boldsymbol{b}}^* = (\det \underset{\approx}{\boldsymbol{q}}_0)\underset{\approx}{\boldsymbol{q}}_0 \bullet (\underset{\sim}{\boldsymbol{a}} \times \underset{\sim}{\boldsymbol{b}}). \tag{19.66}$$

By choosing to effectively change the sign of the permutation tensor under an inversion transformation (so that a right-handed convention is used to evaluate it even though the observer's basis is left-handed), the vorticity vector $\underset{\sim}{\boldsymbol{w}}$ (equal to the angular velocity vector $\underset{\sim}{\boldsymbol{\omega}}$ in rigid motion) will be an axial vector. These considerations are inconsequential except in the extremely rare case of an inverted frame change.

19.3.3. Testing a numerical code for proper invariance under non-evolving frame change

A non-evolving frame change is, with the exception of origin dependence of the position vectors, tantamount to a non-evolving basis change. Therefore, to test a code in this case, the fiducial problem would be re-run with *everything* rotated by some selected *non-evolving* amount. The goal is to check if results are consistent with the fiducial simulation. By the nature of its algorithm, a finite-element method (FEM) momentum solver is likely to pass this test, whereas it could be more difficult for particle methods such as the one in figure 19.5(b), that use a rectilinear background grid and pixilated geometry. Even the FEM test might reveal bugs in its black-box functions. For example, a constitutive source-code typo that accidentally refers to a tensor's xz component where zy is needed would show no symptom of error in a horizontal pulling fiducial problem (because both of these components would be zero), yet it would reveal itself in with angled frame change. If the material is isotropic, then a simple expanding spherical shell test is worthwhile for catching this type of bug. Loss of spherical symmetry would indicate a non-evolving frame bug,

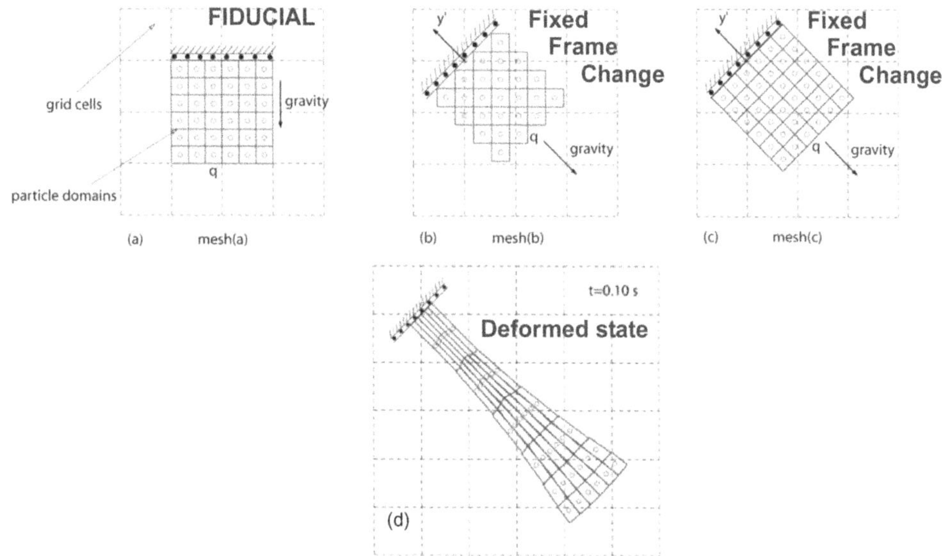

Figure 19.5. A fiducial simulation of vertical stretching of a bar re-run at an angle. Many particle codes discretize a body on a rectilinear grid, producing the stair-stepped boundary in (b) that induces large frame indifference errors in comparison to the vertical stretching of the fiducial problem. The boundary-conforming FEM-style discretization produces far smaller error [4].

which is typically tracked to inadvertently using the wrong components of vectors or tensors.

This section has considered a *non-evolving* frame change, which is effectively just a basis change. As a result, testing a code under non-evolving frame change would entail rotating all geometry and all boundary and initial conditions and all loadings, as in figure 19.5(b). Far more challenging *evolving frame-change* tests are the subject of the next section. This type of test would use figure 19.5(a) as the initial configuration and apply appropriate body forces to induce the same loading as the fiducial problem while *simultaneously* rotating the body and its boundary. To reiterate:

$$\textit{Fixed frame}\text{: } \underset{\approx}{Q} = \underset{\approx}{I},\ \underset{\sim}{k} = \underset{\sim}{0},\ \underset{\approx}{q}_0 \neq \underset{\approx}{I},\ \underset{\sim}{c}_0 \neq \underset{\sim}{0}.$$

$$\textit{Evolving frame}\text{: } \underset{\approx}{Q} = \underset{\approx}{Q}(t),\ \underset{\sim}{k} = \underset{\sim}{k}(t),\ \underset{\approx}{q}_0 = \underset{\approx}{I},\ \underset{\sim}{c}_0 = \underset{\sim}{0}.$$

Thus, for example,

Fixed frame: figure 19.5(c) evolves to become figure 19.5(d).

Evolving frame: figure 19.5(a) evolves to become figure 19.5(d).

The distinction between simple basis change and evolving frame indifference is illustrated yet again in figure 19.6 (where the universal requirement of form invariance of all governing equations goes by the name PMFI, when it is applied to constitutive models).

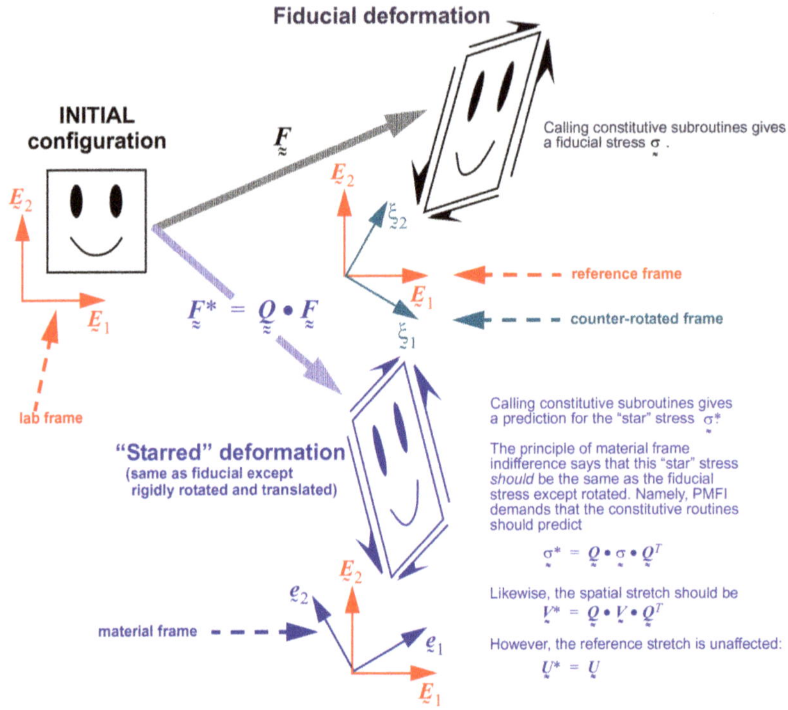

Figure 19.6. The distinction between PMFI and basis rotation. This sketch shows a general deformation and a second (starred) deformation that is identical except for a superimposed counter-clockwise rotation of 30°. Spatial tensors predicted from the constitutive law should be the same as the fiducial predictions except rotated. However, reference tensors such as the right stretch are unaffected by superimposed deformation. When the starred deformation is viewed from the material (co-rotating) frame, it appears identical to the fiducial deformation viewed from the reference (lab) frame. When the fiducial deformation is viewed from the counter-rotated frame, it appears the same as the starred deformation viewed from the reference (lab) frame.

19.3.4 Case 2: evolving frame change (key to frame indifference!)

This section takes the fiducial and star frames to initially coincide so that

$$\underset{\approx}{q}_0 = \underset{\approx}{I} \quad \text{and} \quad \underset{\sim}{c}_0 = \underset{\sim}{0}. \tag{19.67}$$

In this case, the frame-change formulas in (19.58) reduce to[12]

$$\boxed{\underset{\sim}{x}^* = \underset{\approx}{Q} \bullet \underset{\sim}{x} + \underset{\sim}{k}} \quad \text{mixed-type vector,} \tag{19.68a}$$

$$\boxed{\underset{\sim}{X}^* = \underset{\sim}{X}} \quad \text{E-type vector,} \tag{19.68b}$$

$$\boxed{\underset{\sim}{u}^* = \underset{\approx}{Q} \bullet \underset{\sim}{u} + (\underset{\approx}{Q} - \underset{\approx}{I}) \bullet \underset{\sim}{X} + \underset{\sim}{k}} \quad \text{mixed-type vector,} \tag{19.68c}$$

[12] The 'types' are explained in section 19.3.5.

$$\boxed{\underset{\sim}{\mu}^* = \underset{\approx}{Q} \bullet \underset{\sim}{\mu}} \quad \text{e-type vector,} \tag{19.68d}$$

$$\boxed{\underset{\sim}{v}^* = \underset{\approx}{Q} \bullet \underset{\sim}{v} + \underset{\approx}{\Upsilon} \bullet \underset{\approx}{Q} \bullet \underset{\sim}{x} + \dot{\underset{\sim}{k}}} \quad \text{mixed-type vector,} \tag{19.68e}$$

$$\boxed{\underset{\approx}{F}^* = \underset{\approx}{Q} \bullet \underset{\approx}{F}} \quad \text{eE-type tensor,} \tag{19.68f}$$

$$\boxed{\underset{\approx}{B}^* = \underset{\approx}{Q} \bullet \underset{\approx}{B} \bullet \underset{\approx}{Q}^T} \quad \text{ee-type tensor,} \tag{19.68g}$$

$$\boxed{\underset{\approx}{C}^* = \underset{\approx}{C}} \quad \text{EE-type tensor,} \tag{19.68h}$$

$$\boxed{\dot{\underset{\approx}{F}}^* = \underset{\approx}{Q} \bullet \dot{\underset{\approx}{F}} + \underset{\approx}{\Upsilon} \bullet \underset{\approx}{F}} \quad \text{mixed-type tensor,} \tag{19.68i}$$

$$\boxed{\underset{\approx}{L}^* = \underset{\approx}{Q} \bullet \underset{\approx}{L} \bullet \underset{\approx}{Q}^T + \underset{\approx}{\Upsilon}} \quad \text{mixed-type tensor,} \tag{19.68j}$$

$$\boxed{\underset{\approx}{W}^* = \underset{\approx}{Q} \bullet \underset{\approx}{W} \bullet \underset{\approx}{Q}^T + \underset{\approx}{\Upsilon}} \quad \text{mixed-type tensor,} \tag{19.68k}$$

$$\boxed{\underset{\approx}{D}^* = \underset{\approx}{Q} \bullet \underset{\approx}{D} \bullet \underset{\approx}{Q}^T} \quad \text{ee-type tensor,} \tag{19.68l}$$

$$\boxed{\underset{\sim}{a}^* = \underset{\approx}{Q} \bullet \underset{\sim}{a} + \left(\dot{\underset{\approx}{\Upsilon}} + \underset{\approx}{\Upsilon}^2\right) \bullet \underset{\approx}{Q} \bullet \underset{\sim}{x} + 2\underset{\approx}{\Upsilon} \bullet \underset{\approx}{Q} \bullet \underset{\sim}{v} + \ddot{\underset{\sim}{k}}} \tag{19.68m}$$

mixed-type vector.

Note that, for any vector $\underset{\sim}{r}$,

$$\underset{\approx}{\Upsilon} \bullet \underset{\sim}{r} = \underset{\sim}{\varpi} \times \underset{\sim}{r}, \; \dot{\underset{\approx}{\Upsilon}} \bullet \underset{\sim}{r} = \dot{\underset{\sim}{\varpi}} \times \underset{\sim}{r}, \text{ and } \underset{\approx}{\Upsilon} \bullet \underset{\approx}{\Upsilon} \bullet \underset{\sim}{r} = \underset{\sim}{\varpi} \times (\underset{\sim}{\varpi} \times \underset{\sim}{r}) \tag{19.69}$$

where $\underset{\sim}{\varpi}$ is the axial vector associated with superimposed spin. Therefore, these tensor operators in the acceleration formula respectively correspond to angular velocity, tangential acceleration, and centripetal acceleration (reducing respectively to familiar formulas $R\omega$, $R\dot{\omega}$, and $R\omega^2$ for purely circular motion). Because of these terms, a frame-indifference verification test in a numerical code must do more than just rotate the problem inputs (tantamount to a basis change), but it must also add extra body forces that *exactly* counteract the phantom inertial body forces associated with superimposed acceleration. Specifically, the position and velocity fields in the fiducial simulation must be used as part of the computation of extra body forces to add to the test problem of an evolving superimposed rigid motion:

$$\boxed{\underset{\sim}{\boldsymbol{b}}^{\text{extra}} = \ddot{\underset{\approx}{\boldsymbol{Q}}} \bullet \underset{\sim}{\boldsymbol{x}} + 2\dot{\underset{\approx}{\boldsymbol{Q}}} \bullet \underset{\sim}{\boldsymbol{v}} + \ddot{\underset{\sim}{\boldsymbol{k}}}}. \tag{19.70}$$

It is fairly common for frame-change equations to be derived without ever considering the non-evolving frame change that we looked at in the previous section. That is because *non-evolving* frame change relations merely imply that the governing equations must be well-formed tensor equations. As long as the variables appearing in the governing equations are all polar, the more interesting/challenging constraints associated with evolving frame changes can be inferred without loss in generality by assuming that the fiducial and star frames initially coincide as we have done here. All subsequent analyses of frame change in this book will continue with this assumption. In this case,

$$\text{A two-point push tensor transforms by} \quad \underset{\approx}{\boldsymbol{A}}^* = \underset{\approx}{\boldsymbol{Q}} \bullet \underset{\approx}{\boldsymbol{A}}. \tag{19.71a}$$

$$\text{A two-point pull tensor transforms by} \quad \underset{\approx}{\boldsymbol{A}}^* = \underset{\approx}{\boldsymbol{A}} \bullet \underset{\approx}{\boldsymbol{Q}}^T. \tag{19.71b}$$

$$\text{A spatial tensor transforms by} \quad \underset{\approx}{\boldsymbol{A}}^* = \underset{\approx}{\boldsymbol{Q}} \bullet \underset{\approx}{\boldsymbol{A}} \bullet \underset{\approx}{\boldsymbol{Q}}^T. \tag{19.71c}$$

$$\text{A reference tensor transforms by} \quad \underset{\approx}{\boldsymbol{A}}^* = \underset{\approx}{\boldsymbol{A}}. \tag{19.71d}$$

19.3.5 Natural basis (identifying Ee-, eE-, Ee-, and EE-type tensors)

Letting $\underset{\sim}{\boldsymbol{E}}_k$ denote the laboratory basis used in the fiducial problem, we may expand a generic tensor as

$$\underset{\approx}{\boldsymbol{A}} = A_{ij} \underset{\sim}{\boldsymbol{E}}_i \underset{\sim}{\boldsymbol{E}}_j. \tag{19.72}$$

Then, defining a natural basis for the problem with superimposed rotation to be

$$\underset{\sim}{\boldsymbol{e}}_k = \underset{\approx}{\boldsymbol{Q}} \bullet \underset{\sim}{\boldsymbol{E}}_k, \tag{19.73}$$

it follows that:[13]

$$\text{a two-point push tensor expands as} \quad \underset{\approx}{\boldsymbol{A}}^* = A_{ij} \underset{\sim}{\boldsymbol{e}}_i \underset{\sim}{\boldsymbol{E}}_j, \tag{19.74a}$$

$$\text{a two-point pull tensor expands as} \quad \underset{\approx}{\boldsymbol{A}}^* = A_{ij} \underset{\sim}{\boldsymbol{E}}_i \underset{\sim}{\boldsymbol{e}}_j, \tag{19.74b}$$

$$\text{a spatial tensor expands as} \quad \underset{\approx}{\boldsymbol{A}}^* = A_{ij} \underset{\sim}{\boldsymbol{e}}_i \underset{\sim}{\boldsymbol{e}}_j, \tag{19.74c}$$

[13] As pointed out by Bigoni [5], the word 'objective' unfortunately has different definitions in the literature. Therefore, like him, we avoid the term altogether. The term 'objectivity' is often used to refer to what we call frame indifference.

$$\text{a reference tensor expands as} \quad \underset{\approx}{A}^* = A_{ij}\underset{\sim}{E}_i\underset{\sim}{E}_j. \tag{19.74d}$$

In other words, a two-point tensor has the same components as the fiducial tensor, except referenced to a mixed basis. A spatial tensor has the same components as the fiducial tensor except referenced to the rotated basis. A reference tensor has the same components as the fiducial tensor, both to the fiducial laboratory basis. For these reasons, the abbreviations *eE*, *Ee*, *ee*, and *EE* are a shorthand way to refer to the above four types of tensors. Many writers like to emphasize these relationships by using index letters that are capitalized if the associated natural basis is capitalized. For example, components of the deformation gradient, which is an *eE*-type of tensor, might be written F_{iJ}. The problem with this notational choice is that not all tensors can be identified to fall into one of the above four groups.

19.3.6 Scalars are not automatically insensitive to frame change!

Consider the determinant of the four classes of tensor considered so far. Noting that $\det\underset{\approx}{Q} = 1$ for a rotation,

$$\text{if } \underset{\approx}{A} \text{ is a two-point push tensor, then} \quad \det\underset{\approx}{A}^* = \det\underset{\approx}{A}, \tag{19.75a}$$

$$\text{if } \underset{\approx}{A} \text{ is a two-point pull tensor, then} \quad \det\underset{\approx}{A}^* = \det\underset{\approx}{A}, \tag{19.75b}$$

$$\text{if } \underset{\approx}{A} \text{ is a spatial (co-rotational) tensor, then} \quad \det\underset{\approx}{A}^* = \det\underset{\approx}{A}, \tag{19.75c}$$

$$\text{if } \underset{\approx}{A} \text{ is a reference tensor, then} \quad \det\underset{\approx}{A}^* = \det\underset{\approx}{A}. \tag{19.75d}$$

Thus, in all four cases, superimposed rotation has no effect on the determinant. Likewise, another invariant, the magnitude, is unaffected by superimposed rotation. For vectors like the material line segment which transforms according to $\underset{\sim}{\mu}^* = \underset{\approx}{Q} \bullet \underset{\sim}{\mu}$, it follows that superimposed rotation does not affect the vector's length. In fact, the vast majority of scalars are unaffected by superimposed motion, but not all of them. The magnitude of any frame-sensitive vector (like position or displacement) might be frame sensitive. The trace of a spatial or reference tensor will be frame insensitive, but the trace of a two-point tensor will be frame sensitive. For example, the polar rotation angle (found from the trace of the two-point polar rotation tensor) is frame sensitive, which should not be shocking or disconcerting since frame change corresponds to extra rotation. The main point is that you must never declare that a scalar is frame insensitive merely because it is a scalar. That's usually true, but not always.

19.4. Mechanics principles under frame change

Kinematics, the study of motion and deformation, was the sole focus of the previous section. This section on mechanics asserts how forces and stresses that induce motion ought to change under superimposed rigid motion. These assertions are all rooted in one common axiom: as long as appropriate extra body forces in (19.70) are incorporated to provide the acceleration needed for superimposed motion (and hence exactly counteract associated inertial effects), all other actual loads on the body (such as body forces or surface tractions in the fiducial problem) ought to co-rotate under superimposed rotation and be entirely unaffected by superimposed translation[14]. Thus, if $\underset{\sim}{f}$ is a force applied at the current time, any superimposed rigid motion that exists at that time ought to transform it according to

$$\textit{Force:} \quad \boxed{\underset{\sim}{f}^* = \underset{\approx}{Q} \bullet \underset{\sim}{f}}. \tag{19.76}$$

This is the principal mechanics axiom leading to all of the other transformation formulas cited in this section. In continuum mechanics, traction $\underset{\sim}{t}$ is a vector defined to be the current differential force acting on a body from a distributed load per unit differential area at its point of application. Since the differential area *magnitude* is unaffected by superimposed rigid motion, it follows that

$$\textit{Traction:} \quad \boxed{\underset{\sim}{t}^* = \underset{\approx}{Q} \bullet \underset{\sim}{t}}. \tag{19.77}$$

Therefore, any applied traction boundary conditions in the fiducial problem must be transformed according to this equation to become appropriate boundary conditions in any 'star' simulation test for frame indifference. Likewise, in addition to the extra body forces in (19.70), body forces in the fiducial problem must be rotated similarly in a 'star' problem so that the body force in star problem must be

$$\textit{Body force:} \quad \boxed{\underset{\sim}{b}^* = \underset{\approx}{Q} \bullet \underset{\sim}{b} + \ddot{\underset{\approx}{Q}} \bullet \underset{\sim}{x} + 2\dot{\underset{\approx}{Q}} \bullet \underset{\sim}{v} + \ddot{\underset{\sim}{k}}}. \tag{19.78}$$

This is a key equation for setting up engineering simulations in a non-inertial frame: turbine blades, centrifuges, rocket sleds, roller-coaster seats, etc. This result is revisited separately later to write it in a more useful form. If, for example, you are trying to predict tire wear from hard stop-and-go traffic conditions, numerical integration error would be best managed by running the simulation as it is seen by an observer who moves with the bulk motion of the tire. The observer would want to move with a velocity that keeps the rotation hub in the center of the domain. If, for example, the hub moves to the right, it follows that the $\underset{\sim}{k}$ vector would need to point to the *left* by the same amount (since, recall, this vector is the negative of the motion of the observer; if, for example, the wheel moves to the right, then a leftward superimposed translation would be needed to bring it back). Accordingly, if the vehicle brakes hard so that the fiducial (ground-based) observer sees a strong backward-pointing hub acceleration $\underset{\sim}{a}$, then the $\ddot{\underset{\sim}{k}}$ contribution to the extra body force would equal $-\underset{\sim}{a}$, which is a *forward-pointing* phantom body force just like the one you feel pushing you towards the dashboard when riding in a braking car. Similarly, if the wheel is rotating according to some tensor $\underset{\approx}{R}$, then an opposite rotation, $\underset{\approx}{Q} = \underset{\approx}{R}^T$, must be superimposed to create a deformation that looks the same to a stationary observer viewing the star motion as what is seen by a rotating observer viewing the fiducial motion. This effectively turns the inward pointing

[14] A force's point of application co-translates, but the force vector itself is unchanged by translation.

centripetal acceleration in the fiducial problem into an outward-pointing centrifugal body force in the moving frame, as it should! Because of the $2\dot{\underset{\approx}{Q}} \bullet \underset{\sim}{v}$ term, there would also be a Coriolis effect. If the simulation is being run entirely in the rotating frame, the ground-based (non-starred) velocity is not known. Likewise, the non-starred deformed position vector is not known. Recall

$$\underset{\sim}{x}^* = \underset{\approx}{Q} \bullet \underset{\sim}{x} + \underset{\sim}{k} \qquad \text{and} \qquad \underset{\sim}{v}^* = \underset{\approx}{Q} \bullet \underset{\sim}{v} + \underset{\approx}{\Upsilon} \bullet \underset{\approx}{Q} \bullet \underset{\sim}{x} + \dot{\underset{\sim}{k}}. \tag{19.79}$$

Solving these for $\underset{\sim}{x}$ and $\underset{\sim}{v}$ would then allow the body force in the rotating frame to be expressed in terms of 'star' variables being simulated in that frame, allowing evaluation of the transient body force in (19.78) as needed during the simulation.

Recall the axiom that applied forces in the fiducial problem should co-rotate in the 'star' problem. What does this imply about stress? The famous Cauchy tetrahedron argument is used to prove that traction depends linearly on the surface normal $\underset{\sim}{n}$, which implies existence of a tensor, called Cauchy stress $\underset{\approx}{\sigma}$ defined such that

$$\underset{\sim}{t} = \underset{\approx}{\sigma} \bullet \underset{\sim}{n} \text{ so that } d\underset{\sim}{f} = \underset{\approx}{\sigma} \bullet \underset{\sim}{n}\, dA. \tag{19.80}$$

Although differential area dA itself is unaffected by superimposed rigid motion, kinematics tells us that

$$\textit{Current surface normal:} \quad \boxed{\underset{\sim}{n}^* = \underset{\approx}{Q} \bullet \underset{\sim}{n}}. \tag{19.81}$$

Together, the above three equations combine so that

$$\underset{\sim}{t}^* = \underset{\approx}{Q} \bullet \underset{\sim}{t} = \underset{\approx}{Q} \bullet \underset{\approx}{\sigma} \bullet \underset{\sim}{n} = \underset{\approx}{Q} \bullet \underset{\approx}{\sigma} \bullet \underset{\approx}{Q}^T \bullet \underset{\sim}{n}^* = \underset{\approx}{\sigma}^* \bullet \underset{\sim}{n}^*, \tag{19.82}$$

where

$$\textit{Cauchy stress:} \quad \boxed{\underset{\approx}{\sigma}^* = \underset{\approx}{Q} \bullet \underset{\approx}{\sigma} \bullet \underset{\approx}{Q}^T}. \tag{19.83}$$

Because the $\underset{\approx}{Q}$ appears here on both sides in a Rayleigh product, it follows that Cauchy stress is a purely spatial (*ee*-type) tensor. As an operator, it acts on something in the current configuration (namely the current area element vector) to produce something in the current configuration (differential force acting on the area element).

While Cauchy stress gives the actual current force per current area, the first Piola–Kirchhoff (PK1) stress, $\underset{\approx}{P}$, is a nominal stress, meaning it is current the force per *initial* area. In particular, it is defined such that

$$\underset{\approx}{P} \bullet \underset{\sim}{N}\, dA_0 = \underset{\approx}{\sigma} \bullet \underset{\sim}{n}\, dA, \tag{19.84}$$

where $\underset{\sim}{N}$ is the initial unit normal of the surface element, and dA_0 is the initial area. Since frame indifference requires all governing equations (including definitions) be form-invariant, we require that

$$\underset{\approx}{P}^* \bullet \underset{\sim}{N}^*\, dA_0^* = \underset{\approx}{\sigma}^* \bullet \underset{\sim}{n}^*\, dA^*. \tag{19.85}$$

The initial configuration is treated to have no superimposed motion, and therefore:

$$\textit{Initial surface normal:} \quad \boxed{\underset{\sim}{N}^* = \underset{\sim}{N}}. \tag{19.86}$$

Accordingly, because area magnitudes are unaffected by superimposed rigid motion, the previously derived frame-change formulas for $\underset{\approx}{\sigma}^*$ and $\underset{\sim}{n}^*$ show that

$$\textit{First Piola–Kirchhoff stress:} \quad \boxed{\underset{\approx}{P}^* = \underset{\approx}{Q} \bullet \underset{\approx}{P}}. \tag{19.87}$$

Noting that this has only a single $\underset{\approx}{Q}$ on the left side, it follows that the PK1 stress is a two-point push tensor. It acts on something in the initial configuration (namely the initial area element) to give something in the current configuration (the current differential force acting on the current area element).

Traction $\underset{\sim}{t}$ is the current differential force per current area, making $\underset{\sim}{t}\mathrm{d}A$ the force acting on that area, and we can define a 'pull back' of current force by acting on it with the inverse of the deformation,

$$\underset{\approx}{F}^{-1} \bullet \underset{\sim}{t}\,\mathrm{d}A = \underset{\approx}{F}^{-1} \bullet \underset{\approx}{\sigma} \bullet \underset{\sim}{n}\,\mathrm{d}A = \underset{\approx}{F}^{-1} \bullet \underset{\approx}{P} \bullet \underset{\sim}{N}\,\mathrm{d}A_0 = \underset{\approx}{S} \bullet \underset{\sim}{N}\,\mathrm{d}A_0, \tag{19.88}$$

where

$$\underset{\approx}{S} = \underset{\approx}{F}^{-1} \bullet \underset{\approx}{P} \tag{19.89}$$

is called the second Piola–Kirchhoff (PK2) stress and may be interpreted as the *pulled back* force per initial area. The pulled back force is not the initial force; it is merely a distortion of the current force vector, which describes it relative to the initial geometry. A great way to better understand the PK1 and PK2 stresses is to evaluate them in the intuitive fiducial case of uniaxial stress, and to then evaluate them again in the case that the deformed (not initial) configuration is additionally subjected to a 90 degree rotation. Recalling that $\underset{\approx}{F}^* = \underset{\approx}{Q} \bullet \underset{\approx}{F}$ and that $\underset{\approx}{P}^* = \underset{\approx}{Q} \bullet \underset{\approx}{P}$, it follows that

$$\underset{\approx}{S}^* = (\underset{\approx}{F}^*)^{-1} \bullet \underset{\approx}{P}^* = \underset{\approx}{F}^{-1} \bullet \underset{\approx}{Q}^T \bullet \underset{\approx}{Q} \bullet \underset{\approx}{P} = \underset{\approx}{F}^{-1} \bullet \underset{\approx}{P} = \underset{\approx}{S}. \tag{19.90}$$

Therefore, the PK2 stress is a material reference tensor—even though it quantifies something that is happening in the current configuration, it does not change in response to superimposed rigid motion:

$$\textit{Second Piola–Kirchoff stress:} \quad \boxed{\underset{\approx}{S}^* = \underset{\approx}{S}}. \tag{19.91}$$

At this point, we have derived enough transformation formulas to start using them to apply frame indifference constraints to constitutive models. When applied to a constitutive model, the admissibility condition is called the PMFI, and it merely requires that the same constitutive model should be applied by all observers. Inputs to the model must be as they appear to be in the observer's frame. The constitutive model passes frame indifference testing if its predicted output is consistent with the result of merely applying one of the transformation formulas in this section directly to the fiducial output from the same constitutive model. The upcoming examples should clarify this fundamental and fairly intuitive requirement. Imagine a constitutive model that predicts the force required to elongate a rubber band by 2 cm. Frame indifference requires that the same force magnitude would be required to achieve the same elongation even if the experiment is conducted in a state of free-fall where both the rubber band and observer are accelerating as they fall weightless under gravity. It requires the same force magnitude for the same elongation even if the band is being rotated. The force direction must corotate with the material. For either of these non-inertial observer examples, appropriate body forces must be applied to exactly counteract inertial effects. For example, a radially inward pointing body force must be applied to a ball at the end of a band in order to rotate it around a circular path without outwardly stretching the band from centrifugal inertia (and this makes sense when you recall that rotating a stretched band without rotating the observer is intended to emulate oppositely rotating the observer without rotating the band).

19.4.1. Reiteration of key points about superimposed rigid motion

Any deformation always begins at a certain initial configuration and deforms to a new *current* configuration. Under a superimposed rotation, nothing is done to the initial configuration. Superimposed rotation evolves during the fiducial deformation to create a different final 'star' configuration. Both the fiducial and star configurations have *the same initial configuration*. For example, figure 19.7 shows a simple shear fiducial problem and a 'star' that begins from the same initial square configuration, and which experiences the same amount of shear, but is also rotated clockwise by 45°. Depending on the placement of the origin, the star configuration may also include a translation. In physics, the word 'fiducial' means 'regarded or employed as a standard or reference for measurement or calculation'. Here, we are interested in describing how scalars, vectors, and tensors in the rotated 'star' configuration compare with their counterparts in the fiducial configuration.

For an evolving frame change that is initially coincident with the fiducial configuration, a superimposed rigid motion (second situation in figure 19.7) will transform things very differently from the non-evolving frame change. Some scalars, vectors, and tensors will be unaffected by the extra rotation—these so-called *reference* quantities are the same in both the fiducial and 'star' configurations. Other scalars, vectors, and tensors will change in response to the superimposed rotation. Scalar invariants of *most* vectors and tensors (e.g., their magnitudes) remain unchanged, making them invariant under superimposed rotations. Unfortunately, however, some researchers wrongly assert that *any* scalar invariant will be also invariant under superimposed rotation. The fiducial deformation shown

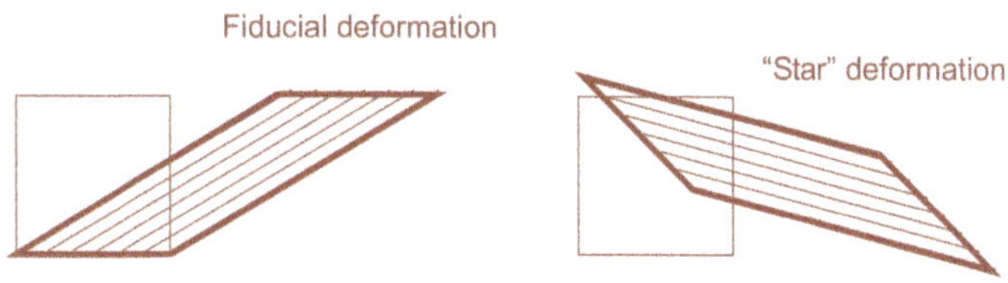

Figure 19.7. Superimposed rotation. The fiducial configuration is simple shear. The 'star' deformation is identical except also subjected to an evolving rotation such that both problems have the same initial state but deviate from each other by increasing rotation as the shearing proceeds.

in figure 19.7 entails a certain amount of material rotation from the shearing. The 'star' deformation entails that same amount of rotation *plus* extra rotation supplied by the superimposed rotation. Thus, the two deformations do *not* have the same rotation angle. Since the rotation angle is obtained from the trace of the polar rotation tensor, this proves that the trace of the rotation tensor is *not* invariant under superimposed rotations even though it *is* invariant under a simple basis change (i.e., a non-evolving change in observer frame).

Under a superimposed rotation, some vectors will change but others will not. For example, the *initial* length and orientation of a material fiber $\underset{\sim}{M}$ will not be affected by a superimposed rotation because (by premise) both the fiducial and 'star' configurations are associated with *the same* initial configuration. Thus,

$$\underset{\sim}{M}^* = \underset{\sim}{M}. \tag{19.92}$$

However, a *deformed* material fiber $\underset{\sim}{m}$ in the fiducial configuration will change upon a superimposed rotation to become a different vector $\underset{\sim}{m}^*$ that is related to its fiducial counterpart by

$$\underset{\sim}{m}^* = \underset{\approx}{Q} \bullet \underset{\sim}{m}, \tag{19.93}$$

where $\underset{\approx}{Q}$ is the superimposed rotation tensor. Whether or not a vector will change upon superimposed rotation depends on the vector's *physical definition.* As a rule, if a vector is defined entirely in terms of the initial state, then it will *not* change upon superimposed rotation. If, on the other hand, a vector is defined entirely in terms of the deformed state, then it will typically transform according to an equation similar to (19.93). Some vectors are defined in a manner that involves *both* the initial and deformed states. For example, displacement $\underset{\sim}{u}$ is defined to be the difference between the deformed and initial position vectors. Configuration-coupling vectors like this will transform in ways unlike a reference vector and unlike a spatial vector:

$$\underset{\sim}{u}^* \neq \underset{\sim}{u} \qquad \text{and} \qquad \underset{\sim}{u}^* \neq \underset{\approx}{Q} \bullet \underset{\sim}{u}. \tag{19.94}$$

Similar statements can be made about how *tensors* will change upon superimposed rotation. If a tensor's definition is cast entirely in terms of the initial state, or if a tensor is defined in a manner that effectively eliminates influence of rotation, then that tensor will be unaffected by a superimposed rotation. For example, the *right* polar stretch tensor $\underset{\approx}{U}$ characterizes the distortional part of the material deformation *before* application of rotation, so no amount of superimposed rotation will affect it. That is, both the fiducial and 'star' deformations have the same right polar stretch:

$$\underset{\approx}{U}^* = \underset{\approx}{U}. \tag{19.95}$$

By contrast, the *left* stretch $\underset{\approx}{V}$, which characterizes distortional deformation *after* application of rotation will transform according to

$$\underset{\approx}{V}^* = \underset{\approx}{Q} \bullet \underset{\approx}{V} \bullet \underset{\approx}{Q}^T. \tag{19.96}$$

Compare this equation with (3.56). The *lab* components of $\underset{\approx}{V}^*$ are related to the *lab* components of $\underset{\approx}{V}$ by an equation that is similar in structure to an ordinary component transformation relation. The Cauchy stress tensor also transforms similarly to (19.96). These spatial tensors always 'co-rotate' with the material:

$$\underset{\approx}{\sigma}^* = \underset{\approx}{Q} \bullet \underset{\approx}{\sigma} \bullet \underset{\approx}{Q}^T. \tag{19.97}$$

Depending on their physical definitions, some other tensors might not transform under superimposed rotation in a manner similar to *either* (19.95) or (19.97). For example, if the deformation gradient tensor for the fiducial deformation is $\underset{\approx}{F}$, then the deformation gradient for the 'star' configuration will be the same except pre-multiplied by $\underset{\approx}{Q}$:

$$\underset{\approx}{F}^* = \underset{\approx}{Q} \bullet \underset{\approx}{F}. \tag{19.98}$$

As an operator, the deformation gradient tensor transforms material line segments in the *initial* configuration into deformed segments in the final configuration. There is no post-multiplication by $\underset{\approx}{q}_0^T$ in the above equation because (recall) this section is a special case for which the fiducial and star problems have *the same* initial configuration ($\underset{\approx}{q}_0 = \underset{\approx}{I}$).

To close, we re-emphasize that evolving superimposed-rotation transformations are fundamentally different from non-evolving basis transformations (which effectively have the same non-evolving superimposed rotation in both the initial and any future configuration). For *spatial* tensors, the transformation formulas appear similar to an ordinary basis change, but other tensors obey structurally different transformation rules. Whereas only isotropic tensors have components that are invariant under a basis transformation, there are many non-isotropic tensors (such as the right stretch in (19.95)) that are unaffected by superimposed rotation. Basis transformation formulas apply in the same way to *all* free scalars, vectors, and tensors regardless of how defined. Evolving superimposed-rotation transformation formulas vary depending on the physical definition of the quantity being transformed. Invariance under non-evolving frame changes merely requires governing equations to be well-formed tensor equations, but invariance under arbitrary frame changes implies far more restriction on admissible constitutive models, which is good news because it typically helps to limit the amount of information that must be sought in laboratory testing. For example, we now explain why laboratory calibration of a constitutive model never needs to include any bulk material rotation. Constitutive response for any (local) material is determined exclusively from knowledge of the initial state and the evolving reference stretch tensor $\underset{\approx}{U}$. The influence of the polar rotation is governed by universal laws that apply to all materials (and thus belong on the outside of constitutive models to allow a single set of routines to be written and maintained by a host code to handle effects of material rotation).

19.4.2 Universal restrictions on anisotropic elastic models

A generally anisotropic elastic material is governed by $\underset{\approx}{\sigma} = g(\underset{\approx}{F})$, where $\underset{\approx}{\sigma}$ is the Cauchy stress, $\underset{\approx}{F}$ is the deformation gradient tensor, and g is an elastic constitutive function. To describe an anisotropic medium, this function implicitly depends on additional material data such as elastic moduli, reinforcing fiber lay-up directions, etc. A key result from frame analysis will be that some of those implicit material arguments must be changed to explicit function arguments in order to pass the requirement of material frame indifference. For software design, this implies that they must never be hard-coded internal parameters, but must instead be actual calling arguments. If a *fiducial deformation* is defined by a deformation gradient $\underset{\approx}{F}$, then the constitutive model predicts a *fiducial stress* $\underset{\approx}{\sigma} = g(\underset{\approx}{F})$. Now consider a 'star' deformation that *begins with the same initial configuration*[15] and is deformed the same as in the fiducial problem except that the deformed state is subjected to additional superimposed rotation defined by an orthogonal tensor $\underset{\approx}{Q}$, as well as possibly a superimposed rigid translation. Then the new 'star' *deformation* has a deformation gradient given by $\underset{\approx}{F}^* = \underset{\approx}{Q} \bullet \underset{\approx}{F}$ with a corresponding predicted 'star' *stress* of $\underset{\approx}{\sigma}^*_{\text{predicted}} = g(\underset{\approx}{F}^*)$. Frame indifference requires that all observers must use exactly the same governing equations, which is why the g-function has no star (think of it as a computer program that is used everyone without any software revisions). Inputs must be whatever they are perceived to be. If both inputs and outputs are known for a fiducial deformation, then inputs and outputs must both change appropriately in the star problem. The intuitive PMFI requires that the stress predicted from the constitutive model using the star input $\underset{\approx}{F}^*$ must be related to the fiducial stress $\underset{\approx}{\sigma}$ (predicted using the exact same model) according to $\underset{\approx}{\sigma}^*_{\text{desired}} = \underset{\approx}{Q} \bullet \underset{\approx}{\sigma} \bullet \underset{\approx}{Q}^T$. In other words, the result of superimposing rotation on the deformed configuration must be a co-rotation of the Cauchy stress. Stated a bit more mathematically, all of the following equations must be true without any restriction whatsoever on the superimposed rotation or translation:

$$\underset{\approx}{\sigma} = g(\underset{\approx}{F}), \tag{19.99a}$$

$$\underset{\approx}{F}^* = \underset{\approx}{Q} \bullet \underset{\approx}{F}, \tag{19.99b}$$

$$\underset{\approx}{\sigma}^*_{\text{predicted}} = g(\underset{\approx}{F}^*), \tag{19.99c}$$

$$\underset{\approx}{\sigma}^*_{\text{desired}} = \underset{\approx}{Q} \bullet \underset{\approx}{\sigma} \bullet \underset{\approx}{Q}^T, \tag{19.99d}$$

[15] Note the distinction here in comparison to group theory that was previously used to identify material symmetries (cubic, tetragonal, etc.) by demonstrating *insensitivity* of material response to rotations applied before the deformation. This section considers how material response must change in response to any superimposed rotation (not necessarily from a symmetry group). There is no assertion that the response must be unaffected by rotation—instead, it must change in the predictable ways that are described in this chapter.

$$\underset{\approx}{\sigma}^*_{\text{predicted}} = \underset{\approx}{\sigma}^*_{\text{desired}}. \tag{19.99e}$$

In essence, a frame-indifference analysis proceeds in the following steps:

STEP 1. Write down the constitutive model twice, once without asterisks and once with asterisks.
STEP 2. Write down the formulas for each of the starred variables in terms of non-starred counterparts, separately considering non-evolving frame changes and superimposed frame changes.
STEP 3. Identify constraints or implications arising from the need for all equations to be true simultaneously for all possible choices of input variables and for all possible choices of the superimposed rotation tensor.

The special case of a non-evolving frame change is merely checking if the constitutive model is comprised of well-formed tensor equations, which can be usually confirmed at a glance and hence is not usually even investigated. The more interesting and subtle observations are drawn from the case of evolving superimposed rotation (defined so that both frames are aligned in the reference configuration, but they become misaligned as the deformation proceeds). For example, these steps are applied to the model $\underset{\approx}{\sigma} = g(\underset{\approx}{F})$ as follows:

STEP 1.

$$\underset{\approx}{\sigma} = g(\underset{\approx}{F}),$$
$$\underset{\approx}{\sigma}^* = g(\underset{\approx}{F}^*).$$

STEP 2a. Non-evolving frame change:

$$\underset{\approx}{\sigma}^* = \underset{\approx}{q} \bullet \underset{\approx}{\sigma} \bullet \underset{\approx}{q}^T,$$

$$\underset{\approx}{F}^* = \underset{\approx}{q} \bullet \underset{\approx}{F} \bullet \underset{\approx}{q}^T.$$

STEP 2b. Evolving frame change:

$$\underset{\approx}{\sigma}^* = \underset{\approx}{Q} \bullet \underset{\approx}{\sigma} \bullet \underset{\approx}{Q}^T,$$
$$\underset{\approx}{F}^* = \underset{\approx}{Q} \bullet \underset{\approx}{F}.$$

STEP 3. Insisting that these equations all be true for any choice of orthogonal and possibly time evolving $\underset{\approx}{Q}$, orthogonal and constant $\underset{\approx}{q}$, and arbitrary deformation gradient $\underset{\approx}{F}$ gives the conditions

$$g(\underset{\approx}{q} \bullet \underset{\approx}{F} \bullet \underset{\approx}{q}^T) = \underset{\approx}{q} \bullet g(\underset{\approx}{F}) \bullet \underset{\approx}{q}^T \quad \text{for all } \underset{\approx}{F} \text{ and all orthogonal } \underset{\approx}{q}, \tag{19.100a}$$

$$g(\underset{\approx}{Q} \bullet \underset{\approx}{F}) = \underset{\approx}{Q} \bullet g(\underset{\approx}{F}) \bullet \underset{\approx}{Q}^T \quad \text{for all } \underset{\approx}{F} \text{ and all orthogonal } \underset{\approx}{Q}. \tag{19.100b}$$

At this point, we must decide if these conditions imply any admissibility constraints on the $g(\underset{\approx}{F})$ function. The basic meanings of these operations are illustrated in figures 19.8 and 19.9. Do not forget that the $g(\underset{\approx}{F})$ must be regarded as a cell-phone

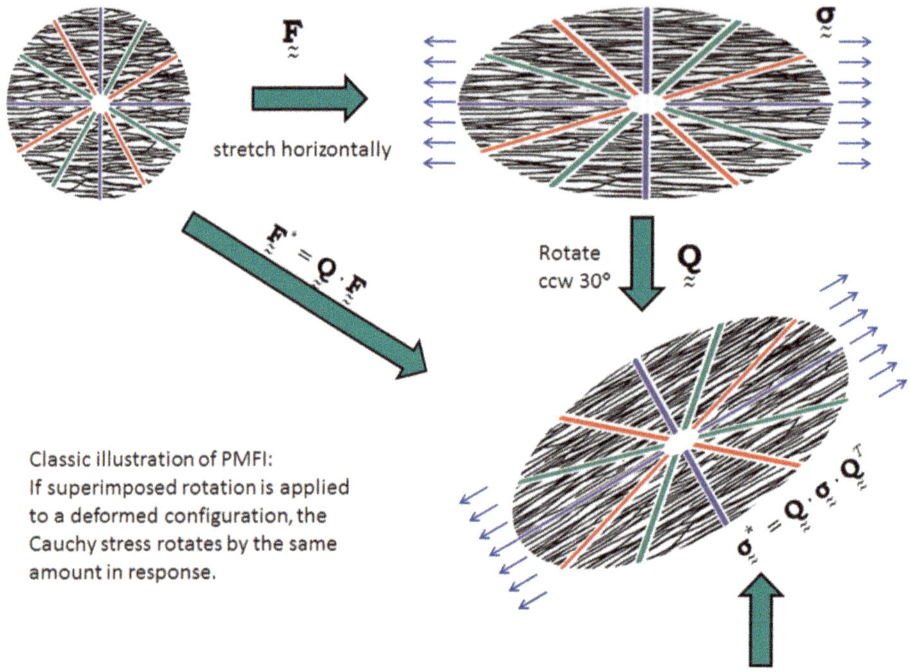

Figure 19.8. The desired change in Cauchy stress induced by superimposing rotation on the deformed configuration. This figure illustrates the classical PMFI restriction in (19.100*b*).

app that is hoped to return the correct answer no matter what frame it is used in. Suppose, for example, that $\underset{\approx}{F}$ is a horizontal stretch. Suppose that $\underset{\approx}{q}$ and $\underset{\approx}{Q}$ are both 90° counter-clockwise rotations. Then $\underset{\approx}{q} \bullet \underset{\approx}{F} \bullet \underset{\approx}{q}^T$ represents a vertical stretch, whereas $\underset{\approx}{Q} \bullet \underset{\approx}{F}$ represents a horizontal stretch followed by a 90° rotation. The first case will elongate initially vertical material fibers, while the second case will elongate initially horizontal fibers and then rigidly rotate them to the vertical. According to the restrictions in (19.100*a*), both of these motions would be required to induce the same stress. That result is not going to happen for a material that is anisotropic with stiff horizontal fibers. The first case would, for such a material, apply elongation in the compliant direction perpendicular to the fibers. The second case would apply elongation in the stiff direction parallel to the fibers, and then rotate the stretched fibers to a vertical orientation. Clearly the second case should produce a much larger vertical stress, but the PMFI restrictions on a model of the form $\underset{\approx}{\sigma} = g(\underset{\approx}{F})$ will not allow that. In fact, this model satisfies PMFI only if the material is isotropic. The only way to get an anisotropic model to satisfy PMFI is to have more calling arguments. Specifically, the function can be generalized to anisotropy only by expanding its arguments to include information giving the material directions. That way, if the material is rotated prior to loading, those material directions can also be rotated appropriately. Then, for example, the act of rotating the material prior to application of loads would include an appropriate rotation of the material fibers as

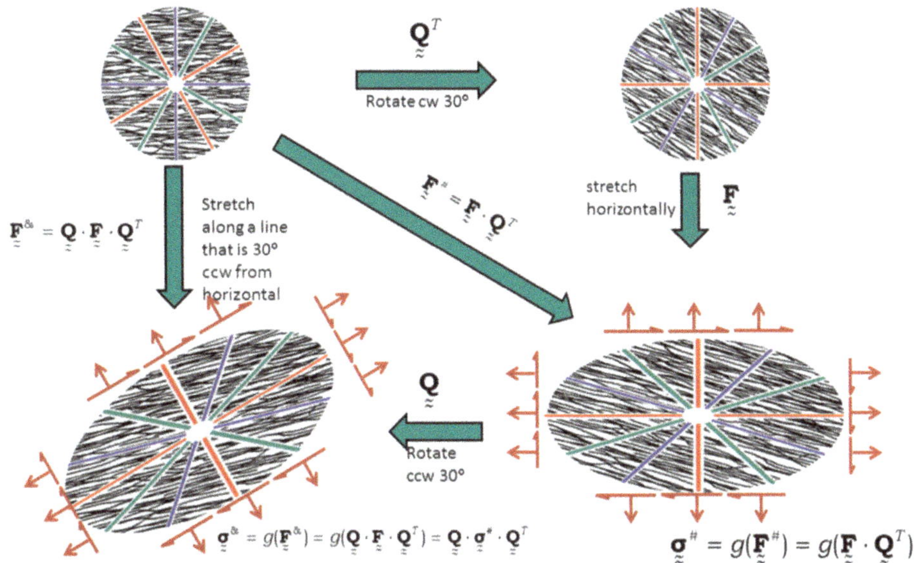

Figure 19.9. A different stress predicted by rotating the material before applying deformation. This figure illustrates (19.101). Rotating the material clockwise and then stretching it horizontally is equivalent to stretching the material along a line oriented counter-clockwise relative to the initial configuration, and then rotating the result clockwise. The deformed state is not the same as that in figure 19.8.

well so that the vertical elongation would then be applied to now vertically oriented fibers to give the required high stress in the fiber direction.

Assuming the material is indeed isotropic, does PMFI tell us anything about the nature of how $g(\underset{\approx}{F})$ depends on $\underset{\approx}{F}$? Yes! Since (19.99*b*) must hold for all orthogonal $\underset{\approx}{Q}$ and all $\underset{\approx}{F}$, it must hold if $\underset{\approx}{Q}$ and $\underset{\approx}{F}$ are replaced with $\underset{\approx}{Q}^T$ and $\underset{\approx}{Q} \bullet \underset{\approx}{F} \bullet \underset{\approx}{Q}^T$, respectively, to give

$$g(\underset{\approx}{F} \bullet \underset{\approx}{Q}^T) = \underset{\approx}{Q}^T \bullet g(\underset{\approx}{Q} \bullet \underset{\approx}{F} \bullet \underset{\approx}{Q}^T) \bullet \underset{\approx}{Q} \quad \text{for all } \underset{\approx}{F} \text{ and all orthogonal } \underset{\approx}{Q}. \quad (19.101)$$

This more convoluted (but equivalent) restatement of PMFI (illustrated in figure 19.9) shows how the answers must change if a superimposed rotation is applied *before* application of the fiducial deformation mapping. It is instructive to specialize each of these PMFI constraints for particular choices of the constitutive *g*-function. For example, a model that takes Cauchy stress to be a multiple of the symmetric part of the deformation gradient will not pass these tests even though it is certainly a well-formed tensor equation. Similarly, a model that takes Cauchy stress to be a multiple of the right Cauchy–Green tensor will fail PMFI constraints, while a model that takes Cauchy stress to be a multiple of the left Cauchy–Greeen tensor will pass all of these tests (despite being a very bad model for other reasons).

Constitutive response comes from deformation of material fibers, while rotation serves a purely kinematic role. This assertion is expressed mathematically by noting that (19.100*b*) may be written as $g(\underset{\approx}{F}) = \underset{\approx}{Q}^T \bullet g(\underset{\approx}{Q} \bullet \underset{\approx}{F}) \bullet \underset{\approx}{Q}$. Since this must hold for all orthogonal $\underset{\approx}{Q}$ it must hold for the particular choice that $\underset{\approx}{Q} = \underset{\approx}{R}^T$, where $\underset{\approx}{R}$ is the polar rotation the polar decomposition $\underset{\approx}{F} = \underset{\approx}{R} \bullet \underset{\approx}{U}$, in which $\underset{\approx}{U}$ is the reference (right) stretch.

Thus, we see that the dependence of Cauchy stress on the deformation gradient $\underset{\approx}{F}$ can be decomposed into multiplicative parts, with the constitutive response determined entirely from the Cauchy stress associated with a pure stretch $\underset{\approx}{U}$:

$$\boxed{g(\underset{\approx}{F}) = \underset{\approx}{R} \bullet g(\underset{\approx}{U}) \bullet \underset{\approx}{R}^T}, \quad \text{where } \underset{\approx}{U} = (\underset{\approx}{F}^T \bullet \underset{\approx}{F})^{1/2} \text{ and } \underset{\approx}{R} = \underset{\approx}{F} \bullet \underset{\approx}{U}^{-1}. \tag{19.102}$$

Recalling that $\underset{\approx}{\sigma} = g(\underset{\approx}{F})$, let us define $\underset{\approx}{\overline{\sigma}} = g(\underset{\approx}{U})$. This is so-called *un-rotated Cauchy stress* predicted under the same deformation without its rotation. Then this result implies that once the constitutive model is applied in the 'un-rotated' (pure stretch) configuration to obtain the un-rotated stress $\underset{\approx}{\overline{\sigma}}$, the result for the actual stress $\underset{\approx}{\sigma}$ is immediately given by:

$$\text{'Re-rotation' operation: } \boxed{\underset{\approx}{\sigma} = \underset{\approx}{R} \bullet \underset{\approx}{\overline{\sigma}} \bullet \underset{\approx}{R}^T} \tag{19.103}$$

this result is valid for any material.

This formula was here shown to be implied by PMFI in (19.100*b*). Conversely, if this holds, it is straightforward to confirm that (19.100*b*) is satisfied. Thus, this formula is necessary and sufficient for an elastic model to satisfy PMFI. Physically, this result implies that the *materials science* (i.e., the physics) of the constitutive model is embodied in the function $g(\underset{\approx}{U})$, while the response to the rotational part of $\underset{\approx}{F}$ is kinematical and therefore handled the same for any elastic material, regardless of the details of the constitutive function g. Though this result was proved here for an elastic material, it applies to any other type of material as well. This is a strong motivation to write *all* constitutive models in the un-rotated frame. Once stress $\underset{\approx}{\overline{\sigma}}$ is computed without rotation, the actual stress $\underset{\approx}{\sigma}$ is immediately found by rotating to the current frame. This approach is called *strong objectivity* (see [6]); it provides an efficient approach to constitutive modeling since any and all constitutive models may presume without loss that $\underset{\approx}{R} = \underset{\approx}{I}$. The Cauchy stress predicted by such models is therefore actually $\underset{\approx}{\overline{\sigma}}$. In numerical implementations, the host code (which invariably has information about the kinematics such as velocity or displacement fields can compute or update the polar rotation tensor $\underset{\approx}{R}$ independently of the constitutive model. Construction of the actual stress is found at any moment when needed by applying (19.103). In doing so, however, it is crucial to evaluate $\underset{\approx}{R}$ at the same moment in time as the time for which $\underset{\approx}{\overline{\sigma}}$ is evaluated. If, for example, the constitutive model is written in an incremental form such that $\underset{\approx}{\overline{\sigma}}$ is the un-rotated stress at the end of a time step, the rotation $\underset{\approx}{R}$ must also be updated to the end of the time step before applying (19.103) to convert $\underset{\approx}{\overline{\sigma}}$ to $\underset{\approx}{\sigma}$. This can be done very accurately by presuming that the rotation axis is stationary during a time step so that only the rotation angle changes. If the rate of the angular velocity vector is treated as constant over the time step, it follows that the rotation axis will be stationary over the step and the rotation angle can be integrated over the step to obtain (via the Euler–Rodrigues formula) the incremental rotation tensor. This approach is far more accurate than merely using a linear update $\underset{\approx}{R}^{\text{new}} = \underset{\approx}{R}^{\text{old}} + \dot{\underset{\approx}{R}}\Delta t$, which will suffer from integration error that causes the updated approximate rotation to not even be orthogonal.

19.4.3 Special restriction on any isotropic elastic material model

An isotropic material has the property that the material may be arbitrarily rotated prior to application of a deformation, and the resulting stress will be the same in either case. In other words, for any orthogonal tensor $\underset{\approx}{Q}$, the stress resulting from a deformation $\underset{\approx}{F} \bullet \underset{\approx}{Q}^T$ is the same as the stress resulting from just $\underset{\approx}{F}$. (Contrast this with the case of the anisotropic material in figure 19.9, where the initial rotation *does* affect the final result.) The transpose has been here added for convenience without loss[16]. Applying an isotropy assertion to the general elastic model, $\underset{\approx}{\sigma} = g(\underset{\approx}{F})$, thus implies that

$$g(\underset{\approx}{F}) = g(\underset{\approx}{F} \bullet \underset{\approx}{Q}^T) \qquad \forall\ \underset{\approx}{Q} \text{ and } \underset{\approx}{F}. \tag{19.104}$$

When combined with (19.101), which must hold for any material (isotropic or not), this condition implies that:

$$\textit{Isotropy condition:}\ \boxed{g(\underset{\approx}{Q} \bullet \underset{\approx}{F} \bullet \underset{\approx}{Q}^T) = \underset{\approx}{Q} \bullet g(\underset{\approx}{F}) \bullet \underset{\approx}{Q}^T} \quad \text{for all } \underset{\approx}{F} \text{ and all orthogonal } \underset{\approx}{Q}. \tag{19.105}$$

Using (19.102), a necessary and sufficient condition for this property to hold is:

$$\textit{Isotropy condition:}\ \boxed{g(\underset{\approx}{Q} \bullet \underset{\approx}{U} \bullet \underset{\approx}{Q}^T) = \underset{\approx}{Q} \bullet g(\underset{\approx}{U}) \bullet \underset{\approx}{Q}^T} \quad \text{for all } \underset{\approx}{U} \text{ and all orthogonal } \underset{\approx}{Q}, \tag{19.106}$$

where $\underset{\approx}{U}$ is the reference polar stretch. Qualitatively, this says that there is no directional dependence of the effort (stress) required to elongate the material. This isotropy condition says, for example, that the two deformed states in figure 18.1 have the same stress magnitudes with the stress directions differing by a rotation that is the same as the rotation of the loading direction[17].The only way to get anisotropic response (such as different stress magnitudes depending on stretch direction) would be to have the constitutive function depend on more than just the deformation as it has here.

---

[16] The assertion must hold for any $\underset{\approx}{Q}$, so it must hold for any $\underset{\approx}{Q}^T$.

[17] Keep in mind that material isotropy does not preclude *induced* anisotropy. As seen in figure 18.1, the transformation of spherical voids into ellipsoids under load reflects an induced material texture. The texture has an anisotropic (directionally dependent) response to new loading increments applied from the deformed state. An additional strain increment in the current loading direction will probably require a larger stress increment than the same strain increment applied transversely.

19.4.4 Symmetry invariants and generators

This chapter's application of PMFI started with the most rudimentary notion of elasticity by assuming that Cauchy stress is a function of the deformation gradient, $\underset{\sim}{\sigma} = g(\underset{\sim}{\boldsymbol{F}})$, from which it was argued that the un-rotated Kirchhoff stress (or un-rotated Cauchy stress) must be an isotropic function of the reference stretch $\underset{\sim}{\bar{\tau}} = f(\underset{\sim}{\boldsymbol{U}})$, meaning that

$$f(\underset{\sim}{\boldsymbol{Q}} \bullet \underset{\sim}{\boldsymbol{U}} \bullet \underset{\sim}{\boldsymbol{Q}}^T) = \underset{\sim}{\boldsymbol{Q}} \bullet f(\underset{\sim}{\boldsymbol{U}}) \bullet \underset{\sim}{\boldsymbol{Q}}^T \quad \text{for all orthogonal } \underset{\sim}{\boldsymbol{Q}}. \tag{19.107}$$

Isotropy further implied that $\underset{\sim}{\bar{\tau}}$ could be written as a linear combination of $\underset{\sim}{\boldsymbol{I}}$, $\underset{\sim}{\boldsymbol{U}}$, and $\underset{\sim}{\boldsymbol{U}}^2$ (called *generators*), with the coefficients in the expansion being functions of the invariants of $\underset{\sim}{\boldsymbol{U}}$. Specifically, it can be shown (see [7]) that a tensor-valued isotropic function, $\underset{\sim}{\bar{\sigma}} = g(\underset{\sim}{\boldsymbol{U}})$, is expressible in the form

$$\underset{\sim}{\bar{\sigma}} = b_0\underset{\sim}{\boldsymbol{I}} + b_1\underset{\sim}{\boldsymbol{U}} + b_2\underset{\sim}{\boldsymbol{U}}^2, \tag{19.108}$$

where b_0, b_1, and b_2 depend only on the scalar invariants $\text{tr}\underset{\sim}{\boldsymbol{U}}$, $\text{tr}\underset{\sim}{\boldsymbol{U}}^2$, and $\text{tr}\underset{\sim}{\boldsymbol{U}}^3$. These invariants may be any *complete and irreducible* invariant triplet (such as the eigenvalues, the trace invariants, the mechanics invariants, etc). A major aim of group theory is to work out similar statements for different types of functions[18]. As we now work through the concepts of material-frame indifference, we will encounter generalized notions of invariants and generators. Any other invariant of this symmetric stretch tensor, $\underset{\sim}{\boldsymbol{U}}$, may be expressed in terms of these three invariants, so we could have simply said that the b_i coefficients are functions of the invariants of $\underset{\sim}{\boldsymbol{U}}$ without naming any particular choice of invariants.

Using (19.102), equation (19.108) may be written

$$\underset{\sim}{\sigma} = b_0\underset{\sim}{\boldsymbol{I}} + b_1\underset{\sim}{\boldsymbol{V}} + b_2\underset{\sim}{\boldsymbol{V}}^2, \quad \text{where } \underset{\sim}{\boldsymbol{V}} = \underset{\sim}{\boldsymbol{R}} \bullet \underset{\sim}{\boldsymbol{U}} \bullet \underset{\sim}{\boldsymbol{R}}^T = \text{ left stretch}. \tag{19.109}$$

Since $\underset{\sim}{\boldsymbol{V}}$ and $\underset{\sim}{\boldsymbol{U}}$ have the same invariants, the coefficients in this expansion are the same functions of the invariants of $\underset{\sim}{\boldsymbol{V}}$. Also, we could multiply both sides of this equation by the Jacobian $J = \det\underset{\sim}{\boldsymbol{F}} = \det\underset{\sim}{\boldsymbol{U}} = \det\underset{\sim}{\boldsymbol{V}}$ to obtain

$$\underset{\sim}{\tau} = b_0^*\underset{\sim}{\boldsymbol{I}} + b_1^*\underset{\sim}{\boldsymbol{V}} + b_2^*\underset{\sim}{\boldsymbol{V}}^2, \quad \text{where } \underset{\sim}{\tau} = J\underset{\sim}{\sigma} = \text{Kirchhoff stress}. \tag{19.110}$$

Since J is an invariant of the stretch, this version of the formula continues to have coefficients that depend only on stretch invariants. As with Cauchy stress, one may define an un-rotated Kirchhoff stress, $\underset{\sim}{\bar{\tau}} = \underset{\sim}{\boldsymbol{R}}^T \bullet \underset{\sim}{\tau} \bullet \underset{\sim}{\boldsymbol{R}}$ so that

$$\underset{\sim}{\bar{\tau}} = b_0^*\underset{\sim}{\boldsymbol{I}} + b_1^*\underset{\sim}{\boldsymbol{U}} + b_2^*\underset{\sim}{\boldsymbol{U}}^2. \tag{19.111}$$

The key result of this discussion was (19.111) which shows that, for an isotropic material, the un-rotated stress (whether Cauchy or Kirchhoff) is expressible as a

[18] See [7] for comprehensive tables of invariants and generators for numerous symmetry groups!

pseudo-linear[19] combination of three tensors: $\underset{\approx}{I}$, $\underset{\approx}{U}$, and U^2. These three tensors are called the *generators* for isotropic materials. Generators for this simple symmetry class (isotropy) as well as for many different symmetry classes are summarized in [7].

Before leaving this section, other general forms for the isotropy class need to be mentioned because of their ubiquitous use in the literature. We have already mentioned that (19.109) through (19.111) are equivalent results. Truth of one implies truth of the others. Another version of the same result is

$$\underset{\approx}{\bar{\tau}} = b_0^{\#}\underset{\approx}{I} + b_1^{\#}\underset{\approx}{C} + b_2^{\#}\underset{\approx}{C}^2, \text{ where } \underset{\approx}{C} = \underset{\approx}{U}^2 = \underset{\approx}{F}^T \bullet \underset{\approx}{F}. \tag{19.112}$$

The $b_i^{\#}$ coefficients in this expansion are functions of the invariants of the right Cauchy–Green tensor $\underset{\approx}{C}$. Equivalence of this result with (19.111)—and hence equivalence with all of the other versions of isotropy—is proved by recalling that the Cayley–Hamilton theorem allows $\underset{\approx}{U}^3$ and higher powers of $\underset{\approx}{U}$ (as well as $\underset{\approx}{U}^{-1}$) to be expressed as linear combinations of the lower-power generators $\underset{\approx}{I}$, $\underset{\approx}{U}$, and $\underset{\approx}{U}^2$ with the coefficients being functions of the invariants of $\underset{\approx}{U}$ or invariants of $\underset{\approx}{C}$ (since invariants of $\underset{\approx}{C}$ are functions of the invariants of $\underset{\approx}{U}$ and vice versa). In other words, generators for isotropy could be alternatively taken as $\underset{\approx}{I}$, $\underset{\approx}{C}$, and $\underset{\approx}{C}^2$. By the Cayley–Hamilton theorem, the generators could even be $\underset{\approx}{C}^{-1}$, $\underset{\approx}{I}$, and $\underset{\approx}{C}$. Note also that these results imply that the un-rotated Cauchy stress $\underset{\approx}{\bar{\sigma}}$ and the un-rotated Kirchhoff stress $\underset{\approx}{\bar{\tau}}$ share a common set of eigenvectors with $\underset{\approx}{C}$ and $\underset{\approx}{U}$, and hence these stresses commute with the stretches. A similar statement applies to the spatial stresses commuting with spatial stretch $\underset{\approx}{V}$ and the left Cauchy–Green tensor $\underset{\approx}{B}$.

This section's main results, equation (19.111) and its various equivalent forms, follow from symmetry and frame indifference arguments, without any reference to the thermodynamically based theory of hyperelasticity, which requires stress to be derivable from a scalar work potential W. Specifically,

$$\underset{\approx}{S} = \frac{\mathrm{d}W}{\mathrm{d}\underset{\approx}{E}}, \tag{19.113}$$

where $\underset{\approx}{S}$ and $\underset{\approx}{E}$ are, respectively, the second Piola–Kirchhoff stress and Green–Lagrange strain, defined by

$$\underset{\approx}{S} = \underset{\approx}{F}^{-1} \bullet \underset{\approx}{\tau} \bullet \underset{\approx}{F}^{-T} = \underset{\approx}{U}^{-1} \bullet \underset{\approx}{\bar{\tau}} \bullet \underset{\approx}{U}^{-1} \tag{19.114}$$

and

$$\underset{\approx}{E} = \tfrac{1}{2}(\underset{\approx}{F}^T \bullet \underset{\approx}{F} - \underset{\approx}{I}) = \tfrac{1}{2}(\underset{\approx}{U}^2 - \underset{\approx}{I}) = \tfrac{1}{2}(\underset{\approx}{C} - \underset{\approx}{I}). \tag{19.115}$$

[19] The combination is pseudo-linear because the coefficients depend on the tensors. Just as one would not claim that a function $(\det \underset{\approx}{A})\underset{\approx}{A}$ is linear with respect to $\underset{\approx}{A}$, one should exercise similar caution when discussing pseudo-linear combinations.

Noting that $\mathrm{d}\underset{\approx}{\boldsymbol{E}} = \frac{1}{2}\mathrm{d}\underset{\approx}{\boldsymbol{C}}$, substituting (19.103) and (19.104) into the hyperelasticity condition of (19.102) thus requires

$$\underset{\approx}{\boldsymbol{\tau}} = 2\underset{\approx}{\boldsymbol{U}} \bullet \frac{\mathrm{d}W}{\mathrm{d}\underset{\approx}{\boldsymbol{C}}} \bullet \underset{\approx}{\boldsymbol{U}} \quad \text{(valid for any hyperelastic material, even anisotropic).} \tag{19.116}$$

As indicated, the above result applies to arbitrary anisotropy provided that the energy function explicitly depends on material texture information (such as fiber orientations) even though this dependence has been shown implicitly here. If the material is known to be isotropic, however, the tensors in the above equation commute with each other, allowing that equation to specialize to

$$\underset{\approx}{\boldsymbol{\tau}} = 2\frac{\mathrm{d}W}{\mathrm{d}\underset{\approx}{\boldsymbol{C}}} \bullet \underset{\approx}{\boldsymbol{C}} \quad \text{(valid for an isotropic hyperelastic material).} \tag{19.117}$$

For large deformations (where $\underset{\approx}{\boldsymbol{C}}$ is not approximately equal to $\underset{\approx}{\boldsymbol{I}}$), note that the Kirchhoff stress itself is not derived directly from a potential—just one factor in its formula is derived from a potential. For isotropy, symmetry arguments may again be invoked to argue that the hyperelastic potential must depend only on the invariants of $\underset{\approx}{\boldsymbol{C}}$. Suppose that these invariants are taken to be

$$T_1 = \mathrm{tr}\underset{\approx}{\boldsymbol{C}}, \quad T_2 = \mathrm{tr}(\underset{\approx}{\boldsymbol{C}}^2), \quad \text{and } T_3 = \mathrm{tr}(\underset{\approx}{\boldsymbol{C}}^3). \tag{19.118}$$

Keeping in mind that the stretch is symmetric, the derivatives are

$$\frac{\mathrm{d}T_1}{\mathrm{d}\underset{\approx}{\boldsymbol{C}}} = \underset{\approx}{\boldsymbol{I}}, \quad \frac{\mathrm{d}T_2}{\mathrm{d}\underset{\approx}{\boldsymbol{C}}} = 2\underset{\approx}{\boldsymbol{C}}, \quad \text{and } \frac{\mathrm{d}T_1}{\mathrm{d}\underset{\approx}{\boldsymbol{C}}} = 3\underset{\approx}{\boldsymbol{C}}^3. \tag{19.119}$$

Then, applying the chain rule to $W(T_1, T_2, T_3)$ gives

$$\frac{\mathrm{d}W}{\mathrm{d}\underset{\approx}{\boldsymbol{C}}} = \frac{\partial W}{\partial T_1}\frac{\mathrm{d}T_1}{\mathrm{d}\underset{\approx}{\boldsymbol{C}}} + \frac{\partial W}{\partial T_2}\frac{\mathrm{d}T_2}{\mathrm{d}\underset{\approx}{\boldsymbol{C}}} + \frac{\partial W}{\partial T_3}\frac{\mathrm{d}T_3}{\mathrm{d}\underset{\approx}{\boldsymbol{C}}}. \tag{19.120}$$

Using (19.119) in (19.120), and then substituting the result back into (19.116) gives

$$\underset{\approx}{\boldsymbol{\tau}} = 2\left[\frac{\partial W}{\partial T_1}\underset{\approx}{\boldsymbol{C}} + 2\frac{\partial W}{\partial T_2}\underset{\approx}{\boldsymbol{C}}^2 + 3\frac{\partial W}{\partial T_3}\underset{\approx}{\boldsymbol{C}}^4\right]. \tag{19.121}$$

Using the Cayley–Hamilton theorem, $\underset{\approx}{\boldsymbol{C}}^4$ may be expressed in terms of $\underset{\approx}{\boldsymbol{I}}$, $\underset{\approx}{\boldsymbol{C}}$, and $\underset{\approx}{\boldsymbol{C}}^2$. Thus, the hyperelastic model in (19.113) does indeed imply (19.112). An interesting question is whether or not the converse is true. If the stress is of the form in (19.112), does it imply hyperelasticity, as defined in (19.113)? The answer is 'no'. Starting with the most rudimentary assumption that $\underset{\approx}{\boldsymbol{\sigma}}$ is a proper function of $\underset{\approx}{\boldsymbol{F}}$, it follows that the PK2 stress $\underset{\approx}{\boldsymbol{S}}$ must be a function of the Green–Lagrange strain $\underset{\approx}{\boldsymbol{E}}$ (this is true even for anisotropic media), but is this enough to guarantee that (19.113) has a solution for the elastic potential W? Using elementary calculus theorems for exact differentials and solvability of partial differential equations, equation (19.113) is solvable for $W(\underset{\approx}{\boldsymbol{E}})$ if and only if the following conditions are satisfied:

i. $\underset{\approx}{\boldsymbol{S}}$ is a function of $\underset{\approx}{\boldsymbol{E}}$ (which is a condition we do satisfy) *and*

ii. the *tangent elastic stiffness* tensor, $\underset{\approx\approx}{\boldsymbol{T}} = \frac{\mathrm{d}\underset{\approx}{\boldsymbol{S}}}{\mathrm{d}\underset{\approx}{\boldsymbol{E}}}$, must be major-symmetric: $T_{ijkl} = T_{klij}$.

Merely satisfying the first of these requirements is not sufficient for the elastic model to be hyperelastic. For hyperelasticity, not only must stress be a path-independent function of strain (which is the first condition), the work $W = \int \underset{\approx}{\boldsymbol{S}}:\mathrm{d}\underset{\approx}{\boldsymbol{E}}$ must also be path-independent.

Even when a material is isotropic, its constitutive tensors, such as stiffness, might not be isotropic. For example, the tangent stiffness $\mathrm{d}\underset{\approx}{\boldsymbol{S}}/\mathrm{d}\underset{\approx}{\boldsymbol{E}}$ corresponding to figure 18.1 is isotropic in the undeformed state, but it acquires anisotropy under compression (typically becoming stiffer in the compression direction). Since anisotropy is intrinsically tied to texture of the microstructure, this fact that isotropic materials generally have anisotropic tangent stiffnesses should come as no surprise in view of the ellipsoidal texture in figure 18.1, and this leads us now into a deeper look at tensor symmetries.

References

[1] Brannon R M 1998 Caveats concerning conjugate stress and strain measures for large distortion of anisotropic materials *Acta Mech.* **129** 107–16

[2] Oberkampf W L and Roy C J 2010 *Verification and Validation in Scientific Computing* (Cambridge: Cambridge University Press)

[3] Kamojjala K C, Brannon R M, Sadeghirad A and Guilkey J 2013 Verification tests in solid mechanics, *Eng. Comput.* at press

[4] Sadeghirad A, Brannon R M and Burghardt J 2011 A convected particle domain interpolation technique to extend applicability of the material point method for problems involving massive deformations, *Int. J. Numer. Methods Eng.* **86** 1435–56

[5] Bigoni D 2012 Nonlinear Solid Mechanics (New York: Cambridge University Press)

[6] Hughes T J R and Winget J 1980 Finite rotation effects in numerical integration of rate constitutive equations arising in large-deformation analysis, *Comput. Methods Appl. Mech. Eng.* **15** 648–55

[7] Zheng Q-S 1994 Theory of representations for tensor functions—a unified invariant approach to constitutive equations *Appl. Math. Rev.* **47** 545–87

Chapter 20

Tensor symmetry (not material symmetry)

Whereas *material* symmetry refers to the insensitivity of the constitutive response to certain transformations (such as rotation) applied to the material before application of a deformation, tensor symmetry refers to insensitivity of a tensor to certain transformations. A *group* is some well-defined set of all transformations of interest. The word 'transformation' means that the output is of the same type as the input. A Laplace transform, for example, takes a function as input and gives a function as output. Rotating a vector is a transformation because the result is also vector. Multiplying a tensor times 3 is a transformation because the result is also a tensor. This chapter looks for special inputs to transformations that will result in no change of the output.

A *symmetry class* under a specified group of transformations is the set of inputs that are unchanged under any choice of transformation from that group. For example, are there any tensors that are unchanged when multiplied by 3? Yes, the zero tensor. In fact, that is the only one, so the symmetry class under that transformation consists of only one member. This chapter seeks symmetry classes under *Rayleigh transformations*. Given a vector $\underset{\sim}{\boldsymbol{a}}$, and tensor $\underset{\approx}{\boldsymbol{A}}$, the Rayleigh transformations under $\underset{\approx}{\boldsymbol{F}}$ are defined

$$\underset{\sim}{\boldsymbol{b}} = \underset{\approx}{\boldsymbol{F}} \bullet \underset{\sim}{\boldsymbol{a}} \qquad \text{or } b_i = F_{ij}a_j, \tag{20.1a}$$

$$\underset{\approx}{\boldsymbol{B}} = \underset{\approx}{\boldsymbol{F}} \bullet \underset{\approx}{\boldsymbol{A}} \bullet \underset{\approx}{\boldsymbol{F}}^T \qquad \text{or } B_{ij} = F_{im}A_{mn}F^T_{nj} = F_{im}F_{jn}A_{mn}. \tag{20.1b}$$

Continuing, Rayleigh transforms of third- and higher-order tensors are

$$B_{ijk} = F_{ip}F_{jq}F_{kr}A_{pqr}, \tag{20.2a}$$

$$B_{ijkl} = F_{ip}F_{jq}F_{kr}F_{ls}A_{pqrs}, \quad \text{etc.} \tag{20.2b}$$

doi:10.1016/978-0-7503-1454-1ch20

Different transformation groups can be defined by requiring choices for $\underset{\approx}{F}$ to belong to some specified set G. Then the problem of finding symmetry classes for first- and second-order tensors would be:

$$\text{find all vectors } \underset{\sim}{a} \text{ such that } \underset{\sim}{a} = \underset{\approx}{F} \bullet \underset{\sim}{a} \qquad \text{for all } \forall \underset{\approx}{F} \in G, \tag{20.3a}$$

$$\text{find all tensors } \underset{\approx}{A} \text{ such that } \underset{\approx}{A} = \underset{\approx}{F} \bullet \underset{\approx}{A} \bullet \underset{\approx}{F}^T \qquad \text{for all } \forall \underset{\approx}{F} \in G. \tag{20.3b}$$

Therefore, the problem is made interesting only when the group G is a restricted set. In symmetry investigations, this group set is often limited to rotations. The isotropy group allows any rotations, while the transverse isotropy group allows only rotations only about a specified axis. The only vector meeting the above condition for the isotropy group is the zero vector, while any vector aligned with the rotation axis will satisfy the condition under the transverse symmetry group. We will soon see that, for 3D, a tensor must be a multiple of the identity to be unchanged under the isotropy group, while the condition is surprisingly different in 2D. This chapter will step through further examples systematically. For now, the main idea is that we are going on a hunt for special vectors or higher-order tensors that are unchanged under certain restricted groups of transformations.

In the first part of this chapter, choices for $\underset{\approx}{F}$ are required to be any orthogonal tensor $\underset{\approx}{Q}$ (i.e., the choices are taken from the isotropy group). When $\underset{\approx}{F}$ is limited to be an orthogonal tensor $\underset{\approx}{Q}$, the Rayleigh transform can be regarded as an orthonormal basis change. Therefore, the symmetry-class problem would be looking for tensors whose components do not change under basis rotations defined by $\underset{\approx}{Q}$. If we allow any choice for orthogonal $\underset{\approx}{Q}$, then we are looking for conditions that must be satisfied by tensor components in order to declare the tensor to be *isotropic*. In that case, the problem statement in (20.3*a*) is looking for any vectors that will not be altered by an arbitrary rotation. The only such vector is the zero vector, so that is the only member of the symmetry class for vectors under the isotropy group, arbitrary rotations. There are infinitely many solutions to the more interesting (20.3*b*); specifically, a single-parameter family of solutions is any multiple of the identity

$$\underset{\approx}{A} = \alpha \underset{\approx}{I} \tag{20.4}$$

for some scalar multiplier α, but a key question is: does this family of solutions cover all possible solutions? Also, are isotropic tensors always going to be multiples of the identity? The answer will be shown in this chapter to depend on the dimension of the space.

If we allow only 90° rotations about the laboratory axes, then we are looking for restrictions on tensor components to declare them to be *cubic*. These sorts of questions are the focus of this chapter, along with exploring properties of tensors that belong to important symmetry classes. Of particular interest is the transverse symmetry class, for which $\underset{\approx}{Q}$ is allowed to be an arbitrary rotation about a given fixed axis. In the case of double-symmetric fourth-order tensors belonging to the transverse symmetry class, we prove that such tensors have five independent parameters, but only four distinct eigenvalues, and we find the corresponding eigenprojectors and explain how this information can be used to resolve ambiguities in Tsai–Wu plasticity models.

20.1 What is isotropy of a tensor?

(i) *Definition* 1: a tensor is *strictly* isotropic if its components are unchanged upon any orthonormal change in basis
(ii) *Definition* 2: a tensor is *proper*-isotropic if its components are unchanged upon any *same-handed* change in basis

Consider a second-order tensor $\underset{\approx}{A}$. According to definition 1,

$$\textit{Strict isotropy:} \quad Q_{ip}Q_{jq}A_{pq} = A_{ij} \text{ for any orthogonal matrix } [Q]. \tag{20.5}$$

According to definition 2,

$$\textit{Proper isotropy:} \quad R_{ip}R_{jq}A_{pq} = A_{ij} \text{ for any proper orthogonal matrix } [R]. \tag{20.6}$$

A proper-orthogonal tensor (i.e., one with determinant +1) is a rotation operation. When a tensor is proper-isotropic, we are saying that it 'looks the same' no matter what orientation you view it from. For proper isotropy, there is no guarantee that the tensor will not look different if you invert it (i.e., if you switch to a left-handed basis or look at the material's tensorial properties indirectly by using a mirror). Strict isotropy insists that the tensor components must be unaffected by both rotations *and reflections.*

Strict isotropy is more restrictive because satisfaction of (20.5) automatically guarantees satisfaction of (20.6), but not vice versa. Which definition is more useful from a practical standpoint? In the vast majority of physics and engineering applications, it is desirable to know when something will be unchanged upon a rotation. Often, one does not even care what happens upon a (often non-physical) reflection. Therefore knowledge of proper isotropy (even when strict isotropy does not hold) is useful enough to be tested first to spotlight important simplifying properties of tensors. Unless otherwise stated, we take the term 'isotropic' to mean 'proper-isotropic'.

This section aims to determine minimal restrictions that must be satisfied by a tensor's components for it to be isotropic (i.e., to satisfy (20.6)). Since (20.6) must hold for *any* rotation matrix, a wise first step in the analysis should first *explore restrictions placed on the components of a tensor for particular choices of the rotation tensor.* Doing this will simplify the general analysis for the set of all possible rotations. Good 'particular' choices for the rotations are 90° rotations about the coordinate axes:

$$\begin{bmatrix} 0 & -1 & 0 \\ 1 & 0 & 0 \\ 0 & 0 & 1 \end{bmatrix}, \quad \begin{bmatrix} 0 & 0 & 1 \\ 0 & 1 & 0 \\ -1 & 0 & 0 \end{bmatrix}, \quad \text{and} \quad \begin{bmatrix} 1 & 0 & 0 \\ 0 & 0 & -1 \\ 0 & 1 & 0 \end{bmatrix}. \tag{20.7}$$

Any component restrictions arising from these special choices will only give you *necessary* conditions for isotropy. Frequently, these necessary conditions turn out to also be sufficient conditions.

For example, to deduce the most general form for an isotropic *vector* $\underset{\sim}{v}$, you must demand that its components be unchanged under a basis transformation; i.e., it must satisfy the equation $R_{ip}v_p = v_i$. In matrix form,

$$\begin{bmatrix} R_{11} & R_{12} & R_{13} \\ R_{21} & R_{22} & R_{23} \\ R_{31} & R_{32} & R_{33} \end{bmatrix} \begin{bmatrix} v_1 \\ v_2 \\ v_3 \end{bmatrix} = \begin{bmatrix} v_1 \\ v_2 \\ v_3 \end{bmatrix}. \tag{20.8}$$

This must hold for all proper-orthogonal matrices $[R]$. Consequently, it must hold for any of the special cases in (20.7). Considering the first case,

$$\begin{bmatrix} 0 & -1 & 0 \\ 1 & 0 & 0 \\ 0 & 0 & 1 \end{bmatrix} \begin{bmatrix} v_1 \\ v_2 \\ v_3 \end{bmatrix} = \begin{bmatrix} v_1 \\ v_2 \\ v_3 \end{bmatrix}, \quad \text{or} \quad \begin{bmatrix} -v_2 \\ v_1 \\ v_3 \end{bmatrix} = \begin{bmatrix} v_1 \\ v_2 \\ v_3 \end{bmatrix}. \tag{20.9}$$

With this simple test, we have already learned an important *necessary* condition for a vector to be isotropic. Namely, v_2 must equal $-v_1$ and v_2 must also equal v_1. The only way to satisfy this constraint is for both of these components to be zero. Thus, an isotropic vector would necessarily have to be of the form $\langle 0, 0, v_3 \rangle$, but (as explained here) this form is not sufficient to say that the vector is isotropic. Knowing that the first two components must be zero does help to simplify further exploration of the general requirement in (20.8). For example, applying (20.8) using the second choice in (20.7) and also using our necessary condition that $v_1 = v_2 = 0$ gives

$$\begin{bmatrix} 0 & 0 & 1 \\ 0 & 1 & 0 \\ -1 & 0 & 0 \end{bmatrix} \begin{bmatrix} 0 \\ 0 \\ v_3 \end{bmatrix} = \begin{bmatrix} 0 \\ 0 \\ v_3 \end{bmatrix}, \quad \text{or} \quad \begin{bmatrix} v_3 \\ 0 \\ 0 \end{bmatrix} = \begin{bmatrix} 0 \\ 0 \\ v_3 \end{bmatrix}. \tag{20.10}$$

This result tells us that v_3 itself also must be zero. Thus, a *necessary* requirement for a vector to be isotropic is that the vector's components must all be zero. Now that these two specialized tests have greatly simplified the nature of vectors that might be isotropic, it is time to use these *necessary* conditions in the general (20.8). Trivially, if a vector is zero, then (20.7) is satisfied for *any* choice of $[R]$. Thus, the zero vector is the *only* isotropic vector.

Looking now at the less trivial case of second-order tensors, note that any linear combination of isotropic tensors is itself isotropic. Specifically, if $\underset{\approx}{A}$ and $\underset{\approx}{B}$ are isotropic, then

$$R_{ip}R_{jq}A_{pq} = A_{ij} \qquad \text{and} \qquad R_{ip}R_{jq}A_{pq} = B_{ij}, \tag{20.11}$$

and it follows that any linear combination $\alpha\underset{\approx}{A} + \beta\underset{\approx}{B}$ will be isotropic because

$$R_{ip}R_{jq}(\alpha A_{pq} + \beta B_{pq}) = \alpha(R_{ip}R_{jq}A_{pq}) + \beta(R_{ip}R_{jq}B_{pq}) = \alpha A_{ij} + \beta B_{ij}. \quad (20.12)$$

The set of all isotropic second-order tensors (of a given tensor order and referenced to a physical space of given dimension) must therefore be a linear space.

20.1.1 Important consequence of linearity of isotropic tensors

Since any linear combination of isotropic tensors will itself be isotropic, it follows that the set of isotropic tensors is a linear subspace (i.e., a *linear manifold*, which is an abstraction of a flat plane through the origin). Thus, the zero tensor will always be an isotropic tensor. More importantly, if there exist *any* non-trivial (i.e., non-zero) isotropic tensors, then there must exist a basis for the subspace of all isotropic tensors. The nature of the basis depends on both the order of the tensor and the dimension of the physical space. For example, below we prove that a second-order tensor in three dimensions (i.e., one whose component matrix is 3×3) is isotropic if and only if it is some multiple of the identity tensor. Consequently, the identity tensor itself is a basis for the subspace of isotropic second-order tensors in 3D. There is only one basis tensor, so this must be a 1D space. Projecting an arbitrary tensor onto this space gives you the *isotropic part* of that tensor. Interestingly, as explained below, second-order tensors in 2D space have a *higher*-dimensional isotropic subspace than second-order tensors in 3D. By this, we mean that there exist *two* distinct proper-isotropic second-order tensors in 2D space (i.e., tensors with 2×2 component matrices), one of which equals the 2D identity tensor and the other equals the 2D permutation tensor. The 2D identity is both proper- and strictly-isotropic. The 2D permutation tensor is not strictly isotropic (its components change sign upon a differently handed change of basis).

20.2 Isotropic second-order tensors in 3D space

Ordinarily, vector and tensor components will change upon a change in basis. Now we ask: what relationships must be satisfied if the components themselves are to remain unchanged upon a change of basis? Such relationships define isotropic tensors. Any linear combination of isotropic tensors is itself an isotropic tensor. Therefore, the set of all isotropic tensors (of a given order and referenced to a given dimension of physical space) forms a linear subspace. Hence, that there exists a finite number of 'primitive' isotropic tensors that serve as a basis. Any isotropic tensor may be expressed as a linear combination of the primitive isotropic basis tensors.

A second-order tensor in 3D space is proper-isotropic if and only if

$$[T] = [R][T][R]^T \qquad \text{for all proper-orthogonal } [R]. \quad (20.13)$$

Since this condition must hold *for all* $[R]$, it must hold for any particular choices for $[R]$. Using the first rotation matrix cited in (20.7) implies

$$\begin{bmatrix} T_{11} & T_{12} & T_{13} \\ T_{21} & T_{22} & T_{23} \\ T_{31} & T_{32} & T_{33} \end{bmatrix} = \begin{bmatrix} 0 & -1 & 0 \\ 1 & 0 & 0 \\ 0 & 0 & 1 \end{bmatrix} \begin{bmatrix} T_{11} & T_{12} & T_{13} \\ T_{21} & T_{22} & T_{23} \\ T_{31} & T_{32} & T_{33} \end{bmatrix} \begin{bmatrix} 0 & 1 & 0 \\ -1 & 0 & 0 \\ 0 & 0 & 1 \end{bmatrix} \tag{20.14}$$

or, multiplying out the right-hand side,

$$\begin{bmatrix} T_{11} & T_{12} & T_{13} \\ T_{21} & T_{22} & T_{23} \\ T_{31} & T_{32} & T_{33} \end{bmatrix} = \begin{bmatrix} T_{22} & -T_{21} & -T_{23} \\ -T_{12} & T_{11} & T_{13} \\ -T_{32} & T_{31} & T_{33} \end{bmatrix}, \tag{20.15}$$

which implies that a necessary condition for a tensor to be isotropic is that $T_{11} = T_{22}$, $T_{12} = -T_{21}$, $T_{13} = 0$, and $T_{23} = 0$. Similar analyses using the other rotation matrices defined in (20.7) eventually implies that all diagonal components must be equal to each other and all off-diagonal components must be zero. In other words, the tensor must be a scalar times the identity tensor $\underset{\approx}{\boldsymbol{I}}$. Of course, since this result was obtained by only three particular choices for the rotation matrix, it is only a necessary condition for isotropy. However, this necessary condition, $\underset{\approx}{\boldsymbol{T}} = \alpha\underset{\approx}{\boldsymbol{I}}$, also satisfies (20.13) *for all* $[R]$, and therefore $\underset{\approx}{\boldsymbol{T}} = \alpha\underset{\approx}{\boldsymbol{I}}$ is both necessary and sufficient for $\underset{\approx}{\boldsymbol{T}}$ to be isotropic. In other words, the most general form for an isotropic tensor referenced to 3D space is

$$\begin{bmatrix} \alpha & 0 & 0 \\ 0 & \alpha & 0 \\ 0 & 0 & \alpha \end{bmatrix}, \tag{20.16}$$

where α is an arbitrary scalar. The set of all isotropic tensors in 3D space is therefore one-dimensional since there is only one arbitrary scalar. The identity tensor $\underset{\approx}{\boldsymbol{I}}$ is a basis for this space. Any general tensor $\underset{\approx}{\boldsymbol{B}}$ may be projected to its isotropic part by the operation

$$\mathrm{iso}\underset{\approx}{\boldsymbol{B}} = \frac{\underset{\approx}{\boldsymbol{I}}(\underset{\approx}{\boldsymbol{I}} : \underset{\approx}{\boldsymbol{B}})}{\underset{\approx}{\boldsymbol{I}} : \underset{\approx}{\boldsymbol{I}}}. \tag{20.17}$$

Note that $\underset{\approx}{\boldsymbol{I}} : \underset{\approx}{\boldsymbol{B}} = \mathrm{tr}\underset{\approx}{\boldsymbol{B}}$ and $\underset{\approx}{\boldsymbol{I}} : \underset{\approx}{\boldsymbol{I}} = \mathrm{tr}\underset{\approx}{\boldsymbol{I}} = 3$. Hence,

$$\mathrm{iso}\underset{\approx}{\boldsymbol{B}} = \tfrac{1}{3}(\mathrm{tr}\underset{\approx}{\boldsymbol{B}})\underset{\approx}{\boldsymbol{I}}. \tag{20.18}$$

This is a very familiar result. The idea of finding the isotropic part by projecting to the space of isotropic tensors becomes less obvious when considering tensors in spaces of different dimensions.

20.3 Isotropic second-order tensors in 2D space

To consider both proper isotropy and strict isotropy of tensors in two dimensions (whose components are therefore 2×2 matrices), we first need to identify the general form for an orthogonal tensor in this space.

$$\text{Let } Q = \begin{bmatrix} a & b \\ c & d \end{bmatrix}. \tag{20.19}$$

We seek restrictions on the components (a, b, c and d) such that

$$[Q]^T[Q] = [I] \tag{20.20}$$

or

$$\begin{bmatrix} a & c \\ b & d \end{bmatrix}\begin{bmatrix} a & b \\ c & d \end{bmatrix} = \begin{bmatrix} 1 & 0 \\ 0 & 1 \end{bmatrix}. \tag{20.21}$$

Multiplying this out shows that the components must satisfy

$$\begin{aligned} a^2 + c^2 &= 1, \\ b^2 + d^2 &= 1, \\ ab + dc &= 0. \end{aligned} \tag{20.22}$$

We can satisfy the first two constraints automatically by setting

$$\begin{aligned} a &= \cos\alpha, \quad c = \sin\alpha, \\ b &= \cos\beta, \quad d = \sin\beta. \end{aligned} \tag{20.23}$$

Then, satisfying the last constraint requires that

$$\cos\alpha\cos\beta + \sin\alpha\sin\beta = 0, \tag{20.24}$$

$$\cos(\beta - \alpha) = 0, \text{ andtherefore } \beta = \alpha + \frac{\pi}{2}. \tag{20.25}$$

The choice $\beta = \alpha + \frac{\pi}{2}$ gives, upon substitution into (20.23),

$$\begin{aligned} a &= \cos\alpha, \quad c = \sin\alpha, \\ b &= -\sin\alpha, \quad d = \cos\alpha. \end{aligned} \tag{20.26}$$

Putting this result back into (20.19) yields a proper-orthogonal matrix, so we will denote it by $[R]$. Namely,

$$[R] = \begin{bmatrix} \cos\alpha & -\sin\alpha \\ \sin\alpha & \cos\alpha \end{bmatrix}. \tag{20.27}$$

On the other hand, the choice $\beta = \alpha - \dfrac{\pi}{2}$ reduces (20.23) to

$$\begin{aligned} a &= \cos\alpha, \quad c = \sin\alpha, \\ b &= \sin\alpha, \quad d = -\cos\alpha. \end{aligned} \tag{20.28}$$

Putting this result back into (20.19) yields an improper-orthogonal matrix,

$$[Q] = \begin{bmatrix} \cos\alpha & \sin\alpha \\ \sin\alpha & -\cos\alpha \end{bmatrix}. \tag{20.29}$$

Equation (20.27) is the most general form for a proper-orthogonal matrix in 2D and (20.29) is the most general form for an improper-orthogonal matrix in 2D. With this result in hand, we are ready to explore the nature of isotropic tensors in 2D. For a second-order 2D tensor to be proper-isotropic its components must satisfy

$$\begin{bmatrix} \cos\alpha & -\sin\alpha \\ \sin\alpha & \cos\alpha \end{bmatrix} \begin{bmatrix} A_{11} & A_{12} \\ A_{21} & A_{22} \end{bmatrix} \begin{bmatrix} \cos\alpha & \sin\alpha \\ -\sin\alpha & \cos\alpha \end{bmatrix} = \begin{bmatrix} A_{11} & A_{12} \\ A_{21} & A_{22} \end{bmatrix}. \tag{20.30}$$

Considering, as a special case, $\alpha = \pi/2$ gives

$$\begin{bmatrix} 0 & -1 \\ 1 & 0 \end{bmatrix} \begin{bmatrix} A_{11} & A_{12} \\ A_{21} & A_{22} \end{bmatrix} \begin{bmatrix} 0 & 1 \\ -1 & 0 \end{bmatrix} = \begin{bmatrix} A_{11} & A_{12} \\ A_{21} & A_{22} \end{bmatrix} \tag{20.31}$$

or

$$A_{11} = A_{22} \text{ and } A_{12} = -A_{21}. \tag{20.32}$$

Since this result was obtained by considering a special rotation, we know it a *necessary* condition for isotropy. However, substituting this condition back into (20.30) shows that it is also a *sufficient* condition. Consequently, the most general form for a proper-isotropic tensor referenced to 2D space is of the form

$$\begin{bmatrix} a & b \\ -b & a \end{bmatrix}, \tag{20.33}$$

where a and b are arbitrary parameters. This final result involves two arbitrary scalars, which implies that the set of proper-isotropic tensors in 2D is two-dimensional. Any tensor in 2D space that is of this proper-isotropic form may be expressed as a linear combination of the following two primitive basis tensors:

$$[\underset{\approx}{I}] = \begin{bmatrix} 1 & 0 \\ 0 & 1 \end{bmatrix} \quad \text{and} \quad [\underset{\approx}{\varepsilon}] = \begin{bmatrix} 0 & 1 \\ -1 & 0 \end{bmatrix}, \tag{20.34}$$

where $\underset{\approx}{\varepsilon}$ is the 2D version of the permutation symbol; namely, ε_{ij} is zero if $i = j$, +1 if $ij = 12$, and –1 if $ij = 21$. In two dimensions, the (proper) isotropic part of a second-order tensor $\underset{\approx}{\boldsymbol{F}}$ would be obtained by projecting the tensor onto the space spanned by the basis in (20.34). This basis is orthogonal (in the sense that the inner product between them is zero), but not normalized (in the sense that the inner product of either basis tensor with itself is not 1), so the appropriate projection operation is

$$\begin{aligned} \text{iso}\underset{\approx}{F} &= \frac{\underset{\approx}{I}(\underset{\approx}{I}:\underset{\approx}{F})}{\underset{\approx}{I}:\underset{\approx}{I}} + \frac{\underset{\approx}{\varepsilon}(\underset{\approx}{\varepsilon}:\underset{\approx}{F})}{\underset{\approx}{\varepsilon}:\underset{\approx}{\varepsilon}} \\ &= \tfrac{1}{2}\underset{\approx}{I}(\underset{\approx}{I}:\underset{\approx}{F}) + \tfrac{1}{2}\underset{\approx}{\varepsilon}(\underset{\approx}{\varepsilon}:\underset{\approx}{F}) \\ &= \frac{\underset{\approx}{I}(F_{11}+F_{22})}{2} + \frac{\underset{\approx}{\varepsilon}(F_{12}-F_{21})}{2}. \end{aligned} \tag{20.35}$$

In component form,

$$\text{iso}\underset{\approx}{F} = \frac{1}{2}\begin{bmatrix} F_{11}+F_{22} & F_{12}-F_{21} \\ F_{21}-F_{12} & F_{11}+F_{22} \end{bmatrix}. \tag{20.36}$$

Incidentally, the 2D polar rotation tensor associated with any 2D $\underset{\approx}{F}$ tensor is a multiple of the isotropic part, with the multiple serving to normalize the rows or columns of iso$\underset{\approx}{F}$. Specifically,

$$\underset{\approx}{R} = \frac{1}{\sqrt{(F_{11}+F_{22})^2 + (F_{12}-F_{21})^2}}\begin{bmatrix} F_{11}+F_{22} & F_{12}-F_{21} \\ F_{21}-F_{12} & F_{11}+F_{22} \end{bmatrix}. \tag{20.37}$$

Recall that proper isotropy is less constraining than strict isotropy. Proper isotropy is a necessary condition for strict isotropy, which means that a strictly isotropic tensor in 2D is necessarily of the form

$$\begin{bmatrix} a & b \\ -b & a \end{bmatrix} \quad \text{(necessary, but not sufficient for } \textit{strict} \text{ isotropy).} \tag{20.38}$$

This result guarantees that the components of the tensor will be unchanged under a proper (same-handed) basis transformation. To be strictly isotropic, it must also satisfy transformations using the matrix $[Q]$ from (20.29). Namely,

$$\begin{bmatrix} \cos\alpha & \sin\alpha \\ \sin\alpha & -\cos\alpha \end{bmatrix}\begin{bmatrix} a & b \\ -b & a \end{bmatrix}\begin{bmatrix} \cos\alpha & \sin\alpha \\ \sin\alpha & -\cos\alpha \end{bmatrix} = \begin{bmatrix} a & b \\ -b & a \end{bmatrix}. \tag{20.39}$$

Considering $\alpha = 0$ shows that b must be zero, which also is easily confirmed to be a sufficient condition to satisfy (20.39). Consequently, for *strict* isotropy, a second-order tensor in 2D must be a multiple of the identity. That is, the permutation tensor $\underset{\approx}{\varepsilon}$ is proper-isotropic, but not strictly isotropic. Its components will change sign upon changing to a differently handed basis. In dyadic form, the 2D permutation tensor may be written

$$\underset{\approx}{\varepsilon} = \underset{\sim}{e}_1\underset{\sim}{e}_2 - \underset{\sim}{e}_2\underset{\sim}{e}_1. \tag{20.40}$$

Changing to a different-handed basis (e.g., exchanging $\underset{\sim}{e}_1$ and $\underset{\sim}{e}_2$) will automatically change the sign of the components. To summarize, if (20.40) is defined to be the permutation tensor for a right-hand basis, then changing to a left-hand basis will change the sign of the *components* of $\underset{\approx}{\varepsilon}$. The tensor $\underset{\approx}{\varepsilon}$ *itself*, being the sum of components times basis dyads, is unaffected by the change of basis. Changing the basis changes the components in a perfectly balancing manner so that the sum of components times basis dyads is unchanged.

20.4 Isotropic fourth-order tensors in 3D

Turning our attention back to tensors in 3D space, methods outlined above for the simpler case of vectors and second-order tensors can be used to prove that the most general form for a *fourth-order* 3D isotropic tensor c_{ijrs} is

$$c_{ijrs} = \alpha(\delta_{ij}\delta_{rs}) + \beta(\delta_{ir}\delta_{js}) + \gamma(\delta_{is}\delta_{jr}). \tag{20.41}$$

For details of the proof, see, for example, [1]. The above expression has three arbitrary parameters, so the space of isotropic fourth-order tensors (referenced to 3D space so that the components are $3 \times 3 \times 3 \times 3$ arrays) is three-dimensional[1], with basis tensors given in the parentheses above. The basis tensors in the parentheses of (20.41) are only one choice for the basis. Any triad of linearly independent combinations of those three tensors is an equally valid basis for the set of all fourth-order isotropic tensors in 3D. In material modeling, a more convenient alternative basis for the space of isotropic fourth-order tensors is

$$P^{\text{iso}}_{ijkl} = \frac{1}{3}\delta_{ij}\delta_{kl}, \tag{20.42a}$$

$$P^{\text{symdev}}_{ijkl} = \frac{1}{2}(\delta_{ik}\delta_{jl} + \delta_{il}\delta_{jk}) + \frac{1}{3}\delta_{ij}\delta_{kl}, \tag{20.42b}$$

$$P^{\text{skew}}_{ijkl} = \frac{1}{2}(\delta_{ik}\delta_{jl} - \delta_{il}\delta_{jk}). \tag{20.42c}$$

These are all constructed from linear combinations of the primitive basis of (20.41). Even though the component formulas for this alternative basis are considerably more complicated, the properties of this basis are irresistible. Specifically, the basis of (20.42) consists of *complementary projectors*! By this we mean

$$\boxed{P^{\text{iso}}_{ijkl}P^{\text{iso}}_{klmn} = P^{\text{iso}}_{ijmn}}, \qquad P^{\text{iso}}_{ijkl}P^{\text{symdev}}_{klmn} = 0, \qquad P^{\text{iso}}_{ijkl}P^{\text{skew}}_{klmn} = 0, \tag{20.43}$$

$$P^{\text{symdev}}_{ijkl}P^{\text{iso}}_{klmn} = 0, \qquad \boxed{P^{\text{symdev}}_{ijkl}P^{\text{symdev}}_{klmn} = P^{\text{symdev}}_{ijmn}}, \qquad P^{\text{symdev}}_{ijkl}P^{\text{skew}}_{klmn} = 0, \tag{20.44}$$

$$P^{\text{skew}}_{ijkl}P^{\text{iso}}_{klmn} = 0, \qquad P^{\text{skew}}_{ijkl}P^{\text{symdev}}_{klmn} = 0, \qquad \boxed{P^{\text{skew}}_{ijkl}P^{\text{skew}}_{klmn} = P^{\text{skew}}_{ijmn}}. \tag{20.45}$$

The physical meanings of these fourth-order tensors become clear by observing how they transform an arbitrary second-order tensor. For any tensor $\underset{\approx}{\boldsymbol{B}}$, note that

[1] The space of isotropic *minor-symmetric* fourth-order tensors in 3D becomes 2D because the minor-symmetry condition ($c_{ijrs} = c_{jirs} = c_{ijsr}$) would require that the coefficient γ in (20.41) to be zero. That is why isotropic elastic stiffness tensors have only *two* independent stiffness moduli.

$$P^{\text{iso}}_{ijkl}B_{kl} = \frac{1}{3}B_{kk}\delta_{ij} = [\text{iso}\underset{\approx}{\boldsymbol{B}}]_{ij}, \tag{20.46a}$$

$$P^{\text{symdev}}_{ijkl}B_{kl} = \frac{1}{2}(B_{ij} + B_{ji}) - \frac{1}{3}B_{kk}\delta_{ij} = [\text{dev}(\text{sym}\underset{\approx}{\boldsymbol{B}})]_{ij} = [\text{sym}(\text{dev}\underset{\approx}{\boldsymbol{B}})]_{ij}, \tag{20.46b}$$

$$P^{\text{skew}}_{ijkl}B_{kl} = \frac{1}{2}(B_{ij} - B_{ji}) = [\text{skw}\underset{\approx}{\boldsymbol{B}}]_{ij}. \tag{20.46c}$$

Thus, as operators, P^{iso}_{ijkl}, P^{symdev}_{ijkl}, and P^{skew}_{ijkl}, respectively, return the isotropic part, the symmetric-deviatoric part, and the skew part of whatever they act on.

20.4.1 Application of isotropic fourth-order tensors to elasticity

Since the tensors in (20.42) comprise an isotropic fourth-order engineering tensor (IFOET) basis, it follows that *any* fourth-order isotropic engineering tensor may be written in the form

$$\alpha P^{\text{iso}}_{ijkl} + \beta P^{\text{symdev}}_{ijkl} + \gamma P^{\text{skew}}_{ijkl}. \tag{20.47}$$

In small deformation elasticity, for example, the fourth-order elastic stiffness tensor E_{ijkl} may be written

$$E_{ijkl} = 3KP^{\text{iso}}_{ijkl} + 2GP^{\text{symdev}}_{ijkl}, \tag{20.48}$$

where K is the bulk modulus and G is the shear modulus. In this case, because the elastic stiffness tensor is minor-symmetric, the coefficient γ in (20.47) is zero. Because the basis in (20.42) consists of complementary projectors, the above expression for the elastic stiffness tensor is a *spectral expansion*. Hence, $3K$ is an eigenvalue of multiplicity 1 with the second-order eigen*tensor* being any multiple of the second-order identity tensor. In other words, the operation $\sigma_{ij} = E_{ijkl}\varepsilon_{kl}$ will produce an isotropic stress σ_{ij} if the strain ε_{kl} is isotropic; specifically, in this case, $\sigma_{ij} = 3K\varepsilon_{ij}$. The product $2G$ is an eigenvalue of multiplicity 5, and any symmetric-deviatoric tensor is the associated eigentensor. That is, $\sigma_{ij} = E_{ijkl}\varepsilon_{kl}$ will predict $\sigma_{ij} = 2G\varepsilon_{ij}$ if $\varepsilon_{kl} = \varepsilon_{lk}$ and $\varepsilon_{kk} = 0$. Finally, an elastic stiffness tensor has an eigenvalue 0 of multiplicity 3 with any skew-symmetric tensor being an eigentensor (i.e., $E_{ijkl}\varepsilon_{kl} = 0$ if $\varepsilon_{kl} = -\varepsilon_{lk}$).

20.4.2 Isotropic elastic materials do not have isotropic fourth-order tangent tensors

A material is said to by hyperelastic if there exists a potential function ϕ, which depends on a strain tensor $\underset{\approx}{e}$ such that the conjugate stress tensor $\underset{\approx}{\sigma}$ is given by[2]

$$\sigma_{ij} = \frac{\partial \phi}{\partial e_{ij}}. \tag{20.49}$$

This implies that

[2] Note that, $\underset{\approx}{\sigma}$ and $\underset{\approx}{e}$ are imprecisely defined as long as they are work conjugate, as defined in classical theories of thermomechanics. For example, the stress and strain in these formulas might be interpreted, respectively, to be the second Piola–Kirchhoff stress $\underset{\approx}{\sigma} = \underset{\approx}{F}^{-1} \bullet (J\underset{\approx}{\sigma}^{\text{cauchy}}) \bullet \underset{\approx}{F}^{-T}$ and Green–Lagrange strain $\underset{\approx}{e} = \frac{1}{2}(\underset{\approx}{F}^T \bullet \underset{\approx}{F} - \underset{\approx}{I})$, where $J = \det \underset{\approx}{F}$ and $\underset{\approx}{\sigma}^{\text{cauchy}}$ is the Cauchy stress.

$$d\sigma_{ij} = E_{ijkl} de_{kl}, \qquad \text{where } E_{ijkl} \equiv \frac{\partial^2 \phi}{\partial e_{ij} \partial e_{kl}}. \tag{20.50}$$

This definition holds for general hyperelasticity. If a *material* is isotropic, then the elastic potential function depends only on the scalar invariants of the strain $\underset{\approx}{e}$:

$$\phi = \phi\left(I_1^e, J_2^e, J_3^e\right), \tag{20.51}$$

where

$$I_1^e = \text{tr}\underset{\approx}{e}, \qquad J_2^e = \frac{1}{2}\text{tr}[(\underset{\approx}{e}^d)^2], \qquad \text{and} \qquad J_3^e = \frac{1}{3}\text{tr}[(\underset{\approx}{e}^d)^3]. \tag{20.52}$$

Here, $\underset{\approx}{e}^d$ denotes the deviatoric part of $\underset{\approx}{e}$. Even though the material is isotropic, corresponding elastic tangent stiffness tensor E_{ijkl} can be easily shown to contain *anisotropic* deformation-induced contributions. In general, the elastic tangent tensor itself is isotropic for an isotropic material only in the limit of vanishing elastic strains. Otherwise, an isotropic material acquires reversible induced anisotropy in the loaded state. Hence, for example, an acoustic wave will travel through a loaded material differently from how it travels through an unstrained medium!

20.5 The isotropic part of a fourth-order tensor

The set of IFOETs[3] is closed under tensor addition and scalar multiplication. This means that any linear combination of IFOET tensors will itself be IFOET. Therefore, the set of all IFOET tensors forms a linear subspace of general fourth-order engineering tensor space.

In the previous section, we showed that any IFOET tensor can be written as a linear combination of P_{ijkl}^{iso}, P_{ijkl}^{symdev}, and P_{ijkl}^{skew}. These three tensors therefore form a basis for the set of all IFOET tensors, and the IFOET subspace must be 3D. This basis is orthogonal in the inner-product sense (e.g., $P_{ijkl}^{\text{iso}} P_{ijkl}^{\text{skew}} = 0$), but it is not normalized. We can define an orthonormal basis for IFOET tensors as

$$\hat{P}_{ijkl}^{\text{iso}} \equiv P_{ijkl}^{\text{iso}}, \tag{20.53a}$$

$$\hat{P}_{ijkl}^{\text{symdev}} \equiv \frac{P_{ijkl}^{\text{symdev}}}{\sqrt{5}}, \tag{20.53b}$$

$$\hat{P}_{ijkl}^{\text{skew}} \equiv \frac{P_{ijkl}^{\text{skew}}}{\sqrt{3}}. \tag{20.53c}$$

The denominators in these equations are the magnitudes of the tensors in the numerators, obtained by taking the square root of the fourth-order inner product of

[3] I.e., isotropic fourth-order engineering tensors (IFOETs) tensors are referenced to 3D physical space so that their components are $3 \times 3 \times 3 \times 3$ arrays (reducible to 9×9 arrays or, if minor-symmetric, 6×6 Voigt or Mandel arrays).

the tensors with themselves. The denominators can be easily remembered because for ordinary second-order engineering tensors isotropic tensors have one independent component, symmetric-deviatoric tensors have five independent components, and skew-symmetric tensors have three independent components.

If three m-dimensional *orthonormal* vectors $\underline{\boldsymbol{n}}_1$, $\underline{\boldsymbol{n}}_2$, and $\underline{\boldsymbol{n}}_3$, form a basis for a 3D subspace embedded in a larger m-dimensional vector space, then any vector $\underline{\boldsymbol{x}}$ in the higher-dimensional space can be projected to the 3D subspace by applying the operation

$$\underline{\boldsymbol{x}}^{\text{projected}} = \underline{\underline{\boldsymbol{P}}} * \underline{\boldsymbol{x}}, \tag{20.54}$$

where

$$\underline{\underline{\boldsymbol{P}}} = \underline{\boldsymbol{n}}_1\underline{\boldsymbol{n}}_1 + \underline{\boldsymbol{n}}_2\underline{\boldsymbol{n}}_2 + \underline{\boldsymbol{n}}_3\underline{\boldsymbol{n}}_3, \tag{20.55}$$

and the * denotes the inner product in the m-dimensional space.

We are interested in finding the IFOET part of a general fourth-order engineering tensor X_{ijkl}. This is accomplished by projecting that tensor to the IFOET subspace. Using (20.54), this operation is found by

$$X_{ijkl}^{\text{IFOET}} = P_{ijklpqrs} X_{pqrs}, \tag{20.56}$$

where the components of the sixth-order (V_3^6) tensor are found by using (20.53) as the orthonormal basis appearing symbolically in (20.55). Namely,

$$P_{ijklpqrs} = P_{ijkl}^{\text{iso}} P_{pqrs}^{\text{iso}} + \frac{1}{5} P_{ijkl}^{\text{symdev}} P_{pqrs}^{\text{symdev}} + \frac{1}{3} P_{ijkl}^{\text{skew}} P_{pqrs}^{\text{skew}}. \tag{20.57}$$

This may look like a tediously complicated operation, but it is actually quite straightforward if the *Mandel* components[4] of the fourth-order tensor X_{pqrs} are available. The operation in (20.56) may be written as

$$X_{ijkl}^{\text{IFOET}} = 3\zeta_1 P_{ijkl}^{\text{iso}} + 2\eta_2 P_{ijkl}^{\text{symdev}} + \eta_3 P_{ijkl}^{\text{skew}}, \tag{20.58}$$

where ζ_1, η_2, η_3 are some invariants of X_{pqrs} defined by

$$\zeta_1 = \frac{1}{3} P_{pqrs}^{\text{iso}} X_{pqrs}, \tag{20.59}$$

$$\eta_2 = \frac{1}{10} P_{pqrs}^{\text{symdev}} X_{pqrs}. \tag{20.60}$$

[4] Mandel components are the same as classical Voigt components except that rows and columns associated with off-diagonal terms are multiplied by $\sqrt{2}$. See chapter 26 or [2] for more details.

$$\eta_3 = \tfrac{1}{3} P^{\text{skew}}_{ijkl} X_{pqrs}. \tag{20.61}$$

If, X_{pqrs} were an elastic stiffness tensor, then ζ_1 would be the bulk modulus and η_2 would be the shear modulus, while η_3 would be zero.

In terms of Mandel components, the 9×9 component matrix for P^{iso}_{ijkl} has zeros everywhere except in the upper-left 3×3 sub-matrix, where the components are all equal to 1/3. The invariant ζ_1 is therefore simply the average of all components in the upper-left 3×3 sub-matrix of the Mandel matrix for X_{pqrs}. Evaluation of this invariant of the fourth-order tensor is somewhat analogous to computing mean stress by the average of diagonal stress components.

In the Mandel system, the 9×9 component matrix for $\tfrac{1}{2}(\delta_{ik}\delta_{jl} + \delta_{il}\delta_{jk})$ has zeros everywhere except that its upper-left 6×6 sub-matrix is the 6×6 identity matrix. In other words, it is the identity operator for symmetric tensors. Similarly, P^{symdev}_{ijkl} is the identity operator for symmetric-deviatoric tensors. To evaluate operation $P^{\text{iso}}_{pqrs} X_{pqrs}$, it is helpful to first compute a scalar ζ_2 by the sum of the first six diagonal components of the Mandel matrix (like a trace of the minor-symmetric part). Then $\eta_2 = \tfrac{1}{10}(\zeta_2 - 3\zeta_1)$.

Finally, the 9×9 Mandel matrix for P^{skew}_{ijkl} is zeros everywhere except the lower-right 3×3 sub-matrix is the 3×3 identity matrix. Hence, η_3 is simply the average of the last three diagonal components of the Mandel matrix for X_{pqrs}.

Suppose, for example, that a general, not necessarily minor- or major-symmetric fourth-order tensor has a 9×9 Mandel component matrix given by

$$\begin{bmatrix} 0 & 0 & 1 & 0 & 5 & 7 & 0 & 0 & 0 \\ 3 & 0 & 6 & 2 & 0 & 0 & 0 & 0 & 0 \\ 0 & 8 & 0 & 0 & 0 & 0 & 0 & 0 & 0 \\ 0 & 0 & 0 & 4 & 0 & 0 & 0 & 0 & 0 \\ 0 & 0 & 0 & 7 & 0 & 0 & 0 & 0 & 0 \\ 0 & 0 & 0 & 0 & 0 & 0 & 0 & 0 & 0 \\ 0 & 0 & 0 & 0 & 0 & 0 & 0 & 0 & 0 \\ 0 & 0 & 0 & 0 & 0 & 0 & 0 & 2 & 0 \\ 0 & 0 & 0 & 0 & 0 & 0 & 9 & 0 & 1 \end{bmatrix}, \tag{20.62}$$

then

$$\zeta_1 = \text{ average } \{0, 0, 1, 3, 0, 6, 0, 8, 0\} = 2, \tag{20.63a}$$

$$\zeta_2 = 0 + 0 + 0 + 4 + 0 + 0 = 4, \tag{20.63b}$$

$$\eta_2 = \tfrac{1}{10}(\zeta_2 - 3\zeta_1) = -0.2, \tag{20.63c}$$

$$\eta_3 = \frac{0+2+1}{3} = 1. \tag{20.63d}$$

These values would then be used in the linear combination defined in (20.58) to obtain the isotropic part of the original tensor.

20.5.1 A scalar measure of anisotropy of fourth-order tensors

To reiterate, the operation $X_{ijkl}^{\mathrm{IFOET}}$ determines the isotropic part of a fourth-order tensor. With this isotropic part in hand, the *deviatoric* part of a fourth-order tensor is simply $X_{ijkl}^{\mathrm{DEV}} = X_{ijkl} - X_{ijkl}^{\mathrm{IFOET}}$. A good scalar quantifier of the degree of anisotropy of a fourth-order tensor is:

$$\text{scalar measure of anisotropy} = \frac{2}{\pi}\ \arctan(X^{\mathrm{DEV}},\ X^{\mathrm{IFOET}}), \tag{20.64}$$

where X^{DEV} is the magnitude of X_{ijkl}^{DEV} and X^{IFOET} is the magnitude of $X_{ijkl}^{\mathrm{IFOET}}$. The two-angle arctangent gives the angle in tensor space that a general tensor X_{ijkl} forms with the linear manifold (plane) of isotropic tensors. The multiplier $2/\pi$ simply normalizes this angle so that this scalar measure of anisotropy varies from 0 to 1. A review of this and other measures of anisotropy may be found in [3]. For example, applying this formula, the tensor in equation (20.62) is found to be 77.7% anisotropic.

20.5.2 Projections and eigenprojectors

The tensors in (20.53) form a useful basis for the set of all isotropic fourth-order tensors. These three fourth-order tensors are very special members of a larger 81-dimensional space (because fourth-order tensors have $3 \times 3 \times 3 \times 3 = 81$ components). The tensors in (20.53) are special because, regarding them as 81-dimensional vectors ($\underset{\sim}{\boldsymbol{N}}_1$, $\underset{\sim}{\boldsymbol{N}}_2$, $\underset{\sim}{\boldsymbol{N}}_3$), they form an orthonormal basis for isotropic fourth-order tensors, meaning that any linear combination of them is itself isotropic and also

$$\underset{\sim}{\boldsymbol{N}}_i \bullet \underset{\sim}{\boldsymbol{N}}_j = \delta_{ij}, \tag{20.65}$$

where the indices run from 1 to 81, making the dot product equivalent to the quadruple dot fourth-order inner product. The act of finding the isotropic part of a generally non-isotropic fourth-order tensor, now also regarded as an 81-dimensional vector $\underset{\sim}{\boldsymbol{T}}$, is equivalent to the operation

$$\underset{\sim}{\boldsymbol{N}}_1(\underset{\sim}{\boldsymbol{N}}_1 \bullet \underset{\sim}{\boldsymbol{T}}) + \underset{\sim}{\boldsymbol{N}}_2(\underset{\sim}{\boldsymbol{N}}_2 \bullet \underset{\sim}{\boldsymbol{T}}) + \underset{\sim}{\boldsymbol{N}}_3(\underset{\sim}{\boldsymbol{N}}_3 \bullet \underset{\sim}{\boldsymbol{T}}) = \underset{\approx}{\boldsymbol{P}} \bullet \underset{\sim}{\boldsymbol{T}}, \tag{20.66}$$

where

$$\underset{\approx}{\boldsymbol{P}} = \underset{\sim}{\boldsymbol{N}}_1\underset{\sim}{\boldsymbol{N}}_1 + \underset{\sim}{\boldsymbol{N}}_2\underset{\sim}{\boldsymbol{N}}_2 + \underset{\sim}{\boldsymbol{N}}_3\underset{\sim}{\boldsymbol{N}}_3. \tag{20.67}$$

Keeping in mind that each 81-dimensional $\underset{\sim}{\boldsymbol{N}}_k$ vector is actually an 81-component fourth-order tensor, it follows that the $\underset{\sim}{\boldsymbol{N}}_k\underset{\sim}{\boldsymbol{N}}_k$ dyads are second-order tensors in 81-dimensional space, representing eighth-order tensors with respect to 3D space. Wow, that is deep, but the intuitive understanding of the $\underset{\approx}{\boldsymbol{P}}$ operator is similar to the sum of orthonormal vectors in the ordinary 3D space where we live. In general, when you multiply one unit vector $\underset{\sim}{\boldsymbol{N}}$ dyadically with another, the resulting operator is a rank-1 projection to the span of $\underset{\sim}{\boldsymbol{N}}$. This means that

$$(\underset{\sim}{\boldsymbol{N}}\underset{\sim}{\boldsymbol{N}}) \bullet \underset{\sim}{\boldsymbol{T}} \tag{20.68}$$

gives the part of $\underset{\sim}{\boldsymbol{T}}$ in the direction of $\underset{\sim}{\boldsymbol{N}}$. Adding two orthonormal vectors to form the operation

$$(\underset{\sim}{\boldsymbol{N}}_1\underset{\sim}{\boldsymbol{N}}_1 + \underset{\sim}{\boldsymbol{N}}_2\underset{\sim}{\boldsymbol{N}}_2) \bullet \underset{\sim}{\boldsymbol{T}} \tag{20.69}$$

will return the projection of $\underset{\sim}{\boldsymbol{T}}$ onto the plane spanned by those orthonormal vectors. Of course, a plane can contain an infinite number of choices of orthonormal basis vectors. If a different orthonormal pair ($\underset{\sim}{\boldsymbol{M}}_1$, $\underset{\sim}{\boldsymbol{M}}_2$) were to be selected as a basis for the plane, then the combination

$$(\underset{\sim}{\boldsymbol{M}}_1\underset{\sim}{\boldsymbol{M}}_1 + \underset{\sim}{\boldsymbol{M}}_2\underset{\sim}{\boldsymbol{M}}_2) \bullet \underset{\sim}{\boldsymbol{T}} \tag{20.70}$$

would again produce the same result: projection of $\underset{\sim}{\boldsymbol{T}}$ onto the plane. The projection tensor in (20.67) is similar, except that it returns the projection of an arbitrary 81-dimensional vector onto the hyper-plane spanned by the three basis tensors for isotropy.

When the two orthonormal vectors correspond to eigenvectors of a symmetric tensor and if the associated eigenvalue has multiplicity 2, then the dyadic summation of these eigenvectors will form a unique *eigenprojector* even though the eigenvectors themselves are not unique. More generally, an eigenprojector for a symmetric tensor is formed by orthonormalizing all eigenvectors associated with a given eigenvalue and then adding the dyads of these vectors[5].

Overall, it is not the individual basis vectors that are physically meaningful—it is the plane itself. An orthonormal basis for the plane is not unique, but the sum of dyads formed from any such basis is unique and hence the physically meaningful. Similarly, eigenvectors are not physically meaningful, but the associated *eigenspaces* (which are defined from the eigenprojectors and geometrically corresponding to a line for multiplicity 1, plane for multiplicity 2, etc) are unique and meaningful. Accordingly, it is highly desirable to obtain closed-form expressions for eigenprojectors. This is done in the next section on transverse isotropy.

20.6 Tensor transverse isotropy

A tensor is said to be *transversely isotropic* about a given unit vector $\underset{\sim}{\boldsymbol{n}}$ if its components are unchanged when the basis is rotated about this vector. If the property is a scalar invariant like mass density, then (by definition) it cannot change under any rotation. The concept of transverse isotropy becomes increasingly complicated for vector and higher-order tensor properties.

[5] This statement can be generalized to non-symmetric tensors by forming dyads of right eigenvectors with left eigenvectors, in which case (for non-symmetry) the projection operation is oblique (like finding a shortened or lengthened shadow when the Sun is not in the 'high noon' position overhead).

Just as there is a distinction between tensor isotropy and material isotropy, we likewise recognize the distinction between *tensor transverse isotropy*, meaning that a tensor's components are unchanged upon basis rotation about a given axis, and *material transverse isotropy*, meaning that a material's response to a given deformation is unchanged if the material is first rotated about a given axis. This section discusses transversely isotropic vectors and tensors. In this discussion, the basis having the 3-direction aligned with the symmetry axis is called the material basis, but the word 'material' is used only to be consistent with terminology of composites engineers. The concept of material transverse symmetry is discussed in section 20.7.

In what follows, we often consider a 'material orthonormal basis' $\{\underset{\sim}{m}_1, \underset{\sim}{m}_2, \underset{\sim}{m}_3\}$ such that $\underset{\sim}{m}_3 = \underset{\sim}{n}$. This basis will not generally coincide with the stationary 'laboratory basis' $\{\underset{\sim}{e}_1, \underset{\sim}{e}_2, \underset{\sim}{e}_3\}$ that is used in general calculations. Components with respect to the material basis are called material components. Components with respect to the laboratory basis are called laboratory components. Our goal in what follows is to first show the structure of the material components of transversely isotropic vectors and tensors. The next goal is to show how those tensors may be expressed *directly* in the laboratory components without having to perform a transformation to the material basis.

A *vector* $\underset{\sim}{u}$ is transversely isotropic if and only if its *material components* are of the form

$$\{\underset{\sim}{u}\} = \begin{Bmatrix} 0 \\ 0 \\ u_3 \end{Bmatrix} \quad \text{in the material basis } \{\underset{\sim}{m}_i\}. \tag{20.71}$$

In other words, the only vector that remains unchanged when rotated about the material symmetry direction $\underset{\sim}{n}$ is a vector that is itself a multiple of $\underset{\sim}{n}$. Thus, since $\underset{\sim}{n}$ is a unit vector, the most general form for a transversely isotropic vector $\underset{\sim}{u}$ is

$$\underset{\sim}{u} = \alpha \underset{\sim}{n}. \tag{20.72}$$

Of course, any vector can be decomposed into the product of magnitude times unit direction, so all vectors are transversely isotropic about their own direction vector. Here, we are requiring symmetry about a particular direction vector, which is the distinction. Thus,

$$\underset{\sim}{u} = \alpha \begin{Bmatrix} n_1 \\ n_2 \\ n_3 \end{Bmatrix} \quad \text{in the laboratory basis } \{\underset{\sim}{e}_i\}. \tag{20.73}$$

Another advantage of (20.72) over (20.71) is that (20.73) can be differentiated if the transverse symmetry axis $\underset{\sim}{n}$ is not constant.

Now consider a general *second-order tensor* $\underset{\approx}{A}$. To be transversely isotropic, the material components of must be of the form

$$[\underset{\approx}{A}] = \begin{bmatrix} A_t & -A_s & 0 \\ A_s & A_t & 0 \\ 0 & 0 & A_a \end{bmatrix} \quad \text{in the material basis} \{\underset{\sim}{m}_i\}. \tag{20.74}$$

Here, the subscripts t and a refer to transverse and axial directions relative to the symmetry axis, and the subscript s indicates skew shear components that represent twisting about the symmetry axis. Referring back to equation (20.33), note that the 'in-plane' (upper 2×2) part of a transversely isotropic 3D tensor is isotropic in that plane. If the tensor is symmetric, then it cannot have skew terms and its most general form then reduces to

$$[\underset{\approx}{A}] = \begin{bmatrix} A_t & 0 & 0 \\ 0 & A_t & 0 \\ 0 & 0 & A_a \end{bmatrix} \quad \text{in the material basis}\{\underset{\sim}{m}_i\}. \tag{20.75}$$

In other words, the tensor must be diagonal in the material basis, with the 11 and 22 transverse material components equal. Thus, a symmetric transversely isotropic 3D tensor must be strictly isotropic in its 2D transverse plane. In dyadic notation, the above equation may be written

$$\underset{\approx}{A} = A_t(\underset{\sim}{m}_1\underset{\sim}{m}_1 + \underset{\sim}{m}_2\underset{\sim}{m}_2) + A_a\underset{\sim}{m}_3\underset{\sim}{m}_3. \tag{20.76}$$

We know that, for 3D space, $\underset{\sim}{E}_1\underset{\sim}{E}_1 + \underset{\sim}{E}_2\underset{\sim}{E}_2 + \underset{\sim}{E}_3\underset{\sim}{E}_3 = \underset{\approx}{I}$ for *any* orthogonal triad $\{\underset{\sim}{E}_1, \underset{\sim}{E}_2, \underset{\sim}{E}_3\}$; therefore it holds for the material basis $\{\underset{\sim}{m}_1, \underset{\sim}{m}_2, \underset{\sim}{m}_3\}$ as well. Therefore, equation (20.76) may be written

$$\underset{\approx}{A} = A_t(\underset{\approx}{I} - \underset{\sim}{m}_3\underset{\sim}{m}_3) + A_a\underset{\sim}{m}_3\underset{\sim}{m}_3. \tag{20.77}$$

This assertion is equivalent to re-writing (20.75) as

$$\begin{bmatrix} A_t & 0 & 0 \\ 0 & A_t & 0 \\ 0 & 0 & A_a \end{bmatrix} = A_t\left(\begin{bmatrix} 1 & 0 & 0 \\ 0 & 1 & 0 \\ 0 & 0 & 1 \end{bmatrix} - \begin{bmatrix} 0 & 0 & 0 \\ 0 & 0 & 0 \\ 0 & 0 & 1 \end{bmatrix}\right) + A_a\begin{bmatrix} 0 & 0 & 0 \\ 0 & 0 & 0 \\ 0 & 0 & 1 \end{bmatrix} \quad \text{in the material basis } \{\underset{\sim}{m}_i\}. \tag{20.78}$$

Since $\underset{\sim}{m}_3 = \underset{\sim}{n}$, equation (20.77) may be written

$$\underset{\approx}{A} = A_t\underset{\approx}{P}_t + A_a\underset{\approx}{P}_a, \qquad \text{where } \underset{\approx}{P}_a = \underset{\sim}{n}\underset{\sim}{n} \qquad \text{and} \qquad \underset{\approx}{P}_t = \underset{\approx}{I} - \underset{\approx}{P}_a. \tag{20.79}$$

This is the most general *direct-notation* form for a symmetric transversely isotropic second-order tensor. When cast in the material basis, this equation merely says that

$$\begin{bmatrix} A_r & 0 & 0 \\ 0 & A_t & 0 \\ 0 & 0 & A_a \end{bmatrix} = A_t[P_a] + A_a[P_a], \quad \text{where } [P_t] = \begin{bmatrix} 1 & 0 & 0 \\ 0 & 1 & 0 \\ 0 & 0 & 0 \end{bmatrix} \quad \text{and} \quad [P_a] = \begin{bmatrix} 0 & 0 & 0 \\ 0 & 0 & 0 \\ 0 & 0 & 1 \end{bmatrix}. \tag{20.80}$$

Equation (20.80) may be alternatively written as

$$\underset{\approx}{\boldsymbol{A}} = A_t \underset{\approx}{\boldsymbol{I}} + (A_a - A_t)\underset{\approx}{\boldsymbol{P}}_a. \tag{20.81}$$

The tremendous advantage of (20.81) over the matrix representation of (20.75) is that (20.81) can be computed *directly* in the laboratory basis without having to perform a coordinate transformation. Namely, in the laboratory basis,

$$\left[\underset{\approx}{\boldsymbol{A}}\right] = A_t \begin{bmatrix} 1 & 0 & 0 \\ 0 & 1 & 0 \\ 0 & 0 & 1 \end{bmatrix} + (A_a - A_t) \begin{bmatrix} n_1 n_1 & n_1 n_2 & n_1 n_3 \\ n_2 n_1 & n_2 n_2 & n_2 n_3 \\ n_3 n_1 & n_3 n_2 & n_3 n_3 \end{bmatrix}. \tag{20.82}$$

Furthermore, equation (20.82) is more useful when the transverse axis is evolving because (20.82) can be differentiated with respect to time.

There is no transversely isotropic *third-order tensor* (other than the isotropic alternating tensor as a degenerate isotropic case).

> We will now present the most general form for a transversely isotropic fourth-order tensor. This discussion requires some advanced terminology and notation, so skip the remainder of this section if you are on your first pass through this book (come back later).

A fourth-order tensor $\underset{\approx}{\boldsymbol{T}}$ is transversely isotropic if and only if

$$T_{ijkl} = R_{ip} R_{jq} R_{kr} R_{ls} T_{pqrs} \tag{20.83}$$

for all proper orthogonal $[R]$ that are rotations about a fixed axis $\underset{\sim}{\boldsymbol{n}}$. If an orthonormal material basis $\{\underset{\sim}{\boldsymbol{m}}_1, \underset{\sim}{\boldsymbol{m}}_2, \underset{\sim}{\boldsymbol{m}}_3\}$ is set up such that $\underset{\sim}{\boldsymbol{m}}_3 = \underset{\sim}{\boldsymbol{n}}$, then the most general form of the 9×9 *Mandel* (see chapter 26) component matrix is

$$[T] = \begin{bmatrix} h_2+h_4 & h_2 & h_3+m_2 & 0 & 0 & m_1 & 0 & 0 & s_5+g_2 \\ h_2 & h_2+h_4 & h_3+m_2 & 0 & 0 & -m_1 & 0 & 0 & s_5+g_2 \\ h_3-m_2 & h_3-m_2 & h_1 & 0 & 0 & 0 & 0 & 0 & s_6+g_3 \\ 0 & 0 & 0 & h_5 & m_3 & 0 & g_4+s_3 & s_4+g_5 & 0 \\ 0 & 0 & 0 & -m_3 & h_5 & 0 & s_4+g_5 & -s_3-g_4 & 0 \\ -m_1 & m_1 & 0 & 0 & 0 & h_4 & 0 & 0 & 0 \\ 0 & 0 & 0 & s_3-g_4 & s_4-g_5 & 0 & s_1 & -g_1 & 0 \\ 0 & 0 & 0 & s_4-g_5 & -s_3+g_4 & 0 & g_1 & s_1 & 0 \\ s_5-g_2 & s_5-g_2 & s_6-g_3 & 0 & 0 & 0 & 0 & 0 & s_2 \end{bmatrix} \tag{20.84}$$

This matrix has a total of 19 independent constants. However, just as the three independent constants in (20.74) were reduced to two independent constants in (20.75) under conditions of symmetry, the above result likewise simplifies considerably with further symmetry assumptions. If the fourth-order tensor $\underset{\approx}{\boldsymbol{T}}$ is major-symmetric, then the five g and three m parameters are zero, leaving only the five h and six s parameters possibly non-zero. If the tensor is minor-symmetric, but not necessarily major-symmetric, then the five g and six s parameters are zero, leaving only the five h and three m parameters possibly non-zero.

In applications, fourth-order tensors are typically *both* major- and minor-symmetric, leaving only the five h parameters. In this case, the 9×9 Mandel component matrix reduces to the following 6×6 component sub-matrix:

$$\begin{bmatrix} h_0 & h_2 & h_3 & 0 & 0 & 0 \\ h_2 & h_0 & h_3 & 0 & 0 & 0 \\ h_2 & h_3 & h_1 & 0 & 0 & 0 \\ 0 & 0 & 0 & h_5 & 0 & 0 \\ 0 & 0 & 0 & 0 & h_5 & 0 \\ 0 & 0 & 0 & 0 & 0 & h_4 \end{bmatrix}, \quad \text{where } h_0 \equiv h_2 + h_4. \tag{20.85}$$

This might, for example, represent the component structure of the elastic stiffness (or compliance) for a material that is made up of laminate layers in the 1–2 plane or a unidirectional fiber composite having the fibers aligned with (or symmetric about) the 3-direction.

Equation (20.85) is the most general form for the Mandel component matrix of a double-symmetric transversely isotropic tensor when the symmetry axis points in the 3-direction. The five parameters $\{h_1, \ldots, h_5\}$ quantify the anisotropy. If the tensor represented by this Mandel matrix is denoted $\underset{\approx}{\boldsymbol{T}}$, then the h_k components are related to conventional four-index T_{ijkl} *material* components as follows:

$$h_1 = T_{3333}, \tag{20.86a}$$

$$h_2 = T_{1122}, \tag{20.86b}$$

$$h_3 = T_{1133}, \tag{20.86c}$$

$$h_4 = 2T_{1212}, \tag{20.86d}$$

$$h_5 = 2T_{3131}. \tag{20.86e}$$

Just as the tensor $\underset{\sim}{\boldsymbol{A}}$ in (20.75) was diagonal only with respect to a basis aligned with the transverse symmetry directions, the component matrix for a fourth-order transversely isotropic stiffness tensor has the structure in (20.85) only when the 3-direction is aligned with the symmetry axis. In (20.79), we upgraded the specialized result of (20.75) by identifying a direct-notation (basis-independent) formulation. Such a formulation is important if, for example, the symmetry axis is not fixed (it might, for example, change orientation in response to rigid-body rotation). Below, we perform a similar analysis in the fourth-order case.

For materials having transverse symmetry about some axis $\underset{\sim}{\boldsymbol{n}}$, a natural basis for the fourth-order elastic stiffness (or compliance) tensor is[6]

$\underset{\approx}{\boldsymbol{b}}_1 =$	$\underset{\approx}{\boldsymbol{b}}_2 =$	$\underset{\approx}{\boldsymbol{b}}_3 =$	$\underset{\approx}{\boldsymbol{b}}_4 =$	$\underset{\approx}{\boldsymbol{b}}_5 =$
$\underset{\approx}{\boldsymbol{P}}\underset{\approx}{\boldsymbol{P}}$	$\underset{\approx}{\boldsymbol{Q}}\underset{\approx}{\boldsymbol{Q}}$	$(\underset{\approx}{\boldsymbol{P}}\underset{\approx}{\boldsymbol{Q}}+\underset{\approx}{\boldsymbol{Q}}\underset{\approx}{\boldsymbol{P}})$	$(\underset{\approx}{\boldsymbol{Q}}\underset{\approx}{\boldsymbol{Q}})^{\Lambda}$	$(\underset{\approx}{\boldsymbol{P}}\underset{\approx}{\boldsymbol{Q}}+\underset{\approx}{\boldsymbol{Q}}\underset{\approx}{\boldsymbol{P}})^{\Lambda}$

, (20.87)

where

$$\underset{\approx}{\boldsymbol{P}} \equiv \underset{\sim}{\boldsymbol{n}}\underset{\sim}{\boldsymbol{n}} = \underset{\approx}{\boldsymbol{P}}_a \qquad \text{and} \qquad \underset{\approx}{\boldsymbol{Q}} = \underset{\approx}{\boldsymbol{I}} - \underset{\sim}{\boldsymbol{n}}\underset{\sim}{\boldsymbol{n}} = \underset{\approx}{\boldsymbol{P}}_t. \tag{20.88}$$

Here, two tensors written adjacent to each other are multiplied dyadically (e.g., the *ijkl* components of $\underset{\approx}{\boldsymbol{P}}\underset{\approx}{\boldsymbol{Q}}$ are $P_{ij}Q_{kl}$). The superscript Λ operator is based on a simpler generalized transpose operator superscripted L which we call a 'leafing' operation because it is a permutation of indices similar to a perfect shuffle (leafing together) a deck of four cards. For any fourth-order tensor with *ijkl* components T_{ijkl}, the *leafing operation* creates a new tensor with *ijkl* given by

$$T^L_{ijkl} = T_{ikjl}. \tag{20.89}$$

Hence, this is an instance of a generalized transpose, effectively swapping the ordering of the middle two basis vectors of the fourth-order tensor. Of particular interest is the leafing operation applied to a tensor–tensor dyad:

$$(\underset{\approx}{\boldsymbol{A}}\underset{\approx}{\boldsymbol{B}})^L_{ijrs} = A_{ir}B_{js}. \tag{20.90}$$

This operator has the symbolic (direct) notation definition that, for any $\underset{\approx}{\boldsymbol{C}}$,

$$(\underset{\approx}{\boldsymbol{A}}\underset{\approx}{\boldsymbol{B}})^L : \underset{\approx}{\boldsymbol{C}} = \underset{\approx}{\boldsymbol{A}} \bullet \underset{\approx}{\boldsymbol{C}} \bullet \underset{\approx}{\boldsymbol{B}}^T. \tag{20.91}$$

The superscript Λ operator is similar except that it also merely imposes minor symmetry on the leafing operation. Specifically, for any two tensors $\underset{\approx}{\boldsymbol{A}}$ and $\underset{\approx}{\boldsymbol{B}}$,

$$(\underset{\approx}{\boldsymbol{A}}\underset{\approx}{\boldsymbol{B}})^{\Lambda}_{ijrs} = \tfrac{1}{4}(A_{ir}B_{js} + A_{is}B_{jr} + A_{jr}B_{is} + A_{js}B_{ir}) \tag{20.92}$$

and hence, for any $\underset{\approx}{\boldsymbol{C}}$,

$$(\underset{\approx}{\boldsymbol{A}}\underset{\approx}{\boldsymbol{B}})^{\Lambda} : \underset{\approx}{\boldsymbol{C}} = \text{sym}[\underset{\approx}{\boldsymbol{A}} \bullet (\text{sym}\underset{\approx}{\boldsymbol{C}}) \bullet \underset{\approx}{\boldsymbol{B}}^T]. \tag{20.93}$$

When forming component matrices for these tensors, we strongly recommend using full 9×9 Mandel components, not the reduced 6×6 forms. The 9×9 matrices allow you to visually confirm when a fourth-order tensor is minor symmetric by noting zeros in the last three rows and last three columns. If not, unexpected nonzero entries are typically handled via an appropriate revision (such as converting from leafing to symmetric leafing) to ensure minor symmetry. With this operator now defined, the components of the five transverse basis tensors are

[6] This basis has some similarity to Walpole's [4].

$$(\underset{\approx}{\boldsymbol{b}}_1)_{ijpq} = n_i n_j n_p n_q, \tag{20.94a}$$

$$(\underset{\approx}{\boldsymbol{b}}_2)_{ijpq} = (\delta_{ij} - n_i n_j)(\delta_{pq} - n_p n_q) = \delta_{ij}\delta_{pq} - (n_i n_j \delta_{pq} + \delta_{ij} n_p n_q) + n_i n_j n_p n_q, \tag{20.94b}$$

$$(\underset{\approx}{\boldsymbol{b}}_3)_{ijpq} = (n_i n_j)(\delta_{pq} - n_p n_q) + (\delta_{ij} - n_i n_j)(n_p n_q) = (n_i n_j \delta_{pq} + \delta_{ij} n_p n_q) - 2 n_i n_j n_p n_q, \tag{20.94c}$$

$$(\underset{\approx}{\boldsymbol{b}}_4)_{ijpq} = \tfrac{1}{2}(\delta_{ip}\delta_{jq} + \delta_{iq}\delta_{jp}) - \tfrac{1}{2}(\delta_{iq} n_j n_p + n_i \delta_{jp} n_q + n_i \delta_{jq} n_p + \delta_{ip} n_j n_q) + n_i n_j n_p n_q, \tag{20.94d}$$

$$(\underset{\approx}{\boldsymbol{b}}_5)_{ijpq} = \tfrac{1}{2}(\delta_{ip} n_j n_q + \delta_{iq} n_j n_q + n_i \delta_{jq} n_p + n_i \delta_{jp} n_p) - 2 n_i n_p n_j n_q. \tag{20.94e}$$

Any transversely isotropic fourth-order minor-symmetric tensor may be written as a linear combination of the five tensors in (20.87). Specifically, if $\underset{\approx}{\boldsymbol{T}}$ is a transversely isotropic double-symmetric fourth-order tensor, then it can be written

$$\underset{\approx}{\boldsymbol{T}} = h_1 \underset{\approx}{\boldsymbol{b}}_1 + h_2 \underset{\approx}{\boldsymbol{b}}_2 + h_3 \underset{\approx}{\boldsymbol{b}}_3 + h_4 \underset{\approx}{\boldsymbol{b}}_4 + h_5 \underset{\approx}{\boldsymbol{b}}_5. \tag{20.95}$$

The h-coefficients are the same as those appearing in (20.85). Accordingly, the Mandel component matrix for $\underset{\approx}{\boldsymbol{b}}_K$ is found by evaluating (20.85) with $h_K = 1$ and the other four independent h-values set to 0. Since the $\boldsymbol{b}$-basis is not orthonormalized (i.e., their mutual inner products are not the Kronecker delta), the h-coefficients are actually contravariant components. An alternative orthonormal basis will be discussed later.

The key advantage of the direct symbolic formulation defined by (20.95) and (20.87) is that it may be applied in any basis, whereas (20.85) applies only in the material basis.

As can be seen by using (20.94a)–(20.94c), we note for future reference that

$$\tfrac{1}{3}(\underset{\approx}{\boldsymbol{b}}_1 + \underset{\approx}{\boldsymbol{b}}_2 + \underset{\approx}{\boldsymbol{b}}_3) = \underset{\approx}{\boldsymbol{P}}^{\text{iso}}, \tag{20.96}$$

where $\underset{\approx}{\boldsymbol{P}}^{\text{iso}}$ is the isotropy projector (which returns the isotropic part of any tensor) that was defined earlier in (20.42a)[7]. Also note that

$$\underset{\approx}{\boldsymbol{b}}_1 + \underset{\approx}{\boldsymbol{b}}_4 + \underset{\approx}{\boldsymbol{b}}_5 = \underset{\approx}{\boldsymbol{P}}^{\text{sym}}, \tag{20.97}$$

where $\underset{\approx}{\boldsymbol{P}}^{\text{sym}}$ is the symmetry projector (which is the identity operator for symmetric tensors). Finally,

$$\tfrac{2}{3}\underset{\approx}{\boldsymbol{b}}_1 - \tfrac{1}{3}\underset{\approx}{\boldsymbol{b}}_2 - \tfrac{1}{3}\underset{\approx}{\boldsymbol{b}}_3 + \underset{\approx}{\boldsymbol{b}}_4 + \underset{\approx}{\boldsymbol{b}}_5 = \underset{\approx}{\boldsymbol{P}}^{\text{symdev}}, \tag{20.98}$$

where $\underset{\approx}{\boldsymbol{P}}^{\text{symdev}}$ is the symmetry-deviator projector (which returns the symmetric-deviatoric part of any tensor), defined earlier in (20.46b). In these particular linear

[7] $P^{\text{iso}}_{ijkl} = \frac{1}{3}\delta_{ij}\delta_{kl}$, $P^{\text{sym}}_{ijkl} = \frac{1}{2}(\delta_{ik}\delta_{jl} + \delta_{il}\delta_{jk})$, $P^{\text{symdev}}_{ijkl} = P^{\text{sym}}_{ijkl} - P^{\text{iso}}_{ijkl}$.

combinations of $\boldsymbol{b}$-tensors, the terms that depend on the symmetry axis $\underset{\sim}{\boldsymbol{n}}$ cancel to produce isotropic tensors that are independent of $\underset{\sim}{\boldsymbol{n}}$. These observations for fourth-order tensors are analogous to the observation for second-order tensors that $\underset{\approx}{\boldsymbol{P}} + \underset{\approx}{\boldsymbol{Q}} = \underset{\approx}{\boldsymbol{I}}$.

The transverse basis tensors in (20.87) become physically meaningful when you consider how they transform an arbitrary second-order tensor. Since the basis tensors in (20.87) are minor-symmetrized, we may consider any tensor $\underset{\approx}{\sigma}$ on which they operate to be symmetric $\sigma_{ij} = \sigma_{ji}$. We will discuss the meaning of the basis in the context of elasticity where $\underset{\approx}{\sigma}$ is the stress tensor and therefore the linear combination for $\underset{\approx}{\boldsymbol{T}}$ in (20.95) is the fourth-order elastic compliance tensor, $\underset{\approx}{\boldsymbol{S}}$,[8] so that $\underset{\approx}{\varepsilon} = \underset{\approx}{\boldsymbol{S}} : \underset{\approx}{\sigma}$ gives the strain. Using the expansion in (20.95) for the compliance,

$$\underset{\approx}{\varepsilon} = h_1 \underset{\approx}{\sigma}_1 + h_2 \underset{\approx}{\sigma}_2 + h_3 \underset{\approx}{\sigma}_3 + h_4 \underset{\approx}{\sigma}_4 + h_5 \underset{\approx}{\sigma}_5, \tag{20.99a}$$

$$\text{where } \underset{\approx}{\sigma}_K \equiv \underset{\approx}{\boldsymbol{b}}_K : \underset{\approx}{\sigma}. \tag{20.99b}$$

In terms of the material basis where the symmetry axis points in the 3-direction, double-dotting the five basis tensors in (20.87) into the arbitrary second-order tensor $\underset{\approx}{\sigma}$ (with a fully populated component matrix) results in the following *material* component matrices for the five $\underset{\approx}{\sigma}_K$ tensors:

$$[\underset{\approx}{\sigma}] = [\underset{\approx}{\boldsymbol{b}}_1 : \underset{\approx}{\sigma}] = \begin{bmatrix} 0 & 0 & 0 \\ 0 & 0 & 0 \\ 0 & 0 & \sigma_{33} \end{bmatrix} = \text{the axial part}, \tag{20.100a}$$

$$[\underset{\approx}{\sigma}_2] = [\underset{\approx}{\boldsymbol{b}}_2 : \underset{\approx}{\sigma}] = \begin{bmatrix} \sigma_{11} + \sigma_{22} & 0 & 0 \\ 0 & \sigma_{11} + \sigma_{22} & 0 \\ 0 & 0 & 0 \end{bmatrix} \sim \text{in-plane isotropic part}, \tag{20.100b}$$

$$[\underset{\approx}{\sigma}_3] = [\underset{\approx}{\boldsymbol{b}}_3 : \underset{\approx}{\sigma}] = \begin{bmatrix} \sigma_{33} & 0 & 0 \\ 0 & \sigma_{33} & 0 \\ 0 & 0 & \sigma_{11} + \sigma_{22} \end{bmatrix}, \text{ which gives a Poisson effect}, \tag{20.100c}$$

$$[\underset{\approx}{\sigma}_4] = [\underset{\approx}{\boldsymbol{b}}_4 : \underset{\approx}{\sigma}] = \begin{bmatrix} \sigma_{11} & \sigma_{12} & 0 \\ \sigma_{21} & \sigma_{22} & 0 \\ 0 & 0 & 0 \end{bmatrix} = \text{in-plane part}, \tag{20.100d}$$

$$[\underset{\approx}{\sigma}_5] = [\underset{\approx}{\boldsymbol{b}}_5 : \underset{\approx}{\sigma}] = \begin{bmatrix} 0 & 0 & \sigma_{13} \\ 0 & 0 & \sigma_{23} \\ \sigma_{31} & \sigma_{32} & 0 \end{bmatrix} = \text{the part that shears the symmetry axis}. \tag{20.100e}$$

[8] Although laughable, it is standard to denote the elastic stiffness tensor by $\underset{\approx}{\boldsymbol{C}}$ and its inverse, the compliance tensor, by $\underset{\approx}{\boldsymbol{S}}$. The fourth-order tensor $\underset{\approx}{\boldsymbol{T}}$ represents either of these.

For linear elasticity, the strain response to an arbitrary stress will be a linear combination of the above tensors. Specifically, the matrix form of (20.99*a*) is

$$\begin{bmatrix}\varepsilon_{11} & \varepsilon_{12} & \varepsilon_{13}\\ \varepsilon_{21} & \varepsilon_{22} & \varepsilon_{23}\\ \varepsilon_{31} & \varepsilon_{32} & \varepsilon_{33}\end{bmatrix} = h_1\begin{bmatrix}0 & 0 & 0\\ 0 & 0 & 0\\ 0 & 0 & \sigma_{33}\end{bmatrix} + h_2\begin{bmatrix}\sigma_{11}+\sigma_{22} & 0 & 0\\ 0 & \sigma_{11}+\sigma_{22} & 0\\ 0 & 0 & 0\end{bmatrix}$$
$$+ h_3\begin{bmatrix}\sigma_{33} & 0 & 0\\ 0 & \sigma_{33} & 0\\ 0 & 0 & \sigma_{11}+\sigma_{22}\end{bmatrix} + h_4\begin{bmatrix}\sigma_{11} & \sigma_{12} & 0\\ \sigma_{21} & \sigma_{22} & 0\\ 0 & 0 & 0\end{bmatrix} + h_5\begin{bmatrix}0 & 0 & \sigma_{13}\\ 0 & 0 & \sigma_{23}\\ \sigma_{31} & \sigma_{32} & 0\end{bmatrix}. \tag{20.101}$$

This result suggests how to design laboratory tests to measure the five h-parameters for transversely isotropic elasticity. One could easily go to the laboratory and subject a transversely isotropic material to a uniaxial stress along the symmetry axis (i.e., $\sigma_{33} \neq 0$, while all other stresses are zero). In this case, the strain response in (20.101) will be

$$\begin{bmatrix}\varepsilon_{11} & \varepsilon_{12} & \varepsilon_{13}\\ \varepsilon_{21} & \varepsilon_{22} & \varepsilon_{23}\\ \varepsilon_{31} & \varepsilon_{32} & \varepsilon_{33}\end{bmatrix} = h_1\begin{bmatrix}0 & 0 & 0\\ 0 & 0 & 0\\ 0 & 0 & \sigma_{33}\end{bmatrix} + h_3\begin{bmatrix}\sigma_{33} & 0 & 0\\ 0 & \sigma_{33} & 0\\ 0 & 0 & 0\end{bmatrix}. \tag{20.102}$$

The above equation implies that, for uniaxial stress loading,

$$\varepsilon_{33} = h_1\sigma_{33} \quad \text{and} \quad \varepsilon_{11} = \varepsilon_{22} = h_3\sigma_{33}. \tag{20.103}$$

No other strain components respond to an axial stress σ_{33}. In other words, an axial stress will produce an axial strain ε_{33} and a Poisson effect (two equal transverse strains ε_{11} and ε_{22}). Transverse isotropy is invalidated if any other independent strain component is observed to be non-negligible in comparison. By measuring these axial and lateral strains, this experiment gives you values for two of the five h-coefficients: $h_1 = \varepsilon_{33}/\sigma_{33}$ and $h_3 = \varepsilon_{11}/\sigma_{33}$.

Under a pure shear in the transverse plane ($\sigma_{11} = -\sigma_{22} = \tau$ with all other stresses zero), the strain response is

$$\begin{bmatrix}\varepsilon_{11} & \varepsilon_{12} & \varepsilon_{13}\\ \varepsilon_{21} & \varepsilon_{22} & \varepsilon_{23}\\ \varepsilon_{31} & \varepsilon_{32} & \varepsilon_{33}\end{bmatrix} = h_4\begin{bmatrix}\tau & 0 & 0\\ 0 & -\tau & 0\\ 0 & 0 & 0\end{bmatrix}, \tag{20.104}$$

which allows measurement of coefficient h_4 and invalidates transverse symmetry if any strain components are inconsistent with this prediction. A simple shear ($\sigma_{12} = \sigma_{21} = \tau$) could be used as an alternative.

Now consider an experiment in which a uniform biaxial stress ($\sigma_{11} = \sigma_{22} = p$) is applied with all other stress components zero. Then the strain will respond in the form

$$\begin{bmatrix}\varepsilon_{11} & \varepsilon_{12} & \varepsilon_{13}\\ \varepsilon_{21} & \varepsilon_{22} & \varepsilon_{23}\\ \varepsilon_{31} & \varepsilon_{32} & \varepsilon_{33}\end{bmatrix} = h_2\begin{bmatrix}2p & 0 & 0\\ 0 & 2p & 0\\ 0 & 0 & 0\end{bmatrix} + h_3\begin{bmatrix}0 & 0 & 0\\ 0 & 0 & 0\\ 0 & 0 & 2p\end{bmatrix} + h_4\begin{bmatrix}p & 0 & 0\\ 0 & p & 0\\ 0 & 0 & 0\end{bmatrix}. \tag{20.105}$$

Measuring ε_{11} and ε_{33} allows re-confirmation of the previously measured coefficient h_3 and determination of the combination $h_2 + h_4$, from which the value of h_2 follows by using the value of h_4 found previously in the transverse shear experiment. Finally, a test in which σ_{13} is non-zero (all other stresses zero) allows determining h_5.

Once the h-components are found, the Mandel compliance is then given by (20.85):

$$[S] = \begin{bmatrix} h_0 & h_2 & h_3 & 0 & 0 & 0 \\ h_2 & h_0 & h_3 & 0 & 0 & 0 \\ h_3 & h_3 & h_1 & 0 & 0 & 0 \\ 0 & 0 & 0 & h_5 & 0 & 0 \\ 0 & 0 & 0 & 0 & h_5 & 0 \\ 0 & 0 & 0 & 0 & 0 & h_4 \end{bmatrix}, \quad \text{where } h_0 \equiv h_2 + h_4. \tag{20.106}$$

The Mandel stiffness matrix is then $[C] = [S]^{-1}$. As explained in chapter 26, the inverse of a Mandel matrix does not require any adjustments by factors of 2 or 4 (which are awkwardly needed when inverting a Voigt matrix).

Keep in mind that it often is not necessary to actually construct and apply the 6×6 Mandel component matrix. Instead, one typically needs only the action of a transverse tensor on a second-order tensor. This fact was first pointed out in (20.99) where the elastic strain formula, $\underset{\approx}{\varepsilon} = \underset{\approx}{\boldsymbol{S}} : \underset{\approx}{\sigma}$ was written as

$$\underset{\approx}{\varepsilon} = h_1 \underset{\approx}{\sigma}_1 + h_2 \underset{\approx}{\sigma}_2 + h_3 \underset{\approx}{\sigma}_3 + h_4 \underset{\approx}{\sigma}_4 + h_5 \underset{\approx}{\sigma}_5, \quad \text{where } \underset{\approx}{\sigma}_K \equiv \underset{\approx}{\boldsymbol{b}}_K : \underset{\approx}{\sigma}. \tag{20.107}$$

The *material* components of the $\underset{\approx}{\sigma}_K$ tensors were given in (20.100). Once the h-moduli are determined from laboratory testing, equation (20.107) can be applied directly in the laboratory basis rather than transforming to material components. Specifically, applying the symbolic expressions in (20.87) gives the following direct formulas:

$$\underset{\approx}{\sigma}_1 = \underset{\approx}{\boldsymbol{b}}_1 : \underset{\approx}{\sigma} = \underset{\approx}{\boldsymbol{P}}\underset{\approx}{\boldsymbol{P}} : \underset{\approx}{\sigma} = (\sigma_n)(\underset{\sim}{\boldsymbol{n}}\underset{\sim}{\boldsymbol{n}}), \tag{20.108a}$$

$$\underset{\approx}{\sigma}_2 = \underset{\approx}{\boldsymbol{b}}_2 : \underset{\approx}{\sigma} = \underset{\approx}{\boldsymbol{Q}}\underset{\approx}{\boldsymbol{Q}} : \underset{\approx}{\sigma} = (\mathrm{tr}\underset{\approx}{\sigma} - \sigma_n)(\underset{\approx}{\boldsymbol{I}} - \underset{\sim}{\boldsymbol{n}}\underset{\sim}{\boldsymbol{n}}), \tag{20.108b}$$

$$\underset{\approx}{\sigma}_3 = \underset{\approx}{\boldsymbol{b}}_3 : \underset{\approx}{\sigma} = (\underset{\approx}{\boldsymbol{P}}\underset{\approx}{\boldsymbol{Q}} : \underset{\approx}{\boldsymbol{Q}}\underset{\approx}{\boldsymbol{P}}) : \underset{\approx}{\sigma} = (\sigma_n)(\underset{\approx}{\boldsymbol{I}}) + (\mathrm{tr}\underset{\approx}{\sigma})(\underset{\sim}{\boldsymbol{n}}\underset{\sim}{\boldsymbol{n}}), \tag{20.108c}$$

$$\underset{\approx}{\sigma}_4 = \underset{\approx}{\boldsymbol{b}}_4 : \underset{\approx}{\sigma} = (\underset{\approx}{\boldsymbol{Q}}\underset{\approx}{\boldsymbol{Q}})^{\Lambda} : \underset{\approx}{\sigma} = \underset{\approx}{\sigma} - (\underset{\sim}{\boldsymbol{t}}\underset{\sim}{\boldsymbol{n}} + \underset{\sim}{\boldsymbol{n}}\underset{\sim}{\boldsymbol{t}}) + (\sigma_n)(\underset{\sim}{\boldsymbol{n}}\underset{\sim}{\boldsymbol{n}}), \tag{20.108d}$$

$$\underset{\approx}{\sigma}_5 = \underset{\approx}{\boldsymbol{b}}_5 : \underset{\approx}{\sigma} = (\underset{\approx}{\boldsymbol{P}}\underset{\approx}{\boldsymbol{Q}} + \underset{\approx}{\boldsymbol{Q}}\underset{\approx}{\boldsymbol{P}})^{\Lambda} : \underset{\approx}{\sigma} = \underset{\sim}{\boldsymbol{t}}\underset{\sim}{\boldsymbol{n}} + \underset{\sim}{\boldsymbol{n}}\underset{\sim}{\boldsymbol{t}} - (\sigma_n)(\underset{\sim}{\boldsymbol{n}}\underset{\sim}{\boldsymbol{n}}), \tag{20.108e}$$

where

$$\underset{\sim}{\boldsymbol{t}} \equiv \underset{\approx}{\sigma} \bullet \underset{\sim}{\boldsymbol{n}} \qquad \text{and} \qquad \sigma_n \equiv \underset{\sim}{\boldsymbol{t}} \bullet \underset{\sim}{\boldsymbol{n}}. \tag{20.109}$$

20.6.1 Eigensystem of a transversely isotropic fourth-order tensor

The eigenproblem for a fourth-order tensor $\underset{\approx}{\underset{\approx}{T}}$ seeks to characterize the set of non-zero tensors $\underset{\sim}{X}$ (called eigentensors) and scalar multipliers λ (eigenvalues) for which

$$\underset{\approx}{\underset{\approx}{T}} : \underset{\sim}{X} = \lambda \underset{\sim}{X}. \tag{20.110}$$

When this condition is expressed in terms of *Mandel*[9] component arrays, it reduces to a standard matrix eigenproblem

$$[T]\{X\} = \lambda\{X\}. \tag{20.111}$$

To understand the importance of using Mandel components for eigen-analysis, refer to [5–7] (which do not use the label 'Mandel') or see the discussion about this topic in chapter 26.

If the tensor $\underset{\approx}{\underset{\approx}{T}}$ is double-symmetric, then only the upper-left 6×6 sub-matrix of the Mandel matrix is non-zero. Hence, the problem reduces to finding the eigensystem of the matrix in (20.106), which can be done analytically because only the fully populated upper-left 3×3 sub-matrix needs to be analyzed, but you must not forget about the coupling of this sub-matrix to the lower-right diagonal sub-matrix through the formula for h_0, which will be important when identifying multiplicities.

Recall from (20.96) that the transversely isotropic tensor may be expanded as

$$\underset{\approx}{\underset{\approx}{T}} = h_1 \underset{\approx}{\underset{\approx}{b}}_1 + h_2 \underset{\approx}{\underset{\approx}{b}}_2 + h_3 \underset{\approx}{\underset{\approx}{b}}_3 + h_4 \underset{\approx}{\underset{\approx}{b}}_4 + h_5 \underset{\approx}{\underset{\approx}{b}}_5. \tag{20.112}$$

This is not the spectral form, which requires expressing a tensor as a linear combination of coefficients times *projectors*; we will show in this section that the eigensystem in fact consists of only four distinct eigenvalues. Closed-form direct-notation expressions for the associated eigenprojectors will be provided.

The eigensystem for the transversely isotropic tensor can be expressed most cleanly by introducing a change of variables as follows[10]:

$$H_1 = h_1 + (2h_2 + h_4) = T_{3333} + 2(T_{1122} + T_{1212}), \tag{20.113a}$$

$$H_2 = h_1 - (2h_2 + h_4) = T_{3333} - 2(T_{1122} + T_{1212}), \tag{20.113b}$$

$$H_3 = 2\sqrt{2}\, h_3 = 2\sqrt{2}\, T_{1133}, \tag{20.113c}$$

$$H_4 = h_4 = 2T_{1212}, \tag{20.113d}$$

[9] Not Voigt! See chapter 26.

[10] Don't you find it annoying when a book introduces a 'helpful change of variables' without explaining *what has motivated it*? We identified this change of variables by using a symbolic math program to find the eigensystem of the original matrix in (20.106). We noted common groupings of the h-components that appeared in this brute-force solution. For example, the grouping $2h_2 + h_4$ appeared frequently, so it would make sense to introduce a change of variables with this grouping. The $2\sqrt{2}$ was introduced in (20.113*c*) because it appeared in our brute-force solution for eigenvalues. Absorbing this factor into our change of variables definition will make the upcoming eigenvalue solution as simply structured as possible.

$$H_5 = h_5 = 2T_{3131}. \tag{20.113e}$$

The four-index T_{ijkl} components are traditional *material* components (i.e., with respect to a transversely isotropic symmetry axis aligned with the 3-direction). The $\underset{\approx}{\mathbf{b}}_I$-tensors in (20.112) are not orthonormal tensors (i.e., $\underset{\approx}{\mathbf{b}}_I :: \underset{\approx}{\mathbf{b}}_J \neq \delta_{IJ}$), so they should be regarded as a covariant basis. Their associated h_i coefficients are properly interpreted as 'contravariant' components (and, as such, would be denoted with superscript indexing in a more advanced non-Cartesian book; see [8]). The H_k in (20.113) merely define a change of variables; they are (contravariant) components in the following alternative (covariant) basis expansion:

$$\underset{\approx}{\mathbf{T}} = H_1\underset{\approx}{\mathbf{B}}_1 + H_2\underset{\approx}{\mathbf{B}}_2 + H_3\underset{\approx}{\mathbf{B}}_3 + H_4\underset{\approx}{\mathbf{B}}_4 + H_5\underset{\approx}{\mathbf{B}}_5. \tag{20.114}$$

The basis tensors implied by the change of variables in (20.113) are

$\underset{\approx}{\mathbf{B}}_1 =$	$\underset{\approx}{\mathbf{B}}_2 =$	$\underset{\approx}{\mathbf{B}}_3 =$	$\underset{\approx}{\mathbf{B}}_4 =$	$\underset{\approx}{\mathbf{B}}_5 =$
$\dfrac{\underset{\sim}{\mathbf{P}}\underset{\sim}{\mathbf{P}} + \frac{1}{2}\underset{\sim}{\mathbf{Q}}\underset{\sim}{\mathbf{Q}}}{2}$	$\dfrac{\underset{\sim}{\mathbf{P}}\underset{\sim}{\mathbf{P}} - \frac{1}{2}\underset{\sim}{\mathbf{Q}}\underset{\sim}{\mathbf{Q}}}{2}$	$\dfrac{(\underset{\sim}{\mathbf{P}}\underset{\sim}{\mathbf{Q}} + \underset{\sim}{\mathbf{Q}}\underset{\sim}{\mathbf{P}})}{2\sqrt{2}}$	$(\underset{\sim}{\mathbf{Q}}\underset{\sim}{\mathbf{Q}})^{\wedge} - \frac{1}{2}\underset{\sim}{\mathbf{Q}}\underset{\sim}{\mathbf{Q}}$	$(\underset{\sim}{\mathbf{P}}\underset{\sim}{\mathbf{Q}} + \underset{\sim}{\mathbf{Q}}\underset{\sim}{\mathbf{P}})^{\wedge}$
$= \dfrac{\underset{\approx}{\mathbf{b}}_1}{2} + \dfrac{\underset{\approx}{\mathbf{b}}_2}{4}$	$= \dfrac{\underset{\approx}{\mathbf{b}}_1}{2} - \dfrac{\underset{\approx}{\mathbf{b}}_2}{4}$	$= \dfrac{\underset{\approx}{\mathbf{b}}_3}{2\sqrt{2}}$	$= \underset{\approx}{\mathbf{b}}_4 - \dfrac{\underset{\approx}{\mathbf{b}}_2}{2}$	$= \underset{\approx}{\mathbf{b}}_5$

. (20.115)

Rather than directly using the moduli H_2 and H_3 in our upcoming expression of the eigensystem, we will introduce one further change of variables to express the relative sizes of these moduli via polar coordinates H and η as

$$H_2 = H \cos\eta \qquad \text{and} \qquad H_3 = H \sin\eta. \tag{20.116}$$

Of course, the values of H and η are found from

$$\boxed{H = +\sqrt{H_2^2 + H_3^2}} \qquad \text{and} \qquad \boxed{\eta = \arctan(H_2, H_3)}. \tag{20.117}$$

We are here emphasizing that a two-argument arctangent must be used. Different computer languages order the arguments differently, so you will need to determine which ordering is appropriate for your software[11].

Above we provided formulas for computing five alternative transverse material parameters (H, η, H_1, H_4, H_5) from the conventional parameters (h_1, h_2, h_3, h_4, h_5). In terms of the new parameters, the eigenvalues of the transversely isotropic fourth-order tensor are

[11] FORTRAN uses ATAN2 (H_3, H_2). Mathematica and C++ use arctan [H_2, H_3].

$$\boxed{\lambda_1 = \frac{H_1 - H}{2} \qquad \text{multiplicity 1 (see discussion below)}}, \tag{20.118a}$$

$$\boxed{\lambda_2 = \frac{H_1 + H}{2} \qquad \text{multiplicity 1 (see discussion below)}}, \tag{20.118b}$$

$$\boxed{\lambda_3 = H_4 \quad \text{multiplicity 2 (see discussion below)}}, \tag{20.118c}$$

$$\boxed{\lambda_4 = H_5 \quad \text{multiplicity 2 (see discussion below)}}, \tag{20.118d}$$

along with a degenerate eigenvalue arising from minor-symmetry,

$$\lambda_5 = 0 \qquad \text{multiplicity 3 (see discussion below).} \tag{20.118e}$$

The corresponding eigenprojectors are

$$\boxed{\begin{aligned} \underset{\approx}{\boldsymbol{P}}_1 &= \underset{\approx}{\boldsymbol{B}}_1 - [\cos\eta \underset{\approx}{\boldsymbol{B}}_2 + \sin\eta \underset{\approx}{\boldsymbol{B}}_3] \\ &= \frac{\underset{\approx}{\boldsymbol{b}}_1}{2} + \frac{\underset{\approx}{\boldsymbol{b}}_2}{4} - \left[\cos\eta\left(\frac{\underset{\approx}{\boldsymbol{b}}_1}{2} - \frac{\underset{\approx}{\boldsymbol{b}}_2}{4}\right) + \sin\eta\frac{\underset{\approx}{\boldsymbol{b}}_3}{2\sqrt{2}}\right] \end{aligned}}, \tag{20.119a}$$

$$\boxed{\underset{\approx}{\boldsymbol{P}}_2 = \underset{\approx}{\boldsymbol{B}}_1 - [\cos\eta \underset{\approx}{\boldsymbol{B}}_2 + \sin\eta \underset{\approx}{\boldsymbol{B}}_3] = \frac{\underset{\approx}{\boldsymbol{b}}_1}{2} + \frac{\underset{\approx}{\boldsymbol{b}}_2}{4} + \left[\cos\eta\left(\frac{\underset{\approx}{\boldsymbol{b}}_1}{2} - \frac{\underset{\approx}{\boldsymbol{b}}_2}{4}\right) + \sin\eta\frac{\underset{\approx}{\boldsymbol{b}}_3}{2\sqrt{2}}\right]}, \tag{20.119b}$$

$$\boxed{\underset{\approx}{\boldsymbol{P}}_3 = \underset{\approx}{\boldsymbol{B}}_4 = \underset{\approx}{\boldsymbol{b}}_4 - \frac{\underset{\approx}{\boldsymbol{b}}_2}{2}}, \tag{20.119c}$$

$$\boxed{\underset{\approx}{\boldsymbol{P}}_4 = \underset{\approx}{\boldsymbol{B}}_5 = \underset{\approx}{\boldsymbol{b}}_5}. \tag{20.119d}$$

When the eigenvalues are all distinct, the tensors in (20.119) are the corresponding unique eigenprojectors. The eigenprojector associated with the zero eigenvalue is

$$\underset{\approx}{\boldsymbol{P}}_5 = \underset{\approx}{\boldsymbol{P}}^{\text{skw}}, \quad \text{where } \underset{\approx}{\boldsymbol{P}}^{\text{skw}} \text{ is defined in (20.43}c\text{).} \tag{20.120}$$

Equations (20.118) and (20.119) allow the transversely isotropic tensor to be expressed as the following spectral expansion:

$$\underset{\approx}{\boldsymbol{T}} = \lambda_1 \underset{\approx}{\boldsymbol{P}}_1 + \lambda_2 \underset{\approx}{\boldsymbol{P}}_2 + \lambda_3 \underset{\approx}{\boldsymbol{P}}_3 + \lambda_4 \underset{\approx}{\boldsymbol{P}}_4. \tag{20.121}$$

Having the spectral expansion allows rapid inversion. Specifically,

$$\underset{\approx}{\boldsymbol{T}}^{-1} = \frac{1}{\lambda_1}\underset{\approx}{\boldsymbol{P}}_1 + \frac{1}{\lambda_2}\underset{\approx}{\boldsymbol{P}}_2 + \frac{1}{\lambda_3}\underset{\approx}{\boldsymbol{P}}_3 + \frac{1}{\lambda_4}\underset{\approx}{\boldsymbol{P}}_4. \tag{20.122}$$

The eigenprojectors and the value of η are the same for both the tensor and its inverse. The eigenvalues are reciprocals of each other.

In elasticity, the operation $\underset{\sim}{\varepsilon} = \underset{\approx}{\boldsymbol{S}} : \underset{\sim}{\sigma}$ can be written as the sum of only four terms instead of the five terms that were required in (20.107). Specifically,

$$\underset{\sim}{\varepsilon} = \lambda_1(\underset{\approx}{\boldsymbol{P}}_1 : \underset{\sim}{\sigma}) + \lambda_2(\underset{\approx}{\boldsymbol{P}}_2 : \underset{\sim}{\sigma}) + \lambda_3(\underset{\approx}{\boldsymbol{P}}_3 : \underset{\sim}{\sigma}) + \lambda_4(\underset{\approx}{\boldsymbol{P}}_4 : \underset{\sim}{\sigma}). \tag{20.123}$$

Let $\alpha = \sigma_{33}$ denote the axial part of the stress, and let $\beta = \frac{1}{2}(\sigma_{11} + \sigma_{22})$ denote the biaxial pressure (i.e., pressure in the transverse plane). Further let $s = \sin\eta$ and $c = \cos\eta$. Then, with respect to the material basis,

$$[\underset{\approx}{\boldsymbol{P}}_1 : \underset{\sim}{\sigma}] = \frac{1}{2}\begin{bmatrix} \beta & 0 & 0 \\ 0 & \beta & 0 \\ 0 & 0 & \alpha \end{bmatrix} - \frac{1}{2\sqrt{2}}\begin{bmatrix} \alpha s - \sqrt{2}\beta c & 0 & 0 \\ 0 & \alpha s - \sqrt{2}\beta c & 0 \\ 0 & 0 & 2\beta s + \sqrt{2}\alpha c \end{bmatrix}, \tag{20.124a}$$

$$[\underset{\approx}{\boldsymbol{P}}_2 : \underset{\sim}{\sigma}] = \frac{1}{2}\begin{bmatrix} \beta & 0 & 0 \\ 0 & \beta & 0 \\ 0 & 0 & \alpha \end{bmatrix} - \frac{1}{2\sqrt{2}}\begin{bmatrix} \alpha s - \sqrt{2}\beta c & 0 & 0 \\ 0 & \alpha s - \sqrt{2}\beta c & 0 \\ 0 & 0 & 2\beta s + \sqrt{2}\alpha c \end{bmatrix}, \tag{20.124b}$$

$$[\underset{\approx}{\boldsymbol{P}}_3 : \underset{\sim}{\sigma}] = \begin{bmatrix} T & \tau & 0 \\ \tau & -T & 0 \\ 0 & 0 & 0 \end{bmatrix}, \quad \text{where } T = \frac{\sigma_{11} - \sigma_{22}}{2} \quad \text{and} \quad \tau = \sigma_{12}, \tag{20.124c}$$

$$[\underset{\approx}{\boldsymbol{P}}_4 : \underset{\sim}{\sigma}] = \begin{bmatrix} 0 & 0 & \sigma_{13} \\ 0 & 0 & \sigma_{23} \\ \sigma_{31} & \sigma_{32} & 0 \end{bmatrix}. \tag{20.124d}$$

The third eigenprojector simply extracts the deviatoric part of the 2×2 in-plane stress matrix. The fourth eigenprojector extracts the part of the stress that shears the transverse symmetry axis. The first two eigenprojectors are more complicated and their physical meanings are less clear because they depend on more than just the symmetry axis.

20.6.2 Discussion: why the zero eigenvalue?

The degenerate fifth eigenvalue (20.118*e*) is zero. Its associated eigenprojector (20.120*e*) is the skew-symmetry projector that was defined earlier in (20.42*c*). *Any skew-symmetric tensor* is an eigentensor of the transversely isotropic double-symmetric fourth-order tensor. The associated eigenvalue is zero. This should come as no surprise because our transversely isotropic tensor has been here defined to be *minor-symmetric*, which means it will give a result of zero whenever it operates on a skew-symmetric tensor. Equation (20.118) states the multiplicity of this zero eigenvalue is 3. That is because the space of skew tensors is three-dimensional (skew tensors have three independent components). The fact that zero is an eigenvalue was not readily apparent from (20.106) because we typeset that matrix in 6×6 form. If we were to use all *nine* Mandel basis tensors (the last three of which are skew-symmetric orthonormalizations of the ordinary $\underset{\sim}{e}_i\underset{\sim}{e}_j$ basis tensors so that these are $(\underset{\sim}{e}_i\underset{\sim}{e}_j - \underset{\sim}{e}_j\underset{\sim}{e}_j)/\sqrt{2}$ for $i \neq j$), then (20.106) would expand with zeros filling out the full 9×9 form. The three additional zeros on the diagonal, corresponding to the three skew-symmetric basis tensors, are the source of the zero eigenvalue of multiplicity 3. The multiplicities cited in (20.118) sum to 9, which is the dimension of the entirety of tensor space. The multiplicities for the non-trivial eigenvalues sum to six, which is the dimension of *symmetric* tensor space. In materials modeling, where most tensors are symmetric, the skew parts of arrays typically are not even shown or considered so that all arrays are written in 6D form (which is a bit risky in practical work, as it is surprisingly easy to inadvertently fail to impose the minor symmetry that is needed to make the last three rows and columns of the full matrix zero).

20.6.3 Discussion: multiplicity

The multiplicities cited in (20.118) are the *lowest possible* multiplicities. The multiplicity of an eigenvalue might actually be higher depending on the values of the original moduli. If, for example, H happens to equal zero, then $\lambda_1 = \lambda_2$, and their multiplicities will therefore combine.

Looking at the matrix in (20.106), you can tell by inspection that h_4 will be an eigenvalue, so it has a multiplicity of *at least* 1. The corresponding eigenvector is

$$\begin{bmatrix} 0 \\ 0 \\ 0 \\ 0 \\ 0 \\ 1 \end{bmatrix}. \tag{20.125}$$

This 6×1 eigenvector *is a Mandel array*. Thus, it actually corresponds to the following *eigentensor:*

$$\begin{bmatrix} 0 & \frac{1}{\sqrt{2}} & 0 \\ \frac{1}{\sqrt{2}} & 0 & 0 \\ 0 & 0 & 0 \end{bmatrix}. \tag{20.126}$$

The $\sqrt{2}$ has appeared through application of the conversion formula in which ordinary components are obtained from Mandel components (namely, the sixth component of a Mandel array is related to the ordinary 12 component of the tensor by $A_6 = \sqrt{2}A_{12}$). This eigentensor is a state of simple shear in the transverse plane. The eigenvalue h_4 may be therefore regarded loosely as a *shear modulus* for this plane.

Above, we explained that h_4 is an obvious eigenvalue of the lower-right 3×3 sub-matrix in (20.106) corresponding to the eigentensor in (20.126). However, our result in (20.118*c*) states that H_4 (which equals h_4) will *always* have a multiplicity of at least *two*. This means that h_4 must also be a less obvious eigenvalue of the non-trivial *upper-left* 3×3 sub-matrix in (20.106). In particular, this second eigenvector associated with h_4 is

$$\begin{bmatrix} \frac{1}{\sqrt{2}} \\ \frac{1}{\sqrt{2}} \\ 0 \\ 0 \\ 0 \\ 0 \end{bmatrix} \text{ which is a pure shear eigentensor } \begin{bmatrix} \frac{1}{\sqrt{2}} & 0 & 0 \\ 0 & \frac{-1}{\sqrt{2}} & 0 \\ 0 & 0 & 0 \end{bmatrix}. \tag{20.127}$$

From the definition of h_0 given in (20.106), the eigenvalue $h_0 - h_2$ identically equals h_4. Thus, we have confirmed that h_4 does indeed have multiplicity of at least 2. Equation (20.126) is a state of simple shear whereas (20.127) is pure shear, each applied in the transverse plane. For a material that is transversely isotropic, these deformations are functionally equivalent, so it makes sense that they both correspond to the same eigenvalue, h_4.

20.6.4 Discussion: why eigenprojectors instead of eigenvectors?

In (20.119*c*), we stated that the eigen*projector* associated with eigenvalue $\lambda_3 = H_4 = h_4$ is

$$\underset{\approx}{\mathbf{P}}_3 = \underset{\approx}{\mathbf{b}}_4 - \frac{\underset{\approx}{\mathbf{b}}_2}{2}. \tag{20.128}$$

Therefore, the Mandel component matrix for $\underset{\approx}{\mathbf{P}}_3$ may be found from (20.106) by setting $h_4 = 1$, $h_2 = -1/2$, and $h_1 = h_3 = h_5 = 0$. Namely,

$$[\underset{\approx}{\mathbf{P}}_3] = \begin{bmatrix} \frac{1}{2} & \frac{1}{2} & 0 & 0 & 0 & 0 \\ -\frac{1}{2} & \frac{1}{2} & 0 & 0 & 0 & 0 \\ 0 & 0 & 0 & 0 & 0 & 0 \\ 0 & 0 & 0 & 0 & 0 & 0 \\ 0 & 0 & 0 & 0 & 0 & 0 \\ 0 & 0 & 0 & 0 & 0 & 1 \end{bmatrix}. \tag{20.129}$$

This is same matrix we would obtain if we had constructed the eigenprojector by summing the dyads (outer products) of the two normalized Mandel eigenvectors in (20.125) and (20.127). The advantage of (20.128) over the component expression in (20.129) is that (20.128) may be cast directly in *any* coordinate system whereas (20.129) holds only when the transverse symmetry axis points in the 3-direction. You can, for example, directly apply rates to (20.128) to obtain time variation associated with rotation of the transversely isotropic axis, but you cannot similarly apply rates to (20.128) because you would miss the part of the rate from rotation of the underlying material basis vectors[12].

The Mandel matrices for any of the other projectors listed in (20.119) can be found similarly. For example, the Mandel matrix for (20.119*a*) can be constructed by applying (20.85) with the h_k coefficients set equal to the coefficients of $\underset{\approx}{\boldsymbol{b}}_k$ in (20.119*a*). Specifically, you would set $h_1 = \dfrac{1 - \cos\eta}{2} = \sin^2\left(\dfrac{\eta}{2}\right)$, $h_2 = \dfrac{1 + \cos\eta}{4} = \frac{1}{2}\cos^2\left(\dfrac{\eta}{2}\right)$, $h_3 = -\left(\dfrac{\sin\eta}{2\sqrt{2}}\right)$, $h_4 = h_5 = 0$ to obtain

$$\left[\underset{\approx}{\boldsymbol{P}}_1\right] = \frac{1}{2}\begin{bmatrix} \hat{c}^2 & \hat{c}^2 & -\sqrt{2}\,\hat{c}\hat{s} & 0 & 0 & 0 \\ \hat{c}^2 & \hat{c}^2 & -\sqrt{2}\,\hat{c}\hat{s} & 0 & 0 & 0 \\ -\sqrt{2}\,\hat{c}\hat{s} & -\sqrt{2}\,\hat{c}\hat{s} & 2\hat{s}^2 & 0 & 0 & 0 \\ 0 & 0 & 0 & 0 & 0 & 0 \\ 0 & 0 & 0 & 0 & 0 & 0 \\ 0 & 0 & 0 & 0 & 0 & 0 \end{bmatrix}, \quad \text{where } \hat{c} \equiv \cos\frac{\eta}{2} \text{ and } \hat{s} \equiv \sin\frac{\eta}{2}. \tag{20.130}$$

If you prefer eigenvectors over eigenprojectors, you can always obtain eigenvectors by Gram–Schmidt orthogonalization (GSO) of the columns of an eigenprojector. In numerical work, potential finite-precision errors can be averted by ensuring that the vectors in the GSO process should be the columns of the projector *ordered from greatest to smallest magnitude*. The number of non-zero vectors resulting from the GSO process will equal the multiplicity cited in (20.118). This multiplicity also equals the trace of the eigenprojector. Therefore, we know that the matrix in (20.130) has only one linearly independent column because its trace equals 1. Because the multiplicity is 1, the projector has at least one non-zero column, and every other column is a multiple of that non-zero column. The multiple might be zero. Consequently, you cannot simply say that the eigenvector will always be a multiple of the *first* column. The first column might be zero, in which case the eigenvector would be a multiple of a *different* non-zero column. Consider, for example, the following limiting cases:

[12] This situation is a like having a vector $\underset{\sim}{\boldsymbol{v}}$ expressed in its decomposed form $\underset{\sim}{\boldsymbol{v}} = s\underset{\sim}{\boldsymbol{\eta}}$, where s is the vector's magnitude and $\underset{\sim}{\boldsymbol{\eta}}$ is its direction. The vector's rate is easily found to be $\dot{\underset{\sim}{\boldsymbol{v}}} = \dot{s}\underset{\sim}{\boldsymbol{\eta}} + s\dot{\underset{\sim}{\boldsymbol{\eta}}}$. With respect to a basis having the 3-direction aligned with the vector itself, the component array for the rate is $\{\dot{\underset{\sim}{\boldsymbol{v}}}\} = \{s\dot{n}_1, s\dot{n}_2, \dot{s}\}$, which is *not* the rate of the vector's aligned component array $\{0, 0, s\}$.

$$\text{if } \eta = 0 \text{ then } [\underset{\approx}{P}_1] = \begin{bmatrix} \frac{1}{2} & \frac{1}{2} & 0 & 0 & 0 & 0 \\ \frac{1}{2} & \frac{1}{2} & 0 & 0 & 0 & 0 \\ 0 & 0 & 0 & 0 & 0 & 0 \\ 0 & 0 & 0 & 0 & 0 & 0 \\ 0 & 0 & 0 & 0 & 0 & 0 \\ 0 & 0 & 0 & 0 & 0 & 0 \end{bmatrix}, \text{ corresponding to eigen vector } \begin{bmatrix} 1/\sqrt{2} \\ 1/\sqrt{2} \\ 0 \\ 0 \\ 0 \\ 0 \end{bmatrix}, \quad (20.131)$$

$$\text{if } \eta = \pi \text{ then } [\underset{\approx}{P}_1] = \begin{bmatrix} 0 & 0 & 0 & 0 & 0 & 0 \\ 0 & 0 & 0 & 0 & 0 & 0 \\ 0 & 0 & 1 & 0 & 0 & 0 \\ 0 & 0 & 0 & 0 & 0 & 0 \\ 0 & 0 & 0 & 0 & 0 & 0 \\ 0 & 0 & 0 & 0 & 0 & 0 \end{bmatrix}, \text{ corresponding to eigen vector } \begin{bmatrix} 0 \\ 0 \\ 1 \\ 0 \\ 0 \\ 0 \end{bmatrix}. \quad (20.132)$$

In the first case, the eigenvector is indeed simply normalization of the first column of the eigenprojector. In the other case, though, the eigenvector is the normalization of the *third* column. For general analytical eigenproblems, eigenprojectors are superior to a collection of eigenvectors. If you claim (mistakenly) that the eigenvector is a multiple of the first column of the matrix in (20.130), then you will be correct *most of the time*, but your solution would break down in the special case that $\eta = \pi$ (and it would suffer from severe round-off errors when η is close to π). The eigenprojector is superior because it is beautifully unbiased—it works every time with uniform numerical accuracy.

20.6.5 Discussion: why four non-zero eigenvalues when transverse isotropy requires five independent *h*-parameters?

Recall that the material component matrix for a transversely isotropic fourth-order double-symmetric tensor has five independent components ($h_1, \ldots, h_5$). However, we found only four distinct eigenvalues. Does this means that the five parameters were not independent after all? Absolutely not. They are still independent.

Consider a related problem in ordinary 3×3 eigen-analyses. If a 3×3 matrix is symmetric, then it has six independent components, but you will never find any more than three distinct eigenvalues. This does not mean that the original six components were not independent after all. Specification of *both* eigenvalues *and* eigenvectors still requires six numbers.

Similarly, a fourth-order transversely isotropic tensor is defined by a symmetry axis $\underset{\sim}{n}$ and *five* numbers ($h_1, \ldots, h_5$). Thus, since the unit vector $\underset{\sim}{n}$ has two independent components, a total of *seven* numbers must be prescribed to fully define a transversely isotropic tensor. We found *four* distinct eigenvalues. The eigenprojectors that go with these eigenvalues are described in terms of the parameter η and the symmetry axis $\underset{\sim}{n}$. Thus, *three* independent numbers are required to specify the eigenprojectors. Taken together, the total number of parameters required to specify both eigenvalues and eigenvectors is therefore still *seven.*

20.6.6 Discussion: positive-definiteness constraints

In elasticity, the fourth-order stiffness tensor must be positive-definite. Therefore, its eigenvalues must be positive. Hence, equation (20.119) implies that *four* conditions must be satisfied for positive-definiteness:

$$H > 0, \qquad H_1 > H, \qquad H_4 > 0, \qquad H_5 > 0. \tag{20.133}$$

Because H is positive, the second condition implies H_1 must be more than merely positive—it must be larger than H. These constraints may be written in terms of conventional four-index material components by using (20.113) and (20.117).

20.6.7 Discussion: do the results reduce properly to isotropy?

This section confirms that the spectral form for transverse isotropy reduces properly when the transverse components are set to values corresponding to isotropy. For isotropy, only two elastic constants (the bulk modulus K and shear modulus G) are required. Specifically, the stiffness is

$$\underset{\approx}{\boldsymbol{C}} = (3K)\underset{\approx}{\boldsymbol{P}}^{\text{iso}} + (2G)\underset{\approx}{\boldsymbol{P}}^{\text{symdev}}, \tag{20.134}$$

or, using (20.96) and (20.98),

$$\underset{\approx}{\boldsymbol{C}} = (K)(\underset{\approx}{\boldsymbol{b}}_1 + \underset{\approx}{\boldsymbol{b}}_2 + \underset{\approx}{\boldsymbol{b}}_3) + (2G)\left(\tfrac{2}{3}\underset{\approx}{\boldsymbol{b}}_1 - \tfrac{1}{3}\underset{\approx}{\boldsymbol{b}}_2 - \tfrac{1}{3}\underset{\approx}{\boldsymbol{b}}_3 + \underset{\approx}{\boldsymbol{b}}_4 + \underset{\approx}{\boldsymbol{b}}_5\right). \tag{20.135}$$

Matching coefficients, the transverse moduli are

$$h_2 = h_3 = K - \tfrac{2}{3}G, \tag{20.136}$$

$$h_1 = K + \tfrac{4}{3}G, \tag{20.137}$$

$$h_4 = h_5 = 2G, \tag{20.138}$$

from which it follows that

$$h_0 \equiv h_2 + h_4 = K + \tfrac{4}{3}G. \tag{20.139}$$

In this case, equations (20.113) give

$$H_1 = 3K + 2G, \tag{20.140a}$$

$$H_2 = \tfrac{2}{3}G - K, \tag{20.140b}$$

$$H_3 = 2\sqrt{2}\left(K - \tfrac{2}{3}G\right), \tag{20.140c}$$

$$H_4 = 2G, \tag{20.140d}$$

$$H_5 = 2G, \tag{20.140e}$$

and (20.117) gives

$$H = |3K - 2G| = \pm(3K - 2G). \tag{20.141}$$

For most materials, $3K > 2G$. However, the $\pm$ allows for the possibility that $3K < 2G$, which corresponds to a material with a negative Poisson's ratio. From (20.116),

$$\cos\eta = \frac{H_2}{H} - \frac{\mp 1}{3} \quad \text{and} \quad \sin\eta = \frac{H_3}{H} = \frac{\pm 2\sqrt{2}}{3}, \tag{20.142}$$

from which it follows that

$$\sin^2\left(\frac{\eta}{2}\right) = \frac{1}{2}\left(1 \pm \frac{1}{3}\right) = \begin{cases} 2/3 \\ 1/3 \end{cases} \quad \text{and} \quad \cos^2\left(\frac{\eta}{2}\right) = \frac{1}{2}\left(1 \mp \frac{1}{3}\right) = \begin{cases} 1/3 \\ 2/3 \end{cases}. \tag{20.143}$$

Equations (20.118) reduce to

$$\lambda_1 = \frac{(3K + 2G) \mp (3K - 2G)}{2} = \begin{cases} 2G \\ 3K \end{cases}, \tag{20.144a}$$

$$\lambda_2 = \frac{(3K + 2G) \pm (3K - 2G)}{2} = \begin{cases} 3K \\ 2G \end{cases}, \tag{20.144b}$$

$$\lambda_3 = H_4, \tag{20.144c}$$

$$\lambda_4 = H_5. \tag{20.144d}$$

Recognizing that the individual 'branches' are functionally equivalent, let us now consider the 'bottom' case:

$$\lambda_1 = 3K, \tag{20.145a}$$

$$\lambda_2 = 2G, \tag{20.145b}$$

$$\lambda_3 = 2G, \tag{20.145c}$$

$$\lambda_4 = 2G. \tag{20.145d}$$

Applying (20.119), the corresponding eigenprojectors are

$$\underset{\approx}{\boldsymbol{P}}_1 = \underset{\approx}{\boldsymbol{B}}_1 - \left[\frac{1}{3}\underset{\approx}{\boldsymbol{B}}_2 - \frac{2\sqrt{2}}{3}\underset{\approx}{\boldsymbol{B}}_3\right] \qquad = \frac{1}{3}(\underset{\approx}{\boldsymbol{b}}_1 + \underset{\approx}{\boldsymbol{b}}_2 + \underset{\approx}{\boldsymbol{b}}_3), \tag{20.146a}$$

$$\underset{\approx}{P}_2 = \underset{\approx}{B}_1 + \left[\tfrac{1}{3}\underset{\approx}{B}_2 - \frac{2\sqrt{2}}{3}\underset{\approx}{B}_3\right] \qquad = \tfrac{2}{3}\underset{\approx}{b}_1 + \tfrac{1}{6}\underset{\approx}{b}_2 - \tfrac{1}{3}\underset{\approx}{b}_3, \tag{20.146b}$$

$$\underset{\approx}{P}_3 = \underset{\approx}{B}_4 \qquad = \underset{\approx}{b}_4 - \frac{\underset{\approx}{b}_2}{2}, \tag{20.146c}$$

$$\underset{\approx}{P}_4 = \underset{\approx}{B}_5 \qquad = \underset{\approx}{b}_5. \tag{20.146d}$$

By comparing (20.146*a*) with (20.96), note that the eigenprojector associated with the eigenvalue $3K$ is $\underset{\approx}{P}^{\text{iso}}$.

In this special case of isotropy, we have $\lambda_2 = \lambda_3 = \lambda_4 = 2G$. Therefore, the unique eigen*space* projector associated with the eigenvalue $2G$ is the sum of the individual (non-unique) eigenprojectors listed above for λ_2, λ_3, and λ_4,

$$\underset{\approx}{P}_2 + \underset{\approx}{P}_3 + \underset{\approx}{P}_4 = \tfrac{2}{3}\underset{\approx}{b}_1 - \tfrac{1}{3}\underset{\approx}{b}_2 - \tfrac{1}{3}\underset{\approx}{b}_3 + \underset{\approx}{b}_4 + \underset{\approx}{b}_5. \tag{20.147}$$

Comparison with (20.98) shows that the eigenspace projector associated with the eigenvalue $2G$ is $\underset{\approx}{P}^{\text{symdev}}$. You can show that the same result is obtained when considering instead the 'top' case for the branch. This proves that the transverse isotropy solution does reduce properly in the limit of isotropy. Namely, any symmetric-deviatoric tensor will be an eigentensor with eigenvalue $2G$. Any isotropic tensor will be an eigentensor with eigenvalue $3K$. To summarize, the spectral expansion in the case of isotropy becomes (as it should)

$$\underset{\approx}{C} = 3K\underset{\approx}{P}^{\text{iso}} + 2G\underset{\approx}{P}^{\text{symdev}}. \tag{20.148}$$

20.6.8 Another special case (jointed materials).

In the previous section, we worked with a stiffness tensor $\underset{\approx}{C}$. Here, we will work with compliance $\underset{\approx}{S}$. Consider a material that has an isotropic compliance tensor,

$$\underset{\approx}{S}^{\text{intact}} = \left(\frac{1}{3K}\right)\underset{\approx}{P}^{\text{iso}} + \left(\frac{1}{2G}\right)\underset{\approx}{P}^{\text{symdev}}, \tag{20.149}$$

or, using (20.96) and (20.98),

$$\underset{\approx}{S}^{\text{intact}} = \left(\frac{1}{9K}\right)\left(\underset{\approx}{b}_1 + \underset{\approx}{b}_2 + \underset{\approx}{b}_3\right) + \left(\frac{1}{2G}\right)\left(\tfrac{2}{3}\underset{\approx}{b}_1 - \tfrac{1}{3}\underset{\approx}{b}_2 - \tfrac{1}{3}\underset{\approx}{b}_3 + \underset{\approx}{b}_4 + \underset{\approx}{b}_5\right). \tag{20.150}$$

Further suppose that this material contains an array of cracks or a collection of joints each having unit normal $\underset{\sim}{n}$. Then a simple model for approximating the compliance of the jointed composite merely enhances the normal and shear compliance. Referring back to (20.101), the jointed compliance should have larger values of ε_{33}, ε_{13}, and ε_{23} than the intact material. Therefore, the compliance in (20.150) should be enhanced by a term of the form

$$\underset{\approx}{S}^{\text{extra}} = \alpha_N \underset{\approx}{b}_1 + \alpha_S \underset{\approx}{b}_5, \tag{20.151}$$

where α_N and α_S are referred to as 'net joint compliances'. With this adjustment, the jointed compliance,

$$\underset{\approx}{S} = \underset{\approx}{S}^{\text{intact}} + \underset{\approx}{S}^{\text{extra}}, \tag{20.152}$$

becomes transversely isotropic with the following h parameters:

$$h_1 = \frac{1}{9K} + \frac{1}{3G} + \alpha_N, \tag{20.153}$$

$$h_2 = h_3 = \frac{1}{9K} - \frac{1}{6G}, \tag{20.153b}$$

$$h_4 = \frac{1}{2G}, \tag{20.153c}$$

$$h_5 = \frac{1}{2G} + \alpha_S, \tag{20.153d}$$

from which it follows that

$$h_0 \equiv h_2 + h_4 = \frac{1}{9K} + \frac{1}{3G}. \tag{20.154}$$

In this case, equations (20.113) reduce to

$$H_1 = h_1 + (2h_2 + h_4) = \frac{1}{3K} + \frac{1}{2G} + \alpha_N, \tag{20.155a}$$

$$H_2 = h_1 - (2h_2 + h_4) = \frac{1}{6G} - \frac{1}{9K} + \alpha_N, \tag{20.155b}$$

$$H_3 = 2\sqrt{2}\,h_3 = 2\sqrt{2}\left(\frac{1}{9K} - \frac{1}{6G}\right), \tag{20.155c}$$

$$H_4 = h_4 = \frac{1}{2G}, \tag{20.155d}$$

$$H_5 = h_5 = \frac{1}{2G} + \alpha_S. \tag{20.155e}$$

Then the eigenvalues are given as usual by (20.117):

$$\lambda_1 = \frac{H_1 - H}{2}, \tag{20.156a}$$

$$\lambda_2 = \frac{H_1 + H}{2}, \tag{20.156b}$$

$$\lambda_3 = H_4, \tag{20.156c}$$

$$\lambda_4 = H_5, \tag{20.156d}$$

where

$$H = +\sqrt{H_2^2 + H_3^2}. \tag{20.157}$$

Recalling (20.133), the positive-definiteness constraints are

$$H_1 > H, H_4 > 0, H_5 > 0. \tag{20.158}$$

Substituting (20.155) into these constraints (assuming K and G are both positive) gives

$$\alpha_S > \frac{-1}{2G} \quad \text{and} \quad \alpha_N > \frac{-1}{K + \frac{4}{3}G}. \tag{20.159}$$

These conditions will always be satisfied in models that take α_S and α_N to be positive. In other words, applying joint compliance enhancements will not destroy positive-definiteness of the net compliance.

20.6.9 Eigenprojector matching in anisotropic plasticity

In plasticity theory, a scalar-valued *yield function* $f(\underset{\sim}{\sigma})$ is generally introduced such that the stress state is elastic if $f(\underset{\sim}{\sigma}) < 0$. The *yield criterion* is the condition that $f(\underset{\sim}{\sigma}) = 0$, and the *yield surface* is the set of stresses that satisfy the yield criterion. As a practical matter, model developers invest considerable time and effort developing isotropic plasticity functions for which the yield function $f(\underset{\sim}{\sigma})$ depends only on invariants of $\underset{\sim}{\sigma}$. A straightforward way to quickly introduce anisotropy is to send an anisotropically transformed stress to the isotropic yield model. Specifically, the generalized yield function is

$$f(\underset{\sim}{\sigma}) = F(\underset{\sim}{\sigma}^*), \tag{20.160}$$

where $F(\underset{\sim}{\sigma}^*)$ is isotropic (i.e., it depends only on invariants of $\underset{\sim}{\sigma}^*$) and

$$\underset{\sim}{\sigma}^* = \underset{\approx}{\boldsymbol{T}}:(\underset{\sim}{\sigma} - \underset{\sim}{\alpha}) \tag{20.161}$$

is a transformed stress. Here, the second-order tensor $\underset{\sim}{\alpha}$ is the back-stress used to model the *Bauschinger effect* in kinematic hardening plasticity theory. If $\underset{\sim}{\alpha}$ has a non-zero isotropic part, it can be used to also model the influence of pressurized fluid in the pore space, in which case $\underset{\sim}{\sigma}^*$ is typically referred to as the *effective stress*. The presence of a nonzero deviatoric part of $\underset{\sim}{\alpha}$ alone makes the function $f(\underset{\sim}{\sigma})$ anisotropic. If the transformation tensor $\underset{\approx}{T}$ is isotropic (and if dev$\underset{\sim}{\alpha} = \underset{\sim}{\mathbf{0}}$), then the yield function in (20.160) will be isotropic. Otherwise, the model will be anisotropic.

As an example, consider the following isotropic function:

$$F(\underset{\sim}{\sigma}^*) = \left\|\underset{\sim}{\sigma}^*\right\|^2 - 1. \tag{20.162}$$

In this case, the equation $F(\underset{\sim}{\sigma}) = 0$ describes a unit sphere in 6D symmetric tensor space. Calling this function using a transformed argument as in (20.160) distorts the sphere into a generalized *ellipsoidal* (Tsai–Wu) yield surface in 6D stress space.

For linear isotropic elasticity, the elastic stiffness has the spectral form

$$\underset{\approx}{C} = 3K\underset{\approx}{\mathbf{P}}^{\text{iso}} + 2G\underset{\approx}{\mathbf{P}}^{\text{symdev}}, \tag{20.163}$$

where K and G are the bulk and shear moduli, respectively. The fourth-order tensors, $\underset{\approx}{\mathbf{P}}^{\text{iso}}$ and $\underset{\approx}{\mathbf{P}}^{\text{symdev}}$ are, respectively, the isotropy and symmetric-deviator projectors[15]. For an isotropic material, the plastic transform tensor must also be isotropic and therefore of the form:

$$\underset{\approx}{T} = \lambda_1\underset{\approx}{\mathbf{P}}^{\text{iso}} + \lambda_2\underset{\approx}{\mathbf{P}}^{\text{symdev}}. \tag{20.164}$$

Choosing $\lambda_1 = 0$ and $\lambda_2 = 1$, gives the von Mises yield function, which presumes that yield occurs at a critical magnitude of the stress deviator independent of pressure, hence generating a yield surface that is a right-circular cylinder in 6D stress space. Choosing both λ_1 and λ_2 to be non-zero would give an approximate model for an isotropic porous material.

For transverse isotropy, the stiffness $\underset{\approx}{C}$ and plastic transformation tensor $\underset{\approx}{T}$ must each be expressible as linear combinations of the basis tensors in (20.87). However, there is no guarantee that these two tensors will necessarily have the same eigenprojectors. Referring to (20.119), they will have the same eigenprojectors only if they both have the same value of η. Since eigenprojectors are characteristic of underlying material symmetries, Schreyer and Zuo [9] point out that it might be useful (especially to deal with a lack of plastic failure data) to *require* eigenprojector matching so that, recalling (20.121) for the case of transverse isotropy,

$$\underset{\approx}{C} = \lambda_1^C\underset{\approx}{\mathbf{P}}_1 + \lambda_2^C\underset{\approx}{\mathbf{P}}_2 + \lambda_3^C\underset{\approx}{\mathbf{P}}_3 + \lambda_4^C\underset{\approx}{\mathbf{P}}_4, \tag{20.165a}$$

$$\underset{\approx}{T} = \lambda_1^T\underset{\approx}{\mathbf{P}}_1 + \lambda_2^T\underset{\approx}{\mathbf{P}}_2 + \lambda_3^T\underset{\approx}{\mathbf{P}}_3 + \lambda_4^T\underset{\approx}{\mathbf{P}}_4. \tag{20.165b}$$

[15] These are defined in (20.43):
$P^{\text{iso}}_{ijkl} = \frac{1}{3}\delta_{ij}\delta_{kl}$, $P^{\text{sym}}_{ijkl} = \frac{1}{2}(\delta_{ik}\delta_{jl} + \delta_{il}\delta_{jk})$, and $P^{\text{symdev}}_{ijkl} = P^{\text{sym}}_{ijkl} - P^{\text{iso}}_{ijkl}$.

Superscripts C and T are used here to distinguish eigenvalues, but both tensors are assumed to have the same eigenprojectors. For $\underset{\approx}{C}$ and $\underset{\approx}{T}$ to have the same eigenprojectors (i.e., the same η), their components must satisfy

$$\tan\eta = \frac{H_2^{(C)}}{H_3^{(C)}} = \frac{H_2^{(T)}}{H_3^{(T)}}, \tag{20.166}$$

where the superscripts, (C) or (T), indicate values for $\underset{\approx}{C}$ or $\underset{\approx}{T}$, respectively. The H-parameters are defined in (20.113b) in terms of h parameters so that this requirement can be written in the following alternative form

$$h_3^{(T)} = \left\{ \frac{h_1^{(T)} - \left(2h_2^{(T)} + h_4^{(T)}\right)}{h_1^{(C)} - \left(2h_2^{(C)} + h_4^{(C)}\right)} \right\} h_3^{(T)}. \tag{20.167}$$

Using (20.86) to write this result in terms of conventional four-index components, the necessary and sufficient condition for two transversely isotropic double-symmetric tensors, $\underset{\approx}{C}$ and $\underset{\approx}{T}$, to have the same eigenprojectors is

$$\boxed{T_{1133} = \left\{ \frac{T_{3333} - 2(T_{1122} + T_{1212})}{C_{3333} - 2(C_{1122} + C_{1212})} \right\} C_{1133}}. \tag{20.168}$$

This formula can be useful to set a reasonable (physically justifiable) value to T_{1133} when it is too difficult to measure it directly.

20.7 Material transverse isotropy

In continuum mechanics, a material is said to be transversely isotropic if there exists a unit axis $\underset{\sim}{\bar{a}}$ such that the material response to a given deformation is unaffected by whether or not the material is first rotated by any angle about the axis before the deformation is applied. This section describes the equations governing transversely isotropic hyperelasticity. As with any hyperelastic material model, the PK2 stress $\underset{\approx}{S}$ is given as the derivative of an energy potential W:

$$\underset{\approx}{S} = 2\frac{\mathrm{d}W}{\mathrm{d}\underset{\approx}{C}}, \tag{20.169}$$

For material transverse isotropy, the potential W depend only on the transverse invariants of the right Cauchy–Green tensor $\underset{\approx}{C}$[14],

[14] Here, we have elected to use the characteristic invariants of $\underset{\approx}{C}$. An alternative invariant triplet, based on the isotropic–isochoric decomposition of $\underset{\approx}{F}$ is often used in practice (see Wikipedia).

$$I_1 = \mathrm{tr}\underset{\approx}{C}, \qquad \text{for which } \frac{\mathrm{d}I_1}{\mathrm{d}\underset{\approx}{C}} = \underset{\approx}{I}, \tag{20.170a}$$

$$I_2 = \mathrm{tr}\underset{\approx}{C}^C = \tfrac{1}{2}\left(I_1^2 - \mathrm{tr}\left(\underset{\approx}{C}^2\right)\right), \qquad \text{for which } \frac{\mathrm{d}I_2}{\mathrm{d}\underset{\approx}{C}} = I_1\underset{\approx}{I} - \underset{\approx}{C}, \tag{20.170b}$$

$$I_3 = \det \underset{\approx}{C}, \qquad \text{for which } \frac{\mathrm{d}I_3}{\mathrm{d}\underset{\approx}{C}} = \underset{\approx}{C}^C = I_3\underset{\approx}{C}^{-1}, \tag{20.170c}$$

$$I_4 = \underset{\sim}{\bar{a}}\underset{\sim}{\bar{a}}\colon \underset{\approx}{C} \qquad \text{for which } \frac{\mathrm{d}I_4}{\mathrm{d}\underset{\approx}{C}} = \underset{\sim}{\bar{a}}\underset{\sim}{\bar{a}}, \tag{20.170d}$$

$$I_5 = \underset{\sim}{\bar{a}}\underset{\sim}{\bar{a}}\colon \underset{\approx}{C}^2 \qquad \text{for which } \frac{\mathrm{d}I_5}{\mathrm{d}\underset{\approx}{C}} = \underset{\sim}{\bar{a}}\underset{\sim}{\bar{a}} \bullet \underset{\approx}{C} + \underset{\approx}{C} \bullet \underset{\sim}{\bar{a}}\underset{\sim}{\bar{a}}. \tag{20.170e}$$

For a transversely isotropic hyperelastic material, the potential function depends only on these invariants. Therefore, applying the chain rule to (20.169) gives

$$\tfrac{1}{2}\underset{\approx}{S} = \frac{\mathrm{d}W}{\mathrm{d}\underset{\approx}{C}} = \frac{\partial W}{\partial I_1}\frac{\mathrm{d}I_1}{\mathrm{d}\underset{\approx}{C}} + \frac{\partial W}{\partial I_2}\frac{\mathrm{d}I_2}{\mathrm{d}\underset{\approx}{C}} + \frac{\partial W}{\partial I_3}\frac{\mathrm{d}I_3}{\mathrm{d}\underset{\approx}{C}} + \frac{\partial W}{\partial I_4}\frac{\mathrm{d}I_4}{\mathrm{d}\underset{\approx}{C}} + \frac{\partial W}{\partial I_5}\frac{\mathrm{d}I_5}{\mathrm{d}\underset{\approx}{C}}. \tag{20.171}$$

Defining $W_k = \partial W/\partial I_k$ and using the derivatives in (20.170),

$$\boxed{\begin{aligned}\tfrac{1}{2}\underset{\approx}{S} = {} & W_1\underset{\approx}{I} + W_2(I_1\underset{\approx}{I} - \underset{\approx}{C}) + W_3I_3\underset{\approx}{C}^{-1} + W_4\underset{\sim}{\bar{a}}\underset{\sim}{\bar{a}} \\ & + W_5(\underset{\sim}{\bar{a}}\underset{\sim}{\bar{a}} \bullet \underset{\approx}{C} + \underset{\approx}{C} \bullet \underset{\sim}{\bar{a}}\underset{\sim}{\bar{a}})\end{aligned}}\ . \tag{20.172}$$

As was done previously in the case of isotropy, the Cauchy stress for transverse isotropy is

$$\boxed{\begin{aligned}\underset{\approx}{\sigma} = {} & \frac{2}{J}[W_3I_3\underset{\approx}{I} + (W_1 + W_2I_1)\underset{\approx}{B} - W_2\underset{\approx}{B}^2 + W_4\underset{\sim}{a}\underset{\sim}{a} \\ & + W_5(\underset{\sim}{a}\underset{\sim}{a} \bullet \underset{\approx}{B}^2 + \underset{\approx}{B}^2 \bullet \underset{\sim}{a}\underset{\sim}{a})]\end{aligned}}\ , \tag{20.173}$$

Here, we have introduced a *non-unit* vector that is the convected version of the initial unit vector,

$$\underset{\sim}{a} = \underset{\approx}{F} \bullet \underset{\sim}{\bar{a}}. \tag{20.174}$$

Note that

$$\underset{\sim}{a} \bullet \underset{\sim}{a} = (\underset{\approx}{F} \bullet \underset{\sim}{\bar{a}})\bullet(\underset{\approx}{F} \bullet \underset{\sim}{\bar{a}}) = (\underset{\sim}{\bar{a}} \bullet \underset{\approx}{F}^T)\bullet(\underset{\approx}{F} \bullet \underset{\sim}{\bar{a}}) = \underset{\sim}{\bar{a}} \bullet \underset{\approx}{C} \bullet \underset{\sim}{\bar{a}} = I_4. \tag{20.175}$$

Therefore,

$$\underset{\sim}{a}\underset{\sim}{a} = I_4\underset{\sim}{\hat{a}}\underset{\sim}{\hat{a}}, \quad \text{where } \underset{\sim}{\hat{a}} \text{ is a unit vector in the direction of } \underset{\sim}{a}. \tag{20.176}$$

For the special case of a bundle of frictionless fibers [10], which is is useful for verifying numerical codes, the energy is determined entirely from the fourth invariant, thus giving

$$\underset{\approx}{S} = 2W_4\underset{\sim}{\bar{a}}\underset{\sim}{\bar{a}} \qquad \text{and} \qquad \underset{\approx}{\sigma} = \frac{2}{J}[W_4 I_4\underset{\sim}{\hat{a}}\underset{\sim}{\hat{a}}]. \tag{20.177}$$

Since $\underset{\sim}{\bar{a}}$ represents the *initial* symmetry axis, it is constant in time. Thus, using the chain rule, the rate of the PK2 stress is

$$\dot{\underset{\approx}{S}} = [2W_{44}\underset{\sim}{\bar{a}}\underset{\sim}{\bar{a}}\underset{\sim}{\bar{a}}\underset{\sim}{\bar{a}}]\colon \dot{\underset{\approx}{C}}, \tag{20.178}$$

where W_{44} is shorthand for $\partial^2 W/\mathrm{d}I_4^2$. This shows that, for the special case of frictionless fiber bundles, the PK2 stiffness has a very intuitive structure of being non-zero only in the fiber direction. The rate of Cauchy stress, on the other hand includes non-intuitive volumetric components due to the rate of J and non-intuitive shear components due to the rate of rotation of $\underset{\sim}{\hat{a}}$. These points are discussed in greater detail in [10].

References

[1] Malvern L E 1969 *Introduction to the Mechanics of a Continuous Medium* (Englewood Cliffs, NJ: Prentice-Hall)

[2] Brannon R M 2003 https://csmbrannon.net/2011/03/03/functional-and-structured-tensor-analysis-for-engineers/

[3] Fuller T J and Brannon R M 2012 On the effects of deformation-induced anisotropy on classical elastic-plastic materials *Int. J. Numer. Anal. Methods Geomech.* **37** 1079–94

[4] Bigoni D 2012 *Nonlinear Solid Mechanics* (New York: Cambridge University Press)

[5] Moraweic A 2004 *Orientations and Rotations: Computations in Crystallographic Textures* (Berlin: Springer)

[6] Schindler W 1997 On the efficient simulation of a particular class of random rotations relevant to computer graphics *J. Comput. Appl. Math.* **81** 107–14

[7] Zheng Q-S 1994 Theory of representations for tensor functions—a unified invariant approach to constitutive equations *Appl. Math. Rev.* **47** 545–87

[8] Cartan E 1966 *The Theory of Spinors* (Paris: Hermann)

[9] Schreyer H L and Zuo Q H 1995 Anisotropic yield surfaces based on elastic projection operators *Trans. ASME* 780–5

[10] Brannon R M 1998 Caveats concerning conjugate stress and strain measures for large distortion of anisotropic materials *Acta Mech.* **129** 107–16

IOP Publishing

Rotation, Reflection, and Frame Changes

Orthogonal tensors in computational engineering mechanics

R M Brannon

Chapter 21

Scalars and invariants

An isotropic invariant is something that remains unchanged upon a general change of orthonormal basis[1]. The components of a vector are *not*, in general, invariant because those components change when we change the basis. However, regardless of the nature of the basis change, the vector itself (being the *sum* of components times basis vectors) will remain unchanged. Thus, in this sense, vectors (velocity, electric field, force, etc) are invariant—these physical quantities themselves do not change simply because we change the basis used to *quantify* them[2]. Likewise, a tensor is invariant under a change of basis; its components will change in a manner that balances the basis change so that the tensor itself (being the sum of components times basis dyads) will not change.

If an invariant is a *single number* (such as temperature or mass density), then it is usually called a *scalar*. The term *scalar invariant* is typically used to denote a single number that is obtained by some operation applied to a directional quantity (vector, tensor, etc) and defined such that its value is unaffected by a (non-evolving) basis change. For example, the magnitude of a vector is a scalar invariant of that vector. The trace of a tensor is a scalar invariant of that tensor. A scalar invariant is a real-valued function of the components of a vector or tensor that will give the same result regardless of what basis is used. For a general (potentially non-orthonormal) basis $\{\underset{\sim}{\boldsymbol{g}}_1, \underset{\sim}{\boldsymbol{g}}_2, \underset{\sim}{\boldsymbol{g}}_3\}$, the component formula for an invariant will typically contain terms involving the metric coefficients $g_{ij} \equiv \underset{\sim}{\boldsymbol{g}}_i \bullet \underset{\sim}{\boldsymbol{g}}_j$. For example, the general definition of the trace of a tensor $\underset{\approx}{\boldsymbol{T}}$ is $\operatorname{tr}\underset{\approx}{\boldsymbol{T}} = T^{ij}g_{ij}$, where T_{ij} are the tensor components in the basis

[1] In this book, the word 'invariant' is to be interpreted as isotropic invariant unless otherwise specified. A different type of invariant can be defined as not changing under whatever symmetry group is pertinent. For example, *transversely invariant* objects are unchanged upon any rotation of the basis about the transverse symmetry axis.

[2] This is analogous to saying that changing language from English to Spanish produces a predicable change from 'c–a–t' to 'g–a–t–o,' but the thing being observed (the actual cat) does not change.

doi:10.1016/978-0-7503-1454-1ch21

expansion $\underset{\approx}{\boldsymbol{T}} = T^{ij}\underset{\sim}{\boldsymbol{g}}_i\underset{\sim}{\boldsymbol{g}}_j$. For an orthonormal basis, $g_{ij} = \delta_{ij}$, and the components T_{ij} are the ordinary Cartesian components where indices are written as subscripts, T_{ij}. Thankfully, for Euclidean spaces such as the three-dimensional (3D) physical world in which engineering problems are solved, determining whether or not a scalar-valued function of components (such as the trace) is a scalar invariant, only requires proving that it is invariant for all *orthonormal* basis changes. If so, then a generalized non-orthonormal component definition of that invariant can always be appropriately defined so that it will also be invariant. We will therefore consider only orthonormal (Cartesian) bases. When using an orthonormal basis, the trace of a tensor $\underset{\approx}{\boldsymbol{T}}$ is defined

$$\sum_{i=1}^{3} T_{ii}. \tag{21.1}$$

To prove that this is a scalar invariant, we must prove that it gives the same result regardless of the (orthonormal) basis used. For example, to prove that the trace is an invariant, we would need to consider the components of the tensor, T_{ij} and $\tilde{T}_{kl}$, with respect to two different orthonormal bases. The components are related by the transformation relation in (3.53),

$$\tilde{T}_{kl} = Q_{ik}Q_{jl}T_{ij}, \tag{21.2}$$

where [Q] is the orthogonal direction-cosine matrix connecting the two systems.

When the trace is taken in the second 'tilde' system, the result is $\tilde{T}_{kk}$, or

$$\tilde{T}_{kk} = Q_{ik}Q_{jk}T_{ij}. \tag{21.3}$$

Because the direction-cosine matrix is orthogonal, we know that $Q_{ik}Q_{jk} = \delta_{ij}$ and therefore, taking repeated indices to be implicitly summed,

$$\tilde{T}_{kk} = \delta_{ij}T_{ij} = T_{ii}. \tag{21.4}$$

Thus, the trace is an invariant because it will give the same result regardless of which system is used to compute it. For a general second-order tensor $\underset{\approx}{\boldsymbol{T}}$ in 3D, there are up to six independent scalar invariants. In practice, the following invariants are typically used:

$$I_1 = \operatorname{tr}\underset{\approx}{\boldsymbol{T}}, \quad J_2^T = \tfrac{1}{2}\operatorname{tr}[(\underset{\approx}{\boldsymbol{T}}^d)^2], \quad J_2^T = \tfrac{1}{3}\operatorname{tr}[(\underset{\approx}{\boldsymbol{T}}^d)^3], \quad \text{and} \quad \left\|\underset{\approx}{\boldsymbol{T}}\right\|. \tag{21.5}$$

Here, $\underset{\approx}{\boldsymbol{T}}^d$ is the deviatoric part of $\underset{\approx}{\boldsymbol{T}}$. The fourth invariant, which is the magnitude of $\underset{\approx}{\boldsymbol{T}}$, given by $\sqrt{T_{ij}T_{ij}}$ is often unfortunately omitted from elementary tensor and continuum mechanics textbooks, but it should be included in the list as an independent invariant because it cannot in general be expressed as a function of the first three invariants *except when the tensor* $\underset{\approx}{\boldsymbol{T}}$ *is symmetric.* In fact, to see that a general real non-symmetric tensor has six invariants, we invoke the definition of an invariant: its value is the same in all bases, making it the same in any convenient basis. In particular, knowing that the component matrix of any real symmetric tensor is diagonal in its principal basis, we know that we can always set up an orthonormal basis aligned with the eigenvectors of $\operatorname{sym}\underset{\approx}{\boldsymbol{T}}$, in which case the decomposition $\underset{\approx}{\boldsymbol{T}} = \operatorname{sym}\underset{\approx}{\boldsymbol{T}} + \operatorname{skw}\underset{\approx}{\boldsymbol{T}}$ becomes

$$\begin{bmatrix} \lambda_1 & -\omega_3 & \omega_2 \\ \omega_3 & \lambda_2 & -\omega_1 \\ -\omega_2 & \omega_1 & \lambda_3 \end{bmatrix}. \tag{21.6}$$

Here, $\lambda_1 \leqslant \lambda_2 \leqslant \lambda_3$ are canonically ordered eigenvalues of $\operatorname{sym}\underset{\approx}{\boldsymbol{T}}$ and ω_k are the components (with respect to the principal basis of $\operatorname{sym}\underset{\approx}{\boldsymbol{T}}$) of the axial vector associated with $\operatorname{skw}\underset{\approx}{\boldsymbol{T}}$. If the eigenvalues of $\operatorname{sym}\underset{\approx}{\boldsymbol{T}}$ are distinct, it follows that the corresponding eigenspaces are mutually perpendicular lines parallel to the eigenvectors. In that case, $|\omega_k|$ are all independent invariants of $\underset{\approx}{\boldsymbol{T}}$. If two eigenvalues are repeated, then the magnitude of the projection of the axial vector onto that eigenspace, as well as the magnitude of the projection onto the line of the remaining eigenvector are independent invariants. If all three λ_k are equal to each other, then the magnitude of the axial vector is still an independent invariant (i.e., its value cannot be determined from the eigenvalues of $\operatorname{sym}\underset{\approx}{\boldsymbol{T}}$).

We have pointed out that a second-order engineering tensor can have up to 6 independent invariants and that these can be identified as components of that tensor with respect to the principal basis of the symmetric part. The principal basis itself can be assigned within a given observer frame by an orthogonal direction cosine matrix, which itself has 3 independent components that are NOT scalar invariants because they will change value upon a change in frame. Thus, we conclude that any second-order engineering tensor may be fully characterized within an observer frame by specification of a total of 9 numbers: 6 invariants that will not change value with a change in frame and 3 frame-dependent numbers needed to construct direction cosines. These 9 numbers are alternatives to the conventional 9 values of the lab component matrix.

IOP Publishing

Rotation, Reflection, and Frame Changes
Orthogonal tensors in computational engineering mechanics
R M Brannon

Chapter 22

PMFI for incremental constitutive models

If an equation is frame-indifferent, is its *rate* also frame-indifferent? In chapter 19, we discussed a simple spring model in its straightforward direct functional form. Determining the forces in a super-structure that has complicated *systems* of interacting springs would entail solving several spring equations simultaneously. Since these equations might be generally nonlinear, a direct analytical solution might not be possible. However, the *rates* of the spring forces will always be related to the *rates* of the spring extensions through a linear system of equations, which is solvable for the rates and can therefore be numerically integrated through time to determine the forces. This is true even when the spring equations are highly nonlinear[1]. Knowing that even simple models (such as elasticity) will generally be solved in conjunction with other more complicated nonlinear models, it is likewise quite common to present material models in rate form. Now we need to explore if doing this has any impact on the frame indifference of the model.

22.1 A frame-indifferent spring rate equation

Consider the rate form of a one-dimensional 'spring' equation

$$\underset{\sim}{\sigma} = E\underset{\sim}{\varepsilon}, \tag{22.1}$$

where, under superimposed rotation,

$$\underset{\sim}{\sigma}^* = \underset{\approx}{\boldsymbol{Q}} \bullet \underset{\sim}{\sigma} \quad \text{and} \quad \underset{\sim}{\varepsilon}^* = \underset{\approx}{\boldsymbol{Q}} \bullet \underset{\sim}{\varepsilon} \tag{22.2}$$

(which we are using to keep the math simple by dealing with vectors instead of tensors). Taking the coefficient E to be constant, the axial 'stress' rate is related to the axial 'strain' rate by

[1] To understand why nonlinear equation becomes linear in rate form, just consider an arbitrarily nonlinear function $y = f(x)$. In rate form, this becomes $\dot{y} = f'(x)\dot{x}$, which is linear with respect to the rates. Numerically integrating through time, of course, only requires knowledge of the starting state.

doi:10.1016/978-0-7503-1454-1ch22

$$\dot{\underset{\sim}{\sigma}} = E\dot{\underset{\sim}{\varepsilon}}. \tag{22.3}$$

If we implement this equation into a computer code as an alternative model for springs will it is required to give consistent results for spring deformations with superimposed rotation. In other words, to be frame-indifferent, it must be true that

$$\dot{\underset{\sim}{\sigma}}^* = E\dot{\underset{\sim}{\varepsilon}}^*. \tag{22.4}$$

If $\dot{\underset{\sim}{\varepsilon}}$ is the spring 'strain' rate for the fiducial motion, what is the spring strain rate for the star motion? Taking rates of (22.2) gives

$$\dot{\underset{\sim}{\varepsilon}}^* = \underset{\approx}{\boldsymbol{Q}} \bullet \dot{\underset{\sim}{\varepsilon}} + \dot{\underset{\approx}{\boldsymbol{Q}}} \bullet \underset{\sim}{\varepsilon} \tag{22.5}$$

or, introducing the superimposed rate of rotation tensor, $\underset{\approx}{\Upsilon} = \dot{\underset{\approx}{\boldsymbol{Q}}} \bullet \dot{\underset{\approx}{\boldsymbol{Q}}}^T$,

$$\dot{\underset{\sim}{\varepsilon}}^* = \underset{\approx}{\boldsymbol{Q}} \bullet \dot{\underset{\sim}{\varepsilon}} + \underset{\approx}{\Upsilon} \bullet \underset{\approx}{\boldsymbol{Q}} \bullet \underset{\sim}{\varepsilon} \tag{22.6}$$

or

$$\dot{\underset{\sim}{\varepsilon}}^* = \underset{\approx}{\boldsymbol{Q}} \bullet \dot{\underset{\sim}{\varepsilon}} + \underset{\approx}{\Upsilon} \bullet \underset{\sim}{\varepsilon}^*. \tag{22.7}$$

Note that

$$\underset{\approx}{\boldsymbol{Q}} \bullet \dot{\underset{\sim}{\varepsilon}} = \dot{\underset{\sim}{\varepsilon}}^* - \underset{\approx}{\Upsilon} \bullet \underset{\sim}{\varepsilon}^*. \tag{22.8}$$

Similarly,

$$\underset{\approx}{\boldsymbol{Q}} \bullet \dot{\underset{\sim}{\sigma}} = \dot{\underset{\sim}{\sigma}}^* - \underset{\approx}{\Upsilon} \bullet \underset{\sim}{\sigma}^*. \tag{22.9}$$

Putting these into (22.3) gives

$$\dot{\underset{\sim}{\sigma}}^* - \underset{\approx}{\Upsilon} \bullet \underset{\sim}{\sigma}^* = E(\dot{\underset{\sim}{\varepsilon}}^* - \underset{\approx}{\Upsilon} \bullet \underset{\sim}{\varepsilon}^*). \tag{22.10}$$

Keep in mind: this is for the *vector* (first-order tensor) spring equation, not a stress–strain relation. In general, an nth-order spatial tensor will involve n terms that subtract away terms involving the superimposed rotation rate. Sometimes, for rates of second-order tensors, it will appear to you that some terms are subtracted while others are added. What is really happening is that a term involving $\underset{\approx}{\Upsilon}$ is being subtracted, but a transpose of $\underset{\approx}{\Upsilon}$ is needed to write it in symbolic (direct) notation. Thus, because $\underset{\approx}{\Upsilon}$ is skew-symmetric (i.e., ($\underset{\approx}{\Upsilon}^T = -\underset{\approx}{\Upsilon}$)), a subtraction involving $\underset{\approx}{\Upsilon}^T$ is the same as an *addition* involving $\underset{\approx}{\Upsilon}$. This commentary will become clearer as we proceed.

For the computer spring model to satisfy the frame indifference requirement of (22.4), the spring must additionally satisfy the original (non-rate) equation,

$$\underset{\sim}{\sigma}^* = E\underset{\sim}{\varepsilon}^*. \tag{22.11}$$

As long as a rate equation is derived from a frame-indifferent *non-rate* formula (as was done here), then this relationship will hold and therefore the principle of material-frame indifference (PMFI) will still be satisfied even without needing any special 'co-rotational rates' that are typically the actual rate minus a term that effectively strips away the part of that rate coming from bulk material rotation. In other words, as long as the non-rate form satisfies frame indifference, the trailing terms on both sides of equation (22.10) will cancel, making the formula just fine when both rates are ordinary rates—no special definitions of weird so-called 'objective' or 'corotational' rates would be needed. Co-rotational rates are introduced to make incremental or rate equations frame-indifferent when those equations were proposed *from the outset* in rate form without knowing with certainty that a frame-indifferent non-rate counterpart relationship even exists. As will be seen later in the context of stress–strain models, not all rate equations are derivable directly from frame-indifferent non-rate equations.

22.2 The PMFI in general

The Cauchy stress measure is not the only stress measure used in the literature. Many researchers employ the *second Piola–Kirchhoff stress* (PK2) tensor, defined by

$$\underset{\approx}{\bar{S}} = J\underset{\approx}{F}^{-1} \bullet \underset{\approx}{\sigma} \bullet \underset{\approx}{F}^{-T}, \tag{22.12}$$

where J is the Jacobian of the deformation,

$$J = \det\underset{\approx}{F}. \tag{22.13}$$

Upon a superimposed rotation,

$$J^* = \det \underset{\approx}{F}^* = \det(\underset{\approx}{Q} \bullet \underset{\approx}{F}) = \det(\underset{\approx}{Q})\det(\underset{\approx}{F}) = J. \tag{22.14}$$

Also

$$\underset{\approx}{\bar{S}}^* = J^*\underset{\approx}{F}^{*-1} \bullet \underset{\approx}{\sigma}^* \bullet \underset{\approx}{F}^{*-T}. \tag{22.15}$$

Substituting the objectivity transformation relations for J, $\underset{\approx}{F}$ and $\underset{\approx}{\sigma}$ reveals that

$$\boxed{\underset{\approx}{\bar{S}}^* = \underset{\approx}{\bar{S}}}. \tag{22.16}$$

In other words, *the PK2 stress is a reference stress tensor*—it is unaffected by a superimposed rigid motion.

The Cauchy stress $\underset{\approx}{\sigma}$ is said to be 'conjugate' to the symmetric part of the velocity gradient $\underset{\approx}{D}$ because the stress power per unit mass is given by

$$P_s = \frac{1}{\rho}\underset{\approx}{\sigma} : \underset{\approx}{D}. \tag{22.17}$$

The symmetric part of the velocity gradient can be shown to be a spatial tensor, i.e.,

$$\underset{\approx}{D}^* = \underset{\approx}{Q} \bullet \underset{\approx}{D} \bullet \underset{\approx}{Q}^T. \tag{22.18}$$

The mass density ρ is a frame-insensitive scalar (i.e., it is unchanged by superimposed rotation). This does *not* mean that the mass density ρ is equal to its initial value ρ_o—it only means that its value with superimposed rotation is the same as its value without it. P_s is itself a frame-insensitive scalar. By using the definition of the PK2 stress, it can be shown that there exists a strain $\underset{\approx}{\varepsilon}$ that is conjugate to the PK2 stress in the sense that

$$P_s = \frac{1}{\rho}\underset{\approx}{\overline{S}} : \dot{\underset{\approx}{\overline{\varepsilon}}}. \tag{22.19}$$

Specifically, this special strain—called the *Green–Lagrange strain*—is defined

$$\underset{\approx}{\overline{\varepsilon}} = \tfrac{1}{2}(\underset{\approx}{C} - \underset{\approx}{I}) = \tfrac{1}{2}(\underset{\approx}{F}^T \bullet \underset{\approx}{F} - \underset{\approx}{I}). \tag{22.20}$$

Recalling that $\underset{\approx}{C}$ is a reference tensor, we immediately see that Green–Lagrange strain $\underset{\approx}{\overline{\varepsilon}}$ is also a reference tensor. That is,

$$\boxed{\underset{\approx}{\overline{\varepsilon}}^* = \underset{\approx}{\overline{\varepsilon}}}. \tag{22.21}$$

Many researchers elect to phrase their constitutive equations in terms of the Green–Lagrange strain and the PK2 stress because they are both reference tensors, and the PMFI is therefore automatically satisfied. Such formulations avoid the complication of moving material directions while having the drawback of needing counter-intuitively nonlinear elastic response functions with respect to those attractively stationary reference material fiber directions [1]. The PMFI has its most dramatic impact on constitutive models that instead opt to use spatial tensors.

22.2.1 Example: elasticity

In its most general sense, a material is usually said to be elastic if the stress in the material depends only on the deformation of the material. This definition is rather sloppy because it violates frame indifference when taken literally. To be more rigorous, a material is elastic if stress can be written as a function of deformation and any number of material constants (moduli), including possibly constant vectors or tensors such as the initial orientation of reinforcing fibers. Let $\underset{\approx}{\sigma}$ denote Cauchy stress and let $\underset{\approx}{F}$ denote the deformation gradient tensor. Any elastic constitutive law may be written

$$\underset{\approx}{\sigma} = f(\underset{\approx}{F}, \{c\}), \tag{22.22}$$

where f is the constitutive function. The term 'constitutive' carries with it an implication that the function f may depend on constant material parameters, here denoted by an array $\{c\}$, some of which might be vectors or tensors. When considering frame indifference, these constants—especially vector/tensor parameters—need to be shown *explicitly*. For example, a transversely isotropic material depends on five elastic constants (c_1, c_2, c_3, c_4, c_5) and on the symmetry axis $\underset{\sim}{n}$. Being a unit vector, the symmetry axis has two independent components, which might be loosely regarded as new material parameters (c_6, c_7). Upon a superimposed rigid rotation $\underset{\approx}{Q}$, the first five constants (scalar elastic moduli) will not change, but the symmetry axis *will* change (at least for small-strain models) from $\underset{\sim}{n}$ to $\underset{\approx}{Q} \bullet \underset{\sim}{n}$. Therefore $c_k^* = c_k$ if $k \leqslant 5$ but $c_k^* \neq c_k$ if $k > 5$. A cubic material depends on two elastic constants (c_1, c_2) and on the orientation of the cubic directions, which may be characterized by an orthogonal tensor $\underset{\approx}{q}$. Physically, we know that application of a superimposed rigid motion *should* change the Cauchy stress to

$$\underset{\approx}{\sigma}^* = \underset{\approx}{Q} \bullet \underset{\approx}{\sigma} \bullet \underset{\approx}{Q}^T. \tag{22.23}$$

Furthermore, when we have a deformation $\underset{\approx}{F}$ and we apply a superimposed rotation $\underset{\approx}{Q}$, then the new deformation gradient is

$$\underset{\approx}{F}^* = \underset{\approx}{Q} \bullet \underset{\approx}{F}. \tag{22.24}$$

When applied under a superimposed rigid motion, PMFI requiires that the constitutive law of (22.22) must still apply:

$$\underset{\approx}{\sigma}^* = f(\underset{\approx}{F}^*, \{c^*\}) \tag{22.25}$$

or

$$\underset{\approx}{Q} \bullet \underset{\approx}{\sigma} \bullet \underset{\approx}{Q}^T = f(\underset{\approx}{Q} \bullet \underset{\approx}{F}, \{c^*\}). \tag{22.26}$$

The PMFI demands that both (22.22) and (22.26) must be true. Thus, we may substitute (22.22) into (22.26) to obtain

$$\underset{\approx}{Q} \bullet f(\underset{\approx}{F}, \{c\}) \bullet \underset{\approx}{Q}^T = f(\underset{\approx}{Q} \bullet \underset{\approx}{F}, \{c^*\}), \tag{22.27}$$

which must hold for all rotations $\underset{\approx}{Q}$ and deformations $\underset{\approx}{F}$.

The above equation represents a restriction on admissible forms for the function *f*. Any proposed elastic constitutive equation of the form (22.22) must satisfy the above restriction. Since the above relation must hold for *all* $\underset{\approx}{Q}$, then it must of course hold for any particular choice of $\underset{\approx}{Q}$. Consider the choice $\underset{\approx}{Q} = \underset{\approx}{R}^T$. Then, noting that $\underset{\approx}{Q} \bullet \underset{\approx}{F} = \underset{\approx}{R}^T \bullet \underset{\approx}{F} = \underset{\approx}{U}$, and recalling from (22.22) that $f(\underset{\approx}{F}) = \underset{\approx}{\sigma}$, PMFI requires that

$$\overline{\underset{\approx}{\sigma}} = f(\underset{\approx}{U}, \{c^*\}), \quad \text{where } \overline{\underset{\approx}{\sigma}} \equiv \underset{\approx}{R}^T \bullet \underset{\approx}{\sigma} \bullet \underset{\approx}{R}. \tag{22.28}$$

This means that the *un-rotated* Cauchy stress must be expressible solely as a constitutive function of the *reference* stretch tensor $\underset{\approx}{U}$ and elastic constants $\{c^*\}$ that are likewise un-rotated to the reference configuration. Consider, for example, a transversely isotropic material that depends on five scalar moduli and one orientation vector $\underset{\sim}{n}$. The five moduli are unaffected by superimposed rotation, so we will not show them explicitly, but the symmetry axis is affected by rotation, so we will show it explicitly. Then (22.28) would imply that

$$\overline{\underset{\approx}{\sigma}} = f(\overline{\underset{\approx}{V}}, \overline{\underset{\sim}{n}}), \tag{22.29}$$

where

$$\overline{\underset{\approx}{\sigma}} \equiv \underset{\approx}{R}^T \bullet \underset{\approx}{\sigma} \bullet \underset{\approx}{R}, \qquad \overline{\underset{\approx}{V}} = \underset{\approx}{R}^T \bullet \underset{\approx}{V} \bullet \underset{\approx}{R} = \underset{\approx}{U}, \qquad \text{and} \qquad \overline{\underset{\sim}{n}} = \underset{\approx}{R}^T \bullet \underset{\sim}{n}. \tag{22.30}$$

We have expressed (22.29) in terms of $\overline{\underset{\approx}{V}}$ to emphasize that all of the independent variables may be regarded as un-rotations of their spatial counterparts. Importantly, an un-rotated reference configuration is not the same thing as the *initial* configuration. It is merely a configuration in which polar rotations have been removed. It can be regarded as a view of the deformation that is seen by an observer tracking with the polar rotation. The un-rotation of a spatial tensor ($\underset{\approx}{\sigma}$ or $\underset{\approx}{V}$) or a spatial vector $\underset{\sim}{n}$ produces a barred tensor or vector that is a *reference* quantity. For example,

$$\bar{\underset{\approx}{\sigma}}^* \equiv \underset{\sim}{R}^{*T} \bullet \underset{\approx}{\sigma}^* \bullet \underset{\sim}{R}^* = (\underset{\sim}{Q} \bullet \underset{\sim}{R})^T \bullet (\underset{\sim}{Q} \bullet \underset{\approx}{\sigma} \bullet \underset{\sim}{Q}^T) \bullet (\underset{\sim}{Q} \bullet \underset{\sim}{R}) = \underset{\approx}{R}^T \bullet \underset{\approx}{\sigma} \bullet \underset{\approx}{R} = \bar{\underset{\approx}{\sigma}}. \quad (22.31)$$

A similar analysis shows that $\bar{\underset{\approx}{V}}^* = \bar{\underset{\approx}{V}}$ and $\bar{\underset{\sim}{n}}^* = \bar{\underset{\sim}{n}}$. In other words, barred quantities are unaffected by any superimposed rotation $\underset{\approx}{Q}$.

Recall that (22.28) and its specialization to (22.29) for transverse isotropy arose from the particular choice $\underset{\approx}{Q} = \underset{\approx}{R}^T$ in the general PMFI requirement of (22.27). Thus, equation (22.28) must, at this stage, be interpreted only as a *necessary* requirement for satisfying PMFI. Now we ask: is it also *sufficient*? The answer is yes! If a material constitutive model can be cast in the form of a functional relationship between barred quantities (i.e., un-rotations of spatial variables), then it automatically satisfies PMFI because barred quantities are unaffected by superimposed rotation and therefore superimposed rotation places no restrictions on the constitutive function f. Therefore, to reiterate:

> Any elastic constitutive model that can be cast as a functional relationship between un-rotated reference-configuration quantities will automatically satisfy PMFI. To use the reference version of the constitutive model in a calculation, you would first replace all input variables with whatever they would be under a superimposed rotation that is the transpose of the polar rotation, and then you would apply the constitutive model to update the dependent variables in the un-rotated reference configuration, and finally re-rotate these results back into the spatial configuration if desired.

To demonstrate that a material constitutive model is frame invariant, you merely need to show that it *can* be phrased in the un-rotated reference. You are not required to evaluate the model in this reference. Suppose, for example, you have an isotropic elasticity model in which the PK2 stress tensor in (22.12) is presumed to be a function of the Green–Lagrange strain $\underset{\approx}{\varepsilon}$ in (22.20):

$$\underset{\approx}{S} = f(\underset{\approx}{\varepsilon}). \quad (22.32)$$

Is this model frame invariant? Yes, because (as shown earlier) $\underset{\approx}{s}^* \equiv \underset{\approx}{s}$ and $\underset{\approx}{\varepsilon}^* = \underset{\approx}{\varepsilon}$. Can $\underset{\approx}{s}$ and $\underset{\approx}{\varepsilon}$ be phrased as polar un-rotations of spatial tensors? Yes. Using the polar decomposition to write $\underset{\approx}{F} = \underset{\approx}{R} \bullet \underset{\approx}{U} = \underset{\approx}{R} \bullet \bar{\underset{\approx}{V}}$, the definition of the PK2 stress (22.12) may be written

$$\underset{\approx}{S} = J\bar{\underset{\approx}{V}}^{-1} \bullet \bar{\underset{\approx}{\sigma}} \bullet \bar{\underset{\approx}{V}}^{-1}, \quad \text{where} \quad \bar{\underset{\approx}{\sigma}} \equiv \underset{\approx}{R}^T \bullet \underset{\approx}{\sigma} \bullet \underset{\approx}{R}. \quad (22.33)$$

Thinking of the overbar as an *operator* such that, for any spatial tensor $\underset{\approx}{A}$,

$$\overline{\underset{\approx}{A}} \equiv \underset{\approx}{R}^T \bullet \underset{\approx}{A} \bullet \underset{\approx}{R}, \tag{22.34}$$

you can readily show that the overbar operation satisfies the following properties for any tensors $\underset{\approx}{A}$ and $\underset{\approx}{B}$ and scalar α:

$$\overline{\alpha \underset{\approx}{A}} = \alpha \overline{\underset{\approx}{A}}, \tag{22.35}$$

$$\overline{\underset{\approx}{A}^{-1}} = \overline{\underset{\approx}{A}}^{-1}, \tag{22.36}$$

$$\overline{\underset{\approx}{A} \bullet \underset{\approx}{B}} = \overline{\underset{\approx}{A}} \bullet \overline{\underset{\approx}{B}}. \tag{22.37}$$

Therefore, equation (22.33) may be written

$$\underset{\approx}{S} = \overline{J \underset{\approx}{V}^{-1} \bullet \underset{\approx}{\sigma} \bullet \underset{\approx}{V}^{-1}}, \tag{22.38}$$

or

$$\overline{\underset{\approx}{S}} = \underset{\approx}{R}^T \bullet \underset{\approx}{S} \bullet \underset{\approx}{R}, \quad \text{where} \quad \underset{\approx}{S} \equiv J \underset{\approx}{V}^{-1} \bullet \underset{\approx}{\sigma} \bullet \underset{\approx}{V}^{-1}. \tag{22.39}$$

The tensor $\underset{\approx}{S}$ can be readily shown to be a *spatial* stress, so this result shows that $\overline{\underset{\approx}{S}}$ can indeed be written as the un-rotation of a spatial quantity. A similar analysis may be used to demonstrate that the Lagrange strain $\overline{\underset{\approx}{\varepsilon}}$ may be likewise written as an unrotation of a spatial strain:

$$\overline{\underset{\approx}{\varepsilon}} = \underset{\approx}{R}^T \bullet \underset{\approx}{\varepsilon} \bullet \underset{\approx}{R}, \quad \text{where} \quad \underset{\approx}{\varepsilon} \equiv \tfrac{1}{2}(\underset{\approx}{V}^2 - \underset{\approx}{I}). \tag{22.40}$$

There are many good reasons to use alternative configurations or alternative stress/strain measures as your master reference for evaluating material models. Consider transverse isotropy. The un-rotated symmetry axis $\overline{\underset{\sim}{n}}$ does *not* generally coincide with the *initial* symmetry axis, which we will denote $\overline{\underset{\sim}{N}}$. The relationship between $\overline{\underset{\sim}{n}}$ and $\overline{\underset{\sim}{N}}$ generally depends on the underlying microstructure, as illustrated in figure 22.1.

In figure 22.1, the symmetry axis for the fiber-dominated material is denoted with the letter 'M' because it deforms with the material. The deformed unit symmetry axis, $\underset{\sim}{m}$, is a unit vector in the direction of $\underset{\sim}{a} = \underset{\approx}{F} \bullet \overline{\underset{\sim}{M}}$. The symmetry axis in the un-rotated configuration is $\overline{\underset{\sim}{m}} = \underset{\approx}{R}^T \bullet \underset{\sim}{m}$, which is a unit vector in the direction of $\overline{\underset{\sim}{a}} = \underset{\approx}{R}^T \bullet \underset{\sim}{a} = \overline{\underset{\approx}{V}} \bullet \overline{\underset{\sim}{M}}$. For the fiber-dominated material, the PK2 stress would be a good choice as the stress to use in a constitutive model because, as seen in (22.33), the PK2 stress essentially 'pulls back' the reference stretch ($\underset{\approx}{U} = \overline{\underset{\approx}{V}}$) from the un-rotated Cauchy stress, which reverses distortion of material fiber direction vectors at the cost of introducing distortion in the fiber stretch scalars (and this is probably a worthwhile trade-off) [1]. For the fiber-dominated material, $\overline{\underset{\sim}{m}}$ characterizes the symmetry axis about which the un-rotated Cauchy stress is centered, but $\overline{\underset{\sim}{M}}$ characterizes the symmetry axis for the PK2 stress. The *initial* symmetry axis $\overline{\underset{\sim}{M}}$ is constant in time, so a material model that uses PK2 stress would not have to track rotations of the symmetry axis (which can be caused by material distortion even if there is no net material rotation). While a PK2 formulation would not be burdened with the need to track changes in symmetry directions for a fiber-dominated material, such a formulation would be required to support *nonlinear*-elastic moduli to avoid compression instabilities [1].

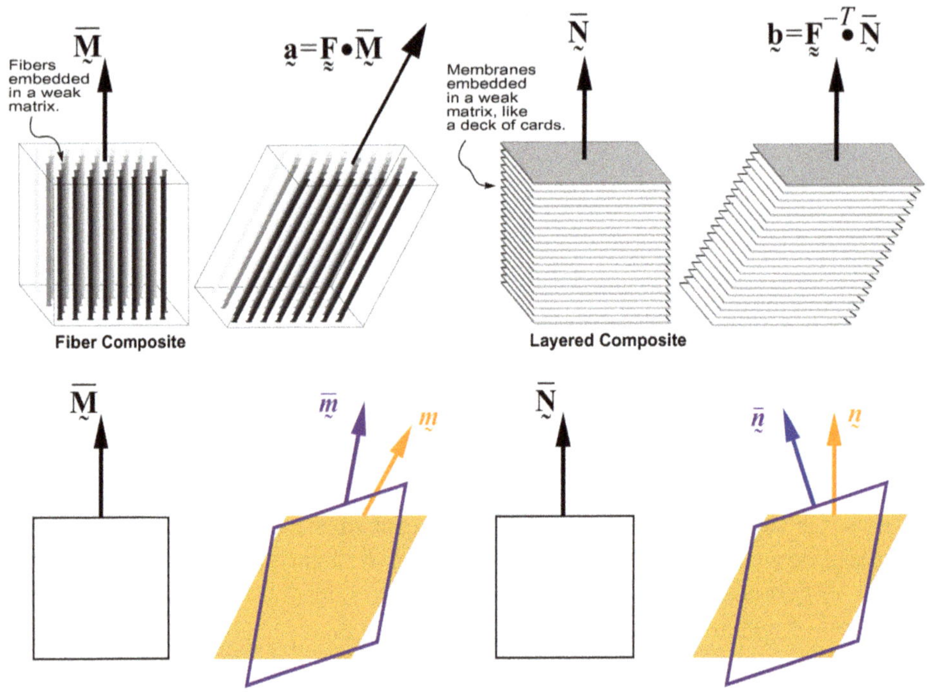

Figure 22.1. Two extreme cases of a transversely isotropic material. For a strongly fiber-dominated material, the transverse symmetry axis tends to co-rotate with the material (in this case, the initial symmetry axis is denoted with the letter 'M' to emphasize that it transforms like a material fiber). For a strongly laminate-dominated structure, the symmetry axis *contra*-rotates (here, the initial normal is denoted with the letter 'N' to emphasize that this axis tends to align with the normal to the laminate planes). The lower figures show the deformation in the un-rotated 'bar' configuration where deformation is a pure stretch. Neither $\bar{\underset{\sim}{m}}$ nor $\bar{\underset{\sim}{n}}$ is coincident with the initial symmetry axis.

The laminate-dominated material figure 22.1 has a symmetry axis that tracks with the laminate *planes*, not with the material fibers. This would imply that neither the un-rotated Cauchy stress *nor* the PK2 stress would be particularly convenient. As was done for the fiber-dominated case, you could always define a new stress that would be appropriate for laminate-dominated microstructures. Transversely isotropic materials that are neither fiber nor laminate-dominated probably would not be conveniently modeled by *any* particular stress–strain definition. Hence, there probably does not exist any 'magic' conjugate stress–strain definition that will be universally appropriate for all types of materials. Choices for stress–strain measures are motivated by *convenience* to make material models as simple in form as possible. You can use any stress–strain definitions that you want, as long as you recognize that your choice might be more complicated than other formulations, therefore possibly making it more difficult to match to experimental data. In the final analysis, however, a complicated stress–strain choice should predict a material response that is consistent with a more elegantly crafted formulation. Model consistency merely requires that a formulation that is cast in terms of the PK2 stress from the outset should give the same result as a model cast in terms of Cauchy stress, once the Cauchy stress is solved to determine the PK2 stress.

No matter what stress-strain choices are adopted, a valid constitutive model should be *accurate and valid—it should be capable of describing real observed experimental data.* Satisfying PMFI is necessary but not sufficient for a model to be a valid. If the fiducial model is inaccurate, then PMFI merely requires the same model to be equally inaccurate (same error) if it is applied with superimposed rotation. Frame indifference is merely a constraint that a model should predict consistent results when applied to a fiducial deformation and a (starred) deformation that is identical but with superimposed rotation. Nothing in PMFI helps to ensure that the *fiducial* model will be predictive in the first place. You could, for example, propose a transverse isotropy model for the un-rotated Cauchy stress in which the symmetry axis does not change in response to deformation. *Such a model would satisfy PMFI,* but it would predict unsatisfactory results in large deformation problems because, as discussed above, the symmetry axis associated with un-rotated Cauchy stress really should distort even in pure stretch conditions (where there is no polar rotation). You are never justified in claiming that your model applies to large deformation problems *simply because it satisfies PMFI.* That principle has absolutely no bearing on the accuracy of a model under large material distortions. Accordingly, any assertion that a model is predictive for large deformation must be justified by comparing its predictions against laboratory data for which the reference stretch tensor not only has one or more eigenvalue significantly different from unity, but also has non-stationary eigenvectors.

22.2.2 Spatial constitutive models

Suppose that an elastic constitutive model is proposed in which the spatial Kirchhoff stress, $\underset{\approx}{\tau} \equiv J\underset{\approx}{\sigma}$ (in which $J = \det \underset{\approx}{F}$ is the Jacobian and $\underset{\approx}{\sigma}$ is the Cauchy stress) is purely a function of some spatial strain measure (and, implicitly, non-directional material properties such as elastic moduli):

$$\underset{\approx}{\tau} = g(\underset{\approx}{\varepsilon}). \tag{22.41}$$

The term 'spatial' is used to describe any tensor that transforms upon a superimposed rigid rotation $\underset{\approx}{Q}$ according to

$$\underset{\approx}{A}^* = \underset{\approx}{Q} \bullet \underset{\approx}{A} \bullet \underset{\approx}{Q}^T. \tag{22.42}$$

The Kirchhoff stress is a spatial tensor, while (22.24) shows that the deformation gradient is *not* a spatial tensor. If the strain measure is spatial, then we know that

$$\underset{\approx}{\varepsilon}^* = \underset{\approx}{Q} \bullet \underset{\approx}{\varepsilon} \bullet \underset{\approx}{Q}^T. \tag{22.43}$$

Equation (22.41) satisfies frame indifference only if

$$\underset{\approx}{\tau}^* = g(\underset{\approx}{\varepsilon}^*). \tag{22.44}$$

In other words,

$$\underset{\approx}{Q} \bullet \underset{\approx}{\tau} \bullet \underset{\approx}{Q}^T = g(\underset{\approx}{Q} \bullet \underset{\approx}{\varepsilon} \bullet \underset{\approx}{Q}^T). \tag{22.45}$$

To be consistent with (22.41), this implies that

$$\underset{\approx}{Q} \bullet g(\underset{\approx}{\varepsilon}) \bullet \underset{\approx}{Q}^T = g(\underset{\approx}{Q} \bullet \underset{\approx}{\varepsilon} \bullet \underset{\approx}{Q}^T), \tag{22.46}$$

which must hold for all rotations $\underset{\approx}{Q}$ and spatial strains $\underset{\approx}{\varepsilon}$.

As a special case, suppose that

$$\tau_{ij} = E_{ijkl}\varepsilon_{kl}, \tag{22.47}$$

where E_{ijkl} are components of a *constant* fourth-order stiffness tensor. With this model, in which the elastic stiffness tensor is constant, substitution of (22.47) into (22.45) gives the restriction

$$E_{ijkl} = Q_{pi}Q_{qj}Q_{rk}Q_{sl}E_{pqrs}. \tag{22.48}$$

This must hold for all $[Q]$, so the only way for this *spatial* constitutive equation to be frame-indifferent would be for the constant stiffness tensor to be isotropic. Hence, the model applies only to material that has no intrinsic anisotropy. To model an anisotropic material, the anisotropic stiffness tensor must be *explicitly* one of the independent variables in the constitutive model. That is, equation (22.41) would need to be revised to read

$$\underset{\approx}{\tau} = g(\underset{\approx}{\varepsilon}, \underset{\approx\approx}{E}). \tag{22.49}$$

Then, under superimposed rotation, $\underset{\approx\approx}{E}^*$ would need to be different from $\underset{\approx\approx}{E}$. Specifically, it would need to transform according to

$$E^*_{ijkl} = Q_{pi}Q_{qj}Q_{rk}Q_{sl}E_{pqrs}. \tag{22.50}$$

To better understand why the spatial formulation of (22.41) requires isotropy of the stiffness tensor, simply consider a fiducial case study that a material with horizontal reinforcing fibers is pulled horizontally, and then compare that to the same deformation with extra rotation that brings the fibers (and loads on them) to be in the vertical direction. If programmed literally, the original constitutive equation of (22.41) has no way to know that the material has rotated, making it fail to predict a stress in the pull-and-rotate scenario that is larger than if the material had been pulled vertically without rotating the fibers. The problem can be rectified by modifying the constitutive equation to include material orientation information as part of its independent variable list. PMFI *requires* such a model modification if the material is anisotropic. If your material is merely transversely isotropic, then PMFI will permit you to propose a spatial constitutive model of the form

$$\underset{\approx}{\tau} = g(\underset{\approx}{\varepsilon}, \underset{\sim}{n}), \tag{22.51}$$

where $\underset{\sim}{n}$ is the axis of transverse symmetry. By including the material direction as an additional independent variable in the constitutive model, you will be able to assert that $\underset{\sim}{n}^* = \underset{\approx}{Q} \bullet \underset{\sim}{n}$, which in turn permits the anisotropic stiffnesses to rotate properly with the material upon superimposed rotation. Solving an anisotropic constitutive problem in the spatial configuration this way is complicated and numerically less efficient than using the un-rotated reference tensor approach described below. Moreover, even if frame indifference issues are handled by including material orientation in the independent variable list, you then face non-PMFI issues of model accuracy resulting from relative distortions of the elastic lattice (see figure 22.1).

22.2.3 Un-rotated constitutive models

Equation (22.46) must hold for all $\underset{\approx}{Q}$, so it must hold for the choice $\underset{\approx}{Q} = \underset{\approx}{R}^T$, so that

$$\underset{\approx}{\overline{\tau}} = g(\underset{\approx}{\overline{\varepsilon}}), \quad \text{where } \underset{\approx}{\overline{\tau}} = \underset{\approx}{R}^T \bullet \underset{\approx}{\tau} \bullet \underset{\approx}{R} \text{ and } \underset{\approx}{\overline{\varepsilon}} = \underset{\approx}{R}^T \bullet \underset{\approx}{\varepsilon} \bullet \underset{\approx}{R}. \tag{22.52}$$

Here, the tensors are said to have been 'un-rotated'. Whereas the equation $\underset{\approx}{\tau} = g(\underset{\approx}{\varepsilon})$ violated PMFI unless the material was isotropic, no such constraint applies to $\underset{\approx}{\overline{\tau}} = g(\underset{\approx}{\overline{\varepsilon}})$. Even though this equation appears to still take only a single variable as input (the un-rotated strain $\underset{\approx}{\overline{\varepsilon}}$), writing a computer program to use this model to update Kirchhoff stress $\underset{\approx}{\tau}$ would actually require bringing in an additional input: the rotation tensor $\underset{\approx}{R}$, which gives sufficient additional information to satisfy PMFI even for an anisotropic material. Un-rotated reference formulations of constitutive models will *always* eliminate PMFI concerns, allowing the modeler to focus on truly constitutive concerns. Consider, for example, a reference-configuration version of (22.49):

$$\underset{\approx}{\overline{\tau}} = g(\underset{\approx}{\overline{\varepsilon}}, \underset{\approx\approx}{\overline{E}}). \tag{22.53}$$

Where the overbars denote Rayleigh product un-rotation pull-back operations using the inverse of the polar rotation tensor. Namely, $\underset{\approx}{\overline{\tau}} = \underset{\approx}{R}^T \bullet \underset{\approx}{\tau} \bullet \underset{\approx}{R} = \underset{\approx}{R}^T > \underset{\approx}{\tau}$, $\underset{\approx}{\overline{\varepsilon}} = \underset{\approx}{R}^T > \underset{\approx}{\varepsilon}$, and $\underset{\approx\approx}{\overline{E}} = \underset{\approx}{R}^T > \underset{\approx\approx}{E}$. Unlike the spatial formulation, the un-rotated stiffness $\underset{\approx\approx}{\overline{E}}$ in this reference formulation *can* permissibly be treated as constant without violating PMFI. In other words, PMFI allows dropping the stiffness as an explicit calling argument, so we can write $\underset{\approx}{\overline{\tau}} = g(\underset{\approx}{\overline{\varepsilon}})$. To validate the model against experimental data, you would probably want to allow the un-rotated stiffness to vary with deformation, but PMFI does not *require* it to vary. Nonlinearities resulting from material shape distortions (i.e., relative rotations of the elastic lattice) are handled independently from PMFI considerations [1].

You might see some formulations that apply a Rayleigh product pull-back using the inverse of the deformation gradient tensor instead of pulling back with only the polar rotation tensor. For example, $\underset{\approx}{F}^{-1} \bullet \underset{\approx}{\tau} \bullet \underset{\approx}{F}^{-T} = \underset{\approx}{F}^{-1} > \underset{\approx}{\tau}$ is the PK2 stress, and the spatial strain and stiffness could be pulled back similarly. Such a formulation, however, is a special case of (22.53), as can be seen by noting, for example, that $\underset{\approx}{F}^{-1} > \underset{\approx}{\tau} = \underset{\approx}{U}^{-1} > \underset{\approx}{\overline{\tau}}$. This shows that a pull-back by $\underset{\approx}{F}^{-1}$ is equivalent to an un-rotation pull-back followed by an *un-distortion* operation by $\underset{\approx}{U}^{-1}$ within the un-rotated frame. The un-distortion operation can be applied without requiring any additional inputs provided to the constitutive model. Thus, since the right stretch $\underset{\approx}{U}$ can be determined uniquely from any un-rotated reference strain, $\underset{\approx}{\bar{\varepsilon}}$, this approach is merely a special case of (22.53).

We consider it to be far more informative to simply observe that PMFI is satisfied by any model that is formulated in the unrotated frame. Any constitutive model that is not already phrased in the un-rotated frame can be *immediately* converted to a model in the un-rotated frame by replacing all tensors with whatever they would be if there were no polar rotation (or, more precisely, what they would be under superimposed rotation equal to the inverse of the polar rotation). This means, for example, that all occurrences of $\underset{\approx}{F}$ would be replaced by the right stretch $\underset{\approx}{U}$. In this way, all nonlinear dependencies on $\underset{\approx}{U}$ in the un-rotated frame are clearly there to improve accuracy of the model against experimental data measured without rotation. Formulating all constitutive models in the un-rotated frame helps to separate PMFI (applicable to any material and automatically satisfied when cast in the un-rotated frame) from the truly *constitutive* (i.e., material-dependent) parts of the model that describe the material response to distortion.

Even though careful PMFI analysis tells you that your model will satisfy PMFI *in principle*, it is nevertheless still important to include PMFI testing in any computer code. For example, to obtain the stress $\underset{\approx}{\tau}^{(n)}$ at a given time step n, the sequence of actions is (1) use the polar rotation at the beginning of the time step to unrotate the *initial* stress, (2) apply the constitutive model to obtain an update of the unrotated stress at the end of the strep (3) convert the updated unrotated stress back to regular stress by re-rotating using the polar rotation at the *end* of the step. Failure to use different rotation tensors will cause a numerical implementation to fail PMFI testing [2]. In numerical modeling, especially when describing motion of fluids, the code designer might find it more useful to adopt an updated Lagrangian formulation (wherein the material state at the beginning of the time step is taken as the reference configuration); this approach accommodates using efficient un-rotated frames wherein the un-rotation operation begins as the identity tensor at the start of the step and finishes as the rotation that occurs over the time step.

22.3 PMFI in rate forms of the constitutive equations

For the purpose of illustration, let us revisit a constitutive law of the following form

$$\tau_{ij} = E_{ijkl}\varepsilon_{kl}, \tag{22.54}$$

in which the stiffness tensor is constant and (to then satisfy PMFI) isotropic. Taking the time rate of both sides (keeping in mind that stiffness is constant) gives

$$\dot{\tau}_{ij} = E_{ijkl}\dot{\varepsilon}_{kl}. \tag{22.55}$$

For satisfying PMFI, we have already asserted that, if the stiffness components E_{ijkl} are constant, then the stiffness tensor must be isotropic. As long as this is true, the constitutive equation defined by (22.55) will satisfy frame indifference because it was obtained by differentiating a frame-indifferent equation. Some researchers assume (incorrectly) that the above equation will violate PMFI because it uses ordinary time rates instead of strange so-called 'co-rotational' rates. In this section, we explain why PMFI is satisfied, and we also explain how co-rotational rates must be introduced to compensate for an error that occurs when a certain approximation for the strain rate is adopted.

Since the strain is spatial, it will transform under a superimposed rotation according to $\varepsilon_{ij}^* = Q_{im}Q_{jn}\varepsilon_{mn}$. Consequently, the rate of the superimposed strain will include contributions from both the rate of ε_{mn} and the rate of $[Q]$. The rate of stress similarly includes such kinematic contributions from material rotation.

Computationally, finding the rate of the strain tensor is sometimes quite cumbersome, so it is common practice to extemporaneously replace the strain rate $\dot{\underset{\approx}{\varepsilon}}$ by the stretching ('rate' of deformation) tensor $\underset{\approx}{\boldsymbol{D}}$, which is just the symmetric part of the velocity gradient:

$$\underset{\approx}{\boldsymbol{D}} = \tfrac{1}{2}(\underset{\approx}{\boldsymbol{L}} + \underset{\approx}{\boldsymbol{L}}^T), \quad \text{where } L_{ij} = \frac{\partial v_i}{\partial x_j}. \tag{22.56}$$

The stretching tensor does have some 'rate-like' qualities. For example, if the principal directions of stretch never change, it equals the rate of the logarithmic strain. However, this tensor is not a true rate for general deformations, so we caution against using it (see chapter 16) because it can lead to *inaccurate* predictions. PMFI does not address issues of accuracy—PMFI merely requires consistency of model predictions associated with material rotation. Because $\underset{\approx}{\boldsymbol{D}}$ *is* frequently used as if it were a strain rate (even though doing so can corrupt accuracy), our task now is to assess the PMFI restrictions that must be placed on rate models that replace an actual strain rate with $\underset{\approx}{\boldsymbol{D}}$.

For elasticity problems, it is common (though sometimes inadvisable) to replace the strain rate by the symmetric part of the deformation gradient so that, *in the absence of rotation*, isotropic constitutive equations are of the form

$$\dot{\underset{\approx}{\sigma}} = g(\underset{\approx}{\boldsymbol{D}}). \tag{22.57}$$

We will show that (22.57) violates PMFI even if the tangent stiffness tensor is isotropic. We will show that the violation of PMFI can be 'repaired' by modifying (22.57) as follows:

$$\overset{\circ}{\underset{\approx}{\sigma}} = g(\underset{\approx}{\boldsymbol{D}}), \tag{22.58}$$

where $\overset{\circ}{\underset{\approx}{\sigma}}$ denotes a special 'co-rotational' rate that *effectively eliminates the part of the stress rate caused by rotation* as discussed below

A fundamental theorem from continuum mechanics states that the velocity gradient tensor $\underset{\approx}{\boldsymbol{L}}$ is related to the deformation gradient tensor $\underset{\approx}{\boldsymbol{F}}$ by

$$\underset{\approx}{\boldsymbol{L}} = \dot{\underset{\approx}{\boldsymbol{F}}} \bullet \underset{\approx}{\boldsymbol{F}}^{-1}. \tag{22.59}$$

Under a superimposed rotation $\underset{\approx}{Q}$, recall that

$$\underset{\approx}{F}^* = \underset{\approx}{Q} \bullet \underset{\approx}{F}. \tag{22.60}$$

The starred velocity gradient is

$$\underset{\approx}{L}^* = \dot{\underset{\approx}{F}}^* \bullet \underset{\approx}{F}^{*-1} = (\dot{\underset{\approx}{Q}} \bullet \underset{\approx}{F} + \underset{\approx}{Q} \bullet \dot{\underset{\approx}{F}}) \bullet (\underset{\approx}{F}^{-1} \bullet \underset{\approx}{Q}^T). \tag{22.61}$$

For convenience, we introduce a superimposed angular rate of rotation tensor $\underset{\approx}{\Upsilon}$ associated with the $\dot{\underset{\approx}{Q}}$ rate of superimposed rotation. Specifically we define

$$\underset{\approx}{\Upsilon} = \dot{\underset{\approx}{Q}} \bullet \underset{\approx}{Q}^T. \tag{22.62}$$

The superimposed angular velocity $\underset{\approx}{\Upsilon}$ acts in addition to any angular velocity that might have been already present in the fiducial deformation. Because $\underset{\approx}{Q}$ is a rotation, the tensor $\underset{\approx}{\Upsilon}$ is skew-symmetric:

$$\underset{\approx}{\Upsilon}^T = -\underset{\approx}{\Upsilon}. \tag{22.63}$$

Using this tensor, the starred velocity gradient of (22.61) simplifies to

$$\underset{\approx}{L}^* = \underset{\approx}{\Upsilon} + \underset{\approx}{Q} \bullet \underset{\approx}{L} \bullet \underset{\approx}{Q}^T. \tag{22.64}$$

The presence of $\underset{\approx}{\Upsilon}$ means that the velocity gradient $\underset{\approx}{L}$ is not a spatial tensor. However, since $\underset{\approx}{\Upsilon}$ is skew-symmetric, the stretching tensor $\underset{\approx}{D} = \text{sym}\underset{\approx}{L}$ is a spatial tensor:

$$\underset{\approx}{D}^* = \underset{\approx}{Q} \bullet \underset{\approx}{D} \bullet \underset{\approx}{Q}^T. \tag{22.65}$$

Recall that the Cauchy stress is a spatial tensor:

$$\underset{\approx}{\sigma}^* = \underset{\approx}{Q} \bullet \underset{\approx}{\sigma} \bullet \underset{\approx}{Q}^T. \tag{22.66}$$

We are now going to prove that this implies that the *rate* of the Cauchy stress is *not* a spatial tensor. The time rate of the starred stress is found by differentiating (22.66) to give

$$\dot{\underset{\approx}{\sigma}}^* = \underset{\approx}{Q} \bullet \dot{\underset{\approx}{\sigma}} \bullet \underset{\approx}{Q}^T + \dot{\underset{\approx}{Q}} \bullet \underset{\approx}{\sigma} \bullet \underset{\approx}{Q}^T + \underset{\approx}{Q} \bullet \underset{\approx}{\sigma} \bullet \dot{\underset{\approx}{Q}}^T. \tag{22.67}$$

With the substitution $\dot{\underset{\approx}{Q}} = \underset{\approx}{\Upsilon} \bullet \underset{\approx}{Q}$, this becomes

$$\dot{\underset{\approx}{\sigma}}^* = \underset{\approx}{Q} \bullet \dot{\underset{\approx}{\sigma}} \bullet \underset{\approx}{Q}^T + \underset{\approx}{\Upsilon} \bullet \underset{\approx}{Q} \bullet \underset{\approx}{\sigma} \bullet \underset{\approx}{Q}^T + \underset{\approx}{Q} \bullet \underset{\approx}{\sigma} \bullet \underset{\approx}{Q}^T \bullet \underset{\approx}{\Upsilon}^T. \tag{22.68}$$

Using (22.67) and recalling that the angular rotation tensor is skew-symmetric, the starred stress *rate* is related to the fiducial stress rate in the following complicated way:

$$\dot{\underset{\approx}{\sigma}}^* = \underset{\approx}{Q} \bullet \dot{\underset{\approx}{\sigma}} \bullet \underset{\approx}{Q}^T + \underset{\approx}{\Upsilon} \bullet \underset{\approx}{\sigma}^* - \underset{\approx}{\sigma}^* \bullet \underset{\approx}{\Upsilon}. \tag{22.69}$$

This equation is the tensor analog of (22.9), which (recall) applied to a *vector* stress. Note that

$$\dot{\underset{\approx}{\sigma}}^* \neq \underset{\approx}{Q} \bullet \dot{\underset{\approx}{\sigma}} \bullet \underset{\approx}{Q}^T. \tag{22.70}$$

Therefore the Cauchy stress *rate* $\dot{\underset{\approx}{\sigma}}$ is *not* spatial even though the Cauchy stress $\underset{\approx}{\sigma}$ is spatial. Consequently, the proposed constitutive equation in (22.57) violates PMFI because it writes a non-spatial tensor $\dot{\underset{\approx}{\sigma}}$ in terms of a spatial tensor $\underset{\approx}{D}$. If we had kept the original strain rate $\dot{\underset{\approx}{\varepsilon}}$ (rather than replacing it with $\underset{\approx}{D}$), then the strain rate would have had non-objective terms involving $\underset{\approx}{\Upsilon}$ similar to those we found for the stress rate. Consequently, using $\dot{\underset{\approx}{\varepsilon}}$ (rather than $\underset{\approx}{D}$) in the rate form of the constitutive relation (22.55), would have cancelled these unruly terms without incident. By using $\underset{\approx}{D}$, no such cancellation occurs and we must alter the constitutive equation appropriately to make these terms effectively disappear (as needed for frame indifference).

22.4 Co-rotational rates (convected, Jaumann, polar, etc)

To rectify the problem with (22.57), it is conventional to propose an alternative constitutive equation of the form

$$\overset{\circ}{\underset{\approx}{\sigma}} = g(\underset{\approx}{D}), \tag{22.71}$$

where $\overset{\circ}{\underset{\approx}{\sigma}}$ denotes a so-called 'co-rotational' rate defined by

$$\overset{\circ}{\underset{\approx}{\sigma}} = \dot{\underset{\approx}{\sigma}} - \underset{\approx}{\Lambda} \bullet \underset{\approx}{\sigma} - \underset{\approx}{\sigma} \bullet \underset{\approx}{\Lambda}^T. \tag{22.72}$$

In this equation, the tensor $\underset{\approx}{\Lambda}$ may be selected to be any convenient physical quantity so long as it transforms under superimposed rotation so that

$$\underset{\approx}{\Lambda}^* = \underset{\approx}{\Upsilon} + \underset{\approx}{Q} \bullet \underset{\approx}{\Lambda} \bullet \underset{\approx}{Q}^T. \tag{22.73}$$

Referring to (22.64), one possible choice for $\underset{\approx}{\Lambda}$ is

$$\underset{\approx}{\Lambda} = \underset{\approx}{L} \leftarrow \text{ this corresponds to the 'convected' rate}. \tag{22.74}$$

This is not the only choice. We could alternatively take the co-rotational tensor to be the vorticity tensor:

$$\underset{\approx}{\Lambda} = \underset{\approx}{W} = \tfrac{1}{2}(\underset{\approx}{L} - \underset{\approx}{L}^T) \leftarrow \text{ this corresponds to the 'Jaumann' rate}. \tag{22.75}$$

We could select the co-rotational tensor to equal the polar spin

$$\underset{\approx}{\Lambda} = \underset{\approx}{\Omega} = \dot{\underset{\approx}{R}} \bullet \underset{\approx}{R}^T \leftarrow \text{ this corresponds to the 'polar' rate}. \tag{22.76}$$

Regardless of the choice for the co-rotational tensor, so long as (22.72) holds, you can demonstrate that

$$\overset{\circ}{\underset{\approx}{\sigma}}{}^* = \underset{\approx}{Q} \bullet \overset{\circ}{\underset{\approx}{\sigma}} \bullet \underset{\approx}{Q}^T. \tag{22.77}$$

In other words, the co-rotational rate is a spatial tensor! Consequently, equation (22.71) satisfies PMFI so long as the constitutive function g in (22.71) is isotropic. Anisotropy can be accommodated by introducing additional 'material orientation' independent variables, as discussed in section 22.2.2.

The constitutive model is implemented by first applying (22.71) to obtain the 'hollow dot' corotational stress rate. In order to update the stress state, the actual stress rate (not the corotational one) can be found at the beginning of the time step by solving for the first term on the right-hand side of (22.72), Multiplying by the timestep gives an estimate of the stress increment needed to update the stress to the end of the step.

Keep in mind that satisfying PMFI has nothing to do with ensuring the model will give accurate or even reasonable results. All of the above choices for the co-rotational rate (i.e., convected, Jaumann, polar, or something else) lead to a constitutive model that satisfies PMFI. The choice for the co-rotational rate *does*, however, influence the model predictions. Therefore, model *accuracy* and model *simplicity* concerns (not PMFI) can guide your choice for a 'good' co-rotational rate.

The choice for the co-rotation tensor is not unique. As long as your choice satisfies (22.73), then your constitutive model will be frame-indifferent. However, *other* physical arguments *besides* invariance may lead you to prefer one co-rotation tensor over another. For example, Dienes [3] demonstrated quite elegantly that the Jaumann rate predicts physically anomalous oscillatory stress when used with *constant linear* isotropic elastic stiffness in simple shear (see figure 22.2 and section 22.6.1), so he recommended the polar rate. Brannon [1] explained that, in principle, the oscillations in a Jaumann model can be removed by using nonlinear elasticity and therefore the use of linear elasticity can be as much to blame for the physically unacceptable predictions as the type of rate used. Implicitly, material modelers judge the goodness of a co-rotational rate by how well it performs for linear elasticity applied at large strain, and by this metric, the Jaumann rate is unacceptable. Even though the polar rate does perform better for simple shear of an isotropic linear-elastic material, Brannon [1] demonstrated that the polar rate is incapable of adequately describing the distortion of directions of material anisotropy for fiber-

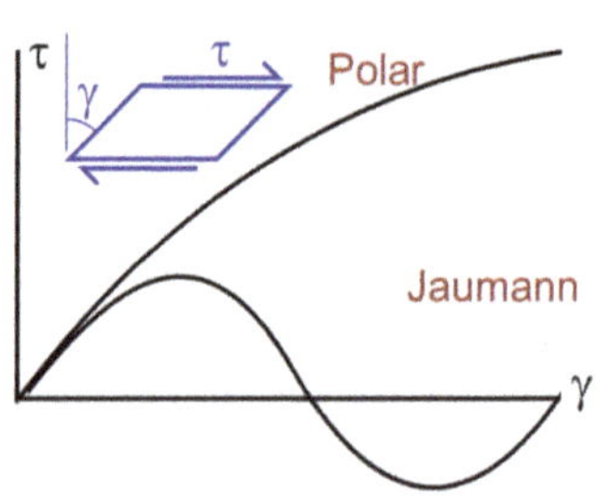

Figure 22.2. A qualitative sketch of Dienes's comparison of Jaumann and polar rates under pure shear using linear elasticity. This oscillatory response shear stress τ versus shear strain γ using the Jaumann rate can be removed by using nonlinear elasticity, proving that a bad corotational rate cannot be blamed per se for non-physical results.

dominated transversely isotropic materials, so she cautiously favored the convected rate (but cast as a Lie derivative—see section 22.5). Brannon also showed that it is not the choice of objective rate *per se* that produces oscillatory stresses. The issue is really rooted in the use of a linear constitutive model. Finally, Brannon recommends abandoning the use of objective rates altogether; instead, recalling (22.52), you can bypass the co-rotation issue by performing all of your constitutive calculations in an un-rotated configuration, then using strong objectivity [4] to handle large material rotations directly in the host code (or in an interface between the host code and the reference-configuration material model). Even in the reference configuration, you must still deal with inherent nonlinearities associated with large stretches and large rotations of a material relative to itself, but at least strong objectivity ensures satisfaction of frame indifference constraints associated with superimposed rotations.

22.4.1 Other objective rates

Dienes observed that the computationally inexpensive Jaumann rate (for which $\underset{\approx}{\Lambda} = \underset{\approx}{W} = \text{skw}\underset{\approx}{L}$) produced oscillatory stress in shear of a linear hypoelastic model, whereas the moderately more expensive polar rate (for which $\underset{\approx}{\Lambda} = \underset{\approx}{\Omega} = \dot{\underset{\approx}{R}} \bullet \underset{\approx}{R}^T$) did not exhibit such undesired behavior. In using the polar rate, the added computational expense arises from the need to compute $\underset{\approx}{\Omega}$, which is constructed from its axial vector in (15.62).The polar spin is related to vorticity by

$$\underset{\approx}{\Omega} = \underset{\approx}{W} + \underset{\approx}{H}, \tag{22.78}$$

where (15.54) defines $\underset{\approx}{H} = \underset{\approx\approx}{T} : \underset{\approx}{D}$ according to

$$\underset{\approx\approx}{T} = \sum_I^3 \sum_J^3 \beta_{IJ}(\underset{\sim}{p}_I \underset{\sim}{p}_J \underset{\sim}{p}_I \underset{\sim}{p}_J) \quad \text{and} \quad \boxed{\underset{\approx}{H} = \sum_I^3 \sum_J^3 \beta_{IJ} D_{IJ}^{\langle P \rangle}\, \underset{\sim}{p}_I \underset{\sim}{p}_J}, \tag{22.79}$$

in which the $\underset{\sim}{p}$ vectors are principal directions of the spatial stretch (with associated eigenvalues λ_K) and the β_{IJ} are skew-symmetric components of a 3×3 matrix (not tensor) defined by

$$\beta_{IJ} = \frac{\lambda_J - \lambda_I}{\lambda_J + \lambda_I} \text{ (with all indices free).} \tag{22.80}$$

After Dienes' influential paper, a virtual cottage industry arose to 'generalize' or broaden these concepts (usually without any clear articulation of the actual goal of the work, which evidently was to make *linear* hypoelasticity as accurate as possible by changing strain definitions, stress definitions, and/or by changing the objective rate definition). Lin [5] succinctly summarizes the various choices for objective rates to be of the same structure as defined above except using different choices for the β_{IJ} coefficients as follows:

$$\text{Polar:} \quad \beta_{IJ} = \frac{\lambda_J - \lambda_I}{\lambda_J + \lambda_I}. \tag{22.81a}$$

$$\text{Eulerian:} \qquad \beta_{IJ} = \frac{\lambda_J^2 + \lambda_I^2}{\lambda_J^2 - \lambda_I^2}. \tag{22.81b}$$

$$\text{Logarithmic:} \qquad \beta_{IJ} = \frac{\lambda_J^2 + \lambda_I^2}{\lambda_J^2 - \lambda_I^2} + \frac{1}{\ln(\lambda_I/\lambda_J)}. \tag{22.81c}$$

Lin also provides a formula for a so-called 'Lagrangian' objective rate. The walk-away lesson is that probably an infinite number of choices exist to satisfy PMFI by replacing the Cauchy stress rate with a co-rotational rate, and the 'favored' choice is the one that gives the most accurate results when the corotational stress rate is linear with respect to the symmetric part of the velocity gradient. Additionally, we reiterate that there is very little value in using co-rotational rates directly as described here. Many years of experience in deploying large-deformation constitutive models in production-quality finite-element codes tell us that a far better approach uses the concept of reference configurations, as now explained.

22.5. Lie derivatives and reference configurations

This section describes a broad class of corotational rates called 'Lie derivatives', which effectively introduce a reference configuration in which the constitutive model can be applied using ordinary (not corotational) rates, after which the results are then 'pushed forward' back to the spatial configuration. Let $\underset{\approx}{\boldsymbol{G}}$ denote a 'push' tensor, which means it transforms under a rigid superimposed rotation $\underset{\approx}{\boldsymbol{Q}}$ according to

$$\underset{\approx}{\boldsymbol{G}}^* = \underset{\approx}{\boldsymbol{Q}} \bullet \underset{\approx}{\boldsymbol{G}}. \tag{22.82}$$

Three particular choices for the $\underset{\approx}{\boldsymbol{G}}$ tensor are quite common:

Co-convected: $\underset{\approx}{\boldsymbol{G}} = \underset{\approx}{\boldsymbol{F}}$, where $\underset{\approx}{\boldsymbol{F}}$ is the deformation gradient tensor.
Contra-convected: $\underset{\approx}{\boldsymbol{G}} = J\underset{\approx}{\boldsymbol{F}}^{-T}$, where $J = \det \underset{\approx}{\boldsymbol{F}}$.
Polar-convected: $\underset{\approx}{\boldsymbol{G}} = \underset{\approx}{\boldsymbol{R}}$, where $\underset{\approx}{\boldsymbol{R}}$ is the polar rotation tensor.

All of these choices transform according to (22.82), as required. Consider any spatial tensor $\underset{\approx}{\boldsymbol{A}}$. Define a generalized 'overbar' pull operation as

$$\overline{\underset{\approx}{\boldsymbol{A}}} = \underset{\approx}{\boldsymbol{G}}^{-1} \bullet \underset{\approx}{\boldsymbol{A}} \bullet \underset{\approx}{\boldsymbol{G}}^{-T}. \tag{22.83}$$

To understand the reason for the transpose on $\underset{\approx}{\boldsymbol{G}}^{-T}$, note that the tensor $\underset{\approx}{\boldsymbol{A}}$ can be written in basis notation as

$$\underset{\approx}{\boldsymbol{A}} = A^{ij} \underset{\sim}{\boldsymbol{g}}_i \underset{\sim}{\boldsymbol{g}}_j, \tag{22.84}$$

where two vectors written side-by-side are multiplied dyadically, and the basis vectors need not be orthonormal. Operating from the left by the 'pull' operator $\underset{\approx}{\boldsymbol{G}}^{-1}$ and from the right by $\underset{\approx}{\boldsymbol{G}}^{-T}$ gives

$$\bar{\underset{\approx}{A}} = A^{ij}\left(\underset{\approx}{G}^{-1} \bullet \underset{\sim}{g}_i\right)\left(\underset{\sim}{g}_j \bullet \underset{\approx}{G}^{-T}\right) = A^{ij}\left(\underset{\approx}{G}^{-1} \bullet \underset{\sim}{g}_i\right)\left(\underset{\approx}{G}^{-1} \bullet \underset{\sim}{g}_j\right). \tag{22.85}$$

In writing the last form, we have noted that $\underset{\sim}{g}_j \bullet \underset{\approx}{G}^{-T}$ is the same thing as $\underset{\approx}{G}^{-1} \bullet \underset{\sim}{g}_j$. By introducing a set of 'helper' vectors defined

$$\bar{\underset{\sim}{g}}_i \equiv \underset{\approx}{G}^{-1} \bullet \underset{\sim}{g}_i. \tag{22.86}$$

Equation (22.85) becomes

$$\bar{\underset{\approx}{A}} = A^{ij}\bar{\underset{\sim}{g}}_i\bar{\underset{\sim}{g}}_j. \tag{22.87}$$

This result shows that the components of $\bar{\underset{\approx}{A}}$ with respect to the pulled-back $\bar{\underset{\sim}{g}}_k$-basis are identical to the components of $\underset{\approx}{A}$ with respect to the spatial $\underset{\sim}{g}_k$-basis. In a very loose sense, $\bar{\underset{\approx}{A}}$ is an 'un-distorted' or 'pulled-back' version of $\underset{\approx}{A}$. For example, when $\underset{\approx}{G} = \underset{\approx}{F}$, the un-distortion operation $\underset{\approx}{G}^{-1}$ takes away material distortion (both rigid rotation and relative rotation and/or stretching of a material lattice). When the $\underset{\approx}{G}$ tensor is a proper rotation, then both the $\bar{\underset{\sim}{g}}_k$ and $\underset{\sim}{g}_k$ bases may be selected to be orthonormal. In this case, $\bar{\underset{\approx}{A}}$ is an *un-rotated* version of $\underset{\approx}{A}$, and the operation $\underset{\approx}{G}^{-1}$ takes away material rotation but does not remove relative rotations associated with distortion of the material lattice. Components of the unrotated tensor with respect to the laboratory basis are identical to components of the spatial tensor with respect to a basis that rotates with the material; phrasing a constitutive model in the unrotated frame is therefore equivalent to using a material frame, but more convenient for using existing linear-algebra packages that implicitly assume a non-evolving basis.

Because $\underset{\approx}{A}$ is presumed to be a spatial tensor, it transforms under superimposed rotation as to

$$\underset{\approx}{A}^* = \underset{\approx}{Q} \bullet \underset{\approx}{A} \bullet \underset{\approx}{Q}^T. \tag{22.88}$$

In light of the assumption of (22.82), the 'barred' tensor $\bar{\underset{\approx}{A}}$ is easily verified to be a 'reference' tensor, meaning that its value is unaffected by a superimposed rigid rotation:

$$\bar{\underset{\approx}{A}}^* = \bar{\underset{\approx}{A}}. \tag{22.89}$$

To describe the rate of the distortion, we will define

$$\underset{\approx}{\Lambda} \equiv \dot{\underset{\approx}{G}} \bullet \underset{\approx}{G}^{-1}. \tag{22.90}$$

By using our assumption of (22.82), it is straightforward to prove that

$$\underset{\approx}{\Lambda}^* = \underset{\approx}{\Upsilon} + \underset{\approx}{Q} \bullet \underset{\approx}{\Lambda} \bullet \underset{\approx}{Q}^T. \tag{22.91}$$

We wish to take rates of $\overline{\underset{\approx}{A}}$. To do this, we need the following helper identity for the rate of an inverse[2]:

$$\frac{d}{dt}\underset{\approx}{G}^{-1} = -\underset{\approx}{G}^{-1} \bullet \dot{\underset{\approx}{G}} \bullet \underset{\approx}{G}^{-1}. \tag{22.92}$$

Using this, you can show that

$$\boxed{\dot{\overline{\underset{\approx}{A}}} = \underset{\approx}{G}^{-1} \bullet \overset{\circ}{\underset{\approx}{A}} \bullet \underset{\approx}{G}^{-T},} \tag{22.93a}$$

$$\text{where } \boxed{\overset{\circ}{\underset{\approx}{A}} = \dot{\underset{\approx}{A}} - \underset{\approx}{\Lambda} \bullet \underset{\approx}{A} - \underset{\approx}{A} \bullet \underset{\approx}{\Lambda}^{T}.} \tag{22.93b}$$

To interpret this equation physically, first let us invert (22.86) to obtain the spatial $\underset{\sim}{g}_i$-basis in terms of the reference $\overline{\underset{\sim}{g}}_i$-basis:

$$\underset{\sim}{g}_i \equiv \underset{\approx}{G} \bullet \overline{\underset{\sim}{g}}_i. \tag{22.94}$$

If the reference basis $\overline{\underset{\sim}{g}}_i$ is constant, then taking rates gives

$$\dot{\underset{\sim}{g}}_i \equiv \dot{\underset{\approx}{G}} \bullet \overline{\underset{\sim}{g}}_i = \underset{\approx}{\Lambda} \bullet \underset{\approx}{G} \bullet \overline{\underset{\sim}{g}}_i = \underset{\approx}{\Lambda} \bullet \underset{\sim}{g}_i. \tag{22.95}$$

Thus, the convected derivative $\overset{\circ}{\underset{\approx}{A}}$ in (22.93*b*) equals the ordinary derivative $\dot{\underset{\approx}{A}}$ minus those parts of the ordinary derivative that arise from rates of the $\underset{\sim}{g}_k$-basis vectors. Recalling that we are regarding the $\underset{\approx}{G}$ as a generic 'push' distortion (or rotation if it is orthogonal), we interpret the convected derivative to be the part of the rate that is *not* caused by this distortion (or rotation). For example, a polar rate is the rate of a tensor from the perspective of an observer who rotates with the material. The co-convected rate is the rate of a tensor from the perspective of an observer who both rotates and deforms with the material fibers. The contra-convected rate corresponds to an observer who rotates and deforms with the material *planes*. From a modeling perspective, these special co-rotational rates are the ones associated with material constitutive response to deformation, stripping away the kinematical (non-constitutive) reorientation. If, for example, you have a material model that is satisfactory for a pure horizontal stretch without material rotation, you can then rotate the material while stretching it and use that same constitutive model in a frame that rotates with the material (so that an observer in that frame perceives only what still seems to be a horizontal stretch); the effect of material rotation can then be incorporated as a purely kinematical contribution in a manner that is universal to any and all possible constitutive models and hence can be developed and implemented as part of a unified computational framework rather than risking simulation errors by having each constitutive model handle its own kinematics.

---

[2] Note the similarity of this identity with the scalar equation, $\frac{d}{dt}\left(\frac{1}{x}\right) = -\dot{x}/x^2 = -(x^{-1})\dot{x}(x^{-1})$.

Importantly, note that (22.93*a*) may be written

$$\dot{\bar{\underset{\approx}{A}}} = \overline{\overset{\circ}{\underset{\approx}{A}}}\,. \tag{22.96}$$

This equation says that the *ordinary* rate of the barred tensor is the same as the bar operation acting on the co-rotational rate. This is an extremely useful observation when using co-rotational rates in constitutive models. Whenever there exists a tensor $\underset{\approx}{G}$ such that the co-rotation tensor $\underset{\approx}{\Lambda}$ is expressible in the form of (22.90), then the co-rotational rate is called a *Lie*[3] *derivative* and you will usually see it written in books in the form

$$\overset{\circ}{\underset{\approx}{A}} = \underset{\approx}{G} \bullet \frac{\mathrm{d}}{\mathrm{d}t}(\underset{\approx}{G}^{-1} \bullet \underset{\approx}{A} \bullet \underset{\approx}{G}^{-T}) \bullet \underset{\approx}{G}^{T}. \tag{22.97}$$

A far more rigorous discussion of Lie derivatives may be found in [6].

22.5.1 Alternative to Lie derivatives in constitutive models

For a mathematically limber reader, the preceding discussion of Lie derivatives probably did shed some light on the physical meaning of a convected derivative. However, the use of such strange time derivatives is foreign to many readers. Furthermore, convected derivatives in numerical computations can be quite cumbersome and sometimes inefficient. Consider, for example, the linear version of (22.71):

$$\overset{\circ}{\underset{\approx}{\sigma}} = \underset{\approx\approx}{E} : \underset{\approx}{D}, \left(\text{in indicial form,} \quad \overset{\circ}{\sigma}_{ij} = E_{ijkl} D_{kl}\right), \tag{22.98}$$

where $\underset{\approx\approx}{E}$ is the fourth-order tangent stiffness tensor. For this equation to be generalized to anisotropic materials, the dependence on $\underset{\approx}{D}$ in (22.71) must be appended to also include dependence on material orientation. In other words, upon a superimposed rigid rotation, not only must $\underset{\approx}{D}$ change, the material stiffness tensor must also change. Specifically, to satisfy PMFI, this spatial model must satisfy

$$\overset{\circ}{\underset{\approx}{\sigma}}{}^{*} = \underset{\approx\approx}{E}{}^{*} : \underset{\approx}{D}{}^{*}. \tag{22.99}$$

Recalling (22.77) and (22.65), both $\overset{\circ}{\underset{\approx}{\sigma}}$ and $\underset{\approx}{D}$ are spatial tensors. Thus, to satisfy PMFI, the stiffness tensor must transform under superimposed rotation such that

$$E^{*}_{ijkl} = Q_{ip} Q_{jq} Q_{kr} Q_{ls} E_{pqrs}. \tag{22.100}$$

In other words, the stiffness components must be referenced to a basis that rotates with the material. Computing the rotated stiffnesses can be quite expensive. An alternative and equivalent technique is to leave the stiffnesses unchanged and instead un-rotate the stress and strain rate. Then, once the constitutive model is applied in this un-rotated frame, the final result may be rotated back to the spatial frame. This

[3] Pronunciation rhymes with 'flea', not 'fly'.

qualitative proposal is derived in mathematical detail below in terms of our generalized distortion/rotation tensor $\underset{\approx}{G}$ and its associated 'bar' operation defined in (22.83).

Barring both sides of (22.71) and applying (22.96) gives

$$\dot{\overline{\underset{\approx}{\sigma}}} = \overline{g(\underset{\approx}{D})}, \quad \text{or in the linear case,} \quad \dot{\overline{\underset{\approx}{\sigma}}} = \overline{\underset{\approx}{E}: \underset{\approx}{D}}. \tag{22.101}$$

We can always define a new constitutive function $\bar{g}$ that depends implicitly on $\underset{\approx}{G}$ such that

$$\overline{g(\underset{\approx}{D})} = \bar{g}(\overline{\underset{\approx}{D}}). \tag{22.102}$$

Hence, we may write the constitutive equation as

$$\dot{\overline{\underset{\approx}{\sigma}}} = \bar{g}(\overline{\underset{\approx}{D}}). \tag{22.103}$$

In the linear case, applying the definition of the 'bar' operation to (22.101) shows that

$$\dot{\overline{\underset{\approx}{\sigma}}} = \overline{\underset{\approx}{E}}: \overline{\underset{\approx}{D}}, \quad \text{where} \quad \overline{E}_{ijkl} = G_{ip}^{-1}G_{iq}^{-1}G_{ir}^{-1}G_{is}^{-1}E_{pqrs}. \tag{22.104}$$

Here we have the *ordinary* rate of an 'un-distorted/un-rotated' stress being a function of the 'un-distorted/un-rotated' stretching. This formulation has several advantages:

- The stress rate is an ordinary rate, not a convected rate, and it is therefore easier to integrate over time.
- After a numerical constitutive model has been developed to give good answers for zero-rotation problems, it can be immediately generalized to permit rotations by simply having the host code perform the 'bar' pull-back operation on all of the arguments to the constitutive model before calling the model. Upon return, the host code then 'pushes forward' the results.
- When expressed in terms of the barred reference configuration, the constitutive equation (22.103) is not restricted to isotropy. For the linear case, the 'bar' stiffness is often regarded as an unchanging (permissibly anisotropic) fourth-order tensor, and the fourth-order push and pull transformations therefore become unnecessary when working exclusively in the 'barred' space.
- For most materials, $\overline{E}_{ijkl}$ will generally vary far less than E_{pqrs}. While the eigenvalues of $\overline{E}_{ijkl}$ are likely to change (nonlinear elasticity), the eigen*vectors*, which represent directions of material asymmetry are unlikely to change very much if the $\underset{\approx}{G}$ tensor is equated with the deformation gradient. By contrast, both the eigenvalues and eigenvectors of E_{pqrs} are likely to change in a highly nonlinear (difficult to model) way.

The two most common choices for the 'bar' push tensor $\underset{\approx}{G}$ are:

- $\underset{\approx}{G} = \underset{\approx}{R}$ from the polar decomposition, in which case, using the Lie derivative in the 'un-rotated' configuration is roughly equivalent to using polar rates in a spatial formulation. By being solved in an unrotated state, the constitutive model may presume without loss that the deformation is always a pure stretch. A disadvantage of this approach is that material directions can still becomed skewed in a pure stretch, implying that the elastic stiffness cannot be held constant in conditions of large distortion.

- $\underset{\approx}{G} = \underset{\approx}{F}$ from the deformation gradient, in which case, using the Lie derivative in the 'un-convected' configuration is equivalent to using convected coordinates. An advantage is that material directions for the stiffness tensor can often be presumed fixed, but the formulation might require special attention to material nonlinearity to avoid instabilities in compression [1].

When you consider thermodynamics, the Cauchy stress is typically abandoned in favor of the Kirchhoff stress $\underset{\approx}{\tau}$, defined to equal the Cauchy stress times the Jacobian of the deformation.

$$\underset{\approx}{\tau} = J\underset{\approx}{\sigma}. \tag{22.105}$$

Unlike the Cauchy stress, using Kirchhoff stress makes the tangent stiffness major symmetric for hyperelasticity [1], so the thermodynamically preferable form for the constitutive law is

$$\overset{+}{\underset{\approx}{\tau}} = \bar{h}\left(\underset{\approx}{\bar{D}}\right). \tag{22.106}$$

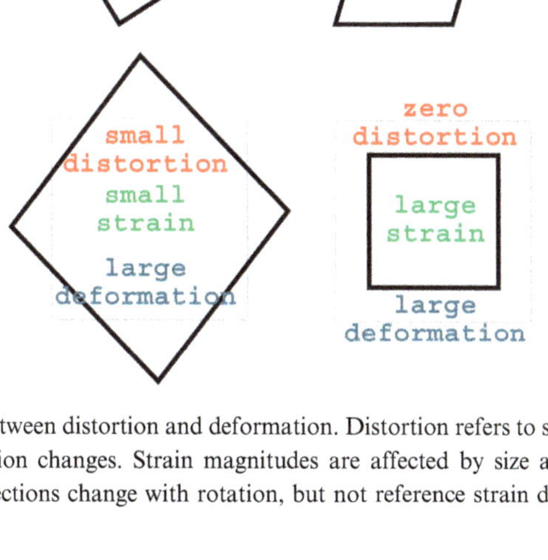

Figure 22.3. Distinction between distortion and deformation. Distortion refers to shape changes. Deformation refers to size and orientation changes. Strain magnitudes are affected by size and shape changes, but not rotation (spatial strain directions change with rotation, but not reference strain directions).

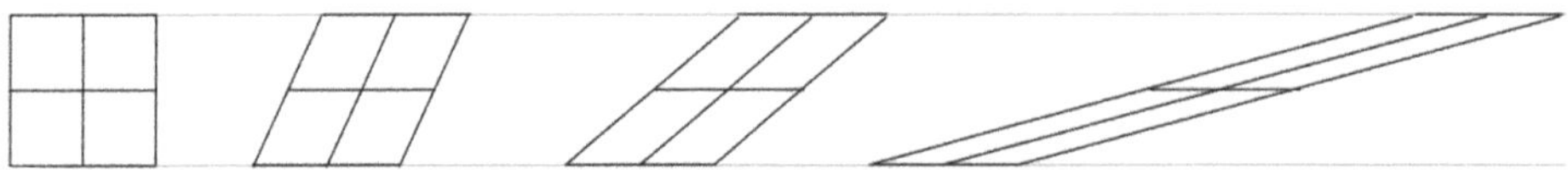

Figure 22.4. Material rotation under simple shear. Originally horizontal fibers do not rotate at all. Originally vertical fibers rotate clockwise, but no more than 90° in the limit as deformation is continued indefinitely.

If we choose $\underset{\sim}{G} = \underset{\sim}{F}$, then $\underset{\sim}{\overline{\tau}}$ equals the PK2 stress and $\underset{\sim}{\overline{D}}$ equals the ordinary rate of Lagrange strain. If we choose $\underset{\sim}{G} = \underset{\sim}{R}$, then $\underset{\sim}{\overline{\tau}}$ is the un-rotated Kirchhoff stress and $\underset{\sim}{\overline{D}}$ is the un-rotated symmetric part of the velocity gradient.

22.6 Frame indifference is only an essential (not final) step

Frame indifference is only a *necessary* self-consistency property that must hold for any general constitutive model. Satisfying PMFI is not *sufficient* to ensure that the constitutive model will perform well under large material distortions. Different constitutive laws can predict widely varying—and sometimes absurd—results even if they satisfy PMFI. The principle of material frame indifference merely requires that results should be as good or lousy under frame change as they are without it.

Recall from (22.72) that a co-rotational stress rate is generally expressible in the following form:

$$\overset{\circ}{\underset{\sim}{\sigma}} = \dot{\underset{\sim}{\sigma}} - \underset{\sim}{\Lambda} \bullet \underset{\sim}{\sigma} - \underset{\sim}{\sigma} \bullet \underset{\sim}{\Lambda}^T. \tag{22.107}$$

For the polar rate, the tensor $\underset{\sim}{\Lambda}$ is equal to the skew-symmetric polar spin $\underset{\sim}{\Omega}$. For the Jaumann rate, $\underset{\sim}{\Lambda}$ is the vorticity $\underset{\sim}{W}$ (i.e., skew-symmetric part of the velocity gradient), but be advised that the Jaumann rate is not a Lie derivative and therefore cannot be integrated numerically using a pulled-back reference formulation. It is straightforward (but not trivial) to demonstrate that both rates are equivalent whenever material distortion is small. By this, we mean that the two rates are nearly equivalent whenever the material has not significantly changed *shape* even if large *size* and/or *orientation* changes have occurred.

Mathematically, a motion involves small distortions if the principal stretches are all nearly equal to each other. For problems with small material distortion, all co-rotational rates predict approximately the same answer, so it makes sense to use the one that is computationally least expensive. For problems with large material distortion, however, special care must be taken to handle the material nonlinearity even if PMFI is satisfied.

22.6.1 Oscillatory stresses under simple shear

To demonstrate that satisfaction of PMFI is not enough to obtain sensible constitutive model predictions, Dienes [3] considered the prediction under simple shear when using an isotropic *linear-elastic* constitutive law

$$\overset{\circ}{\underset{\sim}{\sigma}} = \underset{\approx}{E} : \underset{\sim}{D}, \quad \text{where } \underset{\approx}{E} \text{ is isotropic and constant.} \tag{22.108}$$

Regardless of which objective rate is used, the above equation satisfies PMFI. However, Dienes demonstrated that the *Jaumann* rate predicts anomalous oscillatory shear stresses whereas the polar rate predicts intuitively more appealing monotonically increasing shear stress (see figure 22.2). The Jaumann rate performs so poorly because, for simple shear, the vorticity $\underset{\approx}{\mathbf{W}}$ is constant throughout time, so using the Jaumann objective rate is effectively assuming that a material element tumbles *unceasingly*. Simply drawing a single material element as it deforms demonstrates that such thinking is flawed.

Fibers in figure 22.4 that were originally perpendicular to the shear plane rotate no more than 90° into the shear plane. Fibers in the shear plane never rotate *at all*. If the vorticity $\underset{\approx}{\mathbf{W}}$ were (wrongly) regarded as a measure of overall material rotation rate, then (because the vorticity is constant), it would wrongly imply that material fibers rotate perpetually without ever slowing down as shear progresses. The polar spin $\underset{\approx}{\boldsymbol{\Omega}}$, on the other hand, properly captures material rotation—the polar spin equals the vorticity at first, but it goes to zero as time increases, allowing fiber rotation angles to approach but never exceed 90°. As a rule, all objective rates are equivalent whenever material distortion is small. For small distortions, the material does not significantly change shape, but may permissibly change size and orientation.

The natural formulation for a constitutive model is the one that makes the equations as simple in structure as possible while agreeing (to within tolerable accuracy) with measured data and satisfying applicable physics, kinematics, and thermodynamics constraints[4] both in the measured data domain and in extrapolated domains where the model might be applied despite absence of data[5]. Dienes definitely demonstrated that the Jaumann rate was 'bad' in the context of simple shear with isotropic linear elasticity. However, the fact that the polar rate was 'less bad' does not mean that the polar rate is necessarily *correct*. Brannon [1] demonstrated that even the polar rate gives incorrect predictions when applied to distortion of a fiber-dominant composite material. The heart of the issue is not which objective rate you use. The key to good constitutive modeling is to handle material nonlinearity properly. For fiber-dominant anisotropic materials, Brannon [1] showed that the most simply structured constitutive model capable of predicting nonlinearity arising from distortions of the elastic lattice is one that uses a convected Lie derivative, which guarantees co-rotation of material fibers, but at the cost of needing nonlinear tangent moduli to 'subtract off' unwanted co-elongations of material directions. For laminates, a contra-convected Lie derivative is probably preferable. A composite of fibers and laminates, on the other hand, does not suggest

---

[4] Regardless of which objective stress rate is used, the constitutive model in (22.108) has a serious thermodynamics flaw because the tensor $\underset{\approx}{\mathbf{D}}$ is not a true rate. That equation cannot possibly characterize truly elastic material response, but it is nevertheless acceptably accurate for many engineering applications.

[5] It might seem like a sin to even consider applying a model outside of its calibrated domain. However, important problems (e.g., planetary impact, damage of human tissue, etc) are often known to occur in domains well beyond our moral or technical ability to measure in a lab setting. In such cases, theoretical physics constraints become essential to help formulate reasonable models.

either of these simple choices, but instead points to a 'big data' future for modern constitutive theory that is more 'aware' of underlying microstructure, perhaps using the deformation gradient to evolve fibers and its inverse transpose to evolve planar morphologies like cracks. In that case, a simple unrotated reference (treating all deformations as pure stretches) would suffice to satisfy PMFI.

22.6.2 Handling PMFI in a numerical code

When implementing large-distortion capability into a code, it is essential to permit the constitutive model to retain control of the nonlinear material response associated with large material shape and volume changes. PMFI is a wholly separate issue that demands model prediction consistency under large rotations. PMFI addresses only issues associated with changes in material orientation—PMFI does *not* have any influence on constitutive issues associated with size and/or shape changes of material elements. All material models must satisfy PMFI even if those models are valid only for small strains. Finite-element (and similar) host codes can easily ensure that each material model satisfies PMFI by always sending each model its needed field inputs in the un-rotated reference configuration (e.g., if the model requests the deformation gradient tensor, then the host code instead sends the reference stretch tensor and if the model requests the rotation, the host code sends the identity tensor). With this approach, the host code is responsible for transforming output from the model back to a compatible configuration before proceeding with solution of governing partial differential equations. When correctly implemented, this approach does indeed ensure satisfaction of PMFI and it certainly gives code architects a higher degree of quality control. Furthermore, this strategy permits straightforward implementation of strong objectivity to systems (such as turbine blades) that are subjected to small strains but massive rotations [4]. An alternative (relatively error prone) approach places all responsibilities for PMFI into the realm of the constitutive model. A danger here is that some models might not properly implement frame indifference. Since the host code is not itself ensuring PMFI, the code administrators would be well-advised to implement an automated PMFI checking system that would run *all* model benchmark problems in fiducial (as-submitted) form and under super-imposed rotation. Another disadvantage of putting PMFI in the hands of model developers is that code size might bloat because of redundant un-rotation routines and/or inefficient co-rotational rate calculations in separate models.

Regardless of whether PMFI is handled within the model or host code, large material distortion and dilatation must be addressed within the constitutive model, because this is essentially a nonlinear material response issue, unrelated to PMFI. Special strain measures are often used to simplify the functional form of nonlinear material response. For example, logarithmic (Hencky) strain is often used in conjunction with linear elasticity, which predicts qualitatively appealing *nonlinear* stress–strain response with respect to conventional strain measures under proportional loading. Matching data for non-proportional loading, however, can be extremely difficult with log strain (especially for anisotropic materials).

Alternative reference strain measures such as the Green–Lagrange strain are easier to compute and, unlike log strain, preserve anisotropic material symmetries in their constitutive formulations, but they typically require perversely nonlinear elasticity to match data. Choices for reference strain measures imply associated spatial strains and co-rotational rates when the modeler prefers working in the spatial configuration. Given that different strain definitions and co-rotational rates are typically selected to simplify the structure of constitutive models for different materials, treatment of large material distortion and dilation is properly the responsibility of the material modeler. Rather than concerning themselves with details of constitutive models, host code designers should focus primarily on establishing a useful framework for satisfying and testing PMFI and for preserving integrity of internal variable fields (e.g., using Lagrangian particle tracking to avoid 'mixing' constitutive fields as byproduct of remeshing or advection) [7].

References

[1] Brannon R M 1998 Caveats concerning conjugate stress and strain measures for large distortion of anisotropic materials *Acta Mech.* **129** 107–16

[2] Kamojjala K C, Brannon R M, Sadeghirad A and Guilkey J 2013 Verification tests in solid mechanics, *Eng. Comput.* at press.

[3] Dienes J K 1979 On the analysis of rotation and stress rate in deforming bodies *Acta Mech.* **32** 217–32

[4] Rashid M 1993 Incremental kinematics for finite-element applications *Int. J. Numer. Methods Eng.* **36** 3937–56

[5] Lin R 2003 Hypoelasticity-based analytical stress solutions in the simple shearing process *J. Appl. Math. Mech.* **83** 163–71

[6] Abraham R, Marsden J E and Ratiu T 1988 *Manifolds, Tensor Analysis, and Applications* 2nd edn (Applied Mathematical Sciences vol 75) (New York: Springer)

[7] Sanchez J J, Lucini C B and Strack O E 2016 Lagrangian material tracers (LMT) for simulating material damage in ALEGRA *Sandia National Laboratories Report* 2nd edn #SAND2016-7260

IOP Publishing

Rotation, Reflection, and Frame Changes
Orthogonal tensors in computational engineering mechanics
R M Brannon

Chapter 23

Rigid-body mechanics

The mechanics of rigid bodies can be found in any good dynamics textbook (e.g., [1, 2]). However, few textbooks present the theory in a form that is optimized for readers who are already familiar with advanced tensor analysis. Consequently, such books often provide derivations in awkward component forms. Here we present the theory for readers who already know continuum mechanics and are therefore prepared to appreciate compact tensor formulations.

A body is considered 'rigid' if the distance between any two points in the body remains unchanged, which also implies that angles remain unchanged. By considering an infinitesimal embedded material fiber $d\underset{\sim}{X}$ that transforms under a deformation to $d\underset{\sim}{x} = \underset{\approx}{F} \bullet d\underset{\sim}{X}$, the condition $\|d\underset{\sim}{x}\|=\|d\underset{\sim}{X}\|$ for all material fibers immediately implies that $\underset{\approx}{F}^T \bullet \underset{\approx}{F} = \underset{\approx}{I}$. That is, the deformation gradient tensor must be orthogonal and is usually denoted $\underset{\approx}{R}$ as a reminder that (by also prohibiting material inversion) the motion is a rotation. Also demanding $\underset{\approx}{R}$ be curl-free to satisfy compatibility, it follows that $\underset{\approx}{R}$ must be spatially uniform, implying that the motion is a rotation, possibly with translation.

Referring to figure 23.1, let O_o denote a fixed origin. Let O denote a translating origin that differs from the fixed origin by a vector $\underset{\sim}{k}$. Let $\underset{\sim}{X}$ and $\underset{\sim}{x}$ be, respectively, the *initial* and *current* positions of a particle with respect to O_o, while $\underset{\sim}{r}$ is the current position with respect to the moving O.

By their definitions, both $\underset{\sim}{X}$ and $\underset{\sim}{x}$ are origin-dependent vectors—they will change if a different fixed origin is used. The displacement vector $\underset{\sim}{u}$ points from the initial location of a particle to the current location of the same particle, making it unaffected by the choice of fixed origin. Any vector with this property is called a *free vector*. Displacement, however, is affected by a change in origin that evolves over time, so it is a non-objective vector. Specifically, the vector extending from the initial and final locations of a point will appear different to a moving observer in comparison to a stationary one. According to frame-indifference principles of Newtonian relativity, non-kinematics vectors (such as force, electric field, etc) must be unaffected by a moving origin.

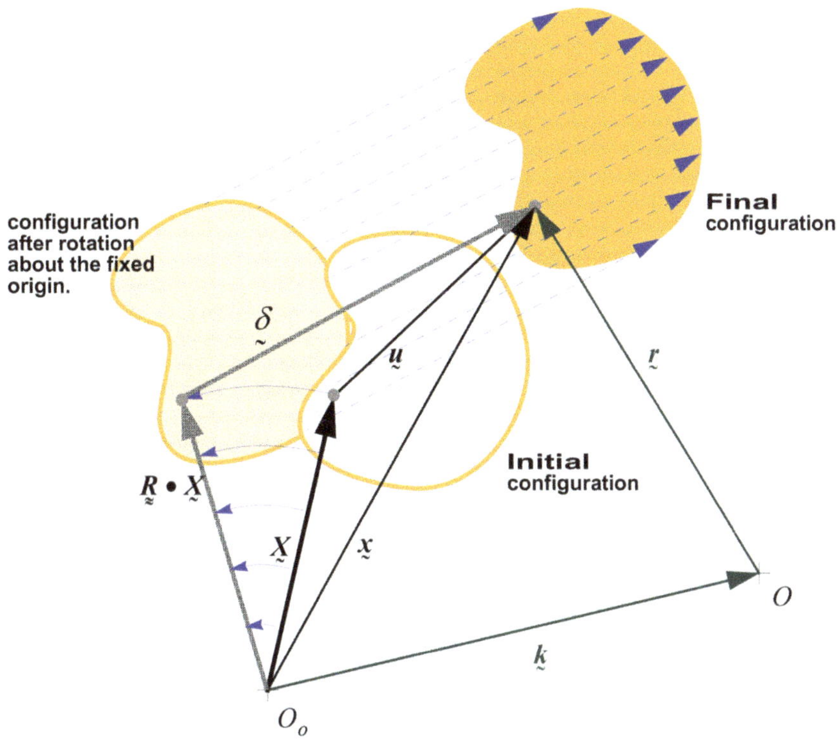

Figure 23.1. Identifying points within a rigid body. The position vector X identifies the *initial* location of the grey particle. The position vectors x and r both point to the *current* location of the same particle, but from different origins.

Figure 23.2 shows how a rigid motion may be regarded as a rotation R about a fixed origin followed by a rigid translation δ. Like position vectors, the translation vector is origin dependent, not free. By contrast, R is a free tensor, making it insensitive to the location of the origin (though, because it is a special case of two-point polar rotation from continuum mechanics, it is sensitive to evolving changes in frame orientation).

23.1. Rate of rotation

The time derivative of the orthogonality condition ($R^T \bullet R = I$) gives $\dot{R}^T \bullet R + R^T \bullet \dot{R} = 0$. Pre-multiplying by R and post-multiplying by R^T gives

$$\Omega^T + \Omega = 0 \text{ where } \boxed{\Omega \equiv \dot{R} \bullet R^T} \text{ ('spin tensor')}. \tag{23.1}$$

The first equation proves that the so-called spin tensor Ω is skew-symmetric. Its axial vector is the angular velocity vector ω. Therefore,

$$\boxed{\Omega \bullet u = \omega \times u} \quad \text{for any vector } u, \quad \text{and hence } [\Omega] = \begin{bmatrix} 0 & -\omega_3 & \omega_2 \\ \omega_3 & 0 & -\omega_1 \\ -\omega_2 & \omega_1 & 0 \end{bmatrix}. \tag{23.2}$$

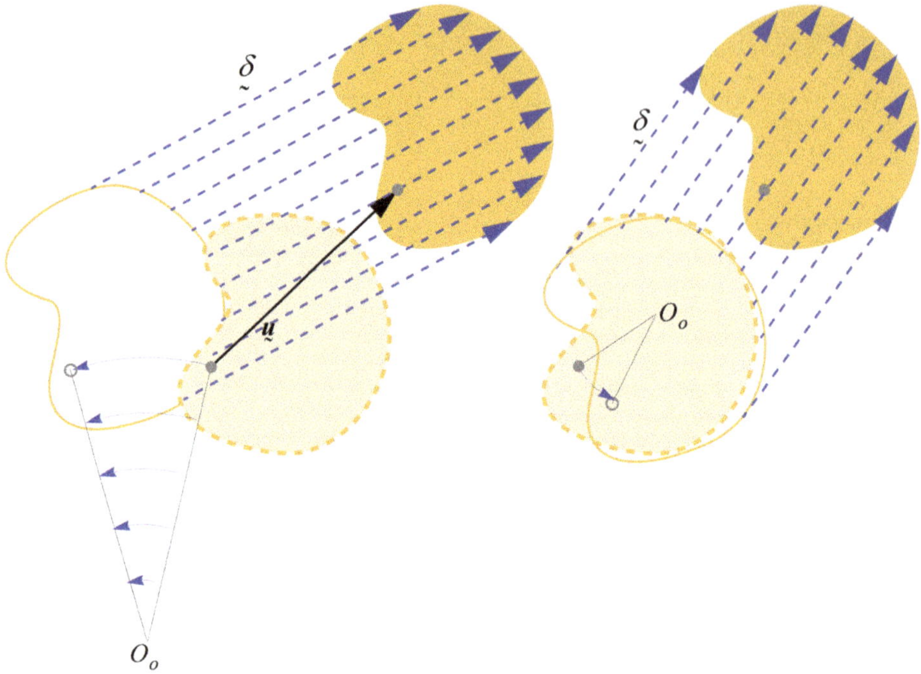

Figure 23.2. Effect of choosing a different fixed origin. These images show the same rigid motion from the dash-boundary body to the dark shaded body. The first stage rotates about the origin to the white-filled intermediate configuration (depicted with sweeping arrows on the left image and without them on the right). The second stage then translates.

This relationship motivates an alternative notation for the axial tensor: $\underset{\approx}{\Omega} = \underset{\sim}{\omega} \times \underset{\approx}{I}$; dotting both sides from the right by $\underset{\sim}{u}$ gives the above equation!

23.2. The slope-intercept description of rigid motion

Under the assumption of rigid motion, the particle's current position $\underset{\sim}{x}$ must be expressible in the form of a rotation plus a translation (analogous to the slope-intercept form of a straight line in which rotation is the slope and displacement is the intercept):

$$\underset{\sim}{x} = \underset{\approx}{R} \bullet \underset{\sim}{X} + \underset{\sim}{\delta}, \tag{23.3}$$

where $\underset{\approx}{R}$ is a rotation tensor and $\underset{\sim}{\delta}$ is a translation vector, both of which depend on time but not position[1]. In shorthand, we indicate this dependence as

$$\underset{\approx}{R} = \underset{\approx}{R}(t), \tag{23.4}$$

$$\underset{\sim}{\delta} = \underset{\sim}{\delta}(t). \tag{23.5}$$

As mentioned earlier, one disadvantage of this description of rotation is that it contains the origin-dependent displacement vector $\underset{\sim}{\delta}$.

[1] We already used compatibility to assert spatial uniformity of rotation.

23.3. The point-slope description of rigid motion

For rigid-motion problems, there is often a special point in the body for which the complete motion is known (or there is a point for which the complete motion can be relatively easily computed). This special point might be the location of a hinge on a rigid robotic arm, the center of mass (CM) of a boomerang, etc. We soon show that the position of the CM is governed by the linear momentum equation, which depends only on the net force on the rigid body. Consequently, *the position of the body's CM is relatively easy to track over time*, making it a good reference point.

A real physical body occupies only a finite amount of space. The *virtual body* is an imagined extension of the rigid body to include all points in space. Suppose, for example, that the real rigid body is a spherical shell. The centroid of this body lies at the sphere center, which is not part of the physical body, but it is part of the virtual body. Points in the infinite virtual body likewise move according to (23.3). Namely,

$$\underset{\sim}{x} = \underset{\approx}{R} \bullet \underset{\sim}{X} + \underset{\sim}{\delta}. \tag{23.6}$$

Let $\underset{\sim}{X}^*$ denote some *particular* convenient point in the virtual body. Let $\underset{\sim}{x}^*$ denote the current position of that reference point. As mentioned, common choices for the reference particle are the body's CM, $\underset{\sim}{x}_c$, are the hinge point on a rigid robotic arm, or any other point that is relatively easy to track over time.

Equation (23.6) holds for *all* points on the rigid body, so it must also hold for the reference point:

$$\underset{\sim}{x}^* = \underset{\approx}{R} \bullet \underset{\sim}{X}^* + \underset{\sim}{\delta}. \tag{23.7}$$

Subtracting (23.7) from (23.6) gives

$$\underset{\sim}{x} - \underset{\sim}{x}^* = \underset{\approx}{R} \bullet (\underset{\sim}{X} - \underset{\sim}{X}^*). \tag{23.8}$$

In this manner, we have separated a rigid motion into two distinct components: the motion of a single reference particle $\underset{\sim}{X}^*$ and motion of all other particles *relative to the reference particle*. In all of the following discussions, we will presume that the reference particle is *Lagrangian* (i.e., *embedded*)—in other words, it always corresponds to the same physical point on the body so that the time rate of $\underset{\sim}{X}^*$ will be zero. As indicated in figure 23.3, some choices for the embedded reference point might travel along simpler paths than others!

23.4. Velocity and angular velocity for rigid motion

Taking the reference point to be the CM, rigid motion may be described by

$$\underset{\sim}{x} = \underset{\sim}{x}_c + \underset{\approx}{R} \bullet (\underset{\sim}{X} - \underset{\sim}{X}_c), \tag{23.9}$$

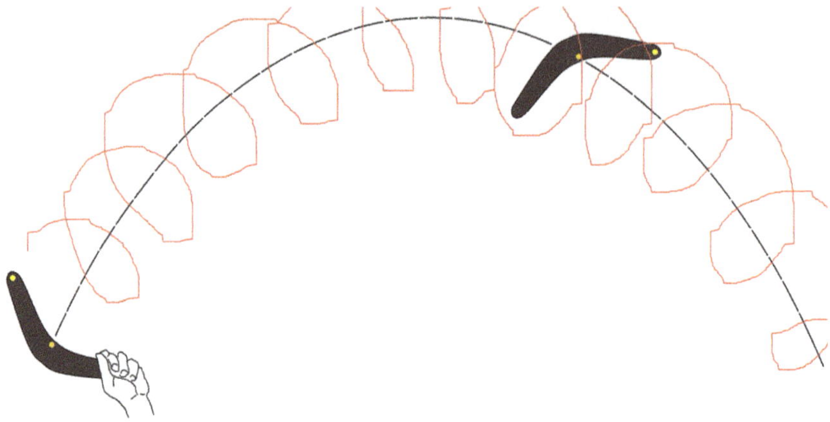

Figure 23.3. Free flight. When the reference point for rigid motion is taken as the CM, the path of the reference point is a simple parabola. When the reference point is taken farther out on the boomerang, its path is a spiral around the path of the CM.

in which the rotation tensor and reference vector are functions of time, not position:

$$\underset{\sim}{\boldsymbol{R}} = \underset{\sim}{\boldsymbol{R}}(t) \quad \text{and} \quad \underset{\sim}{\boldsymbol{x}}_c = \underset{\sim}{\boldsymbol{x}}_c(t). \tag{23.10}$$

In other words, the motion of a rigid body is completely described by specifying the position of the CM $\underset{\sim}{\boldsymbol{x}}_c$ as a function of time and the rotation tensor $\underset{\sim}{\boldsymbol{R}}$ as a function of time. In applications, the net force and torque acting on the body are known, and the resulting motion $\{\underset{\sim}{\boldsymbol{x}}_c(t), \underset{\sim}{\boldsymbol{R}}(t)\}$ is desired (or vice versa).

Material velocity is defined by

$$\dot{\underset{\sim}{\boldsymbol{x}}} \equiv \left(\frac{\partial \underset{\sim}{\boldsymbol{x}}}{\partial t}\right)_{\underset{\sim}{\boldsymbol{X}}} \tag{23.11}$$

or, applying this definition to (23.9),

$$\dot{\underset{\sim}{\boldsymbol{x}}} = \dot{\underset{\sim}{\boldsymbol{x}}}_c + \dot{\underset{\sim}{\boldsymbol{R}}} \bullet (\underset{\sim}{\boldsymbol{X}} - \underset{\sim}{\boldsymbol{X}}_c). \tag{23.12}$$

Noting from (23.9) that $\underset{\sim}{\boldsymbol{X}} - \underset{\sim}{\boldsymbol{X}}_c = \underset{\sim}{\boldsymbol{R}}^T \bullet (\underset{\sim}{\boldsymbol{x}} - \underset{\sim}{\boldsymbol{x}}_c)$, this may be written as

$$\boxed{\dot{\underset{\sim}{\boldsymbol{x}}} = \dot{\underset{\sim}{\boldsymbol{x}}}_c + \underset{\sim}{\Omega} \bullet (\underset{\sim}{\boldsymbol{x}} - \underset{\sim}{\boldsymbol{x}}_c)}, \tag{23.13}$$

where the definition of the angular spin tensor, $\underset{\sim}{\Omega} \equiv \dot{\underset{\sim}{\boldsymbol{R}}} \bullet \underset{\sim}{\boldsymbol{R}}^T$, as well as its relationship to the angular velocity vector $\underset{\sim}{\boldsymbol{\omega}}$, are in equations (23.1) and (23.2). The fact that $\underset{\sim}{\Omega} \bullet \underset{\sim}{\boldsymbol{m}} = \underset{\sim}{\boldsymbol{\omega}} \times \underset{\sim}{\boldsymbol{m}}$ for any vector $\underset{\sim}{\boldsymbol{m}}$ implies that

$$\underset{\sim}{\Omega} = -\underset{\sim}{\boldsymbol{\varepsilon}} \bullet \underset{\sim}{\boldsymbol{\omega}} = \underset{\sim}{\boldsymbol{\omega}} \times \underset{\sim}{\boldsymbol{I}}, \qquad \text{or in indicial form,} \quad \Omega_{ij} = -\sum_{k=1}^{3} \varepsilon_{ijk}\omega_k. \tag{23.14}$$

The component matrix of $\underset{\sim}{\Omega}$ can be constructed from the angular velocity vector:

$$[\underset{\sim}{\Omega}] = \begin{bmatrix} 0 & -\omega_3 & \omega_2 \\ \omega_3 & 0 & -\omega_1 \\ -\omega_2 & \omega_1 & 0 \end{bmatrix}$$

$$\text{or, with respect to an aligned basis, } [\underset{\sim}{\Omega}] = \begin{bmatrix} 0 & -\omega & 0 \\ \omega & 0 & 0 \\ 0 & 0 & 0 \end{bmatrix}. \tag{23.15}$$

Here, the 'aligned basis' is defined to have the 3-direction aligned with the angular velocity vector, and this is useful when deriving identities. If the angular rotation tensor $\underset{\sim}{\Omega}$ is known, then the relationship in (23.14) may be inverted to give the angular velocity vector $\underset{\sim}{\omega}$ by

$$\underset{\sim}{\omega} = -\frac{1}{2}\underset{\approx}{\varepsilon} : \underset{\sim}{\Omega}, \tag{23.16}$$

or, written out explicitly (recalling that $\underset{\sim}{\Omega}$ is skew-symmetric),

$$\omega_1 = \Omega_{32}, \qquad \omega_2 = \Omega_{13}, \qquad \omega_3 = \Omega_{21}. \tag{23.17}$$

As mentioned earlier, axial tensors are defined so that they are related to their axial vectors by the important identity that $\underset{\sim}{\Omega} \bullet \underset{\sim}{m} = \underset{\sim}{\omega} \times \underset{\sim}{u}$ for any vector $\underset{\sim}{u}$. Applying this identity to (23.13) gives

$$\dot{\underset{\sim}{x}} = \dot{\underset{\sim}{x}}_c + \underset{\sim}{\omega} \times (\underset{\sim}{x} - \underset{\sim}{x}_c), \tag{23.18}$$

which is the form of the velocity cited in most elementary dynamics textbooks.

23.5. Time rate of a vector embedded in a rigid body

Define an embedded arrow vector $\underset{\sim}{z}$ to be one that points from one distinct point $\underset{\sim}{x}^{\text{tail}}$ to a second point $\underset{\sim}{x}^{\text{tip}}$, each of which is 'painted' in a rigid body:

$$\underset{\sim}{z} \equiv \underset{\sim}{x}^{\text{tip}} - \underset{\sim}{x}^{\text{tail}}. \tag{23.19}$$

Think of this as an arrow 'tattooed' on the body. Its rate is

$$\dot{\underset{\sim}{z}} \equiv \dot{\underset{\sim}{x}}^{\text{tip}} - \dot{\underset{\sim}{x}}^{\text{tail}}. \tag{23.20}$$

Applying (23.13) each end of the arrow has velocity given by

$$\dot{\underset{\sim}{x}}^{\text{tip}} = \dot{\underset{\sim}{x}}_c + \underset{\sim}{\Omega} \bullet (\underset{\sim}{x}^{\text{tip}} - \underset{\sim}{x}_c) \qquad \text{and} \qquad \dot{\underset{\sim}{x}}^{\text{tail}} = \dot{\underset{\sim}{x}}_c + \underset{\sim}{\Omega} \bullet (\underset{\sim}{x}^{\text{tail}} - \underset{\sim}{x}_c). \tag{23.21}$$

Hence, equation (23.20) becomes

$$\boxed{\dot{\underset{\sim}{z}} \equiv \underset{\sim}{\Omega} \bullet \underset{\sim}{z}}, \qquad \text{or equivalently } \boxed{\dot{\underset{\sim}{z}} = \underset{\sim}{\omega} \times \underset{\sim}{z}}. \tag{23.22}$$

23.6. Acceleration for rigid motion

The second rate of (23.12) gives a very simple expression for the acceleration:

$$\boxed{\ddot{\underset{\sim}{x}} = \ddot{\underset{\sim}{x}}_c + \ddot{\underset{\approx}{R}} \bullet (\underset{\sim}{X} - \underset{\sim}{X}_c)}, \tag{23.23}$$

or, again noting that $\underset{\sim}{X} - \underset{\sim}{X}_c = \underset{\approx}{R}^T \bullet (\underset{\sim}{x} - \underset{\sim}{x}_c)$, the fully spatial expression for the acceleration is

$$\ddot{\underset{\sim}{x}} = \ddot{\underset{\sim}{x}}_c + \ddot{\underset{\approx}{R}} \bullet \underset{\approx}{R}^T \bullet (\underset{\sim}{x} - \underset{\sim}{x}_c). \tag{23.24}$$

Recall that

$$\dot{\underset{\approx}{R}} = \underset{\approx}{\Omega} \bullet \underset{\approx}{R}. \tag{23.25}$$

Therefore

$$\ddot{\underset{\approx}{R}} = \dot{\underset{\approx}{\Omega}} \bullet \underset{\approx}{R} + \underset{\approx}{\Omega} \bullet \dot{\underset{\approx}{R}} = (\dot{\underset{\approx}{\Omega}} + \underset{\approx}{\Omega} \bullet \underset{\approx}{\Omega}) \bullet \underset{\approx}{R}. \tag{23.26}$$

Thus

$$\boxed{\ddot{\underset{\approx}{R}} \bullet \underset{\approx}{R}^T = \dot{\underset{\approx}{\Omega}} + \underset{\approx}{\Omega}^2} \tag{23.27}$$

and

$$\boxed{\ddot{\underset{\sim}{x}} = \ddot{\underset{\sim}{x}}_c + [\dot{\underset{\approx}{\Omega}} + \underset{\approx}{\Omega}^2] \bullet (\underset{\sim}{x} - \underset{\sim}{x}_c)}. \tag{23.28}$$

Applying (23.2) twice shows that

$$\boxed{\underset{\approx}{\Omega}^2 \bullet \underset{\sim}{u} = \underset{\sim}{\omega} \times (\underset{\sim}{\omega} \times \underset{\sim}{u})} \quad \text{for any vector } \underset{\sim}{u}. \tag{23.29}$$

Note from (23.15) that

$$[\underset{\approx}{\Omega}^2] = \begin{bmatrix} 0 & -\omega_3 & \omega_2 \\ \omega_3 & 0 & -\omega_1 \\ -\omega_2 & \omega_1 & 0 \end{bmatrix} \begin{bmatrix} 0 & -\omega_3 & \omega_2 \\ \omega_3 & 0 & -\omega_1 \\ -\omega_2 & \omega_1 & 0 \end{bmatrix}$$

$$= \begin{bmatrix} -\left(\omega_2^2 + \omega_3^2\right) & \omega_1\omega_2 & \omega_1\omega_3 \\ \omega_2\omega_1 & -\left(\omega_3^2 + \omega_1^2\right) & \omega_2\omega_3 \\ \omega_3\omega_1 & \omega_3\omega_2 & -\left(\omega_1^2 + \omega_2^2\right) \end{bmatrix} \tag{23.30}$$

or, rearranging this result,

$$[\underset{\approx}{\Omega}^2] = \begin{bmatrix} \omega_1^2 & \omega_1\omega_2 & \omega_1\omega_3 \\ \omega_2\omega_1 & \omega_2^2 & \omega_2\omega_3 \\ \omega_3\omega_1 & \omega_3\omega_2 & \omega_3^2 \end{bmatrix} - \left(\omega_1^2 + \omega_2^2 + \omega_3^2\right) \begin{bmatrix} 1 & 0 & 0 \\ 0 & 1 & 0 \\ 0 & 0 & 1 \end{bmatrix}. \tag{23.31}$$

In direct notation,

$$\boxed{\underset{\approx}{\Omega}^2 = \underset{\sim}{\omega}\underset{\sim}{\omega} - (\underset{\sim}{\omega} \bullet \underset{\sim}{\omega})\underset{\approx}{I}}. \tag{23.32}$$

Thus, a way of computing $\underset{\approx}{\Omega}^2 \bullet \underset{\sim}{u}$ that is an alternative to (23.29) is

$$\boxed{\underset{\approx}{\Omega}^2 \bullet \underset{\sim}{u} = \underset{\sim}{\omega}(\underset{\sim}{\omega} \bullet \underset{\sim}{u}) - (\underset{\sim}{\omega} \bullet \underset{\sim}{\omega})\underset{\sim}{u}} \text{ for any vector } \underset{\sim}{u}. \tag{23.33}$$

Recall from (23.14) that $\underset{\sim}{\omega}$ is the axial vector associated with the spin tensor $\underset{\approx}{\Omega}$. The alternating tensor $\underset{\approx}{\varepsilon}$ is independent of time. Therefore the time rate of (23.14) gives

$$\dot{\underset{\approx}{\Omega}} = -\underset{\approx}{\varepsilon} \bullet \dot{\underset{\sim}{\omega}}, \tag{23.34}$$

which shows that the *angular acceleration vector* $\dot{\underset{\sim}{\omega}}$ is the axial vector associated with the rate of the spin tensor $\dot{\underset{\approx}{\Omega}}$. Therefore the angular acceleration vector and tensor satisfies the basic identity:

$$\dot{\underset{\approx}{\Omega}} \bullet \underset{\sim}{u} = \dot{\underset{\sim}{\omega}} \times \underset{\sim}{u} \text{ for any vector } \underset{\sim}{u}. \tag{23.35}$$

We may apply (23.35) and (23.33) to (23.28) to write the acceleration strictly in terms of the angular velocity vector:

$$\boxed{\ddot{\underset{\sim}{x}} = \ddot{\underset{\sim}{x}}_c + \dot{\underset{\sim}{\omega}} \times (\underset{\sim}{x} - \underset{\sim}{x}_c) + \underset{\sim}{\omega}[\underset{\sim}{\omega} \bullet (\underset{\sim}{x} - \underset{\sim}{x}_c)] - (\underset{\sim}{\omega} \bullet \underset{\sim}{\omega})(\underset{\sim}{x} - \underset{\sim}{x}_c)}. \tag{23.36}$$

Alternatively, using (23.29) instead of (23.33),

$$\boxed{\ddot{\underset{\sim}{x}} = \ddot{\underset{\sim}{x}}_c + \dot{\underset{\sim}{\omega}} \times (\underset{\sim}{x} - \underset{\sim}{x}_c) + \underset{\sim}{\omega} \times [\underset{\sim}{\omega} \times (\underset{\sim}{x} - \underset{\sim}{x}_c)]}. \tag{23.37}$$

This is the form of the acceleration typically cited in dynamics textbooks. Clearly, however, anyone with even a modicum of skill with tensor analysis would prefer the elegant simplicity of either (23.23) or (23.28).

23.7. Important properties of a rigid body

One of the most profound discoveries in the study of rigid-body dynamics is that we do not need to know the exact mass distribution of the body to completely determine its motion. Rather, it is sufficient to know the total mass M, the location of the CM $\underset{\sim}{x}_c$ and the value of a special second-order 'inertia tensor' $\underset{\approx}{\mu}_c$ defined relative to the CM. These three quantities depend only on the body's geometrical shape and its spatial density distribution, but differently shaped bodies can have the same values of these properties. They may be computed a *priori* at time zero and then easily updated through time if the rotation tensor $\underset{\approx}{R}$ and the position of one point $\underset{\sim}{x}_p$ are known as functions of time.

Suppose that the size and shape of the rigid body B are known. Let ρ denote the mass density[2] of the rigid body. We can define moments of the rigid body as follows:

[2] Spatially varying, but constant (in a Lagrangian sense) with respect to time (e.g., no diffusion).

$$\textit{Zeroth moment (the mass):} \quad M \equiv \int_B \rho \, \mathrm{d}V = \int_{B_o} \rho_o \, \mathrm{d}V_o. \tag{23.38}$$

$$\textit{First moment (the CM):} \quad \underset{\sim}{x}_c \equiv \frac{1}{M} \int_B \underset{\sim}{x} \rho \, \mathrm{d}V. \tag{23.39}$$

$$\textit{Second moment about the CM:} \quad \underset{\approx}{y}_c \equiv \frac{1}{M} \int_B (\underset{\sim}{x} - \underset{\sim}{x}_c)(\underset{\sim}{x} - \underset{\sim}{x}_c) \rho \, \mathrm{d}V. \tag{23.40}$$

Note that the zeroth moment is just the mass of the body. The first moment is the center of mass. For now, the second moment may be regarded as an abstract mathematical entity having no intuitive meaning, but ultimately, you should come to see it as a tensor quantification of the eccentricity of the shape of the body (which is far superior to a ratio of the largest to smallest body dimension, as seen by considering a body that has 'whiskers' extending out large distances without having much mass). Above, the second moment is defined using the position vector relative to the CM. Generalized relative to an arbitrary point,

$$\textit{Second moment about } \underset{\sim}{x}_p\text{:} \quad \underset{\approx}{y}_p \equiv \frac{1}{M} \int_B (\underset{\sim}{x} - \underset{\sim}{x}_p)(\underset{\sim}{x} - \underset{\sim}{x}_p) \rho \, \mathrm{d}V. \tag{23.41}$$

We will later show that the second moment about an arbitrary point $\underset{\sim}{x}_p$ can be computed from the second moment about the CM by the very simple formula:

$$\boxed{\underset{\approx}{y}_p \equiv \underset{\approx}{y}_c + \underset{\sim}{d}_{pc} \underset{\sim}{d}_{pc}}, \quad \text{where } \underset{\sim}{d}_{pc} \equiv \underset{\sim}{x}_p - \underset{\sim}{x}_c. \tag{23.42}$$

Thus, if $\underset{\approx}{y}_c$ and $\underset{\sim}{x}_c$ are known, then the second moment about $\underset{\sim}{x}_p$ can be immediately computed (using a simple dyadic multiplication of the separation vector) without needing to perform any integrations. In other words, while knowledge of $\underset{\approx}{y}_p$ does not provide any additional information, it might simplify certain equations.

We will use a superscript '0' as an alternative notation for the value of a quantity at time zero. Thus, for example,

$$\underset{\sim}{x}^0 \equiv \underset{\sim}{x}|_{t=0} = \underset{\sim}{X}, \tag{23.43}$$

$$\underset{\sim}{x}_c^0 \equiv \underset{\sim}{x}_c|_{t=0} = \frac{1}{M} \int_B \underset{\sim}{x}^0 \rho \, \mathrm{d}V = \frac{1}{M} \int_B \underset{\sim}{X} \rho_o \, \mathrm{d}V_o = \underset{\sim}{X}_c, \tag{23.44}$$

$$\underset{\approx}{y}_c^0 \equiv \underset{\approx}{y}_c|_{t=0} = \frac{1}{M} \int_B (\underset{\sim}{X} - \underset{\sim}{X}_c)(\underset{\sim}{X} - \underset{\sim}{X}_c) \rho_0 \, \mathrm{d}V_0. \tag{23.45}$$

When analyzing the angular momentum equation, we will encounter the expression

$$M \underset{\approx}{\varepsilon} \mathbin{:} (\underset{\approx}{y}_p \bullet \underset{\approx}{\Omega}^T), \tag{23.46}$$

where (recall) M is the total mass of the body, $\underset{\approx}{\boldsymbol{\varepsilon}}$ is the alternating (permutation) tensor, $\underset{\approx}{\boldsymbol{y}}_p$ is the second moment of the body about a point $\underset{\sim}{\boldsymbol{x}}_p$, and $\underset{\approx}{\Omega}$ is the skew-symmetric angular velocity tensor. The expression in (23.46) is linear with respect to $\underset{\approx}{\Omega}$, and it must therefore be linear with respect to its axial vector $\underset{\sim}{\boldsymbol{\omega}}$. In other words, by the representation theorem of tensor analysis, there must exist a tensor $\underset{\approx}{\boldsymbol{\mu}}_p$ that is independent of $\underset{\sim}{\boldsymbol{\omega}}$ such that

$$M\underset{\approx}{\boldsymbol{\varepsilon}}:(\underset{\approx}{\boldsymbol{y}}_p \bullet \underset{\approx}{\Omega}^T) = \underset{\approx}{\boldsymbol{\mu}}_p \bullet \underset{\sim}{\boldsymbol{\omega}}. \tag{23.47}$$

Writing this expression in indicial form reveals that

$$\underset{\approx}{\boldsymbol{\mu}}_p \equiv M[(\mathrm{tr}\underset{\approx}{\boldsymbol{y}}_p)\underset{\approx}{\boldsymbol{I}} - \underset{\approx}{\boldsymbol{y}}_p]. \tag{23.48}$$

The tensor $\underset{\approx}{\boldsymbol{\mu}}_p$ is called the *rotational moment of inertia* for the body. Just as mass inertially resists linear motion, this tensor quantifies the inertial resistance to *rotational* motion. The rotational moment of inertia tensor is both symmetric and positive definite. Consequently, there exists a principal basis in which it is symmetric. If an eigenvalue is small relative to the other eigenvalues, then rotation about the corresponding eigenvector is relatively 'easy' (i.e., very little resistance to rotation about that axis).

23.7.1. Example: sphere

For a sphere of radius R, performing the integrals defining the second moment gives

$$\underset{\approx}{\boldsymbol{y}}_p^{\text{sphere}} = \tfrac{1}{5}R^2\underset{\approx}{\boldsymbol{I}}. \tag{23.49}$$

Applying (23.48), the moment of inertia tensor is therefore

$$\underset{\approx}{\boldsymbol{\mu}}_p^{\text{sphere}} = \tfrac{2}{5}MR^2\underset{\approx}{\boldsymbol{I}}. \tag{23.50}$$

This result can be found in any dynamics textbook.

23.7.2. Example: ellipsoid

Suppose an ellipsoid is defined by three mutually perpendicular vectors $\{\underset{\sim}{\boldsymbol{a}}, \underset{\sim}{\boldsymbol{b}}, \underset{\sim}{\boldsymbol{c}}\}$, which equal the principal axes of the ellipsoid in both orientation and magnitude. We could perform the integrals as was done for the sphere, but the task can be completed more elegantly by thinking of an ellipsoid as a sphere subjected to an affine deformation (and it should be clear that similar analysis applies to any body that could be regarded as a uniformly deformed version of a simpler body). Consider a tensor $\underset{\approx}{\boldsymbol{G}}$, whose component matrix relative to a known fixed laboratory basis is

$$[G] = \begin{bmatrix} a_1 & b_1 & c_1 \\ a_2 & b_2 & c_2 \\ a_3 & b_3 & c_3 \end{bmatrix}. \tag{23.51}$$

In mixed direct-basis notation (with E-vectors denoting the orthonormal laboratory basis), this tensor can be written as

$$\underset{\approx}{G} = \underset{\sim}{a}\underset{\sim}{E}_1 + \underset{\sim}{b}\underset{\sim}{E}_2 + \underset{\sim}{c}\underset{\sim}{E}_3. \tag{23.52}$$

When applied to unit vectors on a *unit sphere*, this tensor will map the sphere to the ellipsoid of interest. Specifically,

$$\underset{\sim}{x}^{\text{ellipsoid}} = \underset{\sim}{x}_p^{\text{ellipsoid}} + \underset{\approx}{G} \bullet (\underset{\sim}{X}^{\text{sphere}} - \underset{\sim}{X}_p^{\text{sphere}}). \tag{23.53}$$

Therefore, using (23.49) for a *unit* sphere gives

$$\underset{\approx}{y}_p^{\text{ellipsoid}} = \underset{\approx}{G} \bullet \underset{\approx}{y}_p^{\text{sphere}} \bullet \underset{\approx}{G}^T = \tfrac{1}{5}(\underset{\approx}{G} \bullet \underset{\approx}{G}^T) \tag{23.54}$$

and

$$\underset{\approx}{\mu}_p^{\text{ellipsoid}} = \tfrac{1}{5}M[(\underset{\approx}{G} : \underset{\approx}{G})\underset{\approx}{I} - \underset{\approx}{G} \bullet \underset{\approx}{G}^T]. \tag{23.55}$$

The tensor $\underset{\approx}{G} \bullet \underset{\approx}{G}^T$ is easily constructed from the sum of axis dyads:

$$\underset{\approx}{G} \bullet \underset{\approx}{G}^T = \underset{\sim}{a}\underset{\sim}{a} + \underset{\sim}{b}\underset{\sim}{b} + \underset{\sim}{c}\underset{\sim}{c}. \tag{23.56}$$

It follows that

$$\underset{\approx}{G} : \underset{\approx}{G} = \operatorname{tr}(\underset{\approx}{G} \bullet \underset{\approx}{G}^T) = a^2 + b^2 + c^2. \tag{23.57}$$

The planar moment of inertia about $\underset{\sim}{a}$ (quantifying how difficult it is to rotate the ellipsoid about that axis) is given by

$$\frac{\underset{\sim}{a} \bullet \underset{\approx}{\mu}_p^{\text{ellipsoid}} \bullet \underset{\sim}{a}}{a^2} = \tfrac{1}{5}M[b^2 + c^2]. \tag{23.58}$$

23.7.3. Switching between the second moment and inertia tensor

The inertia tensor carries the same information as the second moment tensor. As a matter of fact, if $\underset{\approx}{\mu}_p$ is known, then the second moment may be immediately constructed by

$$\underset{\approx}{y}_p = \frac{1}{M}[\tfrac{1}{2}(\operatorname{tr}\underset{\approx}{\mu}_p)\underset{\approx}{I} - \underset{\approx}{\mu}_p]. \tag{23.59}$$

23.7.4. Center-shift for the inertia tensor

Using (23.42), note that $\underset{\approx}{\mu}_p$ can be computed from $\underset{\approx}{\mu}_c$ by

$$\boxed{\underset{\approx}{\mu}_p \equiv \underset{\approx}{\mu}_c + M[(\underset{\sim}{d}_{pc} \bullet \underset{\sim}{d}_{pc})\underset{\approx}{I} - \underset{\sim}{d}_{pc}\underset{\sim}{d}_{pc}]}, \quad \text{where } \underset{\sim}{d}_{pc} \equiv \underset{\sim}{x}_p - \underset{\sim}{x}_c. \tag{23.60}$$

This result extends the parallel axis theorem cited in elementary dynamics textbooks. It is a generalization because we make no requirement of the existence of any parallel axes. Instead, the rotational moment of inertia with respect to an arbitrary point $\underset{\sim}{x}_p$ can be computed if the rotational moment of inertia about the CM is known. The traditional (specialized) *parallel axis theorem* is presented in (23.89).

23.7.5. Moment of inertia about a given axis

Given a plane of unit normal $\underset{\sim}{n}$, the planar moment of inertia is defined by

$$\phi_{cn} \equiv \underset{\sim}{n} \bullet \underset{\approx}{\mu}_c \bullet \underset{\sim}{n} \tag{23.61}$$

or, in terms of the second moment,

$$\phi_{cn} \equiv M[(\mathrm{tr}\underset{\approx}{y}_c) - \underset{\sim}{n} \bullet \underset{\approx}{y}_c \bullet \underset{\sim}{n}]. \tag{23.62}$$

The planar moment of inertia ϕ_{cn} is a measure of how 'difficult' it is to rotate the body about an axis parallel to $\underset{\sim}{n}$ that passes through the CM.

23.7.6. Explicit matrices for the second moment and inertia tensor

To gain an intuitive feeling for the general structure of the various second moments, suppose that we set up a coordinate system with the origin at $\underset{\sim}{x}_p$. Then we can write

$$\underset{\sim}{x} - \underset{\sim}{x}_p = x\underset{\sim}{e}_x + y\underset{\sim}{e}_y + z\underset{\sim}{e}_z. \tag{23.63}$$

The dyad $(\underset{\sim}{x} - \underset{\sim}{x}_p)(\underset{\sim}{x} - \underset{\sim}{x}_p)$ would then have a Cartesian component matrix given by

$$[(\underset{\sim}{x} - \underset{\sim}{x}_p)(\underset{\sim}{x} - \underset{\sim}{x}_p)] = \begin{bmatrix} x^2 & xy & xz \\ yx & y^2 & yz \\ zx & zy & z^2 \end{bmatrix}. \tag{23.64}$$

Thus,

$$[\underset{\approx}{y}_p] \equiv \frac{1}{M}\int_B \begin{bmatrix} x^2 & xy & xz \\ yx & y^2 & yz \\ zx & zy & z^2 \end{bmatrix} \rho \, \mathrm{d}V. \tag{23.65}$$

The moment of inertia tensor is then

$$[\underset{\approx}{\mu}_p] = \int_B \left((x^2 + y^2 + z^2)\begin{bmatrix} 1 & 0 & 0 \\ 0 & 1 & 0 \\ 0 & 0 & 1 \end{bmatrix} - \begin{bmatrix} x^2 & xy & xz \\ yx & y^2 & yz \\ zx & zy & z^2 \end{bmatrix} \right) \rho \, \mathrm{d}V, \tag{23.66}$$

or

$$[\underset{\approx}{\mu}_p] = \int_B \begin{bmatrix} y^2 + z^2 & -xy & -xz \\ -yx & z^2 + x^2 & -yz \\ -zx & -zy & x^2 + y^2 \end{bmatrix} \rho \, \mathrm{d}V. \tag{23.67}$$

The planar moment of inertia about the z-direction would then be

$$\phi_{cz} = \underset{\sim}{e}_Z \bullet \underset{\approx}{\mu}_c \bullet \underset{\sim}{e}_Z = \int_B (x^2 + y^2)\rho \, \mathrm{d}V. \tag{23.68}$$

Thus, ϕ_{cz} is the integral over the body of the distance from the z-axis, when the coordinate system is set up so that the z-axis passes through the CM. The off-diagonal components of the inertia tensor $\underset{\approx}{\mu}_c$ are called the 'products' of inertia. Some books omit the negative sign in their definitions, which is fine only if the negative is re-inserted whenever a change in coordinates is performed. Without the negative, the matrix will not transform like a tensor (so this is a good example to show students the difference between a matrix and a tensor).

We will soon demonstrate that the second moment plays a role in the conservation of angular momentum that is similar to the role played by the mass in the linear momentum equation. Mass is a measure of the body's 'resistance' to changes in linear motion while the inertia tensor measures the body's resistance to changes in angular momentum.

23.7.7 Relationship between the current and initial moments

Conservation of mass guarantees that the zeroth moments are equal. That is, the mass element after rotation equals the mass element before rotation. Of course, since the volume element itself does not change under rigid motion, this shows that mass density of a Lagrangian particle is likewise constant.

$$\rho \, \mathrm{d}V = \rho_o \, \mathrm{d}V_o. \tag{23.69}$$

Recall the definition of the CM:

$$\underset{\sim}{x}_c \equiv \frac{1}{M}\int_B \underset{\sim}{x}\rho \, \mathrm{d}V. \tag{23.70}$$

The second moment about the origin is defined

$$\underset{\approx}{y} \equiv \frac{1}{M}\int_B \underset{\sim}{x}\underset{\sim}{x}\rho \, \mathrm{d}V. \tag{23.71}$$

A rigid motion may be described in terms of the motion of the CM by

$$\underset{\sim}{x} = \underset{\sim}{x}_c + \underset{\approx}{R} \bullet (\underset{\sim}{X} - \underset{\sim}{X}_c). \tag{23.72}$$

Therefore

$$\begin{aligned} \underset{\sim}{x}\underset{\sim}{x} = {} & \underset{\sim}{x}_c\underset{\sim}{x}_c + \underset{\approx}{R} \bullet (\underset{\sim}{X}\underset{\sim}{x}_c - \underset{\sim}{X}_c\underset{\sim}{x}_c) + (\underset{\sim}{x}_c\underset{\sim}{X} - \underset{\sim}{x}_c\underset{\sim}{X}_c) \bullet \underset{\approx}{R}^T \\ & + \underset{\approx}{R} \bullet (\underset{\sim}{X}\underset{\sim}{X} - \underset{\sim}{X}_c\underset{\sim}{X} - \underset{\sim}{X}\underset{\sim}{X}_c + \underset{\sim}{X}_c\underset{\sim}{X}_c) \bullet \underset{\approx}{R}^T. \end{aligned} \tag{23.73}$$

Substituting this into (23.71) and pulling out arguments that do not vary with position gives

$$\underset{\approx}{y} \equiv \underset{\sim}{x}_c \underset{\sim}{x}_c + \underset{\approx}{R} \bullet (\underset{\sim}{X}_c \underset{\sim}{x}_c - \underset{\sim}{X}_c \underset{\sim}{x}_c) + (\underset{\sim}{x}_c \underset{\sim}{X}_c - \underset{\sim}{x}_c \underset{\sim}{X}_c) \bullet \underset{\approx}{R}^T$$
$$+ \underset{\approx}{R} \bullet (\underset{\approx}{y}^0 - \underset{\sim}{X}_c \underset{\sim}{X}_c - \underset{\sim}{X}_c \underset{\sim}{X}_c + \underset{\sim}{X}_c \underset{\sim}{X}_c) \bullet \underset{\approx}{R}^T. \tag{23.74}$$

Simplifying shows that

$$\underset{\approx}{y} \equiv \underset{\sim}{x}_c \underset{\sim}{x}_c + \underset{\approx}{R} \bullet \left(\underset{\approx}{y}^0 - \underset{\sim}{x}^0_c \underset{\sim}{x}^0_c\right) \bullet \underset{\approx}{R}^T. \tag{23.75}$$

From the center-shift theorem (derived below), we know that

$$\underset{\approx}{y}^0_c \equiv \underset{\approx}{y}^0 - \underset{\sim}{x}^0_c \underset{\sim}{x}^0_c. \tag{23.76}$$

Therefore

$$\boxed{\underset{\approx}{y}_c = \underset{\approx}{R} \bullet \underset{\approx}{y}^0_c \bullet \underset{\approx}{R}^T}. \tag{23.77}$$

Now recall that

$$\underset{\approx}{y}_p = \underset{\approx}{y}_c + \underset{\sim}{d}_{pc} \underset{\sim}{d}_{pc} \tag{23.78}$$

and

$$\underset{\sim}{d}_{pc} = \underset{\approx}{R} \bullet \underset{\sim}{d}^0_{pc}. \tag{23.79}$$

Therefore

$$\boxed{\underset{\approx}{y}_p \equiv \underset{\approx}{R} \bullet \underset{\approx}{y}^0_p \bullet \underset{\approx}{R}^T}. \tag{23.80}$$

In other words, the current second moment is related to the initial second moment by a simple rotational ‘push’.

23.7.8. Moment center-shift theorem

Recall the definition of the second moment about some point $\underset{\sim}{x}_p$

$$\underset{\approx}{y}_p \equiv \frac{1}{M} \int_B \underset{\sim}{d}_{xp} \underset{\sim}{d}_{xp} \rho \, \mathrm{d}V, \qquad \text{where } \underset{\sim}{d}_{xp} = \underset{\sim}{x} - \underset{\sim}{x}_p. \tag{23.81}$$

The goal of the moment center-shift theorem is to find a formula that relates $\underset{\approx}{y}_p$ to the (presumably already computed) second moment $\underset{\approx}{y}_c$ which is centered about the CM. Expanded, the integrand of (23.81) is

$$\underset{\sim}{d}_{xp} \underset{\sim}{d}_{xp} = (\underset{\sim}{x} - \underset{\sim}{x}_p)(\underset{\sim}{x} - \underset{\sim}{x}_p) = \underset{\sim}{x}\underset{\sim}{x} - \underset{\sim}{x}\underset{\sim}{x}_p - \underset{\sim}{x}_p \underset{\sim}{x} + \underset{\sim}{x}_p \underset{\sim}{x}_p. \tag{23.82}$$

Noting that $\underset{\sim}{x}_p$ is simply the location of a *particular point* and consequently does not vary with position, it may be pulled out of the integrals in (23.81) to give

$$\underset{\approx}{y}_p \equiv \underset{\approx}{y} - \underset{\sim}{x}_c \underset{\sim}{x}_p - \underset{\sim}{x}_p \underset{\sim}{x}_c + \underset{\sim}{x}_p \underset{\sim}{x}_p. \tag{23.83}$$

As a special case, if $\underset{\sim}{x}_p$ coincides with $\underset{\sim}{x}_c$, then (23.83) becomes

$$\underset{\approx}{y}_c \equiv \underset{\approx}{y} - \underset{\sim}{x}_c\underset{\sim}{x}_c. \tag{23.84}$$

Thus, we may substitute $\underset{\approx}{y} = \underset{\approx}{y}_c + \underset{\sim}{x}_c\underset{\sim}{x}_c$ into (23.83) to obtain

$$\boxed{\underset{\approx}{y}_p \equiv \underset{\approx}{y}_c + \underset{\sim}{d}_{pc}\underset{\sim}{d}_{pc},} \quad \text{where } \underset{\sim}{d}_{pc} \equiv \underset{\sim}{x}_p - \underset{\sim}{x}_c. \tag{23.85}$$

This result shows that $\underset{\approx}{y}_p$ may be immediately constructed without having to perform any new integrals over the body so long as $\underset{\approx}{y}_c$ and $\underset{\sim}{x}_c$ are known. This is why the moments about the centroid are generally considered the 'primitive' moments from which all other moments may be then rapidly constructed.

To remember (23.85), it helps to recognize that $\underset{\approx}{y}_p$ is a measure of the 'resistance' of a body to an angular acceleration. This resistance is the smallest when the axis of rotation passes through the CM. Accordingly, $\underset{\approx}{y}_p$ must always be larger than $\underset{\approx}{y}_c$. Hence, $\underset{\approx}{y}_p$ is obtained from $\underset{\approx}{y}_c$ by *adding* the dyad $\underset{\sim}{d}_{pc}\underset{\sim}{d}_{pc}$.

23.7.9. The parallel axis theorem

Recall the definition of the planar moment of inertia, which measures the resistance of a body to an angular acceleration about an axis parallel to $\underset{\sim}{n}$ passing through the CM $\underset{\sim}{x}_c$:

$$\phi_{cn} = M[(\text{tr}\underset{\approx}{y}_c) - \underset{\sim}{n} \bullet \underset{\approx}{y}_c \bullet \underset{\sim}{n}]. \tag{23.86}$$

We can similarly define the planar moment of inertia about a parallel axis that passes through a different point $\underset{\sim}{x}_p$:

$$\phi_{pn} = M[(\text{tr}\underset{\approx}{y}_p) - \underset{\sim}{n} \bullet \underset{\approx}{y}_p \bullet \underset{\sim}{n}]. \tag{23.87}$$

Substituting (23.85) into this equation,

$$\phi_{pn} = M[(\text{tr}[\underset{\approx}{y}_c + \underset{\sim}{d}_{pc}\underset{\sim}{d}_{pc}]) - \underset{\sim}{n} \bullet (\underset{\approx}{y}_c + \underset{\sim}{d}_{pc}\underset{\sim}{d}_{pc})_p \bullet \underset{\sim}{n}]. \tag{23.88}$$

Rearranging gives the parallel axis theorem:

$$\boxed{\phi_{pn} = \phi_{cn} + r_{pn}^2,} \quad \text{where } r_{pn} \equiv \sqrt{(\underset{\sim}{d}_{pc} \bullet \underset{\sim}{d}_{pc}) - (\underset{\sim}{d}_{pc} \bullet \underset{\sim}{n})^2}\,. \tag{23.89}$$

Note that $\underset{\sim}{d}_{pc} \bullet \underset{\sim}{d}_{pc}$ is the square magnitude of $\underset{\sim}{d}_{pc}$ and $\underset{\sim}{d}_{pc} \bullet \underset{\sim}{n}$ is the magnitude of the part of $\underset{\sim}{d}_{pc}$ that is parallel to $\underset{\sim}{n}$. Sketching these geometrically, the Pythagorean theorem implies that r_{pn} is the separation distance between the two parallel axes.

23.8. Linear momentum of a rigid body

The total linear momentum (sum of mass times velocity) for a body B is defined by

$$\underset{\sim}{L} = \int_B \underset{\sim}{v}\, \mathrm{d}m, \quad \text{where } \underset{\sim}{v} = \dot{\underset{\sim}{x}}, \quad \text{and } \mathrm{d}m = \rho\, \mathrm{d}V. \tag{23.90}$$

Here, ρ is the mass density and $\mathrm{d}V$ is the volume element. Substituting (23.12) into the above definition gives

$$\underset{\sim}{L} = \int_B [\dot{\underset{\sim}{x}}_c + \dot{\underset{\approx}{R}} \bullet (\underset{\sim}{X} - \underset{\sim}{X}_c)]\rho \, \mathrm{d}V. \tag{23.91}$$

Now recall that the rotation tensor $\underset{\approx}{R}$ depends only on time, not position. Likewise, the velocity of the CM depends only on time, not position. Thus, pulling these out of the spatial integral gives

$$\underset{\sim}{L} = M\dot{\underset{\sim}{x}}_c + \dot{\underset{\approx}{R}} \bullet \int_B \underset{\sim}{X}\rho \, \mathrm{d}V - \dot{\underset{\approx}{R}} \bullet (M\underset{\sim}{X}_c). \tag{23.92}$$

The remaining integral equals M times the position $\underset{\sim}{X}_c$ of the center of mass. Therefore, the last two terms cancel leaving only

$$\boxed{\underset{\sim}{L} = M\dot{\underset{\sim}{x}}_c.} \tag{23.93}$$

Thus, the angular spin of a rigid body has no effect on its linear momentum.

23.9. Angular momentum of a rigid body

The total angular momentum about the origin for a body B is defined by

$$H = \int_B \underset{\sim}{x} \times \underset{\sim}{v} \, \mathrm{d}m. \tag{23.94}$$

The cross product $\underset{\sim}{x} \times \underset{\sim}{v}$ may be written as the alternating tensor $\underset{\approx}{\varepsilon}$ operating on the dyad $\underset{\sim}{x}\underset{\sim}{v}$. Consequently, since the alternating tensor does not vary in space (and because the double-dot operation is linear), the above equation may be written

$$\underset{\sim}{H} = \underset{\approx}{\varepsilon}: \int_B \underset{\sim}{x}\underset{\sim}{v} \, \mathrm{d}m. \tag{23.95}$$

Substituting (23.13) into (23.95) gives

$$\begin{aligned}
\int_B \underset{\sim}{x}\underset{\sim}{v} \, \mathrm{d}m &= \int_B \underset{\sim}{x}[\dot{\underset{\sim}{x}}_c + \underset{\approx}{\Omega} \bullet (\underset{\sim}{x} - \underset{\sim}{x}_c)] \, \mathrm{d}m \\
&= M\underset{\sim}{x}_c\dot{\underset{\sim}{x}}_c + \left(\int_B (\underset{\sim}{x}\underset{\sim}{x} - \underset{\sim}{x}\underset{\sim}{x}_c) \, \mathrm{d}m\right) \bullet \underset{\approx}{\Omega}^T \\
&= M\underset{\sim}{x}_c\dot{\underset{\sim}{x}}_c + M(\underset{\approx}{y} - \underset{\sim}{x}_c\underset{\sim}{x}_c) \bullet \underset{\approx}{\Omega}^T.
\end{aligned} \tag{23.96}$$

Thus, noting that $\underset{\approx}{y} - \underset{\sim}{x}_c\underset{\sim}{x}_c = \underset{\approx}{y}_c$,

$$\int_B \underset{\sim}{x}\underset{\sim}{v} \, \mathrm{d}m = M[\underset{\sim}{x}_c\dot{\underset{\sim}{x}}_c + \underset{\approx}{y}_c \bullet \underset{\approx}{\Omega}^T]. \tag{23.97}$$

So the angular momentum about the origin becomes

$$\underset{\sim}{H} = M\underset{\approx}{\varepsilon}: (\underset{\sim}{x}_c\dot{\underset{\sim}{x}}_c + \underset{\approx}{y}_c \bullet \underset{\approx}{\Omega}^T). \tag{23.98}$$

The definition of the inertia tensor about any point $\underset{\sim}{x}_p$ is given in (23.47):

$$M\underset{\approx}{\varepsilon}: (\underset{\approx}{y}_p \bullet \underset{\approx}{\Omega}^T) = \underset{\approx}{\mu}_p \bullet \underset{\sim}{\omega}. \tag{23.99}$$

With this, the angular momentum about the fixed laboratory origin becomes

$$\boxed{\underset{\sim}{H} = M\underset{\sim}{x}_c \times \dot{\underset{\sim}{x}}_c + \underset{\approx}{\mu}_c \bullet \underset{\sim}{\omega}}. \tag{23.100}$$

23.10. Kinetic energy of a rigid body

The total kinetic energy for a body B is defined by

$$\dot{K} = \frac{1}{2}\int_B \underset{\sim}{v} \bullet \underset{\sim}{v}\, \mathrm{d}m \tag{23.101}$$

or, using (23.13) for the velocity,

$$\begin{aligned} \dot{K} &= \frac{1}{2}\int_B [\dot{\underset{\sim}{x}}_c + \underset{\approx}{\Omega} \bullet (\underset{\sim}{x} - \underset{\sim}{x}_c)] \bullet [\dot{\underset{\sim}{x}}_c + \underset{\approx}{\Omega} \bullet (\underset{\sim}{x} - \underset{\sim}{x}_c)]\mathrm{d}m \\ &= \int \left(\frac{1}{2}(\dot{\underset{\sim}{x}}_c \bullet \dot{\underset{\sim}{x}}_c) + \dot{\underset{\sim}{x}}_c \bullet \underset{\approx}{\Omega} \bullet (\underset{\sim}{x} - \underset{\sim}{x}_c) + \frac{1}{2}\mathrm{tr}[\underset{\approx}{\Omega} \bullet (\underset{\sim}{x} - \underset{\sim}{x}_c)(\underset{\sim}{x} - \underset{\sim}{x}_c) \bullet \underset{\approx}{\Omega}^T]\mathrm{d}m\right) \\ &= \frac{1}{2}Mv_c^2 + 0 + \frac{1}{2}M\,\mathrm{tr}[\underset{\approx}{\Omega} \bullet \underset{\approx}{y}_c \bullet \underset{\approx}{\Omega}^T]. \end{aligned} \tag{23.102}$$

It is straightforward to demonstrate that this can be simplified to

$$\boxed{\dot{K} = \frac{1}{2}Mv_c^2 + \frac{1}{2}\underset{\sim}{\omega} \bullet \underset{\approx}{\mu}_c \bullet \underset{\sim}{\omega}}, \quad \text{where } v_c^2 \equiv \dot{\underset{\sim}{x}}_c \bullet \dot{\underset{\sim}{x}}_c. \tag{23.103}$$

Note that the inertia tensor $\underset{\approx}{\mu}_c$ and the angular velocity $\underset{\sim}{\omega}$ in the second term play roles that are, respectively, analogous to mass M and velocity v_c in the first term.

23.11. Newton's equation (balance of linear momentum)

The principle of conservation of linear momentum states that the net force $\underset{\sim}{F}$ acting on a rigid body must equal the rate of change of its linear momentum. That is

$$\underset{\sim}{F} = \dot{\underset{\sim}{L}}. \tag{23.104}$$

Thus using (23.93),

$$\underset{\sim}{F} = M\underset{\sim}{a}_c, \quad \text{where } \underset{\sim}{a}_c \equiv \ddot{\underset{\sim}{x}}_c. \tag{23.105}$$

In other words, the net force equals the total mass times the acceleration of the CM.

23.12. Euler's equation (balance of angular momentum)

The balance of angular momentum requires that the net torque (about the fixed laboratory origin) must equal the rate of change of the angular momentum:

$$\underset{\sim}{T} = \dot{\underset{\sim}{H}}. \tag{23.106}$$

Thus, taking rates of (23.100) gives

$$\underset{\sim}{T} = M\underset{\sim}{x}_c \times \ddot{\underset{\sim}{x}}_c + \underset{\approx}{\mu}_c \bullet \dot{\underset{\sim}{\omega}} + \dot{\underset{\approx}{\mu}}_c \bullet \underset{\sim}{\omega}. \tag{23.107}$$

Recall that

$$\underset{\approx}{\mu}_c = \underset{\approx}{R} \bullet \underset{\approx}{\mu}_c^0 \bullet \underset{\approx}{R}^T. \tag{23.108}$$

Taking rates (noting that $\underset{\approx}{\mu}_c^0$ does not vary with time and $\dot{\underset{\approx}{R}} = \underset{\approx}{\Omega} \bullet \underset{\approx}{R}$) gives

$$\dot{\underset{\approx}{\mu}}_c = \underset{\approx}{\Omega} \bullet \underset{\approx}{\mu}_c - \underset{\approx}{\mu}_c \bullet \underset{\approx}{\Omega}. \tag{23.109}$$

Dotting from the right by $\underset{\sim}{\omega}$, noting that $\underset{\approx}{\Omega} \bullet \underset{\sim}{\omega} = \underset{\sim}{0}$ permits (23.107) to be written

$$\underset{\sim}{T} = M\underset{\sim}{x}_c \times \ddot{\underset{\sim}{x}}_c + \underset{\approx}{\mu}_c \bullet \dot{\underset{\sim}{\omega}} + \underset{\approx}{\Omega} \bullet \underset{\approx}{\mu}_c \bullet \underset{\sim}{\omega} \tag{23.110}$$

or

$$\boxed{\underset{\sim}{T} = M\underset{\sim}{x}_c \times \underset{\sim}{a}_c + \underset{\approx}{\mu}_c \bullet \dot{\underset{\sim}{\omega}} + \underset{\sim}{\omega} \times (\underset{\approx}{\mu}_c \bullet \underset{\sim}{\omega})}. \tag{23.111}$$

This result is the *direct notation* form of Euler's equations of motion. This expression may be applied using any convenient observer frame.

Some texts might use a specialized version of Euler's equation that has no term involving $\underset{\sim}{x}_c$. This might *seem* to mean that the origin is at the CM, but a better interpretation is that the missing term has been absorbed into the torque so that effective torque is $\underset{\sim}{T} - M\underset{\sim}{x}_c \times \underset{\sim}{a}_c$. When evaluated in a frame whose origin is at the CM, this effective torque equals the actual torque. If you wish to take rates of Euler's equation, which would be unusual in practice, the rate of effective torque (as seen by a non-moving observer) is *not* equal to the rate of actual torque. Instead, it is $\dot{\underset{\sim}{T}} - M\dot{\underset{\sim}{x}}_c \times \underset{\sim}{a}_c - M\underset{\sim}{x}_c \times \dot{\underset{\sim}{a}}_c$, which reduces to $\dot{\underset{\sim}{T}} - M\underset{\sim}{v}_c \times \underset{\sim}{a}_c$ if the origin of the non-moving observer happens to instantaneously coincide with the CM. This approach, which uses a non-moving origin that instantaneously coincides with the CM at the instant of evaluation, is not equivalent to using a moving origin that always coincides with the CM and hence is non-inertial and would (according to the principle of frame indifference) require using effective body forces that alter apparent force and torque (see chapter 24). When using a non-moving frame that is instantaneously coincident with the CM, elementary textbooks also typically employ a non-rotating basis (i.e., observer orientation) that happens to be instantaneously aligned with the principal directions of the inertia tensor (so that the products of inertia are zero in that frame). When these special frame choices are adopted, Euler's equation has components given by

$$T_1 = \mu_{11}\dot{\omega}_1 + \omega_3\omega_2(\mu_{33} - \mu_{22}), \tag{23.112a}$$

$$T_2 = \mu_{22}\dot{\omega}_2 + \omega_1\omega_3(\mu_{11} - \mu_{33}), \tag{23.112b}$$

$$T_3 = \mu_{33}\dot{\omega}_3 + \omega_2\omega_1(\mu_{22} - \mu_{11}), \tag{23.112c}$$

which is the form of Euler's equations usually found in elementary dynamics books. The compact, elegant, and general formula in (23.111) is naturally the better choice for anyone who knows tensor analysis.

References

[1] Abraham R, Marsden J E and Ratiu T 1988 *Manifolds, Tensor Analysis, and Applications* 2nd edn (Applied Mathematical Sciences vol 75) (New York: Springer)

[2] Argyris J 1982 An excursion into large rotations *Comput. Methods Appl. Mech. Eng.* 32

Chapter 24

Pseudo-body force for spinning problems

If a body is simultaneously rotating, translating, *and* deforming, then it might be more accurately analyzed using a frame of reference that spins along with the overall rotation of the body. When the body is rigid, material elements would appear to be stationary relative to this rotating frame. When the body deformable, material elements can move and rotate relative to this rotating frame.

In this chapter, we use frame transformations to infer both the the rotating gravity $\underset{\sim}{g}(t)$ and pseudo-body force $\underset{\sim}{b}(\underset{\sim}{x}, t)$ (illustrated figure 24.1) that would be needed in a *non-rotating* problem to produce the same observed or computed stress and displacement fields relative to observer 'B' as those obtained by observer 'A' using the same governing equations. When frame-change formulas are applied to convert A's observations to the frame of observer B, the result is required to agree exactly with what is observed and computed directly by observer B. Students in an introductory physics class might suppose that the correct solution for the body force in figure 24.1 would be $\underset{\sim}{b} = (-R\theta\omega^2)\underset{\sim}{e}_y$, but this would be correct only for steady rotation (where there is no material motion relative to observer B). We seek the more general formula for the pseudo-body force that includes Coriolis effects from motion of the material in the bucket relative to the rotating observer. Such terms could be very important if, for example, an explosive is set off in the rotating bucket [1]. Another application might be the analysis of strain on a human body as it rides in a roller-coaster or fighter jet, where the natural frame is fixed to the passenger seat and is therefore non-inertial, with human flesh thrashing relative to this frame in response to large 'unseen' pseudo body forces.

24.1 Kinematics of superimposed rotation (general analysis)

Figure 24.2 has a detailed caption defining the terminology for this specific discussion of deformation with superimposed rotation. The rotated 'star' (observer A) deformation depicted in figure 24.2 may be regarded as a rigid translation and rotation superimposed upon the 'fiducial' (observer B) deformation shown in figure 24.3.

doi:10.1016/978-0-7503-1454-1ch24

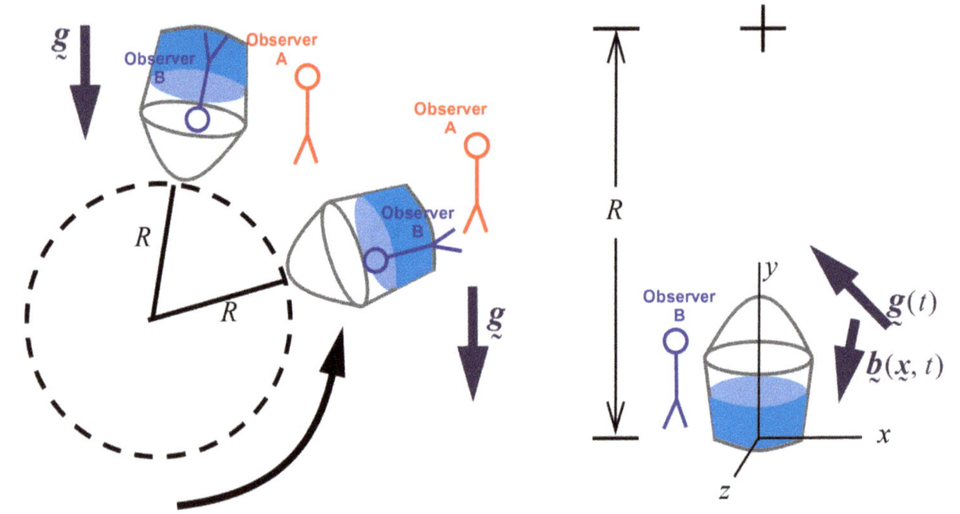

Figure 24.1. A rotating bucket from two perspectives. A simulation performed from the perspective of non-rotating observer A would employ a rotating bucket, and a constant downward gravity vector. A simulation from the perspective of rotating observer B would not only entail a rotating gravity $\underset{\sim}{g}(t)$, but it would also need to include additional pseudo-body force $\underset{\sim}{b}(\underset{\sim}{x}, t)$ to emulate the centrifugal and Coriolis forces of rotation if the fluid in the bucket is sloshing around (as from an explosive).

In numerical simulations of problems with large rotations, it is generally desirable to solve the fiducial (observer B) problem, and then to transform the fiducial results to account for superimposed motion afterward. Our goal is to determine the 'pseudo-body force' $\underset{\sim}{b}(\underset{\sim}{x}, t)$ that must be applied in the more easily analyzed fiducial (observer B) problem in order for its deformations and associated stresses to be identical (after superimposed translation and rotation) to those in the actual (observer A) rotating 'star' problem. In the fiducial problem, the origins O_0 and B_0 (defined generically in figure 24.2) are both assumed stationary without loss:

$$O(t) = O_0, \tag{24.1}$$

$$B(t) = B_0. \tag{24.2}$$

In the 'star' problem, the origin $O^*(t)$ is translating relative to the fiducial origin according to the displacement vector $\underset{\sim}{c}(t)$, and the origin $B^*(t)$ is translating along a path that is always equidistant from $O^*(t)$. Accordingly, the vector connecting these origins in the star problem is simply the rotation of the vector connecting the origins in the initial configuration:

$$\underset{\sim}{R}^* = \underset{\approx}{Q} \bullet \underset{\sim}{R}_0. \tag{24.3}$$

Both the 'star' and the fiducial problems have the same reference configuration. The deformed configurations differ by only a rigid motion. In the fiducial problem, the point $P(t)$ is the time-varying deformed location of P_0. The point $P(t)$ moves exclusively as a result of material deformation with respect to the fiducial frame. The point $P^*(t)$ moves from both this relative motion and from the superimposed translation and rotation.

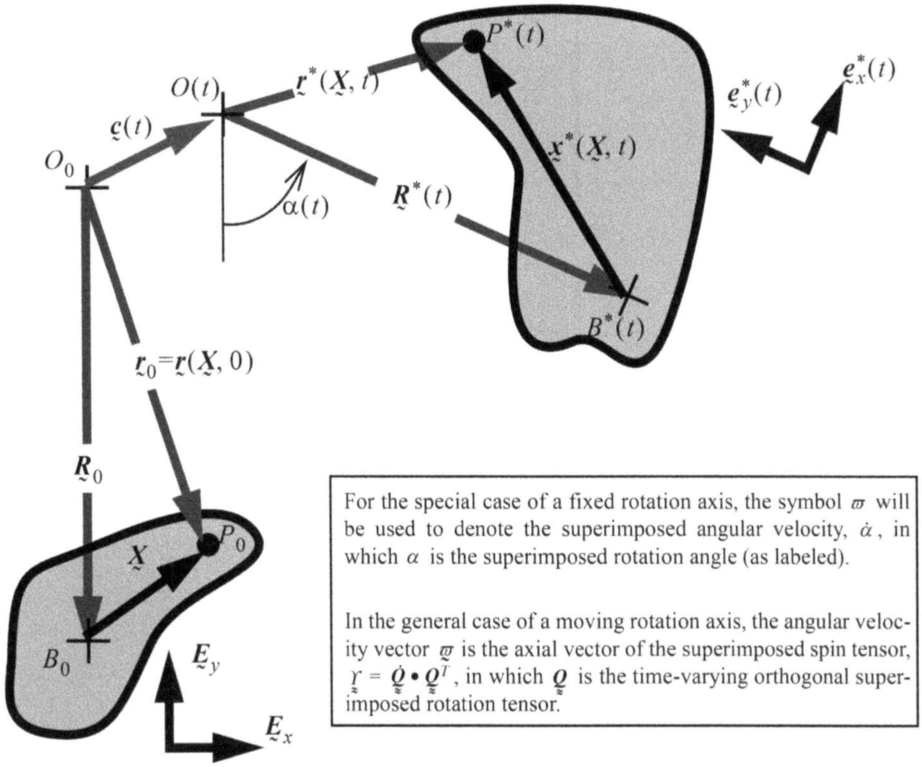

Figure 24.2. A body subjected to simultaneous translation, rotation, and deformation. The origin O_0 is a fixed reference point, which might (for example) be at the initial position of the 'hub' of a wheel if the problem pertains to tire modeling. The moving origin $O(t)$ is displaced from O_0 by a known translation vector $\boldsymbol{c}(t)$; for example, $\boldsymbol{c}(t)$ might be the translation of an automobile (more precisely, the translation of the wheel hub). The origin B_0, which is a secondary point of interest (like the contact point in a rolling problem or the bottom of the bucket in a centrifuge problem), is displaced from origin O_0 by a known vector $\boldsymbol{R}_0$. The moving origin $B^*(t)$ is displaced from the moving origin $O(t)$ by a known vector $\boldsymbol{R}(t) = \boldsymbol{Q}(t) \bullet \boldsymbol{R}_0$, in which $\boldsymbol{Q}(t)$ is a time-varying orthogonal rotation tensor with angular velocity defined in the inset of the figure. The time-varying basis vectors $\boldsymbol{e}_x^*(t)$and $\boldsymbol{e}_y^*(t)$ represent the basis used by the rotated observer $B^*(t)$. The point $P^*(t)$ is the location of some point in the body at time t, and P_0 is the initial location of that same point.

Recall that $\boldsymbol{x}(\boldsymbol{X}, t)$ is the position vector of the point P measured from the moving origin B, while $\boldsymbol{r}(\boldsymbol{X}, t)$ is the position vector of the same point measured from the origin O. Referring to figure 24.3,

$$\boldsymbol{r}(\boldsymbol{X},t) = \boldsymbol{x}(\boldsymbol{X},t) + \boldsymbol{R}_0. \tag{24.4}$$

The vector $\boldsymbol{x}$ connects the origin B to P, whereas the vector $\boldsymbol{x}^*$ connects B^* to P^*. Comparing the rotated 'star' deformation depicted in figure 24.2, with the fiducial deformation in figure 24.3, note that

$$\boldsymbol{x}^* = \boldsymbol{Q} \bullet \boldsymbol{x}, \tag{24.5}$$

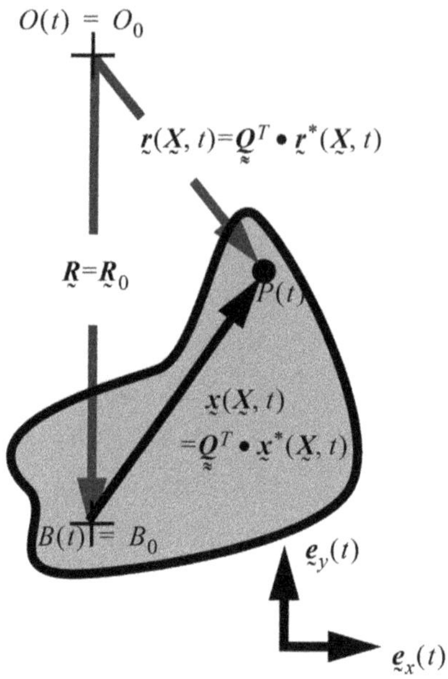

Figure 24.3. Fiducial deformation upon which translation and rotation are superimposed to produce the 'star' deformation depicted in figure 24.2.

$$\boxed{\underset{\sim}{r}^* = \underset{\approx}{Q} \bullet \underset{\sim}{r}}. \tag{24.6}$$

The vector $\underset{\sim}{X}$ connects the origin B_0 to the point P_0. Since both the fiducial and 'star' problems have the same initial configuration, we note that this vector is the same in both problems:

$$\underset{\sim}{X}^* = \underset{\sim}{X}. \tag{24.7}$$

The displacement vector (not shown in the figures to avoid clutter) is defined to connect the initial material point location P_0 to the deformed location of the same material point. Accordingly, the displacement vector in the fiducial problem is

$$\underset{\sim}{u}(\underset{\sim}{X},t) = \underset{\sim}{x}(\underset{\sim}{X},t) - \underset{\sim}{X}. \tag{24.8}$$

Referring to figure 24.2, the displacement vector in the 'star' problem is given by a much more complicated formula:

$$\underset{\sim}{u}^*(\underset{\sim}{X},t) = [\underset{\sim}{c}(t) + \underset{\sim}{R}^*(t) + \underset{\sim}{x}^*(\underset{\sim}{X},t)] - [\underset{\sim}{R}_0 + \underset{\sim}{X}]. \tag{24.9}$$

Using (24.3) and (24.5),

$$\underset{\sim}{u}^*(\underset{\sim}{x},t) = \underset{\sim}{c}(t) + \underset{\approx}{Q} \bullet (\underset{\sim}{x} + \underset{\sim}{R}_0) - (\underset{\sim}{X} + \underset{\sim}{R}_0). \tag{24.10}$$

An alternative, equivalent form is

$$\underset{\sim}{u}^*(\underset{\sim}{x},t) = \underset{\sim}{c}(t) + \underset{\approx}{Q} \bullet (\underset{\sim}{u} + \underset{\sim}{r}_0) - \underset{\sim}{r}_0. \tag{24.11}$$

Let a superposed dot represent the so-called *material rate*, which is the partial derivative of a field with respect to t holding $\boldsymbol{X}$ constant. In light of (24.7), this is also the time derivative holding $\boldsymbol{X}^*$ constant. Material velocity $\boldsymbol{v}$ is defined as the material rate of displacement of the particle P:

$$\boldsymbol{v} = \dot{\boldsymbol{u}}, \tag{24.12a}$$

$$\boldsymbol{v}^* = \dot{\boldsymbol{u}}^*. \tag{24.12b}$$

Thus, using (24.8), (24.4), and (24.10),

$$\boldsymbol{v} = \dot{\boldsymbol{x}} = \dot{\boldsymbol{r}}, \tag{24.13a}$$

$$\boldsymbol{v}^* = \dot{\boldsymbol{c}} + \boldsymbol{Q} \bullet \boldsymbol{v} + \dot{\boldsymbol{Q}} \bullet (\boldsymbol{x} + \boldsymbol{R}_0). \tag{24.13b}$$

Introducing the superimposed spin tensor,

$$\boldsymbol{\Upsilon} := \dot{\boldsymbol{Q}} \bullet \boldsymbol{Q}^T, \tag{24.14}$$

the 'star' material velocity may be written as

$$\boldsymbol{v}^* = \dot{\boldsymbol{c}} + \boldsymbol{Q} \bullet \boldsymbol{v} + \boldsymbol{\Upsilon} \bullet \boldsymbol{Q} \bullet (\boldsymbol{x} + \boldsymbol{R}_0). \tag{24.15}$$

Equivalently, noting that $\boldsymbol{r} = \boldsymbol{x} + \boldsymbol{R}_0$ and also that $\boldsymbol{r}^* = \boldsymbol{Q} \bullet \boldsymbol{r}$,

$$\boldsymbol{v}^* = \dot{\boldsymbol{c}} + \dot{\boldsymbol{r}}^*, \tag{24.16}$$

where

$$\dot{\boldsymbol{r}} = \boldsymbol{Q} \bullet \boldsymbol{v} + \boldsymbol{\Upsilon} \bullet \boldsymbol{r}^*. \tag{24.17}$$

Written out in full,

$$\boxed{\boldsymbol{v}^* = \dot{\boldsymbol{c}} + \boldsymbol{\Upsilon} \bullet \boldsymbol{r}^* + \boldsymbol{Q} \bullet \boldsymbol{v}}. \tag{24.18}$$

Material acceleration is the material rate of velocity. In the fiducial problem, the acceleration is found from the material rate of (24.13*a*):

$$\boldsymbol{a} = \dot{\boldsymbol{v}} = \ddot{\boldsymbol{x}} = \ddot{\boldsymbol{r}}. \tag{24.19}$$

The acceleration $\boldsymbol{a}^*$ in the 'star' problem is found from the material rate of (24.13*b*), which introduces additional terms from the rates of $\boldsymbol{Q}$ and $\boldsymbol{\Upsilon}$:

$$\boldsymbol{a}^* = \ddot{\boldsymbol{c}} + \boldsymbol{\Upsilon} \bullet \dot{\boldsymbol{r}}^* + \dot{\boldsymbol{\Upsilon}} \bullet \boldsymbol{r}^* + \boldsymbol{Q} \bullet \dot{\boldsymbol{v}} + \dot{\boldsymbol{Q}} \bullet \boldsymbol{v}. \tag{24.20}$$

Using (24.17) and (24.14),

$$\boldsymbol{a}^* = \ddot{\boldsymbol{c}} + \boldsymbol{\Upsilon} \bullet (\boldsymbol{Q} \bullet \boldsymbol{v} + \boldsymbol{\Upsilon} \bullet \boldsymbol{r}^*) + \dot{\boldsymbol{\Upsilon}} \bullet \boldsymbol{r}^* + \boldsymbol{Q} \bullet \dot{\boldsymbol{v}} + \boldsymbol{\Upsilon} \bullet \boldsymbol{Q} \bullet \boldsymbol{v}. \tag{24.21}$$

Rearranging,

$$\boldsymbol{a}^* = \ddot{\boldsymbol{c}} + \boldsymbol{\Upsilon} \bullet \boldsymbol{r}^* + \boldsymbol{\Upsilon} \bullet \dot{\boldsymbol{\Upsilon}} \bullet \boldsymbol{r}^* + \boldsymbol{Q} \bullet \boldsymbol{a} + 2\boldsymbol{\Upsilon} \bullet \boldsymbol{Q} \bullet \boldsymbol{v}. \tag{24.22}$$

Letting $\underset{\sim}{\varpi}$ denote the axial vector of $\underset{\approx}{\Upsilon}$, this may be written

this is from superimposed angular acceleration ↓ ; this is from fiducial acceleration (i.e., relative to rotating frame) ↓

$$\underset{\sim}{a}^* = \ddot{\underset{\sim}{c}} + \dot{\underset{\sim}{\varpi}} \times \underset{\sim}{r}^* + \underset{\sim}{\varpi} \times \underset{\sim}{\varpi} \times \underset{\sim}{r}^* + \underset{\approx}{Q} \bullet \underset{\sim}{a} + 2\, \underset{\approx}{\Upsilon} \bullet \underset{\approx}{Q} \bullet \underset{\sim}{v} \tag{24.23}$$

↑ This is from translation ; ↑ This is from superimposed angular velocity ; ↑ Coriolis effect from material velocity in fiducial frame

For any vector $\underset{\sim}{z}$, we define a macron (overbar) operator as the following 'un-rotation' operation:

$$\bar{\underset{\sim}{z}} = \underset{\approx}{Q}^T \bullet \underset{\sim}{z}. \tag{24.24}$$

For any tensor $\underset{\approx}{Z}$, define the macron 'un-rotation' operator as

$$\bar{\underset{\approx}{Z}} = \underset{\approx}{Q}^T \bullet \underset{\approx}{Z} \bullet \underset{\approx}{Q}. \tag{24.25}$$

For higher-order tensors, the macron is defined similarly via the Rayleigh product.

Since $\underset{\sim}{z} = \underset{\approx}{Q} \bullet \bar{\underset{\sim}{z}}$, it follows that

$$\dot{\underset{\sim}{z}} = \underset{\approx}{Q} \bullet \dot{\bar{\underset{\sim}{z}}} + \dot{\underset{\approx}{Q}} \bullet \bar{\underset{\sim}{z}} = \underset{\approx}{Q} \bullet \dot{\bar{\underset{\sim}{z}}} + \underset{\approx}{\Upsilon} \bullet \underset{\approx}{Q} \bullet \bar{\underset{\sim}{z}}. \tag{24.26}$$

Pre-multiplying by $\underset{\approx}{Q}^T$, establishes the following relationship between the un-rotation of the rate and the rate of the un-rotation:

$$\overline{\dot{\underset{\sim}{z}}} = \dot{\bar{\underset{\sim}{z}}} + \bar{\underset{\approx}{\Upsilon}} \bullet \bar{\underset{\sim}{z}}. \tag{24.27}$$

Similarly, for a tensor,

$$\overline{\dot{\underset{\approx}{Z}}} = \dot{\bar{\underset{\approx}{Z}}} + \bar{\underset{\approx}{\Upsilon}} \bullet \bar{\underset{\approx}{Z}} - \bar{\underset{\approx}{Z}} \bullet \bar{\underset{\approx}{\Upsilon}}. \tag{24.28}$$

In terms of the macron un-rotation operator, this section's results for superimposed rotation may be summarized as follows:

Superimposed motion kinematics

$$\overline{\underset{\sim}{R}^*} = R_0, \tag{24.29a}$$

$$\overline{\underset{\sim}{x}^*} = \underset{\sim}{x}, \tag{24.29b}$$

$$\overline{\underset{\sim}{r}^*} = \underset{\sim}{r}, \tag{24.29c}$$

$$\overline{\underset{\sim}{u}^*} = \underset{\sim}{u} + \underset{\sim}{r}_0 - \bar{\underset{\sim}{r}}_0 + \bar{\underset{\sim}{c}}, \tag{24.29d}$$

$$\overline{\underset{\sim}{v}^*} = \underset{\sim}{v} + \bar{\underset{\approx}{\Upsilon}} \bullet \underset{\sim}{r} + \dot{\bar{\underset{\sim}{c}}}, \tag{24.29e}$$

$$\overline{\underset{\sim}{a}^*} = \underset{\sim}{a} + \dot{\bar{\underset{\approx}{\Upsilon}}} \bullet \underset{\sim}{r} + \bar{\underset{\approx}{\Upsilon}} \bullet \bar{\underset{\approx}{\Upsilon}} \bullet \underset{\sim}{r} + 2\bar{\underset{\approx}{\Upsilon}} \bullet \underset{\sim}{v} + \ddot{\bar{\underset{\sim}{c}}}. \tag{24.29f}$$

When analyzing the fiducial problem using a numerical code, the results for the 'star' problem may be obtained by direct application of these equations. The difficult aspect of setting up the fiducial problem, however, is determination of the body force that must be used in the fiducial problem in order to produce the deformation field consistent with the 'star' problem that is of interest. What body forces, for example, would be needed in a simulation that keeps the point of contact of a tire fixed while rolling the road (and gravity) around the tire? The answer is not merely a rotation of the body forces from one perspective to another! The answer must include not only centrifugal forces, but also Coriolis forces. The next section address this question, finishing with examples in the simplified contexts of an accelerating automobile and then a centrifuge.

24.2 Fiducial body force for superimposed rigid motion

The fiducial problem solves the momentum equation given by

$$\underset{\approx}{\sigma} \bullet \overleftarrow{\nabla} + \rho \underset{\sim}{b} = \rho \underset{\sim}{a}, \tag{24.30a}$$

where $\underset{\sim}{b}$ is the body force in the fiducial problem, and the backward divergence is equivalent to a more traditional forward divergence acting on the transpose of stress[1]:

$$\underset{\approx}{\sigma} \bullet \overleftarrow{\nabla} = \frac{d\underset{\approx}{\sigma}}{d\underset{\sim}{x}} : \underset{\approx}{I} = \text{force (per volume) from neighboring material}, \tag{24.30b}$$

$$\underset{\sim}{a} = \left(\frac{\partial \underset{\sim}{x}}{\partial t}\right)_{\underset{\sim}{X}} = \text{material acceleration.} \tag{24.30c}$$

The 'star' problem (e.g., the real centrifuge or tire-rolling problem) solves

$$\underset{\approx}{\sigma}^{*} \bullet \overleftarrow{\nabla}^{*} + \rho^{*} \underset{\sim}{b}^{*} = \rho^{*} \underset{\sim}{a}^{*}, \tag{24.31a}$$

where

$$\underset{\approx}{\sigma}^{*} \bullet \overleftarrow{\nabla}^{*} = \frac{d\underset{\approx}{\sigma}^{*}}{d\underset{\sim}{x}^{*}} : \underset{\approx}{I}, \tag{24.31b}$$

$$\underset{\sim}{a}^{*} = \left(\frac{\partial \underset{\sim}{x}^{*}}{\partial t}\right)_{\underset{\sim}{X}^{*}}. \tag{24.31c}$$

We seek the body force $\underset{\sim}{b}$ that must be used in the fiducial problem (non-starred observer B) in order make the displacement and stress fields differ from the real (starred, observer A) problem only by the transformation appropriate to superimposed rigid motion. Recalling that $\underset{\sim}{x}^{*} = \underset{\approx}{Q} \bullet \underset{\sim}{x}$, the chain rule gives

[1] The ith Cartesian component of the backward divergence is $(\partial \sigma_{ij}/\partial x_j)$, which is the forward divergence of $\underset{\approx}{\sigma}^T$, so the distinction is unimportant in the usual case that stress is symmetric. Although rarely typeset as such, backward gradients and backward divergences are the de facto spatial derivatives used in continuum mechanics (cf. [2]).

$$\frac{d\underset{\approx}{\sigma}^*}{d\underset{\sim}{x}^*} = \frac{d\underset{\approx}{\sigma}^*}{d\underset{\sim}{x}} \bullet \frac{d\underset{\sim}{x}}{d\underset{\sim}{x}^*} = \frac{d\underset{\approx}{\sigma}^*}{d\underset{\sim}{x}} \bullet \underset{\approx}{Q}^T, \tag{24.32}$$

and therefore (24.30*b*) becomes

$$\underset{\approx}{\sigma}^* \bullet \overleftarrow{\nabla}^* = \frac{d\underset{\approx}{\sigma}^*}{d\underset{\sim}{x}} : \underset{\approx}{Q}. \tag{24.32b}$$

To satisfy the principle of material-frame indifference, two deformations that differ by rigid rotation must have Cauchy stress tensors that everywhere are related by

$$\underset{\approx}{\sigma}^* = \underset{\approx}{Q} \bullet \underset{\approx}{\sigma} \bullet \underset{\approx}{Q}^T. \tag{24.33}$$

Noting that $\underset{\approx}{Q}$ does not vary with position, this implies that

$$\underset{\approx}{\sigma}^* \bullet \overleftarrow{\nabla}^* = \underset{\approx}{Q} \bullet (\underset{\approx}{\sigma} \bullet \overleftarrow{\nabla}). \tag{24.34}$$

Therefore, pre-multiplying (24.31*a*) by $\underset{\approx}{Q}^T$ and using the macron (overbar) un-rotation notation defined in the previous section, the 'star' momentum equation may be written as

$$\underset{\approx}{\sigma} \bullet \overleftarrow{\nabla} + \rho \overline{\underset{\sim}{b}^*} = \rho \overline{\underset{\sim}{a}^*}, \tag{24.35}$$

where we have also noted that, for a frame-indifferent solution, $\rho^* = \rho$. Using (24.30*a*) to substitute for the fiducial stress divergence gives the desired solution for the fiducial body force:

$$\underset{\sim}{b} = \overline{\underset{\sim}{b}^*} + \underset{\sim}{a} - \overline{\underset{\sim}{a}^*}. \tag{24.36}$$

Using (24.29*f*), the body force that must be used in the fiducial problem is therefore

Superimposed motion pseudo-body force

$$\boxed{\underset{\sim}{b} = \overline{\underset{\sim}{b}^*} - \underset{\approx}{Q}^T \bullet \ddot{\underset{\sim}{c}} - \dot{\underset{\approx}{\Upsilon}} \bullet \underset{\sim}{r} - \underset{\approx}{\Upsilon} \bullet \underset{\approx}{\Upsilon} \bullet \underset{\sim}{r} - 2\underset{\approx}{\Upsilon} \bullet \underset{\sim}{v}}. \tag{24.37}$$

24.2.1 Special case: pure accelerated translation (without rotation)

For the special case of an experiment being conducted in an accelerating laboratory (e.g., a bucket moving in an automobile accelerating along a straight path), the 'star' frame differs from the fiducial frame only by translation $\underset{\sim}{c}$. In this special case, there is no frame rotation ($\underset{\approx}{Q} = \underset{\approx}{I}$, $\underset{\approx}{\Upsilon} = \underset{\approx}{0}$, and there is no difference between barred and unbarred quantities). Thus, the kinematics relationships in (24.29) reduce unsurprisingly to

Pure translation kinematics

$$\underset{\sim}{R}^* = \underset{\sim}{R}_0, \tag{24.38a}$$

$$\underset{\sim}{x}^* = \underset{\sim}{x}, \tag{24.38b}$$

$$\underset{\sim}{r}^* = \underset{\sim}{r}, \tag{24.38c}$$

$$\underset{\sim}{u}^* = \underset{\sim}{\bar{c}} + \underset{\sim}{u}, \tag{24.38d}$$

$$\underset{\sim}{v}^* = \underset{\sim}{\dot{c}} + \underset{\sim}{v}, \tag{24.38e}$$

$$\underset{\sim}{a}^* = \underset{\sim}{\ddot{c}} + \underset{\sim}{a}. \tag{24.38f}$$

The 'star' problem corresponds to the view of the experiment as seen by a stationary observer (e.g., someone standing on the road watching the accelerating automobile zoom past as someone in the car holds a bucket of water being not only pushed backward from the automobile acceleration but also blown apart from a firecracker exploding inside the bucket). The star problem has a body force $\underset{\sim}{b}^* = \underset{\sim}{g}$ = acceleration of gravity. The star problem also has acceleration from both the firecracker and the automobile. The fiducial problem has acceleration only from the firecracker, while the effect of the automobile's acceleration is accounted for in the fiducial problem by using the following pseudo-body force, given by (24.37) specialized to this problem (i.e., using $\underset{\approx}{Q} = \underset{\approx}{I}$ and $\underset{\approx}{\Upsilon} = \underset{\approx}{0}$):

Pure translation pseudo-body force

$$\boxed{\underset{\sim}{b} = \underset{\sim}{g} - \underset{\sim}{\ddot{c}}}. \tag{24.39}$$

If the car is accelerating forward, this formula correctly predicts a phantom body force that pushes things backward. Alternatively, suppose that the automobile was dropped from a tall building so that its acceleration $\underset{\sim}{\ddot{c}}$ is identical to gravity $\underset{\sim}{g}$. Then this equation says that the pseudo-body force should be zero, which means that the experiment has weightless conditions, as it should.

24.2.2 Special case: centrifuge (steady rotation without translation)

A centrifuge is a special case of the previous results corresponding to no translation ($\underset{\sim}{c} = \underset{\sim}{0}$), a stationary rotation axis (which, using the Euler–Rodrigues formula, implies that $\underset{\approx}{\bar{\Upsilon}} = \underset{\approx}{\Upsilon}$), and a constant superimposed angular velocity (implying that $\underset{\approx}{\dot{\bar{\Upsilon}}} = \underset{\approx}{0}$). These conditions also apply to a merry-go-round or an automobile going around a circular road at a constant speed. In this section, the term 'steady' does not imply that the material points within the centrifuge (or people playing while on the merry-go-round, or drinks spilling inside a turning car, etc) are necessarily stationary with respect to the centrifuge's rotating frame—it just means that the frame's rate of rotation is constant. This section allows transient (i.e., non-steady) material velocity and acceleration relative to the steadily rotating frame, as would be the case of a firecracker being set off inside a rotating bucket of water. We do not even presume that the firecracker explosion does not affect the angular velocity of the centrifuge itself; we only need to treat the observer to continue at a steady pre-explosion rate even if the centrifuge itself slows down.

In this case of steady observer rotation about a fixed (non-translating) axis, the general kinematics relationships in (24.29) reduce to

Centrifuge kinematics:

$$\overline{\underset{\sim}{R}^*} = \underset{\sim}{R}_0, \tag{24.40a}$$

$$\overline{\underset{\sim}{x}^*} = \underset{\sim}{x}, \tag{24.40b}$$

$$\overline{\underset{\sim}{r}^*} = \underset{\sim}{r}, \tag{24.40c}$$

$$\overline{\underset{\sim}{u}^*} = \underset{\sim}{r} - \underset{\approx}{Q}^T \bullet \underset{\sim}{r}_0, \tag{24.40d}$$

$$\overline{\underset{\sim}{v}^*} = \underset{\sim}{v} + \underset{\approx}{\Upsilon} \bullet \underset{\sim}{r}, \tag{24.40e}$$

$$\overline{\underset{\sim}{a}^*} = \underset{\sim}{a} + \underset{\approx}{\Upsilon} \bullet \underset{\approx}{\Upsilon} \bullet \underset{\sim}{r} + 2\underset{\approx}{\Upsilon} \bullet \underset{\sim}{v}. \tag{24.40f}$$

For this special case of a steadily spinning observer, again presuming that the only body force in the 'star' problem is constant gravity ($\underset{\sim}{b}^* = \underset{\sim}{g}$ so that $\bar{\underset{\sim}{b}}^* = \underset{\approx}{Q}^T \bullet \underset{\sim}{g}$), the pseudo-body force in (24.37) reduces to

Centrifuge pseudo-body force

$$\boxed{\underset{\sim}{b} = \underset{\approx}{Q}^T \bullet \underset{\sim}{g} - (\underset{\approx}{\Upsilon} \bullet \underset{\approx}{\Upsilon} \bullet \underset{\sim}{r} + 2\underset{\approx}{\Upsilon} \bullet \underset{\sim}{v})}. \tag{24.41}$$

Letting $\underset{\sim}{\varpi}$ denote the axial vector associated with $\underset{\approx}{\Upsilon}$, this may be written in a more traditional form as

Centrifuge pseudo-body force

$$\boxed{\underset{\sim}{b} = \underset{\approx}{Q}^T \bullet \underset{\sim}{g} - (\underset{\sim}{\varpi} \times \underset{\sim}{\varpi} \times \underset{\sim}{r} + 2\underset{\sim}{\varpi} \times \underset{\sim}{v})}. \tag{24.42}$$

The relative importance of the Coriolis term may be quantified by

$$\eta := \frac{\| 2\underset{\sim}{\varpi} \times \underset{\sim}{v} \|}{\| 2\underset{\sim}{\varpi} \times \underset{\sim}{v} \| + \| \underset{\sim}{\varpi} \times \underset{\sim}{\varpi} \times \underset{\sim}{r} \|}. \tag{24.43}$$

When the centrifugal[2] force, $-\underset{\sim}{\varpi} \times \underset{\sim}{\varpi} \times \underset{\sim}{r}$, is large in comparison to the Coriolis term, $\underset{\sim}{\varpi} \times \underset{\sim}{v}$, this dimensionless Coriolis-importance parameter is small. Otherwise, $\eta \rightarrow 1$ as the Coriolis effects become increasingly important. It is the Coriolis term that makes you feel drunk (i.e., tending to veer off of your path) when you attempt to walk towards the center of a spinning merry-go-round. The faster you walk, the larger your fiducial velocity (i.e., velocity $\underset{\sim}{v}$ relative to the merry-go-round) and hence the larger the value of η.

24.2.3 Buried explosives

A centrifuge can be used to approximate the effects of a large explosive in 1 G of gravity by instead setting off a smaller explosive in higher gravity. For these tests, a bucket of soil is spun in a horizontal centrifuge, making real gravity aligned with the rotation axis.[3] The goal of this section is to determine the pseudo-body force needed in a numerical simulation of the bucket explosion. The numerical simulation sets its origin somewhere in the bucket (such as the bottom of the bucket). This origin for the simulation is the point B in the fiducial problem, taken to be a distance R from the rotation axis once the bucket has begun spinning (as in figure 24.1). The ensuing analysis presumes that the rotation speed is high enough to neglect slight tilt of the bucket from ordinary gravity—the bucket is taken to be extended outward as an extension of the centrifuge arm. If the centrifuge is run at a speed that gives G 'gees' of acceleration, it follows that $R\varpi^2 = Gg$. Thus, if only 'gees' are given, the actual angular velocity may be inferred from

$$\varpi = \sqrt{\frac{Gg}{R}}. \tag{24.44}$$

Taking the z-axis as both the rotation axis and the direction of real gravity, we have

$$\underset{\sim}{\varpi} = \varpi \underset{\sim}{e}_z \quad \text{and} \quad \underset{\sim}{g} = -g\underset{\sim}{e}_z. \tag{24.45}$$

A numerical code solves the fiducial problem using an origin B_0 as indicated in figure 24.3. Let the position vector in the fiducial (numerical) simulation be $\underset{\sim}{x} = x\underset{\sim}{e}_x + ye_y + z\underset{\sim}{e}_z$, thus giving the position vector measured from the hub as

$$\underset{\sim}{r} = \underset{\sim}{R} + \underset{\sim}{x} = x\underset{\sim}{e}_z + (y - R)\underset{\sim}{e}_y + z\underset{\sim}{e}_z. \tag{24.46}$$

Let $\underset{\sim}{v} = v_x\underset{\sim}{e}_x + v_y\underset{\sim}{e}_y + v_z\underset{\sim}{e}_z$ denote the material velocity in the fiducial velocity (i.e., in the non-rotating simulation). Then the pseudo-body force is given by

---

[2] Without the negative, the term would be centripetal. The centripetal force is the force applied to the body to achieve the circular motion, which points toward the center of rotation. The centrifugal force is the pseudo-body force pointing in the opposite direction (i.e., which feels like a force throwing you to the outside of a curve as you move along a circular path).

[3] Our own work in this subject area was supported by ONR MURI grant N00014-11-1-0691.

$$\boxed{\underset{\sim}{b} = -\varpi^2[(R-y)\underset{\sim}{e}_y - x\underset{\sim}{e}_x] - 2(\varpi[v_x\underset{\sim}{e}_y - v_y\underset{\sim}{e}_x])}, \tag{24.47}$$

where the real gravity, $g\underset{\sim}{e}_z$ has been omitted to be consistent with the assumption that the speed is fast enough to neglect a 'dip' in the bucket towards the ground.

For a buried explosive problem, the material in the fiducial problem is expected to have a very large y-component of velocity, implying a large Coriolis force in the positive x-direction if the rotation is positive by the right-hand rule about the z-axis. This implies that high-speed imaging of the explosion should be made with the swing-arm-mounted camera pointing in the negative z-direction so that material motion in the positive x-direction will be clearly visible.

The relative importance of the Coriolis term in this explosive centrifuge problem is

$$\eta: = \frac{v_P}{v_P + \frac{1}{2}v_C}, \tag{24.48}$$

where

$$v_P = \sqrt{v_x^2 + v_y^2} = \text{ fiducial material speed} \perp \text{to rotation axis}, \tag{24.49a}$$

$$v_C = r_P\varpi = \text{speed associated with spinning of centrifuge}, \tag{24.49b}$$

$$r_P = \sqrt{x^2 + (R-y)^2} = \text{distance of the point from the rotation axis}. \tag{24.49c}$$

Our definition of Coriolis importance, η, nicely varies from 0 to 1 as the Coriolis term increases in magnitude relative to the centrifugal term. A more standard (and, in our opinion, obfuscated) measure of importance of the Coriolis term is the Rossby parameter, which has a general definition that simplifies in this case to

$$\zeta = \frac{r_p}{2L} \tag{24.50}$$

in which L is a nebulous length scale. If we define the length scale by $L = v_P/\varpi$, our intuitive η may be expressed in terms of the non-intuitive Rossby parameter ζ by

$$\eta = \frac{1}{1 + \zeta^{-1}}. \tag{24.51}$$

References

[1] Reguerio R *et al* 2013 ONR MURI Project on soil blast modeling and simulation *Dynamic Behavior of Materials* vol 1 ed B Song (Vienna: Springer) pp 341–53

[2] Marris A W *et al* 1967 Two definitions of gradient *J. Appl. Mech.* **34** 789–91

Chapter 25

Computer graphics visualization

Visualization of an object on a computer screen is basically a mapping of three-dimensional (3D) space (in which the body 'virtually' exists) to 2D space (the computer screen). In this chapter, we presume that there exists a single one-eyed observer who is looking through a 'window' onto a 3D body. Given that the observer is on one side of the window and the object is on the other side, a straight line connecting the observer's eye to any particular point on the body must intersect the plane of the window in exactly one point. If this intersection point lies inside the *frame* of the window, then it would correspond to a single point on the computer screen. In this chapter, we derive the equations governing this projection of the body onto the window in a manner consistent with the window orientation and the position of the observer. This mapping onto the window can be regarded as the image shown on a computer screen. What feels to the computer user like manipulation of the body (usually accomplished through click+drag mouse operations), is most efficiently coded as an equivalent change in the window's and/or the observer's position and orientation. Acceleration factors can be introduced so that, for example, a 90° mouse rotation would translate into a 360° apparent rotation of the body. In what follows, we derive the governing equations *without* these acceleration factors.

25.1 Orientation of the body

The orientation of the body can be characterized by three local *embedded* axes, $\{\underset{\sim}{e}_1, \underset{\sim}{e}_2, \underset{\sim}{e}_3\}$. If we let the fixed lab basis be denoted by $\{\underset{\sim}{E}_1, \underset{\sim}{E}_2, \underset{\sim}{E}_3\}$, then (8.9) permits us to fully characterize the body's orientation by

$$\underset{\approx}{R} = \underset{\sim}{e}_1\underset{\sim}{E}_1 + \underset{\sim}{e}_2\underset{\sim}{E}_2 + \underset{\sim}{e}_3\underset{\sim}{E}_3. \tag{25.1}$$

Let us suppose that the current orientation of the body was obtained from some known *initial* orientation. Then the initial orientation tensor is given by

doi:10.1016/B978-0-7503-1454-1ch25

$$\underset{\approx}{R}^0 = \underset{\sim}{e}_1^0 \underset{\sim}{E}_1 + \underset{\sim}{e}_2^0 \underset{\sim}{E}_2 + \underset{\sim}{e}_3^0 \underset{\sim}{E}_3. \tag{25.2}$$

Here and throughout the remainder of this section, a superscript 0 will denote the *initial* value of a quantity.

We can represent the *change* of orientation required to move the triad from $\{\underset{\sim}{e}_1^0, \underset{\sim}{e}_2^0, \underset{\sim}{e}_3^0\}$ to $\{\underset{\sim}{e}_1, \underset{\sim}{e}_2, \underset{\sim}{e}_3\}$ by a tensor $\underset{\approx}{Q}$. There are two ways to compute $\underset{\approx}{Q}$. We can directly apply (8.9) to write

$$\underset{\approx}{Q} = \underset{\sim}{e}_1 \underset{\sim}{e}_1^0 + \underset{\sim}{e}_2 \underset{\sim}{e}_2^0 + \underset{\sim}{e}_3 \underset{\sim}{e}_3^0. \tag{25.3}$$

We can alternatively regard the final orientation $\underset{\approx}{R}$ as a sequential rotation, the first being $\underset{\approx}{R}^0$, and the second being $\underset{\approx}{Q}$ so that (9.8) gives $\underset{\approx}{R} = \underset{\approx}{Q} \bullet \underset{\approx}{R}^0$ from which it follows that

$$\underset{\approx}{Q} = \underset{\approx}{R} \bullet \left(\underset{\approx}{R}^0\right)^T. \tag{25.4}$$

Both (25.3) and (25.4) will give the same result.

25.2 Mapping from the body to the screen

The goal of this section is to find the 'logical coordinates' of a pixel on a computer screen that correspond to any given point on a body. To do this, imagine that you are viewing the body through a rectangular window pane of width $2w$ and height $2h$. Let us permit the window pane to be arbitrarily oriented in space. Consequently, changing the plane of the window—or even rotating the window within its own plane—will result in a change of the appearance of the body's projection onto the window pane.

A pure translation of the body will result in a different perception of the body, as seen through our window. Therefore, in addition to depending on the body's orientation, $\underset{\approx}{R}$, the image of the body in the window will also depend on the location of the body's centroid, which we will denote by $\underset{\sim}{a}$. Another way to change the projection of the body onto the window pane is to change the location of the observer.

We have now defined all quantities that can affect the projection of the body onto the window pane, and we will assign the following symbols to these quantities.

$$\underset{\sim}{c} = \text{the center of the window pane,} \tag{25.5}$$

$$\underset{\sim}{w} = \text{vector from } \underset{\sim}{c} \text{ to the middle of the 'right' edge of the pane,} \tag{25.6}$$

$$\underset{\sim}{h} = \text{vector from } \underset{\sim}{c} \text{ to the middle of the 'top' edge of the pane,} \tag{25.7}$$

$$\underset{\approx}{R} = \text{the orientation of the body,} \tag{25.8}$$

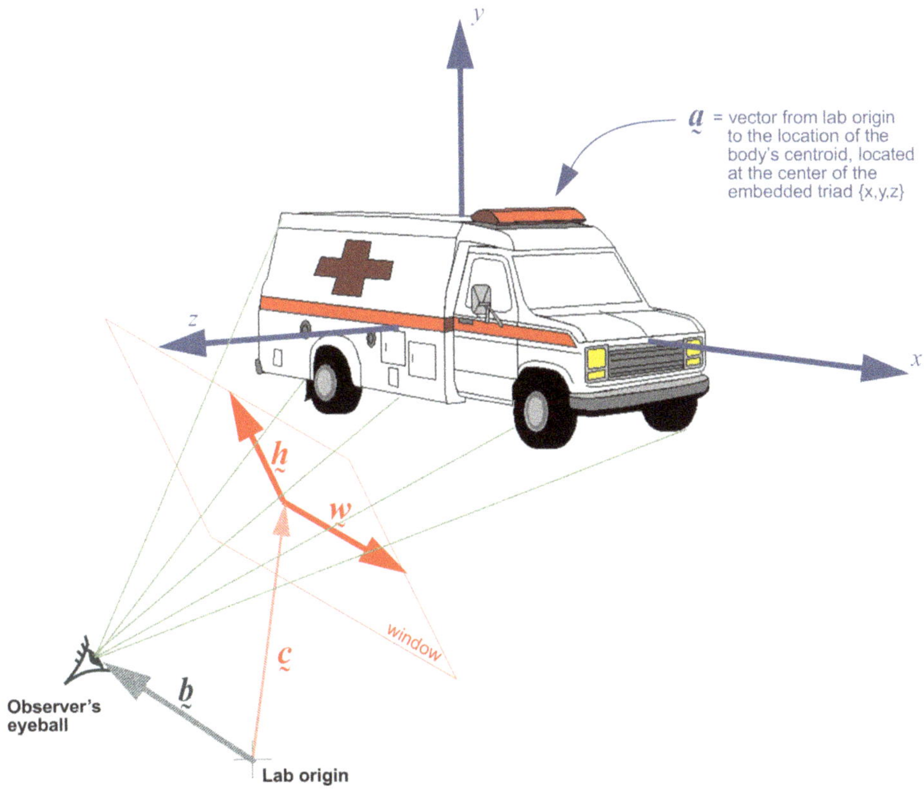

Figure 25.1. A body (the ambulance) being viewed by an observer through a window. The goal of this section is to find the coordinates of points where lines from the eyeball to the object intersect the window. These are then the pixels to light up.

$$\underset{\sim}{a} = \text{location of the body's centroid,} \tag{25.9}$$

$$\underset{\sim}{b} = \text{location of the observer.} \tag{25.10}$$

For the ensuing analysis, we find it useful to introduce symbols for the following additional quantities:

$$\underset{\sim}{n} = \text{normal to window} = \frac{\underset{\sim}{h} \times \underset{\sim}{w}}{\| \underset{\sim}{h} \times \underset{\sim}{w} \|}, \tag{25.11}$$

$$\underset{\sim}{p} = \text{any given point on the body (or its virtual extension),} \tag{25.12}$$

$$\underset{\sim}{P} = \text{image of the point } \underset{\sim}{p} \text{ on the plane of the window,} \tag{25.13}$$

$$\underset{\sim}{s} = \text{`line of sight' vector connecting } \underset{\sim}{p} \text{ to eyeball} = \underset{\sim}{p} - \underset{\sim}{b}. \tag{25.14}$$

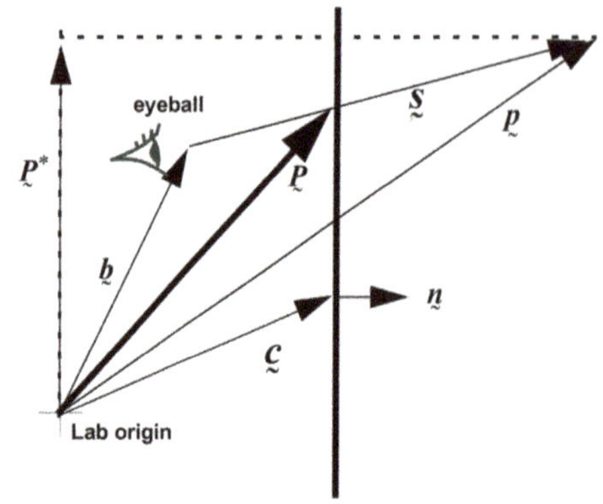

Figure 25.2. Mapping a virtual body to a virtual window plane.

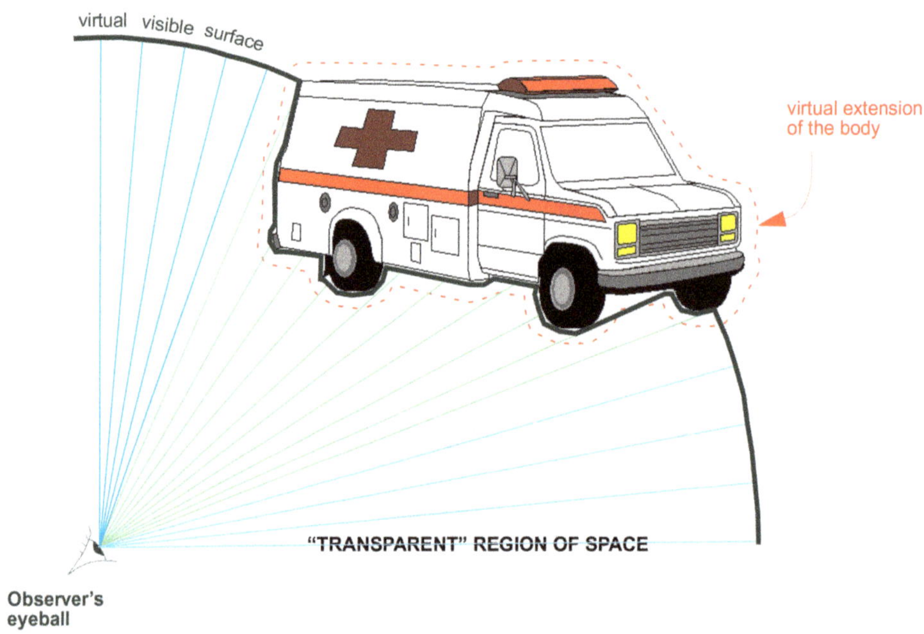

Figure 25.3. The virtual visible surface (VVS) of an object. The opaque region is defined to be the set of points in space that cannot be seen by the observer because those points are blocked from the observer's line of sight. The visible surface of the object is the set of points that *can* be seen by the observer. The VVS is defined by extending the visible surface out to infinity by lines.

Given any point $\underset{\sim}{p}$ in the body (regardless of whether it is visible or hidden—see figure 25.3), we seek a formula for the point $\underset{\sim}{P}$ defined by the intersection of the window's plane and the line of sight (from the eyeball to the point $\underset{\sim}{p}$). Referring to figure 25.2, note that $\underset{\sim}{P}$ equals $\underset{\sim}{b}$ plus some fraction β of the vector $\underset{\sim}{s}$:

$$\underset{\sim}{P} = \underset{\sim}{b} + \beta \underset{\sim}{s}. \tag{25.15}$$

To determine the value of the scalar multiplier β, note that the projection of $\underset{\sim}{\boldsymbol{P}}$ onto $\underset{\sim}{\boldsymbol{n}}$ must equal the projection of $\underset{\sim}{c}$ onto $\underset{\sim}{n}$. Therefore, substituting the above expression for $\underset{\sim}{P}$ into the requirement that $\underset{\sim}{P} \bullet \underset{\sim}{n} = \underset{\sim}{c} \bullet \underset{\sim}{n}$, and then solving for β gives

$$\beta = \frac{\underset{\sim}{c} \bullet \underset{\sim}{n} - \underset{\sim}{b} \bullet \underset{\sim}{n}}{\underset{\sim}{s} \bullet \underset{\sim}{n}}. \tag{25.16}$$

Substituting this result back into (25.15) gives

$$\boxed{\underset{\sim}{P} = \underset{\sim}{b} + \left[\frac{(\underset{\sim}{c} - \underset{\sim}{b}) \bullet \underset{\sim}{n}}{\underset{\sim}{s} \bullet \underset{\sim}{n}}\right] \underset{\sim}{s}} \tag{25.17}$$

or, substituting (25.11) and (25.14),

$$\boxed{\underset{\sim}{P} = \underset{\sim}{b} + \left[\frac{(\underset{\sim}{c} - \underset{\sim}{b}) \bullet (\underset{\sim}{h} \times \underset{\sim}{w})}{\left(\underset{\sim}{p} - \underset{\sim}{b}\right) \bullet (\underset{\sim}{h} \times \underset{\sim}{w})}\right] (\underset{\sim}{p} - \underset{\sim}{b})}. \tag{25.18}$$

These results, which do not presume that the observer's eye is aligned with the window's center would be useful when developing virtual perspective visualizations. The disadvantage of virtual 3D is that it typically requires viewers to wear special goggles, which therefore impede their vision when they need to look away from the screen. A simpler approach, which would involve less constrictive hardware, would consider how a one-eyed person perceives depth: namely by moving their single eye back and forth. A one-eyed 3D visualizer could be constructed by using only a *single* positional sensor that would send a message to the computer about the position of the observer's head relative to the screen. When the observer moves to the right, the image on the screen would adjust naturally to show more of the right-hand side of the object being viewed.

Of course, for the special case that the origin is located at the center of the window ($\underset{\sim}{c} = \underset{\sim}{0}$) and the observer is located a distance d directly away from the center of the window, we have

$$\underset{\sim}{b} = -d\underset{\sim}{n} \quad \text{and} \quad \underset{\sim}{s} = \underset{\sim}{p} + d\underset{\sim}{n} \tag{25.19}$$

and (25.17) becomes[1]

$$\boxed{\underset{\sim}{P} = \frac{\underset{\sim}{P}^*}{\left(\frac{\underset{\sim}{p} \bullet \underset{\sim}{n}}{d}\right) + 1}}, \tag{25.20}$$

where $\underset{\sim}{P}^*$ is the projected location of the point for an observer at infinity. Namely,

$$\boxed{\underset{\sim}{P}^* = \underset{\sim}{p} - \underset{\sim}{p} \bullet \underset{\sim}{n}\underset{\sim}{n}}. \tag{25.21}$$

---

[1] See [1], p 276.

The *logical* coordinates (ξ_w, ξ_h) of the point $\underset{\sim}{P}$ are defined to each range linearly from –1 to 1 if the point $\underset{\sim}{P}$ lies inside the rectangle of the window. Namely,

$$\xi_w \equiv \frac{(\underset{\sim}{P} - \underset{\sim}{c})\bullet \underset{\sim}{w}}{\underset{\sim}{w} \bullet \underset{\sim}{w}} \quad \text{and} \quad \xi_h \equiv \frac{(\underset{\sim}{P} - \underset{\sim}{c})\bullet \underset{\sim}{h}}{\underset{\sim}{h} \bullet \underset{\sim}{h}}. \tag{25.22}$$

25.3 Mapping from the screen to the virtual visible surface

Given the logical coordinates (ξ_w, ξ_h) of a point on the screen, we can reconstruct the point $\underset{\sim}{P}$ by

$$\underset{\sim}{P} = \underset{\sim}{c} + \left(\frac{\xi_w \underset{\sim}{w}}{\sqrt{\underset{\sim}{w} \bullet \underset{\sim}{w}}} + \frac{\xi_h \underset{\sim}{h}}{\sqrt{\underset{\sim}{h} \bullet \underset{\sim}{h}}} \right). \tag{25.23}$$

Since the results from the preceding section correspond to a mapping of 3D space onto a 2D space (the window pane), an inverse mapping from $\underset{\sim}{P}$ back to $\underset{\sim}{p}$ does not exist. However, a partial inverse can be found. Namely, given $\underset{\sim}{P}$ we know that the point $\underset{\sim}{p}$ must equal $\underset{\sim}{P}$ plus some multiple of $\underset{\sim}{P} - \underset{\sim}{b}$:

$$\underset{\sim}{p} = \underset{\sim}{P} + \alpha(\underset{\sim}{P} - \underset{\sim}{b}). \tag{25.24}$$

This equation represents a partial inverse of (25.18) because the scalar α cannot be determined without further information regarding the depth of the point $\underset{\sim}{p}$. However, if we *demand* that the point $\underset{\sim}{p}$ must lie somewhere on the virtual visible image (see figure 25.3), then a unique solution for $\underset{\sim}{p}$ can be found. This point on the virtual visible surface (VVS) will be denoted as

$$\underset{\sim}{q} = \underset{\sim}{P} + \alpha(\underset{\sim}{P} - \underset{\sim}{b}), \quad \text{where } \alpha \text{ is selected so that } \underset{\sim}{q} \text{ lies on the VVS.} \tag{25.25}$$

25.4 Changing the screen image of a body

In light of (25.18), we see that the screen image of a point $\underset{\sim}{p}$ will change whenever any of the following quantities change:

- The location of the observer $\underset{\sim}{b}$.
- The center of the window $\underset{\sim}{c}$.
- The orientation of the window, characterized by $\underset{\sim}{h}$ and $\underset{\sim}{w}$.
- The point itself $\underset{\sim}{p}$.

Existing 3D visualization programs effectively permit any of the above quantities to be altered through various clicking and dragging options. As stated here, of course, we are considering the body to be fixed in space while the observer changes location, orientation, or available viewport (window). The changes in the screen image can be accomplished in such a way that these quantities vary internally within the software, while it appears to the observer that it is the *body* that is being moved. The advantage of programming by changing only the above list is that, computationally, it is cheaper than performing translation and rotation operations on the thousands or even millions of points that define a 3D body[2].

The user interface for achieving a rotation should consider human factors issues. It is possible in principle to obtain a desired reorientation by each clicking and dragging being converted to rotations about an axis perpendicular to the dragging direction. However, this approach does not permit a direct rotation about an axis that is perpendicular to the screen. Well-written and intuitive visualization programs usually support rotation within the plane of the screen when the user clicks and drags along a region that is close to the outer boundary of the screen. Also, to minimize drag distance, most programs support a dragging multiplier so that, for example, dragging around the border by an angle θ results in a body rotation of, say, 4θ.

Reference

[1] Foley J D and Van Dam A 1982 *Fundamentals of Interactive Computer Graphics* (Reading, MA: Addison-Wesley)

[2] This statement applies to geometry mapping, not necessarily to shading. If you have light positions relative to the model, then shadows will change as you move the eye point. If, on the other hand, you move the object, the shadows relative to the light will change differently than moving the eye point. Since many legacy visualization packages do not do shadowing, this is generally not noticeable. Newer graphics cards can now do shadowing in hardware.

IOP Publishing

Rotation, Reflection, and Frame Changes

Orthogonal tensors in computational engineering mechanics

R M Brannon

Chapter 26

Voigt and Mandel components

This chapter defines the Voigt and Mandel (AKA normalized Voigt) systems for saving the 3×3 component matrix for any tensor in an alternative arrangement as a 9×1 array. This alternative nine-dimensional (9D) array is defined so that the last three components will be zero if the original tensor is symmetric, thus making it effectively a 6×1 array in that case. Before discussing this topic in detail, it helps to first talk about a similar problem for ordinary 3D vectors to understand the motivation for the Mandel revision of Voigt notation.

26.1 An introductory 3D example

A 3×1 array of ordinary vector components is meaningless unless you know the *basis* to which those components are referenced. Changing the basis will change the components in a 'contra-balancing' way so that the sum of components times basis vectors remains unchanged. The prefix 'contra' is used because components will change in a way that is contrary to the change in basis vectors. If, for example, you double the length of a basis vector, then the associated component must be *halved* to keep the product of the component times the basis vector unchanged. If more than one basis vector is changed, then all components will change in contra-balancing (and more complicated) ways to ensure that the sum of components times basis vectors does *not* change.

In full basis notation, any vector $\underset{\sim}{v}$ can be expanded in terms of the $\{\underset{\sim}{e}_1, \underset{\sim}{e}_2, \underset{\sim}{e}_3\}$ 'laboratory' basis vectors as

$$\underset{\sim}{v} = v_1\underset{\sim}{e}_1 + v_2\underset{\sim}{e}_2 + v_3\underset{\sim}{e}_3. \tag{26.1}$$

The matrix interpretation of this expansion is

$$\begin{bmatrix} v_1 \\ v_2 \\ v_3 \end{bmatrix} = v_1\begin{bmatrix} 1 \\ 0 \\ 0 \end{bmatrix} + v_2\begin{bmatrix} 0 \\ 1 \\ 0 \end{bmatrix} + v_3\begin{bmatrix} 0 \\ 0 \\ 1 \end{bmatrix}. \tag{26.2}$$

The three arrays on the right-hand side represent the basis vectors. The same array could be expanded alternatively as

$$\begin{bmatrix} v_1 \\ v_2 \\ v_3 \end{bmatrix} = \left(\frac{v_1 + v_2}{2}\right)\begin{bmatrix} 1 \\ 1 \\ 0 \end{bmatrix} + v_3\begin{bmatrix} 0 \\ 0 \\ 1 \end{bmatrix} + \left(\frac{v_2 - v_1}{2}\right)\begin{bmatrix} -1 \\ 1 \\ 0 \end{bmatrix}. \tag{26.3}$$

This expansion uses a different basis that is orthogonal, but not normalized. You could normalize the basis by writing the expansion in a slightly modified way as

$$\begin{bmatrix} v_1 \\ v_2 \\ v_3 \end{bmatrix} = \left(\sqrt{2}\frac{v_1 + v_2}{2}\right)\left\{\frac{1}{\sqrt{2}}\begin{bmatrix} 1 \\ 1 \\ 0 \end{bmatrix}\right\} + (v_3)\left\{\begin{bmatrix} 0 \\ 0 \\ 1 \end{bmatrix}\right\} + \left(\sqrt{2}\frac{v_2 - v_1}{2}\right)\left\{\frac{1}{\sqrt{2}}\begin{bmatrix} -1 \\ 1 \\ 0 \end{bmatrix}\right\}. \tag{26.4}$$

This expansion is identical to equation (26.3) except that the first and last basis vectors have been divided by $\sqrt{2}$ to normalize them and the components have been multiplied by $\sqrt{2}$ to keep the final sum unchanged. In basis form, equation (26.4) may be written

$$\underset{\sim}{v} = v_1^* \underset{\sim}{e}_1^* + v_2^* \underset{\sim}{e}_2^* + v_3^* \underset{\sim}{e}_3^*, \tag{26.5}$$

where

$$\underset{\sim}{e}_1^* \equiv \frac{\underset{\sim}{e}_1 + \underset{\sim}{e}_2}{\sqrt{2}}, \qquad \underset{\sim}{e}_2^* \equiv \underset{\sim}{e}_3, \qquad \underset{\sim}{e}_3^* \equiv \frac{-\underset{\sim}{e}_1 + \underset{\sim}{e}_2}{\sqrt{2}}, \tag{26.6}$$

and

$$v_1^* = \frac{v_1 + v_2}{\sqrt{2}}, \qquad v_2^* = v_3, \qquad v_3^* = \frac{-v_1 + v_2}{\sqrt{2}}. \tag{26.7}$$

If equation (26.6) and equation (26.7) are substituted into equation (26.5), then the original basis expansion of equation (26.1) is recovered.

Basis changes are typically motivated to take advantage of symmetries in a problem of interest. You might consider using the basis in equation (26.6) when vectors in the problem tend to lie in the plane depicted in figure 26.1. When a vector happens to lie in the plane, then its first two *laboratory* components are equal

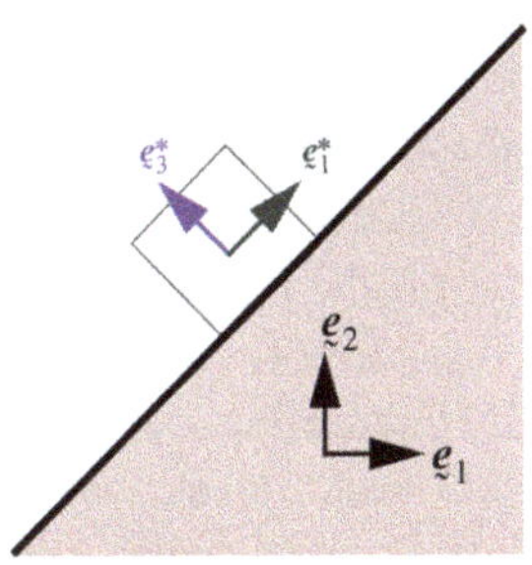

Figure 26.1. A natural basis for problems with some degree of symmetry with respect to a plane oriented at 45° with respect to the lab basis.

($v_1 = v_2$). According to equation (26.7), that means that its components with respect to the new basis are

$$v_1^* = \sqrt{2}\, v_1, \qquad v_2^* = v_3, \qquad v_3^* = 0. \tag{26.8}$$

In other words, only the first two components are non-zero when the vector lies in the plane. Any vector $\underset{\sim}{v}$ lying in this plane has the property that $v_1 = v_2$, so it can be expanded in terms of the *laboratory* basis as $\underset{\sim}{v} = v_1(\underset{\sim}{e}_1 + \underset{\sim}{e}_2) + v_3\underset{\sim}{e}_3$. If a second vector $\underset{\sim}{u}$ also lies in that plane, then the dot product $\underset{\sim}{v} \bullet \underset{\sim}{u}$ is given by

$$\underset{\sim}{v} \bullet \underset{\sim}{u} = 2v_1u_1 + v_3u_3. \tag{26.9}$$

The factor of 2 in front of the first term appears because $\underset{\sim}{e}_1 + \underset{\sim}{e}_2$ dotted with itself has the value of 2. When a vector $\underset{\sim}{v}$ lies in the plane, then its expansion in terms of the new basis is simply $\underset{\sim}{v} = v_1^*\underset{\sim}{e}_1^* + v_2^*\underset{\sim}{e}_2^*$. If a second vector $\underset{\sim}{u}$ also lies in that plane, then the dot product $\underset{\sim}{v} \bullet \underset{\sim}{u}$ is given by

$$\underset{\sim}{v} \bullet \underset{\sim}{u} = v_1^*u_1^* + v_2^*u_2^*. \tag{26.10}$$

Equations (26.9) and (26.10) are using components referenced to two different bases, but the end result for the dot product is the same.

The expansion $\underset{\sim}{v} = v_1(\underset{\sim}{e}_1 + \underset{\sim}{e}_2) + v_3\underset{\sim}{e}_3$ is analogous to equation (26.3) with v_2 set equal to v_1:

$$\begin{bmatrix} v_1 \\ v_2 \\ v_3 \end{bmatrix} = v_1\begin{bmatrix} 1 \\ 1 \\ 0 \end{bmatrix} = v_3\begin{bmatrix} 0 \\ 0 \\ 1 \end{bmatrix} \quad \text{for a vector in the plane.} \tag{26.11}$$

The *new* expansion, $\underset{\sim}{v} = v_1^*\underset{\sim}{e}_1^* + v_2^*\underset{\sim}{e}_2^*$, is analogous to equation (26.4) with v_2 set equal to v_1:

$$\begin{bmatrix} v_1 \\ v_2 \\ v_3 \end{bmatrix} = (\sqrt{2}\, v_1)\left\{\frac{1}{\sqrt{2}}\begin{bmatrix} 1 \\ 1 \\ 0 \end{bmatrix}\right\} + v_3\begin{bmatrix} 0 \\ 0 \\ 1 \end{bmatrix} \quad \text{for a vector in the plane.} \tag{26.12}$$

Both equation (26.11) and equation (26.12) represent expansions for vectors in the plane. Since the plane is 2D, both expansions need only two basis vectors to span the plane. The first basis vector in equation (26.11) is not a unit vector—its dot product with itself gives a value of 2, which accounts for the '2' multiplier in equation (26.9). The basis used in equation (26.12) is identical to that in equation (26.11) except that the basis vectors have been *normalized.* This normalization (division by $\sqrt{2}$) requires the associated component to be *multiplied* by $\sqrt{2}$. Whereas the components with respect to the normalized basis involve an ungainly factor of $\sqrt{2}$, finding these components is a once-only task. Thereafter, operations for vectors in the plane are more similar to formulas for 2D vectors, as was illustrated in equation (26.10), which had no weird factor of 2 in the first term.

This simple example using ordinary 3D vectors nicely sets the stage for similar concepts in basis expansions for tensors. In this simple example, we defined a basis that would naturally break a vector up into two mutually perpendicular parts: one part in the plane and the other part normal to the plane. For tensors, we will define a change in basis that breaks a tensor into its symmetric and skew parts. The set of all symmetric tensors forms a 6D linear manifold that is like the plane in our vector example. The set of all skew tensors forms a 3D linear manifold that is like the normal to the plane.

26.2 Voigt components (inefficient and error prone!)

In full basis notation, any tensor $\underset{\approx}{A}$ (symmetric or not) can be expanded as $A_{ij}\underset{\sim}{e}_i\underset{\sim}{e}_j$, where the $\underset{\sim}{e}_i\underset{\sim}{e}_j$ are basis dyads. The matrix interpretation of this basis expansion is

$$\begin{bmatrix} A_{11} & A_{12} & A_{13} \\ A_{21} & A_{22} & A_{23} \\ A_{31} & A_{32} & A_{33} \end{bmatrix} = A_{11}\begin{bmatrix} 1 & 0 & 0 \\ 0 & 0 & 0 \\ 0 & 0 & 0 \end{bmatrix} + A_{12}\begin{bmatrix} 0 & 1 & 0 \\ 0 & 0 & 0 \\ 0 & 0 & 0 \end{bmatrix} + A_{13}\begin{bmatrix} 0 & 0 & 1 \\ 0 & 0 & 0 \\ 0 & 0 & 0 \end{bmatrix}$$
$$+ A_{21}\begin{bmatrix} 0 & 0 & 0 \\ 1 & 0 & 0 \\ 0 & 0 & 0 \end{bmatrix} + A_{22}\begin{bmatrix} 0 & 0 & 0 \\ 0 & 1 & 0 \\ 0 & 0 & 0 \end{bmatrix} + A_{23}\begin{bmatrix} 0 & 0 & 0 \\ 0 & 0 & 1 \\ 0 & 0 & 0 \end{bmatrix} \tag{26.13}$$
$$+ A_{31}\begin{bmatrix} 0 & 0 & 0 \\ 0 & 0 & 0 \\ 1 & 0 & 0 \end{bmatrix} + A_{32}\begin{bmatrix} 0 & 0 & 0 \\ 0 & 0 & 0 \\ 0 & 1 & 0 \end{bmatrix} + A_{33}\begin{bmatrix} 0 & 0 & 0 \\ 0 & 0 & 0 \\ 0 & 0 & 1 \end{bmatrix}.$$

The six matrices on the right-hand side of this equation can be regarded as a *basis* for any 9 × 9 matrix (they are merely the components of the basis dyads). This expansion is the analog of equation (26.2).

We can alternatively expand the matrix as

$$\begin{bmatrix} A_{11} & A_{12} & A_{13} \\ A_{21} & A_{22} & A_{23} \\ A_{31} & A_{32} & A_{33} \end{bmatrix} = A_{(11)}\begin{bmatrix} 1 & 0 & 0 \\ 0 & 0 & 0 \\ 0 & 0 & 0 \end{bmatrix} + A_{(22)}\begin{bmatrix} 0 & 0 & 0 \\ 0 & 1 & 0 \\ 0 & 0 & 0 \end{bmatrix} + A_{(33)}\begin{bmatrix} 0 & 0 & 0 \\ 0 & 0 & 0 \\ 0 & 0 & 1 \end{bmatrix}$$
$$+ A_{(23)}\begin{bmatrix} 0 & 0 & 0 \\ 0 & 0 & 1 \\ 0 & 1 & 0 \end{bmatrix} + A_{(31)}\begin{bmatrix} 0 & 0 & 1 \\ 0 & 0 & 0 \\ 1 & 0 & 0 \end{bmatrix} + A_{(12)}\begin{bmatrix} 0 & 1 & 0 \\ 1 & 0 & 0 \\ 0 & 0 & 0 \end{bmatrix} \tag{26.14}$$
$$+ A_{[32]}\begin{bmatrix} 0 & 0 & 0 \\ 0 & 0 & -1 \\ 0 & 1 & 0 \end{bmatrix} + A_{[13]}\begin{bmatrix} 0 & 0 & 1 \\ 0 & 0 & 0 \\ -1 & 0 & 0 \end{bmatrix} + A_{[21]}\begin{bmatrix} 0 & -1 & 0 \\ 1 & 0 & 0 \\ 0 & 0 & 0 \end{bmatrix},$$

where

$$A_{(ij)} \equiv \tfrac{1}{2}(A_{ij} + A_{ji}) = \text{the } ij \text{ lab component of sym } \underset{\approx}{A}, \tag{26.15}$$

$$A_{[ij]} \equiv \tfrac{1}{2}(A_{ij} - A_{ji}) = \text{the } ij \text{ lab component of sym } \underset{\sim}{A}. \tag{26.16}$$

The expansion in equation (26.14) is the analog of equation (26.3). The basis tensors in equation (26.14) are called the covariant *Voigt basis.* Note that the last six Voigt basis tensors in equation (26.14) are not unit tensors. They each have magnitude $\sqrt{2}$. The *Mandel basis* (to be discussed later) is identical to the Voigt basis except that the last six basis tensors are normalized by dividing them by $\sqrt{2}$. The associated Mandel components are identical to Voigt components except that the last six are *multiplied* by $\sqrt{2}$. Whereas the Voigt expansion is analogous to equation (26.2), the Mandel expansion would be analogous to equation (26.3). Although Voigt components are themselves easy to obtain, this section will illustrate that subsequent *calculations* are awkward and error-prone because Voigt components correspond to a non-normalized basis. As will be discussed in the next section, normalized Voigt (Mandel) components take a moment or two longer to obtain, but subsequent calculations using them are easier.

The Voigt and Mandel expansions exploit symmetries that often occur in physics. In continuum mechanics, tensors are often either symmetric or skew. Even if they are neither, it is often useful to break a general tensor into the sum of its symmetric and skew parts, which is what the Voigt and Mandel expansions do. The first six terms in these expansions give the symmetric part of $\underset{\sim}{A}$, while the last three terms give the skew part. If, for example, the matrix is symmetric, then equation (26.14) reduces to only six non-zero terms:

$$\begin{aligned}\begin{bmatrix} A_{11} & A_{12} & A_{13} \\ A_{21} & A_{22} & A_{23} \\ A_{31} & A_{32} & A_{33} \end{bmatrix} = {} & A_{11}\begin{bmatrix} 1 & 0 & 0 \\ 0 & 0 & 0 \\ 0 & 0 & 0 \end{bmatrix} + A_{12}\begin{bmatrix} 0 & 0 & 0 \\ 0 & 1 & 0 \\ 0 & 0 & 0 \end{bmatrix} + A_{13}\begin{bmatrix} 0 & 0 & 0 \\ 0 & 0 & 0 \\ 0 & 0 & 1 \end{bmatrix} \\ & + A_{23}\begin{bmatrix} 0 & 0 & 0 \\ 0 & 0 & 1 \\ 0 & 1 & 0 \end{bmatrix} + A_{31}\begin{bmatrix} 0 & 0 & 1 \\ 0 & 0 & 0 \\ 1 & 0 & 0 \end{bmatrix} + A_{12}\begin{bmatrix} 0 & 1 & 0 \\ 1 & 0 & 0 \\ 0 & 0 & 0 \end{bmatrix},\end{aligned} \tag{26.17}$$

which is analogous to equation (26.11). The Mandel expansion for a symmetric tensor similarly reduces a sum of six terms that is related to equation (26.17) in the same way that equation (26.12) is related to equation (26.11).

In the previous section, we considered a plane embedded in ordinary 3D space. Any vector $\underset{\sim}{v}$ in that plane satisfied the constraint $v_1 = v_2$ while vectors normal to the plane satisfied the constraint $v_1 = -v_2$. Now we are considering a 6D linear manifold embedded within general 9D tensor space. Specifically, this manifold is the set of all symmetric tensors. Any symmetric tensor $\underset{\sim}{A}$ satisfies the constraints $A_{12} = A_{21}$, $A_{23} = A_{32}$, and $A_{31} = A_{13}$. Just as our earlier example had a 2D linear manifold (plane) embedded in general 3D vector space, the set of all symmetric tensors is a 6D linear manifold embedded within general 9D tensor space. Just as vectors *normal* to the plane in our vector example satisfied $v_1 = -v_2$, the set of all *skew* tensors satisfy $A_{12} = -A_{21}$, $A_{23} = -A_{32}$, and $A_{31} = -A_{13}$. Just as equation (26.3) broke vectors into their parts in the plane and normal to the plane, the expansion in equation (26.14) breaks up tensors into symmetric and skew parts.

The usual tensor basis expansion $\underset{\approx}{A} = A_{ij}\underset{\sim}{e}_i\underset{\sim}{e}_j$ could be written $\underset{\approx}{A} = A_{ij}\underset{\approx}{E}_{ij}$, where

$$\underset{\approx}{E}_{ij} \equiv \underset{\sim}{e}_i\underset{\sim}{e}_j. \tag{26.18}$$

In terms of $\underset{\approx}{E}_{ij}$, note that $A_{ij} = \underset{\approx}{A}:\underset{\approx}{E}_{ij}$. Since i and j each range from 1 to 3 in implied summations, the expansion $\underset{\approx}{A} = A_{ij}\underset{\approx}{E}_{ij}$ represents a total of nine terms. There is no law demanding a double sum must always be written using a pair of little indices i and j. Henceforth, indices written with CAPITAL letters are be understood to range from 1 to 9, and repeated indices are understood to be summed over this range. With this in mind, the expansion in equation (26.14) could be written compactly as

$$\underset{\approx}{A} = A^K\underset{\approx}{E}_K, \tag{26.19}$$

where

$$A^1 \equiv A_{(11)} \qquad [\underset{\sim}{E}_1] \equiv [\underset{\sim}{e}_1\underset{\sim}{e}_1] = \begin{bmatrix} 1 & 0 & 0 \\ 0 & 0 & 0 \\ 0 & 0 & 0 \end{bmatrix}, \tag{26.20a}$$

$$A^2 \equiv A_{(22)} \qquad [\underset{\sim}{E}_2] \equiv [\underset{\sim}{e}_2\underset{\sim}{e}_2] = \begin{bmatrix} 0 & 0 & 0 \\ 0 & 1 & 0 \\ 0 & 0 & 0 \end{bmatrix}, \tag{26.20b}$$

$$A^3 \equiv A_{(33)} \qquad [\underset{\sim}{E}_3] \equiv [\underset{\sim}{e}_3\underset{\sim}{e}_3] = \begin{bmatrix} 0 & 0 & 0 \\ 0 & 0 & 0 \\ 0 & 0 & 1 \end{bmatrix}, \tag{26.20c}$$

$$A^4 \equiv A_{(23)} \qquad [\underset{\sim}{E}_4] \equiv [\underset{\sim}{e}_2\underset{\sim}{e}_3 + \underset{\sim}{e}_3\underset{\sim}{e}_2] = \begin{bmatrix} 0 & 0 & 0 \\ 0 & 0 & 1 \\ 0 & 1 & 0 \end{bmatrix}, \tag{26.20d}$$

$$A^5 \equiv A_{(31)} \qquad [\underset{\sim}{E}_5] \equiv [\underset{\sim}{e}_3\underset{\sim}{e}_1 + \underset{\sim}{e}_1\underset{\sim}{e}_3] = \begin{bmatrix} 0 & 0 & 1 \\ 0 & 0 & 0 \\ 1 & 0 & 0 \end{bmatrix}, \tag{26.20e}$$

$$A^6 \equiv A_{(12)} \qquad [\underset{\sim}{E}_6] \equiv [\underset{\sim}{e}_1\underset{\sim}{e}_2 + \underset{\sim}{e}_2\underset{\sim}{e}_1] = \begin{bmatrix} 0 & 1 & 0 \\ 1 & 0 & 0 \\ 0 & 0 & 0 \end{bmatrix}, \tag{26.20f}$$

$$A^7 \equiv A_{[32]} \qquad [\underset{\sim}{E}_7] \equiv [\underset{\sim}{e}_3\underset{\sim}{e}_2 - \underset{\sim}{e}_2\underset{\sim}{e}_3] = \begin{bmatrix} 0 & 0 & 0 \\ 0 & 0 & -1 \\ 0 & 1 & 0 \end{bmatrix}, \tag{26.20g}$$

$$A^8 \equiv A_{[13]} \qquad [\underset{\sim}{E}_8] \equiv [\underset{\sim}{e}_1\underset{\sim}{e}_3 - \underset{\sim}{e}_3\underset{\sim}{e}_1] = \begin{bmatrix} 0 & 0 & 1 \\ 0 & 0 & 0 \\ -1 & 0 & 0 \end{bmatrix}, \tag{26.20h}$$

$$A^9 \equiv A_{[21]} \qquad [\underset{\sim}{E}_9] \equiv [\underset{\sim}{e}_2\underset{\sim}{e}_1 - \underset{\sim}{e}_1\underset{\sim}{e}_2] = \begin{bmatrix} 0 & -1 & 0 \\ 1 & 0 & 0 \\ 0 & 0 & 0 \end{bmatrix}. \tag{26.20i}$$

Here, the nine components, $\{A^1, A^2, \ldots, A^9\}$ are called the contravariant *Voigt components* of the tensor. The first six contravariant Voigt components are the lab components of sym $\underset{\approx}{A}$. For example, A^5 is the 31-component of the symmetric part of $\underset{\approx}{A}$. The last three Voigt components are the lab components of skw $\underset{\approx}{A}$; these three skew components are also the components of the axial vector associated with skw$\underset{\approx}{A}$.

Given that we now have two representations (the ordinary 3 × 3 matrix and the 9D array) for tensor components, it is worth summarizing how to go from one format to another.

If the 3 × 3 laboratory component matrix for a tensor is $\begin{bmatrix} A_{11} & A_{12} & A_{13} \\ A_{21} & A_{22} & A_{23} \\ A_{31} & A_{32} & A_{33} \end{bmatrix}$,

then the 9 × 1 Voigt array is $\begin{bmatrix} A_{(11)} \\ A_{(22)} \\ A_{(33)} \\ A_{(23)} \\ A_{(31)} \\ A_{(12)} \\ A_{[32]} \\ A_{[12]} \\ A_{[21]} \end{bmatrix}$.

For example, the 9D contravariant Voigt array for the identity tensor $\begin{bmatrix} 1 & 0 & 0 \\ 0 & 1 & 0 \\ 0 & 0 & 1 \end{bmatrix}$ is $\begin{bmatrix} 1 \\ 1 \\ 1 \\ 0 \\ 0 \\ 0 \\ 0 \\ 0 \\ 0 \end{bmatrix}$.

The 9D contravariant Voigt array for the tensor $\begin{bmatrix} 1 & 6 & 5 \\ 6 & 2 & 4 \\ 5 & 4 & 3 \end{bmatrix}$ is $\begin{bmatrix} 1 \\ 2 \\ 3 \\ 4 \\ 5 \\ 6 \\ 0 \\ 0 \\ 0 \end{bmatrix}$.

The 9D contravariant Voigt array for the tensor $\begin{bmatrix} 0 & -9 & 8 \\ 9 & 0 & -7 \\ -8 & 7 & 0 \end{bmatrix}$ is $\begin{bmatrix} 0 \\ 0 \\ 0 \\ 0 \\ 0 \\ 0 \\ 7 \\ 8 \\ 9 \end{bmatrix}$.

The 9D contravariant Voigt array for the tensor $\begin{bmatrix} 1 & 2 & 3 \\ 4 & 5 & 6 \\ 7 & 8 & 9 \end{bmatrix}$ is found by noting that its symmetric part is $\begin{bmatrix} 1 & 3 & 5 \\ 3 & 5 & 7 \\ 5 & 7 & 9 \end{bmatrix}$ and its skew part is $\begin{bmatrix} 0 & -1 & -2 \\ 1 & 0 & -1 \\ 2 & 1 & 0 \end{bmatrix}$. Therefore, the 9D Voigt array is $[1, 5, 9, 7, 5, 3, 1, -2, 1]^T$.

Following a sequence of calculations using 9D Voigt components, you will eventually wish to re-cast the result in conventional 3×3 Cartesian component form.

If $\begin{bmatrix} A^1 \\ A^2 \\ A^3 \\ A^4 \\ A^5 \\ A^6 \\ A^7 \\ A^8 \\ A^9 \end{bmatrix}$ is the Voigt array for a tensor, then the corresponding 3×3 laboratory component matrix is reconstructed by

$$\begin{bmatrix} A^1 & A^6 - A^9 & A^5 + A^8 \\ A^6 + A^9 & A^2 & A^4 - A^7 \\ A^5 - A^8 & A^4 + A^7 & A^3 \end{bmatrix}. \tag{26.21}$$

We have typeset Voigt components using *superscripts* because the Voigt basis in equation (26.20) *is not normalized.* The last six Voigt basis tensors have a tensor magnitude[1] equal to $\sqrt{2}$, not 1. Since the basis is not orthonormal, Helnwein [1] pointed out that the Voigt components $\{A^1, A^2, \ldots, A^6\}$ are actually *contravariant* components of the tensor $\underset{\approx}{A}$ with respect to the non-normalized *covariant* Voigt basis. Superscript (contravariant) and subscript (covariant) indexing is standard whenever dealing with a non-orthonormal basis. The remainder of this section presumes that the reader is familiar with the terminology and procedures that are standard when working with non-orthonormal bases (i.e., meanings of covariant, contravariant, metrics, etc, and well-known formulas connecting them), but it should be straightforward to follow this section even without such prerequisites.

The covariant metrics for the Voigt basis are

$$[G_{IJ}] = [\underset{\approx}{E}_I : \underset{\approx}{E}_J] = \text{diag}[1, 1, 1, 2, 2, 2, 2, 2, 2]. \tag{26.22}$$

This covariant matrix is not the identity matrix, but at least it is diagonal, which implies that the covariant Voigt basis in equation (26.20) is orthogonal but not normalized. Letting $[G^{IJ}]$ denote the inverse of the $[G_{IJ}]$ matrix, the dual contravariant basis that goes along with the covariant Voigt basis in equation (26.20) is defined

$$\underset{\approx}{E}^I = G^{IJ}\underset{\approx}{E}_J. \tag{26.23}$$

With this 'helper' basis in hand, the contravariant Voigt components can be obtained from

$$A^I = \underset{\approx}{A} : \underset{\approx}{E}^I. \tag{26.24}$$

Whereas equation (26.19) expanded the tensor in terms of the covariant $\underset{\approx}{E}_K$-basis, it is also possible to expand the tensor as a linear combination of the contravariant $\underset{\approx}{E}^K$-basis. Specifically,

$$\underset{\approx}{A} = A_K\underset{\approx}{E}^K, \qquad \text{where} \qquad A_K = \underset{\approx}{A} : \underset{\approx}{E}_K = G_{KL}A^L \tag{26.25}$$

Recalling equation (26.22), this implies that the first three covariant Voigt components identically equal their contravariant counterparts, while the remaining covariant (subscripted) Voigt components are twice as large as their contravariant (superscripted) counterparts:

$$\begin{bmatrix} A_1 \\ A_2 \\ A_3 \\ A_4 \\ A_5 \\ A_6 \\ A_7 \\ A_8 \\ A_9 \end{bmatrix} = \begin{bmatrix} A^1 \\ A^2 \\ A^3 \\ 2A^4 \\ 2A^5 \\ 2A^6 \\ 2A^7 \\ 2A^8 \\ 2A^9 \end{bmatrix} = \begin{bmatrix} A_{(11)} \\ A_{(22)} \\ A_{(33)} \\ 2A_{(23)} \\ 2A_{(31)} \\ 2A_{(12)} \\ 2A_{[32]} \\ 2A_{[12]} \\ 2A_{[21]} \end{bmatrix}. \tag{26.26}$$

[1] The magnitude of any tensor $\underset{\approx}{T}$ (symmetric or not) is $\sqrt{T_{ij}T_{ij}}$.

26.2.1 Specialization for symmetric tensors

Let $\underset{\approx}{\mathbf{A}}$ and $\underset{\approx}{\mathbf{B}}$ be symmetric tensors. In terms of laboratory components, the inner product between them is then

$$\underset{\approx}{\mathbf{A}}:\underset{\approx}{\mathbf{B}} = A_{11}B_{11} + A_{22}B_{22} + A_{33}B_{33} + 2A_{23}B_{23} + 2A_{31}B_{31} + 2A_{12}B_{12}. \tag{26.27}$$

If these symmetric tensors are saved in Voigt form, then

$$\underset{\approx}{\mathbf{A}}:\underset{\approx}{\mathbf{B}} = A^1B^1 + A^2B^2 + A^3B^3 + 2A^4B^4 + 2A^5B^5 + 2A^6B^6. \tag{26.28}$$

The off-diagonal terms contribute twice to the inner product, resulting in the factors of 2 in the last three terms. These factors of 2 appear for reasons similar to the 2 that appeared in equation (26.9). If $\underset{\approx}{\mathbf{A}}$ and $\underset{\approx}{\mathbf{B}}$ are expanded in the form of equation (26.19), then their inner product may be computed directly by

$$\underset{\approx}{\mathbf{A}}:\underset{\approx}{\mathbf{B}} = (A^1\underset{\approx}{\mathbf{E}}_I):(A^J\underset{\approx}{\mathbf{E}}_J) = A^IA^J(\underset{\approx}{\mathbf{E}}_I:\underset{\approx}{\mathbf{E}}_J) \tag{26.29}$$

or, recalling equation (26.22),

$$\underset{\approx}{\mathbf{A}}:\underset{\approx}{\mathbf{B}} = A^IB^JG_{IJ}. \tag{26.30}$$

Expanding this result out explicitly using the metrics in equation (26.22) leads to equation (26.28). Thus, the factors of 2 that appear in the Voigt inner product are merely metrics that arise because the Voigt basis is not normalized.

If $\underset{\approx}{\mathbf{A}}$ is expanded in the form of equation (26.19) but $\underset{\approx}{\mathbf{B}}$ is expanded in the form of equation (26.25), then the inner product may be computed by

$$\underset{\approx}{\mathbf{A}}:\underset{\approx}{\mathbf{B}} = A^IB_I. \tag{26.31}$$

Although this result contains no metrics, it entails the burden of having to find two different types of components for the tensors. In materials modeling, the stress and strain tensors are often stored in contravariant and covariant Voigt forms, respectively, so that the inner product between stress and strain arrays will equal the inner product between stress and strain tensors. This goal, as will soon be demonstrated, is also met using a 'normalized Voigt' (Mandel) system without needing a distinction between 'stress-like' and 'strain-like' arrays.

26.3 Mandel components (nice!)

The Mandel system simply normalizes the Voigt basis, thereby eliminating the need to distinguish between contravariant and covariant components. With the Mandel system, a tensor inner product (:) translates into what looks and feels like an ordinary vector dot product. With the Mandel system, existing linear algebra subroutines may be used without having to modify them to insert ungainly metric factors of 2 or 1/2.

The Mandel basis is the Voigt basis in equation (26.20) with the last six Voigt basis tensors divided by their magnitudes, $\sqrt{2}$. Hence, to keep the combination of components times basis vectors unchanged, the last six Mandel components must equal the Voigt components *multiplied* by $\sqrt{2}$. With the Mandel system, any tensor $\underset{\approx}{A}$ is therefore expanded as

$$\underset{\approx}{A} = A_K^* \underset{\approx}{E}_K^*, \tag{26.32}$$

where

$$A_1^* \equiv A_{11} \qquad [\underset{\approx}{E}_1^*] = [\underset{\sim}{e}_1\underset{\sim}{e}_1] = \begin{bmatrix} 1 & 0 & 0 \\ 0 & 0 & 0 \\ 0 & 0 & 0 \end{bmatrix}, \tag{26.33a}$$

$$A_2^* \equiv A_{22} \qquad [\underset{\approx}{E}_2^*] = [\underset{\sim}{e}_2\underset{\sim}{e}_2] = \begin{bmatrix} 0 & 0 & 0 \\ 0 & 1 & 0 \\ 0 & 0 & 0 \end{bmatrix}, \tag{26.33b}$$

$$A_3^* \equiv A_{33} \qquad [\underset{\approx}{E}_3^*] = [\underset{\sim}{e}_3\underset{\sim}{e}_3] = \begin{bmatrix} 0 & 0 & 0 \\ 0 & 0 & 0 \\ 0 & 0 & 1 \end{bmatrix}, \tag{26.33c}$$

$$A_4^* \equiv \sqrt{2}A_{(23)} \qquad [\underset{\approx}{E}_4^*] = \frac{[\underset{\sim}{e}_2\underset{\sim}{e}_3 + \underset{\sim}{e}_3\underset{\sim}{e}_2]}{\sqrt{2}} = \frac{1}{\sqrt{2}}\begin{bmatrix} 0 & 0 & 0 \\ 0 & 0 & 1 \\ 0 & 1 & 0 \end{bmatrix}, \tag{26.33d}$$

$$A_5^* \equiv \sqrt{2}A_{(31)} \qquad [\underset{\approx}{E}_5^*] = \frac{[\underset{\sim}{e}_3\underset{\sim}{e}_1 + \underset{\sim}{e}_1\underset{\sim}{e}_3]}{\sqrt{2}} = \frac{1}{\sqrt{2}}\begin{bmatrix} 0 & 0 & 1 \\ 0 & 0 & 0 \\ 1 & 0 & 0 \end{bmatrix}, \tag{26.33e}$$

$$A_6^* \equiv \sqrt{2}A_{(12)} \qquad [\underset{\approx}{E}_6^*] = \frac{[\underset{\sim}{e}_1\underset{\sim}{e}_2 + \underset{\sim}{e}_2\underset{\sim}{e}_1]}{\sqrt{2}} = \frac{1}{\sqrt{2}}\begin{bmatrix} 0 & 1 & 0 \\ 1 & 0 & 0 \\ 0 & 0 & 0 \end{bmatrix}, \tag{26.33f}$$

$$A_7^* \equiv \sqrt{2}A_{[32]} \qquad [\underset{\approx}{E}_7^*] = \frac{[\underset{\sim}{e}_3\underset{\sim}{e}_2 - \underset{\sim}{e}_2\underset{\sim}{e}_3]}{\sqrt{2}} = \frac{1}{\sqrt{2}}\begin{bmatrix} 0 & 0 & 0 \\ 0 & 0 & -1 \\ 0 & 1 & 0 \end{bmatrix}, \tag{26.33g}$$

$$A_8^* \equiv \sqrt{2}A_{[13]} \qquad [\underset{\approx}{E}_8^*] = \frac{[\underset{\sim}{e}_1\underset{\sim}{e}_3 - \underset{\sim}{e}_3\underset{\sim}{e}_1]}{\sqrt{2}} = \frac{1}{\sqrt{2}}\begin{bmatrix} 0 & 0 & 1 \\ 0 & 0 & 0 \\ -1 & 0 & 0 \end{bmatrix}, \tag{26.33h}$$

$$A_9^* \equiv \sqrt{2}A_{[21]} \qquad [\underset{\approx}{E}_9^*] = \frac{[\underset{\sim}{e}_2\underset{\sim}{e}_1 - \underset{\sim}{e}_1\underset{\sim}{e}_2]}{\sqrt{2}} = \frac{1}{\sqrt{2}}\begin{bmatrix} 0 & -1 & 0 \\ 1 & 0 & 0 \\ 0 & 0 & 0 \end{bmatrix}. \tag{26.33i}$$

These are called the *Mandel* components and the *Mandel basis*[2]. The introductions of $\sqrt{2}$ are analogous to the $\sqrt{2}$ factors that appear in equation (26.6) and equation (26.7). Because the Mandel basis is ortho*normal*, there is no need to distinguish between contravariant and covariant components. Hence, all Mandel indices are shown simply as subscripts. The Mandel system is marvelously well-suited for computations because all vector (and tensor) operations in this 9D space take indicial and matrix forms that are identical in structure to their counterparts in ordinary 3D space. For example, under the Mandel system, the inner product between two symmetric tensors, $\underset{\approx}{\boldsymbol{A}}$ and $\underset{\approx}{\boldsymbol{B}}$, is given by a formula that is identical in structure to the ordinary dot product between two Cartesian vectors. Namely,

$$\underset{\approx}{\boldsymbol{A}}:\underset{\approx}{\boldsymbol{B}} = A_K^* B_K^*. \tag{26.34}$$

Unlike equation (26.27), this formula involves no ungainly factors of 2. Unlike equation (26.30), this formula requires no metrics (because the metric matrix for an orthonormal basis is simply the identity matrix). Unlike equation (26.31), which required two different types of components (contravariant and covariant), the inner product using Mandel components uses components of the same type (there is no difference between contravariant and covariant components when the basis is orthonormal).

26.4 Voigt components of fourth-order minor-symmetric tensors

A fourth-order tensor $\underset{\approx\approx}{\boldsymbol{C}} = C_{ijkl}\underset{\sim}{\boldsymbol{e}}_i\underset{\sim}{\boldsymbol{e}}_j\underset{\sim}{\boldsymbol{e}}_k\underset{\sim}{\boldsymbol{e}}_l$ can be expanded in terms of the *Voigt* basis in any of the following forms:

$$\underset{\approx\approx}{\boldsymbol{C}} = C^{IJ}\underset{\approx}{\boldsymbol{E}}_I\underset{\approx}{\boldsymbol{E}}_J = C_{IJ}\underset{\approx}{\boldsymbol{E}}^I\underset{\approx}{\boldsymbol{E}}^J = C^I{}_J\underset{\approx}{\boldsymbol{E}}_I\underset{\approx}{\boldsymbol{E}}^J = C_I{}^J\underset{\approx}{\boldsymbol{E}}^I\underset{\approx}{\boldsymbol{E}}_J. \tag{26.35}$$

Two tensors written side-by-side with no multiplier between them (in this case, the basis tensors) are multiplied dyadically[3]. As was the case for second-order tensors, the Voigt components C^{IJ} are actually *contravariant* components.

In mechanics, second-order tensors are often symmetric, making their Voigt array reduce from size 9×1 down to 6×1. Similarly, fourth-order tensors are often minor-symmetric, making their 9×9 Voigt matrix reduce to a 6×6 matrix. For

[2] Mandel extensively used the notion of multiplying the off-diagonal components by a factor of $\sqrt{2}$ as a way to automatically incorporate the necessary factor of 2 in (26.27). The rigorous justification of this approach obtained through deducing the basis tensors and recognizing the distinction as a simple basis change was not, to our knowledge, ever mentioned by Mandel.

[3] Specifically, for any two tensors, $\underset{\sim}{\boldsymbol{A}}$ and $\underset{\sim}{\boldsymbol{B}}$, the expression $\underset{\sim}{\boldsymbol{A}}\underset{\sim}{\boldsymbol{B}}$ denotes a fourth-order tensor whose *ijkl* Cartesian components are $A_{ij}B_{kl}$.

minor-symmetric fourth-order tensors, the four types of Voigt components in equation (26.35) are related to ordinary four-index C_{ijkl} components as follows:

$$[C^{IJ}] = \begin{bmatrix} C_{1111} & C_{1122} & C_{1133} & C_{1123} & C_{1131} & C_{1112} \\ C_{2211} & C_{2222} & C_{2233} & C_{2223} & C_{2231} & C_{2212} \\ C_{3311} & C_{3322} & C_{3333} & C_{3323} & C_{3331} & C_{3312} \\ C_{2311} & C_{2322} & C_{2333} & C_{2323} & C_{2331} & C_{2312} \\ C_{3111} & C_{3122} & C_{3133} & C_{3123} & C_{3131} & C_{3112} \\ C_{1211} & C_{1222} & C_{1233} & C_{1223} & C_{1231} & C_{1212} \end{bmatrix}, \quad \text{symmetric if } \underset{\approx}{\mathbf{C}} \text{ is major symmetric,}$$

$$[C_{IJ}] = \begin{bmatrix} C_{1111} & C_{1122} & C_{1133} & 2C_{1123} & 2C_{1131} & 2C_{1112} \\ C_{2211} & C_{2222} & C_{2233} & 2C_{2223} & 2C_{2231} & 2C_{2212} \\ C_{3311} & C_{3322} & C_{3333} & 2C_{3323} & 2C_{3331} & 2C_{3312} \\ 2C_{2311} & 2C_{2322} & 2C_{2333} & 4C_{2323} & 4C_{2331} & 4C_{2312} \\ 2C_{3111} & 2C_{3122} & 2C_{3133} & 4C_{3123} & 4C_{3131} & 4C_{3112} \\ 2C_{1211} & 2C_{1222} & 2C_{1233} & 4C_{1223} & 4C_{1231} & 4C_{1212} \end{bmatrix}, \quad \text{symmetric if } \underset{\approx}{\mathbf{C}}$$

is major symmetric,

$$\left[C^{I}_{J}\right] = \begin{bmatrix} C_{1111} & C_{1122} & 2C_{1133} & 2C_{1123} & 2C_{1131} & 2C_{1112} \\ C_{2211} & C_{2222} & 2C_{2233} & 2C_{2223} & 2C_{2231} & 2C_{2212} \\ C_{3311} & C_{3322} & 2C_{3333} & 2C_{3323} & 2C_{3331} & 2C_{3312} \\ C_{2311} & C_{2322} & 2C_{2333} & 2C_{2323} & 2C_{2331} & 2C_{2312} \\ C_{3111} & C_{3122} & 2C_{3133} & 2C_{3123} & 2C_{3131} & 2C_{3112} \\ C_{1211} & C_{1222} & 2C_{1233} & 2C_{1223} & 2C_{1231} & 2C_{1212} \end{bmatrix}, \quad \text{not symmetric if } \underset{\approx}{\mathbf{C}}$$

is major symmetric,

$$\left[C^{J}_{I}\right] = \begin{bmatrix} C_{1111} & C_{1122} & C_{1133} & C_{1123} & C_{1131} & C_{1112} \\ C_{2211} & C_{2222} & C_{2233} & C_{2223} & C_{2231} & C_{2212} \\ C_{3311} & C_{3322} & C_{3333} & C_{3323} & C_{3331} & C_{3312} \\ 2C_{2311} & 2C_{2322} & 2C_{2333} & 2C_{2323} & 2C_{2331} & 2C_{2312} \\ 2C_{3111} & 2C_{3122} & 2C_{3133} & 2C_{3123} & 2C_{3131} & 2C_{3112} \\ 2C_{1211} & 2C_{1222} & 2C_{1233} & 2C_{1223} & 2C_{1231} & 2C_{1212} \end{bmatrix}, \quad \text{not symmetric if } \underset{\approx}{\mathbf{C}}$$

is major symmetric.

If the fourth-order tensor $\underset{\approx}{\mathbf{C}}$ is *major*-symmetric, then the contravariant and covariant matrices will be symmetric, but the mixed matrices will not be symmetric. This undesirable property is rectified when using Mandel components for which there is only *one* component matrix for fourth-order tensors, and it is symmetric whenever $\underset{\approx}{\mathbf{C}}$ is major-symmetric.

26.5 Mandel components of fourth-order minor-symmetric tensors

The Voigt system has many inconveniences that do not exist with the Mandel system. A fourth-order minor-symmetric tensor $\underset{\approx\approx}{C} = C_{ijkl}\underset{\sim}{e}_i\underset{\sim}{e}_j\underset{\sim}{e}_k\underset{\sim}{e}_l$ can be expanded in terms of the *Mandel* basis in a very familiar looking Cartesian form:

$$\underset{\approx\approx}{C} = C^*_{IJ}\underset{\approx}{E}^*_I\underset{\approx}{E}^*_J. \tag{26.36}$$

With the Mandel system, this is the only expansion (because, being an orthonormal system, there are no distinctions between covariant and contravariant components). Written as a 6 × 6 matrix, the Mandel components are

$$[C^*_{IJ}] = \begin{bmatrix} C_{1111} & C_{1122} & C_{1133} & \sqrt{2}\,C_{1123} & \sqrt{2}\,C_{1131} & \sqrt{2}\,C_{1112} \\ C_{2211} & C_{2222} & C_{2233} & \sqrt{2}\,C_{2223} & \sqrt{2}\,C_{2231} & \sqrt{2}\,C_{2212} \\ C_{3311} & C_{3322} & C_{3333} & \sqrt{2}\,C_{3323} & \sqrt{2}\,C_{3331} & \sqrt{2}\,C_{3312} \\ \sqrt{2}\,C_{2311} & \sqrt{2}\,C_{2322} & \sqrt{2}\,C_{2333} & 2C_{2323} & 2C_{2331} & 2C_{2312} \\ \sqrt{2}\,C_{3111} & \sqrt{2}\,C_{3122} & \sqrt{2}\,C_{3133} & 2C_{3123} & 2C_{3131} & 2C_{3112} \\ \sqrt{2}\,C_{1211} & \sqrt{2}\,C_{1222} & \sqrt{2}\,C_{1233} & 2C_{1223} & 2C_{1231} & 2C_{1212} \end{bmatrix}. \tag{26.37}$$

The Mandel components have factors of $\sqrt{2}$ in the in the last three rows and last three columns (i.e., locations of off-diagonal components). Consequently, the lower-right 3 × 3 sub-matrix is multiplied *twice* by these factors (once from the rows and once from the columns) to ultimately receive a multiplier of 2. With the Mandel system, if the fourth-order tensor is *major*-symmetric, then the 6 × 6 Mandel matrix will be symmetric. With the Mandel system, the fourth-order minor-symmetric identity tensor is simply the 6 × 6 identity matrix. In ugly contrast to this intuitive result, Voigt's matrix for the fourth-order identity tensor is *not* the 6 × 6 identity matrix. Aside from the strange insertions of $\sqrt{2}$, which you would not even notice when working numerical problems, the Mandel system is far preferable to the Voigt system for this and many other reasons.

Mandel components are especially convenient for tensor operations. For example, the Mandel components of the inverse tensor $\underset{\approx\approx}{C}^{-1}$ are found by simply inverting the Mandel matrix. Contrast this beautiful result with the situation for Voigt components. Taking the inverse of the Voigt contravariant matrix does *not* produce the contravariant Voigt components of the inverse; instead, it produces the *covariant* components of the inverse. Thus, if you use Voigt components, obtaining an inverse requires either pre- or post-adjustments of the matrices with factors or divisors of 2 (the metrics) to obtain the contravariant components. Deducing how to properly perform these adjustments is difficult, tedious, and error-prone.

26.6 Mandel components of fourth-order general tensors

As was done earlier for casting general second-order tensors as 9 × 1 arrays, the Mandel system for casting fourth-order tensors as 9 × 9 matrices applies to general (not necessarily minor-symmetric) tensors. The eighty-one independent components of a general fourth-order tensor may be displayed in terms of a 9×9 Mandel matrix as

$$\left[\begin{array}{ccc|ccc|ccc}
C_{(11)(11)} & C_{(11)(22)} & C_{(11)(33)} & \sqrt{2}C_{(11)(23)} & \sqrt{2}C_{(11)(31)} & \sqrt{2}C_{(11)(12)} & \sqrt{2}C_{(11)[32]} & \sqrt{2}C_{(11)[13]} & \sqrt{2}C_{(11)[21]} \\
C_{(22)(11)} & C_{(22)(22)} & C_{(22)(33)} & \sqrt{2}C_{(22)(23)} & \sqrt{2}C_{(22)(31)} & \sqrt{2}C_{(22)(12)} & \sqrt{2}C_{(22)[32]} & \sqrt{2}C_{(22)[13]} & \sqrt{2}C_{(22)[21]} \\
C_{(33)(11)} & C_{(33)(22)} & C_{(33)(33)} & \sqrt{2}C_{(33)(23)} & \sqrt{2}C_{(33)(31)} & \sqrt{2}C_{(33)(12)} & \sqrt{2}C_{(33)[32]} & \sqrt{2}C_{(33)[13]} & \sqrt{2}C_{(33)[21]} \\
\hline
\sqrt{2}C_{(23)(11)} & \sqrt{2}C_{(23)(22)} & \sqrt{2}C_{(23)(33)} & 2C_{(23)(23)} & 2C_{(23)(31)} & 2C_{(23)(12)} & 2C_{(23)[32]} & 2C_{(23)[13]} & 2C_{(23)[21]} \\
\sqrt{2}C_{(31)(11)} & \sqrt{2}C_{(31)(22)} & \sqrt{2}C_{(31)(33)} & 2C_{(31)(23)} & 2C_{(31)(31)} & 2C_{(31)(12)} & 2C_{(31)[32]} & 2C_{(31)[13]} & 2C_{(31)[21]} \\
\sqrt{2}C_{(12)(11)} & \sqrt{2}C_{(12)(22)} & \sqrt{2}C_{(12)(33)} & 2C_{(12)(23)} & 2C_{(12)(31)} & 2C_{(12)(12)} & 2C_{(12)[32]} & 2C_{(12)[13]} & 2C_{(12)[21]} \\
\hline
\sqrt{2}C_{[32](11)} & \sqrt{2}C_{[32](22)} & \sqrt{2}C_{[32](33)} & 2C_{[32](23)} & 2C_{[32](31)} & 2C_{[32](12)} & 2C_{[32][32]} & 2C_{[32][13]} & 2C_{[32][21]} \\
\sqrt{2}C_{[13](11)} & \sqrt{2}C_{[13](22)} & \sqrt{2}C_{[13](33)} & 2C_{[13](23)} & 2C_{[13](31)} & 2C_{[13](12)} & 2C_{[13][32]} & 2C_{[13][13]} & 2C_{[13][21]} \\
\sqrt{2}C_{[21](11)} & \sqrt{2}C_{[21](22)} & \sqrt{2}C_{[21](33)} & 2C_{[21](23)} & 2C_{[21](31)} & 2C_{[21](12)} & 2C_{[21][32]} & 2C_{[21][13]} & 2C_{[21][21]}
\end{array}\right].$$

Here, round parentheses around any pair of subscripts denote symmetrizing, and square brackets denote skew-symmetrizing. Specifically,

$$C_{(ij)(kl)} = \tfrac{1}{4}(C_{ijkl} + C_{jikl} + C_{ijlk} + C_{jilk}), \tag{26.38}$$

$$C_{(ij)[kl]} = \tfrac{1}{4}(C_{ijkl} + C_{jikl} - C_{ijlk} - C_{jilk}), \tag{26.39}$$

$$C_{[ij](kl)} = \tfrac{1}{4}(C_{ijkl} - C_{jikl} + C_{ijlk} - C_{jilk}), \tag{26.40}$$

$$C_{[ij][kl]} = \tfrac{1}{4}(C_{ijkl} - C_{jikl} - C_{ijlk} + C_{jilk}). \tag{26.41}$$

For example,

$$C_{(11)(22)} = C_{1122}, \tag{26.42}$$

$$C_{(11)(12)} = \tfrac{1}{2}(C_{1112} + C_{1121}), \tag{26.43}$$

$$C_{(11)[12]} = \tfrac{1}{2}(C_{1112} - C_{1121}), \tag{26.44}$$

$$C_{(23)[12]} = \tfrac{1}{4}(C_{2312} + C_{3212} - C_{2321} - C_{3221}), \tag{26.45}$$

$$C_{[21][32]} = \tfrac{1}{4}(C_{2132} - C_{1232} - C_{2123} + C_{1223}), \tag{26.46}$$

and so forth.

The partitioning of the 9×9 Mandel matrix into nine 3×3 submatrices can be identified by zero-valued permutations, positive permutations, and negative permutations of the two indices that go with each row or column. Zero-valued permutation pairs are 11, 22, 33. The positive permutation index pairs (23, 31, 12) are always surrounded by round parentheses to denote symmetrization. Negative permutation pairs [32, 13, 21] are always surrounded by square brackets to denote skew-symmetrization. If we say 'ZP' sub-matrix, then we are referring to the sub-matrix where the first index pairs are zero permutations and the second index pairs are positive permutations. Other submatrices are similar. In other words, the names used to describe the submatrices are

$$\left[\begin{array}{c|c|c} \mathbf{ZZ} & \mathbf{ZP} & \mathbf{ZN} \\ \hline \mathbf{PZ} & \mathbf{PP} & \mathbf{PN} \\ \hline \mathbf{NZ} & \mathbf{NP} & \mathbf{NN} \end{array}\right].$$

The fourth-order identity tensor (with $ijkl$ components $\delta_{ik}\delta_{jl}$) has

$$[\mathbf{ZZ}] = [\mathbf{PP}] = [\mathbf{NN}] = [\mathrm{I}], \text{ with all other submatrices zero.} \tag{26.47}$$

In other words, the Mandel matrix for the fourth-order identity tensor is the 9×9 identity matrix, which is an intuitive result that does not hold for the contravariant Voigt system.

The fourth-order identity on *symmetric* tensor space (for which the $ijkl$ components are $\frac{1}{2}(\delta_{ik}\delta_{jl} + \delta_{il}\delta_{jk})$) has

$$[\mathbf{ZZ}] = [\mathbf{PP}] = [\mathrm{I}], \text{ with all other submatrices zero.} \tag{26.48}$$

When this fourth-order tensor operates on a general second-order tensor, the result is the symmetric part of the input tensor.

The fourth-order identity on *skew* tensor space $\frac{1}{2}(\delta_{ik}\delta_{jl} - \delta_{il}\delta_{jk})$ has

$$[\mathbf{NN}] = [\mathrm{I}], \text{ with all other submatrices zero.} \tag{26.49}$$

When this fourth-order tensor operates on a general second-order tensor, the result is the skew part of the input tensor.

The fourth-order tensor with $ijkl$ components given by $\delta_{ij}\delta_{kl}$ has

$$[\mathbf{ZZ}] = \begin{bmatrix} 1 & 1 & 1 \\ 1 & 1 & 1 \\ 1 & 1 & 1 \end{bmatrix}, \text{ with all other submatrices zero.} \tag{26.50}$$

When this fourth-order tensor operates on a general second-order tensor, the result is the isotropic part of the input tensor times three.

26.7 Fourth-order linear transformations

A linear-elastic stress-strain model

$$\sigma_{ij} = C_{ijkl}\varepsilon_{kl}, \tag{26.51}$$

is a classical example of a fourth-order transformation in which the symmetric strain tensor ε_{kl} is operated on (linearly) by the minor-symmetric stiffness tensor C_{ijkl} to produce the symmetric stress tensor σ_{ij}.

Below are listed three ways to write the above equation in matrix form

$$\begin{bmatrix}\sigma_{11}\\ \sigma_{22}\\ \sigma_{33}\\ \sigma_{23}\\ \sigma_{31}\\ \sigma_{12}\end{bmatrix} = \begin{bmatrix} C_{1111} & C_{1122} & C_{1133} & C_{1123} & C_{1131} & C_{1112}\\ C_{2211} & C_{2222} & C_{2233} & C_{2223} & C_{2231} & C_{2212}\\ C_{3311} & C_{3322} & C_{3333} & C_{3323} & C_{3331} & C_{3312}\\ C_{2311} & C_{2322} & C_{2333} & C_{2323} & C_{2331} & C_{2312}\\ C_{3111} & C_{3122} & C_{3133} & C_{3123} & C_{3131} & C_{3112}\\ C_{1211} & C_{1222} & C_{1233} & C_{1223} & C_{1231} & C_{1212}\end{bmatrix} \begin{bmatrix}\varepsilon_{11}\\ \varepsilon_{22}\\ \varepsilon_{33}\\ \mathbf{2}\varepsilon_{23}\\ \mathbf{2}\varepsilon_{31}\\ \mathbf{2}\varepsilon_{12}\end{bmatrix}, \tag{26.52}$$

this is a Voigt approach.

$$\begin{bmatrix}\sigma_{11}\\ \sigma_{22}\\ \sigma_{33}\\ \sigma_{23}\\ \sigma_{31}\\ \sigma_{12}\end{bmatrix} = \begin{bmatrix} C_{1111} & C_{1122} & C_{1133} & \mathbf{2}C_{1123} & \mathbf{2}C_{1131} & \mathbf{2}C_{1112}\\ C_{2211} & C_{2222} & C_{2233} & \mathbf{2}C_{2223} & \mathbf{2}C_{2231} & \mathbf{2}C_{2212}\\ C_{3311} & C_{3322} & C_{3333} & \mathbf{2}C_{3323} & \mathbf{2}C_{3331} & \mathbf{2}C_{3312}\\ C_{2311} & C_{2322} & C_{2333} & \mathbf{2}C_{2323} & \mathbf{2}C_{2331} & \mathbf{2}C_{2312}\\ C_{3111} & C_{3122} & C_{3133} & \mathbf{2}C_{3123} & \mathbf{2}C_{3131} & \mathbf{2}C_{3112}\\ C_{1211} & C_{1222} & C_{1233} & \mathbf{2}C_{1223} & \mathbf{2}C_{1231} & \mathbf{2}C_{1212}\end{bmatrix} \begin{bmatrix}\varepsilon_{11}\\ \varepsilon_{22}\\ \varepsilon_{33}\\ \varepsilon_{23}\\ \varepsilon_{31}\\ \varepsilon_{12}\end{bmatrix}, \tag{26.53}$$

this is also a Voigt approach.

$$\begin{bmatrix}\sigma_{11}\\ \sigma_{22}\\ \sigma_{33}\\ \sqrt{\mathbf{2}}\sigma_{23}\\ \sqrt{\mathbf{2}}\sigma_{31}\\ \sqrt{\mathbf{2}}\sigma_{12}\end{bmatrix} = \begin{bmatrix} C_{1111} & C_{1122} & C_{1133} & \sqrt{\mathbf{2}}C_{1123} & \sqrt{\mathbf{2}}C_{1131} & \sqrt{\mathbf{2}}C_{1112}\\ C_{2211} & C_{2222} & C_{2233} & \sqrt{\mathbf{2}}C_{2223} & \sqrt{\mathbf{2}}C_{2231} & \sqrt{\mathbf{2}}C_{2212}\\ C_{3311} & C_{3322} & C_{3333} & \sqrt{\mathbf{2}}C_{3323} & \sqrt{\mathbf{2}}C_{3331} & \sqrt{\mathbf{2}}C_{3312}\\ \sqrt{\mathbf{2}}C_{2311} & \sqrt{\mathbf{2}}C_{2322} & \sqrt{\mathbf{2}}C_{2333} & \sqrt{\mathbf{2}}C_{2323} & \sqrt{\mathbf{2}}C_{2331} & \sqrt{\mathbf{2}}C_{2312}\\ \sqrt{\mathbf{2}}C_{3111} & \sqrt{\mathbf{2}}C_{3122} & \sqrt{\mathbf{2}}C_{3133} & \sqrt{\mathbf{2}}C_{3123} & \sqrt{\mathbf{2}}C_{3131} & \sqrt{\mathbf{2}}C_{3112}\\ \sqrt{\mathbf{2}}C_{1211} & \sqrt{\mathbf{2}}C_{1222} & \sqrt{\mathbf{2}}C_{1233} & \sqrt{\mathbf{2}}C_{1223} & \sqrt{\mathbf{2}}C_{1231} & \sqrt{\mathbf{2}}C_{1212}\end{bmatrix} \begin{bmatrix}\varepsilon_{11}\\ \varepsilon_{22}\\ \varepsilon_{33}\\ \sqrt{\mathbf{2}}\varepsilon_{23}\\ \sqrt{\mathbf{2}}\varepsilon_{31}\\ \sqrt{\mathbf{2}}\varepsilon_{12}\end{bmatrix} \tag{26.54}$$

this is the Mandel approach.

These three matrix expressions can be written in terms of Voigt and Mandel basis expansions, respectively, as

$$\sigma^I = C^{IJ}\varepsilon_J, \quad \text{where} \quad \varepsilon_J \equiv G_{JK}\varepsilon^K \quad \text{Voigt}, \tag{26.55}$$

$$\sigma^I = C^I{}_J\varepsilon^J, \quad \text{where} \quad C^I{}_J \equiv C^{IK}G_{KJ} \quad \text{Voigt}, \tag{26.56}$$

$$\sigma_I^* = C_{IJ}^*\varepsilon_J^* \quad \text{Mandel}. \tag{26.57}$$

Only the Mandel formulation, equation (26.57) is computed in a manner that is identical in structure to ordinary Cartesian transformations. The Voigt operations all require computation of covariant or mixed tensor components. For example, equation (26.52) requires insertion of a factor of 2 on the strain array. The indicial equation (26.55) makes it clearer that this adjustment of the contravariant Voigt strain vector is actually creating the *covariant* Voigt strain vector. The second matrix expression, equation (26.53), puts the factor of 2 into the stiffness matrix, thereby converting it to *mixed* Voigt components. That approach permits both the stress and the strain vectors to be of the same type (contravariant), but introduction of factors of 2 into the stiffness produces a matrix that is not symmetric even though the stiffness tensor is major-symmetric. Only the Mandel formulation permits the stress and strain vectors to be of the same type (Mandel) and also permits the stiffness to retain its major symmetry.

Another compelling reason to use the Mandel system arises in the context of finding the eigenvalues and eigentensors of a fourth-order tensor.

26.8 Spectral analysis of fourth-order tensors

The eigenproblem for a fourth-order system (which is important to identify material symmetries in constitutive modeling) seeks eigen*tensors* X_{ij} and eigenvalues λ for which

$$C_{ijkl}X_{kl} = \lambda X_{kl}. \tag{26.58}$$

Referring to equation (26.55), equation (26.56), and equation (26.57), this eigenproblem can be cast in the form of a matrix eigenproblem in 9D space in any of the following ways,

$$[C^{IJ}]\{X_J\} = \lambda[G^{IJ}]\{X_J\} \quad \text{Voigt}, \tag{26.59}$$

$$[C^I{}_J]\{X^J\} = \lambda\{X^I\} \quad \text{Voigt}, \tag{26.60}$$

$$C^*_{IJ}X^*_J = \lambda X^*_I, \quad \text{Mandel}. \tag{26.61}$$

For the first of these expressions, the contravariant metric matrix must be inserted to convert a contravariant component array to its covariant form to ensure that both sides of the equation are working with the same type of 9D vector for $\underset{\sim}{X}$ (rather than covariant on the left-hand side but contravariant on the right). Inserting the metric matrix changes the standard eigenproblem to a general eigenproblem, for which solvers might not be readily available. The second of these equations is a standard eigenproblem. However, if $\underset{\approx}{C}$ is a major-symmetric tensor, then its mixed components in the second equation will *not* be symmetric, hence disallowing a solver that is optimized for symmetric matrices. Only the last of the above three equations—the Mandel statement of the eigenproblem—is free from these drawbacks. In this respect, the Mandel formulation is superior.

If $\underset{\approx}{C}$ is minor-symmetric, then the eigenproblem is cast in Mandel form as

$$\begin{bmatrix} C_{1111} & C_{1122} & C_{1133} & \sqrt{2}C_{1123} & \sqrt{2}C_{1131} & \sqrt{2}C_{1112} \\ C_{2211} & C_{2222} & C_{2233} & \sqrt{2}C_{2223} & \sqrt{2}C_{2231} & \sqrt{2}C_{2212} \\ C_{3311} & C_{3322} & C_{3333} & \sqrt{2}C_{3323} & \sqrt{2}C_{3331} & \sqrt{2}C_{3312} \\ \sqrt{2}C_{2311} & \sqrt{2}C_{2322} & \sqrt{2}C_{2333} & 2C_{2323} & 2C_{2331} & 2C_{2312} \\ \sqrt{2}C_{3111} & \sqrt{2}C_{3122} & \sqrt{2}C_{3133} & 2C_{3123} & 2C_{3131} & 2C_{3112} \\ \sqrt{2}C_{1211} & \sqrt{2}C_{1222} & \sqrt{2}C_{1233} & 2C_{1223} & 2C_{1231} & 2C_{1212} \end{bmatrix} \begin{bmatrix} X_{11} \\ X_{22} \\ X_{33} \\ \sqrt{2}X_{23} \\ \sqrt{2}X_{31} \\ \sqrt{2}X_{12} \end{bmatrix} = \lambda \begin{bmatrix} X_{11} \\ X_{22} \\ X_{33} \\ \sqrt{2}X_{23} \\ \sqrt{2}X_{31} \\ \sqrt{2}X_{12} \end{bmatrix}. \tag{26.62}$$

This is a standard *matrix* eigenvalue problem for which the 6×6 matrix is symmetric whenever $\underset{\approx}{C}$ is major-symmetric. In elasticity theory, the stiffness tensor is positive-definite, making the Mandel matrix positive-definite as well. The Mandel system therefore allows one to use commercially available linear algebra software that is optimized for symmetric positive-definite matrices. Such is not the case with the Voigt system.

Once a 6D Mandel eigenvector is determined, then it can be converted to a 6D contravariant Voigt matrix by dividing the last three components of a Mandel eigenvector by $\sqrt{2}$. Equation (26.21) may then be used to convert the contravariant Voigt matrix to the ordinary 3×3 Cartesian component matrix for the eigen*tensor*.

26.8.1 Example: elastic stiffness

For isotropic linear elasticity, the stiffness tensor has the following Mandel matrix:

$$\begin{bmatrix} \left(K+\frac{4}{3}G\right) & \left(K-\frac{2}{3}G\right) & \left(K-\frac{2}{3}G\right) & 0 & 0 & 0 \\ \left(K-\frac{2}{3}G\right) & \left(K+\frac{4}{3}G\right) & \left(K-\frac{2}{3}G\right) & 0 & 0 & 0 \\ \left(K-\frac{2}{3}G\right) & \left(K-\frac{2}{3}G\right) & \left(K+\frac{4}{3}G\right) & 0 & 0 & 0 \\ 0 & 0 & 0 & 2G & 0 & 0 \\ 0 & 0 & 0 & 0 & 2G & 0 \\ 0 & 0 & 0 & 0 & 0 & 2G \end{bmatrix}, \tag{26.63}$$

where K and G are the bulk and shear moduli, respectively. This matrix has an eigenvalue $3K$ with multiplicity 1 and associated 9D eigenvector $\{1, 1, 1, 0, 0, 0\}$, which is the identity matrix when cast in conventional 3×3 matrix form[4]. In elasticity, this means that an isotropic strain produces an isotropic stress that is $3K$ times that strain. The stiffness matrix also has a second eigenvalue, $2G$ with multiplicity five. The associated 9D orthonormal eigenvectors are

[4] The associated unit eigenvector must be divided by $\sqrt{3}$ to normalize.

$$\frac{1}{\sqrt{2}}\begin{bmatrix}-1\\0\\1\\0\\0\\0\end{bmatrix},\quad \frac{1}{\sqrt{6}}\begin{bmatrix}1\\-2\\1\\0\\0\\0\end{bmatrix},\quad \begin{bmatrix}0\\0\\0\\1\\0\\0\end{bmatrix},\quad \begin{bmatrix}0\\0\\0\\0\\1\\0\end{bmatrix},\quad \begin{bmatrix}0\\0\\0\\0\\0\\1\end{bmatrix}. \tag{26.64}$$

Casting these in conventional 3×3 form gives the unit eigentensors

$$\frac{1}{\sqrt{2}}\begin{bmatrix}-1&0&0\\0&0&0\\0&0&1\end{bmatrix},\ \frac{1}{\sqrt{6}}\begin{bmatrix}1&0&0\\0&-2&0\\0&0&1\end{bmatrix},\ \frac{1}{\sqrt{2}}\begin{bmatrix}0&0&0\\0&0&1\\0&1&0\end{bmatrix},\ \frac{1}{\sqrt{2}}\begin{bmatrix}0&0&1\\0&0&0\\1&0&0\end{bmatrix},\ \frac{1}{\sqrt{2}}\begin{bmatrix}0&1&0\\1&0&0\\0&0&0\end{bmatrix}. \tag{26.65}$$

These five tensors span the set of all tensors that are both symmetric and deviatoric. Therefore any such tensor is an eigentensor with eigenvalue $2G$. Physically, this means that any symmetric deviatoric strain produces a stress that equals $2G$ times the strain.

Incidentally, since the eigenvalues are $3K$ and $2G$, positive definiteness of the linear-elastic stiffness is ensured if and only if K and G are both positive. This implies that Poisson's ratio must fall between -1 and $1/2$. Do an internet search to see that negative Poisson's ratio really is possible—think of the Hoberman sphere, which expands laterally when pulled axially!

Because $2G$ is an eigenvalue of multiplicity 5, the associated eigentensors, shown as 6D vectors in equation (26.64) are not unique. Any linear combination of those vectors is itself an eigenvector. Any vector whose first three components sum to zero is an eigenvector. Even when there is only single multiplicity, as the case for $3K$, the normalized eigentensor is not unique (both $\underset{\approx}{\boldsymbol{I}}$ and $-\underset{\approx}{\boldsymbol{I}}$ can be normalized). What is unique, however, is the *eigenspace* associated with an eigenvalue. For the eigenvalue $3K$, the eigenspace is a hyperline (called the hydrostat) passing through the origin in 9D tensor space such that any tensor on that line is a multiple of $\underset{\approx}{\boldsymbol{I}}$. The eigenspace associated with $2G$ is a hyperplane spanned by the eigenvectors, and this is the set of tensors that are both symmetric and deviatoric. There is an additional eigenvalue of zero, with associated 3D eigenspace being the set of all skew tensors..

Reference

[1] Helnwein P 2001 Some remarks on the compressed matrix representation of symmetric second-order and fourth-order tensors *Comput. Meth. Appl. Mech. Eng.* **190** 2753–70

Chapter 27

Higher-order rotations

In everyday engineering tensor analysis, second-order tensors have 3×3 component matrices and they characterize linear transformations between ordinary (three-component) vectors. This book has focused almost exclusively on rotations in three-dimensional (3D) physical space, in which ordinary three-component vectors are rotated into new orientations by application of a 3×3 component rotation tensor $\underset{\approx}{\boldsymbol{R}}$ (occasionally alternatively denoted $\underset{\approx}{\boldsymbol{Q}}$). When rotating a vector, for example, we wrote $\underset{\sim}{\boldsymbol{v}} = \underset{\approx}{\boldsymbol{R}} \bullet \underset{\sim}{\bar{\boldsymbol{v}}}$. When rotating a tensor we wrote $\underset{\approx}{\boldsymbol{A}} = \underset{\approx}{\boldsymbol{R}} \bullet \underset{\approx}{\overline{\boldsymbol{A}}} \bullet \underset{\approx}{\boldsymbol{R}}^T$, and so forth. These operations relate to ordinary rotations in ordinary physical space. Constitutive models—especially plasticity models—also deal with higher-dimensional spaces which correspond to tensor space itself.

Material constitutive modeling deals with transformations between second-order tensors. For example, the strain tensor is transformed to the stress tensor in constitutive models such as Hooke's law. In most material models, the tensors of interest are symmetric, which means they belong to a 6D subspace of general 9D tensor space. For real materials, constitutive functions are generally nonlinear, but they become linear in incremental or rate form[1]. For example, the stress rate $\dot{\sigma}_{ij}$ might be given by a fourth-order tangent stiffness tensor E_{ijkl} operating on the strain rate $\dot{\varepsilon}_{kl}$. Consequently, fourth-order tensors play a role in constitutive transformations from tensors to tensors that is similar to the role played by ordinary second-order tensors in everyday transformations from vectors to vectors. Just as you can define operations that rotate vectors in ordinary 3D vector space, you can also define

[1] To understand this assertion, consider a nonlinear function $y = f(x)$. In rate form, this becomes $\dot{y} = f(x)\dot{x}$, which is *linear* with respect to the rates $f'(x)$. The function is the slope of the local tangent to the generally nonlinear curve $f(x)$. In constitutive modeling, the analog of $f'(x)$ is a fourth-order tensor typically referred to as the tangent stiffness.

doi:10.1016/978-0-7503-1454-1ch27

operations that rotate *tensors* in 6D symmetric tensor space. Recall that, for ordinary 3D rotations, the rotation operator was generally a fully populated 3 × 3 matrix, which simplified to a simpler structure if the basis was changed from the lab basis to one aligned with the rotation axis. Likewise, a higher-order rotation operation in 6D symmetric tensor space can be cast as fully populated 6 × 6 matrix with respect to the standard basis dyads, but simplifies in structure when cast in terms of some other basis for symmetric tensor space. The purpose of this chapter is to expound on these concepts at an introductory level.

27.1 Rotators: fourth-order rotations in Mandel form

In constitutive modeling, it is very common to generate a tensor $\underset{\approx}{\boldsymbol{B}}$ by rotating a different tensor $\underset{\approx}{\boldsymbol{A}}$ according to

$$\underset{\approx}{\boldsymbol{B}} = \underset{\approx}{\boldsymbol{Q}}^T \bullet \underset{\approx}{\boldsymbol{A}} \bullet \underset{\approx}{\boldsymbol{Q}}, \tag{27.1}$$

where $\underset{\approx}{\boldsymbol{Q}}$ is orthogonal. In indicial notation, this operation may be written

$$B_{ij} = Q_{pi}Q_{qj}A_{pq}, \tag{27.2}$$

or

$$B_{ij} = Y_{pqij}A_{pq}, \quad \text{where} \quad Y_{pqij} = Q_{pi}Q_{qj}. \tag{27.3}$$

The fourth-order tensor Y_{pqij} is called a *rotator* operator. Like an ordinary 3D rotation, it transforms its tensor argument to become a new tensor of the same magnitude but different orientation. Quite frequently, the tensors $\underset{\approx}{\boldsymbol{A}}$ and $\underset{\approx}{\boldsymbol{B}}$ are symmetric, in which case, minor symmetry may be imposed without loss in generality to the tensor Y_{ijpq} so that the above operation may be written

$$B_{ij} = U_{ijpq}A_{pq}, \tag{27.4}$$

where U_{pqij} is a specialized rotator for symmetric tensors defined

$$U_{pqij} = \frac{1}{4}(Q_{pi}Q_{qj} + Q_{qi}Q_{pj} + Q_{pj}Q_{qi} + Q_{qj}Q_{pi}). \tag{27.5}$$

Using Mandel components, this fourth-order tensor may be expressed as a 6 × 6 matrix

$$[U^*] = \begin{bmatrix} Q_{11}^2 & Q_{12}^2 & Q_{13}^2 & \sqrt{2}Q_{12}Q_{13} & \sqrt{2}Q_{11}Q_{13} & \sqrt{2}Q_{11}Q_{12} \\ Q_{21}^2 & Q_{22}^2 & Q_{23}^2 & \sqrt{2}Q_{22}Q_{23} & \sqrt{2}Q_{21}Q_{23} & \sqrt{2}Q_{21}Q_{22} \\ Q_{31}^2 & Q_{32}^2 & Q_{33}^2 & \sqrt{2}Q_{32}Q_{33} & \sqrt{2}Q_{31}Q_{33} & \sqrt{2}Q_{31}Q_{32} \\ \sqrt{2}Q_{21}Q_{31} & \sqrt{2}Q_{22}Q_{32} & \sqrt{2}Q_{23}Q_{33} & (Q_{23}Q_{32}+Q_{22}Q_{33}) & (Q_{23}Q_{31}+Q_{21}Q_{33}) & (Q_{22}Q_{31}+Q_{21}Q_{32}) \\ \sqrt{2}Q_{11}Q_{31} & \sqrt{2}Q_{12}Q_{32} & \sqrt{2}Q_{13}Q_{33} & (Q_{13}Q_{32}+Q_{12}Q_{33}) & (Q_{13}Q_{31}+Q_{11}Q_{33}) & (Q_{12}Q_{31}+Q_{11}Q_{32}) \\ \sqrt{2}Q_{11}Q_{21} & \sqrt{2}Q_{12}Q_{22} & \sqrt{2}Q_{13}Q_{23} & (Q_{13}Q_{22}+Q_{12}Q_{23}) & (Q_{13}Q_{21}+Q_{11}Q_{23}) & (Q_{12}Q_{21}+Q_{11}Q_{22}) \end{bmatrix}. \tag{27.6}$$

In terms of a basis for which the 3-direction is aligned with the rotation axis of $\underset{\approx}{\boldsymbol{Q}}$, this component matrix simplifies to

$$[U^*] = \begin{bmatrix} c^2 & s^2 & 0 & 0 & 0 & -\sqrt{2}\,cs \\ s^2 & c^2 & 0 & 0 & 0 & \sqrt{2}\,cs \\ 0 & 0 & 1 & 0 & 0 & 0 \\ 0 & 0 & 0 & c & s & 0 \\ 0 & 0 & 0 & -s & c & 0 \\ \sqrt{2}\,cs & -\sqrt{2}\,cs & 0 & 0 & 0 & c^2 - s^2 \end{bmatrix}, \tag{27.7}$$

where c and s are, respectively, the cosine and sine of the rotation angle associated with $[Q]$. Note that this matrix involves both the rotation angle α and *twice* the rotation angle (because $cs = \frac{1}{2}\sin 2\alpha$ and $c^2 - s^2 = \cos 2\alpha$). The double angle appears for the same reason double angles appear in Mohr's circle transformation formulas. The sub-matrix in which c and s are *not* squared shows the rotation of the 23 and 31 shear stresses. Unlike the in-plane stress rotations, which ultimately involve 2α, the 23 and 31 shear stresses transform simply by α.

In terms of the 6×1 Mandel arrays for the symmetric tensors $\underset{\approx}{\boldsymbol{A}}$ and $\underset{\approx}{\boldsymbol{B}}$, the operation in (27.4) may be performed via standard matrix manipulations as

$$\{B^*\} = [U^*]^T\{A^*\}. \tag{27.8}$$

Similarly, a *fourth-order* tensor a_{ijkl} (which might, for example, represent an anisotropic tangent stiffness) is often transformed into a different frame (often the material frame) to produce a new tensor b_{ijkl} defined by

$$b_{ijkl} = Q_{pi}Q_{qj}Q_{rk}Q_{sl}a_{pqrs} \tag{27.9}$$

or

$$b_{ijkl} = Y_{pqij}Y_{rskl}a_{pqrs}. \tag{27.10}$$

If the fourth-order tensors are minor-symmetric, this operation may be written

$$b_{ijkl} = U_{pqij}U_{rskl}a_{pqrs}, \tag{27.11}$$

which is computed using Mandel arrays as

$$[b^*] = [U^*]^T[a^*][U^*]. \tag{27.12}$$

Performing frame transformations in this way permits direct computation of Mandel arrays and matrices without having to first revert back to conventional notation, giving a slight boost in computational efficiency.

27.2 Fourth-order 'focused identity' (projection) tensors

For many physical applications areas, a researcher's attention is focused on *subsets* of second-order tensors, rather than on the entire 9D space of every possible tensor.

For example, non-polar[2] constitutive models are rules by which one symmetric tensor (e.g., strain) is transformed into another symmetric tensor (e.g., stress), in which case the modeler's attention is focused on the 6D space of symmetric tensors and constitutive functions (tensor-to-tensor transformations) are characterized by fourth-order *minor*-symmetric tensors. This section discusses the concept of a 'focused identity' tensor corresponding to a linear subset of focused attention (e.g., the set of all symmetric tensors). A 'focused identity' is really just a projection operator. When acting on a member of the focus set, it returns that member unchanged (like an identity operation). When acting on a general member of tensor space, a focused identity operator returns the projection of that tensor onto the focus space. If, for example, the focus is symmetric tensors, then the focused identity for symmetric tensors is the projection operator that will return the symmetric part of any general tensor.

If a tensor subset of interest is closed with respect to addition and scalar multiplication, then the subset is called a *subspace* or *linear manifold.* For example, the set of all unit tensors is *not* a subspace because a linear combination of unit tensors is not itself a unit tensor. The set of all symmetric tensors *is* a subspace because a linear combination of symmetric tensors is itself symmetric. Geometrically, subspaces are analogous to planes or lines that pass through the origin. The term 'hyper-plane' is sometimes used to remind readers that the plane might be embedded within a higher-dimensional space.

Whenever focus is directed toward a subspace, it is always possible (through the projection theorem) to decompose a general 9D tensor into two parts, one contained within the focus space and another normal to the focus space. The part of a tensor contained in the focus space can be obtained through a linear projection operator. In matrix form, projection operators look like 'pieces of the identity', so we will refer to such projection tensors as 'focused identity' tensors.

In the sections that follow, we will presume that there exists a space of focused attention and rotation operations are sought within that focus space. The focused identity for a focus space will always be denoted Δ_{ijkl}, and its components will depend on the focus space. Some examples are listed below.

If focusing on symmetric tensors, then

$$\Delta_{ijkl} = \frac{1}{2}(\delta_{ik}\delta_{jl} + \delta_{il}\delta_{jk}) \tag{27.13}$$

and therefore $\Delta_{ijkl}A_{kl} = A_{ij}$ for any *symmetric* second-order tensor $\underset{\sim}{A}$. Also note that $\Delta_{ijkl}A_{kl} = \frac{1}{2}(A_{ij} + A_{ij}^T)$ for a general tensor and $\Delta_{ijkl}A_{kl} = 0$ for a skew tensor.

---

[2] A material or system is called 'polar' if it potentially supports microscale distributed moments. Examples can include ferroelectrics whose embedded dipoles might twist in an electric field or porous materials whose reentrant ligaments can induce microscopic torques. For polar materials, the stress tensor is not presumed to be symmetric.

If focusing on deviatoric tensors, then

$$\Delta_{ijkl} = \delta_{ik}\delta_{jl} - \tfrac{1}{3}(\delta_{ij}\delta_{kl}) \tag{27.14}$$

and therefore $\Delta_{ijkl}A_{kl} = A_{ij}$ for any *deviatoric* second-order tensor $\underset{\approx}{A}$. Also note that $\Delta_{ijkl}A_{kl} = A_{ij} - \tfrac{1}{3}A_{kk}\delta_{ij}$ for a general tensor and $\Delta_{ijkl}A_{kl} = 0$ for an isotropic tensor.

If focusing on the set of tensors that are both deviatoric and symmetric, then

$$\Delta_{ijkl} = \tfrac{1}{2}(\delta_{ik}\delta_{jl} + \delta_{il}\delta_{jk}) - \tfrac{1}{3}(\delta_{ij}\delta_{kl}) \tag{27.15}$$

and therefore $\Delta_{ijkl}A_{kl} = A_{ij}$ for any symmetric-deviatoric tensor $\underset{\approx}{A}$. Also note that $\Delta_{ijkl}A_{kl} = \tfrac{1}{2}(A_{ij} + A_{ij}^T) - \tfrac{1}{3}A_{kk}\delta_{ij}$ for a general tensor and $\Delta_{ijkl}A_{kl} = 0$ for an isotropic or skew tensor.

In plasticity theory, 'stress space' is a specialized type of focus set. Stress space is defined relative to a (known) stress tensor in a material. Specifically, stress space is also called 'commuting space' because it is the set of all symmetric tensors $\underset{\approx}{X}$ that commute with the stress tensor $\underset{\approx}{\sigma}$ (in other words, $\underset{\approx}{X} \bullet \underset{\approx}{\sigma} = \underset{\approx}{\sigma} \bullet \underset{\approx}{X}$). Characterizing the focused identity is complicated, and readers who are not interested in plasticity should skip ahead now to the next section.

It is straightforward to prove that any member of commuting space will share a common set of eigenvectors with the stress. If the eigenvalues of stress are all distinct, this shared set of eigenvectors will be the same for every member of commuting space. On the other hand, if the stress has any *repeated* eigenvalues, then even though a shared set of eigenvectors exists, the shared eigenvector triads are generally different for different members of commuting space. Consider the degenerate case that the stress is isotropic; then *any* tensor will commute with it and, because any vector is be an eigenvector for the stress, the shared eigenvectors are simply the eigenvectors of each member of commuting space, $\underset{\approx}{X}$ (different members will have different shared eigenvectors). Because the nature of commuting stress space depends on the number of distinct eigenvalues, we will have to define the focused identity tensor appropriate to each possibility.

Suppose the stress has all distinct eigenvalues. Then it has three *unique*[3] rank-1 eigenprojector tensors $\underset{\approx}{P}_k$ formed by the dyadic (outer) product $\underset{\sim}{p}_k\underset{\sim}{p}_k$ (no sum) formed by each of the orthonormalized eigenvectors $\{\underset{\sim}{p}_1, \underset{\sim}{p}_2, \underset{\sim}{p}_3\}$. When the stress has distinct eigenvalues, then the focused identity for commuting space would be

$$\underset{\approx\approx}{\Delta} = \underset{\approx}{P}_1\underset{\approx}{P}_1 + \underset{\approx}{P}_2\underset{\approx}{P}_2 + \underset{\approx}{P}_3\underset{\approx}{P}_3 \tag{27.16}$$

[3] The orthonormalized eigenvectors have unique orientations, but their direction is arbitrary. The eigenprojectors, on the other hand, are truly unique.

and therefore $\Delta_{ijkl}A_{kl} = A_{ij}$ for any second-order tensor $\underset{\approx}{A}$ with the same eigenvectors as the stress.

When the stress has exactly *two* distinct eigenvalues (i.e., when an eigenvalue has a multiplicity of two, making the stress axisymmetric, and confusingly often called triaxial), then the dimension of the associated stress space increases. Letting $\underset{\sim}{n}$ represent the triaxial unit axis (i.e., the normalized eigenvector of the eigenvalue of multiplicity one), the focused identity is then found by minor-symmetrizing the expression $P_{ik}P_{jl} + Q_{ik}Q_{jl}$, where $P_{ij} = n_i n_j$ and $Q_{ij} = \delta_{ij} - n_i n_j$. Thus for triaxial stresses, the focused identity is

$$\Delta_{ijkl} = 2(n_i n_j n_k n_l) + \frac{1}{2}(\delta_{ik}\delta_{jl} + \delta_{il}\delta_{jk}) - \frac{1}{2}(n_i n_k \delta_{jl} + \delta_{ik} n_j n_l + n_j n_k \delta_{il} + \delta_{jk} n_i n_l) \quad (27.17)$$

and therefore $\Delta_{ijkl}A_{kl} = A_{ij}$ for any second-order tensor $\underset{\approx}{A}$ that commutes with the axisymmetric stress.

When the stress has exactly one distinct eigenvalue (which therefore has multiplicity of three), then it is isotropic and *any* tensor will commute with it. Consequently, since we are focused on *symmetric* tensors that commute with the stress, the focused identity for isotropic stress is

$$\Delta_{ijkl} = \frac{1}{2}(\delta_{ik}\delta_{jl} + \delta_{il}\delta_{jk}). \quad (27.18)$$

27.3 Focused rotations

Just as one can define a rotation operation in 3D as a transformation that preserves vector lengths and angles, we can also generalize this concept to higher-dimensional length- and angle-preserving transformations of tensors. For ordinary 3D tensor analysis, a second-order tensor R_{ij} is orthogonal if $R_{ik}R_{jk} = \delta_{ij}$; in this case, the right-hand side is the identity tensor. Analogously, a fourth-order tensor R_{ijkl} is herein said to be *focused orthogonal* if

$$R_{ijmn}R_{klmn} = \Delta_{ijkl}. \quad (27.19)$$

Here, Δ_{ijkl} denotes the *focused identity tensor* appropriate to the second-order tensor space or subspace of interest. If you seek rotation operations that act on the entirety of 9D second-order tensor space, then the focused identity is the regular fourth-order identity.

$$\Delta_{ijkl} = \delta_{ik}\delta_{jl}. \quad (27.20)$$

With this choice, $\Delta_{ijkl}A_{kl} = A_{ij}$ for any second-order tensor $\underset{\approx}{A}$.

In material constitutive modeling, attention is often focused on *symmetric* second-order tensors, in which case one would set the focused identity equal to the symmetry projector:

$$\Delta_{ijkl} = \frac{1}{2}(\delta_{ik}\delta_{jl} + \delta_{il}\delta_{jk}). \quad (27.21)$$

If there is a focus subspace, then there must be an *elided* or null *subspace* (i.e., still present, but ignored because it is not of interest). If you focus on symmetric tensors, then the elided space is, of course, skew tensors. Associated with a *focused* projector Δ_{ijkl}, there will always be a complementary *elided* (null space) projection operator ς_{ijkl} defined as

$$\varsigma_{ijkl} = \delta_{ik}\delta_{jl} - \Delta_{ijkl}. \tag{27.22}$$

This projector will transform an arbitrary tensor into its 'not-of-interest' part. For example, if symmetric tensors are the focus space, then the focused identity is given in (27.21), and the elided projector is $\varsigma_{ijkl} = \frac{1}{2}(\delta_{ik}\delta_{jl} - \delta_{il}\delta_{jk})$; in this case, the operation $\varsigma_{ijkl}A_{kl}$ returns the skew-symmetric part of any tensor $\underset{\approx}{A}$.

27.4 Components of focused identities and elided projectors

Imagine that you are working in a laboratory where a slurry is poured down an incline of 45°. Specifically, suppose that the equation of the plane is given by $x = y$ with respect to Cartesian coordinates aligned with the walls of the laboratory. Then, the normal to the plane is $\underset{\sim}{n} = (\underset{\sim}{e}_2 - \underset{\sim}{e}_1)/\sqrt{2}$. The *focused identity* (also called a *constrained identity*) associated with the plane is the projector $\underset{\approx}{I} - \underset{\sim}{n}\underset{\sim}{n}$. The *laboratory* component matrix of this tensor is

$$[I] - [nn] = \begin{bmatrix} 1 & 0 & 0 \\ 0 & 1 & 0 \\ 0 & 0 & 1 \end{bmatrix} - \frac{1}{2}\begin{bmatrix} 1 & -1 & 0 \\ -1 & 1 & 0 \\ 0 & 0 & 0 \end{bmatrix} = \begin{bmatrix} 1/2 & 1/2 & 0 \\ 1/2 & 1/2 & 0 \\ 0 & 0 & 1 \end{bmatrix} \text{ w.r.t. the lab basis.} \tag{27.23}$$

In this form, this focused identity does not look very much like an identity or even a piece of an identity. If you find yourself frequently tilting your head to line up your view with the slurry plane, then your focus must be on the plane, and it is likely that the equations governing your variables of interest might simplify if you would switch away from the lab basis,

$$\underset{\sim}{e}_1, \quad \underset{\sim}{e}_2, \quad \underset{\sim}{e}_3, \tag{27.24}$$

in favor of an aligned basis

$$\frac{1}{\sqrt{2}}(\underset{\sim}{e}_1 + \underset{\sim}{e}_2), \quad \underset{\sim}{e}_3, \quad \frac{1}{\sqrt{2}}(\underset{\sim}{e}_1 - \underset{\sim}{e}_2). \tag{27.25}$$

The divisor of $\sqrt{2}$ ensures that this basis is not only aligned and orthogonal, but also normalized.

The standard lab-dyadic basis used to describe second-order tensors is

$$\underset{\sim}{e}_1\underset{\sim}{e}_1, \quad \underset{\sim}{e}_1\underset{\sim}{e}_2, \quad \underset{\sim}{e}_1\underset{\sim}{e}_3, \quad \underset{\sim}{e}_2\underset{\sim}{e}_1, \quad \underset{\sim}{e}_2\underset{\sim}{e}_2, \quad \underset{\sim}{e}_2\underset{\sim}{e}_3, \quad \underset{\sim}{e}_3\underset{\sim}{e}_1, \quad \underset{\sim}{e}_3\underset{\sim}{e}_2, \quad \underset{\sim}{e}_3\underset{\sim}{e}_3. \tag{27.26}$$

With respect to this basis, the components of a general tensor $\underset{\approx}{A}$ are typically denoted as

$$A_{11}, A_{12}, A_{13}, A_{21}, A_{22}, A_{23}, A_{31}, A_{32}, A_{33}. \tag{27.27}$$

Of course, these nine components are identified by double subscripting because doing so leads to a natural and intuitive display of the components in tabular form.

Fundamentally, however, when considering tensors as objects themselves, what is really important is that a tensor has nine components each referenced to a companion basis tensor (the lab dyad in this case). You can change the basis tensors if you want to, so long as you change the associated components consistently.

Suppose, for example, that the equations governing your problem become simpler in structure if all tensors are decomposed into symmetric and skew-symmetric parts. Then it is natural to introduce an alternative tensor basis that is aligned with symmetric and skew tensors:

$$
\begin{aligned}
&\underset{\sim}{e}_1\underset{\sim}{e}_1, && \underset{\sim}{e}_2\underset{\sim}{e}_2, && \underset{\sim}{e}_3\underset{\sim}{e}_3,\\
&\frac{1}{\sqrt{2}}(\underset{\sim}{e}_2\underset{\sim}{e}_3 + \underset{\sim}{e}_3\underset{\sim}{e}_2), && \frac{1}{\sqrt{2}}(\underset{\sim}{e}_3\underset{\sim}{e}_1 + \underset{\sim}{e}_1\underset{\sim}{e}_3), && \frac{1}{\sqrt{2}}(\underset{\sim}{e}_1\underset{\sim}{e}_2 + \underset{\sim}{e}_2\underset{\sim}{e}_1),\\
&\frac{1}{\sqrt{2}}(\underset{\sim}{e}_3\underset{\sim}{e}_2 - \underset{\sim}{e}_2\underset{\sim}{e}_3), && \frac{1}{\sqrt{2}}(\underset{\sim}{e}_1\underset{\sim}{e}_3 + \underset{\sim}{e}_3\underset{\sim}{e}_1), && \frac{1}{\sqrt{2}}(\underset{\sim}{e}_2\underset{\sim}{e}_1 + \underset{\sim}{e}_1\underset{\sim}{e}_2).
\end{aligned}
\tag{27.28}
$$

These nine basis tensors are mutually perpendicular to each other (any two different members of this set will have a zero double-dot inner product). The divisors of $\sqrt{2}$ are present to ensure that this basis will be *normalized* (i.e., each basis tensor has unit magnitude). The first six basis tensors are symmetric and the last three are skew-symmetric. The components of a general tensor $\underset{\approx}{A}$ associated with this alternative basis will be simply denoted with a single subscript

$$A_1,\ A_2,\ A_3,\ A_4,\ A_5,\ A_6,\ A_7,\ A_8,\ A_9. \tag{27.29}$$

These are related to the lab-dyadic basis components in (27.27) by

$$
\begin{aligned}
&A_1 = A_{11}, && A_2 = A_{22}, && A_3 = A_{33},\\
&A_4 = \sqrt{2}\,A_{(23)}, && A_5 = \sqrt{2}\,A_{(31)}, && A_6 = \sqrt{2}\,A_{(12)},\\
&A_7 = \sqrt{2}\,A_{[32]}, && A_8 = \sqrt{2}\,A_{[13]}, && A_9 = \sqrt{2}\,A_{[21]}.
\end{aligned}
\tag{27.30}
$$

Here, the notations $A_{(ij)}$ and $A_{[ij]}$ denote the ij component of the symmetric and skew parts of $\underset{\approx}{A}$, respectively[4]. Of course, these components involve multipliers of $\sqrt{2}$ as a result of the normalization of the basis. This ortho*normalized* basis-component system is the *Mandel* system discussed in detail in chapter 26. Researchers who omit the $\sqrt{2}$ adjustments, are using a *Voigt* system. Typically, a Voigt or Mandel array system for describing tensors is used when the tensors of interest are all symmetric, in which case only the first six components are potentially non-zero. Fourth-order linear transformation tensors associated with symmetric tensor space are therefore characterized by 6×6 component matrices.

The principal disadvantage of the Voigt approach is that tensor operations such as the inner product must contain additional factors or divisors of 2 (which are actually metric components that account for the fact that the Voigt basis is not normalized; in fact, properly speaking, Voigt components are actually *contravariant*

[4] For example, $A_{(12)}$ is a shorthand notation for $\frac{1}{2}(A_{12} + A_{21})$, and $A_{[21]}$ is short for $\frac{1}{2}(A_{21} - A_{12})$.

components with respect to the non-normalized tensor basis). The Mandel system generalizes to fourth-order tensors by simply multiplying the last 3 rows and last 3 columns of a 6×6 Voigt matrix by $\sqrt{2}$ (the lower-right 3×3 sub-matrix therefore is multiplied twice by $\sqrt{2}$ so those components equal the Voigt components multiplied by 2). Using the Mandel system, you can readily and efficiently exploit array and matrix manipulation libraries that are built into most modern computing environments. For example, Mandel components of the inverse of a fourth-order tensor are found by simply inverting the tensor's Mandel component matrix—this procedure does not work with the Voigt system. Likewise, with the Mandel system, standard eigensolvers may be used to determine second-order eigen*tensors* associated with fourth-order tensors; with the Voigt system, a more expensive *generalized* eigensolver must be used.

In plasticity (and many other areas of constitutive modeling), it is important to be able to assess the angle between *tensors*. The cosine of the angle between two tensors, $\underset{\approx}{\boldsymbol{A}}$ and $\underset{\approx}{\boldsymbol{B}}$, is defined as

$$\cos\theta = \frac{\underset{\approx}{\boldsymbol{A}} : \underset{\approx}{\boldsymbol{B}}}{(||\underset{\approx}{\boldsymbol{A}}||)(||\underset{\approx}{\boldsymbol{B}}||)}, \quad \text{or} \quad \cos\theta = A_{ij}B_{ij}. \tag{27.31}$$

The mathematical justification for even being able to speak of angles between tensors is that second-order tensors are themselves 9D vectors (in the mathematician's abstract sense of the word), and their double-dot product ':' is an inner product in 9D tensor space, so it plays a role in 9D tensor space that is analogous to the single-dot product for ordinary 3D vectors. It is still the 9D inner product that must be used even if one is limiting attention to the 6D subspace of symmetric tensors. Specifically, if $\underset{\approx}{\boldsymbol{A}}$ and $\underset{\approx}{\boldsymbol{B}}$ happen to be symmetric tensors, their inner product is still given by (27.31), which expands explicitly to

$$\underset{\approx}{\boldsymbol{A}} : \underset{\approx}{\boldsymbol{B}} = A_{11}B_{11} + A_{22}B_{22} + A_{33}B_{33} + 2(A_{23}B_{23} + A_{31}B_{31} + A_{12}B_{12}) \tag{27.32}$$

for symmetric tensors.

When the components of these tensors are saved in *Mandel* form, this operation can be computed by

$$\underset{\approx}{\boldsymbol{A}} : \underset{\approx}{\boldsymbol{B}} = \sum_{K=1}^{6} A_K^* B_K^*. \tag{27.33}$$

Note that the factor of 2 that appears in (27.32) is handled implicitly in this expression because the last three *Mandel* components of each tensor are defined to equal the off-diagonal component times $\sqrt{2}$ which multiply together to give 2.

27.5 Single-plane fourth-order rotations

In the realm of fourth-order tensors, what is the closest analog of ordinary 3D rotations? An ordinary 3D rotation is normally regarded as having an axis of rotation. Any vectors belonging to the plane normal to this axis will remain in that plane after being rotated. For the extension of this concept to higher dimensions, it is the 2D plane—not the 1D axis—that is isomorphic (i.e., analogous) in the conceptual similarity of rotating a vector in a plane (contained in 3D space) and rotating a tensor in a hyperplane (contained in 9D tensor space). Specifically, a single-plane fourth-order rotation operation is one for which there exists a 2D 'active' subspace such that any tensor in this space rotates within this active space. Any tensor in the 'inactive' subspace (i.e., the space orthogonal to the active space) is unchanged upon application of the rotation operation. A general tensor can be decomposed into a part in the inactive space plus a part in the active space (much as an ordinary vector can be expressed as the sum of its part along an axis of rotation plus a part in the plane of rotation, as was illustrated in figure 5.1). Application of a single-plane rotation operation will produce a new tensor that has the same inactive part, while the active part is rotated.

To construct a single-plane rotation operation, you must first obtain an orthonormal basis for the desired active subspace. To do this, you simply need two linearly independent tensors that belong to the active space (the rotation plane). Suppose, for example, these two tensors are denoted $\underset{\approx}{\boldsymbol{m}}$ and $\underset{\approx}{\boldsymbol{n}}$. We will presume here that these tensors are unit tensors (if not, you can always normalize them), but we will not presume that they are orthogonal to each other. You can quickly construct an orthonormal basis $\underset{\approx}{\boldsymbol{M}}$ and $\underset{\approx}{\boldsymbol{N}}$ for the active space (i.e., the space spanned by $\underset{\approx}{\boldsymbol{m}}$ and $\underset{\approx}{\boldsymbol{n}}$) by

$$\underset{\approx}{\boldsymbol{M}} = \frac{\underset{\approx}{\boldsymbol{n}} + \underset{\approx}{\boldsymbol{m}}}{\|\underset{\approx}{\boldsymbol{n}} + \underset{\approx}{\boldsymbol{m}}\|} \quad \text{and} \quad \underset{\approx}{\boldsymbol{N}} = \frac{\underset{\approx}{\boldsymbol{n}} - \underset{\approx}{\boldsymbol{m}}}{\|\underset{\approx}{\boldsymbol{n}} - \underset{\approx}{\boldsymbol{m}}\|}. \tag{27.34}$$

Now that we have an orthonormal basis for the 2D active space, the corresponding single-plane fourth-order rotation tensor may be constructed by

$$R_{ijkl} = (c - 1)(M_{ij}M_{kl} + N_{ij}N_{kl}) + s(N_{ij}M_{kl} - M_{ij}N_{kl}) + \Delta_{ijkl}, \tag{27.35}$$

where c and s are the cosine and sine of the rotation angle, respectively. The tensor Δ_{ijkl} represents the fourth-order focused identity appropriate to the space or subspace you are interested in. Specifically, for rotations defined in full tensor space,

$$\Delta_{ijkl} = \delta_{ik}\delta_{jl}. \tag{27.36}$$

For rotations limited to symmetric tensor space,

$$\Delta_{ijkl} = \frac{1}{2}(\delta_{ik}\delta_{jl} + \delta_{il}\delta_{jk}). \tag{27.37}$$

Note the similarity of (27.34) to (8.71). The expression in (27.34) is valid for any rotation angle in the linear manifold spanned by $\underset{\approx}{\boldsymbol{m}}$ and $\underset{\approx}{\boldsymbol{n}}$. The rotation angle is *not* required to be set so that $\underset{\approx}{\boldsymbol{m}}$ will rotate into $\underset{\approx}{\boldsymbol{n}}$.

27.5.1 Special case: rotating *m* into *n*

If it is desired that $\underset{\sim}{m}$ should rotate into $\underset{\sim}{n}$, then the cosine of the rotation angle is given by $c = m_{ij}n_{ij}$. Back substituting through the definitions in a manner similar to how (8.72) and (8.73) were derived,

$$R_{ijkl} = \frac{(1 + 2c)(n_{ij}m_{kl}) - m_{ij}n_{kl} - (m_{ij}m_{kl} + n_{ij}n_{kl})}{1 + c} + \Delta_{ijkl}. \tag{27.38}$$

This sort of operator might be used, for example, to modify a plasticity constitutive model to support non-associativity (non-alignment of the flow direction with the yield surface normal) in a way that avoids any need for a flow potential surface.

27.5.2 Special case: shared principal space

Suppose, as a special case, that the starting tensors $\underset{\sim}{m}$ and $\underset{\sim}{n}$ share a common set of eigenvectors. Consequently, the tensors $\underset{\sim}{N}$ and $\underset{\sim}{M}$ will also share those eigenvectors. In terms of the shared principal basis, the Voigt (as well as the Mandel) component arrays will be of the form

$$\{\underset{\sim}{M}\} = \{M_1, M_2, M_3, 0, 0, 0\}, \tag{27.39}$$

$$\{\underset{\sim}{N}\} = \{N_1, N_2, N_3, 0, 0, 0\}, \tag{27.40}$$

and, with respect to the shared principal *physical* basis, the Mandel component array for the rotation tensor will be

$$[R] = (c-1)\begin{bmatrix} M_1^2 + N_1^2 & (M_1M_2 + N_1N_2) & (M_1M_3 + N_1N_3) & 0 & 0 & 0 \\ (M_2M_1 + N_2N_1) & M_2^2 + N_2^2 & (M_2M_3 + N_2N_3) & 0 & 0 & 0 \\ (M_3M_1 + N_3N_1) & (M_3M_2 + N_3N_2) & M_3^2 + N_3^2 & 0 & 0 & 0 \\ 0 & 0 & 0 & 0 & 0 & 0 \\ 0 & 0 & 0 & 0 & 0 & 0 \\ 0 & 0 & 0 & 0 & 0 & 0 \end{bmatrix}$$
$$+ s\begin{bmatrix} 0 & (M_1N_2 - M_3N_1) & (M_1N_2 + M_3N_1) & 0 & 0 & 0 \\ (M_2N_1 - M_1N_2) & 0 & (M_2N_3 - M_3N_2) & 0 & 0 & 0 \\ (M_3N_1 - M_1N_3) & (M_3N_2 - M_2N_3) & 0 & 0 & 0 & 0 \\ 0 & 0 & 0 & 0 & 0 & 0 \\ 0 & 0 & 0 & 0 & 0 & 0 \\ 0 & 0 & 0 & 0 & 0 & 0 \end{bmatrix} \tag{27.41}$$
$$+ \begin{bmatrix} 1 & 0 & 0 & 0 & 0 & 0 \\ 0 & 1 & 0 & 0 & 0 & 0 \\ 0 & 0 & 1 & 0 & 0 & 0 \\ 0 & 0 & 0 & 1 & 0 & 0 \\ 0 & 0 & 0 & 0 & 1 & 0 \\ 0 & 0 & 0 & 0 & 0 & 1 \end{bmatrix}.$$

Again, these are components with respect to the shared triad of principal basis vectors in physical space. For example, recalling that Mandel conventions identify the index 1 with the dyad $\underset{\sim}{p}_1\underset{\sim}{p}_1$ and the index 2 with the dyad $\underset{\sim}{p}_2\underset{\sim}{p}_2$, the component R_{12} is cast in conventional 4-index form as R_{1122}. The component R_{44} equals $2R_{2323}$ according to Mandel notation.

27.6 Preferred basis for single-plane rotation

If you choose *not* to use conventional basis tensors (i.e., basis tensors that are formed from dyads of the 3D basis in the usual way), then the component structure of fourth-order tensors can be significantly simplified. Specifically, if you express the components of the above rotation in terms of a 'preferred tensor basis' for which the first basis tensor equals $\underset{\approx}{M}$ and the second basis tensor equals $\underset{\approx}{N}$, then the component matrix for the rotation is simply

$$[R] = \begin{bmatrix} c & -s & 0 & 0 & 0 & 0 \\ s & c & 0 & 0 & 0 & 0 \\ 0 & 0 & 1 & 0 & 0 & 0 \\ 0 & 0 & 0 & 1 & 0 & 0 \\ 0 & 0 & 0 & 0 & 1 & 0 \\ 0 & 0 & 0 & 0 & 0 & 1 \end{bmatrix}. \tag{27.42}$$

The remaining four basis tensors merely have to be orthogonal to the first two 'active' basis tensors. One way to construct these other four basis tensors is to begin with the standard physically referenced basis dyads, $\underset{\sim}{e}_i\underset{\sim}{e}_j$, project them into the inactive space, and then perform Gram–Schmidt orthogonalization. In practice, however, there is not typically a need to do this. In particular, note that (27.35) can be written as the following higher-dimensional version of the Euler–Rodrigues identity, analogous to equation (5.25).

$$\begin{aligned} R_{ijkl} &= c(M_{ij}M_{kl} + N_{ij}N_{kl}) + s(N_{ij}M_{kl} - M_{ij}N_{kl}) \\ &\quad + \Delta_{ijkl} - (M_{ij}M_{kl} + N_{ij}N_{kl}). \end{aligned} \tag{27.43}$$

The component matrix for the first two terms (involving the rotation cosine and sine) is

$$\begin{bmatrix} c & -s & 0 & 0 & 0 & 0 \\ s & c & 0 & 0 & 0 & 0 \\ 0 & 0 & 0 & 0 & 0 & 0 \\ 0 & 0 & 0 & 0 & 0 & 0 \\ 0 & 0 & 0 & 0 & 0 & 0 \\ 0 & 0 & 0 & 0 & 0 & 0 \end{bmatrix}, \tag{27.44}$$

and the component matrix for the remaining terms (last two terms in (27.43)) is

$$\begin{bmatrix} 0 & 0 & 0 & 0 & 0 & 0 \\ 0 & 0 & 0 & 0 & 0 & 0 \\ 0 & 0 & 1 & 0 & 0 & 0 \\ 0 & 0 & 0 & 1 & 0 & 0 \\ 0 & 0 & 0 & 0 & 1 & 0 \\ 0 & 0 & 0 & 0 & 0 & 1 \end{bmatrix}. \tag{27.45}$$

To summarize, the preferred tensor basis permits the rotation tensor to be displayed in matrix form such that it has a single diagonal sub-matrix of the form

$$\begin{bmatrix} c & -s \\ s & c \end{bmatrix} \tag{27.46}$$

corresponding to the active basis tensors, and the remainder of the matrix is the identity. Of course, this example has focused on rotations in symmetric tensor space, which has permitted displaying the rotation tensor as a 6×6 matrix. This rotation operation is still well defined even when it acts on a non-symmetric tensor, but the effect of this tensor on the skew-symmetric part is to simply return zero. Stated differently, the null-space associated with rotations in symmetric tensor space is the set of all skew tensors. If desired, the 6×6 component matrix can be upgraded to a full 9×9 matrix by simply adding zero components everywhere in the last three rows and columns, associated with any tensor basis for the space of skew tensors. With respect to full 9D space, this rotation tensor is not orthogonal—but it is *focused* orthogonal with respect to symmetric tensors.

27.7 Double-plane fourth-order rotations

The single-plane rotation involved a component matrix that, in a preferred basis, was the identity tensor except for a 2×2 embedded rotation sub-matrix. At first, it might seem that the next level of complexity would be the same thing with a 3×3 embedded rotation matrix. However, because a 3×3 rotation matrix must have at least one eigenvalue equal to 1, this possibility degenerates back to the case of a 2×2 rotation embedded in the identity matrix.

The next level of complexity, namely *four* diagonal components of a rotation being identified with cosines of a rotation, corresponds to two independent planes of rotation activated simultaneously so that a tensor basis exists for which the rotation component matrix is of the form

$$\begin{bmatrix} c_1 & -s_1 & 0 & 0 & 0 & 0 \\ s_1 & c_1 & 0 & 0 & 0 & 0 \\ 0 & 0 & c_2 & -s_2 & 0 & 0 \\ 0 & 0 & s_2 & c_2 & 0 & 0 \\ 0 & 0 & 0 & 0 & 1 & 0 \\ 0 & 0 & 0 & 0 & 0 & 1 \end{bmatrix}. \tag{27.47}$$

The subscripts 1 and 2 indicate sines and cosines of two independent rotation angles.

Consider, for example, the tensor $[U]$ given by the Mandel components listed in (27.7). That tensor is a two-plane higher-order rotation operator. It is not immediately apparent to be of the form listed above because the components in (27.7) are with respect to the Mandel basis of (26.33) in which the third lab basis vector $\underset{\sim}{e}_3$ is aligned with the rotation axis for $\underset{\approx}{Q}$. Rather than using the Mandel basis, let us now define a different orthonormal basis for symmetric tensor space as

$$\frac{1}{\sqrt{2}}\begin{bmatrix}0&0&0\\0&0&1\\0&1&0\end{bmatrix},\ \frac{1}{\sqrt{2}}\begin{bmatrix}0&0&-1\\0&0&0\\-1&0&0\end{bmatrix},\ \frac{1}{\sqrt{2}}\begin{bmatrix}1&0&0\\0&-1&0\\0&0&0\end{bmatrix},$$
$$\frac{1}{\sqrt{2}}\begin{bmatrix}0&1&0\\1&0&0\\0&0&0\end{bmatrix},\ \frac{1}{\sqrt{2}}\begin{bmatrix}1&0&0\\0&1&0\\0&0&0\end{bmatrix},\ \frac{1}{\sqrt{2}}\begin{bmatrix}0&0&0\\0&0&0\\0&0&1\end{bmatrix}. \qquad (27.48)$$

Any symmetric tensor may be written as a linear combination of these basis tensors. With respect to this basis, the components of $[U]$ are

$$\begin{bmatrix}\cos\alpha&-\sin\alpha&0&0&0&0\\\sin\alpha&\cos\alpha&0&0&0&0\\0&0&\cos 2\alpha&-\sin 2\alpha&0&0\\0&0&\sin 2\alpha&\cos 2\alpha&0&0\\0&0&0&0&1&0\\0&0&0&0&0&1\end{bmatrix},\quad \text{w.r.t. the specialized basis in (27.48)}, \qquad (27.49)$$

which is precisely of the form in (27.47). This shows that any second-order tensor whose ordinary 3×3 component matrix is a linear combination of the first two basis tensors, so that it is of the form

$$[A]=\begin{bmatrix}0&0&-b\\0&0&a\\-b&a&0\end{bmatrix}, \qquad (27.50)$$

will, upon application of (27.7) rotate into a new tensor that is of a similar form

$$[B]=\begin{bmatrix}0&0&-\bar{b}\\0&0&\bar{a}\\-\bar{b}&\bar{a}&0\end{bmatrix}, \qquad (27.51)$$

and the angle between these two tensors (computed using (27.31)) will equal the rotation angle α associated with $\underset{\approx}{\boldsymbol{Q}}$.

Similarly, any tensor expressible as a linear combination of the second and third basis tensors,

$$[A]=\begin{bmatrix}a&b&0\\b&-a&0\\0&0&0\end{bmatrix}, \qquad (27.52)$$

will rotate into a tensor of the same form separated by an angle (in tensor space) equal to 2α because that is the angle appearing in the second rotation sub-matrix of (27.49). Tensors of this form are deviatoric in the plane perpendicular to the rotation axis. The separation of these tensors by an angle 2α occurs for the same reason that two points on a Mohr's circle will be separated by an angle 2α when the stress itself is rotated by an angle α.

Finally, the presence of the numerals 1 on the last two diagonals in (27.49) implies that any tensor expressible as a linear combination of the last two basis tensors will be completely unaffected by a rotation about the 3-direction, which is an intuitive result given that the last basis tensor is uniaxial about the rotation axis and the second-to-last basis tensor is isotropic in the plane perpendicular to the rotation axis.

27.8 Multi-plane fourth-order rotations

The concept of rotation planes extends as suggested in the previous section so that, when limiting focus to 6D symmetric tensor space, up to three rotation planes can be active at once. For full 9D tensor space, one more rotation plane can be added. A general discussion of rotations in arbitrary dimension spaces can be found in [1].

Reference

[1] Cartan E 1966 *The Theory of Spinors* (Paris: Hermann)

Chapter 28

Closing remarks

This book has offered a comprehensive (although, despite its length, not *exhaustive*) review of theorems and concepts related to rotations and reflections. After beginning this work as a collection of notes created in the early 90s at the University of Wisconsin, most entries in this book were added as a direct outgrowth of applications at Sandia National Laboratories over the course of approximately four subsequent years, followed by further applications in computational solid mechanics research at the University of Utah. As such, this book covers concepts that were found to be of direct value in professional engineering contexts, and they have been presented here using notation and terminology that caters more to engineers than to mathematicians.

Attention has been on proper (physical) rotations in three-dimensional (3D) ordinary engineering space, though we have covered several more advanced excursions to higher dimensions and abstruse applications areas. Much of the text has been devoted to reviewing moderately well-known theorems and identities (such as coordinate transformations) with the aim of improving accessibility of these subjects to practicing engineers. This book has included several bits of original work and theoretical results that are potentially very useful but not well known. Among these topics, the ones we consider to be 'most important' in a research environment and, at the same time, little known (because rarely published or because published for the first time here) are as follows.

- Chapter 5: *converting an axis and angle of rotation into a rotation tensor*. We provided the well-published, but (for some reason) relatively unknown (or misquoted) Euler–Rodrigues formula as well as many other alternative techniques that form the foundation of some numerical integration procedures currently employed in modern finite-element codes.
- Chapter 8: *converting known rotated vectors into a rotation matrix*. This is also a known procedure, but not readily found in engineering reference guides. This section furthermore describes how multi-linear regression (common in

doi:10.1016/978-0-7503-1454-1ch28

digitial image correlation) can be used to assign an overall rotation of an object from knowledge of initial and final locations of points in that body.

- Chapter 12: *polar-decomposition theorem*. Of course, this theorem is very well known, but our more methodical discussion that makes this theorem more intuitive. We provide an apparently new and extraordinarily fast iterative algorithm for computing a polar decomposition (though we acknowledge that the Higham fixed-point iterator remains superior because of its algorithmic simplicity and competitive efficiency). Also, our observation that a deformation gradient tensor can be written down immediately by examining the geometry of the deformation (rather than painstakingly working out a mapping function and taking its derivative) appears to be unknown to many practicing professionals. We explain how to examine a drawing of a deformed material element to immediately infer good estimates for the polar rotation and stretch.
- Chapter 14: *methods to average or interpolate rotations such that the average is itself a rotation*. In computational mechanics, this topic arises frequently when faced with the need to remove advection errors in Eulerian hydrodynamics codes. This chapter surveys various proposals that have been used or considered in that context.
- Chapter 15: derivatives of rotation with respect to axis or angle of rotation, as well as *numerical procedures for integrating angular velocity over time to compute time-varying rotation tensors*. Our discussion points out that several commonly used algorithms are only first-order accurate even in the case that the angular velocity is constant. We provide an alternative that is *exact* in this special case. Additionally, recognizing that rigid motion is ubiquitous in robotics, some insight is provided about why early robots moved in such a 'jerky' unnatural fashion and how, by using a tortuous 'inefficient' movement through control parameter space, a more natural and energy-efficient motion would be achieved in physical space.
- Chapter 16: *time rate (or other derivative) of a principal function* of a symmetric tensor. We reviewed and generalized how to find the time rate of a function that is defined in terms of the principal basis of another tensor. Time rates are complicated by the fact that the principal basis of the independent variable might change in time. One of our contributions was to demonstrate that an existing procedure for finding the rate remains valid even if the independent variable has repeated eigenvalues. As part of the corollary work in this chapter, an important (and we believe original) contribution was the derivation of the error suffered when approximating the log strain rate by the symmetric part of the velocity gradient. Also, we provide an algorithm for computing the log strain rate directly from the symmetric part of the deformation gradient, and show that the difference between these two tensors appears *only* in the off-diagonals when expressed in terms of the principal basis for the stretch, implying that there is zero error in this approximation for the important class of engineering applications that involve no rotation of the reference directions of stretch.

- Chapter 17: *generating a uniformly random rotation.* Having applications in any microscale materials modeling work, as well as many other areas, this capability is of paramount importance. Many researchers, unfortunately, generate random rotations by uniform distributions of Euler angles, and we demonstrated that this strategy does *not* lead to *uniformly* random rotations. Whether seeking uniform or non-uniform distributions, the role of the singular value decomposition is explained (namely what non-uniformities in the distribution are needed in the 3D parameter space in order to compensate for a non-uniform Jacobian in construction of the 9D rotation tensor).
- Chapter 18: *introduction to pole figures.* This chapter introduces simple concepts in crystallography (such as the meaning of an inverse pole figure) with coverage of the mapping functions that transform points on the unit sphere to points on a pole plot.
- Chapter 19: *frame indifference* and *non-inertial (rotating and translating) frames.* The discussion of this important topic is extended more than is usually found in mechanics textbooks specifically to help students and professionals better understand the difference between a simple basis change and a full-fledged evolving frame change, especially addressing why superimposed rotation does not induce extra centrifugal loads on a body—it is because those loads must be exactly counteracted with exactly the same sort of phantom (inertial) body forces that a person feels when thrashed in a roller-coaster. This discussion leads naturally to a derivation of body forces (including Coriolis) one would need to include in simulations of bodies that are deforming by small amounts relative to much larger bulk motion (as with centrifuge experiments, turbine blades, tire analysis, etc.).
- Chapter 20: *distinction between material and tensor symmetry.* This chapter explains brute-force methods for finding tensors belonging to symmetry classes, and emphasizes that an isotropic material will not generally have isotropic tensor properties in the loaded state. For intrinsic transverse isotropy, closed-form solutions are provided for the *four* eigenspaces of a generally five-independent-component fourth-order double-symmetric tensor (such as stiffness), and this solution is applied to show how Tsai–Wu parameters for transversely isotropic plasticity can be assigned to exactly match the eigenspace of the elasticity, which is a physically appealing idea based on the notion that the microscale source of the anisotropy ought to induce identical eigenspaces for elasticity and plasticity (which is a much stronger assertion than saying they ought to both be transversely isotropic).
- Chapter 23: *compact symbolic notation for equations of rigid-body motion.* This chapter summarizes equations for rigid-body mechanics that can be found in a multitude of undergraduate dynamics textbooks, but it states these equations in compact tensor form, not only reducing their size on paper (and revealing structure) but in some cases even generalizing them. For example, in this formulation, the parallel axis theorem is just a special case of a *more general* theorem that takes *less space* to write in compact tensor form!

- Chapter 24: Derivation of the pseudo body forces experienced by a non-inertial observer. These forces are needed to set up non-inertial simulations of these events (e.g., to infer injury risk to a roller coaster passenger, or to explore material deformation in a centrifuge, etc).
- Chapter 25: *equations for 3D perspective projections.* Step-by-step explanations for the derivation of a realistic projection of a point in 3D space (e.g., a point in a finite-element simulation) to its image with proper and fully adjustable perspective for viewing on a computer screen[1].
- Chapter 26: *Voigt versus Mandel components.* The Voigt system is the most common choice for storing a symmetric tensor as a set of six numbers Voigt components are more than just components of an array—they are genuine vector components in a 6D space and are therefore subject to 6D generalizations of basis transformation rules. However, the tensor basis associated with Voigt components is not normalized, which is why manipulations with Voigt components often involve counterintuitive factors of 2. We demonstrated that, by normalizing the Voigt basis, you can work instead with 'Mandel' (also called 'normalized Voigt') components which obey the ordinary matrix algebra, which is crucial when examining the spectral properties of fourth-order tensors.
- Chapter 27: *rotations in higher dimensions.* Being able to interpret symmetric tensors as vectors in 6D space is very important in materials constitutive modeling. Transforming a symmetric tensor to change its orientation in 6D symmetric tensor space is quite different from transforming it in the ordinary 3D way. We demonstrated that, through a change in the *tensor* basis, rotations in 6D symmetric tensor space can be expressed in terms of a fourth-order operator that has one or more submatrices of cosines and sines similar to thc type of submatrices seen in ordinary 3D rotations. We derived an operator that would rotate any given symmetric second-order tensor into a second, specified, tensor of the same magnitude and the same eigenvectors, which could be used (for example) in plasticity models that use non-associativity.
- Appendix A: *FORTRAN (with a bit of Python) subroutines.* Unlike most of the existing literature about rotations, we wrote and tested numerical implementations of the algorithms described in this report, and provided source code. This older programming language is used because the author is an old dog without time to learn new tricks, but the younger readers out there should still be able to decipher the code as her students have done.

This book could be regarded as a companion volume to go along with a lengthy unpublished book *Functional and Structured Tensor Analysis for Engineers*

[1] Side story: that work was originally done to work around a bug in a very old version of Mathematica that prevented proper rendering of 3D spheres. The equations were derived in a panic to prepare graphics for a conference presentation about statistical ceramics failure, in which crack orientations (being unit normals) are depicted as points on a sphere.

(available in a very early draft form online [1]) as well as the much smaller book *Curvilinear Coordinates in a Euclidean Space* which is likewise available in a draft [2] and is being prepped for formal publication soon, though probably with a revised title. Updated/corrected editions are anticipated as this group of books matures, so suggestions for improvement (especially suggestions for expanded citations to the literature) are welcomed. Resource files, such as source code for the various rotation-related operations discussed in this book, may be downloaded from the publisher's website, http://iopscience.iop.org/book/978-0-7503-1454-1. These resource files might be updated as needed to correct errors and/or to improve clarity.

References

[1] Brannon R M 2003 https://csmbrannon.net/2011/03/03/functional-and-structured-tensor-analysis-for-engineers/
[2] Brannon R M 2004 Curvilinear tutorial https://csmbrannon.net/2011/03/07/385/

Appendix

Code listings

It is sometimes easiest to figure out an algorithm by merely looking at actual code. This appendix provides working source code for basic operations involving rotations and/or reflections. If FORTRAN is available, it is shown first, followed (if available) by Mathematica, Python, and C++. These and other resource files may be downloaded from the publisher at http://iopscience.iop.org/book/978-0-7503-1454-1.

A.1 Listing 1: testing if a matrix is a rotation

```
C Fortran
SUBROUTINE CHKROT(DC,IERR)
C      ************************************************************
C      This routine checks if a matrix DC is  a proper rotation.
C      To be orthogonal, the columns (or) rows of DC must form an
C      orthoNORMAL triad. That means that each column dotted with
C      itself must equal 1, and each column dotted with a different
C      column must equal 0. Furthermore, to be a proper rotation,
C      the columns of DC must form a RIGHT-HANDED orthonormal triad.
C      In other words, the determinant of DC must equal +1.
C      ************************************************************
C
C INPUT
C ---
C   DC: candidate rotation matrix (saved as a 3x3)
C
C OUTPUT
C ----
C  IERR = 0 if the matrix is a proper rotation
C
C       = -1 if the matrix is orthogonal but has a negative determinant
C
C       = ij if the dot product between column i and column j is wrong
C          For example, if column 3 is not normalized, then ierr=33.
C          If column 1 is not perpendicular to column 2, then ierr = 12.
```



```
      INCLUDE 'implicit.h'
C For double precison, the "implicit.h" include should contain one
C single line, depending on the desired precision. As appropriate,
C          IMPLICIT REAL(A-H,O-Z)              ! for single precision
C          IMPLICIT DOUBLE PRECISION(A-H,O-Z)  ! for double precision
C ....parameters
C     real*8 pzero,pone
      PARAMETER (PZERO=0.0D0,PONE=0.1D1)
C ....passed
C     real*8 dc
      DIMENSION DC(3,3)
      INTEGER IERR
C
C ....local
      INTEGER I,J,K
C     real*8 dum
ccccccccccccccccccccccccccccccccccccccccccccccccccccccccccccccc end of preamble
      IERR=0
C       The direction cosines are known
C       Check that the columns form an orthonormal set
C
        DO I=1,3
        DO J=1,3
          DUM=PZERO
          DO K=1,3
             DUM=DUM+DC(K,I)*DC(K,J)
          END DO
          IF(I.EQ.J)DUM=DUM-PONE
          IF(ABS(DUM).GT.0.001D0)THEN
             IERR=10*I+J
             RETURN
          END IF
        ENDDO
        ENDDO
C
C       Check if proper rotation by checking if det(dc)=+1
        DUM=DC(1,1)*DC(2,2)*DC(3,3)
     $     +DC(1,2)*DC(2,3)*DC(3,1)
     $     +DC(2,1)*DC(3,2)*DC(1,3)
     $     -DC(3,1)*DC(2,2)*DC(1,3)
     $     -DC(2,1)*DC(1,2)*DC(3,3)
     $     -DC(1,1)*DC(2,3)*DC(3,2)
        DUM=DUM-PONE
        IF(ABS(DUM).GT.0.001D0)THEN
           IERR=-1
           RETURN
        END IF
C
      IERR=0
      RETURN
      END
```

```
************************************
#!/usr/bin/env python
# The MIT License

# Copyright (c) 2017 Brian Leavy

# License for the specific language governing rights and limitations under
# Permission is hereby granted, free of charge, to any person obtaining a
# copy of this software and associated documentation files (the "Software"),
# to deal in the Software without restriction, including without limitation
# the rights to use, copy, modify, merge, publish, distribute, sublicense,
# and/or sell copies of the Software, and to permit persons to whom the
# Software is furnished to do so, subject to the following conditions:

# The above copyright notice and this permission notice shall be included
# in all copies or substantial portions of the Software.

# THE SOFTWARE IS PROVIDED "AS IS", WITHOUT WARRANTY OF ANY KIND, EXPRESS
# OR IMPLIED, INCLUDING BUT NOT LIMITED TO THE WARRANTIES OF MERCHANTABILITY,
# FITNESS FOR A PARTICULAR PURPOSE AND NONINFRINGEMENT. IN NO EVENT SHALL
# THE AUTHORS OR COPYRIGHT HOLDERS BE LIABLE FOR ANY CLAIM, DAMAGES OR OTHER
# LIABILITY, WHETHER IN AN ACTION OF CONTRACT, TORT OR OTHERWISE, ARISING
# FROM, OUT OF OR IN CONNECTION WITH THE SOFTWARE OR THE USE OR OTHER
# DEALINGS IN THE SOFTWARE.

import os,sys
import math

def chk_rot(dc):
    '''
    NAME
        chk_rot

    PURPOSE
Checks if a direction cosine (dc) matrix is a proper rotation.
To be orthogonal, the columns or rows of dc must form an orthonormal triad.
Each column dotted with itself must equal 1, Each column dotted with a different
column must equal 0.
To be a proper rotation, the columns must form a right handed orthonormal triad
(determinant = +1).

    INPUT
        dc:  Candidate rotation matrix

    OUTPUT
        ierr:   0 = Proper rotation
            -1 = Orthogonal, but negative determinant
             ij = Dot product between i and j is wrong
          ie. ierr = 33, column 3 not normalized
       ierr = 12, column 1 not perpendicular to column 2

    AUTHORS
        Rebecca Brannon: theory, algorithm, and code
        Brian Leavy: python implementation
    '''
```

```
    print '\nMatrix rotation tester'
    print '-'*20
    print dc
    # Initialize values
    dim = 3
    tol = 0.001
    ierr = 0
    # Direction cosines are known
    # Check columns form orthonormal set
    for i in range(dim):
      for j in range(dim):
       dum = 0.0
       for k in range(dim):
         dum += dc[k][i]*dc[k][j]
       if i == j:
        dum -= 1
       if abs(dum) > tol:
        ierr = 10 * (i+1) + (j+1)
          print 'Wrong ij:', ierr
        return ierr
    # Check proper rotation
    dum = dc[0][0]*dc[1][1]*dc[2][2] \
        + dc[0][1]*dc[1][2]*dc[2][0] \
        + dc[1][0]*dc[2][1]*dc[0][2] \
        - dc[2][0]*dc[1][1]*dc[0][2] \
        - dc[1][0]*dc[0][1]*dc[2][2] \
        - dc[0][0]*dc[1][2]*dc[2][1]
    dum -= 1.0
    if (abs(dum) > tol):
      ierr -= 1
      print 'Negative determinant:'
      return
    ierr = 0
    print 'Proper rotation:'
    return

if __name__ == '__main__':

    dim = 3
    # Proper rotation
    dc= [[0 for i in range(dim)] for j in range(dim)]
    dc[0][0]=1.0
    dc[1][1]=1.0
    dc[2][2]=1.0
    ierr = chk_rot(dc)

    # Wrong ij
    dc= [[0 for i in range(dim)] for j in range(dim)]
    dc[1][2]=2.0
    dc[1][1]=1.0
    dc[2][2]=1.0
    ierr = chk_rot(dc)

    # Negative determinant
    dc= [[0 for i in range(dim)] for j in range(dim)]
    dc[0][2]=1.0
    dc[1][1]=1.0
    dc[2][0]=1.0
```

```
    ierr = chk_rot(dc)

    sys.exit(0)
```

A.2 Listing 2: converting axis and angle to direction cosines

```
c FORTRAN
      SUBROUTINE AA2DC(axis,angle,dc,ierr)
c     This routine converts axis and angle of rotation to a direction
c     cosine matrix, which is the component matrix of the rotation tensor.
c
c input
c ---
c    AXIS: a unit vector in the direction of the axis of rotation
c    ANGLE: the angle of rotation (by right hand rule around AXIS)
c
c output
c ---
c    DC: the direction cosine matrix defined such that DC(i,j) equals
c        the dot product between the ith laboratory basis vector and the
c        jth rotated basis vector. It is the component matrix for rotation.
c
************************************************************************
      INCLUDE 'implicit.h'
c.....parameters
c     real*8 pone
      parameter (pone=0.1d1)
c.....passed
      integer ierr
c     real*8 axis,angle,dc
      dimension axis(3),dc(3,3)
c.....function (functions instead of subroutines)
c.....external
c.....local (not saved)
c     real*8 c,s,omc
c.....local (saved)
c.....data
c.....statement functions
cccccccccccccccccccccccccccccccccccccccccccccccccccccccccccccccc end of preamble
      ierr=0

      c=cos(angle)
      s=sin(angle)
      omc=pone-c
      dc(1,1)=omc*axis(1)*axis(1)+c
      dc(2,2)=omc*axis(2)*axis(2)+c
      dc(3,3)=omc*axis(3)*axis(3)+c
      dc(1,2)=omc*axis(1)*axis(2)-s*axis(3)
      dc(2,3)=omc*axis(2)*axis(3)-s*axis(1)
      dc(3,1)=omc*axis(3)*axis(1)-s*axis(2)
      dc(2,1)=omc*axis(2)*axis(1)+s*axis(3)
      dc(3,2)=omc*axis(3)*axis(2)+s*axis(1)
      dc(1,3)=omc*axis(1)*axis(3)+s*axis(2)

      RETURN
      END
```

```
(************************* MATHEMATICA ******************************)
AA2DC[axis_, angle_] :=
 Module[{A, axisLength, a, \[Alpha],
   R, \[ScriptCapitalI] = IdentityMatrix[3], aa},
  axisLength = Sqrt[axis.axis]; a = axis/axisLength;
  \[Alpha] = angle;
  A = ( {
     {0, -a[[3]], a[[2]]},
     {a[[3]], 0, -a[[1]]},
     {-a[[2]], a[[1]], 0}
    } );
  aa = Outer[Times, a, a];
  R = Cos[\[Alpha]] (\[ScriptCapitalI] - aa) + aa + Sin[\[Alpha]] A
  ]

*********************************************************
#!/usr/bin/env python
# The MIT License
# Copyright (c) 2017 Brian Leavy
#
# License for the specific language governing rights and limitations under
# Permission is hereby granted, free of charge, to any person obtaining a
# copy of this software and associated documentation files (the "Software"),
# to deal in the Software without restriction, including without limitation
# the rights to use, copy, modify, merge, publish, distribute, sublicense,
# and/or sell copies of the Software, and to permit persons to whom the
# Software is furnished to do so, subject to the following conditions:
# The above copyright notice and this permission notice shall be included
# in all copies or substantial portions of the Software.
# THE SOFTWARE IS PROVIDED "AS IS", WITHOUT WARRANTY OF ANY KIND, EXPRESS
# OR IMPLIED, INCLUDING BUT NOT LIMITED TO THE WARRANTIES OF MERCHANTABILITY,
# FITNESS FOR A PARTICULAR PURPOSE AND NONINFRINGEMENT. IN NO EVENT SHALL
# THE AUTHORS OR COPYRIGHT HOLDERS BE LIABLE FOR ANY CLAIM, DAMAGES OR OTHER
# LIABILITY, WHETHER IN AN ACTION OF CONTRACT, TORT OR OTHERWISE, ARISING
# FROM, OUT OF OR IN CONNECTION WITH THE SOFTWARE OR THE USE OR OTHER
# DEALINGS IN THE SOFTWARE.

import os,sys
from math import *

def aa2dc(axis,angle):
    '''
    NAME
        aa2dc

    PURPOSE
      Converts axis and angle (aa) of rotation to a direction cosine (dc) matrix

    INPUT
       axis:  A unit vector in the diretion of the axis of rotation
       angle: The angle of rotation (right hand rule) around axis

    OUTPUT
        dc:  Direction cosine matrix
```

```
        dc(i,j) = dot product of ith lab base vector and jth rotated base vector

    AUTHORS
        Rebecca Brannon: theory, algorithm, and code
        Brian Leavy: python implementation
    '''
    # Initialize values
    dim = 3
    ierr = 0
    dc= [[0 for i in range(dim)] for j in range(dim)]

    # Trigonometry helper functions
    c = cos(angle)
    s = sin(angle)
    omc = 1.0 - c

    # Create direction cosine matrix
    dc[0][0] = omc * axis[0] * axis[0] + c
    dc[1][1] = omc * axis[1] * axis[1] + c
    dc[2][2] = omc * axis[2] * axis[2] + c
    dc[0][1] = omc * axis[0] * axis[1] - s * axis[2]
    dc[1][2] = omc * axis[1] * axis[2] - s * axis[0]
    dc[2][0] = omc * axis[2] * axis[0] - s * axis[1]
    dc[1][0] = omc * axis[1] * axis[0] + s * axis[2]
    dc[2][1] = omc * axis[2] * axis[1] + s * axis[0]
    dc[0][2] = omc * axis[0] * axis[2] + s * axis[1]

    return dc

if __name__ == '__main__':


    # Euler angle
    axis = [1.0,0.0,0.0]
    angle = 0.0
    print '\nAxis and angle of rotation to direction cosine converter'
    print '-'*40
    print 'axis = ',axis
    print 'angle = ',angle

    # Direction cosine matrix
    dc = aa2dc(axis,angle)
    print 'dc =',dc

    sys.exit(0)
```

A.3 Listing 3: converting direction cosines to axis and angle

The following listing uses preprocessor "ifdef" directives to select the method by which the task is performed. If METHOD1 is defined, then method 1 from section 5.4.4 is used. Otherwise, method 2 from section 5.4.6 is used.

```
c FORTRAN
#ifdef METHOD1
      SUBROUTINE DC2AA(DC,AXIS,ANGLE,ierr)
************************************************************************
      INCLUDE 'implicit.h'
c ....parameters
      parameter (pzero=0.0d0,pone=0.1d1,phalf=0.5d0)
      parameter (puny=0.1d-10)
c ....passed
      integer ierr
      dimension dc(3,3),axis(3)
c ....local
      integer i,j,k
      dimension scr(3,3)
cccccccccccccccccccccccccccccccccccccccccccccccccccccccccccc end of preamble
      ierr=0

c     The axis and angle are desired
      cth=(dc(1,1)+dc(2,2)+dc(3,3)-pone)*phalf
      sth=SQRT(pone-cth*cth)
      angle=acos(cth)
      if(abs(sth).gt.puny)then
          axis(1)=(dc(3,2)-dc(2,3))*phalf/sth
          axis(2)=(dc(1,3)-dc(3,1))*phalf/sth
          axis(3)=(dc(2,1)-dc(1,2))*phalf/sth
      elseif(cth.gt.pzero) then
c         The rotation tensor is the identity
c         and any axis will do.
          axis(1)=pone
          axis(2)=pzero
          axis(3)=pzero
      else
c         The rotation angle is exactly 180 degrees.
c         The rotation axis is therefore parallel to any nonzero
c         column of I+dc, where I is the identity matrix.
          do i=1,3
          do j=1,3
             scr(i,j)=dc(i,j)
          enddo
          enddo

          cth=pzero
          j=-12345
          do k=1,3
             scr(k,k)=scr(k,k)+pone
             sth=scr(1,k)**2+scr(2,k)**2+scr(3,k)**2
             if(sth.gt.cth)then
                cth=SQRT(sth)
                j=k
                scr(1,j)=scr(1,j)/cth
                scr(2,j)=scr(2,j)/cth
                scr(3,j)=scr(3,j)/cth
             endif
          enddo
          axis(1)=scr(1,j)
          axis(2)=scr(2,j)
          axis(3)=scr(3,j)
      endif
```

```
      return
      end

#else
      SUBROUTINE DC2AA(DC,AXIS,ANGLE,ierr)
****************************************************************************
      INCLUDE 'implicit.h'
c ....parameters
      parameter (pzero=0.0d0,pone=0.1d1,ptwo=0.2d1,phalf=0.5d0)
c ....passed
      integer ierr
      dimension dc(3,3),axis(3)
c ....local
      integer i,j,k
      dimension scr(3,3)
cccccccccccccccccccccccccccccccccccccccccccccccccccccccccccccccc end of preamble
      ierr=0

c     The axis and angle are desired
      dum=(dc(1,1)+dc(2,2)+dc(3,3)-pone)*phalf
      angle=acos(dum)
c
c     The rotation axis is  parallel to any nonzero
c     column of R+R^T+(1-trR)I, where I is the identity matrix.
c     and R^T is the tranpose of R. Here R is the rotation matrix DC.
      dum=-ptwo*dum
      do i=1,3
      do j=1,3
         scr(i,j)=dc(i,j)+dc(j,i)
      enddo
      scr(i,i)=scr(i,i)+dum
      enddo

      smag=pzero
      j=-12345
      do k=1,3
         dum=scr(1,k)**2+scr(2,k)**2+scr(3,k)**2
         if(dum.gt.smag)then
            smag=dum
            j=k
         endif
      enddo
      smag=sqrt(smag)
      axis(1)=sign(   scr(1,j)/smag,    dc(3,2)-dc(2,3)    )
      axis(2)=sign(   scr(2,j)/smag,    dc(1,3)-dc(3,1)    )
      axis(3)=sign(   scr(3,j)/smag,    dc(2,1)-dc(1,2)    )

      return
      end

#endif

(* MATHEMATICA *)
DC2AA[dc_] :=
 Module[{ierr, dum, angle, smag, scr, j, axis = ConstantArray[0, {3}]},
  dum = 1/2 (Tr[dc] - 1);
```

```
  angle = ArcCos[dum];
  dum = -2*dum;
  scr = dc + dc\[Transpose] + dum IdentityMatrix[3];
  (*obtain magnitudes of the columns of scr,
  which are also the rows because scr is symmetric*)
  smag = 0;
  j = -12345;
  Do[dum = \[Sqrt](scr[[k]].scr[[k]]);
   If[dum > smag,
    smag = dum;
    j = k];
   , {k, 3}];
  axis[[1]] = sign[1/smag scr[[1, j]], dc[[3, 2]] - dc[[2, 3]]];
  axis[[2]] = sign[1/smag scr[[2, j]], dc[[1, 3]] - dc[[3, 1]]];
  axis[[3]] = sign[1/smag scr[[3, j]], dc[[2, 1]] - dc[[1, 2]]];
  {axis, angle}
  ]


*******************************************
#!/usr/bin/env python
# The MIT License
# Copyright (c) 2017 Brian Leavy
#
# License for the specific language governing rights and limitations under
# Permission is hereby granted, free of charge, to any person obtaining a
# copy of this software and associated documentation files (the "Software"),
# to deal in the Software without restriction, including without limitation
# the rights to use, copy, modify, merge, publish, distribute, sublicense,
# and/or sell copies of the Software, and to permit persons to whom the
# Software is furnished to do so, subject to the following conditions:
# The above copyright notice and this permission notice shall be included
# in all copies or substantial portions of the Software.
# THE SOFTWARE IS PROVIDED "AS IS", WITHOUT WARRANTY OF ANY KIND, EXPRESS
# OR IMPLIED, INCLUDING BUT NOT LIMITED TO THE WARRANTIES OF MERCHANTABILITY,
# FITNESS FOR A PARTICULAR PURPOSE AND NONINFRINGEMENT. IN NO EVENT SHALL
# THE AUTHORS OR COPYRIGHT HOLDERS BE LIABLE FOR ANY CLAIM, DAMAGES OR OTHER
# LIABILITY, WHETHER IN AN ACTION OF CONTRACT, TORT OR OTHERWISE, ARISING
# FROM, OUT OF OR IN CONNECTION WITH THE SOFTWARE OR THE USE OR OTHER
# DEALINGS IN THE SOFTWARE.

import os,sys
from math import *


def dc2aa(dc):
    '''
    NAME
        dc2aa (Method 1)

    PURPOSE
      Converts direction cosine (dc) matrix to an axis and angle (aa) of rotation

    INPUT
        dc:  Direction cosine matrix
          dc(i,j) = dot product of ith lab base vector and jth rotated base vector
```

```
    OUTPUT
       axis:  A unit vector in the diretion of the axis of rotation
       angle: The angle of rotation (right hand rule) around axis

    AUTHORS
        Rebecca Brannon: theory, algorithm, and code
        Brian Leavy: python implementation
    '''
    # Initialize values
    dim = 3
    axis = [0.0,0.0,0.0]
    angle = 0.0
    puny = 1.e-10
    scr = [[0 for i in range(dim)] for j in range(dim)]
    # helper functions
    cth = ( dc[0][0] + dc[1][1] + dc[2][2] -1.0) * 0.5
    sth = sqrt(1.0 - cth * cth)
    angle = acos(cth)


    # Create axis and angle
    if abs(sth) > puny:
      axis[0] = (dc[2][1] - dc[1][2]) * 0.5 / sth
      axis[1] = (dc[0][2] - dc[2][0]) * 0.5 / sth
      axis[2] = (dc[1][0] - dc[0][1]) * 0.5 / sth
    elif cth > 0.0:
      # Rotation tensor is identity (any axis will do)
      axis[0] = 1.0
      axis[1] = 0.0
      axis[2] = 0.0
    else:
      # Rotation angle is 180 degrees, axis is parallel to Identity matrix + dc
      for i in range(dim):
        for j in range(dim):
          scr[i][j] = dc[i][j]

      cth = 0.0
      j = -12345
      for k in range(dim):
        scr[k][k] = scr[k][k] + 1.0
        sth = scr[0][k]**2 + scr[1][k]**2 + scr[2][k]**2
        if sth > cth:
        cth = sqrt(sth)
         j = k
        scr[0][j] = scr[0][j]/cth
        scr[1][j] = scr[1][j]/cth
        scr[2][j] = scr[2][j]/cth

            axis[0] = scr[0][j]
            axis[1] = scr[1][j]
            axis[2] = scr[2][j]

    return axis, angle

def dc2aa2(dc):
    '''
    NAME
        dc2aa (Method 2)
```

```
    PURPOSE
      Converts direction cosine (dc) matrix to an axis and angle (aa) of rotation

    INPUT
        dc:  Direction cosine matrix
          dc(i,j) = dot product of ith lab base vector and jth rotated base vector

    OUTPUT
       axis:  A unit vector in the diretion of the axis of rotation
       angle: The angle of rotation (right hand rule) around axis

    AUTHORS
        Rebecca Brannon: theory, algorithm, and code
        Brian Leavy: python implementation
    '''
    # Initialize values
    dim = 3
    axis = [0.0,0.0,0.0]
    puny = 1.e-10
    scr = [[0 for i in range(dim)] for j in range(dim)]

    # helper functions
    cth = ( dc[0][0] + dc[1][1] + dc[2][2] -1.0) * 0.5
    angle = acos(cth)
    # The rotation axis is parallel to any nonzero column of
    # R+R^T+(1-trR)I
    # ^T = Transpose, I = Identity matrix, R= dc
    cth *= -2.0
    for i in range(dim):
      for j in range(dim):
       scr[i][j] = dc[i][j] + dc[j][i]

      scr[i][i] = dc[i][i] + cth

    smag = 0.0
    j = -12345
    for k in range(dim):
      sth = scr[0][k]**2 + scr[1][k]**2 + scr[2][k]**2
      if sth > smag:
       smag = sth
        j = k

    smag = sqrt(smag)
    axis[0] = copysign(scr[0][j]/smag, dc[2][1] - dc[1][2])
    axis[1] = copysign(scr[1][j]/smag, dc[0][2] - dc[2][0])
    axis[2] = copysign(scr[2][j]/smag, dc[1][0] - dc[0][1])

    return axis, angle

if __name__ == '__main__':

    dim = 3
    # Input direction cosine
    dc= [[0 for i in range(dim)] for j in range(dim)]
    dc[0][0]=1.0
```

```
dc[1][1]=1.0
dc[2][2]=1.0

# Euler angle
axis = [1.0,0.0,0.0]
angle = 0.0
print '\nDirection cosine to axis and angle converter'
print '-'*44
print 'dc = ',dc

# Direction cosine matrix
# Method 1
axis, angle = dc2aa(dc)
print 'axis, angle =',axis,',',angle
# Method 2
axis2, angle2 = dc2aa2(dc)
print 'axis2, angle2 =',axis2,',',angle2

sys.exit(0)
```

A.4 Listing 4: Converting Euler angles to direction cosines

```
c Fortran
      SUBROUTINE EU2DC(eu,dc,ierr)
C
C INPUT
C ———
C   eu: The euler angles defined by the first (zxz) definition
C
C OUTPUT
C ————
c   dc: The direction cosine matrix.
c
      INCLUDE 'implicit.h'
c.....passed
      integer ierr
c     real*8 eu,dc
      dimension eu(3),dc(3,3)
cccccccccccccccccccccccccccccccccccccccccccccccccccccccccccccccccc end of preamble
      ierr=0

      cphi=COS(eu(1))
      sphi=SIN(eu(1))
      cth =COS(eu(2))
      sth =SIN(eu(2))
      cpsi=COS(eu(3))
      spsi=SIN(eu(3))

      dc(1,1)=cphi*cpsi - cth*sphi*spsi
      dc(2,1)=cpsi*sphi + cphi*cth*spsi
      dc(3,1)=spsi*sth

      dc(1,2)=-(cpsi*cth*sphi) - cphi*spsi
      dc(2,2)=cphi*cpsi*cth - sphi*spsi
      dc(3,2)=cpsi*sth
```

```
      dc(1,3)=sphi*sth
      dc(2,3)=-(cphi*sth)
      dc(3,3)=cth

      RETURN
      END

(* MATHEMATICA *)
EU2DC[phi_, theta_, psi_] := ( {
    {Cos[phi], -Sin[phi], 0},
    {Sin[phi], Cos[phi], 0},
    {0, 0, 1}
   } ).( {
    {1, 0, 0},
    {0, Cos[theta], -Sin[theta]},
    {0, Sin[theta], Cos[theta]}
   } ).( {
    {Cos[psi], -Sin[psi], 0},
    {Sin[psi], Cos[psi], 0},
    {0, 0, 1}
   } )

***************************************************************
#!/usr/bin/env python
# The MIT License
# Copyright (c) 2017 Brian Leavy
#
# License for the specific language governing rights and limitations under
# Permission is hereby granted, free of charge, to any person obtaining a
# copy of this software and associated documentation files (the "Software"),
# to deal in the Software without restriction, including without limitation
# the rights to use, copy, modify, merge, publish, distribute, sublicense,
# and/or sell copies of the Software, and to permit persons to whom the
# Software is furnished to do so, subject to the following conditions:
# The above copyright notice and this permission notice shall be included
# in all copies or substantial portions of the Software.
# THE SOFTWARE IS PROVIDED "AS IS", WITHOUT WARRANTY OF ANY KIND, EXPRESS
# OR IMPLIED, INCLUDING BUT NOT LIMITED TO THE WARRANTIES OF MERCHANTABILITY,
# FITNESS FOR A PARTICULAR PURPOSE AND NONINFRINGEMENT. IN NO EVENT SHALL
# THE AUTHORS OR COPYRIGHT HOLDERS BE LIABLE FOR ANY CLAIM, DAMAGES OR OTHER
# LIABILITY, WHETHER IN AN ACTION OF CONTRACT, TORT OR OTHERWISE, ARISING
# FROM, OUT OF OR IN CONNECTION WITH THE SOFTWARE OR THE USE OR OTHER
# DEALINGS IN THE SOFTWARE.

import os,sys
from math import *

def eu2dc(eu):
    '''
    NAME
        eu2dc

    PURPOSE
       Converts Euler angles to a direction cosine (dc) matrix
```

```
    INPUT
        eu:  Euler angles ([phi, theta, psi] or [alpha, beta, gamma]) in range
(2pi, pi, 2pi) radians )

    OUTPUT
        dc:  Direction cosine matrix

    AUTHORS
        Rebecca Brannon: theory, algorithm, and code
        Brian Leavy: python implementation
    '''
    # Initialize values
    dim = 3
    dc= [[0 for i in range(dim)] for j in range(dim)]

    # Trigonometry helper functions
    cphi = cos(eu[0])
    sphi = sin(eu[0])
    ctheta  = cos(eu[1])
    stheta  = sin(eu[1])
    cpsi = cos(eu[2])
    spsi = sin(eu[2])

    # Create direction cosine matrix
    dc[0][0] = cphi * cpsi - ctheta * sphi * spsi
    dc[1][0] = cpsi * sphi + cphi * ctheta * spsi
    dc[2][0] = spsi * stheta
    dc[0][1] = -( cpsi * ctheta * sphi) - cphi * spsi
    dc[1][1] = cphi * cpsi * ctheta - sphi * spsi
    dc[2][1] = cpsi * stheta
    dc[0][2] = sphi * stheta
    dc[1][2] = -( cphi * stheta)
    dc[2][2] = ctheta

    return dc

if __name__ == '__main__':



    # Euler angle
    eu= [0,0,0]
    eu[0]=1.0
    eu[1]=1.0
    eu[2]=1.0
    print '\nEuler angle converter'
    print '-'*20
    print 'eu = ',eu

    # Direction cosine matrix
    dc = eu2dc(eu)
    print 'dc =',dc

    sys.exit(0)
```

A.5 Listing 5: converting direction cosines to Euler angles

```
C FORTRAN
      SUBROUTINE DC2EU(DC,EULER,ierr)
c
c   This routine converts an orthogonal proper rotation matrix to euler
c   angles.
c
c INPUT
c ---
c  DC: direction cosine matrix for the rotation TENSOR
C      DC(i,j) is defined as the inner product between the ith lab
C      basis vector with the jth rotated basis vector.
c
c OUTPUT
c ----
C  EULER: euler angles {phi,theta,psi}, which describe the rotation
C     of the lab triad by the following procedure...First rotate
C     the triad an angle phi about its z-axis. Then rotate the new
C     triad an angle theta about its own new x-axis. Then rotate
C     the newer triad an angle psi about its own z-axis.
c  IERR: =0 if success, =1 otherwise
************************************************************************
      INCLUDE 'implicit.h'
c ....parameters
c     real*8 pzero,pone,puny,small
      parameter (pzero=0.0d0,pone=0.1d1,small=0.1d-8,puny=0.1d-20)
c ....passed
      integer ierr
c     real*8 dc,euler
      dimension dc(3,3), euler(3)
c ....local
c     real*8 dum,dum11,dum12,dum21,dum22,cth,sthsq,theta,psi,phi
ccccccccccccccccccccccccccccccccccccccccccccccccccccccccccccc end of preamble
      cth=dc(3,3)
      sthsq=pone-cth*cth
      if(sthsq.lt.puny)then
        theta=acos(cth)
        psi=pzero
        phi=atan2(dc(2,1),dc(1,1))
        return
      endif


      dum11=dc(1,1)
     $   +((dc(2,3)*dc(3,2) + dc(1,3)*dc(3,1)*dc(3,3))/sthsq)
      dum12=dc(1,2)
     $   -(dc(2,3)*dc(3,1) - dc(1,3)*dc(3,2)*dc(3,3))/sthsq
      dum21=dc(2,1)
     $   -(dc(1,3)*dc(3,2) - dc(2,3)*dc(3,1)*dc(3,3))/sthsq
      dum22=dc(2,2)
     $   +(dc(1,3)*dc(3,1) + dc(2,3)*dc(3,2)*dc(3,3))/sthsq

      dum=sqrt(dum11**2+dum22**2+dum12**2+dum21**2)
      if(dum.gt.small)then
c       The DC matrix is not proper orthogonal
        ierr=1
        return
      endif
      theta=acos(cth)
      phi=atan2(dc(1,3),-dc(2,3))
```

```
      psi=atan2(dc(3,1), dc(3,2))
      euler(1)=phi
      euler(2)=theta
      euler(3)=psi

      ierr=0
      return
      end
```

```
(*********************** MATHEMATICA **************************)
DC2EU[Q_] :=
 Module[{puny = 10^-7, cth, sthsq, dum11, dum12, dum21, dum22, dum,
   theta, phi, psi, euler},
  cth = Q[[3, 3]];
  sthsq = 1 - cth^2;
  theta = ArcCos[cth];
  If[sthsq > puny,
   phi = ArcTan[-Q[[2, 3]], Q[[1, 3]]];
   psi = ArcTan[Q[[3, 2]], Q[[3, 1]]];,
   phi = 0;
   psi = ArcTan[Q[[2, 2]], Q[[2, 1]]];
   ];
  {phi, theta, psi}]
```

```
***************************************************************
#!/usr/bin/env python
# The MIT License
# Copyright (c) 2017 Brian Leavy
#
# License for the specific language governing rights and limitations under
# Permission is hereby granted, free of charge, to any person obtaining a
# copy of this software and associated documentation files (the "Software"),
# to deal in the Software without restriction, including without limitation
# the rights to use, copy, modify, merge, publish, distribute, sublicense,
# and/or sell copies of the Software, and to permit persons to whom the
# Software is furnished to do so, subject to the following conditions:
# The above copyright notice and this permission notice shall be included
# in all copies or substantial portions of the Software.
# THE SOFTWARE IS PROVIDED "AS IS", WITHOUT WARRANTY OF ANY KIND, EXPRESS
# OR IMPLIED, INCLUDING BUT NOT LIMITED TO THE WARRANTIES OF MERCHANTABILITY,
# FITNESS FOR A PARTICULAR PURPOSE AND NONINFRINGEMENT. IN NO EVENT SHALL
# THE AUTHORS OR COPYRIGHT HOLDERS BE LIABLE FOR ANY CLAIM, DAMAGES OR OTHER
# LIABILITY, WHETHER IN AN ACTION OF CONTRACT, TORT OR OTHERWISE, ARISING
# FROM, OUT OF OR IN CONNECTION WITH THE SOFTWARE OR THE USE OR OTHER
# DEALINGS IN THE SOFTWARE.

import os,sys
from math import *


def dc2eu(dc):
    '''
    NAME
        dc2eu

    PURPOSE
       Converts direction cosine (dc) matrix into Euler (eu) angles
```

```
    INPUT
        dc:  Direction cosine matrix

    OUTPUT
       eu:  Euler angles ([phi, theta, psi] or [alpha, beta, gamma]) in range (2pi,
pi, 2pi) radians )
      ierr: 0 = success, 1 = error
    AUTHORS
        Rebecca Brannon: theory, algorithm, and code
        Brian Leavy: python implementation
    '''
    # Initialize values
    dim = 3
    ierr = 0
    eu = [0.0 for i in range(dim)]
    puny = 1.e-20
    small = 1.e-8
    cth = dc[2][2]
    sthsq = 1.0 - cth * cth
    theta = acos(cth)

    # Create Euler angles
    if sthsq < puny:
      psi = 0.0
      phi = atan2(dc[1][0], dc[0][0])
      eu[0] = phi
      eu[1] = theta
      eu[2] = psi
      return eu, ierr

    dum11 = dc[0][0] + (dc[1][2] * dc[2][1] + dc[0][2] * dc[2][0] * dc[2][2]) / sthsq
    dum12 = dc[0][1] - (dc[1][2] * dc[2][0] - dc[0][2] * dc[2][1] * dc[2][2]) / sthsq
    dum21 = dc[1][0] - (dc[0][2] * dc[2][1] - dc[1][2] * dc[2][0] * dc[2][2]) / sthsq
    dum22 = dc[1][1] + (dc[0][2] * dc[2][0] - dc[1][2] * dc[2][1] * dc[2][2]) / sthsq

    dum = sqrt(dum11**2 + dum22**2 + dum12**2 + dum21**2)
    if dum > small:
      # cd matrix is not proper orthogonal
      ierr = 1
      return eu, ierr

    phi = atan2(dc[0][2], -1.0*dc[1][2])
    psi = atan2(dc[2][0], dc[2][1])

    eu[0] = phi
    eu[1] = theta
    eu[2] = psi

    return eu, ierr

if __name__ == '__main__':

    dim = 3
    ierr = 0
    eu = [1.0,0.0,0.0]
```

```
# Direction cosine
dc= [[0 for i in range(dim)] for j in range(dim)]
dc[0][0]=1.0
dc[1][1]=1.0
dc[2][2]=1.0
print dc
print '\nDirection cosine to Euler angle converter'
print '-'*44

# Euler angle
eu, ierr = dc2eu(dc)
print 'eu =',eu
print 'ierr =',ierr

sys.exit(0)
```

A.6 Listing 6: updating rotations by integrating angular velocity

```
C-.—1—.—2—.—3—.—4—.—5—.—6—.—7-
      SUBROUTINE INTEG_ANG_VEL(DT,W,ROLD,RNEW)
C
C     ———
C     Input
C     ———
C     DT: time step
C     ROLD: rotation tensor at the beginning of the step
C     W: angular velocity vector
C             Note: if you have the angular velocity tensor, OMEGA,
C             then you can construct W by
C             W(1) = OMEGA(3,2)
C             W(2) = OMEGA(1,3)
C             W(3) = OMEGA(2,1)
C     ————
C     Output
C     ————
C     Rnew: rotation tensor updated to the end of the step
C
************************************************************************
*
*     PURPOSE:
*     This routine integrates angular velocity over a timestep DT
*     to update a rotation tensor from its "old" value ROLD at the
*     beginning of the step to a new updated value Rnew at the end of
*     the step. This algorithm is EXACT if the angular velocity is
*     constant over the entire time step. It is second-order accurate
*     if the angular velocity has a constant direction but varying
*     magnitude AND the value sent for the angular velocity vector
*     is its value at the half step. The order of accuracy of this
*     algorithm is unknown if the orientation of the angular velocity
*     varies over the time step. Simply sending the angular velocity
*     at the half step might seem like a good strategy, lacking any
*     better ideas. Consider, however, a "toy" problem in which the
*     angular velocity w equals w1 at the beginning of the step and
*     stays equal to this value throughout the first half of the step,
*     then taking a different value w2 for the second half of the step.
*     In this case, the value of w at the half step, (w1+w2)/2 is the
```

```
*     same as it would be if the order of application of these angular
*     velocities were reversed. However, rotating first around
*     "direction 1" and then about "direction 2" produces a very
*     different rotation tensor than the one for rotating first
*     about "direction 2" and then about "direction 1". Because this
*     algorithm does not import information about the angular
*     ACCELERATION, it can have no way of distinguishing between
*     these scenarios. Of course, as the time step is reduced, the
*     problem goes away, but I suspect that this issue will make the
*     algorithm only first-order accurate even if the angular velocity
*     is sent as its value at the half step. Testing is needed.
*
***********************************************************************
C
C  MODIFICATION HISTORY
C  031021:reb:created routine
C
      IMPLICIT NONE
C.....parameters
      REAL*8 ONE,ZERO
      PARAMETER(ZERO=0.0D0,ONE=0.1D1)
C.....passed
      REAL*8 DT
      REAL*8 W,ROLD, RNEW
      DIMENSION W(3),ROLD(3,3),RNEW(3,3)
C.....local (not saved)
      INTEGER I,J
      REAL*8 WSQMAG,WMAG,S,C
      REAL*8 Q,A
      DIMENSION Q(3,3),A(3)
ccccccccccccccccccccccccccccccccccccccccccccccccccccccccccccc end of preamble
C     Get the square magnitude of the angular velocity vector
      WSQMAG = W(1)*W(1) + W(2)*W(2) + W(3)*W(3)
C
C     The rotation doesn't change if the angular velocity is zero
      IF(WSQMAG.EQ.ZERO)THEN
         DO I=1,3
         DO J=1,3
           RNEW(I,J) = ROLD(I,J)
         ENDDO
         ENDDO
         RETURN
      ENDIF
C
      WMAG=SQRT(WSQMAG) ! magnitude of the angular velocity vector
C
C     Incremental rotation axis:
      A(1)=W(1)/WMAG
      A(2)=W(2)/WMAG
      A(3)=W(3)/WMAG
C
C     Cosine and sine of the incrmental rotation angle
      C=COS(WMAG*DT)
      S=SIN(WMAG*DT)
C
C     Assemble the incrmental rotation by the Euler-Rodrigues formula
      Q(1,1)=(ONE-C)*A(1)*A(1) + C
      Q(2,2)=(ONE-C)*A(2)*A(2) + C
      Q(3,3)=(ONE-C)*A(3)*A(3) + C
```

```
C
      Q(3,2)=(ONE-C)*A(3)*A(2)  + S*A(1)
      Q(1,3)=(ONE-C)*A(1)*A(3)  + S*A(2)
      Q(2,1)=(ONE-C)*A(2)*A(1)  + S*A(3)
C
      Q(2,3)=(ONE-C)*A(2)*A(3)  - S*A(1)
      Q(3,1)=(ONE-C)*A(3)*A(1)  - S*A(2)
      Q(1,2)=(ONE-C)*A(1)*A(2)  - S*A(3)
C
C     Update the rotation
      RNEW(1,1)  = Q(1,1)*ROLD(1,1)  + Q(1,2)*ROLD(2,1)  + Q(1,3)*ROLD(3,1)
      RNEW(2,1)  = Q(2,1)*ROLD(1,1)  + Q(2,2)*ROLD(2,1)  + Q(2,3)*ROLD(3,1)
      RNEW(3,1)  = Q(3,1)*ROLD(1,1)  + Q(3,2)*ROLD(2,1)  + Q(3,3)*ROLD(3,1)
      RNEW(1,2)  = Q(1,1)*ROLD(1,2)  + Q(1,2)*ROLD(2,2)  + Q(1,3)*ROLD(3,2)
      RNEW(2,2)  = Q(2,1)*ROLD(1,2)  + Q(2,2)*ROLD(2,2)  + Q(2,3)*ROLD(3,2)
      RNEW(3,2)  = Q(3,1)*ROLD(1,2)  + Q(3,2)*ROLD(2,2)  + Q(3,3)*ROLD(3,2)
      RNEW(1,3)  = Q(1,1)*ROLD(1,3)  + Q(1,2)*ROLD(2,3)  + Q(1,3)*ROLD(3,3)
      RNEW(2,3)  = Q(2,1)*ROLD(1,3)  + Q(2,2)*ROLD(2,3)  + Q(2,3)*ROLD(3,3)
      RNEW(3,3)  = Q(3,1)*ROLD(1,3)  + Q(3,2)*ROLD(2,3)  + Q(3,3)*ROLD(3,3)
C
      RETURN
      END
```

A.7 Listing 7: Generating a uniformly random unit normal

```
      SUBROUTINE RANDA(AXIS)
c     This routine outputs a random unit normal
c
c OUTPUT
c ———
C    AXIS: The random unit axis.
C
************************************************************************
C
      INCLUDE 'implicit.h'
c ....parameters
c     real*8 pi
      parameter (pi=0.3141592653589793d1)
c     real*8 pone,ptwo
      parameter (pone=0.1d1,ptwo=0.2d1)
c ....passed
c     real*8 axis
      dimension axis(3)
c ....functions
c     real*8 rand
c ....local
c     real*8 phi, theta
c ....local (SAVED)
c     real*8 seed
c     save seed
c     data seed/0.0/
c ....statement functions
      include "randnum.h"
cccccccccccccccccccccccccccccccccccccccccccccccccccccccccccccc end of preamble
```

```
      phi=ptwo*pi*randum()
      theta=ACOS(pone-ptwo*randum())
      axis(1)=sin(theta)*cos(phi)
      axis(2)=sin(theta)*sin(phi)
      axis(3)=cos(theta)
      return
      end
```

```
******************************************************************
#!/usr/bin/env python
# The MIT License
# Copyright (c) 2017 Brian Leavy
#
# License for the specific language governing rights and limitations under
# Permission is hereby granted, free of charge, to any person obtaining a
# copy of this software and associated documentation files (the "Software"),
# to deal in the Software without restriction, including without limitation
# the rights to use, copy, modify, merge, publish, distribute, sublicense,
# and/or sell copies of the Software, and to permit persons to whom the
# Software is furnished to do so, subject to the following conditions:
# The above copyright notice and this permission notice shall be included
# in all copies or substantial portions of the Software.
# THE SOFTWARE IS PROVIDED "AS IS", WITHOUT WARRANTY OF ANY KIND, EXPRESS
# OR IMPLIED, INCLUDING BUT NOT LIMITED TO THE WARRANTIES OF MERCHANTABILITY,
# FITNESS FOR A PARTICULAR PURPOSE AND NONINFRINGEMENT. IN NO EVENT SHALL
# THE AUTHORS OR COPYRIGHT HOLDERS BE LIABLE FOR ANY CLAIM, DAMAGES OR OTHER
# LIABILITY, WHETHER IN AN ACTION OF CONTRACT, TORT OR OTHERWISE, ARISING
# FROM, OUT OF OR IN CONNECTION WITH THE SOFTWARE OR THE USE OR OTHER
# DEALINGS IN THE SOFTWARE.

import os,sys
import random
from math import *

def randa():
    '''
    NAME
        randa

    PURPOSE
       Output a random unit normal

    INPUT

    OUTPUT
        axis:  A random unit axis

    AUTHORS
        Rebecca Brannon: theory, algorithm, and code
        Brian Leavy: python implementation
    '''
    # Initialize values
    dim = 3
    axis= [0.0 for i in range(dim)]

    # Trigonometry helper functions
```

```
    phi = 2.0 * pi * random.random()
    theta = acos(1.0 - 2.0 * random.random())

    # Create random axis
    axis[0] = sin(theta) * cos(phi)
    axis[1] = sin(theta) * sin(phi)
    axis[2] = cos(theta)

    return axis

if __name__ == '__main__':

    print '\nRandom axis generator'
    print '-'*44

    # Direction cosine matrix
    axis = randa()
    print 'axis =', axis

    sys.exit(0)
```

A.8 Listing 8: Generating a uniformly random rigid rotation

```
C This file contains two subroutines, RANDC and RNDC, both of
C which compute a uniformly random rotation tensor. The only difference
C between them is the way that they do it.
C
C The routine randc computes the rotation by generating two
C uniformly random unit vectors, orthonormalizing them,
C and then generating the third vector by a right-hand cross product.
C
C The routine rndc computes the rotation via Shoemake's algorithm.
C
      SUBROUTINE RANDDC(R)
C This routine calls either RANDC or RNDC, depending on how the
C programmer desires the rotation to be generated.
      INCLUDE 'implicit.h'
C     real*8 r
      DIMENSION R(3,3)
!     CALL RANDC_QUATERN(R)
      CALL RANDC_E1E2(R)
      RETURN
      END
C
C
C
************************************************************************
      SUBROUTINE RANDC_E1E2(R)
C This routine computes a uniformly random rigid rotation tensor.
C
C This routine calls randa to generate a uniformly random
C direction for the rotated E1 axis.  Then randa is called again
C to generate a second random axis. The rotated E2 is taken to
C equal the projection of the second random direction to the
C plane with normal equal to e1. Finally, the third direction is
```

```
C given by the right hand cross product of E1 x E2.
C
C OUTPUT
C ----
C  R: The direction cosine tensor components.  The ij component
C     of R equals the ith lab basis vector dotted with the jth rotated
C     basis vector.
************************************************************************
      INCLUDE 'implicit.h'
C     real*8 small,r,dum
      PARAMETER (SMALL=.1D0)
      DIMENSION R(3,3)
      EXTERNAL RANDA
cccccccccccccccccccccccccccccccccccccccccccccccccccccccccccc end of preamble
      CALL RANDA(R(1,1))
    7 CALL RANDA(R(1,2))
C     subtract off the part of e2 that is in the direction of e1
      DUM=R(1,1)*R(1,2)+R(2,1)*R(2,2)+R(3,1)*R(3,2)
      R(1,2)=R(1,2)-DUM*R(1,1)
      R(2,2)=R(2,2)-DUM*R(2,1)
      R(3,2)=R(3,2)-DUM*R(3,1)
      DUM=SQRT(R(1,2)**2+R(2,2)**2+R(3,2)**2)
C
C     make certain that the two vectors are independent.
C     If dum is small, then the second random vector is nearly
C     parallel to the first, and we need to sample again.
      IF(DUM.LT.SMALL)GO TO 7
C
      R(1,2)=R(1,2)/DUM
      R(2,2)=R(2,2)/DUM
      R(3,2)=R(3,2)/DUM
C
      R(1,3)=R(2,1)*R(3,2)-R(3,1)*R(2,2)
      R(2,3)=R(3,1)*R(1,2)-R(1,1)*R(3,2)
      R(3,3)=R(1,1)*R(2,2)-R(2,1)*R(1,2)
C
      RETURN
      END
************************************************************************
      SUBROUTINE RANDC_QUATERN(R)   !previously called RNDC
C This routine computes a uniformly random rigid rotation tensor
C using Shoemake's method. A uniformly random rotation corresponds to
c a uniform distribution of quaternions on the surface of a
c FOUR-dimensional unit sphere. Shoemake's algorithm creates realizations
c for the spherical coordinates (Y1,Y2,Y3) on the 4D unit sphere surface.
c Specifically, Shoemake's coordinates are defined such that the
c quaternion is
c                u0 = cos(Y2) cos(Y3)
c                u1 = sin(Y1) sin(Y3)
c                u2 = cos(Y1) cos(Y3)
c                u3 = sin(Y2) sin(Y3)
c
c The area element in 4D is
c random rotation. Of the many algorthims in this file, this one
c appears to be the most efficient.
C
C OUTPUT
C ----
C  R: The direction cosine tensor components.  The ij component
C     of R equals the ith lab basis vector dotted with the jth rotated
```

```
C     basis vector.
*************************************************************************
      INCLUDE 'implicit.h'
      PARAMETER (ONE=0.1D1,TWO=0.2D1,
     & TWOPI=6.2831853071795864769252867665590057683943387987502d0)
C     real*8 r
      DIMENSION R(3,3)
c ....statement functions
      include "randnum.h"
cccccccccccccccccccccccccccccccccccccccccccccccccccccccccccccccc end of preamble
      r1=randum()
      r2=randum()
      r3=randum()

      Y1=TWOPI*r1
      c1=Cos(Y1)
      s1=Sin(y1)

      Y2=TWOPI*r2
      c2=cos(y2)
      s2=sin(y2)
C
      c3=SQRT(ONE-r3)
      s3=SQRT(r3)
c
      U1=s1*c3
      U2=c1*c3
      U3=s2*s3
      U0=c2*s3
C
c     Now that the quaternion components are known, we can construct
c     the rotation. The formula is
c      R = COEFI [I] + COEFUU [UU] + COEFE [axU]
c     where [I] is the identity tensor, [UU] is the dyad of U with
c     itself, and axU is the axial tensor associated with U.
c
      COEFI = TWO*U0*U0-ONE
      COEFUU= TWO
      COEFE = TWO*U0
C
      R(1,1)=COEFI+COEFUU*U1*U1
      R(2,2)=COEFI+COEFUU*U2*U2
      R(3,3)=COEFI+COEFUU*U3*U3
C
      R(2,3)=COEFUU*U2*U3-COEFE*U1
      R(3,1)=COEFUU*U3*U1-COEFE*U2
      R(1,2)=COEFUU*U1*U2-COEFE*U3
C
      R(3,2)=COEFUU*U3*U2+COEFE*U1
      R(1,3)=COEFUU*U1*U3+COEFE*U2
      R(2,1)=COEFUU*U2*U1+COEFE*U3
C
      RETURN
      END
C
C
*************************************************************************
      SUBROUTINE RANDC_AA(R)
```

```
C This routine computes a uniformly random rigid rotation tensor
C This routine calls randa to generate a uniformly random
C direction for the rotation axis. Then it solves the equation
c angle-sin(angle)=Pi*rnd, where rnd is a random number on [0,1),
c for the rotation angle. From there, the rotation is constructed by
c using the Euler-Rodrigues formula
C
C OUTPUT
C ----
C  R: The direction cosine tensor components.  The ij component
C     of R equals the ith lab basis vector dotted with the jth rotated
C     basis vector.
**************************************************************************
      INCLUDE 'implicit.h'
C     real*8 small,r,dum
      DIMENSION R(3,3),axis(3)
      parameter (pi=0.31415926535897932384626433832795028841971d1)
      parameter (sixpi=0.18849555921538759430775860299677017305l8d2)
      parameter (third=0.33333333333333333333333333333333333333333d0)
      parameter (puny=0.1d-11)
      parameter (maxknt=100)
      EXTERNAL RANDA
      INCLUDE 'randnum.h'
cccccccccccccccccccccccccccccccccccccccccccccccccccccccccccc end of preamble
      CALL RANDA(axis)

      rnd = randum()
c
c  ...Numerically solve the equation angle-sin(angle)=Pi*rnd
c     for the rotation angle. The following is a Newton fixed-point
c     iterator.
      angle= (sixpi*rnd)**third
      do knt=1,maxknt
        err=abs((angle-sin(angle))/pi-rnd)
        if(err.lt.puny)go to 8
        angle=(pi*rnd-angle*cos(angle)+sin(angle))/(one-cos(angle))
      enddo
      call bombed('iterator for equal area angle failed')



    8 continue
      call aa2dc(axis,angle,R,ierr)

      RETURN
      END

*****************************************
#!/usr/bin/env python
# The MIT License
# Copyright (c) 2017 Brian Leavy
#
# License for the specific language governing rights and limitations under
# Permission is hereby granted, free of charge, to any person obtaining a
# copy of this software and associated documentation files (the "Software"),
# to deal in the Software without restriction, including without limitation
# the rights to use, copy, modify, merge, publish, distribute, sublicense,
# and/or sell copies of the Software, and to permit persons to whom the
# Software is furnished to do so, subject to the following conditions:
```

```
# The above copyright notice and this permission notice shall be included
# in all copies or substantial portions of the Software.
# THE SOFTWARE IS PROVIDED "AS IS", WITHOUT WARRANTY OF ANY KIND, EXPRESS
# OR IMPLIED, INCLUDING BUT NOT LIMITED TO THE WARRANTIES OF MERCHANTABILITY,
# FITNESS FOR A PARTICULAR PURPOSE AND NONINFRINGEMENT. IN NO EVENT SHALL
# THE AUTHORS OR COPYRIGHT HOLDERS BE LIABLE FOR ANY CLAIM, DAMAGES OR OTHER
# LIABILITY, WHETHER IN AN ACTION OF CONTRACT, TORT OR OTHERWISE, ARISING
# FROM, OUT OF OR IN CONNECTION WITH THE SOFTWARE OR THE USE OR OTHER
# DEALINGS IN THE SOFTWARE.

import os,sys
import random
import numpy
from math import *
def randa():
    '''
    NAME
        randa

    PURPOSE
        Output a random unit normal

    INPUT

    OUTPUT
        axis:  A random unit axis

    AUTHORS
        Rebecca Brannon: theory, algorithm, and code
        Brian Leavy: python implementation
    '''
    # Initialize values
    dim = 3
    axis= numpy.zeros(dim)

    # Trigonometry helper functions
    phi = 2.0 * pi * random.random()
    theta = acos(1.0 - 2.0 * random.random())

    # Create random axis
    axis[0] = sin(theta) * cos(phi)
    axis[1] = sin(theta) * sin(phi)
    axis[2] = cos(theta)
    print 'axis = ',axis
    return axis

def randc():
    '''
    NAME
        randc

    PURPOSE
       Computa a uniformly random rigid rotation tensor

    INPUT
```

```
    OUTPUT
        r:  random rotation tensor

    AUTHORS
        Rebecca Brannon: theory, algorithm, and code
        Brian Leavy: python implementation
    '''
    # Initialize values
    dim = 3
    dum = 0.0
    small = 0.1
    i=0
    #r = [[0 for i in range(dim)] for j in range(dim)]
    r = numpy.matrix('0. 0. 0.; 0. 0. 0.; 0. 0. 0.')
    #r = numpy.matrix([[0., 0., 0.],[0., 0., 0.],[0., 0., 0.]])
    print 'r = ', r
    # Trigonometry helper functions

    # Create rotation tensor
    r[0][0] = randa()
    #print 'r00 = ',randa()
    while dum < small:
      r[0][1] = randa()
      # Subtract part of e2 in direction of e1
      #dum = r[0][0] * r[0][1] + r[1][0] * r[1][1] + r[2][0] * r[2][1]
      dum = numpy.dot(r[0][0], r[0][1]) + numpy.dot(r[1][0], r[1][1]) + numpy.dot
(r[2][0], r[2][1])
      #print 'dum1 = ',dum
      r[0][1] -= dum * r[0][0]
      r[1][1] -= dum * r[1][0]
      r[2][1] -= dum * r[2][0]
      dum = numpy.sqrt(r[0][1]**2 + r[1][1]**2 + r[2][1]**2)
      #print 'dum2 = ',dum
      # Check vectors are independent
      # if dum is small, vector2 ~ parallel to vector 1
      print 'r = ',r

    r[0][1] /= dum
    r[1][1] /= dum
    r[2][1] /= dum

    r[0][2] = r[1][0] * r[2][1] - r[2][0] * r[1][1]
    r[1][2] = r[2][0] * r[0][1] - r[0][0] * r[2][1]
    r[2][2] = r[0][0] * r[1][1] - r[1][0] * r[0][1]

    return r
if __name__ == '__main__':

    dim = 3
    # Random rigid rotation
    #R= [[0.0 for i in range(dim)] for j in range(dim)]
    #R = numpy.zeros((dim,dim))
    #R = numpy.matrix([[0., 0., 0.],[0., 0., 0.],[0., 0., 0.]])
    R = numpy.matrix('0. 0. 0.; 0. 0. 0.; 0. 0. 0.')
```

```
    print '\nRandom rigid rotation'
    print '-'*44

    # Direction cosine matrix
    R = randc()
    print 'R =',R

    sys.exit(0)
****************************************************************
#!/usr/bin/env python
# The MIT License
# Copyright (c) 2017 Brian Leavy
#
# License for the specific language governing rights and limitations under
# Permission is hereby granted, free of charge, to any person obtaining a
# copy of this software and associated documentation files (the "Software"),
# to deal in the Software without restriction, including without limitation
# the rights to use, copy, modify, merge, publish, distribute, sublicense,
# and/or sell copies of the Software, and to permit persons to whom the
# Software is furnished to do so, subject to the following conditions:
# The above copyright notice and this permission notice shall be included
# in all copies or substantial portions of the Software.
# THE SOFTWARE IS PROVIDED "AS IS", WITHOUT WARRANTY OF ANY KIND, EXPRESS
# OR IMPLIED, INCLUDING BUT NOT LIMITED TO THE WARRANTIES OF MERCHANTABILITY,
# FITNESS FOR A PARTICULAR PURPOSE AND NONINFRINGEMENT. IN NO EVENT SHALL
# THE AUTHORS OR COPYRIGHT HOLDERS BE LIABLE FOR ANY CLAIM, DAMAGES OR OTHER
# LIABILITY, WHETHER IN AN ACTION OF CONTRACT, TORT OR OTHERWISE, ARISING
# FROM, OUT OF OR IN CONNECTION WITH THE SOFTWARE OR THE USE OR OTHER
# DEALINGS IN THE SOFTWARE.

import os,sys
from math import *


def rndc():
    '''
    NAME
        rndc

    PURPOSE
Random rigid rotation using Shoemake's method

    INPUT

    OUTPUT
        r:  random rigid rotation tensor

    AUTHORS
        Rebecca Brannon: theory, algorithm, and code
        Brian Leavy: python implementation
    '''
    # Initialize values
    dim = 3
    small = 0.1
    r = [[0 for i in range(dim)] for j in range(dim)]
```

```
    # Trigonometry helper functions

    # Create direction cosine matrix
    r[0][0] = 0.0

    return r

if __name__ == '__main__':

    dim = 3
    # Random rigid rotation
    R= [[0.0 for i in range(dim)] for j in range(dim)]

    print '\nShoemake random rigid rotation'
    print '-'*44

    # Direction cosine matrix
    R = rndc()
    print 'R =',R

    sys.exit(0)
```

A.9 Listing 9: polar decomposition in 3D and 2D

The following listing shows the 3 × 3 polar decomposition first, followed by the extremely simple 2 × 2 decomposition routine at the end.

```
C
C
C
C---.---1---.---2---.---3---.---4---.---5---.---6---.---7--
      SUBROUTINE POLARDECOMP(F,R,U)                  !very well tested
C*********************************************************************
C     PURPOSE: This routine computes the polar decompositon,
C
C            F = RU
C
C     for a 3x3 matrix F. Here, R is an orthogonal matrix
C     and U is a symmetric positive definite matrix.
C
C     This routine is limited to 3x3 matrices specifically to
C     optimize its performance.
C
C     This routine returns a proper rotation if det[F]>0,
C     or an improper orthogonal tensor if det[F]<0.
C
C     The host code is responsible for NOT calling this routine
C     if det[F]=0. The closer that [F] is to being singular, the
C     longer this routine will take (it might give up).
C
C input
C ---
C
```

```
C     F: the 3x3 matrix to be decomposed into the form F=RU
C
C output
C ---
C     R: polar rotation tensor (3x3 orthogonal matrix)
C     U: polar stretch tensor stored as a single-index Voigt array
C        with components ordered 11, 22, 33, 12, 23, 31
C        Permission is granted to modify this subroutine to send U as a
C        less-efficiently stored 3x3 matrix. Likewise, permission is
C        granted to modify this routine to return the left stretch
C        [V] instead of the right stretch.
C
      INCLUDE 'implicit.h'
C.....passed arguments
      DIMENSION F(3,3), R(3,3), U(6)
cccccccccccccccccccccccccccccccccccccccccccccccccccccccccccccc end of preamble
C
C     PHASE 1: GET THE POLAR ROTATION TENSOR
      CALL POLARROTATION(F,R)
C
C
C     PHASE 2: GET THE STRETCH
C
C The right stretch is found by [U] = [R]^T  [F]
      U(1) = R(1,1)*F(1,1) + R(2,1)*F(2,1) + R(3,1)*F(3,1)        ! U11
      U(2) = R(1,2)*F(1,2) + R(2,2)*F(2,2) + R(3,2)*F(3,2)        ! U22
      U(3) = R(1,3)*F(1,3) + R(2,3)*F(2,3) + R(3,3)*F(3,3)        ! U33
      U(4) = R(1,1)*F(1,2) + R(2,1)*F(2,2) + R(3,1)*F(3,2)        ! U12
      U(5) = R(1,2)*F(1,3) + R(2,2)*F(2,3) + R(3,2)*F(3,3)        ! U23
      U(6) = R(1,3)*F(1,1) + R(2,3)*F(2,1) + R(3,3)*F(3,1)        ! U31
C
C The left stretch in the polar decomposition F = VR can be computed
C by [V] = [F] [R]^T. Just uncomment the following lines if you want
C the left stretch V (and, of course, add it to the calling arguments).
C
C     V(1) = F(1,1)*R(1,1) + F(1,2)*R(1,2) + F(1,3)*R(1,3)        ! V11
C     V(2) = F(2,1)*R(2,1) + F(2,2)*R(2,2) + F(2,3)*R(2,3)        ! V22
C     V(3) = F(3,1)*R(3,1) + F(3,2)*R(3,2) + F(3,3)*R(3,3)        ! V33
C     V(4) = F(1,1)*R(2,1) + F(1,2)*R(2,2) + F(1,3)*R(2,3)        ! V12
C     V(5) = F(2,1)*R(3,1) + F(2,2)*R(3,2) + F(2,3)*R(3,3)        ! V23
C     V(6) = F(3,1)*R(1,1) + F(3,2)*R(1,2) + F(3,3)*R(1,3)        ! V31
C
C
      RETURN
      END ! SUBROUTINE POLARDECOMP
C
C
C---.---1---.---2---.---3---.---4---.---5---.---6---.---7--
      SUBROUTINE POLARROTATION(F,R)                 !very well tested
C     PURPOSE: This routine computes the tensor [R] from the polar
C     decompositon (a special case of singular value decomposition),
C
C            F = RU = VR
C
C     for a 3x3 invertible matrix F. Here, R is an orthogonal matrix
C     and U and V are symmetric positive definite matrices.
C
C     This routine determines only [R]. After calling this routine, you
```

```
C     can obtain [U] or [V] by  [U] = [R]^T [F]    and  [V]=[F][R]^T
C     where "^T" denotes the transpose. The Higham algorithm is easier
C     and equally accurate, so this is mostly an interesting exercise.
C
C     This routine is limited to 3x3 matrices specifically to
C     optimize its performance. You will notice below that all matrix
C     operations are written out long-hand because explicit
C     (no do-loop) programming avoids concerns about some compilers
C     not optimizing loop and other array operations as well as they
C     could for small arrays.
C
C     This routine returns a proper rotation if det[F]>0,
C     or an improper orthogonal tensor if det[F]<0.
C
C     The host code is responsible for NOT calling this routine
C     if det[F]=0. The closer that [F] is to being singular, the
C     longer this routine will take.
C
C     This routine uses an iterative algorithm, but the iterations are
C     continued until the error is minimized relative to machine
C     precision. Therefore, this algorithm should be as accurate as
C     any purely analytical method. In fact, this algorithm has been
C     demonstrated to be MORE accurate than analytical solutions because
C     it is less vulnerable to round-off errors. It is only slightly
C     better than than the algorithmically simpler HIGHAM algorithm,
C     which simply computes R by a fixed point operation that averages
C     a tensor with its own inverse transpose (starting with F, the
C     result is then piped recursively to the same operator until
C     it doesn't change. That's the rotation. .
C
C     Reference for scaling method:
C     Brannon, R.M. (2004) "Rotation and Reflection" (draft version).
C
C     Reference for fixed point iterator:
C     Bjorck, A. and Bowie, C. (1971) "An iterative algorithm for
C     computing the best estimate of an orthogonal matrix." SIAM J.
C     Numer. Anal., vol 8, pp. 358-364.
C
C input
C ----
C    F: the 3x3 matrix to be decomposed into the form F=RU
C
C output
C ----
C    R: polar rotation tensor (3x3 orthogonal matrix)
C
C  030701|Rebecca Brannon| devised algorithm and created routine
C  030829|Rebecca Brannon| used Farnsworth's multiply reduction idea
C  050524|Rebecca Brannon| fixed gnu compiler optimizer problem
C       INCLUDE 'implicit.h'
C.....parameters
      PARAMETER (ONE=0.1D1, TWO=0.2D1, THREE=0.3D1 )
      PARAMETER (HALF=0.5D0)
C
C.....passed arguments
      DIMENSION F(3,3), R(3,3)
C.....local (helper) variables
      INTEGER KNT
```

```
cccccccccccccccccccccccccccccccccccccccccccccccccccccccccccccc end of preamble
C Step 1: Compute [C] = Transpose[F] . [F] (Save into E for now)
      E11 = F(1,1)*F(1,1) + F(2,1)*F(2,1) + F(3,1)*F(3,1)
      E22 = F(1,2)*F(1,2) + F(2,2)*F(2,2) + F(3,2)*F(3,2)
      E33 = F(1,3)*F(1,3) + F(2,3)*F(2,3) + F(3,3)*F(3,3)
      E23 = F(1,2)*F(1,3) + F(2,2)*F(2,3) + F(3,2)*F(3,3)
      E31 = F(1,3)*F(1,1) + F(2,3)*F(2,1) + F(3,3)*F(3,1)
      E12 = F(1,1)*F(1,2) + F(2,1)*F(2,2) + F(3,1)*F(3,2)
C
C*************************************************************************
C
C Step 2: To guarantee convergence, scale [F] by multiplying it by
C         Sqrt[3]/magnitude[F]. This is allowable because this routine
C         finds ONLY the rotation tensor [R]. The rotation for any
C         positive multiple of [F] is the same as the rotation for [F]
C         itself.
C         Scaling [F] by a factor sqrt(3)/mag[F] requires replacing the
C         previously computed [C] matrix by a factor 3/squareMag[F],
C         where squareMag[F] is most efficiently computed by trace[C].
C         Complete computation of [E]=(1/2)([C]-[I]) with [C] now
C         being scaled.
                S=THREE/(E11+E22+E33)
                E11 = HALF*(S*E11-ONE)
                E22 = HALF*(S*E22-ONE)
                E33 = HALF*(S*E33-ONE)
                E23 = HALF*(S*E23)
                E31 = HALF*(S*E31)
                E12 = HALF*(S*E12)
C
C Step 3: Replace S with Sqrt(S) and set the first guess for [R] equal
C         to the scaled [F] matrix,   [A]=Sqrt[3]F/magnitude[F]
                S=SQRT(S)
                A11 = S*F(1,1)
                A21 = S*F(2,1)
                A31 = S*F(3,1)
                A12 = S*F(1,2)
                A22 = S*F(2,2)
                A32 = S*F(3,2)
                A13 = S*F(1,3)
                A23 = S*F(2,3)
                A33 = S*F(3,3)
C
C Step 4. Compute error of this first guess.
C     The matrix [A] equals the rotation if and only if [E] equals [0]
      ERRZ= E11*E11 + E22*E22 + E33*E33 + TWO*(E12*E12+E23*E23+E31*E31)
C
C Step 5.  Check if scaling ALONE was sufficient to get the rotation.
C     This occurs whenever the stretch tensor is isotropic.
C     A number X is zero to machine precision if (X+1.0)-1.0 evaluates
C     to zero. Typically, machine precision is around 1.e-20.
      IF  ( ERRZ + ONE .EQ.ONE )   GO TO 10
C
C*************************************************************************
      DO 6 KNT=1,2000
C
C Step 6: Improve the solution.
C
C     Compute a helper matrix, [X]=[A][I-E].
```

```
C     Do *not* overwrite [A] at this point.
C     There is *NOT* a programming error with the indices here.
C     The following expression reflects the fact that
C     E21=E12,  E32=E23, and E13=E31.
C
      X11 = A11 - (A11*E11 + A12*E12 + A13*E31)
      X21 = A21 - (A21*E11 + A22*E12 + A23*E31)
      X31 = A31 - (A31*E11 + A32*E12 + A33*E31)
      X12 = A12 - (A12*E22 + A13*E23 + A11*E12)
      X22 = A22 - (A22*E22 + A23*E23 + A21*E12)
      X32 = A32 - (A32*E22 + A33*E23 + A31*E12)
      X13 = A13 - (A13*E33 + A11*E31 + A12*E23)
      X23 = A23 - (A23*E33 + A21*E31 + A22*E23)
      X33 = A33 - (A33*E33 + A31*E31 + A32*E23)
C
C     Transfer result to [A]
C
      A11 = X11
      A21 = X21
      A31 = X31
      A12 = X12
      A22 = X22
      A32 = X32
      A13 = X13
      A23 = X23
      A33 = X33
C
C Step 7: Using the improved solution compute the new
C         error tensor, [E] = (1/2) (Transpose[A].[A]-[I])
      E11 =  HALF*(A11*A11 + A21*A21 + A31*A31 - ONE)
      E22 =  HALF*(A12*A12 + A22*A22 + A32*A32 - ONE)
      E33 =  HALF*(A13*A13 + A23*A23 + A33*A33 - ONE)
      E23 =  HALF*(A12*A13 + A22*A23 + A32*A33)
      E31 =  HALF*(A13*A11 + A23*A21 + A33*A31)
      E12 =  HALF*(A11*A12 + A21*A22 + A31*A32)
C
C Step 8: compute new error
      ERR=  E11*E11 + E22*E22 + E33*E33 + TWO*(E12*E12+E23*E23+E31*E31)
C
C Step 9: If new error is smaller than old error, then keep iterating.
C Otherwise, we have reached machine precision accuracy.
      IF(ERR.GE.ERRZ)GO TO 10
      ERRZ=ERR
    6 CONTINUE
C
C***********************************************************************
C
C Step 10: Reaching here means convergence!
C
C         Save rotation and return
C
   10 CONTINUE
C     print*,'———————————————— knt=',knt
      R(1,1) = (one+A11)-one
      R(2,1) = (one+A21)-one
      R(3,1) = (one+A31)-one
```

```
      R(1,2) = (one+A12)-one
      R(2,2) = (one+A22)-one
      R(3,2) = (one+A32)-one
      R(1,3) = (one+A13)-one
      R(2,3) = (one+A23)-one
      R(3,3) = (one+A33)-one
C
      RETURN
      END ! SUBROUTINE POLARROTATION
```

The following routine is specialized for the polar decomposition of a 2×2 matrix.

```
C---.---1---.---2---.---3---.---4---.---5---.---6---.---7--
      SUBROUTINE POLARDECOMP2X2(F,R,U)
C***********************************************************************
C     PURPOSE: This routine computes the polar decompositon,
C
C             F = RU
C
C     for a 2x2 matrix F. Here, R is a proper orthogonal matrix
C     (i.e., a rotation) and U is a symmetric positive definite matrix.
C
C     This routine is limited to 2x2 matrices specifically to
C     optimize its performance.
C
C     This routine will return a proper rotation if det[F]>0,
C     or an improper orthogonal tensor if det[F]<0.
C
C input
C ---
C
C    F: the 2x2 matrix to be decomposed into the form F=RU
C
C output
C ---
C    R: polar rotation tensor (2x2 orthogonal matrix)
C    U: polar stretch tensor (2x2 positive definite symmetric)
C
C
C   date |  person        | modification history:
C  ------|----------------|------------------------------------
C  030701|Rebecca Brannon| devised algorithm and created routine
C
      INCLUDE 'implicit.h'
C.....passed arguments
      DIMENSION F(2,2), R(2,2), U(2,2)
C.....local (helper) variables
cccccccccccccccccccccccccccccccccccccccccccccccccccccc end of preamble
C
      C=F(1,1)+F(2,2)
      S=F(2,1)-F(1,2)
      D = SQRT(C*C + S*S)
      C = C/D
      S = S/D
C
      R(1,1) =  C
      R(2,2) =  C
```

```
      R(1,2) = -S
      R(2,1) =  S
C
      U(1,1) =  C*F(1,1) + S*F(2,1)
      U(2,1) =  C*F(2,1) - S*F(1,1)
      U(1,2) =  C*F(1,2) + S*F(2,2)
      U(2,2) =  C*F(2,2) - S*F(1,2)
C
C     If you want left stretch [V] from the decomposition F=VR,
C     uncomment these lines (and add [V] to the calling arguments).
C
C     V(1,1) =  C*F(1,1) - S*F(1,2)
C     V(2,1) =  C*F(2,1) - S*F(2,2)
C     V(1,2) =  C*F(1,2) + S*F(1,1)
C     V(2,2) =  C*F(2,2) + S*F(2,1)
C
      RETURN
      END ! SUBROUTINE POLARDECOMP2X2
```

A.10 Listing 10: principal function time rate (PFTR) in FORTRAN

Source code for demo

```
  PROGRAM PrinFunctionDemo
C
C     The following program is a special purpose demonstration of the
C     principal function time rate (PFTR) algorithm for the special
C     [A] and [Adot] matrices cited in the example of the Rotation
C     book.  The [A] and [Adot] matrices are here hard-coded
C     at the evaluation time T=1. The function f(x) and its derivative
C     fprime(x) are also hardcoded in this stand-alone demo.
C
      IMPLICIT NONE
      REAL*8 T    ! time
      REAL*8 A,Q,LAM
      DIMENSION A(3,3),Q(3,3),LAM(3)
      REAL*8 ADOT,YDOT
      DIMENSION ADOT(3,3),YDOT(3,3)
cccccccccccccccccccccccccccccccccccccccccccccccccccccccccccccccc end of preamble
      OPEN (66,FILE="driver.out", STATUS="UNKNOWN")
C
C
C     lab components of [A] evaluated at T=1...
      T=1.0
      A(1,1) = 41.0*EXP(T-1.0)
      A(2,2) = 34.0*T*T*T
      A(3,3) = 50.0*T*T
      A(1,2) = -12.0*T
      A(2,3) = LOG(T)
      A(3,1) = 0.0
      A(2,1) = A(1,2)
      A(3,2) = A(2,3)
      A(1,3) = A(3,1)
      CALL PRETTY_PRINT_MATRIX(66,"THE MATRIX A",A)
      CALL JACOBI(A,Q)
      LAM(1)=A(1,1)
```

```
      LAM(2)=A(2,2)
      LAM(3)=A(3,3)
      WRITE(66,*)'lam1=',LAM(1)
      WRITE(66,*)'lam2=',LAM(2)
      WRITE(66,*)'lam3=',LAM(3)
      CALL PRETTY_PRINT_MATRIX(66,"DIRECTION COSINES",Q)
C
C     lab components of [Adot] evaluated at T=1
      T=1.0
      ADOT(1,1) = 41.0*EXP(T-1.0)
      ADOT(2,2) = 102.0*T*T
      ADOT(3,3) = 100.0*T
      ADOT(1,2) = -12.0
      ADOT(2,3) = 1.0/T
      ADOT(3,1) = 0.0
      ADOT(2,1) = ADOT(1,2)
      ADOT(3,2) = ADOT(2,3)
      ADOT(1,3) = ADOT(3,1)
      CALL PRETTY_PRINT_MATRIX(66,"THE MATRIX ADOT",ADOT)
C
C
      CALL PRINRATE(LAM, Q, ADOT,     YDOT )
      CALL PRETTY_PRINT_MATRIX(66,"THE MATRIX YDOT",YDOT)
C
      STOP
      END
C
C
C
C
C
C
C
C---.----1----.----2----.----3----.----4----.----5----.----6----.----7--
      SUBROUTINE PRETTY_PRINT_MATRIX(IOU,MATRIX_NAME,A)
      IMPLICIT NONE
      CHARACTER*(*) MATRIX_NAME
      CHARACTER*80 JNKSTR
      INTEGER IOU, J
      REAL*8 A(3,3)
cccccccccccccccccccccccccccccccccccccccccccccccccccccccccccccc end of preamble
      WRITE(IOU,8001)(A(1,J),J=1,3)
      IF(MATRIX_NAME.NE.' ') THEN
         JNKSTR=MATRIX_NAME//' = '
         WRITE(IOU,8002)JNKSTR,(A(2,J),J=1,3)
      ELSE
         WRITE(IOU,8002)MATRIX_NAME,(A(2,J),J=1,3)
      END IF
      WRITE(IOU,8003)(A(3,J),J=1,3)
      WRITE(IOU,*)
C
      RETURN
C
 8001 FORMAT(T20,     '_',                  T71, '_',       /,
     &       T19,    '|    ',  3(G13.6,3X),  T72,  '|'       )
 8002 FORMAT(A18,    '|    ',  3(G13.6,3X),  T72,  '|'       )
 8003 FORMAT(T19,    '|_   ',  3(G13.6,3X),  T71, '_|'       )
      END
```

```
C--.---1---.---2---.---3---.---4---.---5---.---6---.---7--
      SUBROUTINE JACOBI(A,V)
C This routine gets eigenvalues and orthonormalized eigenvectors
      IMPLICIT NONE
      INTEGER I,IH,J,K,NEQ
      REAL*8 A,V,SUM,SSUM,SI,TT,CO,TL,T1,TOL,AA,AMAX,ERR
      REAL*8 PZERO,PONE
      PARAMETER (PZERO=0.0D0,PONE=0.1D1)
      REAL*8 TX
      PARAMETER (TX=1.E-4)
      DIMENSION A(3,3), V(3,3)
Ccccccccccccccccccccccccccccccccccccccccccccccccccccccccccccccccccccccc
C
      NEQ = 3
      SUM = PZERO
      TL = 20.
      TOL = 1.0E-5
      DO 200 I=1,NEQ
         DO 190 J=1,NEQ
            IF(TL.GT.PZERO) V(I,J) = PZERO
            SUM = SUM + DABS(A(I,J))
 190     CONTINUE
 200  CONTINUE
      DO 189 I=1,NEQ
         IF(TL.GT.PZERO) V(I,I)=1.0
 189  CONTINUE
      IF(SUM.LE.PZERO) RETURN
      SUM = SUM/DFLOAT(NEQ*NEQ)
 400  SSUM = PZERO
      AMAX = PZERO
      DO 700 J=2,NEQ
         IH = J - 1
      DO 700 I=1,IH
         AA = DABS(A(I,J))
         IF(AA.GT.AMAX) AMAX = AA
         SSUM = SSUM + AA
         IF(AA.LT.0.1*AMAX) GO TO 700
         AA=ATAN2(2.0*A(I,J),A(I,I)-A(J,J)+1.E-10)/2.0
         SI = DSIN(AA)
         CO = DCOS(AA)
C
        DO 500 K=1,NEQ
          TT = A(K,I)
          A(K,I) = CO*TT + SI*A(K,J)
          A(K,J) = -SI*TT + CO*A(K,J)
          TT = V(K,I)
          V(K,I) = CO*TT + SI*V(K,J)
          V(K,J) = -SI*TT + CO*V(K,J)
 500    CONTINUE
!
        A(I,I) = CO*A(I,I) + SI*A(J,I)
        A(J,J) = -SI*A(I,J) + CO*A(J,J)
        A(I,J) = PZERO
        DO 600 K=1,NEQ
          A(I,K) = A(K,I)
          A(J,K) = A(K,J)
 600    CONTINUE
 700  CONTINUE
      ERR=DABS(SSUM)/SUM
C     IF(ERR.GT. TOL) GO TO 400
```

```
C
C       The following checks if ERR is smaller than machine precision
        IF(ERR*TX+PONE.GT.PONE)GO TO 400
C
C       To reach here, the algorithm has converged.
C       Repair small round off errors by replacing numbers smaller
C       than 1.e-15 by zero.
        DO I=1,3
        DO J=1,3
          IF(ABS(V(I,J)).LT.1.E-15)V(I,J)=PZERO
        ENDDO
        ENDDO
C
C
C
        RETURN
        END
C
C—-.——1——.——2——.——3——.——4——.——5——.——6——.——7—
      FUNCTION F(X)
      IMPLICIT NONE
      REAL*8 X,F
cccccccccccccccccccc end of preamble
      F=X*X
      RETURN
      END
C—-.——1——.——2——.——3——.——4——.——5——.——6——.——7—
      FUNCTION FPRIME(X)
      IMPLICIT NONE
      REAL*8 X,FPRIME
cccccccccccccccccccc end of preamble
      FPRIME=2.0*X
      RETURN
      END
C
C—-.——1——.——2——.——3——.——4——.——5——.——6——.——7—
      SUBROUTINE PrinRate(LAM, Q, ADOT,     YDOT )
C     PURPOSE: This routine computes the rate of a principal function
C                 Y = f(A)
C
C     The principal function is computed by applying the function f
C     in the principal basis for [A] (i.e., in A-space).
C
C     Prior to calling this function, an eigensolver be called so that
C     the eigenvectors of [A] are stored in the columns of [Q].
C
C input
C ——
C  lam: array containing eigenvalues of [A]
C    Q: matrix whose columns contain corresponding orthonormalized
C       eigenvectors of [A]
C Adot: lab components of the time rate of [A]
C
C output
C ——
C Ydot: lab components of the time rate of Y=f(A)
C
C  MODIFICATION HISTORY
```

```
C  030914|Rebecca Brannon| created routine
      IMPLICIT NONE
C.....parameters
      REAL*8 ZERO
      PARAMETER (ZERO=0.0D0)
C.....passed
      REAL*8 LAM,Q,ADOT,YDOT
      DIMENSION LAM(3),Q(3,3),ADOT(3,3),YDOT(3,3)
C.....function (functions instead of subroutines)
      REAL*8 F,FPRIME
      EXTERNAL F,FPRIME
C.....local (not saved)
      INTEGER I,J,M,N,R,S,K,L
      REAL*8 YD,GAM(3,3)
cccccccccccccccccccccccccccccccccccccccccccccccccccccc end of preamble
C     first get the gamma coefficients
      DO I=1,3
      DO J=1,3
         IF(LAM(I).EQ.LAM(J))THEN
           GAM(I,J)=FPRIME(LAM(I))
         ELSE
           GAM(I,J)=(F(LAM(I))-F(LAM(J)))/(LAM(I)-LAM(J))
         ENDIF
      ENDDO
      ENDDO
C
C     for debugging purposes, write the resulting gam matrix
C     This line would, of course, be deleted for regular applications
      CALL PRETTY_PRINT_MATRIX(66,"THE GAMMA COEFS",GAM)
C
      DO M=1,3
      DO N=1,3
C
      YD=ZERO
      DO R=1,3
      DO S=1,3
      DO K=1,3
      DO L=1,3
        YD = Q(M,K)*Q(N,L)*Q(R,K)*Q(S,L)*GAM(K,L)*ADOT(R,S)           + YD
C
      ENDDO
      ENDDO
      ENDDO
      ENDDO
C
      YDOT(M,N)=YD
      ENDDO
      ENDDO
      RETURN
      END
```

Output from demo

```
|‾    41.0000        -12.0000        0.000000E+00     ‾|
```

```
The matrix A =       |     -12.0000          34.0000          0.000000E+00     |
                     |_    0.000000E+00     0.000000E+00      50.0000         _|

 lam1=    50.00000000000000
 lam2=    50.00000000000000
 lam3=    25.00000000000000
                      _                                                       _
                     |      0.800000        0.000000E+00    -0.600000          |
direction cosines    |     -0.600000        0.000000E+00    -0.800000          |
                     |_    0.000000E+00      1.00000         0.000000E+00     _|

                      _                                                       _
                     |       41.0000         -12.0000         0.000000E+00     |
The matrix Adot =    |      -12.0000          102.000          1.00000         |
                     |_    0.000000E+00       1.00000          100.000        _|

                      _                                                       _
                     |       100.000          100.000          75.0000         |
The gamma coefs =    |       100.000          100.000          75.0000         |
                     |_      75.0000          75.0000          50.0000        _|

                      _                                                       _
                     |       3650.00         -2616.00         -12.0000         |
The matrix Ydot =    |      -2616.00          7224.00          84.0000         |
                     |_     -12.0000          84.0000          10000.0        _|
```

A.11 Listing 11: Mathematica demo (rate of a principal function)

Rates of principal functions

In[1]:=
```
ClearAll["Global`*"];
```

Note to non-*Mathematica* users: Text like this contains explanatory comments. Inside actual commands, which are rendered in black bold on separate lines, comments are enclosed in (* *) delimiters. This file has been formatted so that both the command and its output (when not suppressed by a semicolon) are contained in black boxes.

The purpose of this notebook is to illustrate how to find the rate of a principal function of a symmetric tensor. Laboratory components of the tensor $\underset{\approx}{A}$ that vary with time will be saved in a sample matrix called [A].

We might also write $[A^{LL}]$ where the label "LL" is added as a reminder that these are referenced to the laboratory basis

Case study:

In[2]:=
```
A = ALL = ( 41 e^(t-1)   -12 t     0
            -12 t        34 t^3    Log[t]
            0            Log[t]    50 t^2 );
```

The tensor is constructed by pairing up lab components with lab basis dyads: $\underset{\approx}{A} = A^{LL}_{ij}\, \underset{\sim}{e}^L_i\, \underset{\sim}{e}^L_j$.

With respect to the principal basis, labeled with a "P", the expansion is $\underset{\approx}{A} = A^{PP}_{ij}\, \underset{\sim}{e}^P_i\, \underset{\sim}{e}^P_j$

The $[A^{PP}]$ matrix is diagonal: $[A^{PP}] = \begin{bmatrix} \lambda_1 & 0 & 0 \\ 0 & \lambda_2 & 0 \\ 0 & 0 & \lambda_3 \end{bmatrix}$, and therefore

$$\underset{\approx}{A} = \lambda_1\, \underset{\sim}{e}^P_1\, \underset{\sim}{e}^P_1 + \lambda_2\, \underset{\sim}{e}^P_2\, \underset{\sim}{e}^P_2 + \lambda_3\, \underset{\sim}{e}^P_3\, \underset{\sim}{e}^P_3$$

A principal function, $f[\underset{\approx}{A}]$, is computed by first performing a change of basis to write the tensor's components in the principal basis. Then the operation $\underset{\approx}{Y} = f\left[\underset{\approx}{A}\right]$ is of the same form except with the function applied to the eigenvalues:

$$\underset{\approx}{Y} = f[\lambda_1]\, \underset{\sim}{e}^P_1\, \underset{\sim}{e}^P_1 + f[\lambda_2]\, \underset{\sim}{e}^P_2\, \underset{\sim}{e}^P_2 + f[\lambda_3]\, \underset{\sim}{e}^P_3\, \underset{\sim}{e}^P_3 \text{ and therefore}$$

$$[Y^{PP}] = \begin{bmatrix} f[\lambda_1] & 0 & 0 \\ 0 & f[\lambda_2] & 0 \\ 0 & 0 & f[\lambda_3] \end{bmatrix}.$$

While this algorithm for evaluating a principal function is straightforward, finding the rate of a principal function is a far more daunting task. We can't simply take the rate of the components in the principal basis. The rate must include terms accounting for rates of the eigenvectors as well, and this job is further complicated by the fact that eigenvectors are never unique. An algorithm for finding this rate is illustrated below in the special case that the the function is so simple that the correct answer can be found by alternative methods. Specifically, we will use $f(x) = x^2$ because then $f\left(\underset{\approx}{A}\right) = \underset{\approx}{A} \cdot \underset{\approx}{A}$ can be computed directly using laboratory components, with the rate found directly using laboratory components. The algorithm is numerical, so we must specify an evaluation time. For our case study, we use $t = 1$

Exact solution for verification testing:

```
In[3]:= fCaseStudy[x_] := x²; Print["function:  ", "f(x)=x²"]
AoftCaseStudy = A; Print["[A]=", MatrixForm[A]];
YoftCaseStudy = A.A; Print["[Y]=A.A=", MatrixForm[YoftCaseStudy]];
tCaseStudy = 1;
Print["At the evaluation time, t = ", tCaseStudy, ", lab components are"]
AcaseStudy = A /. t → tCaseStudy;
Print["the tensor [A] = ", MatrixForm[AcaseStudy]]
AdotCaseStudy = D[AoftCaseStudy, t] /. t → tCaseStudy;
Print["the rate of the tensor [Ȧ] = ", MatrixForm[AdotCaseStudy]]
Y = YcaseStudy = YoftCaseStudy /. t → tCaseStudy;
Print["the principal function [Y] = ", MatrixForm[YcaseStudy]]
YdotCaseStudy = D[YoftCaseStudy, t] /. t → tCaseStudy;
Print["the rate of the principal function [Ẏ] = ", MatrixForm[YdotCaseStudy]]
```

function: $f(x) = x^2$

$$[A] = \begin{pmatrix} 41 e^{-1+t} & -12 t & 0 \\ -12 t & 34 t^3 & \mathrm{Log}[t] \\ 0 & \mathrm{Log}[t] & 50 t^2 \end{pmatrix}$$

$$[Y] = A.A = \begin{pmatrix} 1681 e^{-2+2t} + 144 t^2 & -492 e^{-1+t} t - 408 t^4 & -12 t\,\mathrm{Log}[t] \\ -492 e^{-1+t} t - 408 t^4 & 144 t^2 + 1156 t^6 + \mathrm{Log}[t]^2 & 50 t^2 \mathrm{Log}[t] + 34 t^3 \mathrm{Log}[t] \\ -12 t\,\mathrm{Log}[t] & 50 t^2 \mathrm{Log}[t] + 34 t^3 \mathrm{Log}[t] & 2500 t^4 + \mathrm{Log}[t]^2 \end{pmatrix}$$

At the evaluation time, t = 1, lab components are

$$\text{the tensor } [A] = \begin{pmatrix} 41 & -12 & 0 \\ -12 & 34 & 0 \\ 0 & 0 & 50 \end{pmatrix}$$

$$\text{the rate of the tensor } [\dot{A}] = \begin{pmatrix} 41 & -12 & 0 \\ -12 & 102 & 1 \\ 0 & 1 & 100 \end{pmatrix}$$

$$\text{the principal function } [Y] = \begin{pmatrix} 1825 & -900 & 0 \\ -900 & 1300 & 0 \\ 0 & 0 & 2500 \end{pmatrix}$$

$$\text{the rate of the principal function } [\dot{Y}] = \begin{pmatrix} 3650 & -2616 & -12 \\ -2616 & 7224 & 84 \\ -12 & 84 & 10000 \end{pmatrix}$$

User input

```
In[12]:= f[x_] := fCaseStudy[x];
A = AcaseStudy;
Adot = AdotCaseStudy;
```

This notebook is written in a manner that the above user inputs can be changed to something else besides what was defined above in the verification case study. The following flag will be used to tell the notebook whether or not to check answers against the case-study's direct solution

```
In[15]:= checkAnswer = True
Out[15]= True
```

Finding the principal function

In Mathematica, the eigenvectors for matrices are not necessarily normalized to be unit vectors, so we apply the Orthogonalize command to ensure that we are working with an orthonormal set of eigenvectors.

In[19]:=
```
{eval, evecs} = Eigensystem[A]; evec = Orthogonalize[evecs];
Print[
  "The spectral decomposition, [A^LL]=[q^LP][A^PP][q^PL],\nwhich is traditionally
    written [A]=[Q][Λ][Q]^T, is..."];
Q = qLP = evec^T; APP = DiagonalMatrix[eval]; qPL = evec;
Print[MatrixForm[A], " = ", MatrixForm[qLP], MatrixForm[APP], MatrixForm[qPL]]
```

The spectral decomposition, $[A^{LL}]=[q^{LP}][A^{PP}][q^{PL}]$,
which is traditionally written $[A]=[Q][\Lambda][Q]^T$, is...

$$\begin{pmatrix} 41 & -12 & 0 \\ -12 & 34 & 0 \\ 0 & 0 & 50 \end{pmatrix} = \begin{pmatrix} 0 & -\frac{4}{5} & \frac{3}{5} \\ 0 & \frac{3}{5} & \frac{4}{5} \\ 1 & 0 & 0 \end{pmatrix}\begin{pmatrix} 50 & 0 & 0 \\ 0 & 50 & 0 \\ 0 & 0 & 25 \end{pmatrix}\begin{pmatrix} 0 & 0 & 1 \\ -\frac{4}{5} & \frac{3}{5} & 0 \\ \frac{3}{5} & \frac{4}{5} & 0 \end{pmatrix}$$

Check that the decomposition is correct.

In[20]:=
```
{A == qLP.APP.qPL, A == Q.DiagonalMatrix[eval].Q^T} // Simplify
```

Out[20]= {True, True}

The following exploits Mathematica's symbolic math capabilities to verify that ambiguity in the eigenvectors does not affect the final result.

The case study problem has been designed so that, as seen, one of the eigenvalues is repeated. Accordingly, the first two eigenvectors are arbitrary in direction as long as they are both perpendicular to each other and in the same spanning plane. To verify that this ambiguity will not influence the final results, we replace the eigenvectors with an arbitrary rotation as follows

In[21]:=
```
ev = evec;
If[checkAnswer, ev = {{Cos[θ], -Sin[θ], 0}, {Sin[θ], Cos[θ], 0}, {0, 0, 1}}.evec; qLP = ev^T; qPL = ev;
  A == Simplify[qLP.APP.qPL]]
Q = ev^T;
Print["Direction cosine matrix:    [Q] = ", MatrixForm[Q]];
```

Out[22]= True

Direction cosine matrix: $[Q] = \begin{pmatrix} \frac{4\,\mathrm{Sin}[\theta]}{5} & -\frac{4\,\mathrm{Cos}[\theta]}{5} & \frac{3}{5} \\ -\frac{3\,\mathrm{Sin}[\theta]}{5} & \frac{3\,\mathrm{Cos}[\theta]}{5} & \frac{4}{5} \\ \mathrm{Cos}[\theta] & \mathrm{Sin}[\theta] & 0 \end{pmatrix}$

The principal function is then

In[24]:=
```
Y = Simplify[qLP.DiagonalMatrix[f[eval]].qPL];
Print["Lab components of Y=f(A) are ", MatrixForm[Y]]
If[checkAnswer, Simplify[Y == YcaseStudy]]
```

Lab components of Y=f(A) are $\begin{pmatrix} 1825 & -900 & 0 \\ -900 & 1300 & 0 \\ 0 & 0 & 2500 \end{pmatrix}$

Out[26]= True

For the case study, the replacement of the first two eigenvectors by an arbitrary rotation in the eigenplane made the direction cosine matrix depend on that rotation angle, but the final result for Y=f(A) was seen here to be unaffected by that ambiguity (i.e., not dependent on θ), which is as it should be!

Step 1: Create the G(x,y) function

```
In[27]:= G[x_, y_] := If[x == y, f'[x], (f[x] - f[y])/(x - y)];
```

Step 2: Use this function to evaluate $\gamma_{ij} = G(\lambda_i, \lambda_j)$

```
In[28]:= γ = Table[G[eval[[i]], eval[[j]]], {i, 3}, {j, 3}];
         Print["[γ] = ", MatrixForm[γ]]

[γ] = (100 100 75
       100 100 75
       75  75  50)
```

Step 3: Compute the laboratory components of the rate of Y=f(A).

```
In[30]:= Ydot = Simplify[Table[
           Sum_{p=1}^{3} Sum_{q=1}^{3} Sum_{k=1}^{3} Sum_{l=1}^{3} Q[[m, k]] Q[[n, l]] Q[[p, k]] Q[[q, l]] Adot[[p, q]] γ[[k, l]]
           , {m, 3}, {n, 3}]];
         Print["lab components of rate of Y: ", MatrixForm[Ydot]]
         If[checkAnswer, Ydot == YdotCaseStudy]

lab components of rate of Y: (3650  -2616  -12
                              -2616  7224   84
                              -12    84     10000)

Out[32]= True
```

Tip: after using the `Sum[...]` command, the beautiful formatting as nice summation symbols is achieved by clicking anywhere in the line and typing CTRL–SHIFT–N.

www.ingramcontent.com/pod-product-compliance
Ingram Content Group UK Ltd.
Pitfield, Milton Keynes, MK11 3LW, UK
UKHW051910100926
8556IPUK00005B/838

9 780750 314527